Distribution and ecology of vascular plants in a tropical rain forest

Geobotany 1

Series Editor
M. J. A. WERGER

DR W. JUNK PUBLISHERS THE HAGUE – BOSTON – LONDON 1981

J. B. HALL AND M. D. SWAINE
Formerly of Department of Botany, University of Ghana, Accra

Distribution and ecology of vascular plants in a tropical rain forest

Forest vegetation in Ghana

DR W. JUNK PUBLISHERS THE HAGUE – BOSTON – LONDON 1981

Distributors

for the United States and Canada

Kluwer Boston, Inc.
190 Old Derby Street
Hingham, MA 02043
U.S.A.

for all other countries

Kluwer Academic Publishers Group
Distribution Center
P.O. Box 322
3300 AH Dordrecht
The Netherlands

This volume is listed in the Library of Congress Cataloging in Publication Data

ISBN 90 6193 681 0 (this volume)
ISBN 90 6193 895 3 (series)

Cover design: Max Velthuijs

PRINTED IN THE NETHERLANDS

Dedicated to Albert Adai Enti, who was our first guide in the Ghanaian forest, and whose understanding and knowledge of its plants have been a constant source of information and inspiration.

Foreword

It is a privilege to be asked and a pleasurable duty for me to write the foreword of this book.

The conservation and wise utilisation of the humid tropical forests, a unique biome, are matters of great concern and importance to millions living within and around these forests and, perhaps, less directly, to the totality of mankind.

These forests provide many essential products and services for mankind. The list is lengthy and need not be repeated here. Suffice it to say that there are not many aspects of human activity which do not utilise some of these products, services or derivatives therefrom. Yet it is the view of those most closely associated with the study of these forests that what is known is but a minuscule portion of what there is to know. The products and services now utilised, are perhaps some infinitesimal part of the full potential.

All over the tropical world, however, these forests are being destroyed. At first, slowly, but now surely gathering tempo. This is true also of Ghana. Tracts of forest land are converted to other uses, often ephemeral and not sustained. Irreversible changes take place in our environment. The gains are shortlived, the losses unobtrusively accumulate and stay forever. The accelerating rate of deforestation, in the face of our relatively scanty knowledge of this biome, is indeed a sad reflection of the state of human affairs.

It is in this setting that one welcomes this book by Messrs. J. B. Hall and M. D. Swaine.

John Hall came to Ghana in 1955 as a Biology master at Mfantsipim School, Cape Coast, moved in 1964 to the University of Cape Coast (then a University College) as a Botany Lecturer, and worked from 1966 to 1980 as Lecturer and Senior Lecturer at the Department of Botany, University of Ghana, Legon. He has collected plants in all parts of Ghana, was in charge of the Ghana Herbarium from 1969 to 1980 and has published widely.

Mike Swaine joined the Department of Botany, University of Ghana, Legon in 1971 and worked as Lecturer and Senior Lecturer till 1979. His degrees in Forestry and Botany at the University College of North Wales prepared him for the association with John Hall. Together they roamed the remotest parts of Ghana, collected, described and wrote. Mike Swaine, as I write, is now with the Institute of South-East Asian Biology in the Department of Botany, University of Aberdeen, Scotland.

This book is the result of a survey of the floristic composition of the closed forest in Ghana. It introduces the ecology and utilisation of Ghanaian forest in relation to work on tropical forest elsewhere. Full descriptions of forest types are given and concentration is on the more detailed species distribution, including herbs and shrubs, from which the forest classification is derived.

Excellent maps show the distribution of species most commonly found in the plots assessed, the localities of herbarium specimens, the plots the authors found the species in, and notes on the life form, reproductive morphology and ecology of each species together with references of taxonomic treatments.

I consider this a very important contribution to knowledge of the humid tropical forest in Ghana.

In a sense this book should update and complement earlier workers; Hutchinson and Dalziel 'Flora of West

Tropical Africa', Taylor 'Synecology and Silviculture in Ghana' and Irvine 'Woody Plants of Ghana'. I pay tribute to those earlier workers and also to these latest.

This book should be of great interest to a wide variety of people: botanists, ecologists, foresters, geographers, pharmacists, taxonomists, teachers, students and all users of forests.

I hope it improves their perception of overall forest composition, of species regeneration, of areas where conservation is most necessary, and spurs them on, one way or the other, to save this invaluable asset.

J. H. François

Chief Conservator of Forests
Department of Forestry, Accra

Preface

Of the world's major ecosystems, tropical forest is the most complex both architecturally and floristically and, largely because of this complexity, the least understood. In comparison with most tropical forests, those of Ghana are well known, at least taxonomically. Herbarium collections span 200 years. Botanists and foresters have visited every part of the forest zone. Its plant species are described in a comprehensive Flora of West Tropical Africa which has gone through two editions. Although new species continue to be found and described, they are encountered very seldom. The taxonomic foundation has therefore been laid for an account not merely of the occurrence of forest species in Ghana, but also of their distribution within the country.

Information on species distribution can be gleaned from the localities of herbarium specimens cited in Floras, or from the labels of specimens in large herbaria, such as the Ghana Herbarium in the University of Ghana, Legon. Such information will be reliable as far as it goes, but very incomplete. Botanical collectors, naturally enough, have concentrated their attention on accessible places near main roads and towns, and on those low-growing species with accessible flowers, and especially on species which flower for much of the year. Foresters get off the beaten track, but their main interest is in large trees. The smaller trees, and especially the large climbers, have been generally neglected.

Between 1971 and 1975, we undertook a survey of the floristic composition of Ghanaian closed-canopy forest. We listed all vascular plant species present on 155 sample plots in mature forest, each plot measuring 25 m × 25 m, distributed throughout the forest zone; the number of plant species on each plot ranged from 20 to 200. Species were also listed on a further 130 plotless samples. Of a total 1 300 species recorded (out of a total forest flora of 2 000 species), the 749 which occurred four or more times were used for ordination. Our new classification of Ghanaian forest (Hall & Swaine 1976) was derived from this ordination. A byproduct of the survey was much data on species distribution, and the presentation of this data is a major aim of this book.

Of the 749 species ordinated, we have not mapped those whose distribution was very sparse and uninformative, and were therefore not adequately represented in our survey records. Our 624 maps therefore show the distributions of that third of Ghana's forest species which are most commonly encountered in mature, non-swampy forest. It is unfortunate in some ways that secondary forest species have had to be omitted, as this group includes many important medicinal plants such as *Rauvolfia vomitoria,* and *Voacanga africana,* and the most promising of the sources of sweeteners, *Thaumatococcus daniellii.*

Each map shows our survey records by solid dots; because sample plots were scattered fairly uniformly through the forest zone, and all species present were recorded, it is possible from the distribution of these dots to draw conclusions as to relative abundance and dispersion of the species. Some species are seen to be evenly spread throughout the forest types in which they occur, whilst the distribution of others seems to be influenced by historical rather than by ecological factors. From the reference maps, the plot numbers in which any species occurs can be identified, and the exact location of each plot is given in an appendix. Using this information it should therefore be possible to find the species on the ground. The histograms accompanying the maps are derived from sample plot data, and indicate the apparent ecological preference of each species. Herbarium records are generally shown by circles. Where the locality for a herbarium specimen coincides with that of a survey record, the former is generally omitted, though it is included, slightly displaced, in the case of some rare species.

The brief notes accompanying the maps are intended to bring the taxonomy of the species up to date, to indicate species in need of further taxonomic work, and to summarise life forms and phenology concisely and consistently.

Distribution maps of African plants are listed by Lebrun & Stork (1977), and maps continue to appear in the Distributiones Plantarum Africanarum series published by the Jardin Botanique National de Belgique. The overwhelming majority of these maps show distribution on a continental, or regional scale, and it has been suggested to us that the time is not yet ripe for attempting to map distributions within a single country. Although we are only too well aware of the possibilities for error in field determinations, we can only plead that we have done all we can to minimise these, by collecting specimens for verification in the herbarium, by crosschecking between plots, and so on. Our maps are certainly imperfect, but it is equally certain that before they can be substantially improved, much of the remaining forest will have given way to farms and timber plantations.

Janzen (1975) begins his text on Tropical Plant Ecology with the words: 'It is my opinion that the study of plant biology in general, and the biology of tropical plants in particular, has suffered greatly from over emphasis on descriptive morphology, phytosociology, and systematics.' As our book deals with precisely these three aspects of botany, perhaps we should apologise for inflicting yet more suffering. We hope nevertheless that even the most adventurous ecologists may find here at least some raw material for their daring generalisations.

Acknowledgements

Our greatest debt is to Mr Daniel Abbiw who accompanied us on almost all our forays: tirelessly organising the meals, cutting the plot boundaries, identifying trees and, back in the herbarium, placing records onto the maps. We were assisted in the field especially by Mr R. S. Amponsah, Mr E. B. Hansen, Mr A. K. Mensah, and Dr J. M. Lock. Much of the rather tedious business of transferring records from field sheets to cards, and from computer print-outs to maps was undertaken by Mr R. S. Amponsah and Miss Alberta Banson, assisted during vacations by some of our students. Mr Eric Ofei helped with photography. We are grateful to the Head of the Department of Botany, University of Ghana, for unstinting support, to the Director of the Royal Botanic Gardens at Kew, and the Silviculturist of the Forestry Department in Kumasi for access to their respective herbaria, and to many other members of the Forestry Department and Game & Wildlife Department for their assistance both in the field and in tracing forest records. The manuscript was ably typed and retyped by Miss Mispah Torgbetse and Mr Thompson Akoto. Among those who read the draft manuscript, and whose comments and suggestions have been so useful, should be mentioned Dr Julius Amatekpor, Dr Kojo Attoh, Dr Emmanuel Asibey, Mr John François, Prof. P. Greig-Smith, Dr P. J. Grubb, Dr Nigel Hepper, Prof. G. W. Lawson, Dr A. J. M. Leeuwenberg, Dr Diana Lieberman, Dr J. M. Lock, Dr Jean Maley, Prof. P. W. Richards, Dr Michael Talbot, Dr Edmund Tanner, Dr M. J. A. Werger, Mr Frank White, Dr G. E. Wickens, and two anonymous referees. The Publications Board of the University of Ghana, and the Forestry Department of the Ghana Government contributed towards the cost of publication. Dr Aké Assi has been of inestimable assistance in solving many knotty problems of identification both in the field and in the herbarium. Finally we should like to acknowledge our debt to those early intrepid foresters H. N. Thompson, T. F. Chipp, C. Vigne, J. E. Andoh and A. A. Enti, whose pioneering made our project possible.

Acknowledgement is made for the following figures:
Bureau de Recherches géologiques et minières, Paris (3.4.).
Government of Ghana (1.4.).
Journal of Ecology (8.3, 8.13, 8.16).
Longman Group Ltd. & Dr J. Jeník (8.18).
National Aeronautics and Space Administration, U.S. Government (1.1, 1.2, 3.3, 8.6).
Organisation of African Unity (3.7a).

Contents

PART ONE : INTRODUCTION TO GHANAIAN FOREST

List of abbreviations*

A.R.S. Agricultural Research Station.

A-sample Sample plot measuring 25 m × 25 m on which all vascular plant species were enumerated during our survey (Hall & Swaine 1976).

asl Above sea level.

B-sample Plotless sample in which the first 30–60 species encountered were listed during our survey (Hall & Swaine 1976).

C Large climber capable of reaching the canopy.

c Small understorey climber.

c. circa (approximately).

CEC Cation exchange capacity (measure of the ability of soil colloids to attract cations).

CT Chorological transgressor (a species occurring widely in more than one chorological region).

dbh Diameter at breast height (1.3 m).

DS Dry Semi-deciduous forest-type.

DSFZ Fire Zone subtype of Dry Semi-deciduous forest.

DSIZ Inner Zone subtype of Dry Semi-deciduous forest.

Ep Epiphyte.

FF Ferns and Fern Allies.

Fl Flower.

F.R. Forest reserve.

Fr Fruit.

FWTA Flora of West Tropical Africa (second edition).

gbh Girth at breast height (1.3 m).

GC Guineo-Congolian Wide (a species occurring in all three subcentres of the Guineo-Congolian Region).

G.P.R. Game Production Reserve.

GW Guinea Wide (a species occurring in West African forest on both sides of the Dahomey Gap, but absent from the Congo basin).

H Herb.

ht. Height.

Lvs Leaves.

m-equiv Milli-equivalents.

MS Moist Semi-deciduous forest-type.

MSNW North-west subtype of Moist Semi-deciduous forest.

MSSE South-east subtype of Moist Semi-deciduous forest.

N.P. National Park.

pH Hydrogen ion concentration (measure of acidity).

Pt Pteridophyte.

S Shrub or pygmy tree (less than 2 m high).

SM Southern Marginal forest-type.

SO South-east Outlier forest-type.

sp. Species (singular).

sp. aff. Species related to.

spp. Species (plural).

subsp. Subspecies.

T Tree capable of exceeding 30 m height (megaphanerophyte).

t Tree 2–30 m high.

TEB Total exchangeable bases (metallic cations adsorbed to soil colloids).

UG Upper Guinea Endemic (a species restricted to West African forest west of the Dahomey Gap).

var. Variety.

WE Wet Evergreen forest-type.

W.S. Wildlife Sanctuary.

* SI units and points of the compass are excluded from this list.

Upland Evergreen forest in Tano Ofin Forest Reserve.

PART ONE

Introduction to Ghanaian forest

CHAPTER 1

What is forest?

Two main kinds of vegetation cover the surface of Ghana: forest and savanna; these, and the various formations into which they may be subdivided were defined at a Pan-African conference at Yangambi (C.S.A. 1956), and more recently by UNESCO (1973). Forest is dominated by trees at least 5 m high, whose crowns interlock, and lacks a continuous grassy ground layer. The crowns of trees in savanna are separated, or trees may be absent; there is always a seasonally dense understorey of grasses. Taxonomic distinctions are also important. Very few plant species occur naturally in both kinds of vegetation – notable exceptions are the trees *Afzelia africana* and *Diospyros mespiliformis.* The grasses which flourish in forest – such as *Leptaspis cochleata* and *Olyra latifolia* – are not found in savanna, and the grass tribe Andropogoneae which is so abundant in savanna is absent from forest. Both the Yangambi and the UNESCO classifications recognise a third formation-type – woodland – as intermediate between forest and savanna. The crowns of trees in woodland may almost touch, and the grass layer is rather sparse. Woodland is not extensive in Ghana, however, being restricted to isolated fringes and patches along the forest-savanna boundary; floristically, Ghanaian woodland is usually closer to savanna than to forest. Further features of forest are the prevalence of large woody climbers – lianes – and the complexity of the canopy, composed of crowns at different heights above the ground. Buttressing and cauliflory occur in some forest trees, but never in savanna. A thick woody or corky outer bark is characteristic of many savanna trees, whereas in forest trees the outer bark is generally thin.

This latter character is related to the incidence of fire. Occasional ground fires seem to be an important factor in certain Ghanaian forests (see p. 78), but in most savanna fierce fires are an annual occurrence.

1.1 The extent of Ghanaian forest

The boundary between forest and savanna is usually abrupt, being sharpened by the fires which run through the savanna, but may be complex. Because the forest is much richer in chlorophyll during the dry season than is savanna, it reflects much less red light. Thus the two kinds of vegetation are readily distinguished in the imagery (Fig. 1.1) obtained from the Landsat satellites, the first of which was launched in 1972. The outline of the forest zone in Ghana (Fig. 1.3 & 2.4) was obtained from such satellite images (see also Swaine, Hall & Lock 1976). The area of the forest zone in Ghana is about 82 000 km^2, representing 34% of the total area of Ghana.

The forest reserves and wildlife reserves shown in Fig. 1.3 are listed in Appendix 1. They have a total area of about 18 000 km^2, and thus cover 22% of the forest zone.

The integrity of reserves can be well seen by comparing Fig. 1.2 with 1.3 – the forest reserves stand out clearly in the Landsat image. Regular inspection of Landsat images would, in fact, be an excellent way of monitoring illegal forest clearing in inaccessible areas.

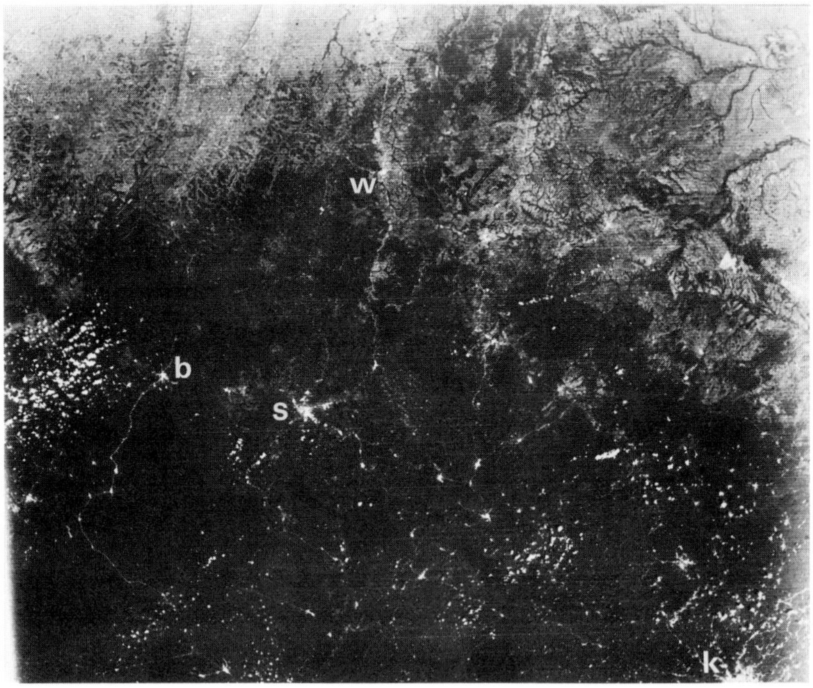

Fig. 1.1 LANDSAT-1 image (band 5 – red) of the forest-savanna boundary north-west of Kumasi on 25 November 1973. Forest appears darker than savanna. Note the complexity of the boundary – forest may penetrate savanna along streams, and savanna may penetrate forest along mid-slopes. Built-up areas and roads in forest appear white. b = Berekum; k = Kumasi; s = Sunyani; w = Wenchi.

Fig. 1.2 LANDSAT-1 image (band 7 – infra-red) of part of Brong-Ahafo and western Ashanti regions. Forest Reserves (compare with Fig. 1.3a) appear darker than surrounding forest. k = Kumasi.

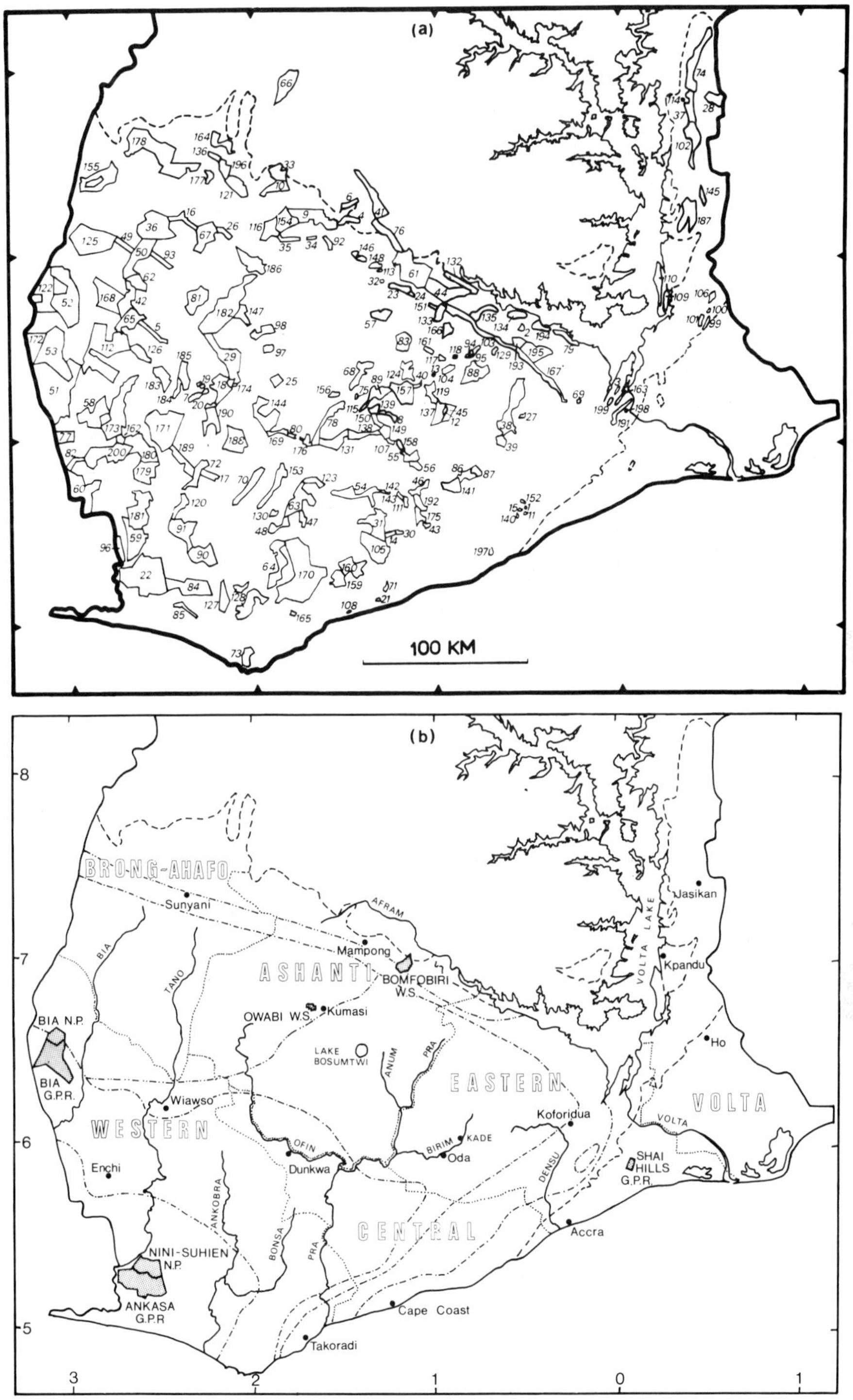

Fig. 1.3 Forest zone of Ghana. Forest-savanna boundary indicated by broken line.

(a) Forest Reserves – for key to numbers see Appendix 1(a).

(b) Wildlife Reserves (shaded): National Parks (N.P.), Game Production Reserves (G.P.R.), Wildlife Sanctuaries (W.S.) – see Appendix 1(b) for details.

Names of administrative regions in open capitals; boundaries dotted. Boundaries of forest-types – · · –; of subtypes – · · · –.

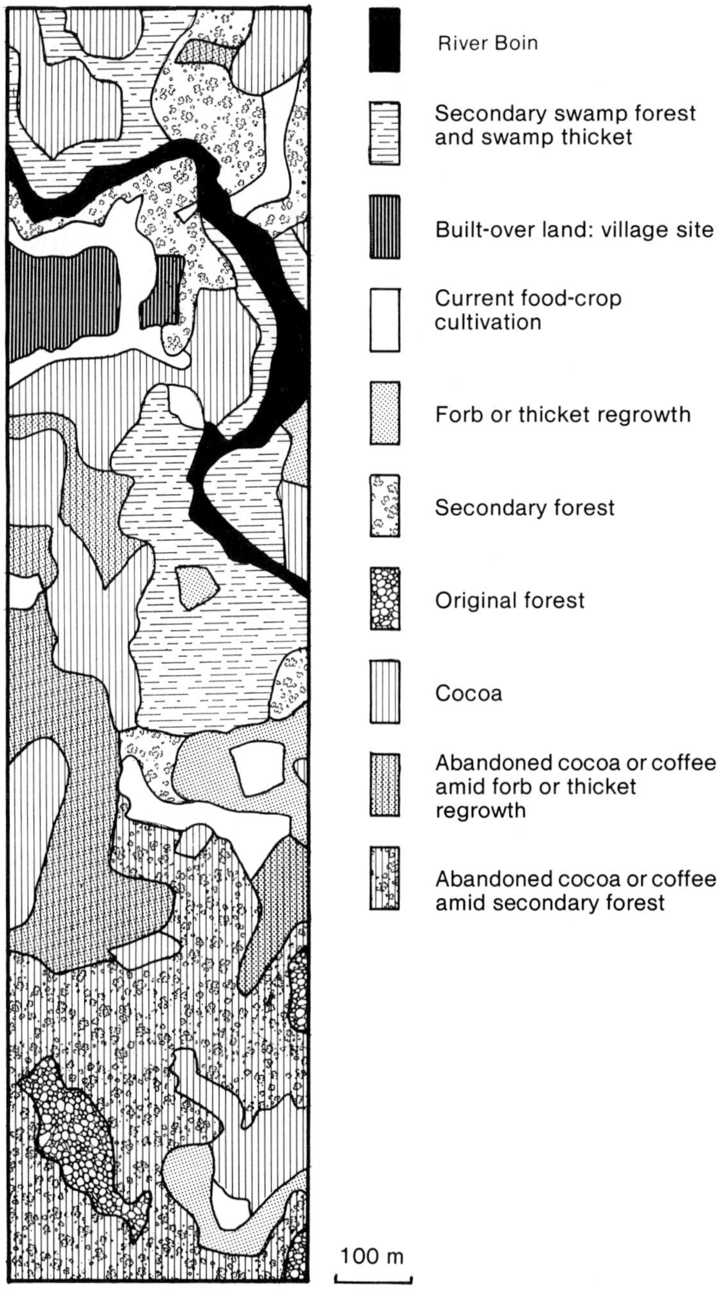

Fig. 1.4 Vegetation mosaic near Asemkrom just north-west of Nini-Suhien National Park (Fig. 8.15) in south-west Ghana (after Ahn 1958).

1.2 Secondary and swamp forest, and the concept of primary forest

The rate of forest clearing outside reserves is such that intact forest will soon have virtually disappeared except within reserves and in small patches of sacred forest near villages (p. 85). The proportion of forest outside reserves declined from 20% in 1955 to 5% in 1972 according to estimates in the annual reports of the Forestry Department for those years.

The countryside of the unreserved three-quarters of the forest zone is now a patchwork of man-influenced (secondary) vegetation types (Fig. 1.4). When forest is cleared for food farming (Fig. 1.5), a few large isolated trees may be left unfelled, but they are usually damaged when the brushwood is burnt and may gradually die and disintegrate. After a couple of years, the farm (Fig. 1.6) is usually abandoned, and the crop is swamped by coppice from cut trees together with tall herbaceous or sub-shrubby weeds (Fig. 1.7) such as *Solanum* spp.; various Marantaceae (e.g. *Marantochloa leucantha* (K. Schum.) Milne-Redh., *Thaumatococcus daniellii* (Benn) Benth.), Compositae such as *Melanthera scandens* (Schum. & Thonn.) Roberty, *Aspilia africana* (Pers.) C.D. Adams and *Eupatorium odoratum* Linn. (Hall, Kumar & Enti 1972), and grasses (e.g. *Setaria* spp., *Panicum maximum* Jacq., *P. laxum* Sw. and *Paspalum* spp.). Ahn (1958) distinguishes three stages in regrowth on old farms. The above stage ('forb regrowth') is followed by thicket with abundant woody climbers such as *Adenia* sp., *Acacia* spp., and *Entada scelerata* A. Chev., and young trees which shade out and kill the light-demanding weeds. The pioneer trees are quick-growing, short-lived species such as *Musanga cecropioides, Trema orientalis, Harungana madagascariensis, Ficus exasperata, Rauvolfia vomitoria* and *Macaranga* spp. (Fig. 1.8) which form the early phase of the third-stage secondary forest. Taller, longer-lived species such as *Funtumia africana, Albizia* spp., *Chlorophora* spp., *Terminalia ivorensis* and *Nauclea diderrichii* take over after 10–20 years to give older secondary forest (Fig. 1.10) from which the mature forest gradually develops. Ahn indicates differences between regrowth on newly-cleared forest and on land which has been repeatedly farmed. The numbers of secondary species within the major life-form categories are shown in Table 1.1.

Table 1.1 Species restricted to secondary forest and farms (i.e. 'secondary species').

	Non-climbing herbs	Climbers	Phanerophytes micro- & nano-	Phanerophytes mega- & meso-	Total
Secondary species	257	214	88	37	596
Total forest species in category	660	605	441	414	2 120
Percent secondary species	39	35	20	9	28

More shade trees are usually left standing over cocoa farms than over food farms; from a distance it may, in fact, be difficult to distinguish cocoa farms from intact forest. Much of the forest on hill slopes in the Volta Region, which appears intact from the valley, proves on closer inspection to be plantations of cocoa.

Swamp vegetation is also described by Ahn. *Raphia* palms, often dominant in swamps (Fig. 1.9), are a source of building material and of palm wine. Swamp is sometimes cleared for rice-farming.

Although less heterogeneous than the anthropogenic communities sketched above, the vegetation within forest reserves is by no means uniform. The accessible parts of most forest reserves have been lumbered at some time, and old logging tracks and the gaps left by felled trees may bear tangled thicket or secondary forest.

Gaps also result from natural tree fall, and these may be especially numerous where the soil is shallow (as on rocky hill summits) or unstable (as in swamps). In the drier forests, regrowth may be

Fig. 1.5 Forest cleared and burned for farming west of Bogoso, south-west Ghana; April 1977.

Fig. 1.6 New plantain farm close to site of Fig. 1.5.

Fig. 1.7 Shrubs and herbs growing on abandoned farm: the broad-leaved grass is *Setaria megaphylla;* climbers include *Melanthera scandens* and *Centrosema pubescens.*

Fig. 1.8 Musanga cecropioides, Alchornea cordifolia, and other trees of young secondary forest.

Fig. 1.9 Raphia swamp near Tarkwa, with *Ficus eriobotryoides* and *Anthocleista vogelii.*

Fig. 1.10 Older secondary forest with *Terminalia ivorensis. Ricinodendron heudelotii, Musanga cecropioides* and *Funtumia africana.*

slow, and thickety gaps consequently cover a high proportion of the forest area. The term 'primary forest' is often taken to imply forest which has never been subject to human interference. Two difficulties are presented, however, by this usage. Firstly, it is unlikely that such forest exists, at least in Africa; Jones (1955–56), for example, found abundant pottery in a Nigerian forest which had previously been thought of as primary. Secondly, and more fundamentally, the effects of natural disturbance by tree fall may not be readily distinguishable from the effects of disturbance by man. We therefore use the term 'primary' merely to imply only that the canopy is high, and more or less closed ('mature phase' forest in the sense of Whitmore 1975 and 1978). In contrast, 'secondary' forest has a very broken canopy, mostly low, and tangled undergrowth ('gap' and 'building' phases of Whitmore; 'chablis' phase of Hallé, Oldeman & Tomlinson 1978). By 'secondary' species, we mean species that are normally absent from 'primary' forest; 'primary' species are best developed in 'primary' forest, but may also be present in 'secondary' forest.

Forest regeneration is discussed more fully in Chapter 6.

CHAPTER 2

Variation and classification of forest

Seral and swamp communities such as those mentioned above were named and briefly described by Chipp (1927) and Taylor (1960). Such small-scale local variation is relatively easy to recognise and explain. Even the earliest forest investigations, however, revealed the existence of differences from one part of the forest zone to another in the structure and floristic composition of mature high forest. It is with this kind of variation that this book is principally concerned.

2.1 Previous classifications of closed-canopy forest

H. N. Thompson (1910) produced a sketch-map of forest zones (Taylor 1960, p. 34) based entirely on his own travels – several hundred miles, largely on foot during the first six months of 1908. He distinguished a wetter 'rain forest' from a drier 'monsoon forest', and a marginal zone said to be transitional between monsoon and savanna forest. Although the lines of demarcation are – understandably – incorrect, it is remarkable that he was able in so short a time to lay a good foundation for subsequent classification.

T. F. Chipp served as a forest officer in Ghana from 1910 to 1914 and again in 1921; much of his time was spent in botanical survey and collecting work. Chipp attempted to apply current Clements-Tansley (Tansley & Chipp 1926) notions of climax and succession to Ghana's forests, and devised a scheme (Chipp 1927) in which the wettest forest (Thompson's 'rain forest') was described as the climax: *Cynometra-Lophira* Association. Three successively drier forest types were named the *Lophira-Entandrophragma, Entandrophragma-Khaya,* and *Triplochiton-Piptadenia* preclimaxes respectively. In the text and in a sectional diagram, Chipp indicated that these types occur in sequence from the sea inland, in accordance with decreasing rainfall. Unfortunately his map (Chipp 1927, p. 49; and reproduced without comment by Taylor 1960, p. 35) bears very little relation to his text, and one can only presume that it was printed in error from some hasty and mis-copied diagram.

Chipp's successor as forest botanist was Chidlow Vigne who worked in Ghana from 1922–1948. He never produced a classification of forests, but his incomparable knowledge of the forest flora was used by St. Clair Thompson (1936), who divided the forest as follows: *Cynometra-Tarrietia* association (rain or evergreen forest); *Lovoa-Guarea-Pentadesma* association (wet mixed deciduous forest); *Khaya-Entandrophragma* association (moist mixed deciduous forest); *Triplochiton-Cistanthera* (i.e. *Nesogordonia*) association (dry mixed deciduous forest). St. Clair Thompson did not attempt to map these associations, but he did provide lists of indicator species.

Taylor's (1952) classification (Fig. 2.1a), based on the ideas of Chipp, Vigne and St. Clair Thompson, found general acceptance, although it was criticised by Foggie (1957) on the grounds that Taylor's associations are not sufficiently discrete. Further support for the classification is provided in a book on Ghana's forest trees (Taylor 1960). It has provided a useful framework for planning forest

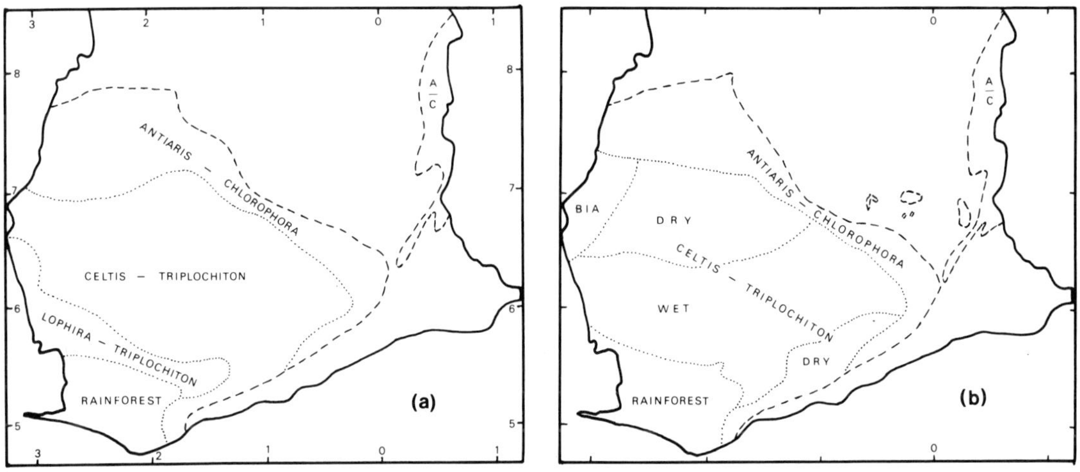

Fig. 2.1 Earlier classifications of Ghanaian forest:

(a) TAYLOR 1952

Rainforest
Cynometra-Lophira-Tarrietia Association

Moist Semi-deciduous forest
Lophira-Triplochiton Association
Celtis-Triplochiton Association
Antiaris-Chlorophora Association

(b) MOONEY 1959

Rainforest
Moist Semi-deciduous forest
Celtis-Triplochiton Association
(i) Wet Facies
(ii) Dry Facies
(iii) Bia Association
Antiaris-Chlorophora Association

– – – – = forest-savanna boundary; . . . . . . = forest-type boundaries.

management. A slightly different scheme (Fig. 2.1b) was presented by Mooney (1961) but this seems to have been largely ignored.

The main defect of Taylor's (and *a fortiori* Mooney's) scheme is its apparent arbitrariness. The only mention of the method by which it was reached is a note (Taylor 1960, p. 35): 'This has been prepared from personal reconnaissances and from Forestry Department records, mainly enumeration surveys'. Because neither his methods of treating data nor his criteria for classification are defined, it is not possible to suggest improvements or to incorporate new data such as might be produced by subsequent enumerations. Taylor neglected the driest forests, perhaps because they had not been enumerated, and one cannot say where they should be placed in his scheme. The reason for this failure of definition is Taylor's adherence to the Anglo-American Clements-Tansley school of phytosociology, in which communities are classified according to 'dominant' species. In the mixed tropical forest of Ghana the canopy is usually formed by a fluctuating mixture of species none of which is dominant.

The continental Zürich-Montpellier or Braun-Blanquet approach to vegetation classification (see Westhoff & van der Maarel 1973 and Werger 1974a, 1974b, for up-to-date accounts), in which associations are defined by 'characteristic' species which may or may not be dominant, was used in Ivory Coast by Mangenot and his associates (Mangenot 1955; Emberger, Mangenot & Miège 1950). They made lists of all species present on small, fairly homogeneous sample plots, usually 100 m^2, and grouped these lists to define their associations. They probably exaggerated the distinctness of these communities, but the criticisms of Aubréville (1950–1) seem too severe in view of the comparative objectivity of their approach. Subsequent phytosociological work in Ivory Coast (Guillaumet & Adjanohoun 1971), has extended the scheme of Mangenot without fundamentally altering it.

Nevertheless, there are good grounds for doubting that the Zürich-Montpellier methods are appropriate to species-rich tropical forest (Schnell 1976, 1:170). Merely listing all the species present in a stand that is anywhere near a 'minimum area', however defined, will usually take several hours and if importance and sociability are also to be meaningfully assessed stratum by stratum, hours will stretch into days. Species lists (relevés) have to be juggled in species/stand tables, and this presents great difficulties where the number of species runs into several hundreds. It is probably significant that Mangenot never presented any organised phytosociological tables, on which Zürich-Montpellier associations and higher categories should properly be based.

2.2 Our classification

For our classification of Ghanaian forest (Hall & Swaine 1976), we followed Mangenot in using small sample plots. Our 'A' samples measured 25 m by 25 m, and on these all species present were listed: shrubs, climbers, herbs, saplings and seedlings. Being interested primarily in variation on a regional rather than local scale, we sited the plots in forest as mature and intact as could be found in the area to be sampled. The canopy had to be closed; vegetation on swampy ground, or with an abundance of secondary species, or containing crop plants, or with savanna grasses, was avoided. Most samples were located in forest reserves, but we took advantage of the small plot size to sample sacred groves and small forest fragments, especially in areas lacking reserves. In addition to the species lists, we recorded for each plot the names and numbers of trees and lianes equal to or exceeding 10 cm diameter at breast height, the height and name of the tallest tree in or near the plot and the basal area (using a Spiegel Relaskop). Greig-Smith (1969) has shown that, in species-rich vegetation, presence alone gives as much information as can usefully be employed in the numerical analysis of the lists which we intended to use. Accordingly we made no attempt to assess the abundance of anything other than the woody stems. Aspect and slope were recorded, notes were made on the general appearance of the vegetation, and fifteen soil samples were collected from the upper 15 cm of soil, bulked and mixed.

A total of 155 A-sample plots was recorded during the course of the survey. These were supplemented by 130 B-samples, much quicker to record, in which we merely listed the first 30–60 species encountered; (see Fig. 2.2 and Appendix 2 for detailed locations of samples).

Altogether 1 248 species were recorded in the 155 A-sample plots. The number found in each plot ranged from about 20 in the driest to about 200 in the wettest forests. The aim of analysis of such data is to arrange the plots in such a way that plots with similar floristic composition are placed close together; species should also be arranged so that those with similar distributions are placed close together. This process is known as ordination. Species found in only one plot clearly cannot be used in comparisons between plots; those found only in two or three plots are scarcely more useful. After eliminating all such species we were left with 749 species that were found in four or more plots. It was this matrix of 155 plots by 749 species that was subjected to ordination.

Various methods are available for ordination; we chose 'reciprocal averaging' (Hill 1973) because it is little influenced by rare species and because the species and plot ordinations are directly comparable; the value of this latter property is explained later (p. 16). The computer program (which was run in the Computing Laboratory of the University College of North Wales) was allowed to proceed until six axes for plots and six for species were extracted. Values on each axis are scaled to range from 0 to 100.

Up to this point the analysis is completely objective. Subsequent use of the axes depends on the aims and experience of the ecologist. Fig. 2.3 is the diagram obtained when plots are graphed according to their ordination scores on axes 1 and 6. We selected this combination of axes because it gave the best scatter of points, and the points were arranged in a meaningful way. Neither axis reflects change in a single environmental variable. Axis 1 is mainly responsive to the environmental moisture gradient: the forest plots under the highest rainfall have the lowest axis 1 values (mean species score less than 20) and the driest forests have the highest axis 1 values, (mean species score more than 65). Between axis 1 values of 20 to 40, however, there is no clear relationship with moisture. In this range of axis 1, axis 6 distinguishes three groups of plots: the 'wettest' lowland plots with low values on axis 6; lowland plots of lower rainfall with intermediate axis 6 values; upland plots with high axis 6 values. In the axis 1 range of 45–60 there is a marked

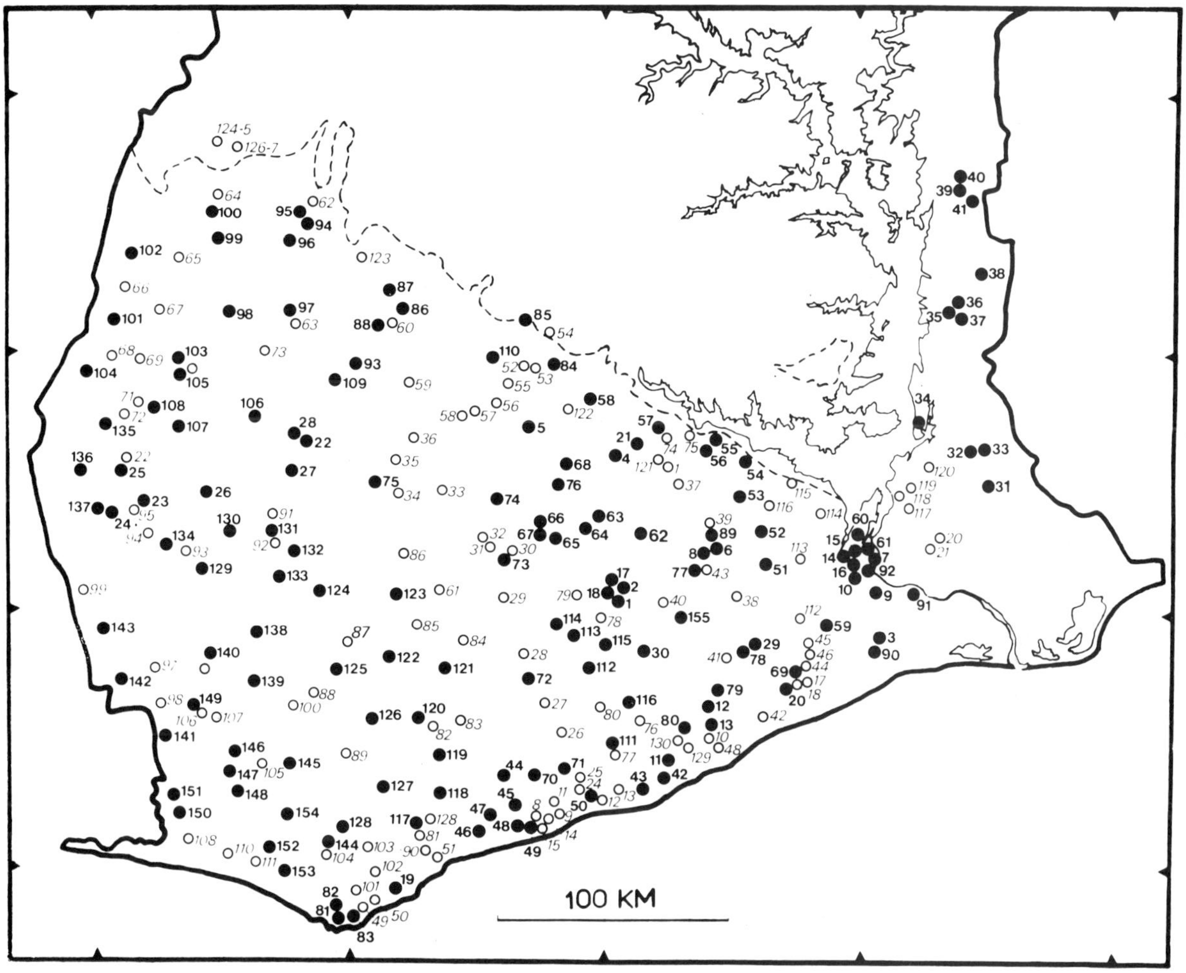

Fg. 2.2 Locations of plots recorded in survey of forest by Hall & Swaine (1976); A-samples – bold numbers and solid circles; B-samples – italic numbers and open circles. For details of plot positions see Appendix 2.

tendency for forests on granite to have higher axis 6 values than forests on shales and quartzites.

A glance at the diagram will show that the points do not, however, always form discrete clusters, suggesting that variation in floristic composition of forests is more or less continuous. It is now widely recognised, in fact, that such continuous variation is the norm in vegetation. It does not follow, however, that division of the ordination to yield a classification must be pointless, or entirely arbitrary. For practical purposes it is often necessary to employ a limited number of types, about which generalisations can be made, even when intermediates between types are recognised to exist. The manner in which division is made will depend partly on the purpose of classification, and partly on the subjective features of point distribution.

In dividing the area of the ordination diagram (Fig. 2.3) into seven portions, each defining one forest type, we used the following criteria: compact clusters of points should not be divided; lines should pass through gaps between point clusters; lines should be straight; types should occupy geographically coherent areas, and correlate with ecological factors, such as rainfall, which are known to be of ecological importance. The distribution of these seven types in the Ghana forest zone is shown in Fig. 1.3b & 2.4.

The B-samples were not used in the analysis which yielded the ordination diagram, but were

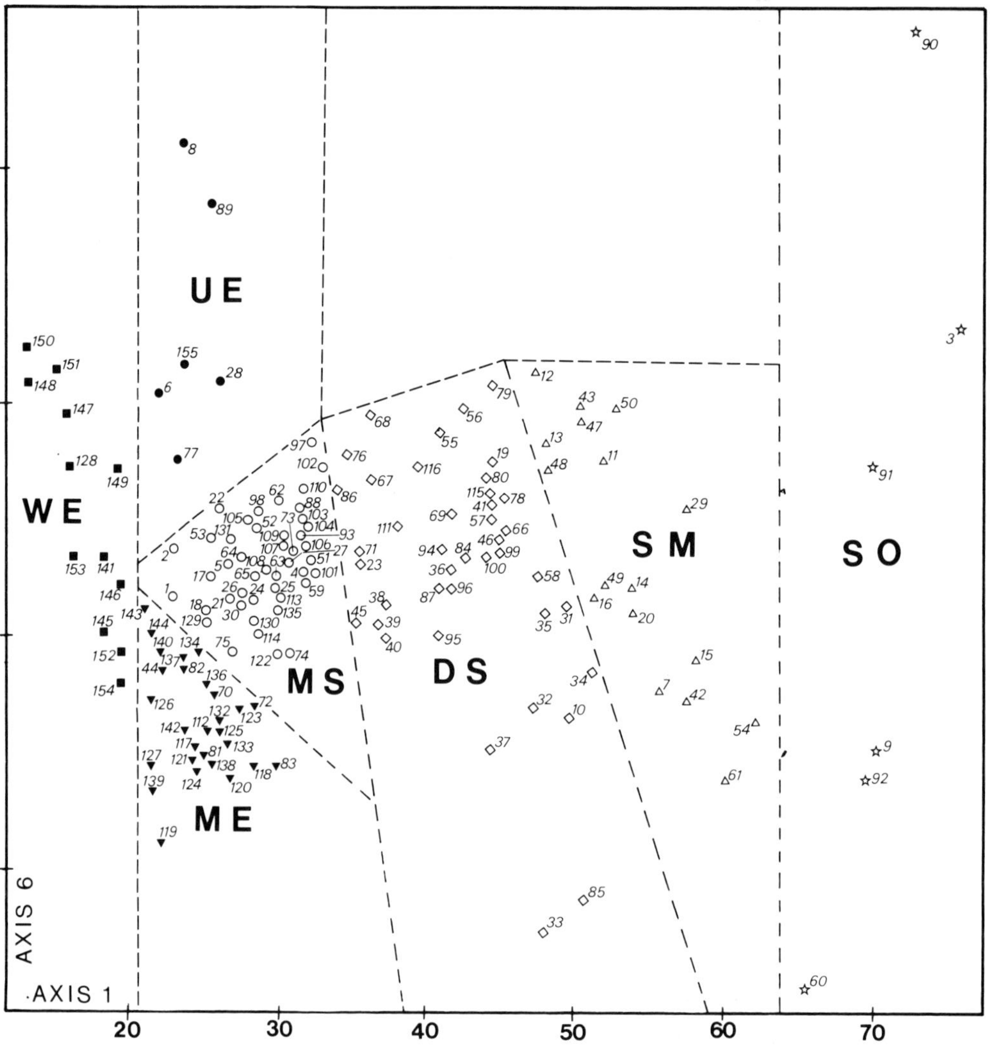

Fig. 2.3 Ordination of A-samples partitioned into forest types: WE = Wet Evergreen; ME = Moist Evergreen; UE = Upland Evergreen; MS = Moist Semi-deciduous; DS = Dry Semi-deciduous; SM = Southern Marginal; SO = South-east Outlier. Sample numbers refer to Fig. 2.2 and Appendix 2. Axis values are expressed as means of species ordination scores; (see Appendix 4 for explanation).

identified (Swaine & Hall 1976) by finding the mean of the species ordination scores (listed in Appendix 3), on axes 1 and 6. These mean scores are related linearly to plot ordination scores (because 'reciprocal averaging' was used), and therefore give the appropriate plot position on the ordination diagram. A worked example of plot identification from species scores is given in Appendix 4.

Comparison of Fig. 1.3b with Fig. 2.1a shows that our map of forest-types is not greatly different from that of Taylor, except that we have been able to distinguish two kinds of forest in the zone which Taylor calls 'Coastal scrub and grassland', and subtypes have been recognised in Taylor's *'Antiaris-Chlorophora'* and *'Celtis-Triplochiton'* associations. The main difference is that our methods are defined and repeatable, and the map can be improved as further plots are included. The only way in which a forest can be identified to association in Taylor's scheme is by determination of its geographical position. This may be satisfactory to a forester, who is interested in large areas of average forest, but a biologist may wish to identify small patches of forest which seem to differ from that in the surrounding area.

Although most of the plots shown in Fig. 2.2 are

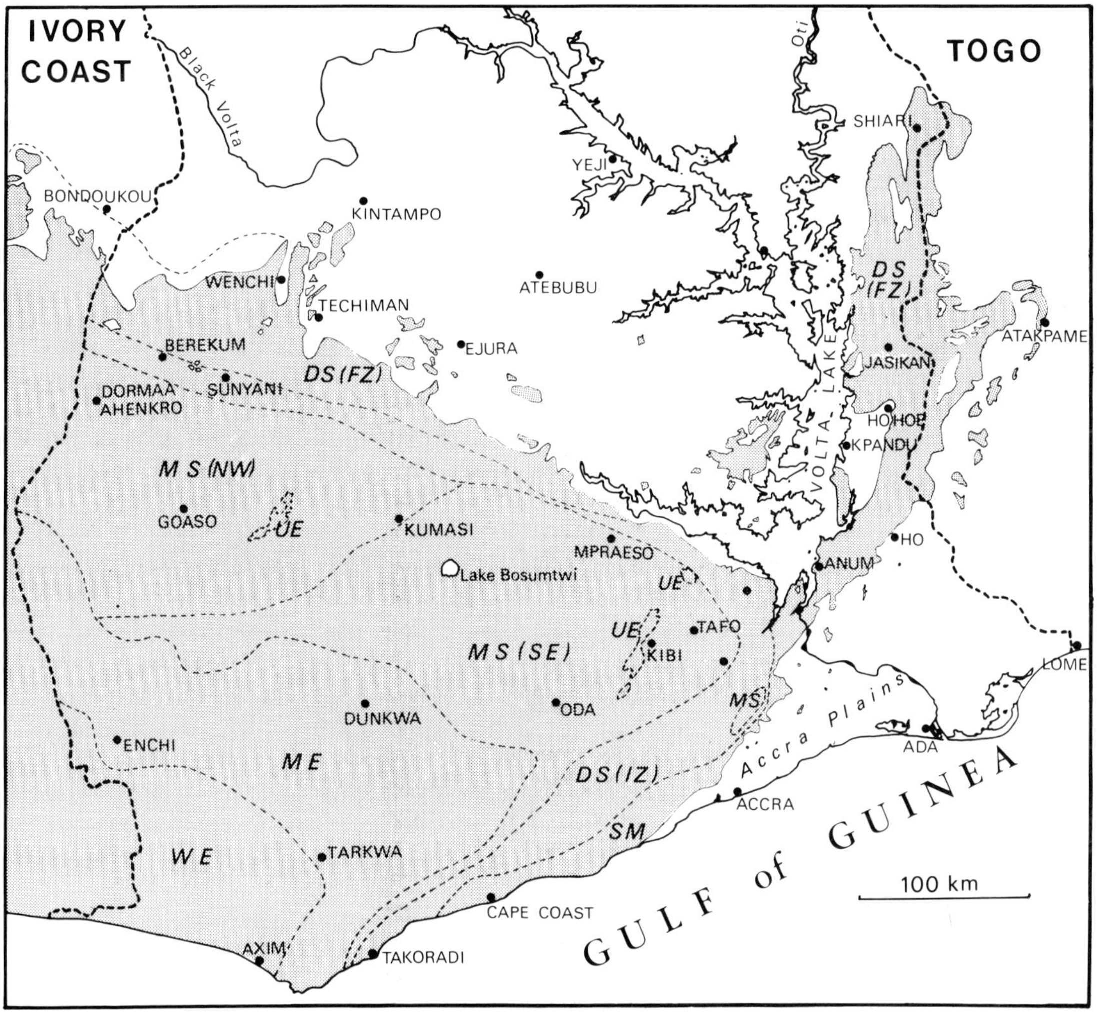

Fig. 2.4 Distribution of forest types (Hall & Swaine 1976) in southern Ghana. Forest is stippled. Forest-type and subtype boundaries shown by broken line (- - -). Forest-type abbreviations: WE = Wet Evergreen; ME = Moist Evergreen; MS = Moist Semi-deciduous (NW = North-west subtype; SE = South-east subtype); DS = Dry Semi-deciduous (FZ = Fire Zone subtype; IZ = Inner Zone subtype); SM = Southern Marginal. Limits of the Volta Lake and forest margin mapped from LANDSAT imagery.

of the type appropriate to the zone in which they lie, there are interesting exceptions. A-samples 66 and 67, situated in forest on shallow soils on the rocky top of Gyedemoa Hill, belong to a type drier than that of the surrounding lowland. A-sample 29, on the side of a granite inselberg, is of a drier type than A-sample 78 at its foot, for the same reason.

Two kinds of name have been widely used for tropical forest-types – those incorporating important physiognomic and environmental features, and those based on characteristic species. Both have disadvantages. The drawback of the first system is the limited number and relative nature of available terms. A forest with rainfall considered high enough to justify the epithet 'wet' in one part of the tropics may be considered 'dry' in another. Deciduousness is difficult to assess, and varies in its completeness. The Clements-Tansley school named associations according to dominant species. Foggie (1957) is thus justified in criticising Taylor (1952), who adhered to this school, for naming his associations after trees which were neither dominant nor, in general, restricted to the associations so named. The Zürich-Montpellier school defines associations by the table of species and stands, rearranged to bring together characteristic species. Frequently the name for the association is obtained by hyphenating the roots of the generic names of two of the species occurring in the relevés classified in that association and adding the suffix '-etum'. The suffix can alternatively be added to the root of just one generic name. These species usually (but not necessarily) are characteristic or conspicuous species. They need not be present in any particular stand of the association, though often they will have been recorded in the type relevé. An alliance (grouping several associations) is named by attaching the suffix '-ion' to the root of the generic name of a species, again usually a characteristic or conspicuous one, and in many instances the same species used in the type association's name. Order and class names (superior ranks in the hierarchy) are built up in the same way, using the suffixes '-etalia' and '-etea' respectively (Barkman, Moravec & Rauschert 1976). This system has been extensively used in francophone Africa for naming forest-types (e.g. Schnell 1952; Lebrun & Gilbert 1954; Mangenot 1955). Our species-poor South-east Outlier forest-type was named the *Diospyrion abyssinicae* by Jeník & Hall 1976. The Zürich-Montpellier system has been almost universally accepted in continental Europe, but we feel for the reasons given above (p. 13) that it cannot be successfully applied to species-rich mixed tropical forest. Our forest-types, if they were to be named by this system, would certainly deserve a hierarchical position above that of association, but debate as to whether they represent alliances or orders would seem rather futile, especially in the light of their admitted inter-gradation.

Somewhat reluctantly, then, we have decided to use physiognomic-environmental names for our types even though the types are defined floristically. It should be borne in mind, however, that each is understood to be preceded by 'Ghanaian'. Thus 'Ghanaian Wet Evergreen forest' may have certain features in common with 'Costa Rican Wet Evergreen Forest', but no close similarity is implied. The connotations of the epithets used in some of the names are explained below – 'wet' (p. 19), 'upland' (p. 95), 'evergreen' and deciduous' (p. 51). These five types (Wet Evergreen, Moist Evergreen, Upland Evergreen, Moist Semi-deciduous, Dry Semi-deciduous) are all well-known in West African countries other than Ghana. The remaining two types – Southern Marginal and South-east Outlier – have been given geographical names because they have not yet been recognised outside Ghana.

These types are described in turn in Chapter 8 (p. 71). In the intervening chapters we consider the various aspects of the forest as a whole.

CHAPTER 3

Forest environment

The distribution of forest-types is largely determined by a complex of interacting environmental factors of which climate, geology and soils are the most important. Rainfall has direct effects on moisture availability and indirect effects on soil nutrient levels. Soil properties also depend directly on the nature of the parent rock from which the soil is derived by weathering, and indirectly on geology through its influence on relief. It is therefore not to be expected that any single factor in isolation would show perfect correlation with forest-type limits.

3.1 Climate

A glance at the ordination diagram (Fig. 2.3) shows that axis 1 of the ordination is mainly concerned with moisture; the relationship is clearly demonstrated in Fig. 3.1. At the extreme left, with low axis 1 values, is the Wet Evergreen type, while the very dry Outlier type has the highest axis 1 values. Fig. 3.2 shows mean annual rainfall in and near the forest zone, based on the 170 stations from which complete records were obtainable for at least the five years preceding 1963. Comparison with Fig. 1.3 shows a fairly good correlation of the 1 750 mm isohyet with the limit of the Wet Evergreen forest-type, and of the 1 250 mm and 1 000 mm isohyets with the limits of the Southern Marginal type. Another correlation is of the 1 500 mm isohyet with the boundary between the South-east and North-west subtypes of the Moist Semi-deciduous forest-type.

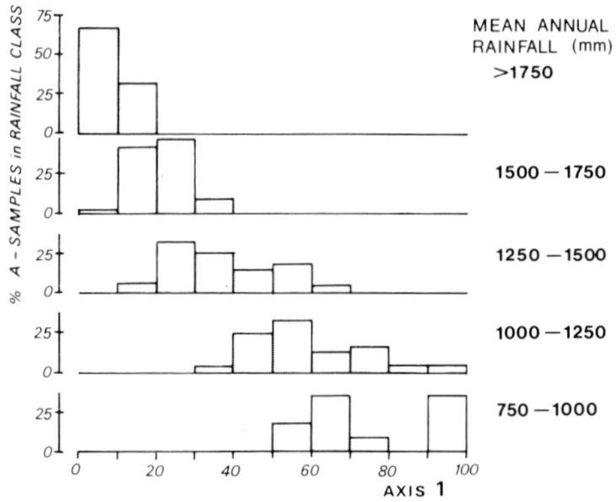

Fig. 3.1 Relation of axis 1 of the ordination (Fig. 2.3) with mean annual rainfall. A-samples were allocated to rainfall classes from their positions on Fig. 3.2.

The high total rainfall in south-west Ghana results from higher monthly totals in the rainy season, but the total length of the two dry seasons is similar (4–5 months with mean rainfall <100 mm) to that in the Moist Evergreen and South-east subtype of the Moist Semi-deciduous forest-types. Throughout the forest zone, the rainfall pattern is of the two-peak type with maxima in May-June and September-October. Potential evapotranspiration is also important, but the number of stations yielding data from which it may be calculated is much less than that for rainfall. Of the various formulae for the calculation of potential evapotranspiration, that of Papadakis (1966) is the simplest, being a measure of mean saturation deficit multiplied by a constant

$$E = 0.5625\ (e_{ma} - e_d)$$

where E is monthly (Papadakis) evapotranspira-

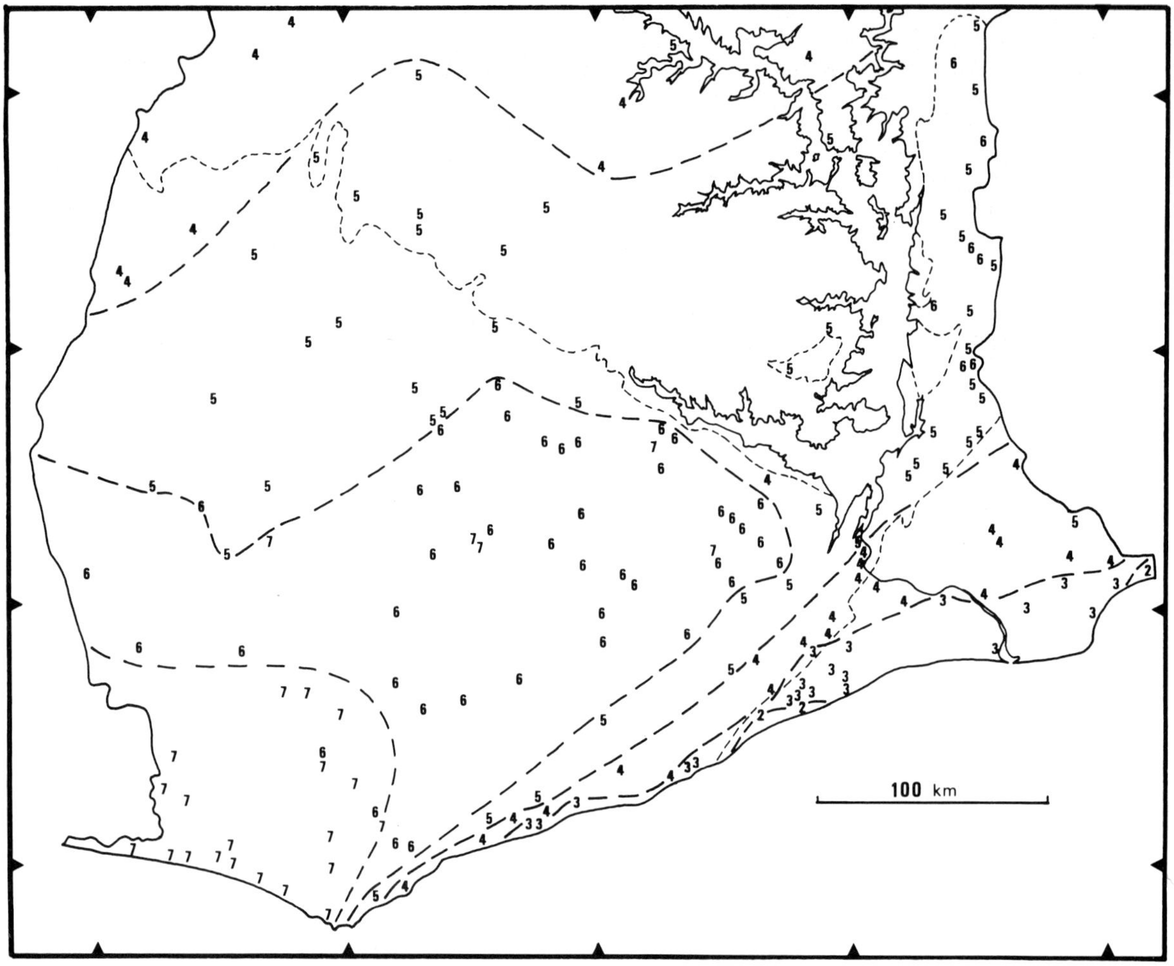

Fig. 3.2 Mean annual rainfall for all stations with five or more consecutive records prior to 1963. Isohyets (bold broken lines) are shown at 250 mm intervals. Numbers indicate mean annual rainfall within the following ranges: *2:* 500–750 mm; *3:* 750–1 000 mm; *4:* 1 000–1 250 mm; *5:* 1 250–1 500 mm; *6:* 1 500–1 750 mm; *7:* >1 750 mm. Forest-savanna boundary shown by fainter broken lines.

tion in centimetres; e_{ma} is saturation vapour pressure at the monthly mean daily maximum, and e_d is monthly mean vapour pressure, both in millibars.

Figures obtained by this method are too low in humid areas and too high in dry areas, as compared to values for pan evaporation, but nevertheless give differences from and ratios with precipitation that correlate quite well with vegetation. Papadakis defines a dry month as one in whch precipitation is less than half of potential evapotranspiration. By this criterion stations in Moist Evergreen forest have no dry months, and the forest zone as a whole has two or fewer dry months and an annual precipitation : potential evapotranspiration ratio greater than unity. Papadakis potential evapotranspiration is 750 to 1 000 mm along the coast, 1 000 to 1 250 mm in most of the forest zone, 1 250 to 2 000 mm in the Guinea Savanna zone and up to 2 250 in the Sudan savanna zone near the northern border, where the harmattan is most intense.

Temperature variation in the forest zone is rather slight; the mean monthly maximum in the hottest month (February or March) is 31–33 °C, and the mean monthly minimum in the coldest month (December–January in the northern part and August in the south) is 19–21 °C. The average daily range is 8–9 °C, and the seasonal range of daily mean temperature is 3–4.5 °C. Mean temperatures on the summits of the highest hills (700 m) in the forest zone may be 3–5 °C lower than those of the surrounding plains.

Anyone who has flown from Accra to Tamale

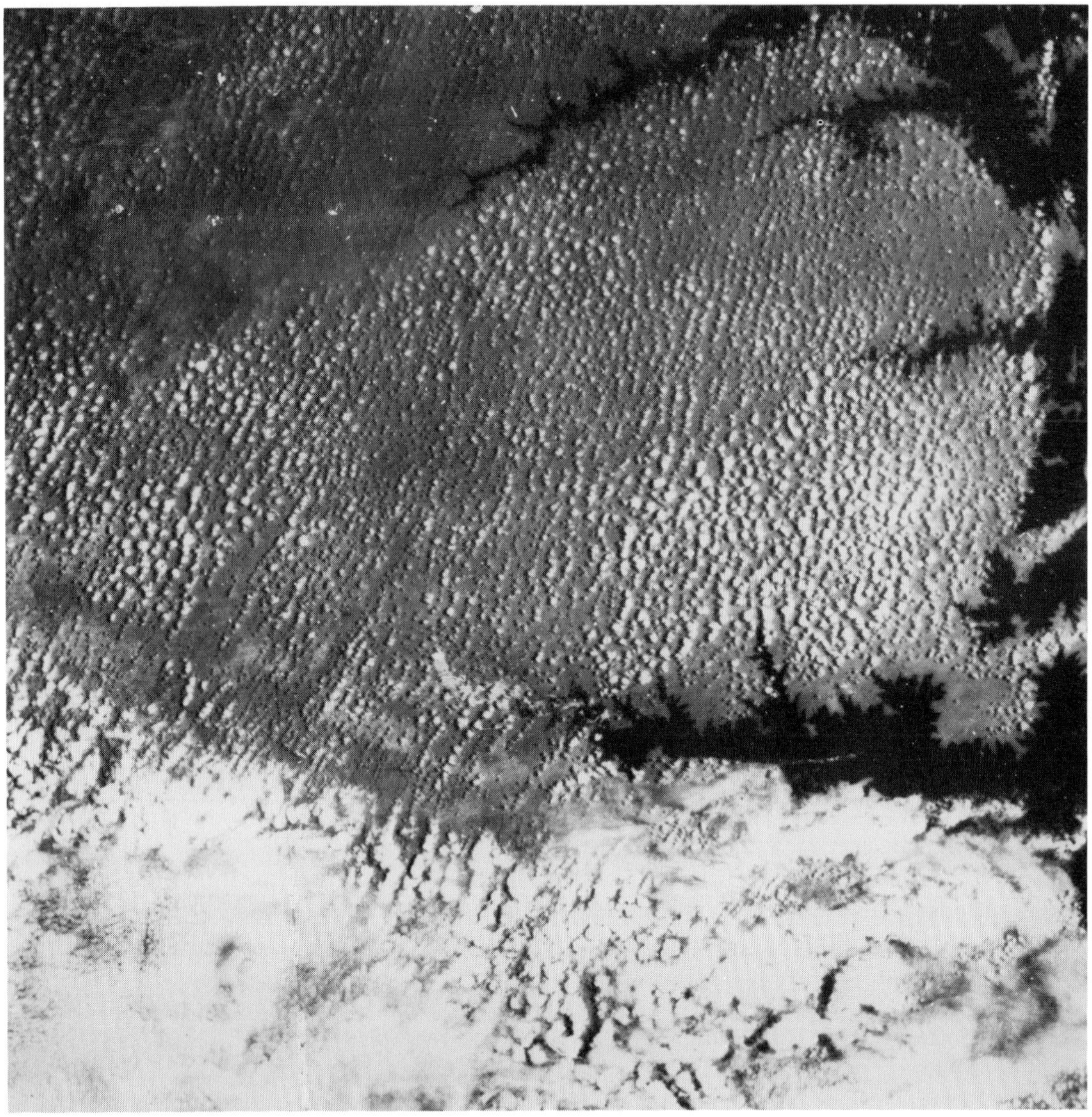

Fig. 3.3 LANDSAT-1 image (band 7 – infra-red) of 18 September 1972 at 0930 hours. The Volta Lake appears black; the Afram arm is seen at lower right. Clouds are white. Note deep cumulus over forest zone (cf. Fig. 2.4), virtual absence of cloud over open water, thin cloud (aligned in the direction of prevailing wind) over savanna, and heavier cloud over forest outlier in the south-east Afram Plains.

will realise that cloudiness is much higher (Fig. 3.3) and hours of bright sunshine much lower in the forest zone than in savanna. Seasonal variation in these parameters in relation to forest phenology is discussed by Gibbs & Leston (1970). Cloudiness appears to be greater over hill ranges, although exact data are not available.

3.2 Geology and relief

The main geological features of southern Ghana (Bates 1962) are shown in Fig. 3.4. Two major zones may be distinguished: the folded or titled Middle Precambrian rocks in the south, west and eastern margin which trend from south-west to north-east

Fig. 3.4 Main geological formations of southern Ghana. See Fig. 2.4 for distribution of forest vegetation.

and the level-bedded late Precambrian to early Palaeozoic (Voltaian) rocks of the northern centre, which lie unconformably on the Precambrian basement. Smaller areas of more recent sediments cover the extreme south-east and south-west corners of Ghana.

The relationships of the Precambrian rocks are still a matter of dispute (Brunnschweiler, Dempster & Kusnir 1972), but from a descriptive point of view the Birrimian and Tarkwaian rocks are strongly folded and more or less metamorphosed sediments and volcanics associated with granitic rocks, while the Dahomeyan consists of tilted gneisses. The Buem and Togo rocks are folded and metamorphosed sediments and volcanics forming a narrow strip between the Birrimian and Dahomeyan provinces. Relief is shown in Fig. 3.5. Most of the Birrimian province bears low, undulating country, but narrow south-west to north-east trending ranges of hills occur, up to 700 m high, mostly formed by the volcanic-rich Upper Birrimian rocks and often capped with bauxite. The hills of the Atewa Range and Tano Ofin forest reserves, which bear the Upland Evergreen forest-type, are of this kind. Hills also occur in the granite; these do not reach elevations as great as the Upper Birrimian hills, but they may be steep-sided, and, especially in the drier parts of the forest zone, soils may be shallow and the rock partly exposed. Nyanao Hill near Nsawam and the hills of the Dome River Forest Reserve are good examples.

The Dahomeyan rocks in Ghana comprise two main types. A belt of dark-coloured basic gneiss runs up the centre of the Dahomeyan area, and abrupt rocky hills up to 400 m high (see Fig. 8.10) – Shai, Krobo, Osudoku, and Adaklu – are constructed of this rock. The first three of these hills, in areas with mean annual rainfall of 800 to 1 100 mm, bear forest of South-east Outlier type, whereas the fourth, with rainfall of about 1 300 mm, is capped with Dry Semi-deciduous forest. Belts of paler, acidic gneiss, forming a flatter landscape which is devoid of forest, lie on either side of the basic gneiss belt.

The landscape developed from Buem-Togo rocks consists of steep-sided ridges ranging in altitude from 300 m in the south to 800 m in the north, dominated by quartzites and sandstones, alternat-

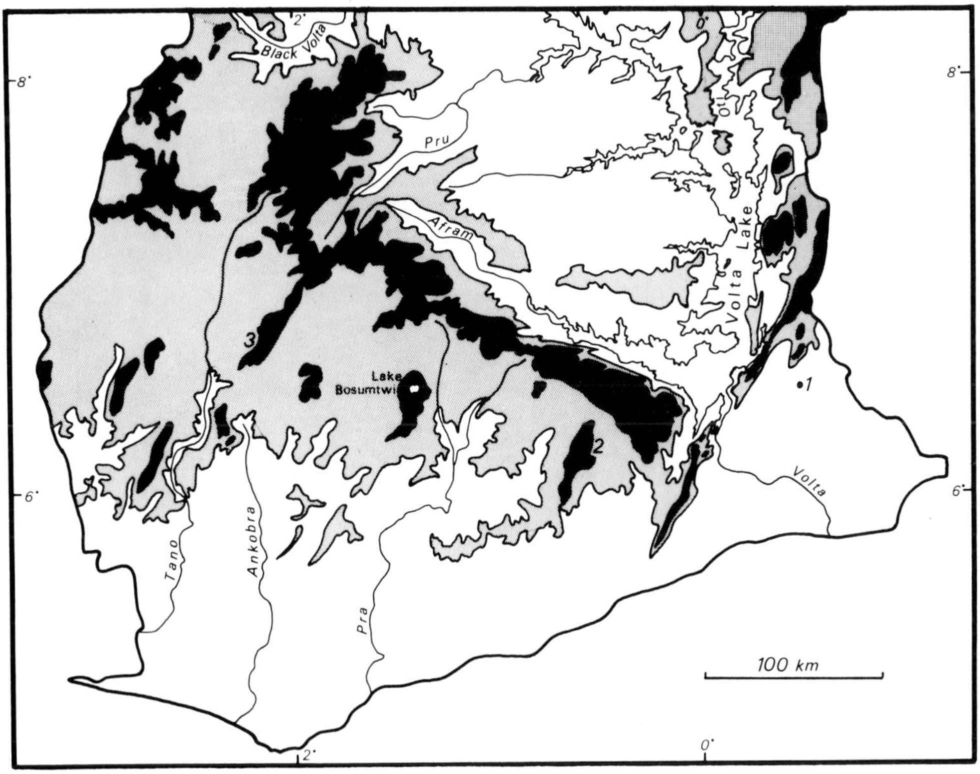

Fig. 3.5 Relief and major rivers in southern Ghana. Land above 300 m altitude black, 150–300 m stippled, below 150 m white. 1: Adaklu Hill; 2: Atewa Range F.R.; 3: Tano Ofin F.R.

ing with narrow valleys cut in the softer shales. The hill-slopes and small valleys on the ranges to the east of the Volta Lake bear Dry Semi-deciduous forest; the broader valleys are usually covered by savanna, as are some of the hill-summits (Jeník & Hall 1966). The range south of the Volta Lake is subject to a very steep rainfall gradient. The drier slopes abutting the Dahomeyan bear Southern Marginal or even South-east Outlier forest, while the highest part of the ridge, with higher rainfall and much fog near Aburi, bears Moist Semi-deciduous forest. Dry Semi-deciduous forest occupies sites intermediate in moisture. The road from Koforidua to Kumasi runs below the impressive sandstone cliffs which form the south-western edge of the Voltaian basin. Above the cliffs are isolated bauxite-capped peaks up to 700 m high, and a plateau at about 500 m which falls on its north-eastern side to the low plains of the central Voltaian basin. This high sandstone rim of the Voltaian bears forest of Moist Semi-deciduous type on the south-west flank, which faces the rain-bearing winds, and Dry Semi-deciduous forest on the north-east flank, in the rain shadow. The plains, underlain by Voltaian mudstones and shales, are savanna-covered.

Within most of the forest-zone, parent rock is buried under great depths of weathered material and seems to have no recognisable influence on forest-type. At the driest end of the forest spectrum, South-east Outlier forest occurs both on base-rich gneiss (under low rainfall) and on base-poor Togo series quartzites and shales (under higher rainfall). Within the Dry Semi-deciduous and Southern Marginal types, however, there is evidence that floristic composition of the forest may be influenced by rock substrate; for further discussion see p. 85. Where rock reaches the surface, and soil is consequently very shallow, forest may be of a type drier than that on surrounding deeper soils (p. 39).

3.3 Soils

Like vegetation, soil type is determined by a complex of environmental factors of which climate, nature of bed-rock, topography and drainage, and vegetation itself are the most important, and also by the length of time during which it has developed. The soils of Ghana have been classified in a hierarchical system devised by C. F. Charter (Brammer 1962) according to which the orders – the highest level in the hierarchy – are named by the two leading genetic factors. Most forest soils belong to the Climatophytic Earths order, for which climate and vegetation are the dominant factors. The other orders are Topohydric Earths (topography and drainage), and Lithochronic Earths (parent rock and age). This classification thus accepts the belief, expressed in the preceding paragraph, that the effects of climate on forest soils generally override those of parent rock except where soils are shallow and immature.

The lowest level in the hierarchy is the soil series, consisting of profiles which are essentially similar in their constituent horizons as defined by colour, humus content, texture, nature of clay fraction, pH, and parent rock. In most parts of the forest zone where the land is undulating or hilly, the soil series change with position in the topography, forming a catena. Corresponding to the soil may be vegetation catenas. These are well-known in savanna, and have also been recognised in forest (Lawson, Armstrong-Mensah & Hall 1970), but the differences between forest at different levels on a well-drained slope are slight compared to the difference between these and the swampy valley bottom (Fig. 3.6). Hall & Swaine (1976) found that the upper- and middle-slope forests compared by Lawson et al. 1970 were placed close together in their ordination of the whole forest zone.

The group of series which together form a catena under the same conditions of climate and parent rock constitute a soil association – the next highest level in the hierarchy. Series occupy small scattered

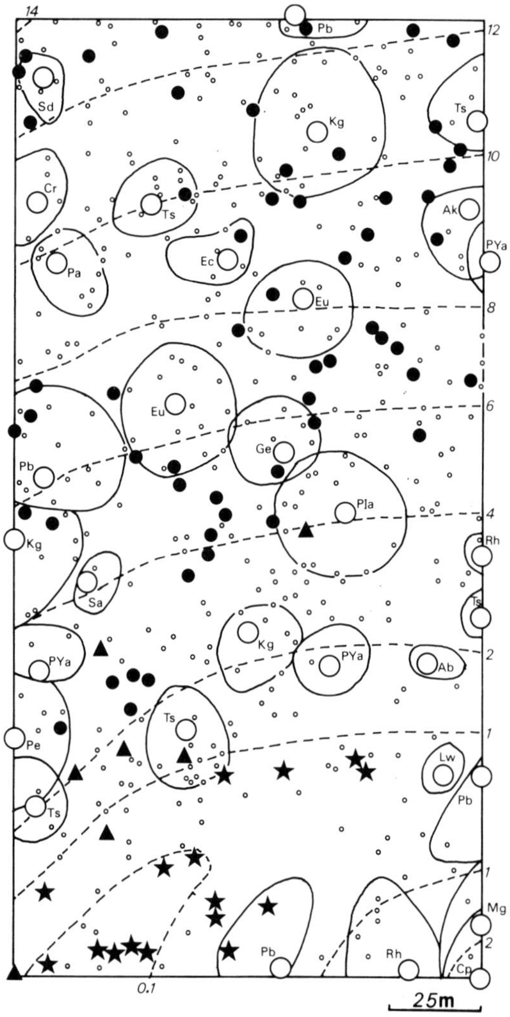

Fig. 3.6 Distribution of trees ⩾0.5 m gbh in 2 ha of primary forest including a swampy depression and the hillside above it (contours in metres) in Bia National Park (see Fig. 1.3b for position). Catena effects are demonstrated by *Raphia hookeri* palms (black stars) in swamp, *Hexalobus crispiflorus* (black triangles) on lower slopes, and *Corynanthe pachyceras* (black dots) on middle and upper slopes. The crowns of all trees ⩾2 m gbh (large open circles) are mapped and named with the following abbreviations.

Abbreviations
Ab = Alstonia boonei
Ak = Anopyxis klaineana
Cp = Ceiba pentandra
Cr = Chlorophora regia
Ec = Entandrophragma cylindricum
Eu = E. utile
Ge = Guibourtia ehie
Kg = Klainedoxa gabonensis
Lw = Lannea welwitschii
Mg = Maranthes glabra
Pb = Parkia bicolor
Pe = Parinari excelsa
Pla = Piptadeniastrum africanum
PYa = Pycnanthus angolensis
Rh = Ricinodendron heudelotii
Sa = Stereospermum acuminatissimum
Sd = Samanea dinklagei
Ts = Terminalia superba
All other trees shown by small open circles.

patches or belts along the contours and can only be mapped at large scales of about 1:10 000. The association is the smallest unit that can be mapped at the more usual scales of 1:100 000 or 1:250 000. Ahn 1961 presents a map of soil associations in south-west Ghana at a scale of 1:250 000. Although such maps are doubtless useful to agriculturalists, they have the great disadvantage to ecologists that boundaries between associations must necessarily follow geological boundaries even though differences in parent rock may be of negligible importance for vegetation.

It is at the next level in the hierarchy – the great soil group – that useful comparisons can first be made with vegetation mapped at a small scale. The great soil groups are mapped for the whole of Ghana by Brammer (1962 – and copied by Ahn 1970, p. 205). Two groups account for almost all forest soils: forest oxysols which are highly leached, yellowish, very acid and nutrient-poor: and forest ochrosols which are less leached, reddish, slightly acid to neutral and better supplied with nutrients. Brammer's map shows a region of forest oxysols in the south-west corner of Ghana more or less coincident with the Wet Evergreen forest type, a region of forest ochrosols underlying most of the Moist and Dry Semi-deciduous forest, and an intervening area of oxysol-ochrosol intergrades corresponding to Moist Evergreen forest.

Oxysols and ochrosols together form the Great Soil Group Family of Latosols – the level immediately below order (with which we started this discussion).

This Ghana system of classification has proved useful in Ghana and parts of Nigeria, but has not been adopted elsewhere in Africa – presumably because it was never adequately published. The system of Aubert 1963 has attracted more followers, initially in francophone countries only but more recently, with the publication of the soil maps of the pan-African CCTA (D'Hoore 1964) and IAOA (Boulet, Fauck, Kaloga, Leprun, Riquier & Vieillefon 1971), closely based on Aubert's scheme, many anglophone countries have also adopted it.

In Fig. 3.7 are presented a map of soils in southern Ghana according to the Charter system

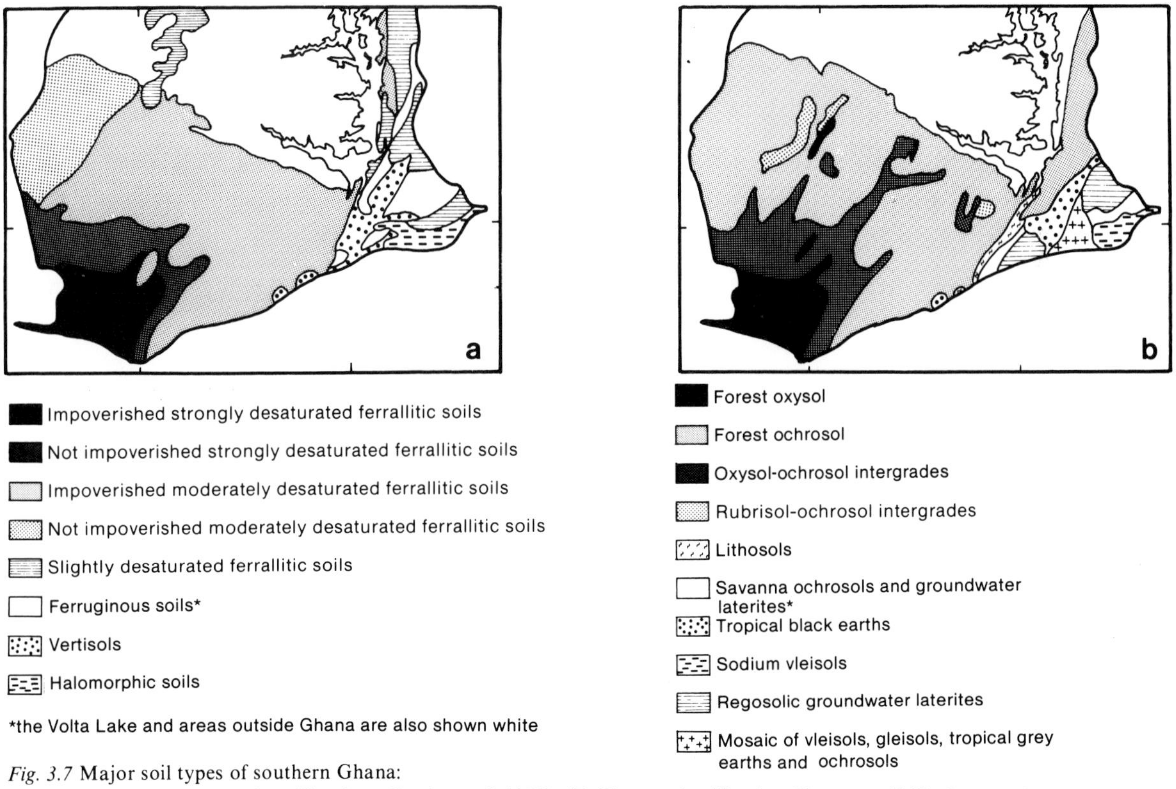

Fig. 3.7 Major soil types of southern Ghana:
(a) IAOA classification (simplified from Boulet *et al.* 1971); (b) Charter classification (Brammer 1962). See text for explanation of terms, and Table 3.2 for comparison with forest-type map.

and a simplified version of the IAOA map for comparison. It is apparent that the oxysols and forest ochrosols together correspond to the ferrallitic soils of the IAOA system, and the savanna ochrosols correspond to ferruginuous soils. These are the two major groups of tropical sesquioxide-rich soils in the Aubert system; differences between them are summarised in Table 3.1. Ferrallitic soils are subdivided in the IAOA map according to the degree of desaturation of the cation exchange complex, and the degree of 'impoverishment' – i.e. removal of clay by leaching from upper horizons.

Table 3.1 Major differences between ferruginous and ferrallitic soils.

Character	Ferruginous soils	Ferrallitic soils
Weatherable minerals	often present	usually absent
Clay minerals	some 2:1 lattice clays often present but kaolinitic (1:1) clays dominant.	2:1 lattice clays generally absent; kaolinitic clays dominant.
Gibbsite (Aluminium oxide)	usually absent	usually present
$SiO_2 : Al_2O_3$ ratio	>2	⩽2
Percent saturation of cation exchange capacity	⩾40	<40
Colour	reddish	yellowish
Depth of profile	<250 cm	usually several metres

After D'Hoore (1964).

Apart from the appeal of its international acceptance, we prefer the IAOA map because of its better correspondence to our forest type (Table 3.2). Lithosols are mapped separately on the IAOA map, but in view of their patchy, limited extent we have preferred to group them with the associated ferruginous or ferrallitic soil. The occurrences of South-east Outlier forest on ferrallitic soils are restricted, however, to lithosols.

The remaining major soil type on which forest occurs is the vertisols (tropical black earths), with abundant 2:1 lattice clays and high cation status; South-east Outlier forest is extensive on basic gneiss hills having this kind of soil.

Any classification must be built around some chosen aim; soil classifications are generally meant to summarise information relevant to limitations on plant growth – whether of agricultural crops or natural vegetation. Plant roots are concentrated in the upper 30 cm or so of the profile (Okali, Hall & Lawson 1973; Bernhard-Reversat, Huttel & Lemée 1978); the properties of the topsoil should thus be of great interest in soil classification. Unfortunately the topsoil is especially subject to disturbance by agricultural activities such as land clearance, burning, cultivation, harvesting and so on. For a stable classification, applicable to farmed and fallow soils alike, it is thus necessary to take subsoil as well as topsoil characteristics into account. For the ecologist who is interested in natural or semi-natural vegetation, however, topsoil characters alone may well be adequate – at least for correlation with broad vegetation types.

Table 3.2 Correspondence between soil type and forest type.

Soil type[a]	Forest type[b]
Impoverished, strongly desaturated ferrallitic	WE
Not impoverished, strongly desaturated ferrallitic	ME
Impoverished, moderately desaturated ferrallitic	MSSE, DS (most)
Not impoverished, moderately desaturated ferrallitic	SM, SO (part), MSNW, DS (part)
Slightly desaturated ferrallitic	DS (part)
Ferruginous	DS (part)
Vertisol	SO (most)

[a] Soil type according to the IAOA classification of Boulet *et al.* (1971).
[b] Forest type abbreviations as in Appendix 1 (p. 328).

As mentioned above (p. 14) we collected topsoil from each forest A-sample recorded, and the results of analysis of these soil samples provided good correlations with forest-type (Fig. 3.8), better than could be obtained from any of the formal classifica-

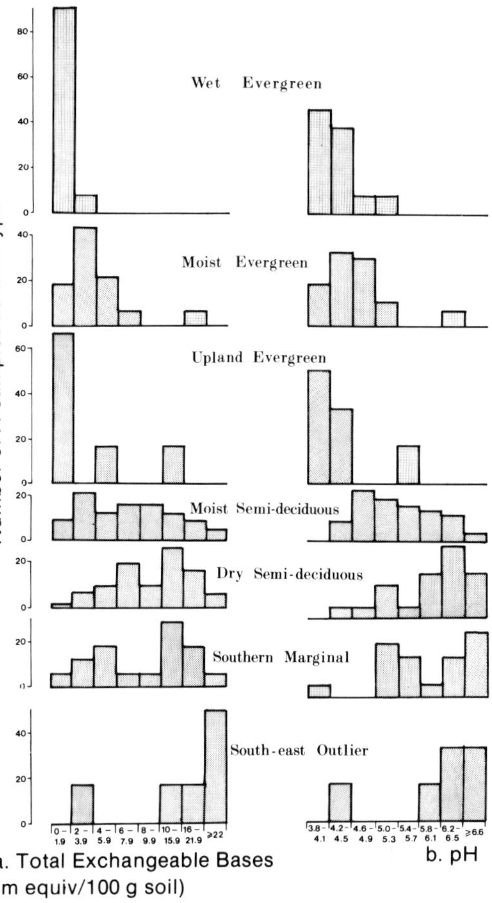

Fig. 3.8 Soil characteristics of forest types:
(a) Total Exchangeable Bases (TEB) determined by flame photometry on ammonium acetate extract; (b) pH by glass electrode. Soil was taken from top 15 cm in A-samples.

tions discussed above. As expected, the soils of the Wet Evergreen forest have consistently very low levels of total exchangeable bases, low saturation, and therefore low pH. The soils of Upland Evergreen forest are very distinct (see section 8.11, p. 98).

Fig. 3.9 shows our results for percentage saturation of cation exchange capacity plotted on a map of the Ghana forest zone. It can be seen that saturation of less than 40% (ferrallitic soils *sensu* D'Hoore see Table 3.1) is typical of the Evergreen forest-types, while saturation above 40% (ferruginous soils *sensu* D'Hoore) is found in the other forest-types. In Fig. 3.7, however, ferrallitic soils appear much more extensive, possibly because more of the profile was used in assessing saturation.

3.4 The forest-savanna boundary in relation to environmental factors

Comparison of Fig. 3.2 with Fig. 1.3 shows that the 1 250 mm isohyet cuts across the northern forest-savanna boundary at right angles, so that the north-west corner of the forest zone receives less than 1 250 mm while the south-central part of the savanna receives more. The fact that rainfall may be higher in some parts of the southern savanna than in some parts of the northern forest has led certain authors (e.g. Chipp 1922) to suppose that the boundary is unnatural, and mainly determined by the incidence of grass fires started by farmers. Swaine, Hall & Lock (1976), however, show that the boundary position can be explained in terms of a combined effect of geology, topography and rainfall. Forest extends farthest to the north, into the area receiving less than 1 250 mm rainfall, along the western edge of the country underlain by Birrimian rock, while it retreats farthest into the area with over 1 250 mm rainfall on the lower Voltaian rocks. It seems that Birrimian rocks (greywackes and lavas) perhaps because of more favourable cation yield, give rise to a soil more favourable to forest growth than do the Voltaian rocks, which comprise sandstones and shales. Portères 1950 also shows how the position of the forest-savanna boundary in Ivory Coast may be influenced by geology. It seems probable that geology and relief are important determinants of the northern forest boundary; a glance at Fig. 3.4 and 1.3 shows that this boundary follows the scarp bordering the Voltaian basin more closely than it does the isohyets. The existence of forest under low rainfall along the Voltaian rim may be explained by the effects of higher elevation in reducing evapotranspiration and of steeper topography in improving drainage and preventing iron-crust formation. The forest outliers in the Afram Plains occupy low hills (compare Fig. 3.5 and Fig. 1.3). Peltre 1977 likewise concludes for the Baoulé region of the Ivory Coast that the position of the forest-savanna boundary cannot be explained as a consequence of farming, but he considers palaeoclimatic factors to be more important than present-day climate, geology, relief or soils.

The stability of the boundary is enhanced by annual savanna fires which tend to prevent the spread of forest. Recently, De Rham (1970) has

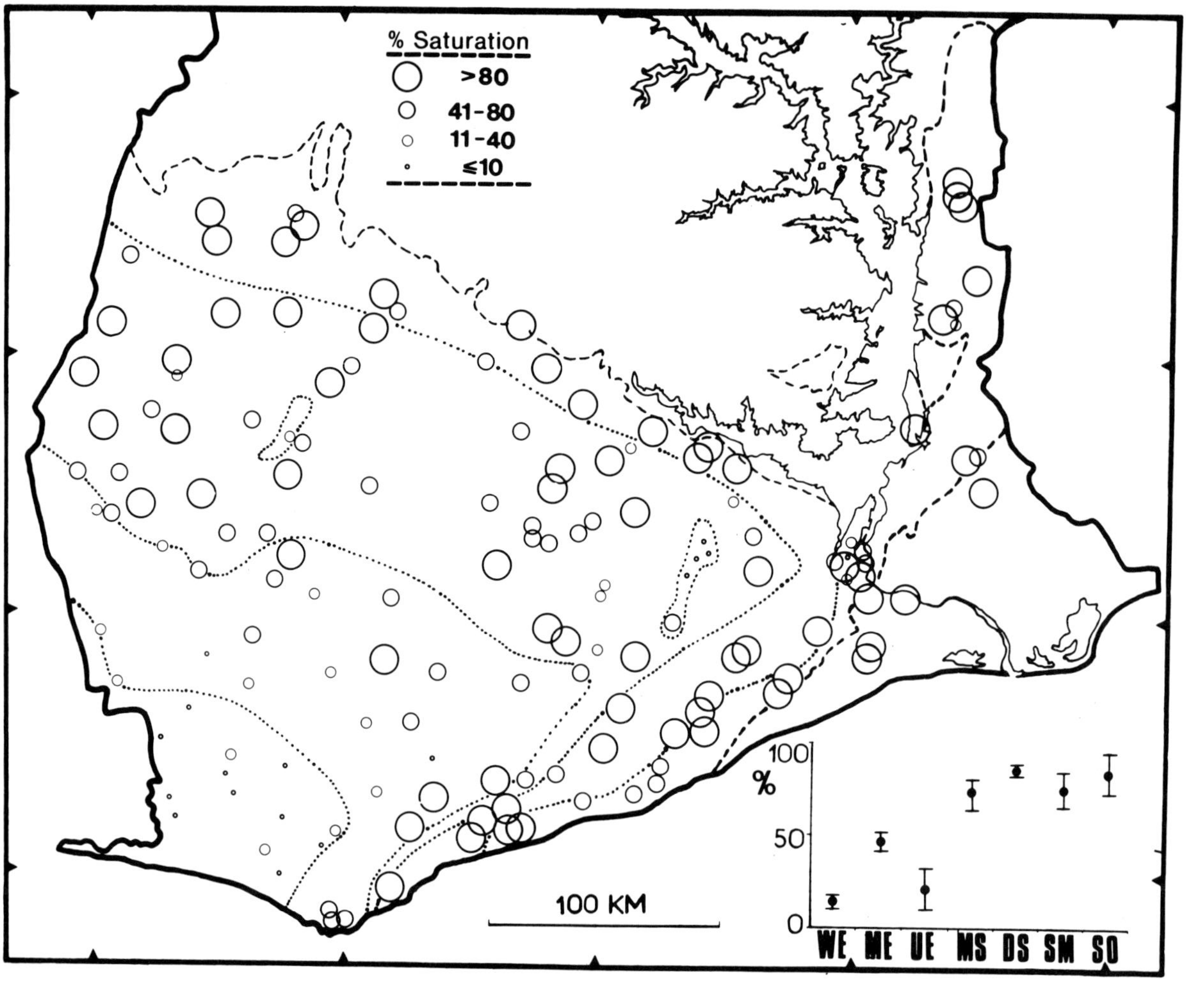

Fig. 3.9 Percent saturation of Cation Exchange Capacity (CEC) in A-samples. CEC was assessed by measuring ammonium leached with KCl. Boundaries: forest-savanna, broken line; forest-types, dotted. Graph shows means and standard errors of CEC saturation in each type.

suggested that low rates of nitrogen mineralization and the consequent extremely low levels of inorganic nitrogen in savanna soils may be as important as fire in retarding the advance of forest species into savanna areas which, in terms of climate, seem suitable for forest vegetation.

For a recent description of soil conditions across the forest-savanna boundary in Ghana see Markham & Babbedge (1979).

CHAPTER 4

Forest flora

4.1 State of knowledge of the forest flora

The first extensive collections of plants in the forest zone of Ghana were made about 180 years ago in the Akwapim district by the Danes: Isert & Thonning. Akwapim (together with adjacent parts of the Accra Plains) is thus the type locality for many of the species described from their collections by Schumacher, Vahl and Willdenow (Hepper 1976). The first botanist to penetrate deeply into the forest was Cummins, who accompanied the Ashanti Expedition from Cape Coast to Kumasi in 1895–96. The earliest collectors in the Wet Evergreen forest were Burton & Cameron (1883), who looked for gold and collected a few plant specimens as a hobby. H. N. Thompson, whose report (Thompson 1910) paved the way for the establishment of a Forestry Department in Ghana (p. 68), collected some specimens of trees during his exploratory journeys through the forest. The most thorough forest collections, however, have been made by the staff of the Forestry Department: Chipp's collections provided the basis for the Forest Herbarium, Kumasi which was formally established in 1923. Chipp was followed by a distinguished line of curators: Vigne, Andoh and Enti. All discovered new species which were named after them – *Saba thompsonii, Tabernaemontana chippii, Tarenna vignei, Vernonia andohii, Dissotis entii* among many others. The first check-lists of the forest flora were published by Chipp (1913–1914) and included 579 species. Burtt Davy & Hoyle (1937) published a list of 792 species from Ghana forest, but included few herbs and no ferns or orchids. The second edition of the Flora of West Tropical Africa (Hutchinson & Dalziell 1954–1972) included about 1 800 species of Ghana forest plants; recent collecting has brought the figure to over 2 100. Some of these new records are reported by Morton (1963, 1964 & 1966); others, in Cucurbitaceae by Chakravarty (1968), Orchidaceae by Hall & Bowling (1969), pteridophytes by Hall & Bigger (1974). The majority of new records are, however, reported for the first time in the present book. New species continue to be described from Ghana (e.g. Hall 1972, 1974a, 1974b, 1976, 1980, Hall & Leeuwenberg 1979, Lock 1978, Lock & Hall 1973, Robbrecht 1975).

West Africa is fortunate in being one of the very few large tropical areas to have a Flora (Flora of West Tropical Africa; FWTA) which is both recent and complete. But continuing revision for other Floras, especially of East Tropical Africa, Cameroun, Gabon and Zaire, (and the production of monographs) frequently shows the need for changes in the circumscription and names of taxa occurring in our area. We have tried to make our treatment of the Ghana forest flora an improvement over that in FWTA, but have not incorporated suggested name changes that are controversial, or which ignore the differences between what to us seem to be good taxa. In certain instances – for example *Napoleonaea* (Map 399) – we have drawn attention to a taxonomic problem by mapping two recognisable entities without being able to give them acceptable names. No new names, species or combinations appear for the first time in this book.

Our maps show the known distribution of 643 common species, including 384 trees and shrubs, 64 herbs and 195 climbers. A checklist of a further 350 species of forest trees (including 95 species not recorded for Ghana in FWTA) is provided in Appendix 5. It was decided not to include a checklist of additional shrubs, herbs and climbers, owing to the greater difficulty of checking names.

Although the forest flora of Ghana as a whole is well-known, knowledge of the distribution of species within Ghana is scanty. The shortcomings of herbarium records are discussed on p. 103. Except for the best-known timber species (see Chapter 11), Forestry Department stock survey records are taxonomically unreliable: all trees encountered are given a name, but many species which we know to occur are never recorded; the practice of converting to scientific names the vernacular names called out by tree spotters is obviously open to several kinds of error; voucher specimens are very rarely collected; little time is spent in the field checking uncertainties.

The present book can make no claim to finality – the total area of all our 285 samples only amounts to about 15 ha – but it goes some way towards improving our knowledge of distributions. We may well be accused by conservative taxonomists of foolhardiness in attempting to produce maps from such limited data. Our answer to such a charge of prematurity must be that, in view of the rate at which forest is disappearing (p. 7), there may never be sufficient time for the thorough collection of data which a really complete treatment would demand.

4.2 Composition of the flora

For each family in the Ghana forest flora, Appendix 6 gives the number of species by which it is represented, categorised by life form, habitat and broad geographic distribution.

The largest family is Rubiaceae with 218 species, almost all of which are woody, mostly in primary forest. The second largest is Orchidaceae with 148 species, most of which are epiphytes. The pteridophytes taken together (family limits are notoriously changeable) amount to 126 species, almost equally divided between ground herbs and epiphytes. The largest genus of trees is *Ficus* (33 spp., many of which are epiphytic), of herbaceous epiphytes *Bulbophyllum* (30 spp.), and climbers *Salacia* (24 spp.). In marked contrast with Malaysia, with enormous genera such as *Shorea* (180 spp.), there are no large genera of tall trees in Ghana: the largest genus in this category is *Entandrophragma* with only 4 species. Larger genera exist among understorey trees – *Diospyros* with 17 species, *Drypetes* with 13 and *Memecylon* with 11. All the larger genera and families are typical of primary forest. The families of the large lianes (Celastraceae, Combretaceae, Connaraceae, Dichapetalaceae and Loganiaceae), with a total of 115 climbing species, contain virtually no secondary forest species (defined by their absence from primary forest rather than by their presence in secondary forest). Some relatively large families consist mainly of trees: Anacardiaceae, Annonaceae, Ebenaceae, Sapotaceae, Sterculiaceae, Tiliaceae and Violaceae have a combined total of about 180 species of which scarcely any are secondary forest species. Families consisting mainly of secondary forest species and weeds include Acanthaceae, Amaranthaceae, Asclepiadaceae, Compositae, Cucurbitaceae, Cyperaceae, Gramineae, Labiatae, Solanaceae, Urticaceae & Vitaceae, together with many smaller ones. It may be noted that the 'primary' families are in general lower on the phylogenetic scale of such recent authors as Cronquist (1968) than are the 'secondary' families. Also noteworthy is the contrasted ecological faithfulness of some pairs of related taxa. Apocynaceae, for example, is 87% 'primary', while Asclepiadaceae is only 47% 'primary'. In the Melastomataceae, *Memecylon* is 100% 'primary', while *Tristemma* (5 spp.) is absent from primary forest. In Euphorbiaceae, *Drypetes* is 'primary' and *Macaranga* 'secondary'.

Finally we may note (see also Richards 1973) the very small number of palms.

4.3 Diversity

The poverty of the African forest flora in comparison with those of Malaysia and Amazonia has often been remarked (Richards 1973). Whitmore 1975 estimates at 7 900 species the total vascular flora of Malaya, which has an area (formerly almost entirely forested) of 132 000 km^2. In comparison with

Ghana, Malaya thus has nearly four times the number of species for about one and a half times the area. Axelrod & Raven (1978) suggest that post-Oligocene aridity caused widespread extinction in a previously rich forest flora.

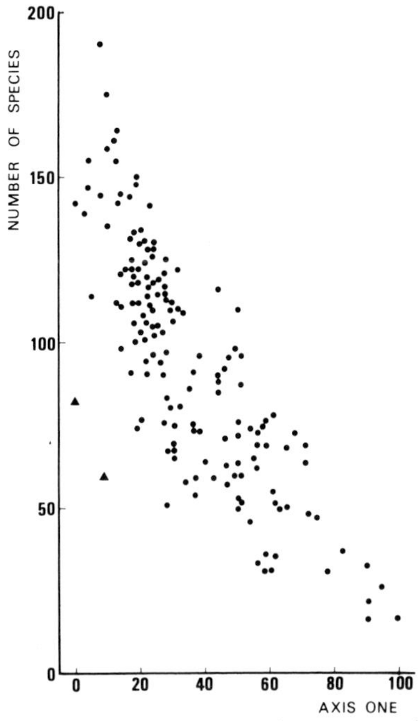

Fig. 4.1 Relation between diversity (expressed as number of species per A-sample) and score on axis 1 of ordination (Fig. 2.3). A-samples with *Scaphopetalum amoenum* dominant in understorey are shown by triangles.

Fig. 4.1 shows the relation between diversity, expressed as total number of species per A-sample, and axis 1 value of the A-sample in the ordination (Fig. 2.3). As explained on p. 19, axis 1 is closely correlated with the moisture gradient in the forest zone. It is thus clear that diversity, so defined, is correlated with the moisture status of the forest (swamp and secondary forest are here excluded from consideration). In the Wet Evergreen forest, up to 200 species of vascular plants were recorded per 0.0625 ha plot, while in the dry South-east Outlier forest-type, the number of species was as low as 20.

Species-area curves for plots in Moist Semi-deciduous and Wet Evergreen forest are shown in Fig. 4.2; the slope is significantly steeper in the wetter forest. Whitmore (1975, p. 5) gives species-area curves for trees but not for all species. From Fig. 4.3 it can be seen that rate of increase in number of tree species with increasing area of plot is much greater in Malaya than in any West African forest.

May (1976) summarises much recent theoretical work which argues that diversity is less in plant communities exposed to 'perturbation' (i.e. severe disturbance) – in which term one could include a dry season. The lower diversity of forests in which the dry season is more severe could thus be anticipated. Even our Wet Evergreen forest is

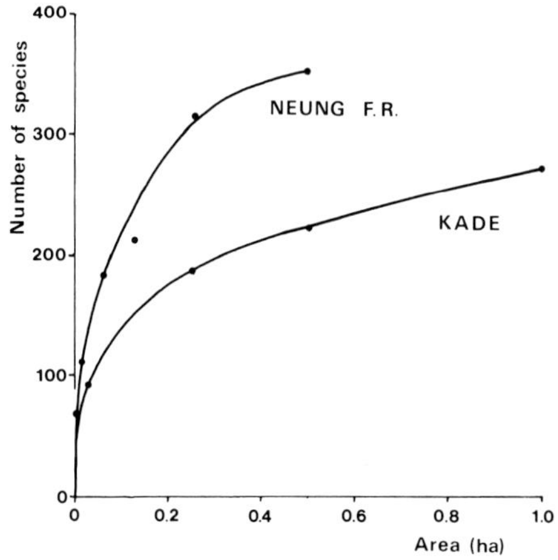

Fig. 4.2 Species-area curves (all species) for Ghanaian forest: Kade (Moist Semi-deciduous South-east subtype) and Neung F.R. (Wet Evergreen type).

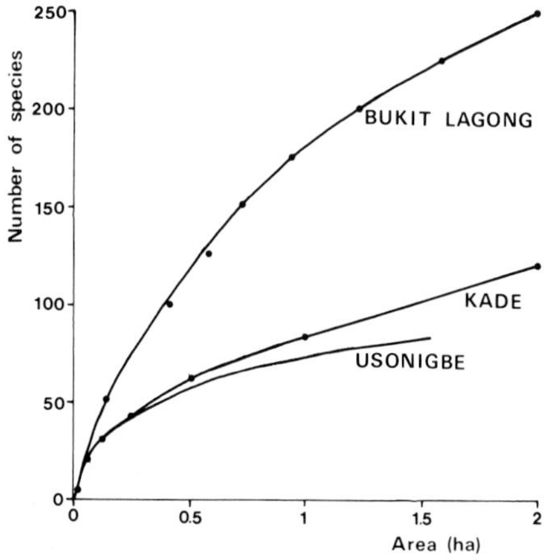

Fig. 4.3 Species area curves for trees ⩾10 cm dbh. Curves for Bukit Lagong F.R. (Malaya) from Whitmore (1975) and Usonigbe F.R. (Nigeria) from Baur (1964). Curve for Kade (Ghana) from our own data.

subject to a fairly long dry season (p. 19), whereas Malayan forest is less subject to drought (Whitmore 1975). If the horizontal axis in Fig. 4.1 were extended to the left, corresponding to forests with less moisture stress, and a regression line through the plotted points were similarly to be extrapolated leftwards, one could perhaps interpret the greater diversity of Malayan forest in purely climatic terms.

Connell (1978) favours the view that diversity will be greatest where perturbation is fairly frequent, and that the most stable forests will have lower diversity. His view is based on the apparent stability of single-dominant forests, dominated by various species of Caesalpiniaceae (Letouzey 1968, p. 126). In Ghana *Talbotiella gentii* (Map 576) forms such forests (Hall & Swaine 1980); there is no evidence, or even likelihood, however, that they occur in localities peculiarly free from perturbation.

4.4 Chorological considerations

It is a major aim of that part of descriptive plant geography called chorology to divide the map of the world into areas of relatively homogeneous flora. Within such an area a high proportion of species will be endemic. The first attempt at a comprehensive chorological map of Africa was made by Lebrun (1947), who recognised the African tropical forest as constituting the 'Région guinéenne'. Monod (1957) renamed this area the 'Région guinéo-congolaise, and White (1965) anglicised the term to 'Guineo-Congolian Region'.

The authors of chorological maps (e.g. Aubréville 1962, Léonard 1965) have generally favoured hierarchical schemes in which the area under consideration is completely divided into Regions, these into Domaines, and the Domaines into Sectors. White (1976a, 1978, 1979) pointed out that endemics are not, in general, evenly spread, but rather concentrated in limited areas which he calls Major Centres of Specific Endemism, (or 'Region' for the sake of brevity). He provisionally defines a phytogeographical Region as having higher than 50% endemism, and more than 1 000 endemic species. On these criteria, the forested Guineo-Congolian region, for example, is very distinct, perhaps not surprisingly when the adjacent regions, the Sudanian and Zambezian Regions, consist of savanna. The 'Regions' are separated by 'transition zones', in which endemism is normally very much less than 50%. The Guinea-Congolian/Sudanian, and Guinea-Congolian/Zambezian transition zones consist of forest-savanna mosaic, in which species characteristic of the respective adjacent Regions occur intermingled.

Within the Guineo-Congolian Region White (1979) recognises three Subcentres of Specific Endemism. These are the Upper Guinea Subcentre (corresponding to Aubréville's 1962 'Domaine libéro-ivoréen), the Lower Guinea Subcentre (corresponding to Aubréville's 'Domaine camerouno-gabonais'), and a Congolian Subcentre (corresponding to Aubréville's 'Domaine congolais'). Each of these subcentres possesses a considerable number of endemics, though the proportion is in each case less than 50% of the total flora. The subcentres are separated by 'intervals' with relatively few endemics. Between the Upper Guinea and Lower Guinea subcentres is the 'Dahomey Interval' (or 'Dahomey Gap' as it is generally called), consisting of a corridor of grassland vegetation reaching the coast from the northern savannas. The vegetation of the 'Sangha River Interval', separating the Lower Guinea and Congolian subcentres is not well known, but includes much swamp forest. The geographical positions of the Guineo-Congolian Region, its subdivisions, and adjacent phytochoria, are shown in Fig. 4.4.

Once the map has been divided into phytochoria on the basis of endemism, the flora can be divided into corresponding 'elements', and each species assigned to an element according to its distribution. The following elements can be recognised in the Ghana forest flora; we have assigned all our mapped species to one or other of them.

1. Guineo-Congolian Wides. These species occur in all three subcentres, but not (or scarcely) beyond the Guineo-Congolian Region.
2. Guinea Wides. These occur in both Upper and Lower Guinea Subcentres, but not in the Congolian Subcentre.
3. Upper Guinea Endemics. These are restricted to the Upper Guinea Subcentre. Near-endemics occur widely in the Upper Subcentre, but also have restricted populations elsewhere.
4. Satellites are restricted populations within Ghana of species whose major occurrence is in a subcentre other than the Upper Guinea.

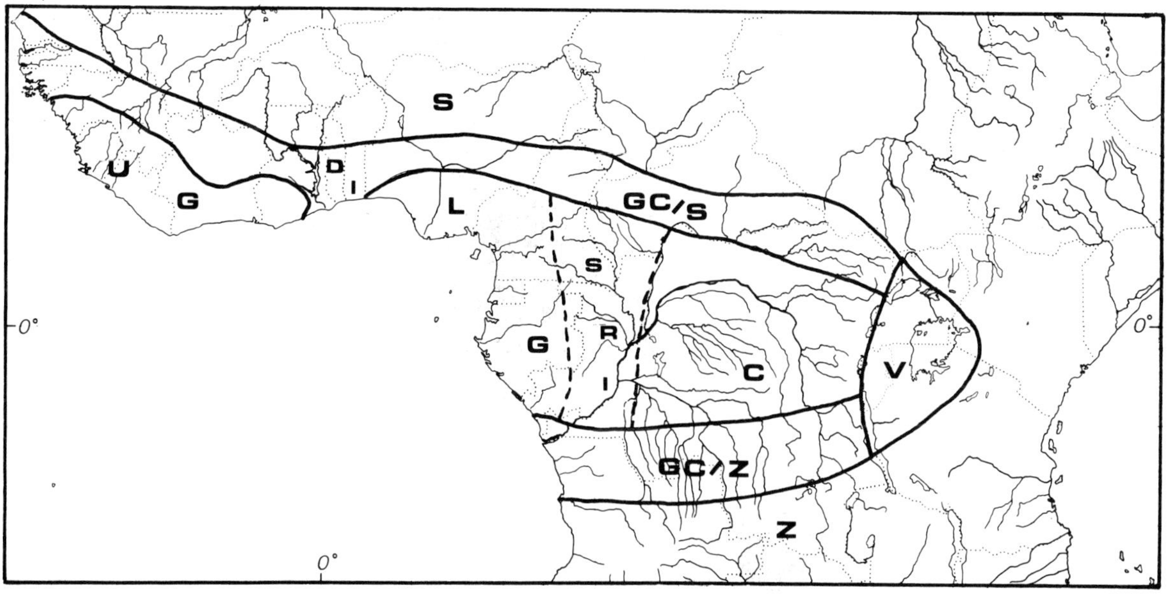

Fig. 4.4 The Guineo-Congolian Region and adjacent phytochoria (after White 1979). The Guineo-Congolian Region comprises the following three subcentres of Specific Endemism: Upper Guinea (UG), Lower Guinea (LG) and Congolian (C). Separating the subcentres are the Dahomey Interval (DI) and the Sangha River Interval (SRI). Other phytochoria shown are the Sudanian Region (S), the Zambezian Region (Z), the Lake Victoria Regional Mosaic (V), the Guinea-Congolian/Sudania Transition Zone (GC/S) and the Guinea-Congolian/Zambezia Transition Zone (GC/Z).

5. Chorological transgressors are widely distributed outside the Guineo-Congolian region. Some of these are also ecological transgressors in that they occur widely in vegetation other than lowland forest.

Of the 643 species described in the notes accompanying our maps, (the species most frequently encountered in Ghanaian closed-canopy forest, see p. 104), 88% are Guineo-Congolian, and only 12% are transgressors (Table 4.1).

It should be noted that the term 'element' may be used in a different sense when the possible history rather than the present composition of the flora is being investigated. *Sacoglottis gabonensis* (Baill.) Urb. for example, is the only African representative of the otherwise American family Humiriaceae (Thorne 1973), and may for this reason be assigned to an American, or Afro-American element. White 1979 points out that unnecessary confusion may result from the failure to distinguish clearly between 'descriptive chorology', whose geographical elements are based simply on the facts of present-day distribution, and 'interpretative chorology', distinguishing genetical elements according to hypotheses as to their historical origin. *Sacoglottis gabonensis* is, descriptively, a Guineo-Congolian

Table 4.1 Distribution within chorological groups (White 1979) of our 643 mapped species.

	Chorological group							
	Upper Guinea endemics	Upper Guinea near-endemics	Guinea Wides	Lower Guinea satellite	Congolian satellite	Guineo-Congolian Wides	Chorological transgressors	Ecological-chorological transgressors
Number of species	103	8	209	1	1	245	50	26
Percent of total	16	1	32	0.2	0.2	38	8	4

species, though it can be considered, interpretatively, to belong to an American genetical element.

For the purpose of analysing the whole Ghanaian forest flora, including species of secondary and swamp forest, and ruderals, we have adopted a classification into slightly different elements (Tables 4.3 and 4.4). The West African element is defined as those species restricted to the area of the Flora of West Tropical Africa (FWTA), or which extend into Cameroun beyond the eastern border of the FWTA area. This delimitation has the advantage for our purpose that the areas of forest to east and to west of Ghana are approximately equal. From Table 4.3 it can be seen that 85% of the ruderal species of farms and paths have areas extending beyond Africa, as compared to 3% for primary forest species. The largest distributional category of primary, secondary and swamp species is 'widespread African'. In terms of chorological elements, a few included species, such as *Rubus pinnatus* var. *afrotropicus* are Afro-montane satellites, a few such as *Diospyros mespiliformis* are 'ecological transgressors', occurring both in forest and savanna, but the vast majority of the species in the first four distributional categories (endemic to Africa), comprising about 85% i.e. 1 800 of the total species list, are certainly Guineo-Congolian. 340 species (16% of total) occur only west of the Dahomey gap, as compared to 430 species (20% of total) which are endemic to West Africa, (*sensu* FWTA) but which extend on both sides of the gap. Of these 430, only 51 are not known to occur in Ivory Coast – thus they reach their western limit in Ghana.

We can use these figures as follows to assess the importance of the Dahomey Gap as a phytogeographical barrier. Because the areas of forest to west and east of Ghana are similar, and do not differ greatly in species richness (Lebrun 1976), we may propose the null hypothesis that of the 368 species extending into the Ghana forest zone from only one direction, the contribution from the east should not differ significantly from the western contribution, if the Dahomey Gap is not of phytogeographical importance. In fact the respective numbers are 51 and 317 as against an expectation of 368/2 = 184. Thus:

$$\chi^2 = \frac{(51 - 184)^2}{184} + \frac{(317 - 184)^2}{184} = 192.26$$

which is very highly significant. The null hypothesis must therefore be rejected – clearly the contribution from the west is very much greater. The difference is most reasonably attributed to the presence of the Dahomey Gap on Ghana's eastern border.

It is also of interest to examine Aubréville's (1962) distinction between a 'domaine périphérique septentrional' running across the Dahomey gap in the drier northern part of the forest, and a 'domaine libéro-ivoréen' in the wetter forests. The data and calculations presented in Table 4.5 support Aubréville in showing that the Dahomey Gap is much more important for the 'wet' than for the 'dry' species, but it also shows that even the 'dry' species are significantly restricted.

Table 4.2 Comparison of distribution-type of species in matched plots in forest reserves from Ivory Coast, Ghana and Nigeria.

Forest-type and position	Sample plot	Total species	Number of species in sample plot absent from total flora: Ivory Coast (and further west)	Ghana (and beyond)	Nigeria (and further east)
(a) Dry Semi-deciduous forest Gh. plot 520 km from each of others	Maraoué (Iv.C.)	116	1	2 (2)	12 (10)
	Bomfoum (Gh.)	98	0	1	7 (7)
	Ijaiye (Nig.)	129	6 (5)	5 (4)	1
(b) Lowland evergreen forest; Gh. plot 640 km from Iv.C. plot and 860 km from Nig. plot.	Sakré (Iv.C.)	139	1	10 (7)	44 (32)
	Subri (Gh.)	112	0	1	14 (13)
	Sapoba (Nig.)	85	23 (27)	22 (26)	1

Percentages of plot totals in brackets. Plots are A-samples each of 0.0625 ha, except for Sapoba which was a B-sample (see p. 14).

Table 4 3 Relation between *successional* position and global distribution of Ghana forest species.

Habitat	Distribution						
	West Africa (Senegal to Cameroun)			Wider distribution in Africa	Extends beyond Africa	Unclassified	Total
	West of Dahomey Gap only		Both sides of Dahomey Gap				
	Ghana only	Extending West of Ghana					
Primary forest	(2) 23 (92)	(18) 243 (77)	(22) 302 (70)	(54) 733 (70)	(3) 43 (15)	(1) 11 (65)	1 355 (64)
Secondary forest	0	(12) 56 (18)	(20) 91 (21)	(45) 207 (20)	(22) 99 (34)	(1) 3 (18)	456 (22)
Farm & paths	0	0	(2) 3 (1)	(13) 18 (2)	(85) 119 (41)	0	140 (7)
Swamp & riverain	(1) 2 (8)	(10) 14 (4)	(18) 27 (6)	(50) 73 (7)	(21) 30 (10)	0	146 (7)
Unclassified	0	(17) 4 (1)	(30) 7 (2)	(39) 9 (1)	0	(13) 3 (18)	23 (1)
Total	(1) 25 (100)	(15) 317 (100)	(20) 430 (100)	(49) 1 040 (100)	(14) 291 (100)	(1) 17 (100)	2 120 (100)

In any cell of the table the number of species *n* is not bracketed, the preceding number is *n* as a percentage of the row total, and the succeeding bracketed number is *n* as a percentage of the column total.

Table 4.4 Relation between *ecological* preference and global distribution of Ghana forest species.

Ecological preference	Distribution						
	West Africa only (Senegal to Cameroun)			Wider distribution in Africa	Extends beyond Africa	Unclassified	Total
	West of Dahomey Gap only		Both sides of Dahomey Gap				
	Ghana only	Extending West of Ghana					
Evergreen forest	(2) 9 (36)	(40) 155 (49)	(18) 69 (16)	(35) 136 (13)	(3) 12 (4)	(2) 8 (47)	389 (18)
Dry forest	(1) 5 (20)	(8) 28 (9)	(25) 91 (21)	(52) 190 (18)	(14) 52 (18)	0	366 (17)
Mesic forest, or ubiquitous	(1) 6 (24)	(8) 89 (28)	(19) 216 (50)	(52) 585 (56)	(19) 212 (73)	(1) 7 (41)	1 115 (53)
Upland forest	(3) 4 (16)	(18) 25 (8)	(17) 24 (6)	(56) 79 (8)	(6) 9 (3)	(1) 1 (6)	142 (7)
Unclassified	(1) 1 (4)	(19) 20 (6)	(28) 30 (7)	(46) 50 (5)	(6) 6 (2)	(1) 1 (6)	108 (5)
Total	(1) 25 (100)	(15) 317 (100)	(20) 430 (100)	(49) 1 040 (100)	(13) 291 (100)	(1) 17 (100)	2 120 (100)

In any cell of the table the number of species *n* is not bracketed, the preceding number is *n* as a percentage of the row total, and the succeeding bracketed number is *n* as a percentage of the column total.

Table 4.5 Test of null hypothesis that the Ghana forest zone receives equal contributions of species from east and west.

	Species preference	
	Evergreen forest	Dry forest
West African endemic species extending into Ghana from west only (W)	155	28
West African endemic species extending into Ghana from east only (E)	8	10
Total species (T)	163	38
$\chi^2 = \frac{(W - \frac{1}{2}T)^2}{\frac{1}{2}T} + \frac{(E - \frac{1}{2}T)^2}{\frac{1}{2}T}$	132.6	8.5
Probability	$\ll 0.001$	< 0.01

Table 4.6 Contingency table analyses of species in forest plots in Ivory Coast and Nigeria which do not occur in the total forest flora of both countries, classified according to whether or not they occur in the Ghana forest flora (data from Table 4.2).

	In Nigeria but not Iv. Coast	In Iv. Coast but not Nigeria	Total
(a) Dry semi-deciduous forest			
Present in Ghana flora	1	10	11
Absent from Ghana flora	5	2	7
Total	6	12	18
χ^2 (Yates' correction) = 4.9			
$0.05 > P > 0.01$			
(b) Lowland evergreen forest			
Present in Ghana flora	1	34	35
Absent from Ghana flora	22	10	32
Total	23	44	67
$\chi^2 = 29.3$			
$P \ll 0.001$			

Table 4.7 Forest species thought to be endemic in Ghana.

Name	Family	Habitat	Forest-type (abbrev. see Appendix 1 p. 328)	Life form	Neo- or palaeo-endemic*	Map number
Alsodeiopsis chippii Hutch.	Icacinaceae	lry forest	WE	pygmy tree	neo	–
Bonamia vignei Hoyle	Convolvulaceae	lry forest	MS	large climber	palaeo	–
Bowringia discolor J. B. Hall	Papilionaceae	riverain forest	WE	large climber	palaeo	–
Cola umbratilis Brenan & Keay	Sterculiaceae	lry forest	WE	small tree	neo	154
Commiphora dalzielii Hutch.	Burseraceae	lry forest	SO	shrub	palaeo	–
Dalbergia setifera Hutch. & Dalz.	Papilionaceae	lry forest	SM	small tree	neo	–
Diaphananthe suborbicularis Summerh.	Orchidaceae	lry forest	DS	epiphytic herb	neo	–
Grewia megalocarpa Juss.	Tiliaceae	lry forest	SM-SO	shrub	palaeo	–
Hymenostegia gracilipes Hutch. & Dalz.	Caesalpiniaceae	riverain forest	WE	medium tree	neo	–
Ledermanniella bowlingii (J. B. Hall) C. Cusset	Podostemaceae	river bed	DS	aquatic herb	palaeo	–
Monocyclanthus vignei Keay	Annonaceae	lry forest	WE	pygmy tree	palaeo	–
Nephthytis swainei Bogner	Araceae	lry forest	ME-WE	ground herb	palaeo	–
Oleandra ejurana C. D. Adams	Davalliaceae	lry forest	DS	epiphytic fern	palaeo	–
Psychotria ankasensis J. B. Hall	Rubiaceae	lry forest	WE	ground herb	palaeo	–
Talbotiella gentii Hutch. & Greenway	Caesalpiniaceae	lry forest	SO-SM	medium tree	palaeo	–
Turraea ghanensis J. B. Hall	Meliaceae	lry forest	SM	pygmy tree	palaeo	–
Uvariopsis globiflora Keay	Annonaceae	lry forest	MS	small tree	neo	616
Virectaria tenella J. B. Hall	Rubiaceae	lry forest	DS	ground herb	palaeo	–

Several other imperfectly known species have been omitted.

* The term 'neo-endemic' is here applied to species which are closely related to some other West African species; they appear to have speciated within fairly recent times. 'Palaeo-endemics' lack extant close relatives, or their nearest relatives are absent from West Africa.

The objection can be made to the above reasoning that because the gap between Ghana forest and Nigerian forest is greater than that between Ghana forest and Ivory Coast forest, we would in fact expect a rather lower representation of eastern species in Ghana. Another approach to assessing the phytogeographical importance of the Dahomey Gap is to compare species lists from forest samples in Ghana with those in similar forest, and similarly remote, to the west. The results of such a comparison are represented in Table 4.2. All of the species recorded in the Ghana plots occur in Ivory Coast, but 7% and 13% of the species in the 'dry' and 'wet' Ghana plots respectively are not known from Nigeria. Table 4.6 shows that a significantly lower proportion of the Ivory Coast than of the Nigerian plot species are absent from Ghana, and that the difference is greater for 'wet' than for 'dry' plots. Tables 4.2 and 4.5 thus provide ample confirmation of the importance of the Dahomey Gap.

Forest species endemic to Ghana are tabulated in Table 4.7. Further prospecting may reveal that some of these occur in neighboring countries, but it may also discover other, at present unknown species. Of these 18 species, 7 are known from Wet Evergreen forest – the type richest in species (p. 90) – and it is here also that Ghana's only endemic genus (*Monocyclanthus*) occurs. The other end of the moisture spectrum, Southern Marginal and South-east Outlier forest, supports a further five endemic species – a high number in view of the limited extent and floristic poverty of these types, but explicable when it is remembered that comparable vegetation seems to be virtually absent from neighbouring countries. Only three of the endemic species are sufficiently abundant to justify mapping – *Cola umbratilis* (Map 155), *Uvariopsis globiflora* (Map 615), and *Talbotiella gentii* (Map 576). Two are neo-endemics, *Cola umbratilis* being closely related to *C. buntingii* Bak.f. of 'wet' forest in Liberia and western Ivory Coast, and *Uvariopsis globiflora* being related to *U. guineensis* Keay, which occurs from Guinée to western Ivory Coast. Both of these occur so close to the Ivory Coast that it seems inevitable that they must in places transgress the border. The third mapped endemic, *Talbotiella gentii,* is not very closely related to the other two species of *Talbotiella,* in E. Nigeria and Cameroun. The report (Hall 1976) of the discovery of *T. gentii* in N. Cameroun seems to have been in error.

4.5 Species distributions within Ghana

Our forest-types are derived from species distributions within Ghana (p. 14). It is thus to be expected that some relation should be found between species ranges and forest-type boundaries. Very few species, however, are restricted to a single forest-type (see tables of characteristic species in Chapter 8). Each of our maps is accompanied by a histogram showing relative abundance in the types.

It has been suggested to us that the distributions shown in our maps could be classified into a limited number of types. Curtis (1959) showed that a high proportion of Wisconsin plant species reach their geographical limits within a comparatively narrow zone running across the state, which corresponds approximately to the limit of glaciation. We have not attempted to plot all species, but from Fig. 4.5, showing the limits of areas within Ghana of *Diospyros* species, it seems unlikely that such an attempt would produce a simple pattern of distinct distribution-types, although White (1962) has successfully classified the total distributions of *Diospyros* species in Africa.

Broad distribution patterns are determined by two kinds of factors – the environmental and the

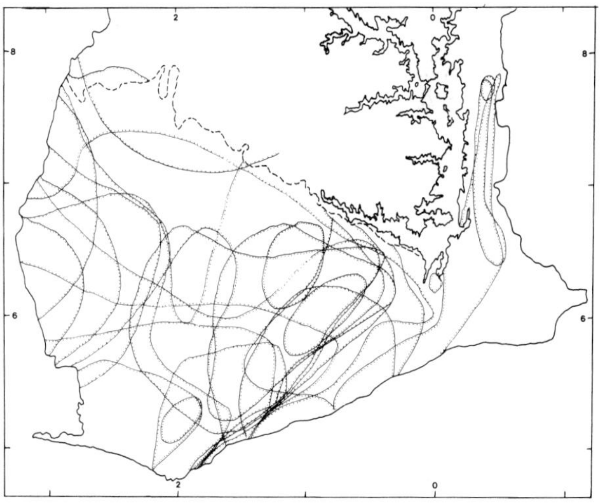

Fig. 4.5 Distribution limits of *Diospyros* species with the forest zone of Ghana. Dots show the side of the line on which the species occurs. Note the coincidence of many limits with the south-eastern boundary of Moist Semi-deciduous forest (cf. Fig. 2.4), and the limits of species centred on the Ivory Coast which run across type boundaries.

historical. Where the area of distribution of a species coincides with that of one of our forest-types, or its limits are parallel to those between forest-types, it is likely that environmental factors are paramount. Thus the area of *Diospyros chevalieri* (Map 211) is almost identical with that of Wet Evergreen forest, through which it is evenly spread, and adaption to the prevailing environmental conditions of low soil nutrient status and high rainfall provides an adequate explanation for its distribution. *Diospyros abyssinica* (Map 209) is not confined to the Southern Marginal type, but its other occurrences are in the adjacent types, and again its distribution can be explained environmentally, (though in Malawi its distribution is less predictable and White (1970, p. 103) classifies is as an 'ecological transgressor'). Other 'congruent' species (i.e. those distributed throughout an ecological zone) are *D. mespiliformis* (whose main area is savanna), *D. monbuttensis*, *D. sanza-minika*, and *D. vignei*.

Distributions of the remaining species of *Diospyros* are inconsistent with a purely environmental explanation, being patchy, and failing to fill evenly any environmental zone however defined. In this they resemble *Turraeanthus africanus* in Ivory Coast, discussed by Aubréville (1950–51); (*Turraeanthus* is fairly 'congruent' in Ghana, see Map 605). *Diospyros cooperi* (Map 212) only occurs close to the Ivory Coast border, in three forest types. *Diospyros piscatoria* (Map 220) has a distribution in the centre of the forest zone, in the Moist Evergreen and Moist Semi-deciduous types. *D. gabunensis* (Map 214) is almost confined to Moist Evergreen forest, but is restricted to the eastern part of the type. *D. heudelotii* (Map 215) is spread through most of the main forest block, apart from the Upland Evergreen and the two driest types – but does not occur east of the River Volta (it is absent from Nigeria). Other species have a fragmented distribution. *D. mannii* (Map 217), for example, has one area similar to that of *D. cooperi*, another similar to *D. gabunensis*, and a third area in the east-central part of the forest zone. *D. kamerunensis* (Map 216) has a continuous distribution, but is unexpectedly rare or absent in the north-west quarter of the forest zone. (F. White tells us that in Cameroun *D. kamerunensis* is restricted to the wettest forests).

Species such as these appear not to be spread throughout the area which, from available evidence, would seem suitable for them. It is possible that they are limited by complex biological interactions, for example with competitor species and predators. The most straightforward explanation (not entirely incompatible with the foregoing), is a historical one – either they are becoming extinct in parts of their formerly more extensive areas, or they are gradually spreading from one or more centres, or refuges.

4.6 History of West African forest vegetation

Recent years have seen an upsurge of interest in Quaternary climatic fluctuations in Tropical Africa, and their effects on flora and vegetation (Livingstone 1975, Wickens 1975, Hamilton 1976). Climatic change has been invoked to help explain several related phytogeographical problems, especially patterns of endemism, levels of diversity, and patterns of disjunction and vicariance.

Two explanations may be put forward for the occurrence of a taxon in two disjunct areas: either it has become extinct in the intervening zone, or it has accomplished long-distance dispersal from one area to the other. In the former case extinction could have resulted from a failure to compete with more aggressive species, or from climatic change; in practice these possibilities are not usually separable. Long-distance dispersal will be the favoured hypothesis where only one or two species show a particular pattern of disjunction, and where the propagules of the species in question seem capable of such dispersal. Climatic disruption of a previously continuous area will be suspected where many species are involved, some with heavy seeds of short viability.

In the forest flora of Ghana the principal disjunctions are between the species of evergreen forests of south-west Ghana and similar vegetation in south-west Ivory Coast ('espèces sassandriennes' of Mangenot 1956 and Guillaumet 1967), and south-east Nigeria, Cameroun and Gabon. Table 4.4 shows that of 389 evergreen forest species, 380 occur in other countries, and therefore show such disjunction. Aubréville (1949 & 1962) is followed by Guillaumet (1967) in suggesting that these forests must at some remote period have been in continuity, but subsequent desiccation has divided them into their present isolated blocks. 42% (see Table 4.4) of these 'wet' species are 'western Guinea' species, endemic to the area west of the Dahomey Gap, and it may be presumed that the majority of these have evolved since the 'savanisation' of the gap. For some, a close vicariant relative is known in the Cameroun-Gabon forests: White (1962) cites *Diospyros vignei* occurring from Liberia to Ghana, and its closest relative *D. cinnabarina* in Cameroun. The acaulous palm *Sclerosperma mannii* Wendl. is represented in south-west Ghana by an entire-leaved form, while the typical form in south-east Nigeria to Gabon has pinnate leaves – speciation may here be still in progress.

The evergreen forests of south-west Ghana, and south-west Ivory Coast & Liberia have also been separated sufficiently long for some speciation to occur – *Cola umbratilis* in Ghana and *Cola buntingii* Bak.f. in Ivory Coast are obvious vicariants. Populations of *Diospyros vignei* in south-west Ghana differ in several characteristics from those of south-west Ivory Coast and Liberia, and may prove to be distinct (White 1978).

Distribution patterns of 'wet' species thus provide persuasive evidence for previous wetter periods. Aubréville (1949 & 1962) and Guillaumet (1967) also suppose that a period of greater desiccation than the present can be deduced from distributions, but the evidence which they cite is much less strong. Guillaumet classifies the 'wet' species of south-west Ivory Coast (espèces sassandriennes) into the following five groups: I – SW Ivory Coast only; II – SW Ivory Coast and Liberia only; III – areas of I and/or II plus SW Ghana; IV – areas of I and/or II plus SE Nigeria-Zaire; V – areas of I to IV combined. He lists all the species in these five groups, and includes the following numbers of species in each: I – 17; II – 78; III – 21; IV – 23; V – 24. Because he finds more species in IV than in III, he concludes that the flora of SW Ghana is impoverished, and that this must be due to a past dry period. 'Une étude numérique de la flora des diverses régions du massif forestier occidental devrait montrer la relative pauvreté du Ghana'. In fact recent collecting has revealed the presence in Ghana of substantial proportions of Guillaumet's groups I, II & IV – a revised list would be: I – 13; II – 59; III – 43; IV – 15; V – 32. SW Ghana thus has almost as many endemics as SW Ivory Coast (9 as against 13), and of the 58 SW Ivory Coast and Liberia species which occur in only one other area, 43 are in Ghana and only 15 in SE Nigeria to Zaire. The greater number of species in groups I and II as compared to the SW Ghana endemics can be explained by the greater size and higher present-day rainfall of the SW Ivory Coast and Liberia evergreen forest block and there is no need to postulate a past period of desiccation, on this distributional evidence. There are, in any case, several 'wet' species known from SW Ghana and the SE Nigeria to Gabon block which do not occur further west: these include *Allexis cauliflora* (Jeník & Enti 1968), *Sclerosperma mannii* Wendl., and *Didelotia unifoliolata.*

On the other hand, the small patches of 'drier' vegetation which occur on shallow soils here and there within moist forest may well be relicts from a time when the climate was drier and such vegetation was more extensive. Some examples are mentioned on p. 87. Bia National Park (Fig. 1.3b) lies partly in the Moist Semi-deciduous and partly in the Moist Evergreen forest zones. A granite outcrop within the Park, however, bears forest of Southern Marginal type, with *Afzelia africana, Dennettia tripetala, Diospyros abyssinica, Grewia carpinifolia, Markhamia tomentosa* and *Strophanthus hispidus.* The bauxite-covered summits in the Tano Ofin and Krokosua Hills forest reserves are dotted with patches of natural savanna which could also be relicts.

There is also abundant evidence of other kinds for periods of great desiccation in other parts of West and Tropical Africa (Raven & Axelrod 1974, Livingstone 1975). This includes the belt of fixed sand dunes from Senegal to Northern Nigeria extending far south of the present southern limit of the mobile Saharan dunes; reworked Kalahari Sands underlying much of the present forest area in Zaire; evidence of past low lake-levels, and so on. What is lacking is direct evidence from the West African forest zone itself. One may well ask, if forest in the western Guinea block was really restricted by drought to very small refugia, as represented by Aubréville and Guillaumet, why these refugia harboured such large numbers of species which are at present restricted to high rainfall regions, rather than species characteristic of the drier kinds of forest which now grow nearer the savanna boundary. Further hypotheses can, of course, be devised to meet this objection. It cannot be assumed that high rainfall is essential for the species which now grow in the 'wet' forest areas – possibly they are merely tolerant of the very low levels of exchangeable cations which result from prolonged leaching (p. 27). If the soil were to be impoverished by a long wet spell, 'dry' species might be unable to invade the resulting poor soils even if the rainfall decreased greatly. Perhaps the low diversity of the western forest block as compared to the Cameroun-Gabon block is in fact the result of extinction by drought of species which are moisture-demanding. Obvious hygrophytes, such as the filmy fern *Trichomanes guineense* (Map 596) might have re-invaded the area by long-distance dispersal when moister conditions returned. We really know very little about the ecological tolerances and requirements of any forest species. *Pentadesma butyracea* is a typical Wet Evergreen species in the main part of the forest of Ghana (see Map 451) – but it also occurs commonly in fringing forest in mountain valleys near the Togo border above the northern edge of our distribution maps! In Ghana *Sloetiopsis usambarensis* (Map 539) is restricted to dry forest near the coast, but in Ivory Coast it occupies much wetter, even evergreen forest. *Angylocalyx oligophyllus* (Map 41) in Ghana is one of the most characteristic species of lowland evergreen forest, but in Nigeria it is widespread in Dry Semi-deciduous forest.

Distributional data are of only limited value in assessing past climatic change because they can give us no indication of time-scale. It is known that many families and genera now confined to tropical forest extended into Libya during Tertiary times 30–40 million years ago (Aubréville 1970, 1971). Was this the period when the evergreen forests of Ivory Coast, Ghana and Cameroun were last in continuity? Or were the species common to these forests spread right across West Africa during the most recent pluvial period, which commenced about 12 000 years ago? Recent detailed studies of deep ocean sediments (Emiliani 1966, Shackleton & Opdyke 1977, Pastouret *et al.* 1978) show that major oscillations in world climate during the Pleistocene were much more numerous than had been suspected from terrestrial geomorphological and pollen-analysis studies. Any one of them might have been responsible for a species becoming locally extinct, spreading to a new area, or speciating. The modern distribution map may be regarded as a palimpsest on which are superposed the 'writings' of innumerable climatic changes. It would be naive for us to suggest, as phytogeographers have sometimes done, a simple scheme of two or three 'dry' and 'wet' phases to 'explain' distribution patterns.

Disjunctions within the areas of montane and sub-montane species have sometimes been attributed to the dissection of once-continuous areas by climatic change. Moreau (1966) suggested that presently isolated blocks of African Montane forest could have been continuous during cool pluvial periods corresponding to northern glaciations, and Morton (1962) used a similar argument to explain disjunctions in the West African montane flora. Livingstone (1975) and Williams (1975) point out, however, that Pleistocene cool periods were dry, not wet, in Tropical Africa, and it is very unlikely that montane forest has ever been much more widespread than at present. And as pointed out by Jones in the discussion following Morton's (1961) paper, montane habitats are often inherently disjunct. To take an example from the flora of Ghana, *Impatiens kamerunensis* Warb. var. *obanensis* (Keay) Grey-Wilson, which is disjunct between hills in SE Nigeria and the Ghana-Togo border, is confined to waterfalls and cataracts – a habitat which could never conceivably be continuous. Long-distance dispersal must be invoked here, and might therefore also be feasible for some other montane species. The fern *Asplenium schnellii* Tard. occurs only in Ghanaian Upland Evergreen forest on the Atewa Range and in mountains along the western Ivory Coast border, but long-distance dispersal of fern spores is easily imaginable. The fern *Pteris vittata* Linn. has its only West African mainland stations on and around gold-mine slag heaps in Ghana (Hall 1970); it must have reached these by long-distance dispersal during the past century. 'Baker's Law' (Baker 1955) states that a population resulting from long-distance dispersal is likely to be self-compatible even when the parent population is self-incompatible. South-east Outlier forest in the Shai Hills of Ghana includes several disjunct species (Jeník & Hall 1976). One of these, *Coccinia grandis* (Linn.) J. O. Voigt, is normally dioecious (in fact dioecy is a generic characteristic of *Coccinia*) but is monoecious at Shai. *Vepris heterophylla* has hermaphrodite flowers and polyembryonous seeds (suggesting apomixy) at Shai, but is dioecious in its Mali and Cameroun stations.

It seems likely therefore, that these two species have reached Shai by long distance dispersal.

Direct evidence for past vegetation and flora can only come from dated fossils – pollen, leaves and wood. Although pollen analysis has yielded much information about vegetation change in East Africa (see Livingstone 1975 for references) there is as yet little data from the West African forest zone. Pollen has been identified in cores through peat in the seabed just off Ivory Coast (Assémien, Filleron, Martin & Tastet 1971; Frédoux & Tastet 1976; Frédoux 1977), but data are too scanty for firm conclusions to be drawn concerning vegetation change. The studies of Hall, Talbot & Swaine (1978) on fossil leaf floras from Lake Bosumtwi in Ghana reveal that about 10 000 B.P. a vegetation of semi-deciduous forest similar to that of the present day occupied the area, thus casting doubt on the applicability to this area of Livingstone's (1975, p. 274) remark that evergreen forest was much more extensive at that time. Knowledge of past climates and vegetation in the tropics is summarised by UNESCO/UNEP/FAO (1978).

To summarise – we should have liked to be able to use the distribution maps in this book to support some convincing account of vegetational history, but this would be premature in the present state of knowledge. The maps show many intriguing patterns, but these may not all be explicable until we know much more about past climates and vegetation in the West African forest zone.

CHAPTER 5

Forest structure and physiognomy

5.1 Trees and shrubs

The quantity of trees on a forest plot can be assessed in various ways, notably as tree density (number of trees per unit area), basal area (total cross-sectional area at breast height – 1.3 m above the ground), as maximum height of trees, maximum girth, or as maximum crown diameter.

Tree density and basal area vary considerably from place to place in forest, but show surprisingly little systematic difference in average between types (Table 5.1). Dawkins (1959), after examining data from many parts of the world, concluded that mature tropical forest has a mean basal area of 32 m^2ha^{-1}. The generally lower figures for Ghana forest may be a consequence of the high incidence of disturbance by farmers and loggers.

Maximum tree height, on the other hand, correlates well with forest-type (Fig. 5.1). The tallest trees are found in Moist Semi-deciduous forest. The lower maximum height in drier forest-types is to be expected as a consequence of greater moisture stress. Reduction of maximum height in the Evergreen forests is more unexpected – in Malaysia the

Table 5.1 Basal area and density of trees in Ghanaian forest-types.

Type	Mean basal area[a] (m^2ha^{-1})	Standard error of mean	Mean density[b] (trees ha^{-1})	Standard error of mean
Wet Evergreen	25.5	1.1	445	30
Moist Evergreen	23.6	0.6	505	19
Upland Evergreen	26.8	1.9	485	78
Moist Semi-deciduous				
South-east subtype	24.2	0.8	480	29
North-west subtype	24.3	1.1	514	29
Dry Semi-deciduous				
Inner Zone subtype	24.2	1.1	504	32
Fire Zone subtype	22.3	1.2	474	32
Southern Marginal	22.2	2.0	469	37
South-east Outlier	21.0	2.3	456	29
Average	23.8	1.3	487	11

[a] Measured by Spiegel Relaskop.
[b] Trees $\geqslant$0.1 m dbh.
Data from forest survey (see p. 14)
Analysis of variance revealed no significant difference between forest-types for either measure.

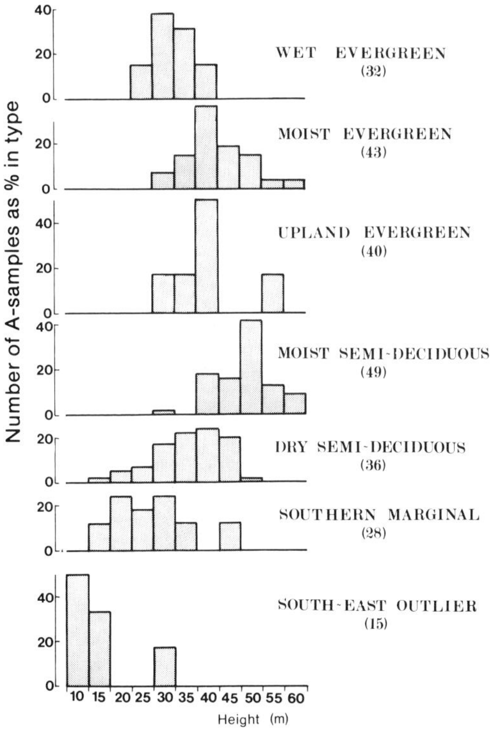

Fig. 5.1 Height (m) of tallest tree in or near each A-sample; mean for each type in brackets.

wettest forests appear to have the biggest trees (Whitmore 1975, p. 122) – but may perhaps be ascribed to limiting nutrient levels (Hall & Swaine 1976).

Tree species may be classified (Raunkiaer 1934) according to their maximum attainable height. Megaphanerophytes can exceed a height of 30 m, at which size they have on average a breast-height girth of about 100 cm (Rollet 1974). Heights for other classes are: mesophanerophytes (8–30 m), microphanerophytes (2–8 m), and nanophanerophytes (0.25–2 m). Most of the nanophanerophytes are better described as treelets or 'pygmy' trees than as shrubs, because they have a single main stem, but true shrubs, branching at or near ground-level do also occur, especially in the drier forest. Basal sprouting of trees is a conspicuous feature of South-east Outlier and Southern Marginal forest.

Branching characteristics have long been used by foresters as aids to identification in the field, and have been analysed and classified by Hallé & Oldeman (1970), and Hallé, Oldeman & Tomlinson (1978). We have assigned many of our mapped species to their models (see notes accompanying the maps). The architectural patterns (models) described by these authors are based largely on studies of saplings and understorey species, although it is sometimes possible to determine the model of a mature tree from the form of vertical branches which reiterate the pattern seen in saplings. There remains a need for a descriptive terminology (even if less exact) for mature crown form of large trees. Emergents such as *Entandrophragma utile, Klainedoxa gabonensis, Piptadeniastrum africanum* (Fig. 3.6), and *Cylicodiscus gabunensis* (Fig. 5.2) commonly bear crowns 30–40 m in diameter, but some tall trees, such as *Celtis mildbraedii,* have strikingly small crowns, seldom exceeding 15 m diameter.

The boles of most megaphanerophytes are buttressed at the base, but the degree of buttressing varies from fairly small, rounded buttresses that are little more than root-spurs, to the tall plank-buttresses stretching to 10 m or more up the trunk that are a feature of such trees as *Ceiba pentandra* (Fig. 5.3), *Piptadeniastrum africanum* and *Terminalia superba.* On the other hand some large trees, such as *Daniellia ogea,* have more or less cylindrical trunks with no trace of buttressing. It is widely supposed that buttresses offer support to the trunk (Henwood 1973); a review of 19 alternative theories have been presented by Smith (1972, see also Smith 1979). One could hypothesize that the need for buttresses could be related to other characteristics of the species such as mechanical properties of the wood, depth and extent of root system, and crown diameter (McMahon 1975), and to such environmental factors as wind force and direction, and soil depth. Unfortunately modelling is hampered by the great difficulty of investigating the root systems of mature trees.

The numbers of non-climbing phanerophyte species within the different life-form categories (Raunkiaer 1934) are shown in Table 5.2, in relation to their global distributions. It is clear (see also Fig. 5.4) that, on the whole, the larger trees are more widely distributed than the smaller ones.

5.2 Climbers

Climbers are abundant in all forest-types, comonly comprising about 40% of the total flora on small sample plots. They vary greatly in size (Fig. 5.5),

Fig. 5.2 Crown of *Cylicodiscus gabunensis:* emergent tree 50 m high near Kade.

Fig. 5.3 Ceiba pentandra near Kade showing buttresses 15 m high. Maize farm in foreground.

Table 5.2 Global distribution of non-climbing phanerophytes in the Ghana forest flora.

	Mega-	Meso-	Micro-	Nano-	Unclassified	Total
Endemic to Ghana	0	1	11	2	0	14
Non-endemics west of Dahomey Gap only	11	58	62	39	0	170
West Africa only, both sides of Dahomey Gap	14	68	75	38	1	196
More widely distributed in Africa	72	173	112	66	0	423
Occurring outside Africa	1	10	10	23	0	44
Unclassified	0	6	2	0	0	8
Total	98	316	272	168	1	855

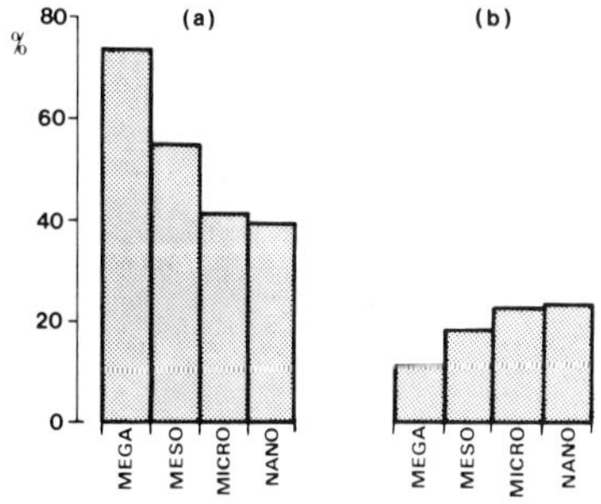

Fig. 5.4 Global distribution of Ghanaian forest phanerophytes. Each bar represents the percentage of species in a phanerophyte class which fall into one of the following distribution types:

(a) restricted to Africa but extending outside West Africa (423 species); (b) west of the Dahomey Gap only, excluding Ghanaian endemics (170 species). Data from Table 5.2.

Fig. 5.5 Mixture of different sizes of climbers in forest understorey in Atewa Range F.R.

ranging from slender-stemmed plants twining in the understorey, such as *Pyrenacantha* spp., to giant lianes in genera such as *Agelaea, Combretum, Salacia* and *Strychnos* whose stems may reach 30 cm diameter. Although the contribution of climbers to the basal area of forest is quite small, they may form a disproportionately large part of the canopy (Kira & Ogawa 1971). Their stem structure is frequently anomalous (Obaton 1960), and stem anatomy is often useful in identification (Hall & Lock 1975). The numbers of species of large and small climbers are shown in Table 5.3, together with their successional position. Small climbers are especially characteristic of recently disturbed forest, only 43% occurring in primary forest as against 87% for large climbers.

Table 5.3 Successional preferences of climber species.

	Large climbers	Small climbers	Unclassified	Total
Occurring in primary forest	212	153	1	366
Restricted to secondary forest	22	179	0	201
Restricted to farms and paths	1	12	1	14
Restricted to swamp forest	9	13	0	22
Unclassified	1	2	0	3
Total	245	359	2	606

5.3 Epiphytes

Vascular epiphytes are much less abundant and widespread than climbers, being absent completely from South-east Outlier forest, very rare in Southern Marginal forest, and abundant only in Upland Evergreen forest. The two main groups of epiphytes are the orchids and ferns, comprising respectively 121 (see also Hall & Bowling 1969) and 66 epiphytic species in Ghana, and thus amounting between them to 9% of the Ghana forest flora. Most of these species are, however, quite rare, and moreover, under-sampled in our forest survey because they are often hidden in the canopy. *Solenangis scandens* (Schltr.) Schltr. was the only epiphytic orchid recorded as many as 5 times during our forest survey. The only completely epiphytic ferns that may be regarded as common are *Asplenium africanum* (Map 59) and *Microsorium punctatum* (Map 383 & Fig. 5.7), encountered in respectively 23 and 43 A-samples during our survey. Some root climbers on tree boles, especially members of Araceae, may or may not be rooted in the soil; they may thus be facultative epiphytes.

Among woody epiphytes, species of *Ficus* are frequent. These may remain epiphytic, as its usual with *F. sagittifolia,* or send roots to the ground and eventually 'strangle' the host (Fig. 5.6). Hemiparasitic epiphytic shrubs in the family Loranthaceae seem to be much more abundant on isolated trees left by farming, or on the crowns of cocoa and cola plantations, than in mature forest; none were recorded during our survey. Other kinds of woody epiphyte are largely restricted to Upland Evergreen forest (Chap. 8.11).

Table 5.4 shows that 22% of epiphytes are restricted to Upland forest (above about 500 m) as compared to only 5% of ground-rooted species.

Table 5.4 Ecological restrictions of epiphytic and parasitic species as compared to ground-rooted holophytes*.

	Epiphytes	Parasites	Ground-living holophytes	Unclassified	Total
'Wet' forest	31	2	347	5	385
'Dry' forest	7	2	351	6	366
Upland forest	46	0	93	2	141
Mesic forest, or widespread	112	7	981	14	1 114
Unclassified	11	0	95	0	106
Total	207	11	1 867	27	2 112

* Holophytes are plants with normal photosynthetic metabolism, and not obviously parasitic or saprophytic.

Fig. 5.6 A strangling fig. (*Ficus thonningii*) on *Adansonia digitata* in the Shai Hills G.P.R.

Fig. 5.7 Microsorium punctatum, an epiphytic fern in forest understorey of Atewa Range F.R.

Johannson (1974) gives a valuable account of West African vascular epiphytes, largely based on his own studies in Liberia.

5.4 Phanerophyte leaves

Leaves and leaflets of woody forest plants are predominantly notophyll (area 30 to 45 cm^2), entire-margined, glabrous and acuminate. Large leaves – macrophyll and megaphyll – are mostly confined to the forest understorey. The leaves of many species, in fact, are much larger, more strongly toothed, more acuminate and hairier in the young sapling stage than in the canopy (Fig. 5.8). Small leaves are commoner and large ones rarer in the drier forest types (Fig. 5.9). Givnish & Vermeij (1976) present convincing physiological reasons as to why small leaves are adaptive for moisture conservation and management in drier environments – either in the upper layers of the canopy, or in regions with lower rainfall and longer dry seasons. In another paper Givnish (1978) proposes that compound leaves may be adaptive in seasonally arid environments in which deciduous trees are common, and in forest gaps; the compound leaf may be considered as a 'cheap throwaway branch'. Compound leaves are by no means confined to these environments in Ghana, and several species of canopy tree start life in the understorey with compound leaves which are replaced by simple leaves as the tree reaches the more arid canopy.

Mc Key *et al.* (1978) report that trees in forest on a nutrient-poor site have a leaf phenolic content twice as high as that of trees on richer soils, and suggests that phenolics may discourage predation by monkeys. Comparable data on phenolic or other secondary compounds in Ghanaian forests are lacking, but such differences are to be expected between our nutrient-poor evergreen forest types and the drier types with richer soils.

Small (1972) produces evidence from temperate vegetation that evergreenness may be advantageous in nutrient-poor habitats. Addicott (1978) discusses

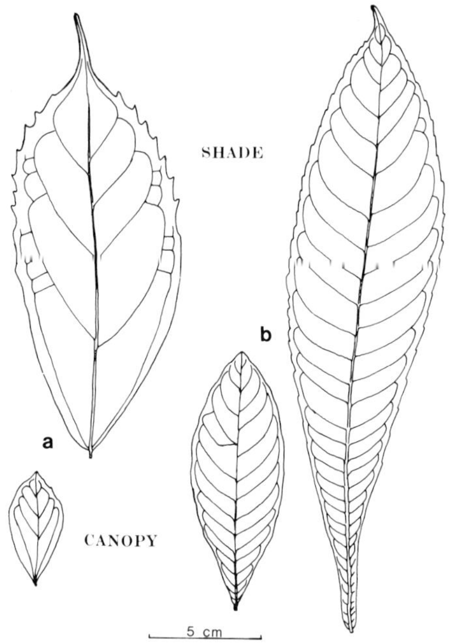

Fig. 5.8 Understorey and canopy leaves of (a) *Celtis mildbraedii*, a large tree, and (b) *Hugonia platysepala*, a liane.

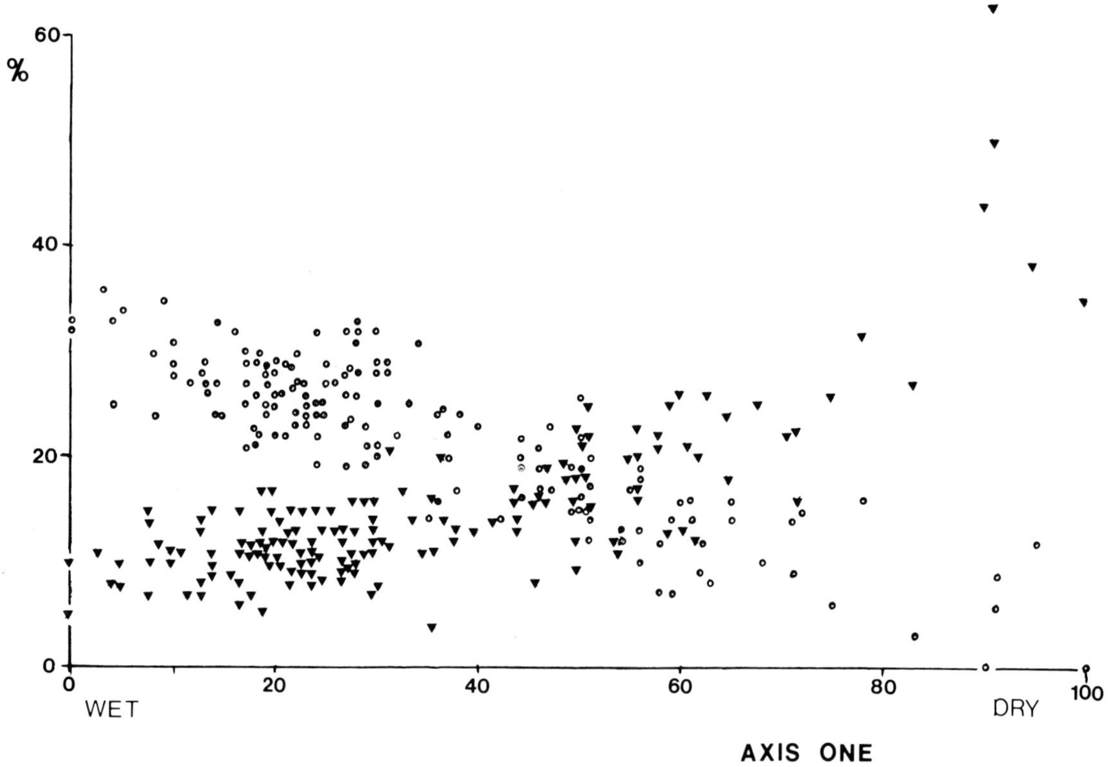

Fig. 5.9 Relation between leaf size and A-sample ordination score on axis 1 (Fig. 2.3). Percentages of species in each A-sample shown for the following categories: microphylls (inverted triangles), and sum of macro- and megaphylls (open circles).

the mechanism of leaf abscission with particular reference to seasonal drought, but suggests that leaf abscission can provide several benefits besides reducing moisture loss in dry seasons.

We have used degree of deciduousness to name most of our forest-types. Deciduousness is not, in fact, an easy parameter to assess; a complete range is found (Longman & Jenik 1974) from species which are always fully-foliaged *(Strombosia glaucescens)*, through those which change in leafiness from season to season without ever becoming quite leafless *(Anogeissus leiocarpus)*, those which flush new leaves just as the old ones are shed *(Lophira alata)*, to those which have a well-marked deciduous period *(Terminalia superba)*. There is no forest-type in Ghana which is completely deciduous at any time of year. Even in the driest type – South-east Outlier forest – complete leaflessness occurs only in some of the canopy species. Deciduousness is commonest in large trees (Table 5.5 & Fig. 5.10), and our assessment of forest types was therefore based on them. Let us use the index:

number of trees of deciduous large-tree species × 100 ÷ total trees of large-tree species.

Table 5.5 Deciduousness of species in the various classes of non-climbing phanerophytes.

	Mega-	Meso-	Micro-	Nano-	Unclassified	Total
Evergreen	51	290	262	166	0	769
Deciduous	43	24	8	2	0	77
Unclassified	4	2	2	0	1	9
Total	98	316	272	168	1	855

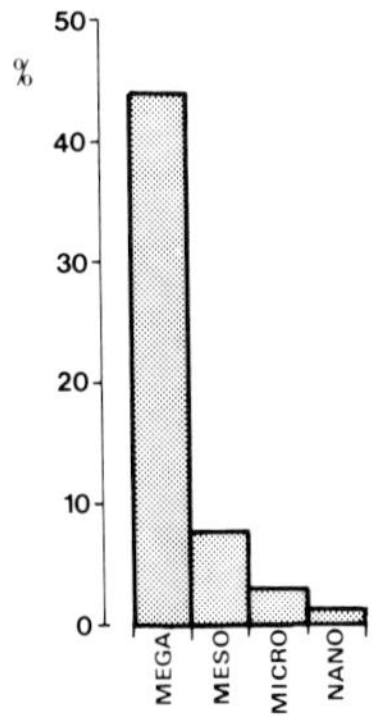

Fig. 5.10 Percentage of deciduous species in the various non-climbing phanerophyte classes in the Ghana forest flora (data from Table 5.5).

For our A-sample plots, we found the three evergreen forest-types to have about 10–20% deciduousness, and the two Semi-deciduous types to have 30–70% deciduousness. The Fire Zone sub-type of the Dry Semi-deciduous type is the driest forest which contains many megaphanerophytes and is the most strikingly deciduous (up to 100%) on this scale (Fig. 5.11). An exceptional fully-deciduous understorey tree is *Schumanniophyton problematicum* (Fig. 5.12), a rare species of Moist and Wet Evergreen forest. A few climbers are deciduous, for example *Chasmanthera dependens* Hochst., as are some epiphytic orchids, such as *Polystachya reflexa* Lindl.

Fig. 5.11 Deciduous canopy of Fire Zone subtype of Dry Semi-deciduous forest in Dede F.R.

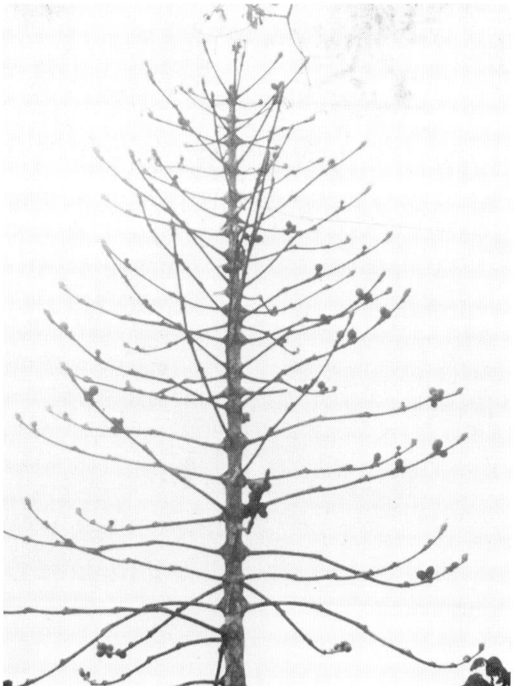

Fig. 5.12 Schumanniophyton problematicum – a deciduous understorey tree of evergreen forest.

Fig. 5.13 Herbs on the forest floor of Atewa Range F.R.
(a) *Selaginella versicolor* (Selaginellaceae); (b) *Costus engleranus* (Zingiberaceae);
(c) *Sarcophrynium brachystachys* (Marantaceae); (d) *Leptaspis cochleata* (Gramineae).

Fig. 5.14 Habenaria macrandra, a ground orchid on the forest floor near Kibi.

5.5 Ground herbs

The most abundant erect herbs (Fig. 5.13) of the forest floor are ferns, such as *Selaginella versicolor* Spring, various species of *Asplenium, Bolbitis* and *Pteris,* and also members of the Marantaceae such as *Ataenidia, Marantochloa* and *Sarcophrynium.* There are also specialised broad-leaved grasses, such as *Leptaspis* and the broad-leaved sedges *Hypolytrum* and *Mapania.* Creeping herbs include the peculiar *Costus engleranus* (Zingiberaceae), with succulent leaves, a very low rate of transpiration (Lawson *et al.* 1970) and a poorly-developed root system. *Geophila* and *Hymenocoleus* species (Rubiaceae) and various Acanthaceae are also members of this synusia. The herb-layer is best developed where the canopy is light and ground moisture abundant; such conditions are characteristic of Upland Evergreen forest, low-lying forest near swamp margins, the Fire Zone of the Dry Semi-deciduous forests, and in gaps (whether natural or man-made) in other forest types.

Herbs of unusual life-form include a single widespread holoparasite *Thonningia sanguinea* (Map 585), and a few rare holosaprophytic species including orchids, e.g. *Auxopus,* and the Burmanniaceae: *Gymnosiphon* and *Burmannia.* Geophytic herbs include orchids of the genera *Corymborkis, Eulophia, Oeceoclades (Eulophidium)* and *Habenaria* (Fig. 5.14), and the fireball lilies, *Scadoxus (Haemanthus).* Annuals are completely lacking.

CHAPTER 6

Reproduction

6.1 Vegetative reproduction

Many forest herbs reproduce mainly vegetatively from creeping stolons as in *Geophila* spp. and *Costus engleranus,* from buds on the inflorescences as in *Chlorophytum sparsiflorum* Bak., or from buds on the leaves as in *Scadoxus cinnabarinus* (Hall & Tomlinson 1973) and many ferns, such as *Bolbitis gemmifera.* A patch of a particular herbaceous species on the forest floor may thus be clonal. Vegetative reproduction of some climbers, such as *Urera* spp., is effected by development of adventitious roots from stems which fall to the ground. The stem of the climber *Cissus aralioides* may break up into separate internodes, each of which can regenerate.

Forest trees, in general, show little capacity for vegetative reproduction, and this seems especially true of secondary species such as *Musanga cecropioides* (Fig. 1.8) and *Trema orientalis.* Some small trees, however, typically reproduce vegetatively either from root suckers *(Millettia thonningii)* or by the layering of drooping branches as in *Sloetiopsis usambarensis* and *Scaphopetalum amoenum* (Jeník 1969). The last two species, growing in dry and wet forest respectively, can completely dominate the understorey over considerable areas, their dense shade inhibiting the regeneration of other trees and thus reducing floristic diversity (Fig. 4.1). Root suckering seems to be particularly common in Fire Zone forest.

6.2 Flowers and pollination

The floral biology of West African forest plants present an almost unexplored territory to the observer. Reasons for the paucity of knowledge include the uncommonness of flowers in the forest understorey, and the inaccessibility of the canopy. For those species with a characteristic 'syndrome' (Faegri & Van der Pijl 1971) of floral characters, it is possible to guess at the pollinator with little fear of error. Rubiaceous genera such as *Oxyanthus,* and apocynaceous genera such as *Tabernaemontana,* with a long white, tubular corolla that is sweetly scented at night will almost certainly be pollinated by very long-tongued hawk-months. Trees and climbers with red flowers, such as *Bombax, Combretum, Erythrina,* and *Hildegardia* are in many cases known to be pollinated by sunbirds. Bees may be expected to favour Papilionaceae and Zingiberaceae. Bat-pollination has been demonstrated in *Ceiba* (Baker & Harris 1959), but as yet in no other Ghanaian forest tree. Flowers of *Pararistolochia, Amorphophallus* and *Bulbophyllum* have the purplish colour, long bristles and offensive odour that are attractive to flies.

The great majority of species however, even including the orchids, possess small or medium-sized flowers, without any obvious adaptation to a particular type of pollinator. Some of the canopy trees in this category are probably wind-pollinated. *Celtis* is widely dispersed in modern pollen rain, presumably by wind (Hamilton 1974); *Chlorophora* and *Antiaris* are almost certainly also wind-pollinated. Recent observations on the small understorey and forest-margin shrub *Mallotus oppositifolius* (Lock & Hall 1981) illustrate the difficulty that may be encountered in assessing the probability of wind-pollination. *Mallotus* is dioecious, has an inconspicuous perianth, lacks nectar, and the male flowers produce large quantities of dry, powdery, pollen – all features suggesting wind pollination. But both sexes of flower have a sweet scent; male flowers are visited by many bees and syrphid flies of many species, but only the syrphid flies have been seen to alight on female flowers.

Many tropical forest grasses and sedges are known to be insect-pollinated, including *Hypolytrum* (Lorougnon 1973). Bawa & Opler (1975) have pointed out that many plants of tropical forest have unisexual flowers. Of the large families in West African forest, for example, Ebenaceae and Euphorbiaceae are invariably unisexual – the former being dioecious and the latter monoecious or dioecious. In Rubiaceae, the largest forest family, flowers are hermaphrodite but in many species are heterostyled (Baker 1958). Many Connaraceae (Baker 1962) and *Mostuea* in the Loganiaceae (Leeuwenberg 1961) are also heterostyled. In the Sapindaceae, there is the peculiar situation that flower sex may be determined at a relatively late stage of development; all flowers are potentially hermaphrodite, but flowers, inflorescences and even individuals at any one time, may be unisexual (Leenhouts 1973). A *Deinbollia* sp. (Sapindaceae) grown in a greenhouse at Wageningen, was observed to produce in successive flowering episodes male flowers only, then female flowers only, then male, and so on (Leeuwenberg, pers. comm.). Although Meliaceae have generally been considered hermaphrodite, it is now known that hermaphrodite flowers occur in only 8 out of 51 genera, the remaining genera having monoecious, polygamous or dioecious flowers (Styles 1972, Pennington & Styles 1975, Styles & Khosla 1976).

All these features promote out-crossing, and their prevalence casts doubt (Ashton 1969) on the assertion that self-fertilization is the rule in tropical forest species (Fedorov 1966). No careful study has been attempted on the breeding systems of Ghanaian forest trees comparable to those initiated in Malaya (Ashton 1976, Gan *et al.* 1977, Kaur *et al.* 1978).

Phenological observations on some Ghanaian forest trees, including data on flowering, are presented by Taylor (1960). For most species, however, labels on herbarium specimens provide the only record of phenology. In the notes accompanying our mapped species we have indicated seasons where these appear to be reasonably coherent and consistent. Where records are lacking or very few we omit any mention of seasonality. Species whose records of flowering and fruiting are scattered through much of the year are described as 'not clearly seasonal'; some of these might, in fact, have several distinct flowering seasons during the year, but careful field observation over several years would be needed to establish the fact. We have indicated fruiting season for fewer species than we have for flowering season, because fruits generally persist for much longer than flowers, and fruiting specimens are less often collected than flowering specimens. Widespread synchronization of flowering within a species is commonly observed. In mid-April 1977, for example, all the mature specimens of the understorey treelet *Oxyanthus pallidus* observed in forest near Kade were just coming into flower; anthesis was completed within a week. Four weeks later, in the same forest, *Oxyanthus formosus* (which had been vegetative a month previously) was opening its flowers, while *O. pallidus* was in young fruit. A few days later, flowering *O. formosus* was seen several hundred kilometres to the west, near Samreboi.

The factors responsible for this synchronization have seldom been investigated experimentally. It is known, however, that some *Coffea* species respond to heavy rainfall (Rees 1964) and that *Hildegardia barteri* is responsive to photoperiod (Hall, Okali & Lock 1975).

Semi-continuous flowering and fruiting is unusual in the forest, being unknown in the canopy and found only among a few understorey species, such as *Heisteria parvifolia, Microdesmis puberula* and *Salacia uregaensis,* and in secondary forest species, such as *Trema orientalis* and *Harungana madagascariensis.*

6.3 Fruits and dispersal

Peak fruiting of forest trees takes place in the dry season (Taylor 1960; Longman & Jeník 1974). Not all forest trees fruit regularly every year – *Triplochiton scleroxylon* is well-known for its erratic fruiting, and attempts have been made (Mackenzie 1959) to correlate heaviness of crop with the previous year's weather.

About 70% of tall trees and climbers have fleshy fruits, and this proportion scarcely varies between forest-types. The proportion of species with fleshy fruits (Table 6.1) is higher among understorey trees (about 80%), and lower among ground herbs (about 40%). Some kinds of fleshy fruit appear to be adapted to dispersal by particular kinds of animal. Large, fibrous fruits, such as those of *Desplatsia* are favoured by elephants, as are fruits with large woody pyrenes such as *Balanites wilsoniana* and *Panda oleosa*. Seedlings of all these species have been seen growing in elephant dung. Our attempts to grow *Panda* seedlings from ripe fruits collected below a parent tree were completely unsuccessful. No other forest animal can swallow the pyrene, so the extinction of elephants may well result in the disappearance of *Panda* (Alexandre 1977).

The most thorough study of fleshy-fruit dispersal yet made in Ghana was concerned with baboons in the Outlier forest of the Shai Hills Game Production Reserve on the Accra Plains. These animals deposit dung on flat rocks, whence it was collected every month for 15 months; the seedlings germinated from the dung were identified and counted (Lieberman, Hall, Swaine & Lieberman 1979). Of 100 fleshy-fruited species (including some savanna species) which occur in the Shai Hills the fruits of 36 are known to be eaten by baboons. Oil palm fruits are cracked and the kernels consumed; the seeds of many other species pass safely through the gut, their germination often being enhanced in the process. Fleshy fruits of all types are taken, ranging

Table 6.1 Relation of fruit type to tree size.

Fruit type	Phanerophyte type	
	Mega-	Meso- & Micro-
Fleshy	60	78
Wind-dispersed	29	6
Other	11	16
Number of species	45	63

Percentage representation among trees ⩾30 cm gbh (1 153 individuals of 115 species) in 2 ha of Moist Semi-deciduous forest near Kade.

from figs *(Ficus)* and *Nauclea* with very numerous tiny seeds embedded in the flesh, to *Vitex* with a single large pyrene, and the baobab *(Adansonia)* with many large seeds embedded in rather dry, sweet pulp within a large, brittle fruit.

Of the common primates in moist forest, the *Cercopithecus* monkeys seem to be very largely frugivorous, but a quantitative study of their diet is hampered by the impracticability of collecting dung samples and difficulties in observing the animals (P. Curtin, pers. comm.). *Colobus* monkeys are mainly leaf-eating; their effect on fruit is destructive (Mc Key 1978); they prefer it green, and are particularly fond of immature wind-dispersed fruits of genera such as *Mansonia, Hippocratea, Calycobolus, Pterygota* and *Triplochiton* (D. Olson, pers. comm.).

The other important forest frugivores are birds (especially hornbills, plantain-eaters and barbets), and fruit-bats. No systematic observations appear to have been made on these in Ghana but the small fruits of *Zanthoxylum xanthoxyloides* are popular with bulbuls, pigeons and starlings. Van der Pijl (1969) defines fruit-classes adapted to different dispersers. Mc Key (1975) distinguishes between specialist and non-specialist frugivorous birds. The former live more or less exclusively on fruit, favouring fruits with high nutrient content, and have a 'gentle' digestive system, allowing toxic seeds without a thickened testa or endocarp to pass quickly through the gut unharmed. Non-specialists have gizzards, which would rupture the seeds favoured by specialists, thus poisoning the birds; they prefer seeds which are well-protected mechanically or too small to be crushed, even when the fruits are not highly nutritious, as their diet includes other sources of nutrients. Much more work will be needed to determine whether the kinds of fruits taken by specialized as against non-specialized frugivorous birds do indeed differ in the ways predicted by Mc Key, and whether comparable relationships can be discerned among mammals.

Adhesive fruits are rare in forest, being found in some forest grasses such as *Leptaspis* and, surprisingly, in the large liane *Pisonia aculeata* Linn. Some epiphytes, especially those in Loranthaceae, and *Rhipsalis*, have 1-seeded berries with flesh that remains viscid even when passed through a bird's gut, so the seed adheres to the tree-branch on which it is defaecated.

Wind-dispersed propagules (Fig. 6.1) are of two main types: winged or plumed, and minute (dust-like). The former are found in about 20–30% of tall trees and lianes, 5% of understorey trees, and not at all in ferns or epiphytes. Dust-like propagules are, of course, found in all pteridophytes (except *Selaginella*) and in orchids; this propagule-type is consequently common in ground herbs (about 30%) and overwhelmingly predominant among epiphytes; it is not found in trees. Burrows (1975) discusses the aerodynamic features of wind-dispersed propagules and demonstrates the great importance of turbulent air-flow for dispersal of dust seeds.

Keay (1957) found for trees in a Nigerian forest a similar relation between height of crown and incidence of wind-dispersal. Clearly, for a given angle of descent the distance by which a seed is dispersed from the parent will be roughly proportional to its starting height. The crowns of emergent trees, moreover, are much better exposed to the wind than are crowns in the understorey. Although it is true that winged and plumed seeds may drop straight down in calm weather, spectacular dispersal effects may sometimes be observed in the strong gusting winds which precede the late dry-season storms: the fleece-invested seeds of *Ceiba pentandra* may be blown from tree crowns in great clouds, reminiscent of snow.

The family Celastraceae is represented in forest by three abundant genera: *Campylostemon, Hippocratea* and *Salacia,* the first two having capsular fruits and winged seeds, the last with fleshy berries. The 20 species of *Hippocratea* and *Campylostemon* known in Ghana are all lianes, none of which flower in the understorey though some, such as *H. indica* Willd., tend to be weedy and to flower in exposed situations along roadsides and forest margins. *Salacia* is represented in Ghana by 25 species, very similar to *Hippocratea* in flower and shoot morphology. Although many of these are also large lianes which flower in the canopy, some, such as *S. camerunensis,* are understorey climbers, and two are understorey shrubs. *Salacia* is never weedy.

Explosive fruits are found in 10–15% of species. The percentage does not vary greatly with stratum, this observation being in accordance with the finding of Beer & Swaine (1977) that distance dispersed is scarcely affected by height of release. Swaine & Beer (1977) investigated explosive dispersal in the exotic rain forest tree *Hura crepitans* Linn., and found modal and maximum dispersal distances to be 30 m and 45 m respectively. It is possible that thick-walled pods of such species as *Bussea occidentalis* may have comparable dispersing ability. The thick pod wall may give some protection to seeds against destructive predators (though *Colobus* monkeys open *Bussea* fruits and eat the seeds), and seeds may be larger than those capable of wind dispersal. Nevertheless, explosive dispersal is not a very 'popular' method in rain-forest, even in the Leguminosae. Seventeen species of leguminous trees are included in Table 6.1, but of these, only six rely on explosive dispersal; the remainder includes 5 wind-dispersed, 5 animal-dispersed, and 1 with no obvious mechanism.

Rinorea includes 11 species of understorey trees and shrubs, and its fruit is an explosive capsule containing dry seeds. Closely related to *Rinorea* is *Decorsella paradoxa* (syn. *Gymnorinorea abidjanensis*), which is remarkable in that the ovary dehisces shortly after fertilization, and the seeds develop exposed on the placentas (Hutchinson & Dalziel 1954; vol. 1 p. 105). The seeds have a fleshy orange testa, and the fruit bears a strong superficial resemblance to that of *Uvaria.* It is perhaps more reasonable to regard *Decorsella* as having evolved a more adaptive propagule type, than as demonstrating atavistic tendencies to gymnospermy.

6.4 Germination and establishment

The seeds of tropical forest trees are notorious for the briefness of their viability. Storage of seed is a major difficulty in attempts to propagate certain desirable species such as *Triplochiton scleroxylon* in plantations (Jones 1973). But while it is generally true that the seed of most trees and lianes of mature forest lacks dormancy, and germinates at once on the damp forest floor, there are many forest species which behave quite differently. If forest is cleared for farming, or a tree-fall exposes a patch of bare soil, seedlings quickly appear of species that are never seen in mature forest. It is commonly believed that the seeds of these plants arrive after clearing, but in fact most of them have lain dormant in the shaded soil, and germinated only after the canopy has been removed and light let in. Keay (1960) exposed topsoil from closed-canopy forest in boxes

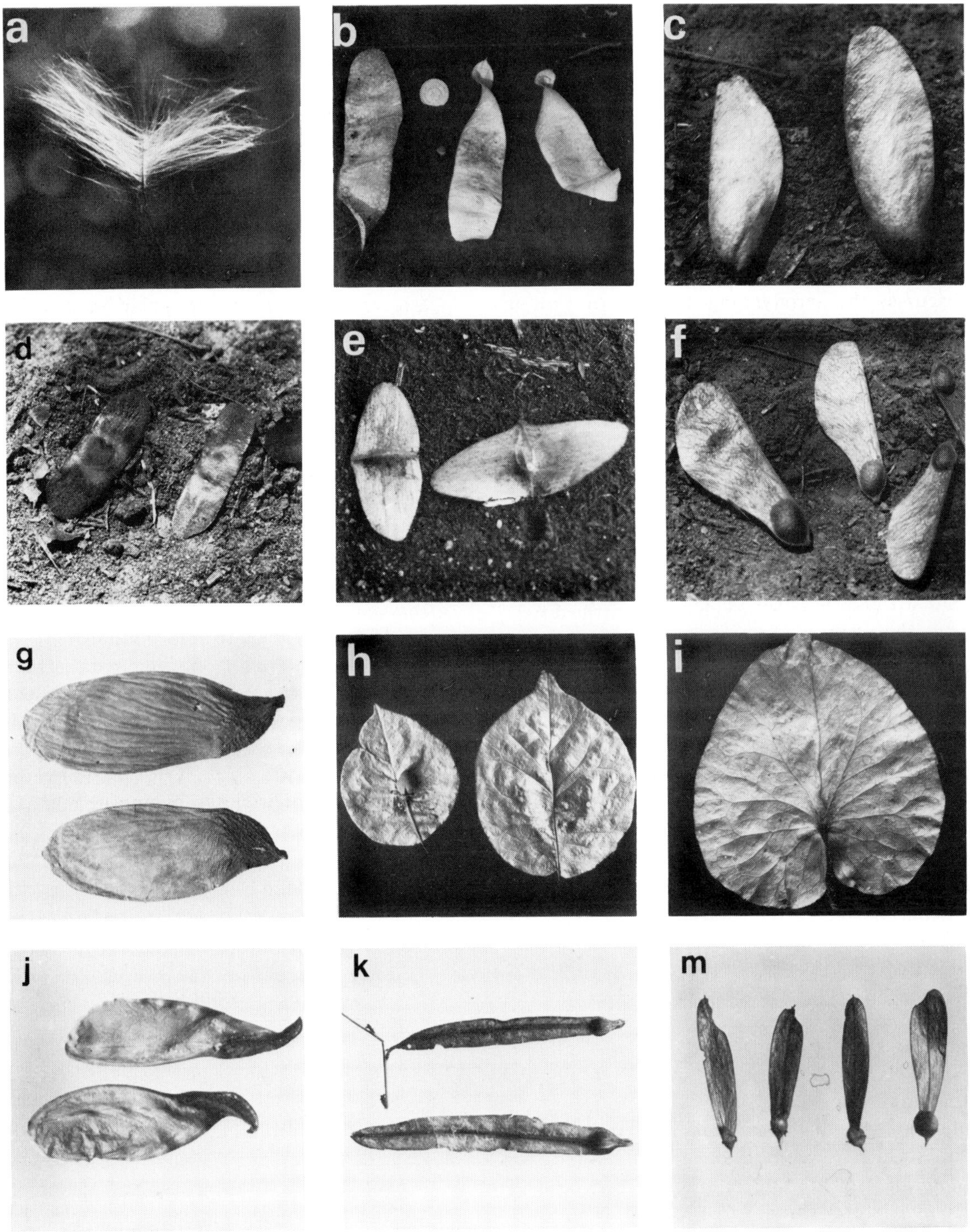

Fig. 6.1 Wind-dispersed propagules.

Trees: (a) *Funtumia africana,* plumed seed, in flight; (b) *Aubrevillea platycarpa,* thin pod; (c) *Hildegardia barteri,* fruit with inflated wing; (d) *Dalbergia* sp., fruit with elongated wing; (e) *Terminalia superba,* fruit with transverse wing; (f) *Mansonia altissima,* winged fruit; (g) *Pterygota macrocarpa,* winged seed.

Climbers: (h) *Neuropeltis acuminata,* leafy bract; (i) *Calycobolus heudelotii,* accrescent calyx; (j) *Hippocratea africana,* winged seed; (k) *Afrobrunnichia erecta,* winged pedicel; (m) *Ventilago africana,* fruit with terminal wing.

and counted the seedlings that appeared. We (Hall & Swaine 1981) have repeated Keay's experiment using soil from different forest-types, and compared germination success in boxes kept in semi-shade, under a black polythene open tent, with germination in open sunlight. Some of the results are summarised in Table 6.2. Enormous numbers of seedlings grew in some of the sun-lit boxes (Fig. 6.2); the boxes in partial shade usually had only about one-tenth as many, most of which showed very poor establishment. Only 10–25% of the species found in the sun-lit box were also found in the shaded box of the same sample pair; of the remainder, the majority presumably require light or wider fluctuations of temperature (Longman 1969) for germination. On the other hand, the few species which were found only in shaded boxes were uncommon species that were probably so restricted by chance.

Most of the species which germinated were found in a wide variety of primary forest sites. This suggests that the factors which influence the distribution of primary species (and hence the forest type) are less important to secondary species.

The most abundant species, *Solanum erianthum (S. verbascifolium),* occurred in all except the Outlier sample, and accounted for 48% of the total 1 907 seedlings recorded. Most widespread was *Trema orientalis* (see also Alexandre 1978), occurring in all samples and contributing 16% of the seedlings. *Musanga cecropioides* was the most abundant species in the Wet Evergreen forest, contributing 61% of the total in that sample. All three species are typical pioneers of forest gaps: shrub, small tree and medium tree respectively. All these species are difficult or impossible to germinate from fresh seeds, and are all likely to be dispersed by birds or fruitbats. The extraordinary density of seedlings arising from topsoil samples taken far from seed-bearers suggests that (a) dispersal is efficient, (b) passage through a gut may enhance germination, (c) viability is protracted. Experimental work is needed to investigate these points.

Grasses, forbs and slender understorey climbers were also well-represented among species needing exposure for germination, and a particularly interesting group is that of large trees with a similar requirement. Of these, *Chlorophora* spp. and *Nauclea diderrichii* have tiny seeds, about 1 mm diam, similar in size to those of *Musanga* and *Ficus* spp. *Alstonia boonei* and *Zanthoxylum* spp. have rather larger seeds, 2–3 mm in diameter, similar in size to those of *Trema*. The seeds of *Anthocleista* spp., *Ricinodendron heudelotii* and *Terminalia ivorensis* are quite large, up to 10 mm in diameter (as are those of *Lophira alata,* an ecologically similar species not found in our boxes). These species can therefore regenerate only in rather large gaps. In closed-canopy forest, they can be found only as big trees, and as these are not likely to occur within a small 25 m × 25 m plot, we undersampled them or missed them.

A succession of secondary tree species in a large forest gap in Upland Evergreen forest of the Atewa Range F. R. is shown in Fig. 6.3, which presents

Table 6.2 Numbers of seedlings and species germinating in topsoil[a] collected from 1 m² quadrats in closed-canopy forest of different types.

Treatment of seed-boxes	Forest-type									
	Wet Evergreen		Moist Semi-deciduous		Dry Semi-deciduous		Southern Marginal		South-east Outlier	
	Seedl.	Spp.	Seedl.	Spp.	Seedl.	Spp.	Seedl.	Spp.	Seedl.	Spp.
Full sunlight	171	21	339	31	693	36	394	42	114	25
Shade[b]	12	6	37	4	29	8	21	4	13	4

[a] Topsoil was removed to a depth of c. 3 cm, and mixed thoroughly before being spread 3 cm deep over 3 cm of heat-sterilized soil in wooden trays. Soil was regularly watered, and counting continued for about 100 days.

[b] Shade was provided by a black polythene canopy which reduced illumination to about 0.25% of full sunlight.

Fig. 6.2 Seedlings of secondary forest species germinating in topsoil removed from Dry Semi-deciduous forest at Odoben (see Fig. 8.8), exposed and regularly watered. The sedge is *Mariscus alternifolius;* the young shrub with broad, pinnately-lobed leaves is *Solanum torvum; Carica papaya* has palmately-lobed leaves; the tall flowering herb is *Crassocephalum crepidioides;* the young tree with ovate leaves with basal nerves is *Trema orientalis.*

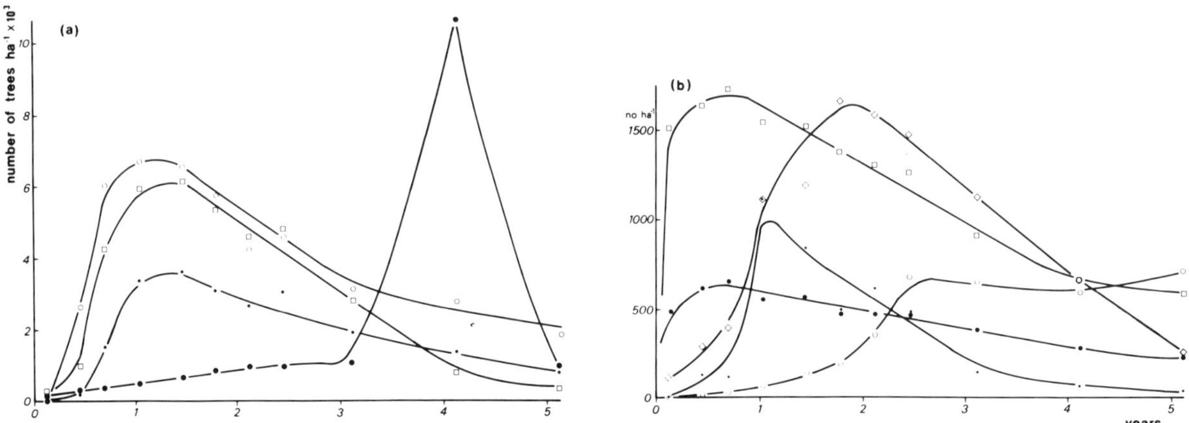

Fig. 6.3 Changes in population density of trees colonising from seed a large clearing in Upland Evergreen forest in the Atewa Range F.R. (see text for discussion).

Key to symbols:

(a) *Albizia zygia* ●, *Harungana madagascariensis* ·, *Musanga cecropioides* □, *Terminalia ivorensis* ○.

(b) *Funtumia africana* ○, *Margaritaria discoidea* ◇, *Nauclea diderrichii* ·, *Rauvolfia vomitoria* □, *Trema orientalis* ●.

repeated counts of tree seedlings and saplings on a permanent transect established a month after clearing. *Rauvolfia vomitoria* and *Trema orientalis* germinated most quickly, almost reaching their maximum density within the first month. *Harungana madagascariensis, Musanga cecropioides, Nauclea diderrichii* and *Terminalia ivorensis* germinated more slowly, but had all reached maximum density within a year. *Terminalia* seedlings developed from winged fruits blown in from trees around the gap shortly after clearance; the seeds of other species were presumably lying dormant in the soil as where those of *Margaritaria discoidea.* The second wave of trees – *Albizia* spp. and *Funtumia africana* – arose, like *Terminalia,* from efficient, wind-dispersed propagules which are capable of germinating under the shade of young secondary growth, but rarely become established under a mature canopy. *Pericopsis elata* (absent from Atewa) appears to belong to the 'second wave' (Ampofo & Lawson 1972).

Finally, the trees of mature forest will invade: these include such species as *Guarea cedrata* and *Strombosia glaucescens.* Their seedlings survive on the floor of mature forest, and respond with faster growth to slight gaps in the canopy. It is this category of tree which was most adequately sampled during our survey.

CHAPTER 7

Utilisation and conservation of Ghanaian forest

7.1 History of forest exploitation for non-timber products

Archaeological evidence shows that man has lived in the forest zone of Ghana for many thousands of years (Dickson 1971). Stone Age populations must, however, have been quite small and have had little impact on the forest. There is no information on the date of introduction of agriculture, but the transition to a 'Neolithic' pottery-using culture has been dated at about 3 000 B.C. in the northern part of the forest zone near Abetifi (Smith 1975). Stone tools were apparently still being used in this area as late as 1 300 A.D. (Posnansky & McIntosh 1976), but just north of the forest boundary there is evidence for iron-working as early as 200 A.D. (Posnansky & McIntosh 1976); it is likely that these iron-workers were agriculturalists. Fairly large towns are known to have been founded near the forest-savanna boundary at least as early as 1 100 A.D. (Posnansky & McIntosh 1976) and are suspected to have traded in gold obtained from the forest zone. The first of a series of important forest states, Akwamu, arose about 1 500 A.D. Until the mid-nineteenth century, however, disease and warfare combined to keep the population low. In 1853 the population of Southern Ghana was roughly estimated at 250 000. The first true census in 1891 gave about 800 000 for Southern Ghana. The figure increased to 1 million by 1920, and was about 4 million in the most recent 1970 Census (Gaisie & De Graft-Johnson 1976).

Before 1850 exploitation of the forest was largely restricted to subsistence agriculture. At any given time, only a small part of the forest zone would have been under cultivation. About the middle of the nineteenth century there was an upsurge in the production of palm oil for export (Dickson 1971), but the areas of cultivation were concentrated along the southern and eastern fringes of the forest zone. Gum copal was also exported in some quantity (up to 500 tons in 1876), but was collected from *Daniellia ogea* trees growing wild in the forest. Rubber was tapped from wild *Funtumia elastica* trees and *Landolphia owariensis* lianes; annual production reached 3 000 tons by the end of the nineteenth century. Kola *(Cola nitida)* has been exported from the Ghana forest zone to the northern savanna country since prehistoric times, though the volume of this inland trade cannot be determined. The kola tree is indigenous in the forest, but was also cultivated, especially when the trade from the coast increased – reaching an annual level of 14 000 tons in 1921.

The first introduced export crop of importance to be grown in the forest zone was coffee, annual exports peaking at about 70 tons in the 1890's. But all the foregoing crops are insignificant in comparison with cocoa *(Theobroma cacao* Linn.). From its introduction by Tetteh Quarshie in 1878 cocoa (or cacao as it may be called) was adopted eagerly by farmers in the eastern part of the forest zone, and by 1911 Ghana was the leading producer of the crop – a position it held until 1978. Between 1909 and 1923, annual cocoa exports rose from 20 000 to 200 000 tons. Cocoa farming techniques were soon carried into forest in Akim and Ashanti by migrant farmers (Hill 1963). From 1923 to 1959 cocoa exports ranged between 200 and 250 thousand tons. By 1965 production had increased to 550 thousand tons; then there was a gradual decline to about 300 thousand tons in 1978.

Although the volume of timber removed from Ghanaian forest is great, the volume of fuelwood is vastly greater. Foggie and Piasecki (1962) estimated annual consumption as 4.5 million m^3 used directly as firewood and 1.5 million m^3 converted to charcoal (some of this being derived from the savanna zone). Recent Forestry Department annual reports state only the quantities known to be taken from forest reserves – a trifling amount of about 5 000 m^3. Virtually all fuelwood is obtained from trees felled when land is cleared for farming. The population of Ghana has almost doubled since Foggie and Piasecki reported their estimates and, if fuelwood consumption has increased proportionately, it seems likely that at least twice as much wood is now burnt as is used for timber. Fuelwood is, in fact, by far the largest source of domestic energy in Ghana.

Apart from fuelwood, the villager obtains from indigenous forest plants a wide range of other commodities, which are exhaustively listed by Irvine (1961). These include rope, washing sponge, poles and thatch for house-construction, wild fruits, medicines, domestic utensils and so on. Most of the plants used for these purposes are relatively abundant, harvesting scarcely affects their numbers. The same is not true, however, of the trees most used as chewing sticks: *Garcinia epunctata* and *Garcinia kola* (Adu-Tutu *et al.* 1979). These trees are already rare, and could well become extinct in Ghana if cutting continues at the present rate.

Recently there has been interest in exploiting as alternatives to synthetic sweeteners the intense sweetness of the fruits of three indigenous forest plants: *Thaumatococcus daniellii* (Benn) Benth., *Synsepalum dulcificum* (Schum. & Thonn.) Daniell, and *Dioscoreophyllum cumminsii* (Stapf) Diels (Cagan 1973). Attempts are being made to establish plantations, and it is possible that they may be developed as crops.

7.2 The Ghana timber industry

Foggie and Piasecki (1962) report that the earliest recorded export of Ghana timber was of *Lophira alata* which was used for the keels of Elizabethan ships in the sixteenth century. Be that as it may, the timber in the oldest houses and castles along the coast of Ghana is imported pitch pine, and we may assume that there was no local timber industry before the time of the first exports of West African mahogany (*Khaya* spp.) in 1888. An exception was the use of wawa logs (*Triplochiton scleroxylon*) for hollowed-out fishing canoes. Presumably these must have been dragged to the coast.

Accounts of the development of the Ghana timber industry are given by Logan (1946 and 1947), and Dickson (1971) as well as by Foggie & Piasecki (1962). The figures for timber exports shown in Fig. 7.1 are derived from these, and from annual reports of the Ghana (Gold Coast) Forestry Department. The most striking feature of Fig. 7.1 is the enormous increase in exports immediately after the Second World War. Until this time mahogany accounted for about 90% of exports; Fig. 7.2 shows that the rapid rise in popularity of *Triplochiton* (wawa, obeche) as an export timber was mainly responsible for the overall increase. *Triplochiton* was first exported in 1948, but by 1953 it had

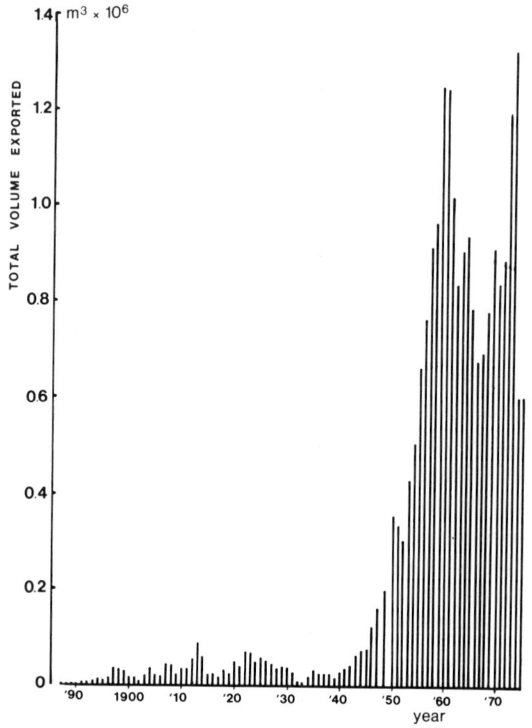

Fig. 7.1 Total volume of sawn timber and logs exported from Ghana between 1887 and 1975.

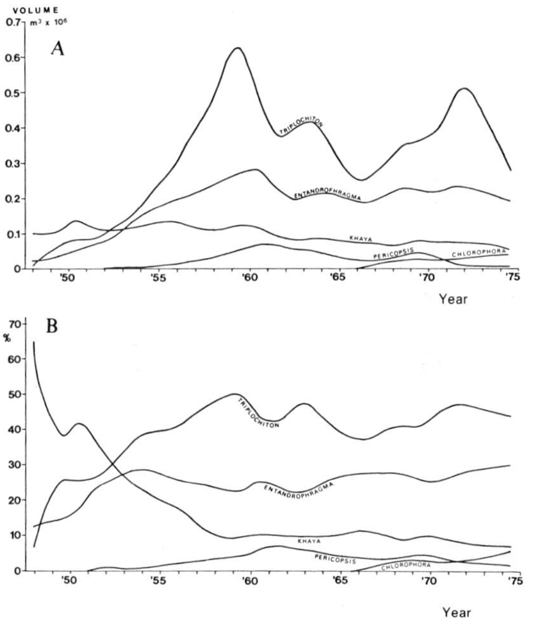

Fig. 7.2 Export volume trends (smoothed moving averages of two years) in five major timbers (*Chlorophora, Entandrophragma* and *Khaya* each comprise several species).
(a) Volume; (b) Volume as percentage of annual total.

overtaken *Khaya* in volume exported and since 1955 has generally provided more than 40% of total exports. *Entandrophragma* spp. (sapele, edinam, utile, omu) provide fine timber very similar to mahogany – it is likely, in fact, that part of the exports of 'mahogany' before the Second World War were in fact *Entandrophragma* – and since the mid-1950's they have held second place to *Triplochiton.* Fig. 7.2a shows that *Khaya* exports have declined somewhat since 1956, presumably because of increasing rarity of timber-sized trees in the forest.

Pericopsis elata (kokrodua, afrormosia), which yields the most expensive Ghanaian timber, was first exported from the north-western part of the forest zone (to which it is restricted, Fig. 11.6) in 1952. Its decline since 1961 again reflects the exhaustion of stocks in the forest.

Chlorophora (odum, iroko) is abundant and well-known in many parts of the forest zone (Fig. 11.1), but until 1966 export was forbidden in order to preserve supplies for local use. Subsequent increase is due entirely to the export of sawn timber; the export of logs is still forbidden.

The proportion of total exports derived from species other than the above has not altered greatly since the war, though the volume rose during the post-war expansion of the timber trade.

The Ghana Forestry Department classifies timber by value (Appendix 7). Class 1 includes 14 species, class 2 has 12 species and class 3 has 23 species. All the species graphed in Fig. 7.2 belong to class 1. All the class 2 timbers are exported but, like class 3 have not always been exported in quantities commensurate with their qualities and abundance. *Piptadeniastrum africanum* (2), *Celtis* spp. (3), and *Pyranthus angolensis* (3) are striking examples of under-utilisation.

In recent years *Guibourtia ehie* (shedua), though formerly unclassified, has been successfully promoted as an export timber and now has Class 2 rating.

Total annual timber production (including local use) has averaged about 1.85 million m³ since 1958; during this time between 50% and 70% has been exported. Thus about 23 m³ of timber has been extracted annually from each square kilometre of the forest zone. Since the average extracted volume of a tree is 15 m³ (Annual Report of Ghana Forestry Department 1972), slightly more than one tree per square kilometre is extracted each year.

Because of the great variation within the forest zone in density of human settlement, and consequently also in forest cover, such a figure is difficult to interpret; data from forest reserves are more meaningful. Fig. 7.3 shows the rapid increase since 1958 in the volume of timber extracted from forest reserves. The contribution to Ghana's timber production made by the reserves rose dramatically from 13% in 1958 to 78% in 1975. Table 7.1 suggests that the rate of recruitment in the forest reserves as a whole is somewhat less than the rate at which trees are being removed for timber. It is not possible, however, to construct an exact balance sheet. Firstly, mean growth rates differ greatly from one species to another; Baur's graph (1964, p. 148) shows figures for average time of passage from 2.7 to 3 m girth ranging from 4 years for *Triplochiton scleroxylon* to 19 years for *Lophira alata.* The subdivision of the timber classes as (a), (b) or (c) (Appendix 7) is, in fact, made according to growth rate (Baidoe 1976). In the same paper, Baidoe states that recruitment rates are inversely correlated with rainfall, and thus differ between forest types. Sec-

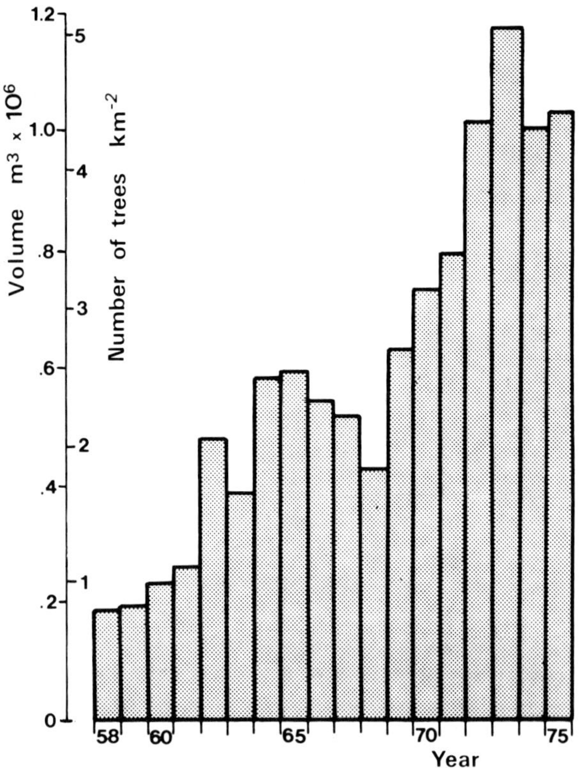

Fig. 7.3 Volume of timber extracted from forest reserves 1958–1975; exported and locally consumed timber included. The values for number of trees per km^2 are based on the total area of forest reserves in the forest zone, and the mean volume per extracted tree (see Table 7.1).

ondly, growth rates are determined from repeated measurements of trees in girth increment sample plots, and variation between individual trees in such plots is commonly found to be very large. Two considerations not taken into account in Table 7.1 mean that the useful recruitment is less than the table suggests: (1) not all of the forest reserves can be exploited for timber – in many the terrain is too steep and rocky (p. 69) – and (2) not all potentially marketable trees are in fact cut by timber firms. Thus our estimate may be taken to lead to the conclusion reached by the more detailed analysis of Baidoe (1976): rates of cutting exceed recruitment rates.

Ghanaian forest has largely been managed under a selection system (Baidoe 1970), with a felling cycle of 25 years, reduced to 15 years in 1971. Before felling, treatment, combined with stock survey ('combined operations'), involved the arsenite poisoning of uneconomic large trees and trees of bad form. Promising saplings were freed from competition by weeding and climber cutting. 4 300 km^2 (more than a quarter of the total reserved area) was treated in this way up to 1971 (Britwum 1976) but doubt as to the value of the treatment has now led to its abandonment.

Although selection felling seldom extracts more than 2 trees per ha, many more trees are destroyed

Table 7.1 Estimated balance sheet for removal by cutting, and natural recruitment of marketable tree species in Ghana forest 1975.

	Vol. (m^3)	Vol./area[a] (m^3 km^{-2})	Number of trees[b]/area (trees km^{-2} yr^{-1})
A. *Removals*			
Total for forest zone	1.85 × 10^6	23	1.5
Total for forest reserves only	1.44 × 10^6	96	6.4
B. *Recruitment to 3 m girth* (approx. exploitable size)			
Mean density of all trees in 2.70 – 2.99 m girth class[c] (trees km^{-2})			150
Average duration of growth from 2.70 to 3.00 m gbh[d]			20 years
Recruitment rate for all trees (trees km^{-2} yr^{-1})			10
Proportion of marketable trees in 2.70 – 2.99 m girth class			0.3–0.5
Recruitment rate of marketable trees (trees km^{-2} yr^{-1})			3–5

[a] Area of forest zone c. 82 000 km^2; Area of forest reserves 15 000 km^2.
[b] Mean volume of extracted trees c. 15 m^3 (Forestry Dept. Annual Report 1972).
[c] Rollet (1974 p. 120).
[d] Baur (1964 p. 148).

incidentally as the tree falls, and as a result of the construction of access roads. Young trees and saplings are also damaged: Mensah (1966) found about 10% of established trees to be damaged after normal selection felling in the 77 ha forest compartment which he investigated. Dawkins (1959) cites felling damage as a major constraint on the productivity of the selection system. The large gaps in the canopy are rapidly covered by climbers, which tend to inhibit tree growth, and the remaining isolated trees are more liable to be blown down or to lose large branches during storms. The forest at Kade described by Lawson *et al.* (1970) was cut over for timber about 1920; sixty years later the hauling tracks are still plainly visible in the undergrowth, and the upper canopy has many gaps. Forest that is very rich in economic species, for example Bia-Tano Forest Reserve, may have scarcely any canopy left; in such forest we had difficulty in finding enough intact forest to contain a 25 m × 25 m plot during our survey.

The Nigerian Forestry Department tried to convert much of their forest estate to a uniform condition by the Tropical Shelterwood system (TSS), but this system has now been abandoned since it proved both expensive and generally unable to provide an adequate stocking of valuable young trees (Adeyoju 1975, p. 76–77 and Lawton 1978). In Ghana, TSS was tried, on an experimental basis, in Bobiri and Asenanyo Forest Reserves, but is not likely to be more widely applied (Osafo 1969).

Up to 1967 the Forestry Department made little attempt to establish extensive plantations within the forest zone. In 1971 an ambitious annual planting target was set of 65 km^2 (mainly of teak, *Terminalia ivorensis, Gmelina, Eucalyptus, Cedrela* and pine), mean annual planting from 1966–1971 having been 19.2 km^2. In 1972 79.6 km^2 were planted, but by 1975 clearing costs and problems of seed supply had reduced the figure to 28.5 km^2 per annum. Even at the 1972 rate, it would take about 170 years completely to replant the forest reserves. For many years to come, the Ghana timber industry will continue to depend mainly on natural regeneration. Kio (1979) warns of the great risks in over-reliance on plantations of exotics in the tropical forest, and recommends continued research into the ecology of natural forest communities to provide a firmer basis for their scientific management.

In the long run, however, it seems inevitable that forest reserve exploitation must become more intensive. The main reason for the sharp increase in the contribution of reserves to total production since 1958 (Fig. 7.3) is simply that closed forest outside the reserves has almost all been destroyed (p. 7). If Ghana's export timber trade is to survive, it may well become necessary to reduce the minimum permissible girth for extraction from the present limits (according to species) of 210 to 330 cm and to promote the utilization of a wider range of species. Conversion to plantations will be concentrated in areas where stocking of desirable species is at present lowest; these are the higher rainfall areas (Wet and Moist Evergreen forest-types), and in the drier forest (Fire Zone subtype of the Dry Semi-deciduous type).

Pulp mills which could process a mixture of all tree species obtained by clear felling have been planned for Gabon and Ivory Coast (Goudet 1975). Part of the revenue from this process could be used to establish plantations of fast-growing species such as eucalypts, *Gmelina* and pines for continued operation of a pulp industry. The government hopes to establish a similar mill with an annual capacity of 60 000 tons of paper pulp at Daboase on the River Pra, drawing supplies from 100 km^2, mainly plantations, in the nearby Subri River Forest Reserve.

Another government project envisages an iron and steel complex to be established near Takoradi, using iron ore from Opon Mansi Forest Reserve. The annual requirement of charcoal would be 120 000 tons, equivalent to 150 km^2 of forest, or a much larger area of plantations.

It is clear from the foregoing that very little of forest as we know it today is likely to survive beyond the next century. This book will be a valuable record for the 22nd century botanist of the distributions which forest plants used to have, but is unlikely to be of much direct help in his attempts to find them.

7.3 Forest and wildlife conservation

Before 1880 farmed land was allowed to revert to forest after two or three years of cropping. Because the population was quite small (p. 63), almost all of the forest zone was in fact covered by mature forest. By 1900, however, forest was being rapidly cleared for cocoa farms – from which forest is permanently excluded.

Alarm at the increasing rate of destruction of forest led administrators to feel the need for legislation to conserve at least part of the forest. In consequence H. N. Thompson, a forester of wide experience in the Indian and Nigerian colonial forest services, was invited to inspect the forests of Ghana and to make recommendations for their administration. Thompson made an extensive tour of Ghana in 1908. On his advice, a Forestry Department was established and the first Conservator (N. C. Mc Leod) appointed in 1909, even before the publication of his report (Thompson 1910). A Forest Bill, proposing the establishment of forest reserves, was drafted in 1911, but opposition from local people prevented it from becoming law. Between 1909 and the beginning of the First World War in 1914, the small Forestry Department staff toured the forest zone, improving knowledge of the flora and looking for areas suitable for eventual reservation. After the war, unavailing efforts were made to persuade local authorities to reserve part of their forests voluntarily. Eventually a Forest Ordinance was passed in 1927, giving the Government power to order forest reservation where local authorities persisted in their refusal. A target was set of 6 000 square miles (15 500 km^2) of reserve, that is about 20% of the forest zone; at the beginning of the Second World War in 1939, by dint of courage and persistence in the face of much opposition, this goal had been virtually achieved (Logan 1946). Since 1939, the size of the permanent forest has continued to increase, though more slowly (Fig. 7.4).

Cocoa farms existing within forest reserves at the time of demarcation have generally been allowed to persist. Especially in the Volta Region, the area of allowed farms has tended to increase, and some of the ostensible 'reserves' there now consist of more farms than forest. On the whole, however, the Department has been very successful in checking illegal farming; 525 illegal farmers were successfully prosecuted in the courts during 1972, for example. The creation of the artificial Volta Lake in 1964 resulted in the partial flooding and abandonment of six forest reserves in the Afram and Pawmpawm valleys, this being the single largest loss to reservation.

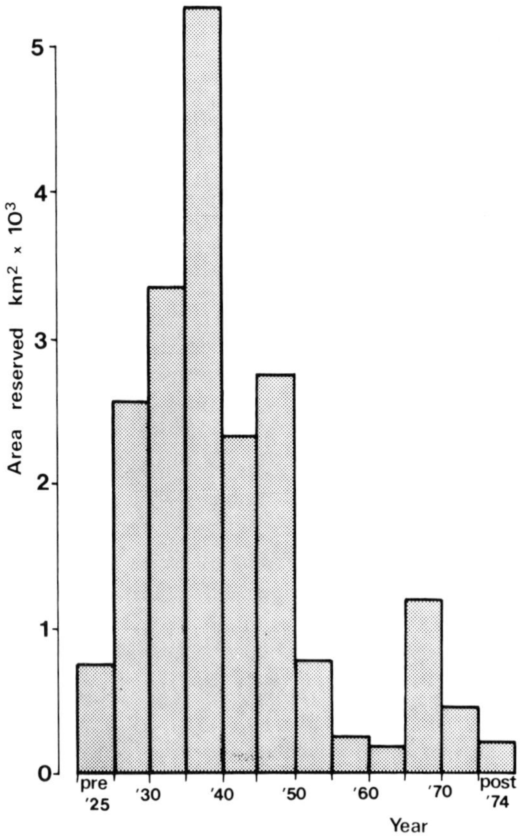

Fig. 7.4 Estimated area of forest reserves added to permanent forest estate in 5-year periods.

Although the forest reserves now supply nearly all of Ghana's timber (Fig. 7.3), Thompson and the other early foresters were as much concerned to prevent environmental deterioration as to safeguard the future of a timber industry which was then, in any case, on a very small scale. Thompson (1910, p. 11) described sombrely the danger that savanna will creep into the forest from the Accra Plains, and urged large-scale planting of *Casuarina* and *Cedrela* as a remedy. He also worried (*op. cit.* p. 97) about the little patches of savanna within the northern forest margin – which we now regard (p. 81) as ancient relicts. The priorities of the early Forestry Department were, therefore, to ring the forest zone with reserves that could stop the savanna from advancing, to reserve hills in order to safeguard water supplied and prevent erosion, and to keep the atmosphere sufficiently moist for forest crops such as cocoa by scattering reserves throughout the forest zone. Because of this concern for the environment, one quarter of the forest estate consists of 'protection reserves', too hilly or rocky to be easily worked for timber; the remaining three-quarters comprises 'production reserves' (this classification is, however, informal and without legal implication).

Reserves for the preservation of game animals in Ghana were legally constituted as far back as 1909, but unfortunately it was not until 1965 that a Department of Game & Wildlife was established to make the law effective. During the interim hunting continued unchecked. (Asibey 1970 & 1972). Since 1965 the Department has organised effective administration of many of the previously existing game reserves and wildlife sanctuaries, and is pursuing a policy of securing wildlife reserves in all the major ecological regions of the country. All the Department's reserves (wildlife conservation areas) are administered according to the provisions of the Wild Animals Preservation (Amendment) Ordinance of 1952. Within the limits imposed by the Ordinance the Department can, however, formulate a management policy for each reserve according to its size, richness in animal life and degree of freedom from disturbance. The various kinds of reserve are titled in accordance with the definitions accepted in the African Convention on the Conservation of Nature and Natural Resources (Organisation of African Unity 1970). Commercial lumbering, for example, is incompatible with the status of 'National Park', but controlled cutting of timber (provided that animal life is not threatened) may be permitted in Game Reserves (including Wild Life Sanctuaries and Game Production Reserves).

By far the largest Wildlife Reserves lie in the northern savanna. Fig. 1.3b and Appendix 1(b) show the positions of those within the forest zone; many of them were formerly forest reserves or parts of forest reserves, and these are indicated in Appendix 1(b).

It is a major aim of conservation policy to preserve intact adequate samples of forest of each of the major types. Table 7.2 shows the proportions of the areas of these types which lie within forest or wildlife reserves. All of the Upland Evergreen type is so protected, and about 30% of the other evergreen types. Most South-east Outlier forest is protected within the Shai Hills G.P.R. Least protected, therefore, are the Inner Zone of the Dry Semi-deciduous type and the Southern Marginal type. Most of the surviving specimens of these types consist, in fact, of forest patches preserved by villagers as sacred groves. As traditional beliefs weaken in the face of western education and religion, these groves will be increasingly endangered, and it is to be hoped that ways may be found to ensure their continued conservation.

Table 7.2 Number and areas (km^2) of forest and wildlife reserves in the various forest types.

	Forest type or subtype									
	WE	ME	UE	MSSE	MSNW	DSIZ	DSFZ	SM	SO	Total
Number of reserves entirely within the type (or subtype)	9	38	0	52	25	10	29	10	1	174
Area of entirely-included reserves	1 374	4 400	0	1 859	3 567	136	2 592	62	0.8	13 991
Number of reserves lying in more than one type (or subtype)	6	12	4	13	6	12	6	2	1	31
Area of partly-included reserves	536	1 131	292	632	474	557	290	31	5.0	3 948
Total reserved area in type	1 910	5 531	292	2 491	4 041	693	2 882	93	5.8	17 939
Total area of type	6 570	17 770	292	18 460	14 430	8 630	12 810	2 360	20	81 341
Percentage of type reserved	29	31	100	13	28	8	23	4	29	22
Percentage of total reserved area	11	21	1.5	14	23	4	16	0.5	0.03	100

Data from Appendix 1(a) and (b), and Table 8.1.
Abbreviations for forest-types as in Appendix 1.

We must not assume, however, that forest is assured of adequate conservation because it lies within a reserve. At present the best forest is within 'protection forest reserves', but as the demand for timber rises, these will be increasingly cut over. The production forest reserves on comparatively level ground are in imminent danger of profound alteration (p. 67).

It is for this reason that forest National Parks, though at present small in area, are so important. Unfortunately, their future is at present doubtful. Bia National Park was established in 1974 with an area of 305 km^2. Within 2 years this had been reduced to 230 km^2, and the following year further reduced to only 80 km^2. The park was set up largely to provide an ecosystem in which endangered fauna such as various primates and forest elephants can survive. Elephants are very important in the regeneration of many tree species (Alexandre 1977), but it is unlikely that the present reduced area of the Park will indefinitely allow the persistence of elephants. It is to be hoped that the Nini-Suhien National Park will be better treated.

Many reasons may be adduced for conservation of representative intact forest, some cultural and some economic. The forest environment has set the scene for the historical development of most of the peoples of southern Ghana. It is our duty to ensure that such an important part of the country's heritage is preserved for future generations. Experience elsewhere in the world has shown that urbanised, industrialised man needs wilderness areas into which, from time to time, he can escape from his artificial working environment. It can also be argued that forest animals – elephants, monkeys, antelopes, birds – have a right to continued existence. The tropical ecosystem is the most complex that has ever existed. If therefore provides the best natural laboratory for the study of the relations between organisms, and their interactions with the environment. If it is destroyed, man will lose many scientific opportunities – and an ever-deepening understanding of the world around him must be regarded as one of man's most important and fundamental aims. On the material side, foreign tourists will want to include on their itinerary a visit to 'tropical jungle'; tourists bring foreign exchange. Of the hundreds of species in tropical forest, many have as yet no known use. But it is impossible to predict what important uses might be found in the future and extinct species can never be replaced. Hedberg (1979) lists many rare Ghanaian plants (lists provided by J. B. Hall); none of these are in imminent danger of extinction, but many will be endangered as forest continues to be cleared. It is increasingly obvious that the timber trees which we already use are genetically variable (Gan *et al.* 1977) and that selection programmes can greatly increase timber yields (Tufuor 1976). Germ plasm is sought by geneticists from unusually well-grown trees, and from populations growing in unusual sites. Nature reserves may serve, then, as gene banks, by drawing on which the genotypes of plantation species can be improved.

CHAPTER 8

Descriptions of forest types

This chapter compares, in some detail, the forest types defined by our classification (Hall & Swaine 1976). Table 8.1 gives the approximate area of land occupied by each type. The comparisons given in Table 8.2 between our types and those described in Ivory Coast by Guillaumet & Adjanohoun (1971) are based on data obtained during our tour of Ivory Coast forest in August 1975.

For each type an attempt is made to summarise those floristic features which have separated it from other types. Because of the continuous nature of floristic variation between types, the distinction is necessarily complex and cannot be fully described in a few words. For each type, however, there are generally a few species which seem characteristic either by being especially common, or by their scarcity in other areas. These species are tabulated for each type, together with data on frequency within the type, occurrence in other types, chorological group, and abbreviations for life-form.

Throughout this section the use of the term 'frequency' applies to the percentage of A-sample plots from our own survey in which the species was present. 'Density' is a term reserved for number per unit area; it is possible for a species to have high frequency but low density over a particular area,

Table 8.1 Areas occupied by the various forest-types in Ghana.

Types and subtypes	Area (km^2) subtype	Area (km^2) type	Percent of total forest area
Wet Evergreen		6 570	8.1
Moist Evergreen		17 770	21.9
Upland Evergreen		292	0.3
Moist Semi-deciduous		21 890	40.4
South-east	18 460		
North-west	14 430		
Dry Semi-deciduous		32 890	26.4
Inner Zone	8 630		
Fire Zone	12 810		
Southern Marginal		2 360	2.9
South-east Outlier		20	0.03
Total forest area		81 342	100.03

Table 8.2 Comparison between our forest-types and those recognised in Ivory Coast by Guillaumet & Adjanohoun (1971).

Ghana type	Ivory Coast type (approximate equivalent)
Wet Evergreen	Forêt à *Diospyros* spp. et *Mapania* spp. Forêt à *Eremospatha* et *Diospyros mannii* (partly)
Moist Evergreen	Forêt à *Eremospatha* et *Diospyros mannii* (partly) Forêt à *Turraeanthus* et *Heisteria* Variante à *Nesogordonia* et *Khaya ivorensis*
Upland Evergreen	Forêt à *Parinari excelsa*
Moist Semi-deciduous	Forêt à *Celtis* spp. et *Triplochiton* (partly)
Dry Semi-deciduous (inner zone)	Forêt à *Celtis* spp. et *Triplochiton* (partly)
Dry Semi-deciduous (fire zone)	Forêt à *Aubrevillea kerstingii* et *Khaya grandifoliola*
Southern Marginal	–
South-east Outlier	–

and in another area to have local patches of high density but an overall low frequency. The relationship between frequency and density is fully discussed by Greig-Smith (1964).

In the tables of characteristic species, the expression: 'Number of other types' means: 'Number of other Ghanaian forest-types where the species was found in an A-sample'.

Chorological groups follow White (1979), and are the same as those used in the notes accompanying the maps (see p. 108), except that near-endemics are grouped with endemics, and ecological transgressors are not distinguished.

UG – Upper Guinea Endemic (West of Dahomey Gap only)
GW – Guinea Wide (East and West of Dahomey Gap; West of Sangha River Interval)
GC – Guineo-Congolian Wide (Upper and Lower Guinea and Congolia)
CT – Chorological Transgressor (occurring in other regions as well as the Guineo-Congolian Region).

Abbreviations for life-forms are as follows:
T – large tree capable of exceeding 30 m height (megaphanerophyte)
t – smaller tree
C – large climber capable of reaching the canopy
c – small understorey climber
S – shrub or pygmy tree
H – herb
Pt – pteridophyte
Ep – epiphyte

A successful classification based on floristics will show a correlated segregation of other characteristics of forest vegetation; these are discussed under 'structure'. Environmental conditions correlated with the floristic differences are discussed, and a final section attempts to assess the economic productivity of the type in relation to conservation of vegetation, and includes mention of particular localities which seem good examples of the type.

Moist Semi-deciduous forest occupies by far the largest area and may be considered as a floristic nodum from which the other types differ to a greater or lesser extent by the absence of some species and the presence of additional ones. No other type possesses such a large share of the total flora and it is therefore with Moist Semi-deciduous forest that we begin our comparison between types.

8.1 Moist Semi-deciduous type

This most extensive of the forest types in Ghana agrees fairly closely in range with Taylor's (1960) *Celtis-Triplochiton* Association. Forest very like it occurs widely in West Africa.

8.1.1 Floristics

Very few species are confined to this type. Of all the species used in the ordination analysis 76% occur here. Many of the commonest species in Ghana achieve their greatest frequency here (Table 8.3).

It is notable that one of Taylor's defining associates, *Triplochiton scleroxylon* is less frequent here (37% of samples) than in the Dry Semi-

Table 8.3 Characteristic species of Moist Semi-deciduous forest.

Species	Percent frequency	Occurrence elsewhere (number of types)	Chorological group	Life-form
Leptaspis cochleata	98	4	CT	H
Celtis mildbraedii	93	4	GC	T
Nesogordonia papaverifera	91	4	GW	T
Culcasia angolensis	88	4	GC	C/EP
Griffonia simplicifolia	93	5	GW	C
Calycobolus africanus	91	5	GC	C
Microdesmis puberula	88	5	GC	t
Baphia nitida	95	6	GW	t

deciduous type (56%). Two species of *Sterculia* might be considered somewhat characteristic: *S. rhinopetala* (65%) and *S. oblonga* (60%) are each only recorded in 3 other types, all with much lower frequency. The majority of A-sample plots contain over 100 species of vascular plant.

8.1.2 Structure

Trees in this type become taller than in any other: heights often exceed 50 m and sometimes 60 m. Fig. 8.1 is a profile diagram prepared in Bobiri F.R. (see Fig. 1.3) on 0.032 ha where, although the upper canopy is discontinuous, two trees exceed 50 m in height. This upper canopy consists of a mixture of deciduous and evergreen species in varying proportions, but over the whole type, the two occur in about equal proportions – hence the term 'Semi-deciduous'. The understorey trees are sometimes gregarious; for example, population density of *Cleidion gabonicum* in Bobiri reaches 280 trees (⩾10 cm gbh) per hectare. In contrast the upper canopy is generally composed of a few species from a possible list of about sixty.

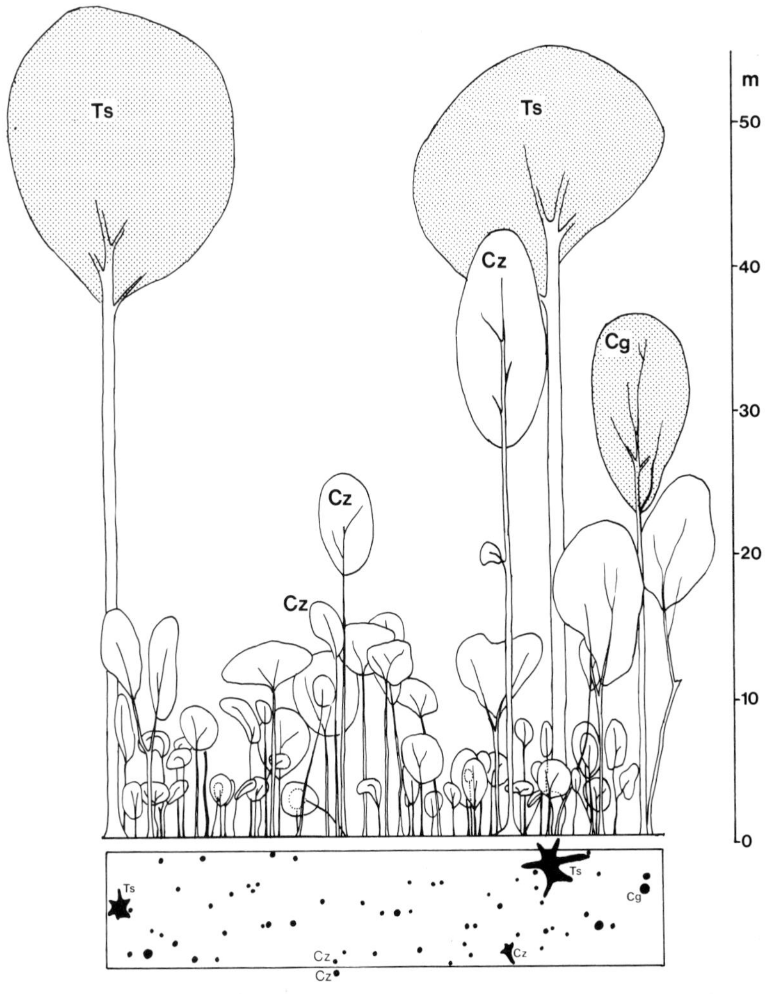

Fig. 8.1 Profile diagram of strip 8 m × 40 m in Moist Semi-deciduous forest in Bobiri F.R. Species with stippled crowns are not represented by seedlings or saplings ⩽3 m high on the strip. All individuals of species represented in the canopy are named. Abbreviations: Cg = *Cylicodiscus gabunensis,* Ts = *Triplochiton scleroxylon,* Cz = *Celtis zenkeri.*

8.1.3 Environment

The annual rainfall within the type varies between 1 200 and 1 800 mm. This moderate rainfall leads to more depletion of soil nutrients than in types of lower rainfall. Base saturation is generally high, however, (60 to 80%) providing a pH of about 5 to 6. Total exchangeable bases (TEB) are generally below 10 m – equiv/100 g soil, but this appears adequate for the considerable tree growth characteristic of the type. The type as a whole has only a moderate elevation (150–600 m), higher elevations within the area carrying forest of a distinct type (Upland Evergreen q.v.).

8.1.4 Conservation and utilisation

The type is the most productive in the forest zone. The soil is ideal for most of the forest zone crops, including cocoa, which was first introduced in the south-east corner (Akwapim) and later spread to Ashanti and Brong-Ahafo (p. 63). Cocoa and other farms are very much in evidence all over the Moist Semi-deciduous type, and there is a great demand for more farmland. Although the type is well supplied with forest reserves, most have been heavily exploited for timber, and pressure on these reserves from both timber contractors and farmers is likely to increase. The least disturbed forest is found on steep hills where difficulties of access have made logging operations uneconomic. The type is especially rich in economic timber species, about half those listed in Appendix 7 showing greatest abundance here. Since it is the type of largest area, bears the largest trees, and probably has the highest growth rates, it undoubtedly contributes the major proportion of forest produce from Ghana. Jack (1959) gives a detailed account of forest within this type from a forester's point of view.

Bia National Park (p. 70) lies across the southern limit of the type in the west and contains undisturbed forest on a gently undulating topography. It was originally the largest of the forest National Parks and carries a high stocking of valuable timber species. Despite its potential as a reserve of genetic material for the major economic species, most of its former area has recently been converted to a Game Production Reserve which permits timber extraction. The reserve is also noted for the abundance of elephants *(Loxodonta africana cyclotis).*

8.1.5 Subdivision of type

The 1 500 mm mean annual rainfall isohyet (Fig. 3.2) which runs NE-SW through Kumasi, approximately coincides with the boundary separating two floristically distinct subtypes within Moist Semi-deciduous forest (Fig. 1.3). Mooney (1961) made a similar subdivision of Taylor's *Celtis-Triplochiton* association into a 'dry' facies and 'wet' facies (Fig. 2.1b). He also distinguished a further subdivision, the 'Trans Bia' (for which there seems little floristic justification) on the doubtful basis of forest structure.

Our subdivision is based on a re-ordination of those A-samples comprising the Moist Semi-deciduous type (Hall & Swaine 1976); the subtypes are well characterised by indicator species, and differ in average rainfall and also soil richness (Table 8.4).

Table 8.4 Environmental differences between the North-west and South-east subtypes of Moist Semi-deciduous forest.

	Subtype	
	North-west	South-east
	Number of A-samples	
Mean annual rainfall (mm)		
1 000–1 250	0	1
1 250–1 500	19	3
1 500–1 750	2	20
	Mean value for A-samples	
Total Exchangeable Bases (m-equiv/100 g soil)	9.2	7.6
Percent saturation	77	69
pH	5.6	5.3

8.2 North-west subtype of Moist Semi-deciduous forest

8.2.1 Floristics

The drier North-west subtype contains several species found only rarely elsewhere in Ghana, but which are widespread in African forest (Table 8.5).

8.2.2 Conservation and utilisation

Thompson (1910, p. 62–63) describes the forests in the western part of the area of this subtype as

Table 8.5 Characteristic species of Moist Semi-deciduous forest North-west subtype.

Species	Percent frequency		Occurrence elsewhere (number of types)	Chorological group	Life-form
	NW	SE			
Argomuellera macrophylla	38	0	0	CT	S
Chlamydocarya thomsoniana	29	4	1	GC	c
Entandrophragma utile	33	4	2	GC	T
Guibourtia ehie	71	8	4	GW	T
Pericopsis elata	5	0	0	GC	T
Khaya anthotheca	19	0	1	GC	T

NW = North-west subtype; SE = South-east subtype.

largely uninhabited; the same was true up to the Second World War, and the Forestry Department was therefore able to obtain numbers of big reserves (Table 7.2). Two large and well-organised timber companies, Mim Timber Company (Mim) and Gliksten (Sefwi-Wiawso), have exploited the area since the war, and the farmers who settled in the cut-over forest have now transformed most of the forest outside reserves to cocoa farms. Recent attempts to expel farmers from new reserves along the Ivory Coast border have met severe opposition.

The Mim Company has been particularly active; it was largely responsible for developing the market for 'afrormosia' *(Pericopsis elata),* and has so effectively exploited this species that marketable specimens are increasingly difficult to find (Fig. 7.2). the company is now marketing *Guibourtia ehie,* a previously unclassified timber but which was twelfth by volume on the list of exported species in the Forestry Department Annual Report for 1972.

Because the area is (or was) unusually well-stocked with valuable timber, logging is severe, and rather little canopy remains in exploited reserves.

The North-west subtype harbours far more elephants than any other part of the Ghana forest. At least two large populations are known, one ranging from near Mim south-westwards to the Krokosua Hills F.R., and the other centred in Bia National Park and Bia Tawya F.R. west of the Bia River (C. Martin, *pers. comm.*). Elephants are particularly sensitive to disturbance in their home range, and around Mim and Goaso have become serious pests of farmland and are commonly shot. Across the Bia, the elephants have remained undisturbed until recently, when logging operations began, but are already more in evidence as they are forced from their usual home range.

There are a number of interesting hills within forest reserves in the subtype which have not been exploited commercially. Bosumkese and Kwamisa hills both give their names to forest reserves, and carry good examples of untouched NW subtype forest.

8.3 South-east subtype of Moist Semi-deciduous forest

8.3.1 Floristics and structure

There are fewer species characteristic of the South-east subtype and although rare in the North-west subtype, all of these occur commonly in other types (Table 8.6).

Stocking density figures from Forestry Department enumerations suggest that *Khaya ivorensis* (Fig. 11.2) and *Terminalia ivorensis* (Fig. 11.7) achieve their greatest density in forest reserves within this subtype. In our survey, however (Map 324). *K. ivorensis* appears more frequently (30%) in the Moist Evergreen forest.

All of the trees listed in Table 8.6 are of economic importance; *Cola nitida* is, of course, the edible kola, which here grows wild.

8.3.2 Conservation and utilisation

The South-east subtype contains a number of localities which have been important centres for forest research. Bobiri F.R., just east of Kumasi,

Table 8.6 Characteristic species of Moist Semi-deciduous forest South-east subtype.

Species	Percent frequency SE	Percent frequency NW	Occurrence elsewhere (number of types)	Chorological group	Life-form
Turraeanthus africanus	38	5	2	GC	t
Daniellia ogea	29	5	2	GW	T
Khaya ivorensis	21	5	2	GW	T
Illigera pentaphylla	33	0	3	GC	C
Pteris togoensis	29	5	3	GW	Pt
Chytranthus macrobotrys	25	10	3	GC	t
Cola nitida	21	10	4	UG	t

SE = South-east subtype; NW = North-west subtype.

has been the site of many silvicultural experiments and ecological observations (Foggie 1947), and Pra Anum F.R., north of Akim Oda, includes a number of pioneer plantations. The University of Ghana has an Agricultural Research Station at Kade containing a small reserve of closed-canopy forest which has been the subject of several ecological publications: (Greenland & Kowal 1960; Lawson, Armstrong-Mensah & Hall 1970; John 1973).

Because it has been settled longer than the North-west subtype, the South-east subtype had less forest available for reservation. The percentages reserved are respectively 28 and 13 for these subtypes (Table 7.2).

8.4 Dry Semi-deciduous type

This type of forest exists under a wider range of environmental conditions than does any other type. It forms a peripheral band around the moister forest types and is adjacent, in the north, to the Guinea savanna zone. Forest of this kind appears to be widespread in west Africa; species lists from Casamance (Senegal), Ivory Coast and Nigeria all give satisfactory identification with this type.

The type includes most of the area of Taylor's *Antiaris-Chlorophora* association, and only differs notably in the western extension of its southern arm Figs. 1.3b and 2.1a).

8.4.1 Floristics

Antiaris and *Chlorophora* do indeed achieve their greatest abundance here. *Antiaris toxicaria* (syn. *A. africana*) is present in 70% of our A-samples in this type, but is also commonly found in all other types (Map 48). *Chlorophora* species (Map 124 and Fig. 11.1) show a similar distribution, but they are found less commonly in closed forest because the seedlings require high light levels for germination and establishment (p. 60).

The type is characterised by several species (Table 8.7), including *Triplochiton scleroxylon* which is more frequent in our samples of Dry than of Moist Semi-deciduous forest (Taylor's *Celtis-Triplochiton*) – see Map 603. Few species are confined to this type and only two exceed 80% frequency: *Lecaniodiscus cupanioides* (Table 8.7), and *Griffonia simplicifolia* which is very common in all forest types (Map 284). *Trichilia prieuriana,* although common in other types, reaches especially high densities in Dry Semi-deciduous forest (Map 594): in our survey all plots containing 10 or more trees (⩾10 cm dbh) of this species fell within this type, yielding a total of 156 trees in 41 plots.

Almost all of the A-samples contained between 40 and 100 species.

8.4.2 Structure

The vertical structure of those forests is generally well-developed with a more or less continuous upper canopy, though at a generally lower height than in Moist Semi-deciduous forest. The tallest trees are mostly between 30 and 45 m in height.

8.4.3 Environment

This type occurs under a wide range of annual

Table 8.7 Characteristic species of Dry Semi-deciduous forest.

Species	Percent frequency	Occurrence elsewhere (number of types)	Chorological group	Life-form
Zanthoxylum leprieurii	21	0	CT	t
Dicranolepis grandiflora	16	0	GW	t
Baissea zygodioides	44	1	UG	C
Sterculia tragacantha	44	2	GC	t
Diospyros monbuttensis	37	2	GC	t
Motandra guineensis	72	3	GC	C
Trilepisium madagascariense	63	3	CT	t
Cola millenii	67	3	GW	t
Celtis zenkeri	60	3	GC	T
Triplochiton scleroxylon	56	3	GC	T
Lecaniodiscus cupanioides	88	4	GC	t

rainfall. In the Volta Region mean annual rainfall often exceeds 1 500 mm, especially near the higher hills. In contrast, Dry Semi-deciduous forest near Berekum (Brong-Ahafo Region) receives little over 1 000 mm each year. Most of the type, however, receives between 1 250 and 1 500 mm.

8.4.4 Subdivision of the type

The geology of the underlying rock is also quite varied (Fig. 3.4) and may well have an important influence on the flora in different areas, particularly on rocky hills where soils are shallow. A re-ordination of the type (Fig. 8.2) shows that A-samples on Togo and Buem rocks tend to have higher axis 2 values than those on Basement Complex rocks.

Re-ordination, however, showed the clearest floristic differences between forest adjacent to savanna, and forest adjacent to other forest types (Fig. 8.3). Neither rainfall nor soil conditions correlate with this distinction, but many of the plots close

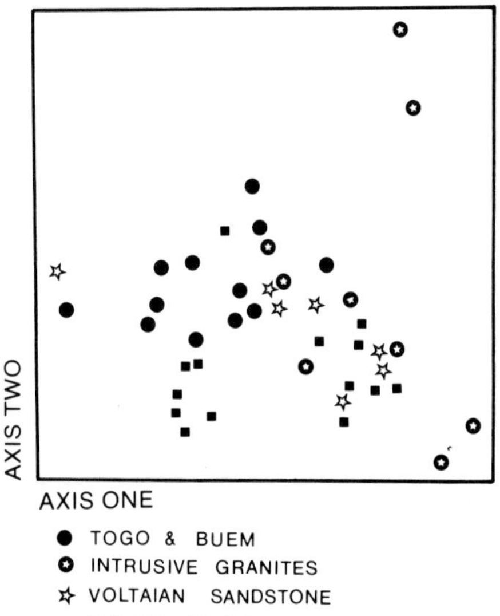

Fig. 8.2 Reordination of Dry Semi-deciduous A-samples showing variation of vegetation in relation to rock type (Fig. 3.4).

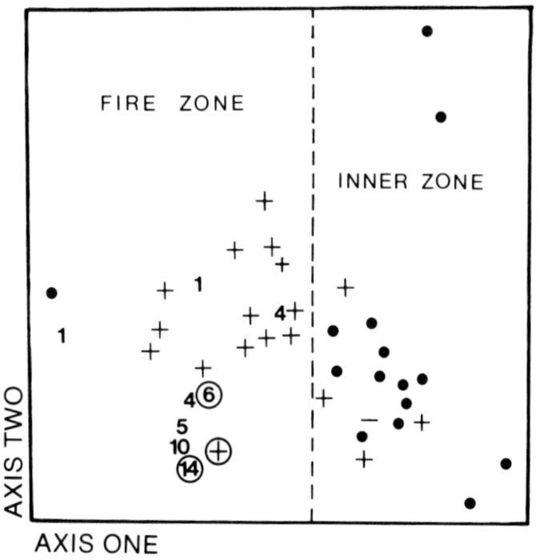

Fig. 8.3 Partition of reordination (Fig. 8.2) into Fire Zone and Inner Zone subtypes. Fig. 2.4 shows that Fire Zone plots are closer to the northern forest-savanna boundary. Presence (+), absence (●) or number of trees of *Elaeis guineensis* in A-samples is shown. A-samples in which ground fires have been reported (see text) are ringed.

to Guinea savanna bore signs of fire in the understorey and we received several reports of ground fires in such forest. We have concluded that occasional ground fires are the major factor determining the nature of these 'Fire Zone' forests (Hall & Swaine 1976). The distinction between 'Fire Zone' and 'Inner Zone' subtypes had not previously been recognised.

8.5 Fire Zone subtype of Dry Semi-deciduous forest

8.5.1 Floristics, structure and environment

The forests along the boundary with savanna have many peculiarities, both structural and floristic. The woody understorey is sparse, and there may be few large lianes. Consequently the forest floor is unusually well illuminated, and often carries a dense weedy growth of marantaceous forbs such as *Marantochloa cuspidata,* various herbaceous Acanthaceae and slender climbers. Characteristic species are listed in Table 8.8; it will be noted that two of them – *Afzelia africana* and *Diospyros mespiliformis* are among the very few trees which grow both in forest and savanna. *Spondias mombin* has edible fruits, and has almost certainly been introduced from tropical America. Other rarer species of tree more or less restricted to this subtype are *Anogeissus leiocarpus* (another savanna-living species), *Aubrevillea kerstingii, Celtis africana, Detarium senegalense, Khaya grandifoliola, Zanha golungensis,* and *Parkia filicoidea* (if this latter is recognised as a distinct species).

Aubréville (1938 & 1949) was struck with the distinctness of this subtype in the Ivory Coast, calling it 'forêt dense sèche semi-décidue'. He surmised that it represents the remnants of a forest which, before the advent of agriculture, stretched far to the north, covering what is now Guinea savanna. He also supposed that many secondary forest species originated in this kind of vegetation. Guillaumet & Adjanohoun (1971) named it 'Forêt à *Aubrevillea kerstingii* et *Khaya grandifoliola*' after two of its characteristic species.

Aubréville (1938) recognised that fire sometimes runs into these forests, but regarded this as exceptional. It is our belief that, on the contrary, fire is an important factor here, and is responsible for many of the distinctive characteristics. In Tain Tributaries Block I F.R. (Fig. 1.3a) where we took A-sample number 96 (Appendix 2 and Fig. 2.2) we were told by the forest ranger than a ground fire had occurred 15 years previously, after a period of drought, and that A-sample 94 (Yaya F.R.) had been burnt at the same time. There had been a fire in Nsemre F.R., near A-sample 95, just prior to our visit in 1973. Subsequently we saw forest in Dede F.R. that had experienced a fire in 1972.

It was clear that such forest-floor fires do not necessarily kill all the trees. In a 50 m × 50 m plot at Dede F.R. we found that only 15% (13 out of 84) of the trees with girth greater than 30 cm had died, although 54% (77 out of 143) of the trees with girth

Table 8.8 Characteristic species of Dry Semi-deciduous forest Fire Zone subtype.

Species	Percent frequency FZ	Percent frequency IZ	Occurrence elsewhere (number of types)	Chorological group	Life-form
Spondias mombin	20	0	0	CT	t
Afzelia africana	45	5	1	CT	t
Tylophora conspicua	30	5	1	GC	c
Diospyros mespiliformis	20	0	1	CT	t
Sterculia tragacantha	40	23	2	GC	t
Cola millenii	65	27	3	GW	t
Canthium vulgare	35	18	3	CT	t
Cola gigantea var. *glabrescens*	50	18	4	GW	T
Elaeis guineensis	95	29	4	CT	t

FZ = Fire Zone subtype; IZ = Inner Zone subtype.

10–30 cm had been killed. Fires may thus cause the thinning of the understorey which, as stated above, is a feature of the subtype (Fig. 8.4). There was no evidence, at Dede or Nsemre, for invasion by savanna grasses, but the removal of undergrowth by burning had favoured the germination of many weedy forest species such as *Momordica charantia* Linn. and *Palisota hirsuta* (Thunb.) K. Schum., and secondary forest trees such as *Ceiba pentandra* and *Ricinodendron heudelotii* (Fig. 8.5). The plots which had burned 15 years previously seemed typical, structurally and floristically, of the subtype. Clearly, regeneration of forest trees would be completely prevented by very frequent burning, but if the interval between fires is sufficient to allow some young trees to reach a girth of 20 cm or so, there seems no reason why forest should not persist indefinitely. Fire Zone forest, however, has a very patchy canopy, with many tangled gaps, in which the introduced pawpaw tree (*Carica papaya* Linn.) is often conspicuous.

Table 8.8 indicates that the oil palm *Elaeis guineensis* is characteristic of the Fire Zone. The abundance of oil palms is shown more clearly when the number of trees per A-sample is plotted (Fig. 8.2). In fact 45 out of the 66 oil palm trees recorded in all our Ghana A-samples came from Fire Zone forest. Although it is possible that a long history of agriculture is responsible, it seems likely that the resistance of oil palms to burning and the known high-temperature requirement for germination of their seeds may also play a part.

Deciduousness is higher in Fire Zone forest than in any other type or subtype (p. 51). Perhaps the heavy crop of leaf litter in the dry season predisposes to fire.

Forests of the Volta Region all lie within the Fire Zone (Fig. 2.2). In addition to the species characteristic of the subtype in general, these forests harbour a few lowland species which here reach their western limit (e.g. *Maranthes kerstingii*), and disjunct sub-montane species on the mountains (e.g. *Carex neochevalieri* Kuk., *Maesa lanceolata, Rhus longipes,* and *Sorindeia zenkeri*).

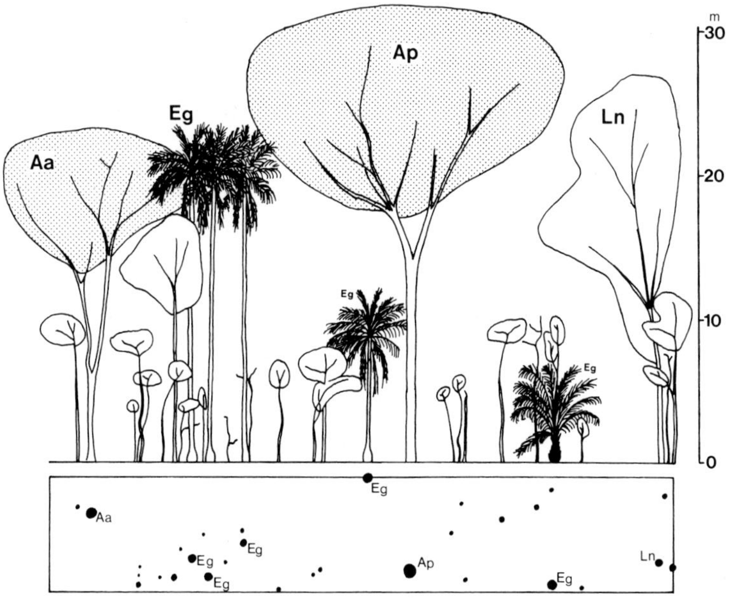

Fig. 8.4 Profile diagram of strip 8 m × 45 m in Dry Semi-deciduous forest Fire Zone subtype in Nsemre F.R.; stippling as in Fig. 8.1.
Abbreviations:
Aa = *Afzelia africana* Ln = *Lannea nigritana*
Ap = *Aubrevillea platycarpa* Eg = *Elaeis guineensis*

Fig. 8.5 Nsemre F.R. in June 1973 showing regrowth of Marantaceae following recent ground fire. The strangling fig in the foreground is epiphytic on *Elaeis guineensis;* other palms are also this species.

8.5.2 Conservation and utilisation

The subtype is well supplied with forest reserves (Table 7.2), most of which are classified as protection reserves. These have received little attention from timber contractors; economic species are rare here, with the exception of odum (*Chlorophora* spp.). Many of the reserves were established for the purpose of retarding the southern advance of savanna, a policy forcefully advocated by Thompson (1910, p. 104).

The modern view is sceptical of his dire warnings, as there is no evidence that the forest-savanna boundary has greatly changed its position, at least during historical times (Swaine, Hall & Lock 1976). The natural occurrence of fires, however, suggests that the balance between forest and some other kind of vegetation is critical; incautious and intensive farming in this zone has indeed allowed grasses to establish to the exclusion of woody plants, forming patches of 'derived savanna'. Some of the grassland patches cited by Thompson (1910, p. 97) as evidence of savanna advance are now thought, however, to be natural features of the vegetation. Such patches are found throughout the Fire Zone where the soil is rocky or shallow. Grassy sandstone pavements, with a few savanna trees, may be seen at various points on the Kwahu ridge, for example near Kwahu Tafo, but their flora includes so many rare and relict species (Hall 1966) that they cannot possibly be of recent origin. The summits of the hills in the Volta Region are typically grassland and similarly carry an interesting disjunct flora (Morton 1962).

Fig. 8.6 (a) and (b) are satellite images of part of the Volta Region. Relief is well shown in Fig. 8.6a (infra-red) and closed-canopy forest appears pale. The distinction between forest and savanna is best seen in the red wavelength (Fig. 8.6b) when forest is dark, savanna pale and roads and towns almost white. Fig. 8.7 is an interpretative map of the same area.

Forest land is undoubtedly better than savanna for agriculture, and there has always been heavy pressure for the release of forest reserves for farming, especially in the Volta Region where forest is almost entirely confined to the steep hills that have been reserved to prevent erosion, but where little logging has been done. The Forestry Department has gone some way toward meeting this demand by encouraging the 'taungya' system. This means that farmers are allowed to grow crops inside reserves on condition that they plant and care for young trees (often teak – *Tectona grandis* Linn.f.) that can form plantations when farming stops.

Some of Ghana's most spectacular scenery is found in the Volta Region. Mount Adaklu (570 m elevation), just south of Ho, bears an impréssive 200 m crag of basic gneiss, and the summit is covered by well-preserved forest of low stature which contrasts with the savanna covering the surrounding plains.

Further north, beyond Kpandu, the Togo Plateau Forest Reserve occupies a mountain system remarkable for its variety of vegetation and for a number of disjunct plant species typical of submontane grassland. These two localities thus seem worth special protection. We have not surveyed the forest outlier on hills near Odonkokrom on the Afram Plains (Fig. 2.4), but it is most likely to be of the Fire Zone subtype.

8.6 Inner Zone subtype of the Dry Semi-deciduous forest

8.6.1 Floristics and structure

The species which help to distinguish between this subtype and the Fire Zone subtype are mostly common also in other forest types (Table 8.9). It is likely that their absence from the Fire Zone is due to their intolerance of even light fires. All are fairly small woody plants and have thin bark.

Structurally, these forests closely resemble Moist Semi-deciduous forest, the tallest trees being only a little shorter (compare Fig. 8.1 and Fig. 8.8).

8.6.2 Conservation and utilisation

The subtype has rather few forest reserves (Table 7.2), especially in the southern arm which is only protected in the tiny Opimbo F.R. (1 km^2). In the north, fairly large portions of Pamu-Berekum, Kwamisa, Opro River and Bomfoum Forest Reserves fall within the subtype. The north-eastern part of the subtype, in the Yilo Krobo area between Huhunya and Asesewa, has been farmed very intensively, and the forest has been largely replaced by a kind of secondary grassland in which the grass

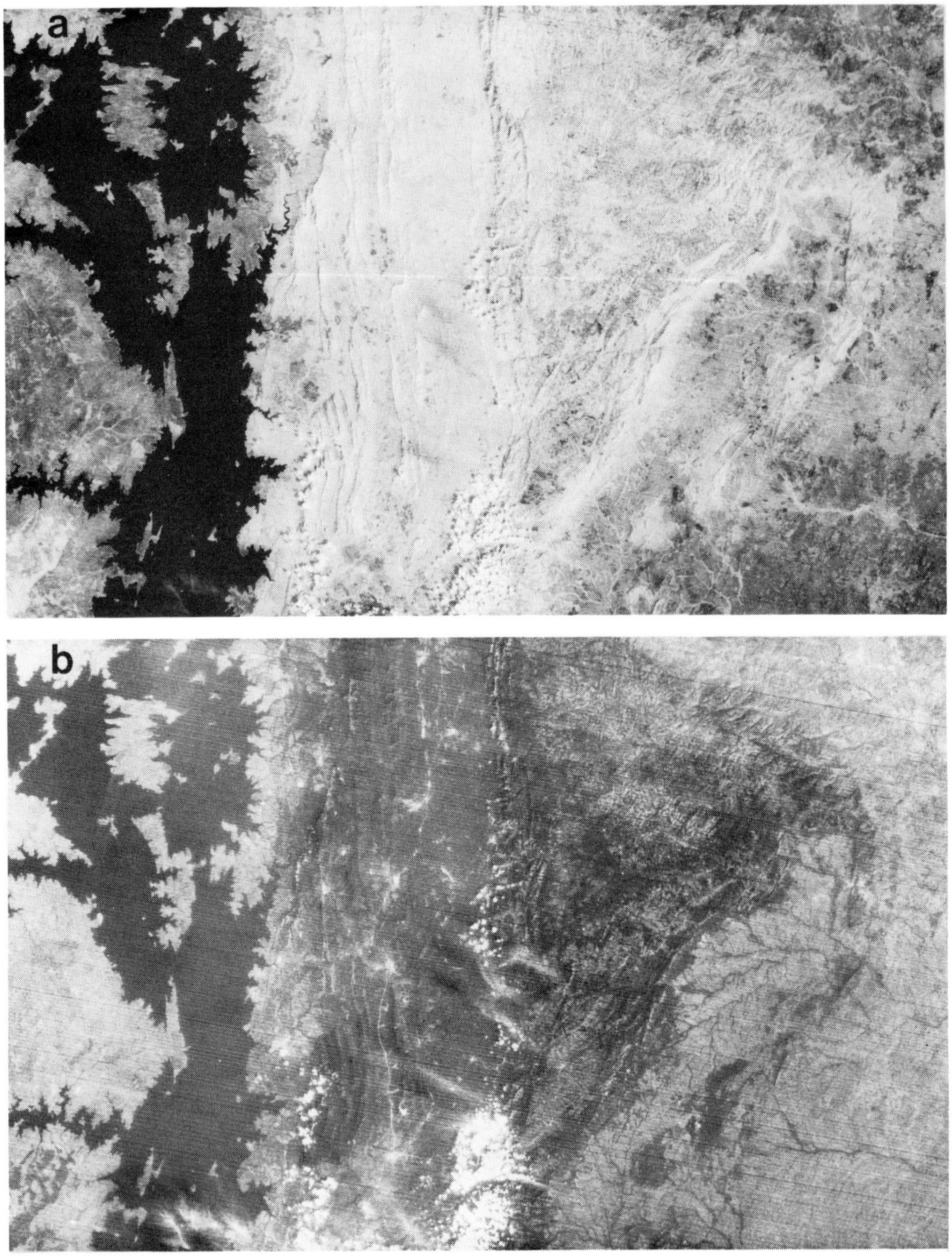

Fig. 8.6 LANDSAT-2 images of part of the Volta Region of Ghana and the adjacent part of the Republic of Togo on 2 Feb. 1976 at 0930 hours (see Fig. 8.7 for interpretation).

(a) Band 7 (infra-red). Open water appears black, as does the most recently burnt savanna, and cloud shadow. Pale areas are mostly forest: fringing forest contrasts with burned savanna and open water. Unburned savanna has a range of tones, and may approach forest. Hill ridges are thrown into relief because of shadows cast by the morning sun. Forest reserves do not show as clearly as in Fig. 1.2 (see p. 68 for discussion of encroachment on forest reserves in Volta Region).

(b) Band 5 (red). Open water and forest are black: fringing forest is therefore difficult to distinguish from water. Savanna is pale, the complex forest-savanna boundary is better shown than in Fig. 8.6a, and transient savanna burns are less obtrusive. Towns and roads are almost white, but sometimes difficult to distinguish in savanna.

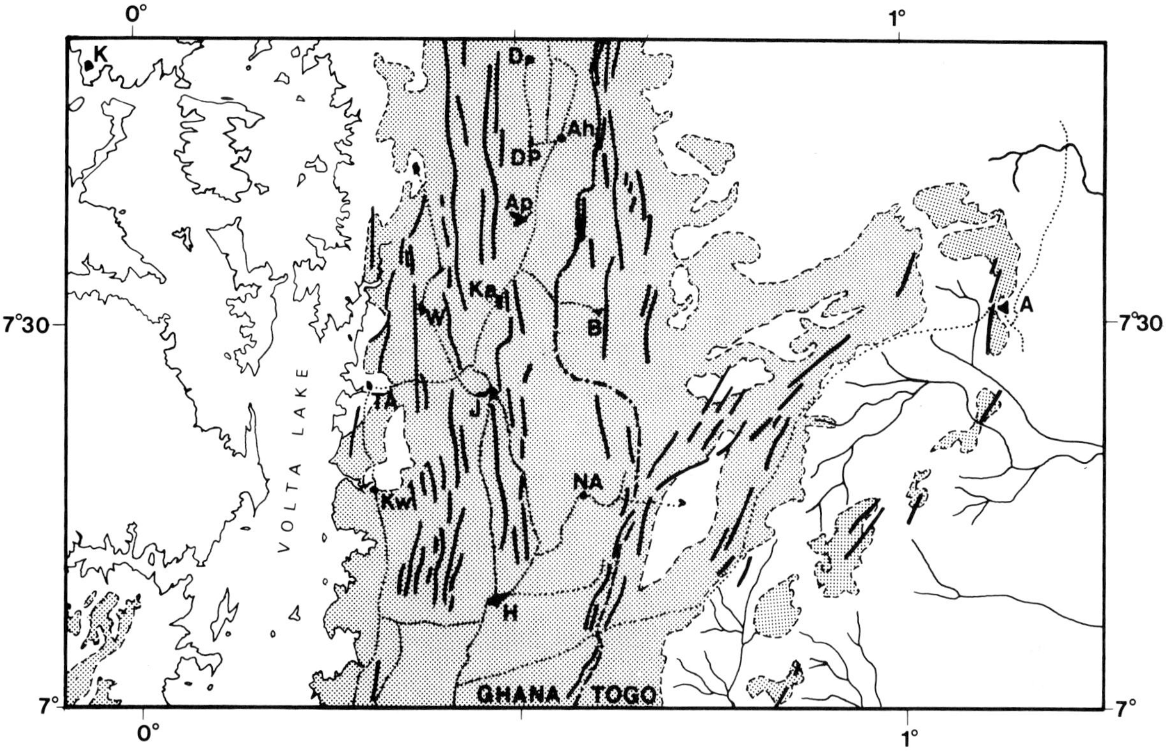

Fig. 8.7 Map compiled with the aid of the images in Fig. 8.6. Forest-savanna boundary: broken line; approximate international boundary: broken-dotted line; topographic lineaments: heavy line; lake boundary and rivers: fine line; forest land: stippled; roads: dotted. Key to marked towns: A = Atakpame; Ah = Ahamansu; Ap = Apesokubi; B = Badou; D = Dodo; DP = Dodi Papase; H = Hohoe; J = Jasikan; K = Kete Krachi; Ka = Kadjebi; Kw = Kwamekrom; NA = New Ayoma; TA = Tapa Abotoase; W = Worawora.

Table 8.9 Characteristic species of Dry Semi-deciduous forst Inner Zone subtype.

Species	Percent frequency		Occurrence elsewhere (number of types)	Chorological group	Life-form
	IZ	FZ			
Drypetes singroboensis	27	0	1	UG	t
Dracaena arborea	36	5	3	GW	t
Hymenostegia afzelii	64	0	5	GW	t
Sphenocentrum jollyanum	41	0	6	GW	S

IZ = Inner Zone subtype; FZ = Fire Zone subtype.

Panicum maximum is dominant. Few indigenous trees are spared by farmers, except for *Chlorophora excelsa, Newbouldia laevis* and *Ceiba pentandra.* Firewood plantations of the small introduced tree *Cassia siamea* Lam. are, however, widespread.

8.7 Southern Marginal type

There is no equivalent for this forest type in any previous classification; most of it falls within the area of Taylor's (1960) 'Coastal Thicket and Scrub'.

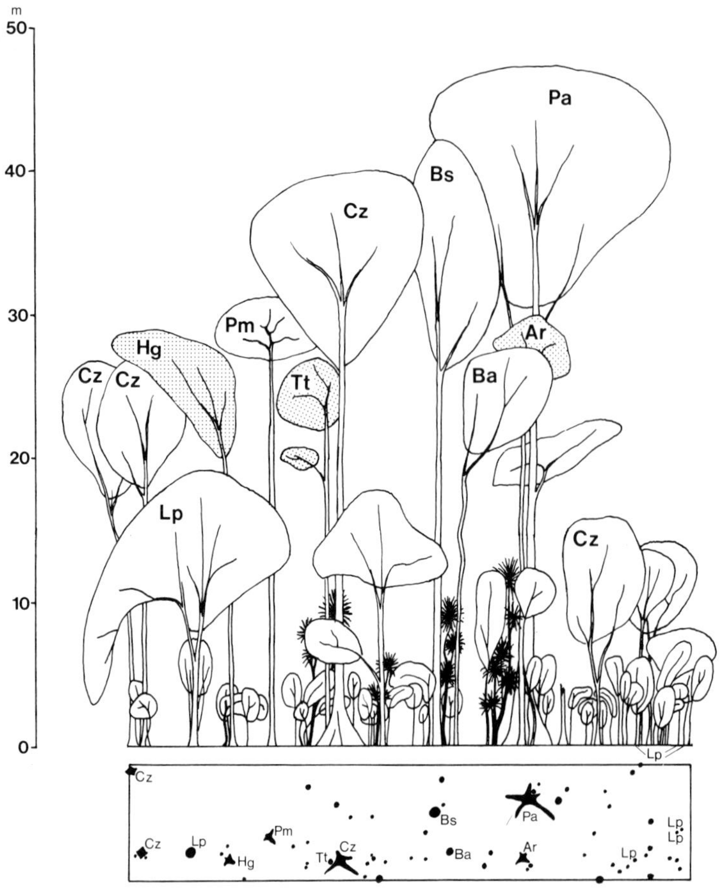

Fig. 8.8 Profile diagram of a strip 8 m × 40 m in forest relict near Odoben (A-sample 116) – Dry Semi-deciduous forest Inner Zone subtype. Stippling as in Fig. 8.1.

Abbreviations:
Ar = *Aningeria robusta*
Bs = *Blighia sapida*
Cz = *Celtis zenkeri*
Hg = *Holoptelea grandis*
Lp = *Lecaniodiscus punctatus* (source of holotype)
Pa = *Piptadeniastrum africanum*
Pm = *Pterygota macrocarpa*
Tm = *Trilepisium madagascariense*
Tt = *Trichilia tessmannii*

The trees shown with long leaves are *Dracaena arborea.*

8.7.1 Floristics

The type is well characterised by species which are uncommon elsewhere (Table 8.10). It is notable that most of these are trees, though mostly of small stature. *Nesogordonia papaverifera* is the only tree species of importance to the timber trade, and although it never achieves great size in this type it is locally very abundant. Over a third of all the trees (⩾10 cm dbh) of this species recorded in our survey occurred in eight Southern Marginal plots.

Gregariousnes is, in fact, a pronounced feature of this forest-type. *Hildegardia barteri, Hymenostegia afzelii, Manilkara obovata* and *Talbotiella gentii* may all achieve densities exceeding 20 trees (⩾10 cm dbh) per 0.0625 ha A-sample. In the understorey, the shrub *Sloetiopsis usambarensis* (p. 55) may be completely dominant. The most remarkable of these gregarious species is the Ghanaian endemic *Talbotiella gentii,* which may completely dominate not only the canopy but also the understorey and ground layers (Hall & Swaine 1981). Like *Hymenostegia,* and most other tropical African trees which form essentially monospecific stands, it is a member of Caesalpiniaceae.

Table 8.10 Characteristic species of Southern Marginal forest.

Species	Percent frequency	Occurrence elsewhere (number of types)	Chorological group	Life-form
Teclea verdoorniana	59	1	GC	t
Cynometra megalophylla	24	1	GW	t
Artabotrys velutinus	65	1	GW	C
Drypetes floribunda	76	2	GC	t
Drypetes parvifolia	59	2	GW	t
Diospyros abyssinica	59	2	CT	t
Salacia leptoclada	53	2	CT	C
Dennettia tripetala	47	2	GW	t
Dialium guineense	76	3	GW	t
Hildegardia barteri	30	3	GW	t
Nesogordonia papaverifera	71	4	GW	t
Campylostemon angolense	71	4	GC	C
Lecaniodiscus cupanioides	88	4	GC	t

8.7.2 Structure

These forests mostly occur today as small fragments but are true forest in the sense that they have a closed canopy formed by trees and a fairly dense undergrowth. Although the trees are rather short, rarely exceeding 35 m, the vertical structure is fairly well-developed (Fig. 8.9). Species which are normally understorey trees may form the main canopy (e.g. *Hymenostegia afzelii* in Fig. 8.9) with emergents of *Antiaris* and *Ceiba* forming a discontinuous upper canopy. Some trees may attain considerably girth. The forest patches are mostly surrounded by thicket which probably affords some protection from fires in nearby farms or grassland.

8.7.3 Environment

The type ranges from west of Cape Coast to Akosombo and varies in annual rainfall and rock type. Annual precipitation is relatively low, however (1 000–1 250 mm). Soil conditions are greatly influenced by the underlying rock, since most of the forest occurs on hills with shallow soil. Two main subtypes may be recognised, corresponding to the major rock types (Fig. 8.10). The north-easterly portion lies on the Togo series rocks of the Akwapim ridge – mostly quarzites and shales – and bears relatively acidic soil, compared with the south-westerly portion which is principally on soils of granitic origin. Fig. 8.11 demonstrates the floristic differences between the forest on the two rock types. *Cynometra megalophylla* (normally a riverain species but here growing in upland situations) seems confined to the granitic sites, and *Talbotiella gentii* to acidic rocks.

8.7.4 Conservation and utilisation

Relatively little of the 2 360 km² of this type is today covered by forest. The land is mostly covered by thicket, farms and savanna (partly derived from more woody vegetation by farming). Forest occurs as small patches mostly on rocky hills not favoured for farming. There are very few forest reserves (Table 7.2) and most of our samples were taken in sacred groves which are of major importance in preserving widely scattered examples of the type. Notable amongst them is Amoawa Hill between Winneba and Mankesim (Fig. 8.10) which contains a remarkable stand of *Nesogordonia*. The preponderance of sacred groves (often used as cemeteries) amongst our samples may account for the usually high levels of phosphorus recorded in the soils from this type.

Timber species are virtually absent from these forests. The associated paucity of forest reserves no doubt explains why such a distinct type has been ignored in the past. Forest produce is mostly

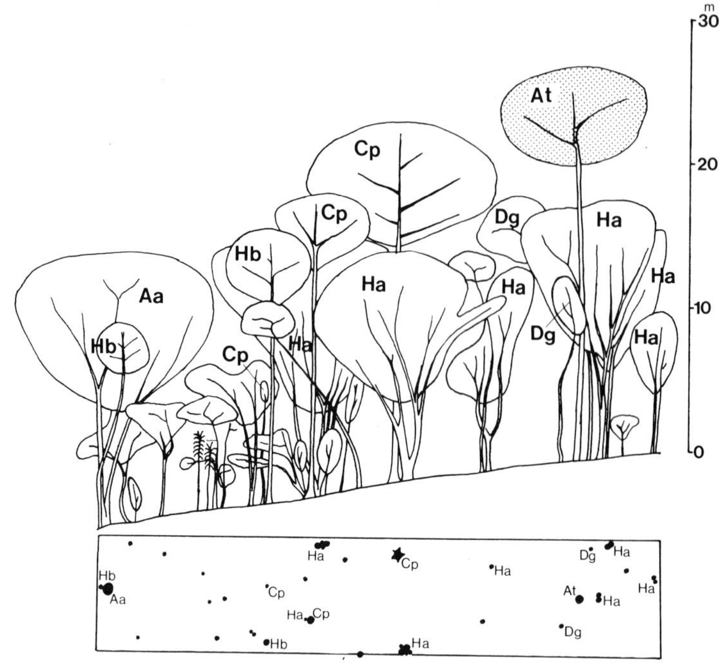

Fig. 8.9 Profile diagram of strip 8 m × 40 m in Southern Marginal forest in Sapawsu F.R. Stippling as in Fig. 8.1.
Abbreviations:

Aa = *Afzelia africana*	Cp = *Ceiba pentandra*	Ha = *Hymenostegia afzelii*
At = *Antiaris toxicaria*	Dg = *Dialium guineense*	Hb = *Hildegardia barteri*

Fig. 8.10 Map showing distribution of all A-samples of Southern Marginal (●) and South-east Outlier (◀) type. A few Dry Semi-deciduous samples (■) are also shown. Savanna is stippled. South-east Outlier forest occurs west of the R. Volta between the broken line and the forest-savanna boundary. Major geological formations (see also Fig. 3.4) with dotted boundaries: To = Togo and Buem (horizontal hatching); D = Dahomeyan; GB = Granites and Birrimian; S = Sekondian; TQ = Tertiary and Quaternary sediments. The boundary between the Dahomeyan and Togo series is coincident for much of its length with the forest-savanna boundary.

The major physiographic features were drawn from a LANDSAT-2 image of 28 December 1975.

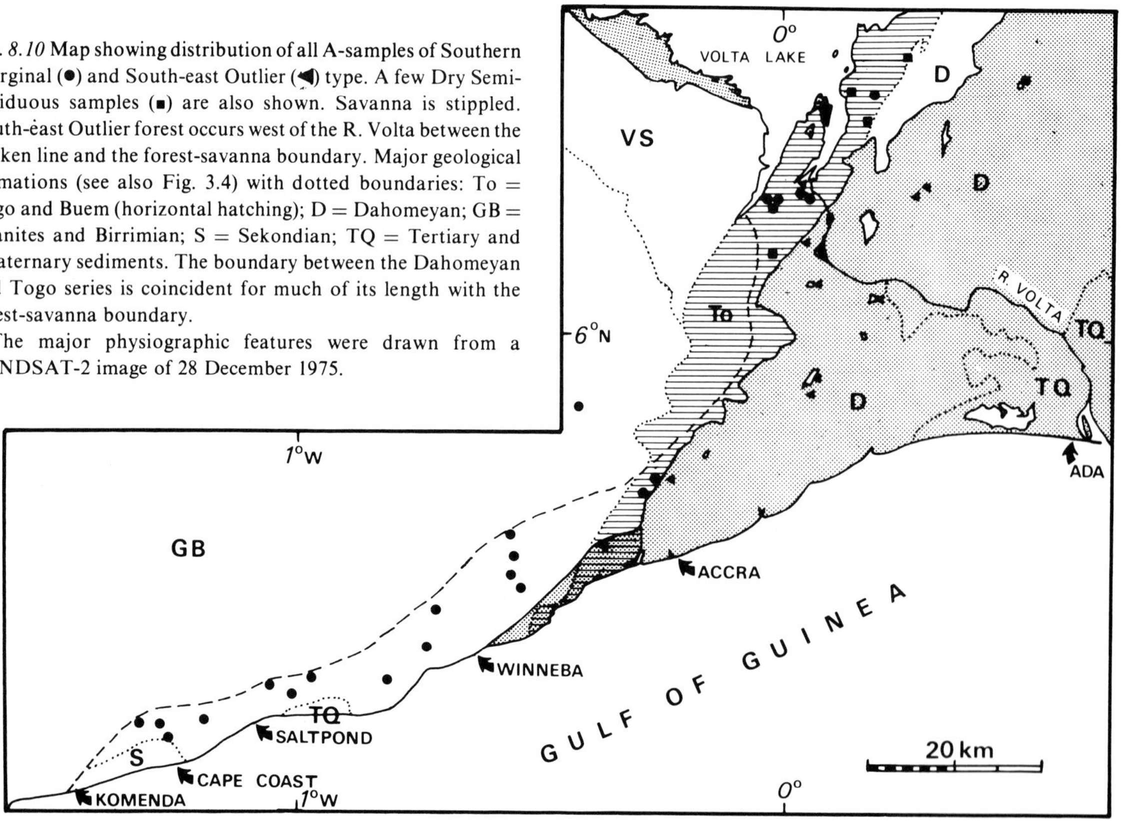

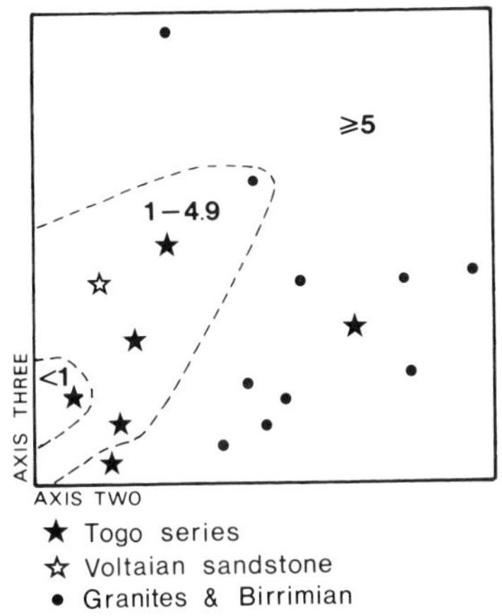

Fig. 8.11 Reordination of Southern Marginal A-samples; showing variation of vegetation in relation to rock type (Fig. 3.4). Broken lines separate plots with soils of the following calcium concentrations (m-equiv./100 g dry soils): <1; 1–4.9; ≥5. The isolated Togo series site occupies a valley bottom on rich alluvial soil with no rock exposure; all other Togo series sites are on rocky slopes.

firewood; as the trees are often close-grained, the wood makes good charcoal. These forests are the nearest to many of the large coastal towns and may be expected to have suffered continued pressure for farm land and firewood for many centuries. It is probably because the trees are slow-growing, and forest takes such a long time to recover from clearing, that thicket, rather than secondary forest, is the predominant vegetation of the zone. Thicket may, however, be the climax vegetation on maritime slopes exposed to salt-laden winds.

Forest similar to Southern Marginal forest is found in isolated localities in various parts of the forest zone with higher rainfall. Rocky granite outcrops near Mim (Ebaw Rock), Otoase (Koran hill), Achiase and in Bia National Park (Apeketewa Rock) bear such forest presumably as a result of the unusually shallow soil. In the gorge of the Black Volta in Bui National Park, riverain forest gives way upslope to forest of the Southern Marginal type.

8.8 South-east Outlier type

This is by far the least extensive of our forest-types, occupying scarcely more than 20 km^2 in small scattered patches. It is, however, one of the most distinctive, both floristically and structurally. Previous classifications have ignored it because of its negligible contribution to the total area of forest in Ghana. Jeník & Hall (1976), however, recognise it as true forest in their account of the vegetation of the Accra plains, erecting for it the alliance *Diospyrion*.

8.8.1 Floristics

The most striking floristic feature of these forests is their paucity of species. Some A-samples contained fewer than 20 species and none more than 40. Characteristic species are nevertheless plentiful (Table 8.11). *Vepris heterophylla*, which may be locally very abundant, has an interesting disjunct distribution, its only other known stations being in Mali and northern Cameroun (Waterman, Meshal, Hall & Swaine 1978). Other less common disjuncts include the small tree *Ochna ovata* and the undershrub *Crossandra nilotica* Oliv. which are not known elsewhere in West Africa.

The low diversity of the flora leads to a high degree of gregariousness amongst the trees. As much as 90% of the stocking may be of one species.

8.8.2 Structure

Canopy trees in these forests are short (usually 10–15 m) although occasional scattered emergents of *Ceiba pentandra* and *Adansonia digitata* Linn. occur. On the Shai Hills, the forest has two distinct strata (Fig. 8.13): an upper canopy of *Millettia thonningii*, *Diopyros abyssinica* and *D. mespiliformis*, with an understorey of *Drypetes parvifolia* (Fig. 8.12). Elsewhere, *Talbotiella gentii* (Krobo mountain, Sapawsu F.R.) forms local patches with a *Drypetes parvifolia* understorey; *Hymenostegia afzelii* with some *Diospyros abyssinica* forms a single layered canopy on Osodoku mountain. The small tree *Maytenus undata* is also locally abundant in the latter locality.

Basal sprouting and profuse epicormics (Fig. 8.12b) are a feature of many of these species (except *Diospyros*) and basal sprouting is no doubt

Fig. 8.12 South-east Outlier forest.
(a) Forest in Shai Hills G.P.R.; dry season; (b) *Eugenia coronata* trunk showing basal branching; Shai Hills G.P.R.

Table 8.11 Characteristic species of South-east Outlier forest.

Species	Percent frequency	Occurrence elsewhere (number of types)	Chorological group	Life-form
Vepris heterophylla	33	0	GW	t
Tiliacora funifera	67	1	CT	C
Uvaria ovata	67	1	UG	c
Millettia thonningii	50	1	GW	t
Premna quadrifolia	50	1	GW	C
Ritchiea reflexa	50	1	UG	c
Diospyros abyssinica	100	2	CT	t
Drypetes parvifolia	100	2	GW	t
Erythroxylum emarginatum	67	2	CT	S
Drypetes floribunda	50	2	GC	t
Salacia leptoclada	50	2	CT	C
Asparagus warneckei	33	2	UG	c

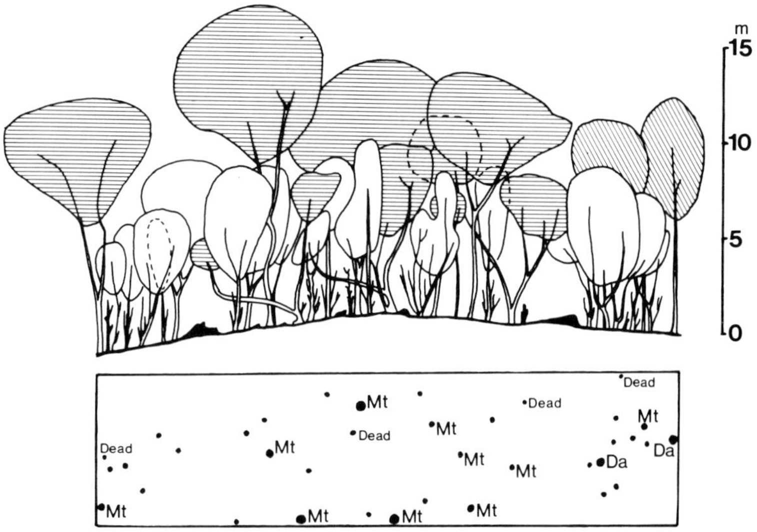

Fig. 8.13 Profile diagram of strip 8 m × 30 m in South-east Outlier forest in the Shai Hills. *Millettia thonningii* (Mt = horizontal shading) and *Diospyros abyssinica* (Da = oblique shading) form the upper canopy; *Drypetes parvifolia* (unshaded) forms the understorey. All species regenerating.

important as means of vegetative reproduction, or at least replacement.

8.8.3 Environment

South-east Outlier is forest at its driest extreme in Ghana. Mean annual rainfall in the Accra Plains ranges from about 750 to 1 000 mm. In 1976 7 rainfall at the Shai Hills was a mere 650 mm.

As in other areas of low rainfall, the parent rock has an important influence on soil conditions. Most of the examples of this type are associated with rocky hills and outcrops which may provide local pockets of soil with enhanced water-retention. The rock type has a controlling influence on soil chemistry; Table 8.12 gives environmental data for each of the six A-samples in the type.

The higher rainfall and more acidic rock on the shales and quartzites of the Togo series yield a soil poorer in nutrients than soil developed over basic

Table 8.12 Summary of environmental conditions in six South-east Outlier forest samples.

Site	Rock type	Approx annual rainfall (mm)	Soil factors					
			Ca^{++}	TEB	CEC	Percent sat. of CEC	pH	P
Mampong Quarry (Shai)	Basic gneiss	800	32	38	36	100	6.5	603
Shai Hills	Basic gneiss	800	17	21	19	100	6.9	54
Osodoku Mountain	Basic gneiss	1 100	19	26	27	97	6.4	10
Krobo Mountain	Basic gneiss	1 100	18	22	23	96	6.6	21
Yogaga Mountain (Kpong)	Shale/Quartzite	1 250	9	13	14	89	5.9	13
Sapawsu F.R.	Shale/Quartzite	1 400	1	2	11	20	4.2	14

CA^{++} Calcium ion concentration (m-equiv/100 g dry soil)
TEB Total exchangeable Bases (m-equiv/100 g dry soil)
CEC Cation Exchange Capacity (m-equiv/100 g dry soil)
P Available phosphorus (ppm)

gneiss under a lower rainfall. In the type as a whole, the soils are among the most nutrient-rich in the forest zone and this may explain why forest exists under such low rainfall.

8.8.4 Conservation and utilisation

This unique kind of forest, which has not been reported from other countries, occupies such a small area of land that it is vulnerable and could easily be eliminated entirely. Fortunately, some of the largest examples are already protected by one means or another. The Shai Hills Game Production Reserve includes a sizeable tract in the central valley of the hills; Sapawsu F.R. protects some of the better examples of *Talbotiella* forest, and local customs associated with the Battle of Dodowa at Katamanso have effectively preserved about 1 km^2 of thicket and forest (Pinkwae) near Amrahia on the Legon-Dodowa road.

Since economic timber species are absent, the only threat to the type presently is from firewood cutting, which is already encroaching on parts of the Krobo Mountain site.

Similar forest has been seen to the north of the forest zone on a rocky hillside above the Black volta gorge at Bui (8° 16′N 2° 16′W). The principal tree there is the rare *Pteleopsis habeensis* (it also occurs at Sapawsu) and the remaining species include many typical of the Shai Hills.

8.9 Wet Evergreen type

This forest type, restricted to the highest rainfall zone of Ghana, is perhaps the most easily recognised and corresponds closely with Taylor's *Lophira-Tarrietia* (i.e. *Heritiera)-Cynometra* 'Rainforest' Association.

8.9.1 Floristics

These forests are floristically very rich (up to 200 species per A-sample), and contain more characteristic species than any other type (Table 8.13), partly because they share a border with only one other type. For this reason it is much easier to name this type than other types after suitable species. All three of the species so used by Taylor are quite characteristic, although *Lophira alata* was seldom included in our A-samples as its seeds require high light intensities for germination.

Table 8.13 also shows that the global distribution of characteristic species is generally more restricted than is the case for other types. *Cola umbratilis* is endemic to Ghana, and half of the others do not occur east of the Dahomey Gap.

The chorological relationships of the Wet Evergreen flora are discussed on p. 38.

As a consequence of high diversity, the density of individual species tends to be low, but certain tree species are usually present. Among these, *Dacry-*

Table 8.13 Characteristic species of Wet Evergreen forest.

Species	Percent frequency	Occurrence elsewhere (number of types)	Chorological group	Life-form
Soyauxia grandifolia	62	0	UG	t
Trichomanes guineense	54	0	GW	Pt
Agelaea trifolia	54	0	GW	C
Cola umbratilis	50	0	UG	t
Scaphopetalum amoenum	42	0	UG	S
Guaduella macrostachys	38	0	GW	H
Cynometra ananta	38	0	UG	T
Coula edulis	31	0	GW	t
Scytopetalum tieghemii	69	1	UG	t
Placodiscus oblongifolius	62	1	UG	t
Pentadesma butyracea	62	1	GW	T
Heritiera utilis	54	1	UG	T
Diospyros chevalieri	54	1	UG	S
Chytranthus cauliflorus	54	1	UG	t
Strephonema pseudocola	54	1	GW	T
Heisteria parvifolia	92	2	GC	S
Aganope leucobotrya	83	2	GW	t
Cola chlamydantha	77	2	GW	t
Ctenitis jenseniae	77	2	GC	Pt

odes klaineana (92% frequency), *Strombosia glaucescens* (85%), *Diospyros sanza-minika* (85%) and *Dialium aubrevillei* (85%) are found widely in other types, but achieve their greatest frequency here.

8.9.2 Structure

The structure of Wet Evergreen forest is essentially the same as that of the Moist Semi-deciduous type (Fig. 8.14), but vertically compressed, canopy trees rarely exceeding 40 m.

Although notably shorter (mean for the type is 32 m), the largest trees may have massive boles and the basal area per hectare of the type is not significantly different from that in more productive areas. The forest is termed evergreen because of the relative scarcity of deciduous trees in the canopy (<20%).

In places, the large shrub *Scaphopetalum amoenum* locally forms a dense, virtually monospecific, understorey by means of vegetative layering (Jeník 1969). The exclusion of all other species which would normally form the understory, and of many of the ground herbs, leads to species lists less than half as long as those of neighbouring samples (Fig. 4.1).

The moist conditions of Wet Evergreen forests support life-forms which are absent or rare elsewhere. In other forest types filmy ferns are confined to tree boles or moist river banks; here *Trichomanes guineense* is a frequent element of the herbaceous ground flora. Several species of saprophytic angiosperms are only recorded from this type: they are *Sciaphila africana* A. Chev., *Voyria primuloides* Bak., *Gymnosiphon longistylus* (Benth.) Hutch. and *Burmannia congesta* (Wright) Jonker. Epiphytes are less noticeably abundant than in Upland Evergreen forest, but include many species not known from other types, e.g. *Polystachya fractiflexa* Summerh. and *P. monolenis* Summerh.

8.9.3 Environment

Annual rainfall is higher in this type than anywhere else in Ghana. Virtually all stations within the type record mean annual falls in excess of 1 750 mm and in places higher than 2 000 mm. Consequently, the soils show the most severe effects of leaching; total exchangeable bases in our samples never exceeded 2 m-equiv/100 g soil and base saturation was rarely greater than 15%, giving pH values almost entirely within the range 3.8 to 4.3.

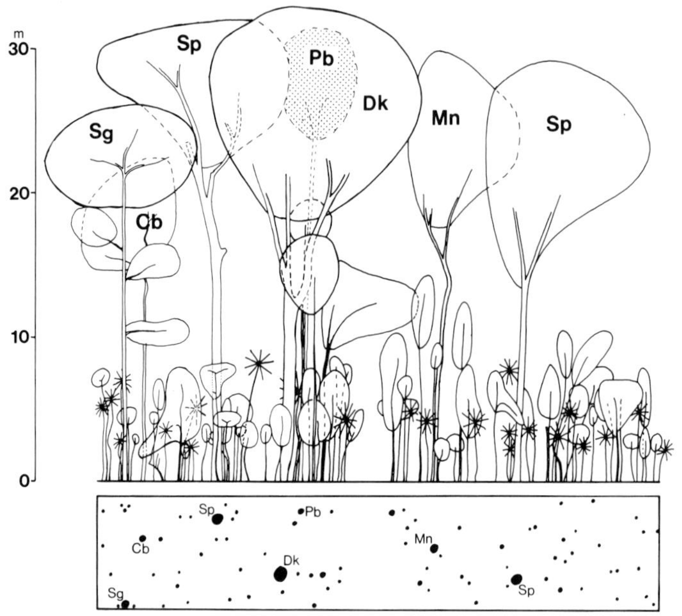

Fig. 8.14 Profile diagram of strip 8 m × 40 m in Wet Evergreen forest in Neung F.R. Stippling as in Fig. 8.1.
Abbreviations:
Cb = *Calpocalyx brevibracteatus*
Dk = *Dacryodes klaineana*
Mn = *Memecylon normandii*
Pb = *Pentadesma butyracea*
Sg = *Strombosia glaucescens*
Sp = *Strephonema pseudocola*
The long-leaved trees in the understorey are *Cola chlamydantha* and *Cola umbratilis*.

Such scarcity of free nutrients may well explain the reduced stature of the forest, and it has been suggested (Monk 1966, Small 1972) that evergreenness may be an adaptation for the conservation of absorbed nutrients. Decomposition rates are high, leaf litter and humus never abundant; the bulk of the plant nutrients at any time is probably held within the plants.

Fig. 8.15 displays the main features of the Wet Evergreen area in south-western Ghana. Most of this forest-type is underlain by basement complex rocks, but there is a small area of land formed of Tertiary sediments of a kind which under high rainfall in Ivory Coast and Nigeria give rise to floristically distinct vegetation (Mangenot 1950; Hall 1977). In Ghana the sediments are mostly swampy or farmed and little closed forest remains. Of our plots only two B-samples fell within this area, and we are thus unable to assess the influence of the sediments on the vegetation. Quaternary sands behind the coast have given rise to distinctive very highly leached soils (Ahn 1959) which are podzolic and support only short grassland (containing many herbaceous species not known elsewhere in Ghana) in areas subject to seasonal freshwater flooding. Well-developed *Raphia* swamps occupy permanently flooded areas, and secondary forest is found on the more elevated parts of the topography.

8.9.4 Conservation and utilisation

Wet Evergreen forest contains relatively few of the presently valuable timber species. Of class I timbers, only *Heritiera utilis* and *Lovoa trichilioides* may be considered fairly abundant. Fairly common timbers in class 2 include (*Guarea cedrata* and *Lophira alata*. There is a notable absence, however, of Ghana's chief export species *Triplochiton scleroxylon*. The value of these reserves is further reduced because the useable bole length of most exploitable trees is relatively short.

The poor soils of the area have, in the past, repelled cocoa or food farmers and the population has been low for a forest area. Such poor returns from both farming and forestry have allowed a larger proportion of undisturbed forest to remain here than in other types. The best examples of the

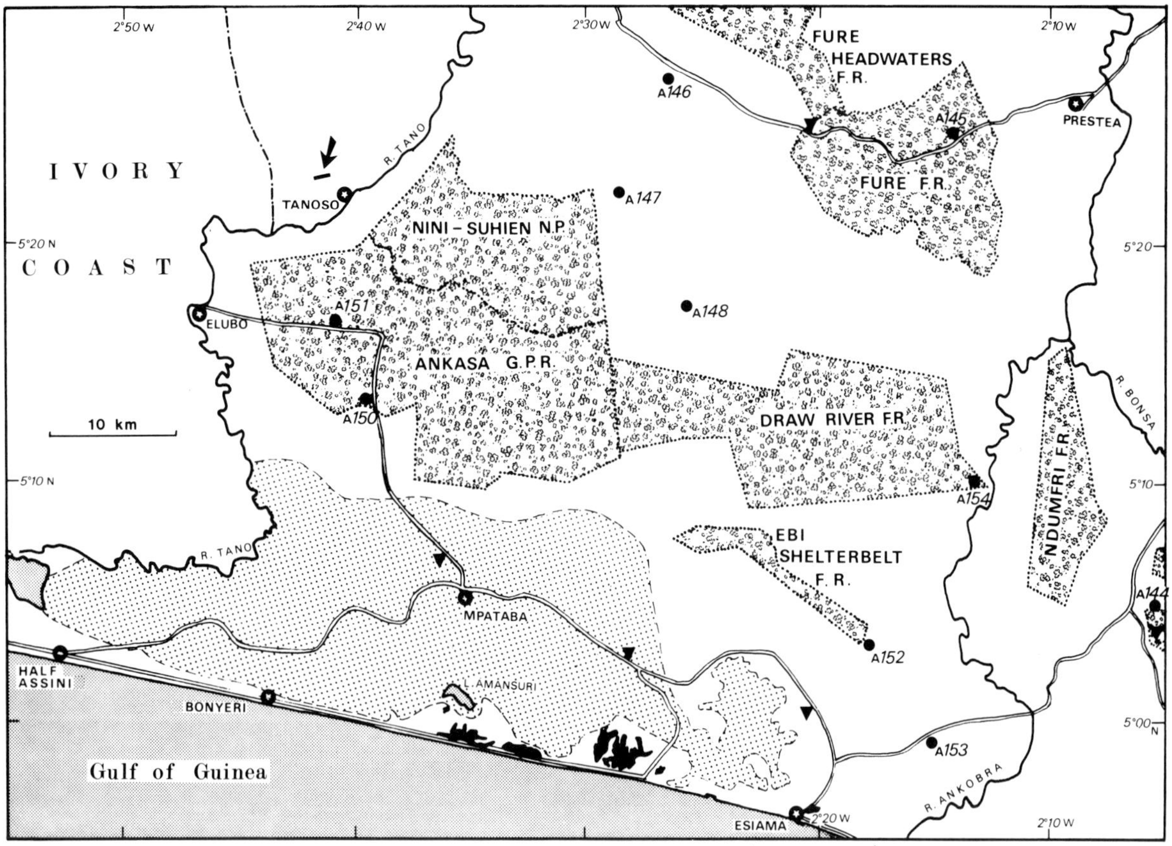

Fig. 8.15 Map of south-west Ghana. Reserved forest (all Wet Evergreen) shown by symbolic tree crowns, Tertiary sands by coarse stipple and open water by fine stipple. Irregular black patches are grassland on leached Quaternary podzolic sands. A-samples (●) and B-samples (▼) are shown. The position of the vegetation mosaic (Fig. 1.4) is arrowed. After Ahn 1961.

type are to be found (Fig. 8.15) in the Ankasa G.P.R. and Nini-Suhien N.P. (together occupying the area of the former Ankasa River F.R.), and in the Draw River F.R. It is greatly to be hoped that the Nini-Suhien National Park (170 km^2) will be preserved intact and free from timber-cutting.

Ankasa is also one of only two known localities for Ghana's sole endemic forest genus – *Monocyclanthus* (Annonaceae) and the other locality also falls within the type at Benso.

8.10 Moist Evergreen type

Much of Taylor's *Triplochiton-Lophira* 'transition zone' falls within this type. Although the type is somewhat intermediate between Wet Evergreen and Moist Semi-deciduous, it seems nonetheless worthy of type status.

8.10.1 Floristics

Although less well supplied with species confined to one type than is Wet Evergreen forest, the Moist Evergreen type has more clearly characteristic species than does the Moist Semi-deciduous forest (Table 8.14). Many species present in Wet Evergreen forest are absent here; others are more or less confined to the two lowland evergreen types (e.g. *Angylocalyx oligophyllus,* Map 41); other species, notably absent from wetter forest, make their appearance here (e.g. *Triplochiton scleroxylon,* Map 603).

Although not as rich as Wet Evergreen, the type has great floristic diversity, with up to 170 species per A-sample.

Table 8.14 Characteristic species of Moist Evergreen forest.

Species	Percent frequency	Occurrence elsewhere (number of types)	Chorological group	Life-form
Placodiscus bancoensis	19	0	UG	t
Rinorea breviracemosa	15	0	GW	t
Petersianthus macrocarpus	56	1	GC	T
Diospyros gabunensis	33	1	GC	t
Airyantha schweinfurthiii subsp. *confusa*	15	1	GC	C
Dichapetalum toxicarium	67	2	UG	C
Maranthes glabra	30	2	GC	t
Angylocalyx oligophyllus	93	3	GC	S
Acridocarpus smeathmannii	78	3	GW	c
Rinorea oblongifolia	70	3	GC	t

8.10.2 Structure

Although very similar to Moist Semi-deciduous forest in appearance (Fig. 8.16), the tallest trees in Moist Evergreen forests are slightly shorter on average (43 m). Deciduous trees form only a small proportion of the canopy (<20%), and the application of the term 'evergreen' is thus appropriate.

8.10.3 Environment

Annual rainfall within the type seems very close to that of the South-east subtype of the Moist Semi-deciduous forest. The soil, however, is clearly poorer in nutrients (Fig. 3.8) and the vegetation distinct, suggesting that there might well be climatic differences which are insufficiently expressed by recent data on annual precipitation.

Although never achieving great altitude, steep-sided hills are a feature of some parts of the type (Bowiye Range F.R., Opon Mansi F.R.) and are mostly of Upper Birrimian rocks.

8.10.4 Conservation and utilisation

The type is rich in economic species, being second only to Moist Semi-deciduous forest in this respect. Table 7.2 shows that 31% of the type is included in forest reserves and must contribute heavily to the country's timber production. Subri River F.R., Ghana's largest forest reserve (c. 600 km²), lies entirely within the type. Much of the Subri reserve and those on the Birrimian hills are at present

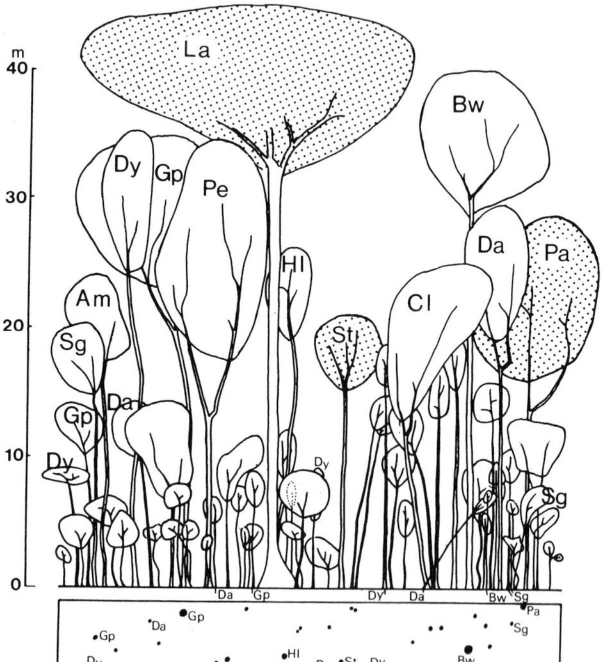

Fig. 8.16 Profile diagram of strip 8 m × 40 m in Moist Evergreen forest in Subri River F.R. Stippling as in Fig. 8.1. All individuals of species represented in the canopy are named.
Abbreviations:

Am = *Anthonotha macrophylla*
Bw = *Blighia welwitschii*
Cl = *Cola lateritia*
Hl = *Homalium letestui*
La = *Lophira alata*
Pa = *Phyllocosmus africanus*
Da = *Dialium aubrevillei*
Dy = *Drypetes aylmeri*
Gp = *Gilbertiodendron preussii*
Pe = *Parinari excelsa*
Sg = *Strombosia glaucescens*
St = *Scytopetalum tieghemii*

largely unexploited because of the difficulties of access. In the southern part of Subri, however, a large planting programme is under way replacing the natural forest by pure stands mainly of *Gmelina arborea*, intended for a pulp mill to be built near Daboase (p. 67). Opon Mansi F.R. is threatened by exploitation of its iron ore deposits.

The large African Timber and Plywood Sawmill and Veneer factory at Samreboi, established at the end of the second World War, lies just within the northern edge of the Wet Evergreen forest-type, but most of their timber has come from Moist Evergreen forest.

A large part of the Bia Game Production Reserve (p. 70) falls within this type. Previously Bia Tributaries South Forest Reserve, it was little exploited until recently, when many concessions have been granted within the area, and a rich harvest of economic Meliaceae is expected.

8.11 Upland Evergreen type

Upland Evergreen forests (Swaine & Hall 1977) occur on isolated hill ranges (500–750 m elevation) within the area of the Moist Semi-deciduous type. Taylor (1960) included them in this *Triplochiton-Lophira* association (Moist Evergreen type) because in both of these forest-types occurs a mixture of trees typical of Moist Semi-deciduous and of Wet Evergreen forests. Our ordination (Fig. 2.3), however, showed a complete separation of Upland and Moist Evergreen.

Upland Evergreen forest is quite similar to the 'Forêt dense humide à *Parinari excelsa*' (Guillaumet & Adjanohoun 1971) of hills in western Ivory Coast, despite the greater elevation (>1 000 m) and much more abundant *Parinari excelsa* in the latter.

8.11.1 Floristics

Table 8.15 shows that, in contrast with other types, the most characteristic species of Upland Evergreen forest are herbaceous rather than woody. Ferns, both epiphytic and ground-living, are both abundant and diverse (Adams & Alston 1955; Hall & Bigger 1974). The tree fern *Alsophila manniana (Cyathea manniana)* is common in swampy valleys in the Atewa Range F.R., which includes the best example of the type. Numerous disjunct species which have their nearest stations on other West African mountains indicate the submontane nature of the type (Hall, Brookman-Amissah, Leston & Dodoo 1973; Swaine & Hall 1977). These include a member of the temperate genus *Rubus pinnatus* Willd. var. *afrotropicus* (Engl.) C. E. Gust., and the recently described *Hymenocoleus multinervis*

Table 8.15 Characteristic species of Upland Evergreen forest.

Species	Percent frequency	Occurrence elsewhere (number of types)	Chorological group	Life-form
Peperomia fernandopoiana	50	0	GW	Ep
Asplenium dregeanum	33	0	CT	Ep;Pt
Arthropteris monocarpa	33	0	CT	Ep;Pt
Angraecum birrimense	33	0	GW	Ep
Lonchitis currori	33	0	CT	Pt
Lasianthus repens	33	0	GC	H
Chidlowia sanguinea	83	1	UG	t
Alchornea floribunda	67	1	GC	t
Sarcophrynium prionogonium	67	1	GC	H
Lophira alata	33	1	GC	T
Piptostigma fugax	67	2	UG	t
Uvariopsis globiflora	67	2	UG	t
Rinorea welwitschii	50	2	GC	t
Costus engleranus	83	3	GW	H
Tabernaemontana pachysiphon	83	3	GW	t
Strombosia glaucescens	100	4	GC	T
Leptaspis cochleata	100	4	CT	H

Robbrecht which is definitely known elsewhere only from forest on the mountains of Dans and Nimba near the Ivory Coast-Liberia border (Robbrecht 1977).

Sample plots in Upland Evergreen forest generally provide longer species lists than those in the surrounding Moist Semi-deciduous forest, and the flora is further enlarged by the diverse topography of the hill summits.

8.11.2 Structure

Vertical structure is similar to that of Moist Semi-deciduous forest but, as in the other evergreen types, vertically compressed (Fig. 8.17), the tallest trees rarely exceeding 45 m.

Deciduous canopy trees seem rarer here than in any other forest type. Although based on only a small sample of trees (31) only 6% were deciduous. The great diversity of epiphyte species and life-forms is characteristic. Although epiphytes are most abundant on some emergent canopy trees (Fig. 8.18), understorey trees carry an unusually large epiphytic flora. Not all the largest trees are heavily adorned, but extreme examples may support 40 species of vascular epiphytes. Among these are many woody epiphytic species, mostly rare or unknown elsewhere in Ghana: these include *Hymenodictyon floribundum* (Steud. & Hochst.) B. L. Robinson, *Preussiella kamerunensis* Gilg, *Medinilla mannii* Hook.f., *Schefflera barteri* and *Anthocleista microphylla*.

8.11.3 Environment

The hills on which Upland Evergreen forests occur are generally isolated and steep-sided with more or less flat summits which are the last remains of the

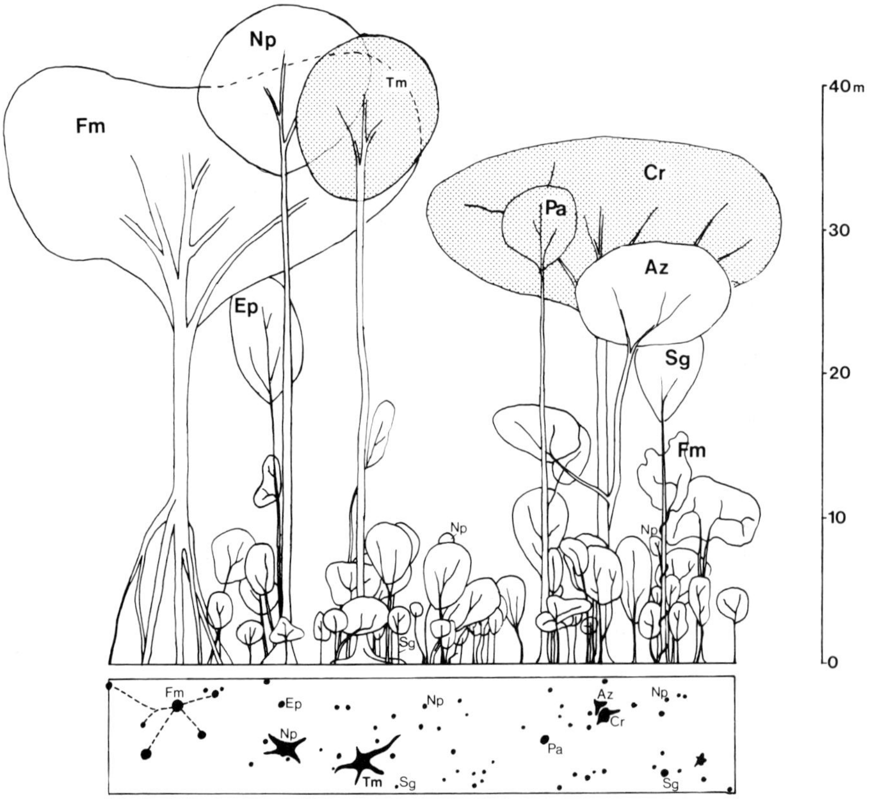

Fig. 8.17 Profile diagram of strip 8 m × 45 m in Upland Evergreen forest in Atewa Range F.R. Stippling as in Fig. 8.1. All individuals of species represented in the canopy are named. Abbreviations:
Az = *Albizia zygia* Ep = *Enantia polycarpa* Np = *Nesogordonia papaverifera* Sg = *Strombosia glaucescens*
Cr = *Chlorophora regia* Fm = *Ficus macrosperma* Pa = *Pycnanthus angolensis* Tm = *Trilepisium madagascariense*

Fig. 8.18 Dense growth of canopy epiphytes including the ferns *Drynaria laurentii* and *Oleandra distenta;* Atewa Range F.R.

Tertiary peneplain which once covered southern Ghana. The soils which have developed are thus very ancient and often bauxitic. They are rich in clays which have been aggregated by humus to form a surprisingly light-textured soil (Fig. 8.19). The elevation of the hill summits (mostly >500 m) leads to reduced temperatures, increased rainfall and mistiness (Fig. 8.20), and probably a reduction in annual incoming radiation. These conditions favour leaching (with associated low base saturation and pH), and reduced rates of organic decomposition which are demonstrated by high levels of soil organic matter and nitrogen (Fig. 8.19).

The forests are very uneven, patches of closed canopy alternating with thickets, swamps and even grassland. This may be attributed to various causes: bauxite occurs at variable depths and may be exposed at the surface; the soil is thus sometimes too shallow to support large trees and is covered by dense thicket, or in drier areas, by grassland. Cooper (1958) presents maps of bauxite deposits in the Tano Ofin F.R. near Nyinahin and in the

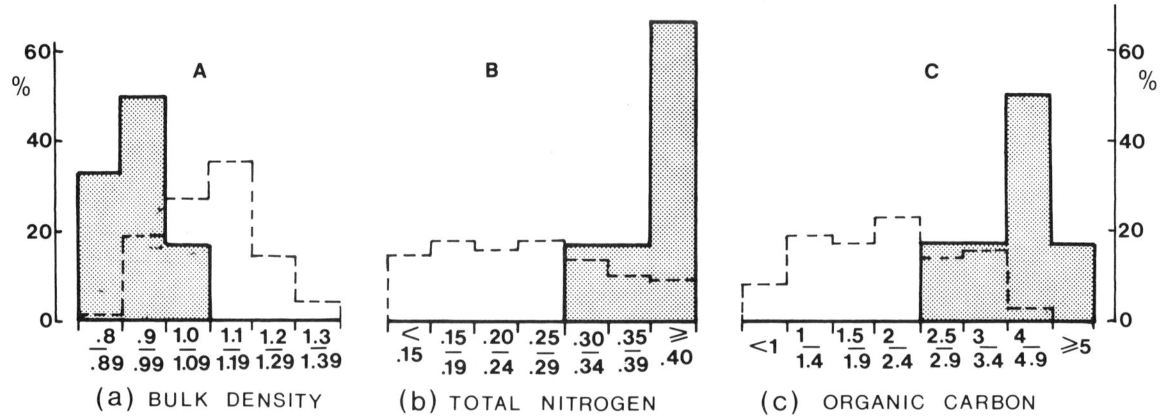

Fig. 8.19 Distinctive features of Upland Evergreen forest soils. Data from analysis of topsoil in A-samples: total nitrogen by Kjeldahl, and organic carbon by Walkley & Black. Shaded histograms: Upland Evergreen samples; unshaded histograms with broken outlines: all other samples. (a) Bulk density (g cm^{-3}). (b) Total nitrogen (percent dry soil). (c) Organic carbon (percent dry soil).

Fig. 8.20 Summit clouds on Mt. Ejuanema near Nkawkaw.

Krokosua Hills F.R. which show numerous small meadows within the forest. The flora of these meadows includes several rare species, including the herb *Aeschynomene deightonii* Hepper, found in Ghana only at Nyinahin and Krokosua, and outside Ghana in Guinée and Sierra Leone. The meadows are clearly relict rather than the result of human interference.

The broken nature of the forest cover is probably contributed to by increased wind-throw of large trees. Shallow, clayey soil, steep slopes, and swamps do not provide the best anchorage for large trees, and the elevated position must expose trees to high winds at times.

8.11.4 Conservation and utilisation

Valuable timbers are uncommon in Upland Evergreen forests, and difficulty of access has until recently prevented exploitation. The best examples of the type (at Tano Ofin and Atewa) are both within forest reserves and recent bauxite surveys have established roads to the summits. This has lead to some timber extraction at high elevation at least on the Atewa Range.

The reserves of bauxite, although of relatively low grade, are a potential threat to the survival of Upland Evergreen forest. Useable ore, however, occurs only in local patches, and it should be possible to exploit it without completely destroying the intervening forest.

Another locality for the type has been recorded on Mt. Ejuanema, above Nkawkaw. The forest on the Afao Hills above Awaso, where bauxite is currently mined, shows some similarity with Upland Evergreen forest, but was included with Moist Semi-deciduous forest by the ordination.

We are here concerned principally with closed-canopy forest. But these uplands include other kinds of interesting vegetation. The bovals (meadows on bauxite outcrops), swamps and thickets are probably unique in Ghana and deserve protection in their own right.

Conversion of natural Moist Evergreen forest to plantations of *Pinus caribaea*, Subri River Forest Reserve.

PART TWO

The maps

CHAPTER 9

Mapping methods and map format

9.1 Mapping plant distributions

Maps of plant distribution fall into two main categories. The first type comprises maps showing localities where herbarium specimens of a particular taxon have been collected. Such maps are of very great value on a continental scale. Those prepared in the series 'Distributiones Plantarum Africanarum' published by the Belgian National Botanic Garden use a dot size corresponding to one degree squares, and give an excellent impression of distribution in Africa as a whole (e.g. White 1976b). For mapping of smaller areas, use of herbarium records has the disadvantage that patchiness of any map, and differences in dot density between maps, is as likely to be the result of the vagaries of collectors as to reflect real distributional features. It is for this reason that distribution maps of this kind have seldom been prepared for individual tropical African countries and those attempted have not been very informative (e.g. Chakravarty 1968).

The second type of map collates the observations, within a national grid, of large numbers of local naturalists, who fill in forms and reports for the grid square with which they are concerned, but will not usually make collections. Collecting of the rarer species will, in fact, be actively discouraged by the mapping organisers. The Atlas of the British Flora (Perring & Walters 1962) provides an excellent example – the grid used was of 10 km × 10 km squares. This kind of map is only feasible where knowledgeable amateurs are abundant, and where the flora is well known and described.

Mapping for this book has adopted a hybrid procedure. Our 285 sample sites are well scattered through the forest zone of Ghana (Fig. 2.2) but, for practical reasons, are neither systematic nor random in arrangement. Judging from species-area curves (Fig. 4.2) each A-sample included about half and each B-sample about a quarter of the species which could have been found in a 1 ha plot. Little significance, therefore, can be attached to the absence of a species from any particular sample, but absence from several adjacent plots may be significant. Because the 'A' plot size was constant and all species were listed, the proportion of A-samples in which a species was recorded gives a fairly objective picture of its relative abundance. Density of herbarium collections, in contrast, is much more closely related to conspicuousness, length of flowering season, accessibility of flowering shoots, and accessibility and attractiveness of the locality. The most frequently-collected forest plants in Ghana include *Microdesmis puberula,* a small tree which flowers more or less continuously, and *Oxyanthus* spp. which have large flowers. We found *Griffonia simplicifolia* (Map 284) to be the most widespread and abundant forest species, but perusal of herbarium specimens would lead to the impression that this species is largely restricted to thickets on the Accra Plains and forest margins (where it is readily collected in flower or fruit). Some of the most abundant species, especially of large climbers – for example *Leptoderris fasciculata* (Map 344) – had never even been recorded from Ghana prior to our survey.

The objection may be made that our method places too heavy reliance on relevé records, necessarily based mainly on the field recognition of leafy shoots and the boles of trees, although the flora is still imperfectly known. Mistakes in identification, the critics rightly claim, are inevitable – and without voucher specimens, accuracy cannot be checked by specialist taxonomists. Our defence must be that the collection of a voucher specimen for each identification would have made the survey impossibly time-consuming: more than 20 000 specimens would have been required. For difficult and critical species we did in fact collect many vouchers, which are kept in the Ghana Herbarium of the Department of Botany, University of Ghana, but whose collecting positions we have not mapped because many are too scrappy – leaves catapulted from tall trees for example – to be included with good herbarium specimens. The herbarium specimens included on our maps will give a reasonable picture of our species concepts. In cases where we could not always distinguish between a group of two or three species, perhaps because flowers are essential for identification, we have pooled our survey records, and distinguished only the herbarium records. *Cola gigantea* var. *glabrescens* and *C. lateritia* var. *maclaudii,* for example, are quite distinct in fruit, usually distinguishable by tree form, but not separable as saplings – for this reason they share Map 150.

Of the more than 2 000 species of vascular plants in Ghanaian forest, only about 60% were recorded during our survey. Since we confined our sampling to closed-canopy forest on well-drained sites, secondary and swamp forest species and ruderals were seldom recorded, and are mostly excluded from the maps. Epiphytes were undersampled (p. 47) and very few have been mapped. Because of the small plot size, rare species were usually missed and have only been mapped if they show an interesting distribution pattern, perhaps being restricted to one type.

Certain important timber species were rarely seen in our survey either because they occur in swamp (e.g. *Mitragyna* spp.), are rare (e.g. *Pericopsis elata*), or do not regenerate under the forest canopy (e.g. *Terminalia ivorensis*). Because it seemed absurd to omit the most valuable forest plants from a book about the forest, we have prepared special maps for nine such species from Forestry Department enumeration data (Chapter 11, p. 318).

9.2 Explanation of symbols and text

Each of the following 624 maps plots the known localities for one or more species within the forest zone of Ghana. The records derive principally from our own survey of 285 sample sites (see Fig. 2.2) and these are shown by solid dots (except where otherwise stated in the notes accompanying the map). Each of these localities is provided with latitude and longitude to the nearest 10 sec in Appendix 2.

Additional records, shown by open symbols (except where otherwise stated), are of specimens in the Ghana Herbarium (Legon) or in the Forest Herbarium (Kumasi), or of localities given in the Flora of West Tropical Africa (Hutchinson & Dalziel 1954–72).

The distribution of the species amongst the forest types is shown on each map as a histogram giving the percentage of the A-samples which contain the species.

The species are arranged in alphabetical order by generic and specific name.

On each double page, notes are given for the species whose corresponding map is indicated by a number in bold face. These include life-form, biology, ecology and occasionally chemistry of the species, with numerical referencing (p. 362) of relevant publications.

The notes are arranged in the following sequence:

1. Map number and species name, including authority.
2. Volume and page reference to the revised edition (1954–72) of the Flora of West Tropical Africa (FWTA), where the name is used by FWTA in the same sense as by us.
3. Numerical reference (p. 362) to names not used in FWTA: these include species published since 1972, new combinations, and names which must now be used in a wider, narrower, or different sense following taxonomic revision. References are sometimes to the original publication, and sometimes to a Flora or other book if this is likely to be more readily available, and uses the name in the same sense as that accepted by us.
4. In the case of name changes, the FWTA synonym is given in brackets. 'Incl.' ('In-

cluding') precedes names of species cited by FWTA as district, and occurring in the FWTA area, but not accepted as distinct by us.

5. Where we have mapped together two or more species, or two or more varieties, these additional names are given without brackets.
6. Family name.
7. Vernacular name, preceded by language, only where the name is widely used, in a fairly unambiguous way.
8. Type locality, for species described from specimens collected in Ghana.
9. Plant size and life-form, and distinctive morphological features. Each species is placed in one of the following categories:
   Tree – large (>30 m max. ht.), medium (10–30 m), small (2–10 m), pygmy (<2 m). Buttresses are mentioned if they occur; no mention means no buttresses.
   Shrub – erect woody plant branching from the base, less than 5 m tall.
   Climber – small (slender stem, usually confined to the understorey); large (woody lianes capable of reaching the crowns of large trees); root climber (attached to support by adventitious roots, sometimes becoming epiphytic). Anatomical anomalies in the stems of climbers are briefly indicated.
   Ground herb – little or no secondary thickening, usually less than 2 m tall.
   Epiphyte – vascular plant which is normally rooted on the trunk or branch of a woody plant, and not in the ground.
   Unless otherwise stated, plants lack latex or strongly aromatic bark.
10. If no statement is made to the contrary, leaves are evergreen, simple, lamina devoid of glands, lamina margin entire, lamina and twigs glabrous or subglabrous. Lamina dimensions apply to leaves of mature plants, not saplings; categories (according to Raunkiaer 1934) are as follows:

Leptophyll	<25 mm^2
Nanophyll	25–225 mm^2
Microphyll	225–2 025 mm^2
Notophyll	2 025–4 500 mm^2
Mesophyll	4 500–18 225 mm^2
Macrophyll	18 225–164 025 mm^2
Megaphyll	>164 025 mm^2

11. Architectural model. Names and definitions of models are according to Hallé, Oldeman & Tomlinson (1978). The main diagnostic features of each model are given below, less exactly, as we have tried to avoid using all the technical terms employed by Hallé *et al.*

A. Easily distinguished models

Holttum:	Trunk strictly unbranched, inflorescence terminal.
Corner:	Trunk strictly unbranched, inflorescence lateral.
Tomlinson:	Trunk branched only at ground level, inflorescences terminal or lateral.
Chamberlain:	Trunk apparently unbranched, but stem in fact sympodial, overtopping terminal inflorescences.
Schoute:	Dichotomous branching occurs.
McClure:	Aerial branching pattern determinate, new aerial shoots initiated by ground-level branching.
Leeuwenberg:	No main trunk above lowest branch; branching pattern sympodial, 2–3 equal branches initiated below inflorescences which are terminal on shoots.

B. Other models

The remaining models (other than Stone's model, which has not been recognised in any of our mapped species) are grouped in Table 9.1. Characters used to distinguish each model from others in the same cell of Table 9.1 are given below.

Koriba:	Similar to Leeuwenberg's model at first, but one branch becoming erect, and continuing the vertical growth of the main trunk.
Prévost:	Similar to Koriba's model, except that trunk-forming branch is distinct from the start.
Scarrone:	Main trunk monopodial, branches pseudo-dichotomous, similar to Leeuwenberg's model.

Table 9.1 Synopsis of the characteristics of the less-easily distinguishable architectural models of Hallé, Oldeman & Tomlinson (1978).

		Trunk sympodial	Trunk monopodial
Branching of branches sympodial	Leaves on branchlets spirally arranged, not distichous	Koriba, Prévost	Scarrone, Aubréville
	Leaves on branchlets distichous		Fagerlind, Petit
Branching of branches monopodial	Leaves on branchlets spiral	Champagnat	Rauh, Attims
	Leaves on branchlets distichous	Nozeran, Mangenot, Troll	Massart, Roux, Cook

Aubréville: '*Terminalia*-type' branching; branching of branches obviously sympodial, leaves and lateral inflorescences crowded at the upturned ends of branchlets; further branchlets originate below the leaf-tuft; main branches apparently whorled.

Fagerlind: Branching of branches less obviously sympodial, inflorescences terminal on branchlets; branches apparently whorled.

Petit: Very similar to Fagerlind's model, but branches not whorled.

Champagnat: Axes bend over secondarily.

Rauh: Branches are apparently whorled.

Attims: Branches not whorled.

Nozeran: 'Cocoa-type' branching; new erect axis with spiral leaves originates laterally from older axis which is topped by horizontal axes with distichous leaves.

Mangenot: Axes initially erect with spiral leaves, then bending over, and with distichous leaves; new erect axes originating near point of bending of older axes.

Troll: Axes all with distichous leaves, becoming erect secondarily.

Massart: Branches apparently whorled.

Roux: Branches not whorled.

Cook: Branches or branchlets strongly resembling compound leaves.

Hallé *et al.* are referred to after the name of the model in the case of those species which they cite; where no reference is given, the identification of the model has been made by us.

12. Floral morphology and biology. Flower types, defined by Faegri & van der Pijl (1971) are illustrated in Fig. 9.1. Flower colour and sometimes pollination mechanism (where known) are mentioned. Period of flowering (months numbered 1–12) are given where herbarium data and Floras indicate that the flowering season is limited, and fairly consistent from year to year (see p. 56). 'Not clearly seasonal' means flowering has been recorded through much of the year. No mention means that available data are inadequate. Flower size is according to linear dimensions of longest perianth member:

    Large (>20 mm)
    Medium (6–20 mm)
    Small (<6 mm)

If flower sexuality, and cauliflory are not mentioned, flowers are hermaphrodite, and among the leaves.

13. Fruit morphology, colour, and dispersal mechanism, where known. By 'fruit' we mean the biological unit of the infructescence, rather than the structure derived from the gynoecium of one flower. Seed lengths (excluding wings and parachutes) are as follows:

    Very large (>20 mm)
    Large (11–20 mm)
    Medium (2–10 mm)
    Small (<2 mm)

Weight would have been a better measure

Fig. 9.1 Blossom types (after Faegri & Van Der Pijl 1971).
(a) Funnel: *Rothmannia longiflora*
(b) dish: *Mussaenda chippii*
(c) bowl: *Monocyclanthus vignei*
(d) bell: *Cola reticulata*
(e) trap: *Ceropegia gemmifera*
(f) brush: *Mimosa pudica*
(g) gullet: *Leonotis nepetifolia*
(h) flag: *Berlinia tomentella*

than length, but data are not available for more than a few species. Fruiting seasons are indicated according to the same principles as given above for flowering season.

14. Ecology. The phrase 'regenerates in shade' means that seeds germinate in shade, and that healthy saplings (or other young stages in the case of herbs) are commonly seen on the floor of closed-canopy forest; we do not imply that the species can necessarily continue to grow to maturity under shade. Gregariousness is mentioned when it is frequent.

15. Soil preferences. These were examined for each soil character by a computer program which calculated the 'G' statistic (Sokal & Rohlf 1969) for a contingency table of species occurrence against soil character class, both obtained from our A-samples (p. 14). Soil characters and their classes were as follows:

	Class							
	1	2	3	4	5	6	7	8
(a) Clay content as percent of oven-dry soil by weight	0–9	10–14	15–19	20–24	25–34	≥35		
(b) Total exchangeable bases (m-equiv. per 100 g oven-dry soil)	0–1.9	2.0–3.9	4.0–5.9	6.0–7.9	8.0–9.9	10.–15.9	16.–21.9	≥22.0
(c) Acidity pH	3.8–4.1	4.2–4.5	4.6–4.9	5.0–5.4	5.4–5.7	5.8–6.1	6.2–6.5	≥6.6

A 'G' statistic figure was regarded as 'very significant' if the probability of obtaining this or a higher value was less than 0.001. Because this level of significance could not usually be obtained if the number of records of occurrence was less than 10, we have not indicated soil preference for any rare species, even though it is likely that some of them are in fact restricted to a particular range of soil classes.

16. Distribution outside Ghana. We have used distribution to place each species into one of the chorological classes described by White (1969); these are described on p. 72. All classes can be distinguished by a perusal of FWTA with the exception of Guinea Wides and Guineo-Congolian Wides, the limit between which (the Sangha River Interval) cuts through Zaire. White (1969) gives the chorological class for all our mapped species of *Diospyros, Maranthes, Psychotria, Strychnos,* Meliaceae and Lecythidaceae, and a few other taxa. In other cases we have classed nontransgressive species extending to Zaire as Guineo-Congolian Wides, though some may in fact be Guinea Wides.

17. Utilisation. For timber species we provide Ghana Forestry Department classification and trade name (see Appendix 7), and reference to Fergusson (1960) who describes timber properties and provides coloured plates of each timber. Occasionally note is made of some other usage (e.g. chewing stick, or medicine).

18. Plant Chemistry. Some recent work is referenced.

19. Taxonomic notes. Reasons are given for some taxonomic decisions that may be controversial.

CHAPTER 10

Distribution maps with notes on morphology, taxonomy, chorology, ecology and utilisation

Key to maps

● Records from the survey of Hall & Swaine (1976).
○ Herbarium and Flora records.

(Where two or more taxa are mapped together, other symbols may be used; these are explained in the accompanying note).

Broken line – northern forest limit.

Thin solid line – margin of Volta River and Lake.

Thick solid line – Ghana coastline & political boundary.

Marginal marks indicate 1° divisions of longitude and latitude.

The number at the top-right corner of the map relates the map to the notes, with the same number, on the same double page.

The histogram on each map shows the percentage of A-samples (see p. 14 for definition) in each forest-type which contain the species:

Abbreviation	Forest-type	Number of A-samples
WE	Wet Evergreen forest	12
ME	Moist Evergreen forest	28
UE	Upland Evergreen forest	6
MS	Moist Semi-deciduous forest	45
DS	Dry Semi-deciduous forest	41
SM	Southern Marginal forest	17
SO	South-east Outlier forest	6

Notes for the important timber species which are mapped separately in Chap. 11 (p. 318) are included in alphabetical sequence with the other notes.

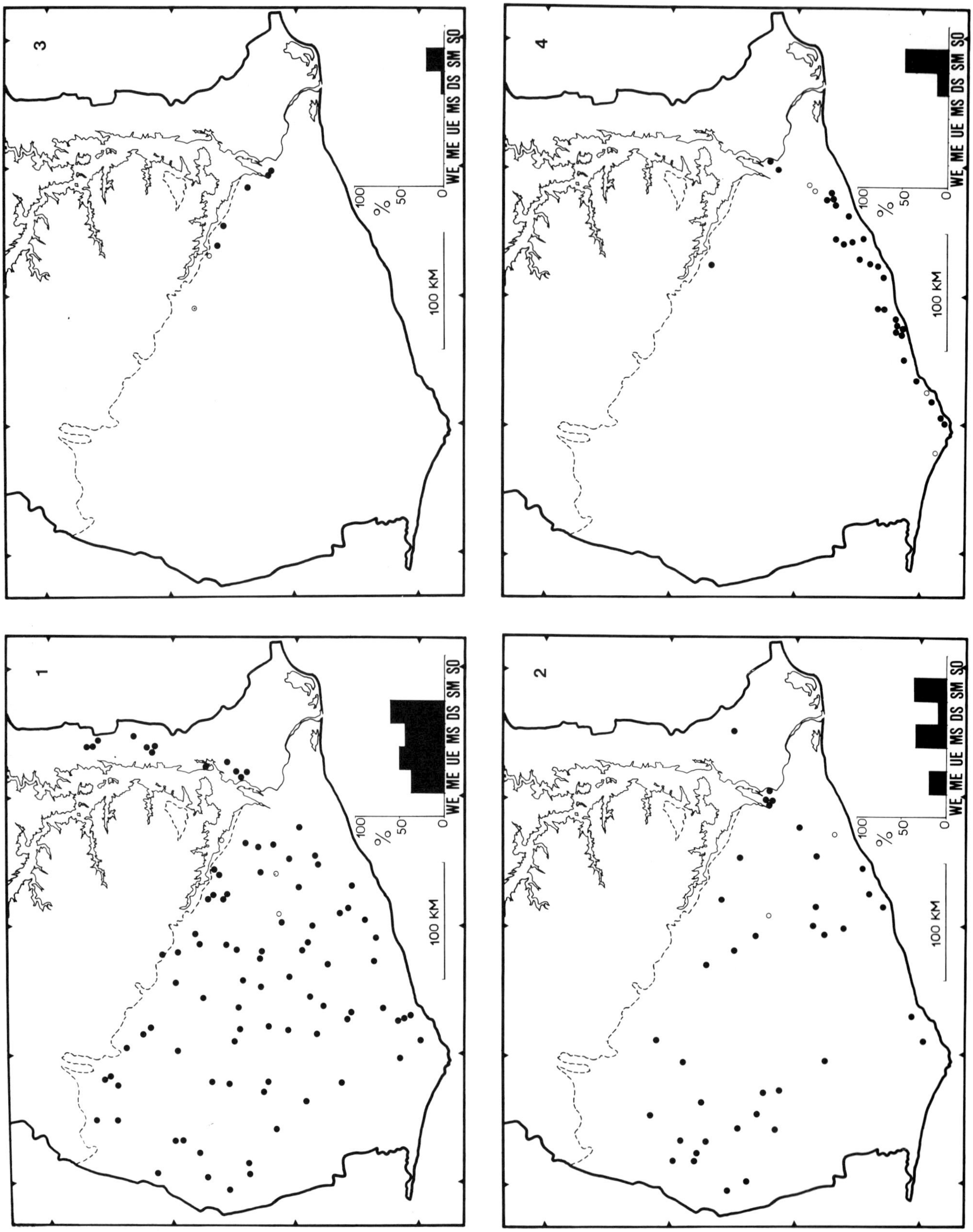
1
2
3
4
100 KM
%
100
50
0
WE ME UE MS DS SM SO

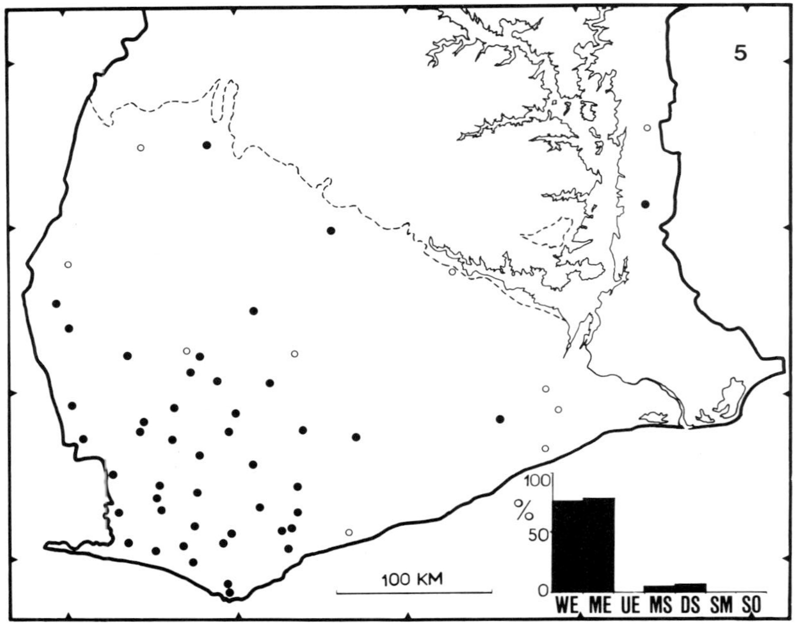

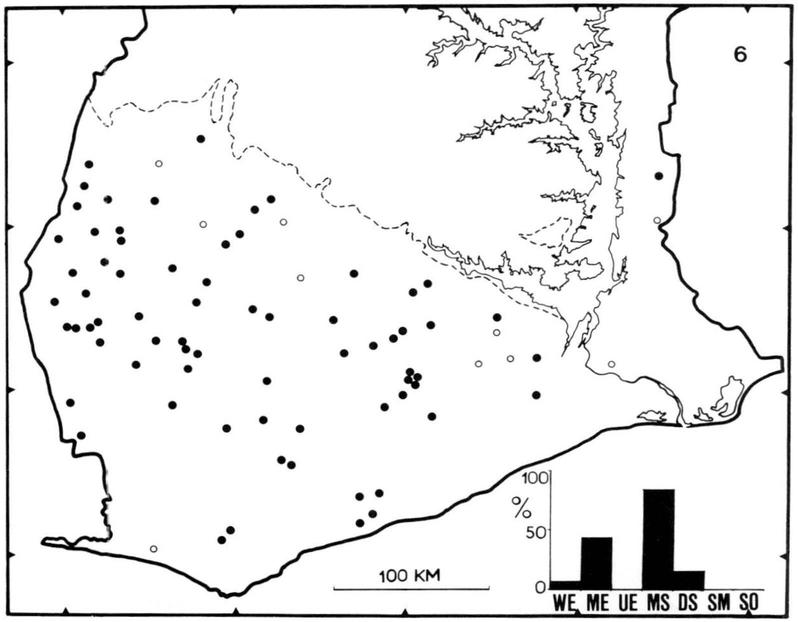

1 ACACIA KAMERUNENSIS Gandoger (28), (*A. pennata* of FWTA 1:500, partly, not of (L.) Willd.). Mimosaceae. Akan: nwere.
Large woody climber, thorny, xylem normal, cylindrical; lvs bipinnate, leaflets leptophyll. Fl small, cream, crowded in brush inflorescence, 8–9. Fr flat, thin dehiscent pod, wind-dispersed; seeds medium. Extrafloral nectaries on petiole. Seedlings found in shade, but rapid growth in gaps. Senegal to Congo; (Guineo-Congolian wide). Woody stem commonly crushed for chewing sponge. (Akan: sawe) (1).

2 ACACIA PENTAGONA (Schumach. & Thonn.) Hook. f. (28), (*A. pennata* of FWTA 1:500, partly, not of (L.) Willd.). Mimosaceae. Akan: nwere. (Type from Oyarifa near Accra).
Differs from *A. kamerunensis* in slightly broader leaflets and dehiscent pods. Stout, thorny climber, xylem deeply five-lobed in section; lvs bipinnate, leaflets leptophyll. Fl small, cream, brush, 8–9. Fr flat, thin dehiscent pod, wind-dispersed; seeds medium. Extra-floral nectaries on petiole. Seedlings found in shade, but rapid growth in gaps. Senegal to Congo; (Guineo-Congolian wide). Uses and local names as for *A. kamerunensis*.

3 ACALYPHA NEPTUNICA Muell. Arg., FWTA 1:410. Euphorbiaceae.
Large, somewhat hairy shrub, lvs notophyll. Fl monoecious small, inconspicuous, probably wind-pollinated, not clearly seasonal, Fr small explosive capsule; seed medium. Gregarious in understorey of dry forest, confined to Upper Voltaian rocks. Not known elsewhere in West Africa but widespread in Central and East Africa; (Congolian satellite).

4 ACRIDOCARPUS ALTERNIFOLIUS (Schumach. & Thonn.) Niedenzu, FWTA 1:352. Malpighiaceae. (Type probably from near Accra).
Slender understorey climber; lvs notophyll. Fl yellow, medium, bowl, 7–11. Fr samara, wind-dispersed; seed medium. Regenerates in shade. Glandular patches on leaf underside. S. Leone to Nigeria; (Guinea wide).

5 ACRIDOCARPUS SMEATHMANNII (DC.) Guill. & Perr., FWTA 1:352. Malpighiaceae.
Understorey climber in forest, but large shrub in thicket; lvs notophyll. Fl yellow, medium, bowl, mostly 12–3. Fr samara, wind-dispersed; seed medium. Glandular patches on leaf underside. Regenerates in shade. Mainly on poor soils in wetter part of forest zone, but flowers mainly in openings and at drier margins of the range, as in thickets at edge of the Accra Plains. Guinea Bissau to Nigeria; (Guinea wide).

6 ADIANTUM VOGELII Mett. ex Keys. FWTA, FF: 39. Adiantaceae.
Ground fern; lf bipinnate, leaflets nanophyll. Spores very small, wind-dispersed. More than half of plots from which the species was recorded had soil pH 4.6–5.3. Grows in deep shade. Guinée to Zanzibar; (chorological transgressor).

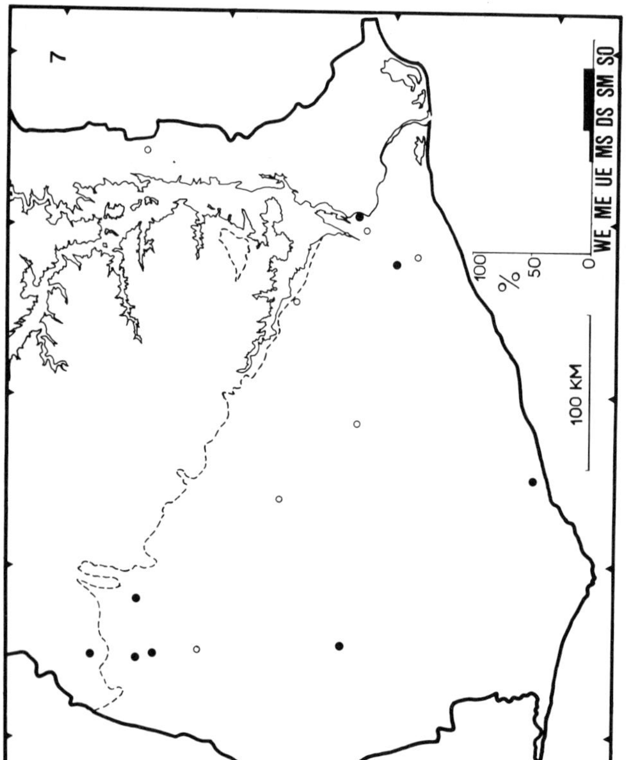
7
100 KM
100
%
50
0
WE ME UE MS DS SM SO

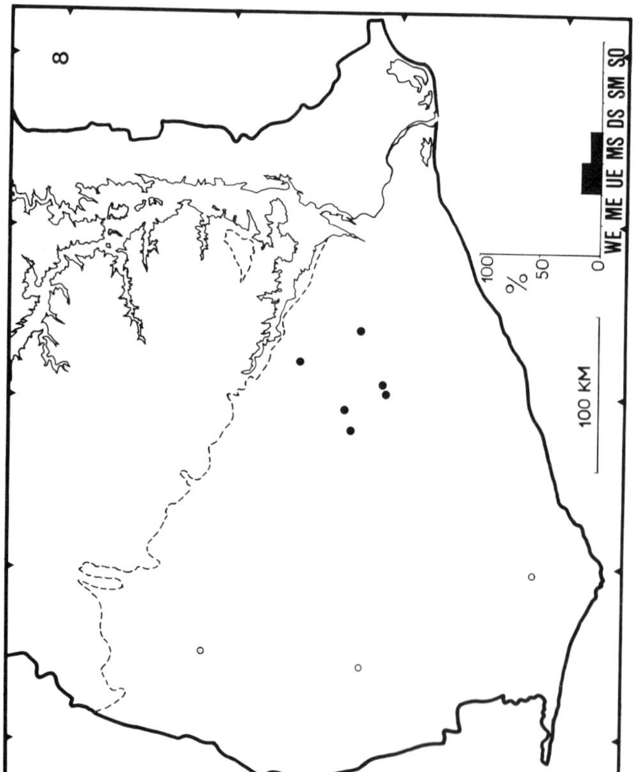
8
100 KM
100
%
50
0
WE ME UE MS DS SM SO

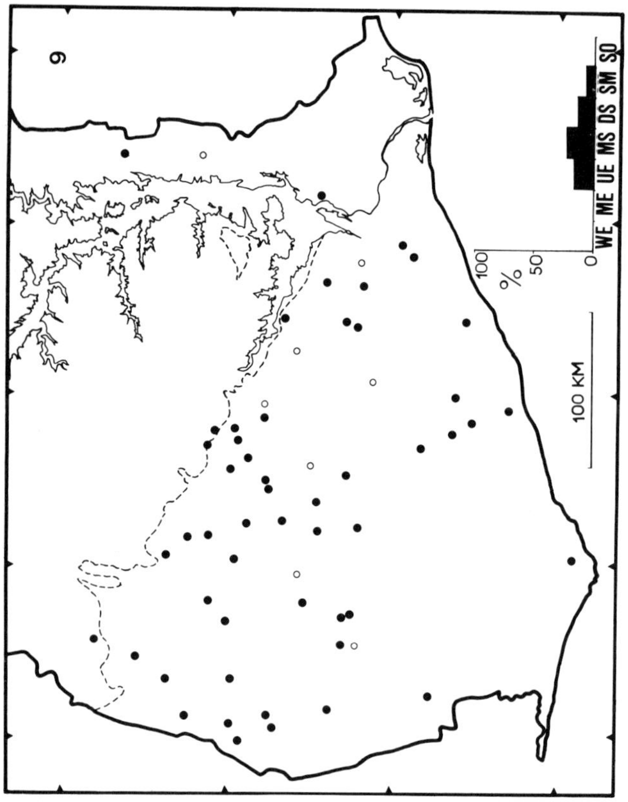
9
100 KM
100
%
50
0
WE ME UE MS DS SM SO

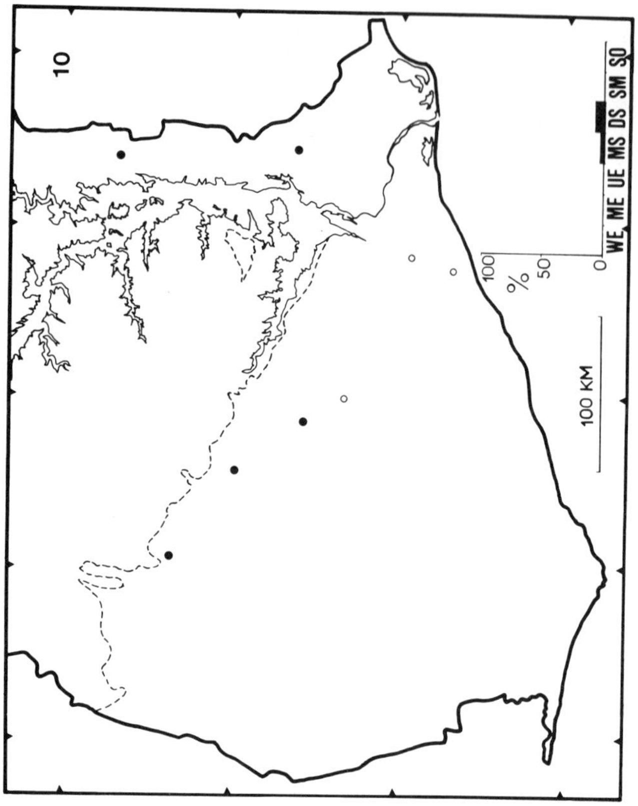
10
100 KM
100
%
50
0
WE ME UE MS DS SM SO

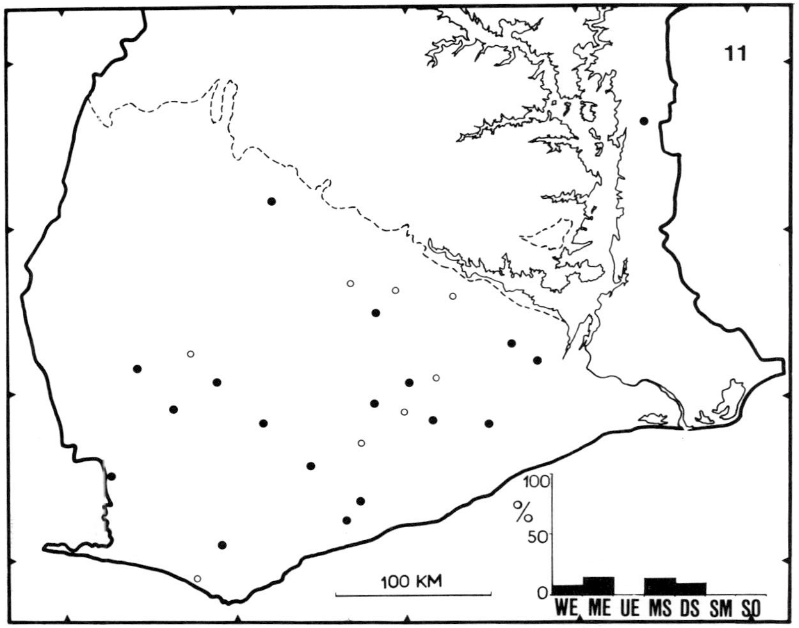

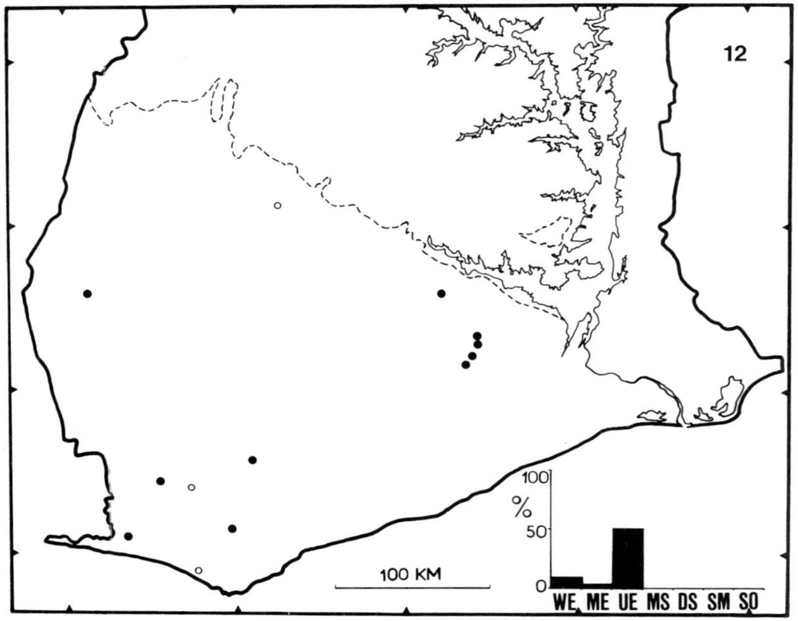

7 AEGLOPSIS CHEVALIERI Swingle, FWTA 1:687. Rutaceae. Akan: kwae-akenka.

Large aromatic spiny shrub; lvs notophyll, but may be microphyll in dry forest. Fl white, medium, dish, 2–4. Fr large berry, gummy flesh around large seeds. Liberia to Togo; (Upper Guinea endemic).

8 AFRAMOMUM CHRYSANTHUM Lock (82). Zingiberaceae. (*A. daniellii* of FWTA 3:76, partly, not of (Hook. f.) K. Schum.). Akan: adowa wisa. (Type from Atewa Range F.R.).

Large aromatic stoloniferous herb; lvs macrophyll. Tomlinson's model. Fl yellow, large, funnel, inflorescences arising directly from rhizome and from base of erect leafy shoots, probably pollinated by bees, mostly 7–11. Fr red, grooved berry. Intolerant of deep shade. S. Leone to Togo; (Upper Guinea endemic). Seeds medium, peppery, smooth and shiny, less widely used in traditional medicine than those of *A. melegueta* K. Schum. (Akan: famu wisa).

9 AFRAMOMUM GEOCARPUM Lock & Hall (83). Zingiberaceae. (*A. leptolepis* of FWTA 3:76, partly, not of (K. Schum.) K. Schum.). Akan: sensam. (Type from Atewa Range F.R.).

Large aromatic stoloniferous herb; lvs macrophyll. Tomlinson's model. Fl pale mauve, large, funnel, inflorescences arising directly from rhizome away from erect leafy shoots, probably pollinated by bees, 1–5. Fr geocarpic berry. Seeds medium, scarcely peppery. Intolerant of deep shade. S. Leone to Togo; probably also in Cameroun; (probably Guinea wide).

10 AFRAMOMUM STANFIELDII Hepper, FWTA 3:76. Zingiberaceae.

Medium-sized aromatic stoloniferous herb; lvs mesophyll. Tomlinson's model. Fl white, large, funnel, 3–4, inflorescences arising from base of erect leafy shoots, Fr red, smooth berry. Seeds medium, slightly peppery. Intolerant of deep shade. Iv. Coast to Nigeria; (Guinea wide).

11 AFROBRUNNICHIA ERECTA (Asch.) Hutch. & Dalz., FWTA 1:139. Polygonaceae.

Slender tendril climber; wood cylindrical, normal, releasing drinkable water when cut; lvs notophyll. Undescribed model (59). Fl small, white, dish, 12–2. Fr samara (Fig. 6.1k), wind-dispersed; seed medium. Mainly at edges of forest gaps. S. Leone to Zaire; (Guineo-Congolian wide).

12 AFROSERSALISIA AFZELII (Engl.) A. Chev., FWTA 2:30. Sapotaceae. Akan: bakunini.

Large tree with buttresses; lvs mesophyll to notophyll, with white latex. Aubréville's model. Fl small, greenish-yellow, brush, 2–5. Fr red, one-seeded berry with edible pulp; seed large. Seedlings (26) abundant in shade, often below parent tree. Outside the wetter forest apparently confined to uplands above 500 m, and there sometimes gregarious. S. Leone to Gabon; (Guinea wide). Wood pale red, hard, unclassified, used for charcoal.

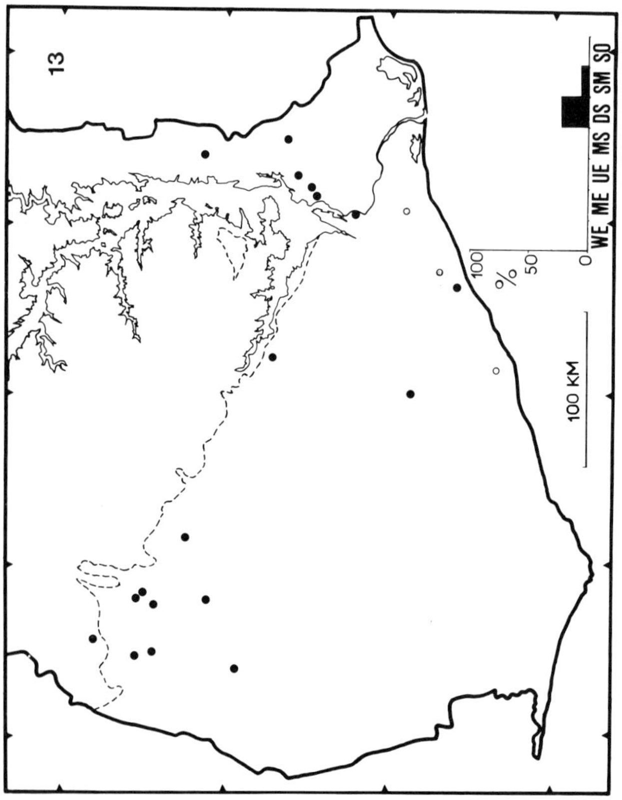
13
100
%
50
0
WE ME UE MS DS SM SO
100 KM

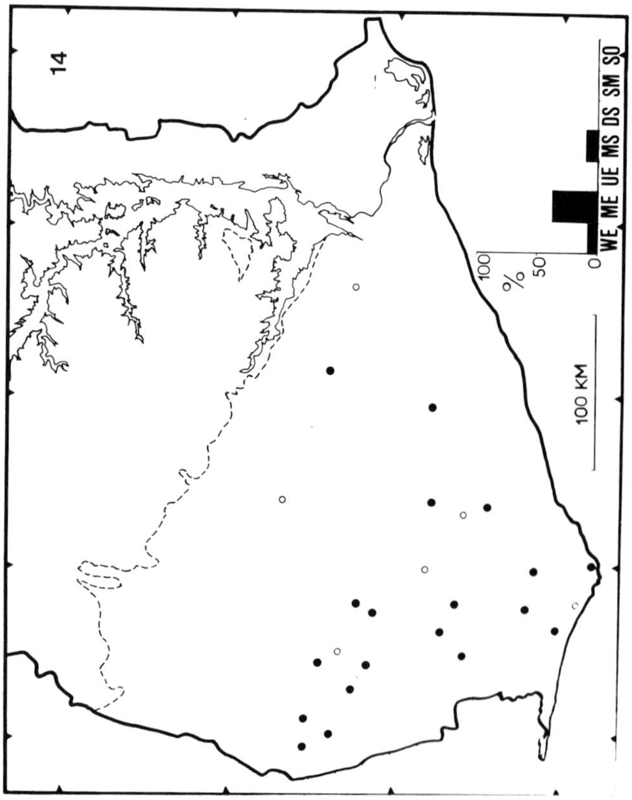
14
100
%
50
0
WE ME UE MS DS SM SO
100 KM

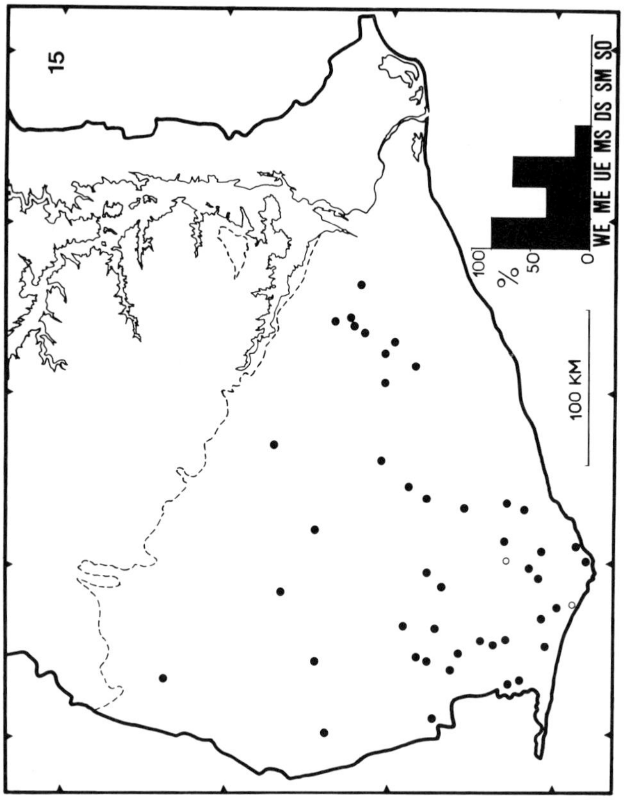
15
100
%
50
0
WE ME UE MS DS SM SO
100 KM

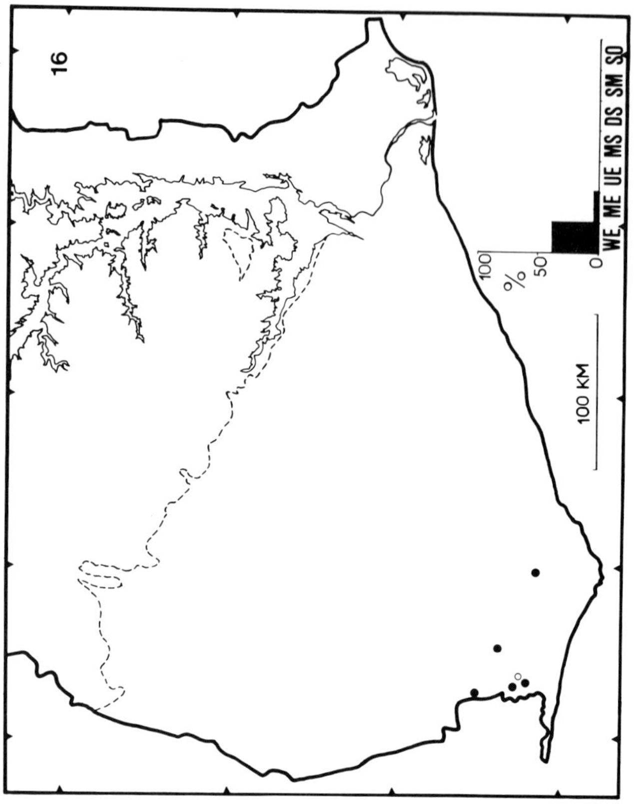
16
100
%
50
0
WE ME UE MS DS SM SO
100 KM

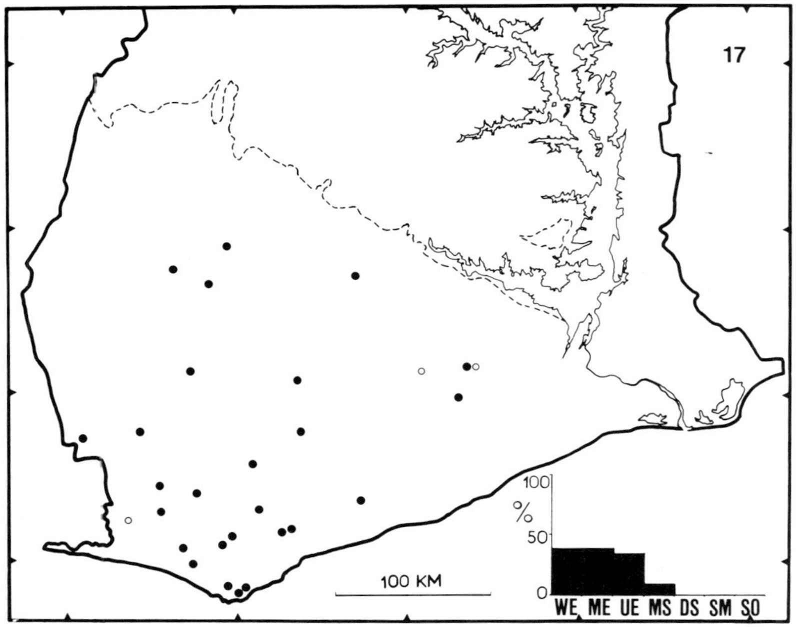

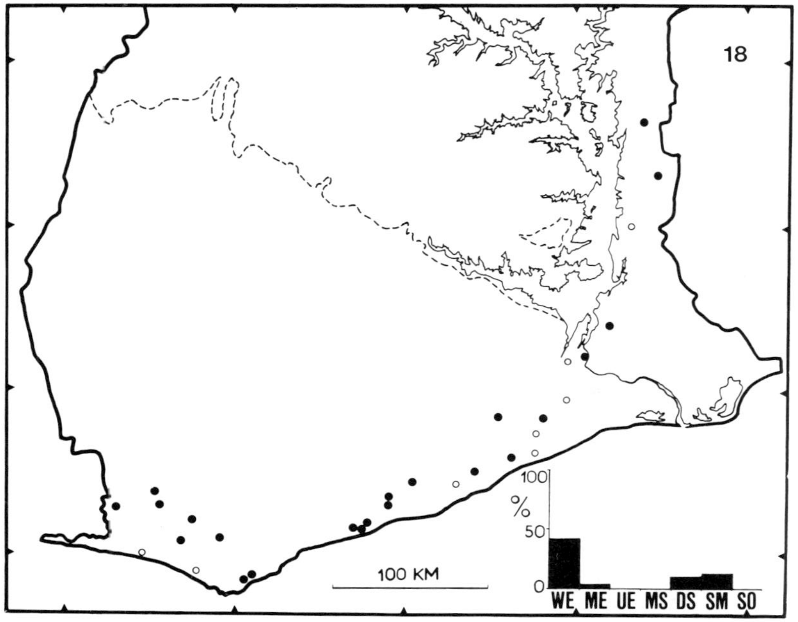

13 AFZELIA AFRICANA Sm., FWTA 1:459. Caesalpiniaceae. Akan: papao.
Medium to large tree; lvs pinnate, leaflets notophyll, deciduous. Troll's model. Fl greenish-white, medium, scented, flag, 2–3. Fr woody explosive pod; seeds very large, black with orange aril. Regenerates in moderate shade. Dry forest and also in adjacent savanna; fire-resistant. Recorded from moist forest only on rocky hills. Roots with ectotrophic mycorrhiza (69a). Senegal to Sudan and Tanzania; (ecological and chorological transgressor). Class 3 timber, trade name: afzelia (40), sometimes planted but slow-growing.

14 AFZELIA BELLA Harms var. *gracilior* Keay, FWTA 1:461. Caesalpiniaceae. Akan: papaonua.
Often confused with *A. africana* by foresters. Large tree; lvs pinnate, leaflets notophyll, deciduous. Troll's model. Fl white, large, scented, flag, 4–9. Fr woody, explosive pod; seeds very large, black with orange aril; regenerates in shade. Iv. Coast to Ghana; species is a Guineo-Congolian wide. Class 3 timber.

15 AGANOPE LEUCOBOTRYA (Dunn) Polhill (92), (*Ostryoderris leucobotrya* Dunn, FWTA 1:522). Papilionaceae.
Small, straggling tree; lvs pinnate, leaflets notophyll. Fl white, flag, medium-sized, 10–3. Fr flat, thin indehiscent pod, wind-dispersed; seeds large. Regenerates in shade, but flowers only at edge of clearings. Plot records very significantly associated with acid, base-poor soils of high clay content. S. Leone to Nigeria; (Guinea wide).

16 AGELAEA sp. aff. DUCHESNEI De Wild. & Th. Dur. (99). Connaraceae.
This species is distinguished by the presence on young leaves of dense, white, cottony hairs which disappear during maturation. It does not match perfectly any known species, but cannot be described until the flowers and fruits are discovered. Woody climber; lvs trifoliolate, leaflets notophyll. Fl and Fr unknown. Seedlings in deep shade. Iv. Coast to Ghana; (Upper Guinea endemic).

17 AGELAEA NITIDA Soland. ex Planch., FWTA 1:746. Connaraceae.
Stout woody climber with deeply-furrowed stem, and wood with included phloem rings; lvs trifoliolate, leaflets notophyll. Fl small, white, dish, entomophilous, but seldom seen. Fr one-seeded red follicle; seed medium to large, black with yellow aril. Regenerates in deep shade. Iv. Coast to Zaire; (Guineo-Congolian wide). Many of the named species of *Agelaea* are, in fact, microspecies connected by intermediate forms; all specimens here called *A. nitida* have minute appressed stellate hairs on the leaves, but some approach *A. grisea* Schellenb., *A. dewevrei* De Wild. & Th. Dur., and *A. pseudobliqua* Schellenb.

18 AGELAEA OBLIQUA (P. Beauv.) Baill., FWTA 1:745. Connaraceae.
Stout woody climber similar anatomically to *A. nitida*, (or scrambling shrub in thickets); lvs trifoliolate, leaflets notophyll. Fl and Fr as in *A. nitida*, but collected much more often as they are formed when the species grows as a low shrub in thickets. Regenerates in deep shade. Guinea Bissau to Cameroun; (Guinea wide). All *Agelaea* specimens with glabrous lamina and glabrous to pubescent petiole have been included in *A. obliqua*, but the distribution histrogram suggests that at least two taxa are probably involved.

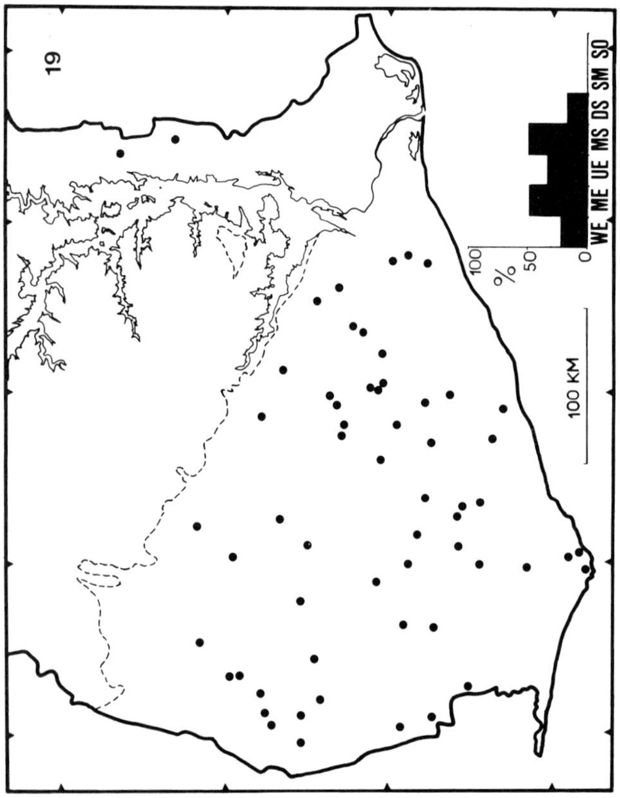
19
100 KM
%
100
50
0
WE ME UE MS DS SM SO

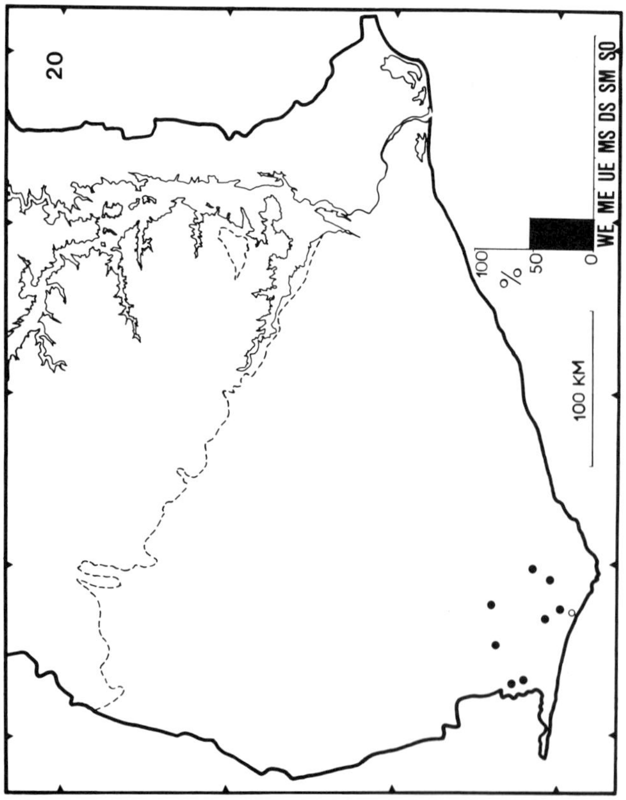
20
100 KM
%
100
50
0
WE ME UE MS DS SM SO

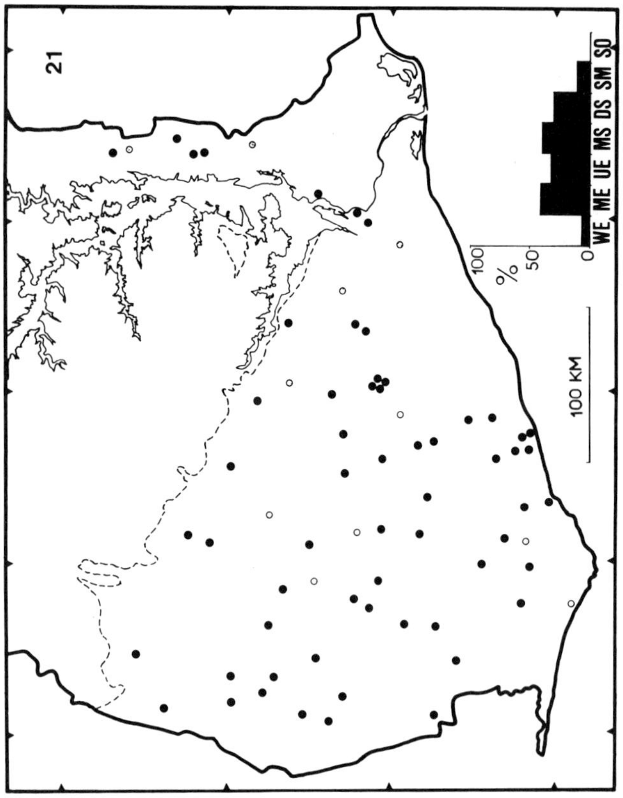
21
100 KM
%
100
50
0
WE ME UE MS DS SM SO

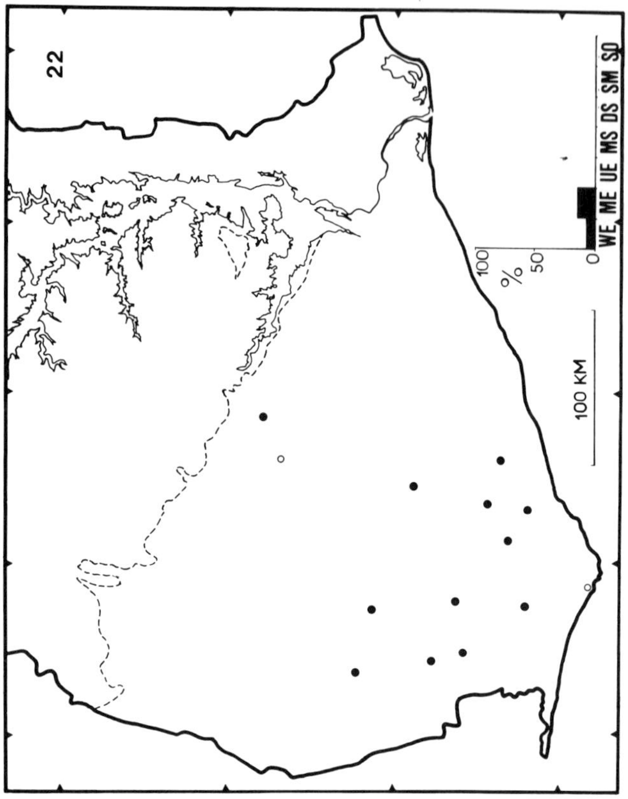
22
100 KM
%
100
50
0
WE ME UE MS DS SM SO

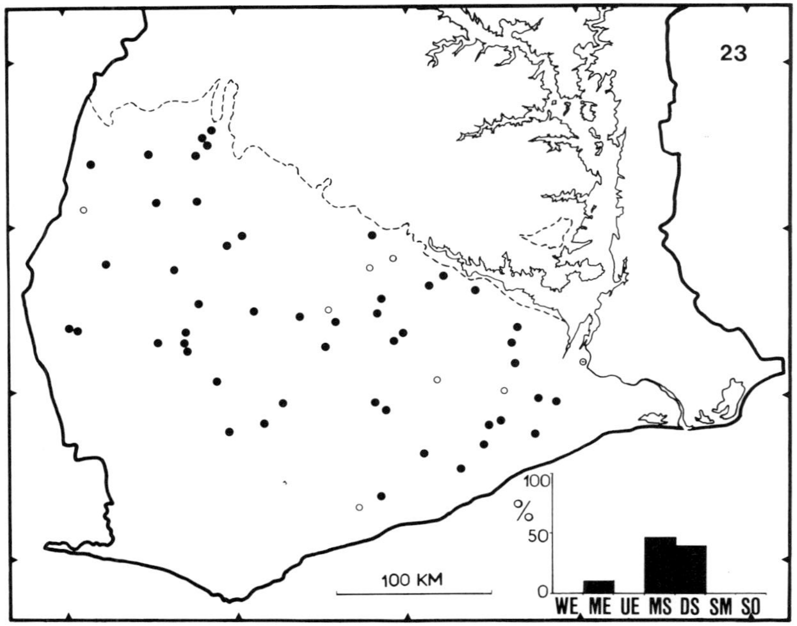

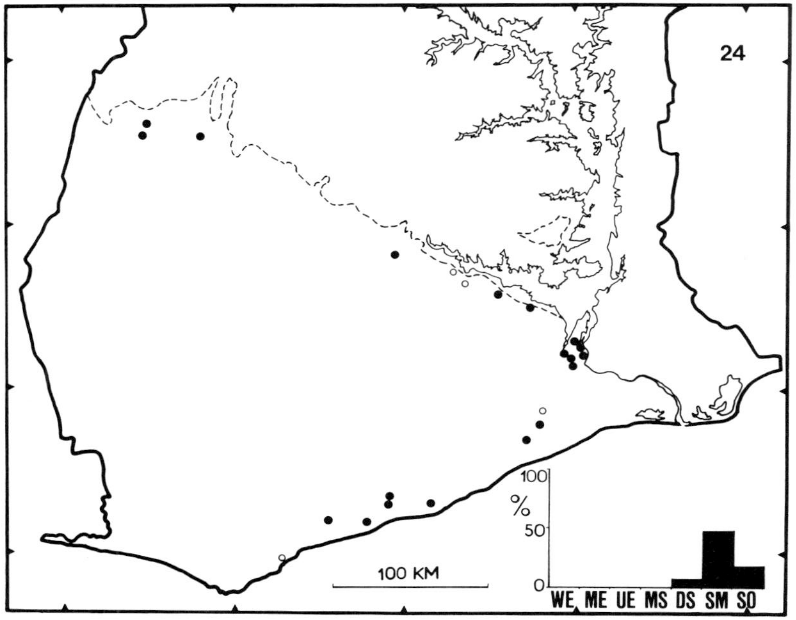

19 AGELAEA cf. PILOSA Schellenb. FWTA 1:745. Connaraceae.
Woody climber; lvs trifoliolate, leaflets notophyll, with sparse spreading hairs on lamina. Fl and Fr as in *A. nitida,* but not yet collected in Ghana. Regenerates in shade. 70% of plot records were from plots with soil pH 4.2–5.3. Iv. Coast to Nigeria; (Guinea wide).

20 AGELAEA TRIFOLIA (Lam.) Gilg. FWTA 1:746. Connaraceae
Woody climber; lvs trifoliolate, leaflets notophyll when mature but larger in sapling, densely spreading-hairy especially when young. Fl and Fr as in *A. nitida,* seldom collected. Seedlings in shade. All plot records from soils with total exchangeable bases of <2 m-equiv./100 g soil. Guinea Bissau to Nigeria; (Guinea wide).

21 AIDIA GENIPIFLORA (DC.) Dandy, FWTA 2:114. Rubiaceae. Akan: otwensono.
Medium tree (or shrub in coastal thickets); lvs notophyll. Petit's model. Fl white, large, scented, dish with tube, 12–4. Fr berry with thin flesh; seeds medium. Regenerates in shade. Guinea Bissau to Cameroun; (Guinea wide).

22 AIRYANTHA SCHWEINFURTHII (Taub.) Brummitt subsp. *confusa* (Hutch. & Dalz.) Brummitt (33). (*Baphiastrum confusum* (Hutch. & Dalz.) Pellegr., FWTA 1:513). Papilionaceae.
Stout, woody climber; lvs notophyll. Fl white and yellow, medium to large, flag. Fr small pod, seeds medium-size, red. Iv. Coast to Zaire; (Guineo-Congolian wide).

23 ALAFIA BARTERI Oliv., FWTA 2:73. Apocynaceae.
Woody climber with white latex; lvs notophyll. Fl white, medium, scented, dish with tube, 3–8. Fr follicle; seeds medium, each with plume of long hairs, wind-dispersed. S. Leone to Cameroun; (Guinea wide).

24 ALAFIA SCANDENS (Thonn.) De Wild., FWTA 2:73. Apocynaceae. (Type from Akwapim).
Woody climber with white latex; lvs notophyll, more acuminate than *A. barteri.* Fl white and red, medium, strongly scented, dish with tube. Fr and seed as in *A. barteri.* Senegal to Zaire; (Guineo-Congolian wide).

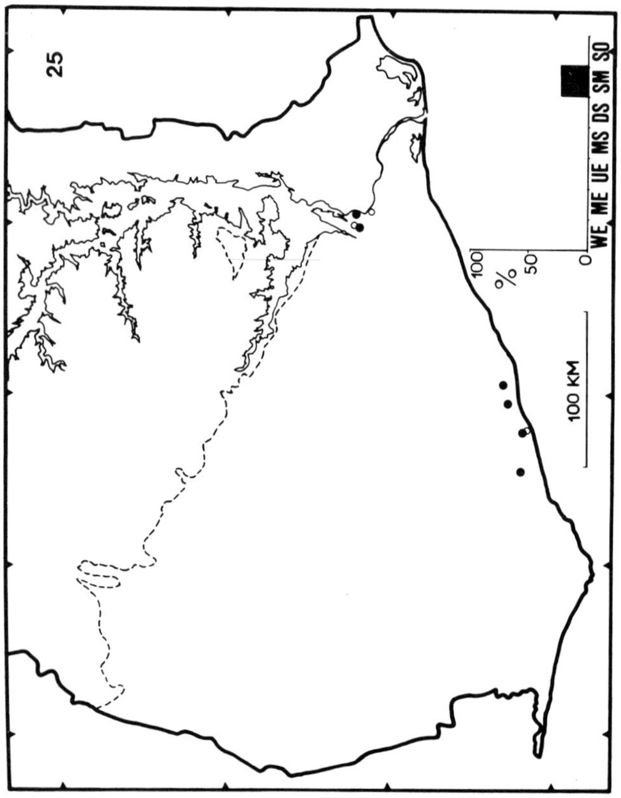
25
100 KM
%
100
50
0
WE ME UE MS DS SM SO

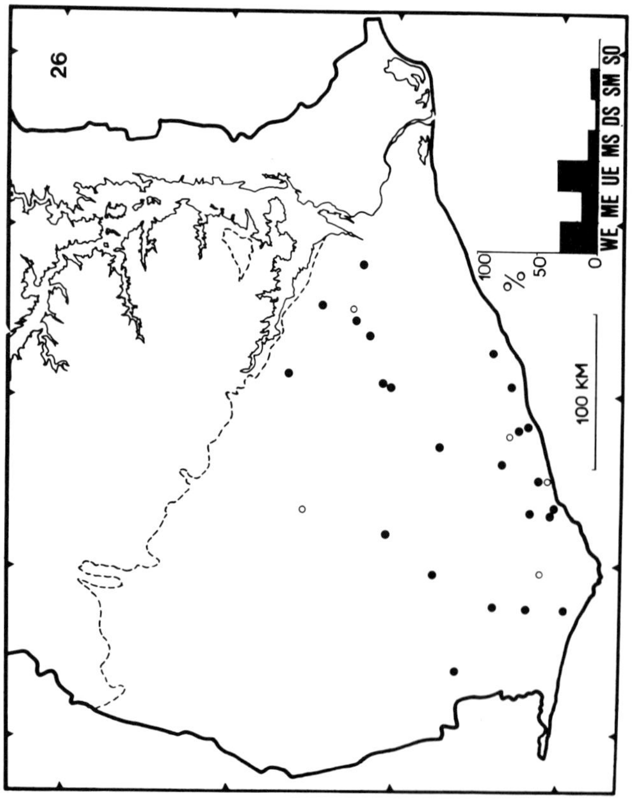
26
100 KM
%
100
50
0
WE ME UE MS DS SM SO

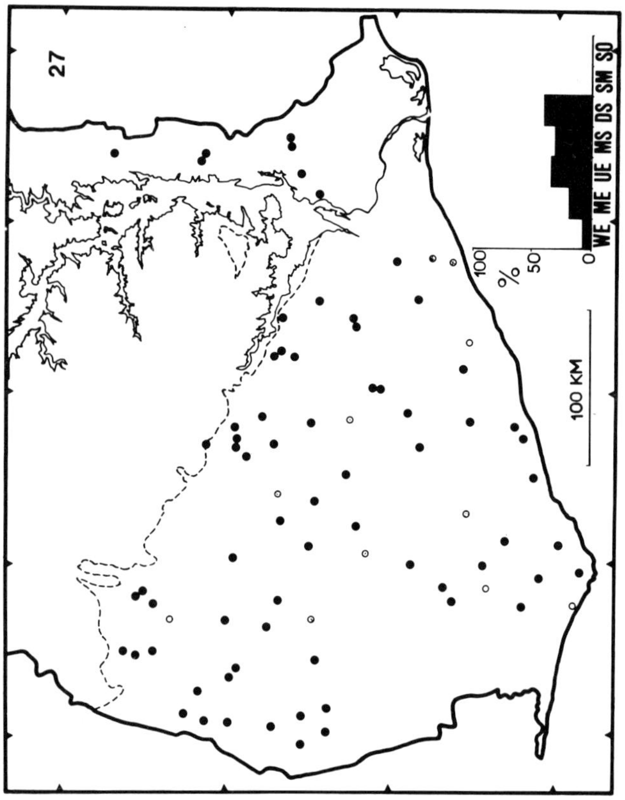
27
100 KM
%
100
50
0
WE ME UE MS DS SM SO

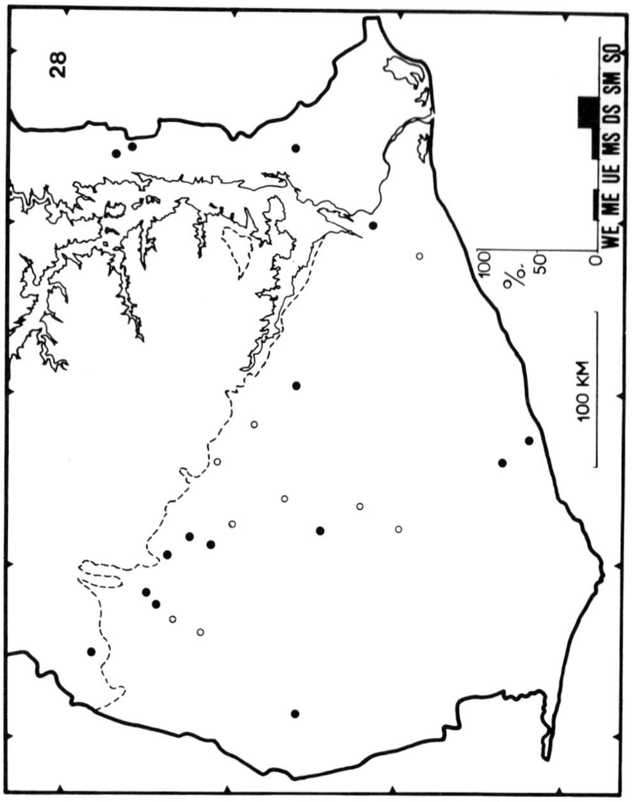
28
100 KM
%
100
50
0
WE ME UE MS DS SM SO

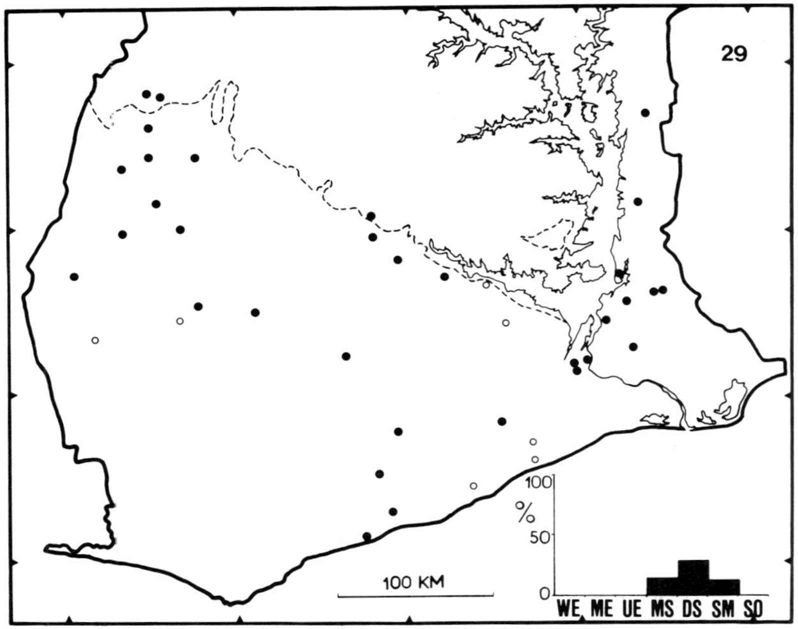

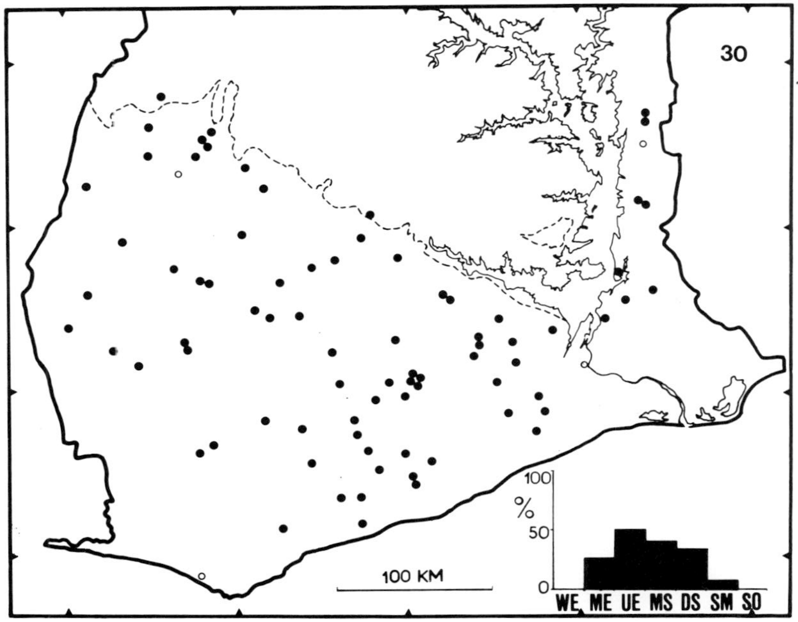

25 ALBERTISIA CUNEATA (Keay) Forman (44). (*Epinetrum cuneatum* Keay, FWTA 1:70). Menispermaceae. (Type from Yamoransa near Cape Coast).
Slender glabrous understorey climber; lvs microphyll. Fl small, dioecious, inconspicuous, yellowish, 3–12. Fr not known. Locally abundant in dry forest. Described from Ghana and subsequently discovered in Senegal; (Upper Guinea endemic).

26 ALBERTISIA SCANDENS (Mangenot & Miège) Forman (44). (*Epinetrum scandens* Mangenot & Miège, FWTA 1:70). Menispermaceae.
Slender hairy understorey climber; lvs notophyll. Fl small, dioecious, inconspicuous, yellowish. Fr beaked, bristly, 1-seeded orange berry, 11–7; seed large. Iv. Coast to Ghana; (Upper Guinea endemic).

27 ALBIZIA ADIANTHIFOLIA (Schumach.) W. F. Wight, FWTA 1:502. Mimosaceae. Akan: pampena. (Type from Berekuso).
Large tree; lvs bipinnate, leaflets nanophyll, deciduous. Troll's model (59). Fl greenish-white and red, small, clustered in brush inflorescence, 11–4. Fr flat, thin pod, wind-dispersed, 12–5; seeds medium. Seedlings found in shade but rapid growth only in gaps; common in secondary forest. Senegal to South Africa; (Chorological transgressor). Class 3 timber.

28 ALBIZIA FERRUGINEA (Guill. & Perr.) Benth., FWTA 1:502. Mimosaceae. Akan: awiemfosamina.
Large tree; lvs bipinnate, leaflets nanophyll, deciduous. Troll's model. Fl small, greenish-white, brush inflorescence, 3–4. Fr flat, thin, pod, wind-dispersed; seeds medium to large. Seedlings found in shade but rapid growth only in gaps. Senegal to Angola and Uganda; (Guineo-Congolian wide). Class 3 timber, trade name: albizzia (40).

29 ALBIZIA GLABERRIMA (Schumach. & Thonn.) Benth., FWTA 1:502. Mimosaceae. Akan: okoro-akoa. (Type from Oyarifa near Accra).
Medium tree; lvs bipinnate, leaflets microphyll, deciduous. Troll's model. Fl small, white, brush inflorescence, 1–3. Fr flat, thin pod, wind-dispersed; seeds large. Seedlings found in shade, but rapid growth only in gaps. Guinée to Rhodesia; (Chorological transgressor).

30 ALBIZIA ZYGIA (DC.) J. F. Macbr., FWTA 1:502. Mimosaceae. Akan: okoro.
Large tree; lvs bipinnate, leaflets microphyll, deciduous. Troll's model. Fl white and red, small, brush inflorescence, 2–4. Fr flat, thin pod, wind-dispersed; seeds medium to large. Seedlings found in shade, but rapid growth only in gaps; common in secondary forest. Senegal to Tanzania; (Chorological transgressor). Class 3 timber, trade name: okuro (40).

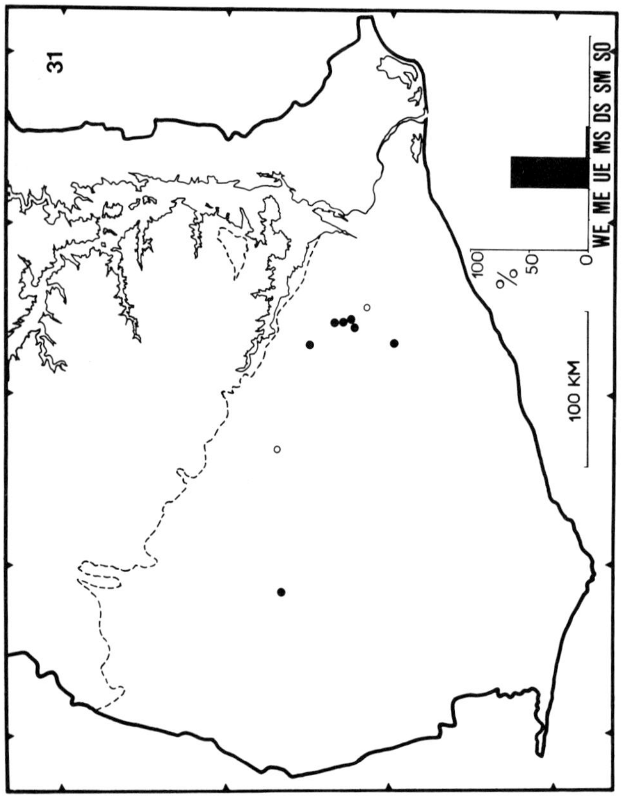
31
100 KM
%
100
50
0
WE ME UE MS DS SM SO

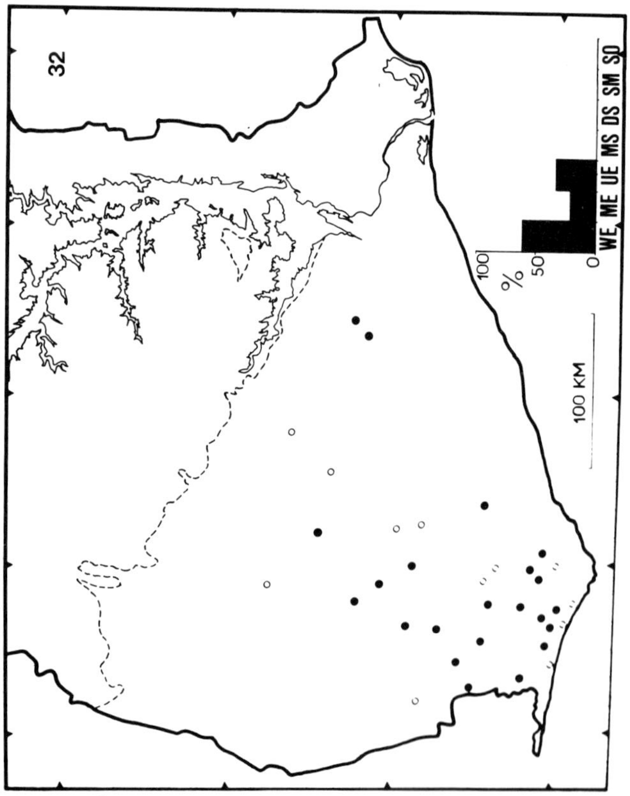
32
100 KM
%
100
50
0
WE ME UE MS DS SM SO

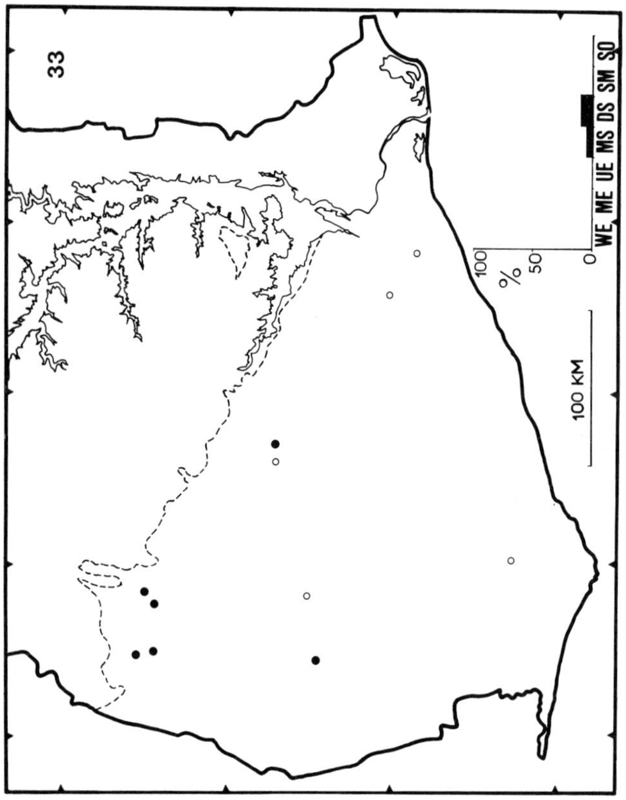
33
100 KM
%
100
50
0
WE ME UE MS DS SM SO

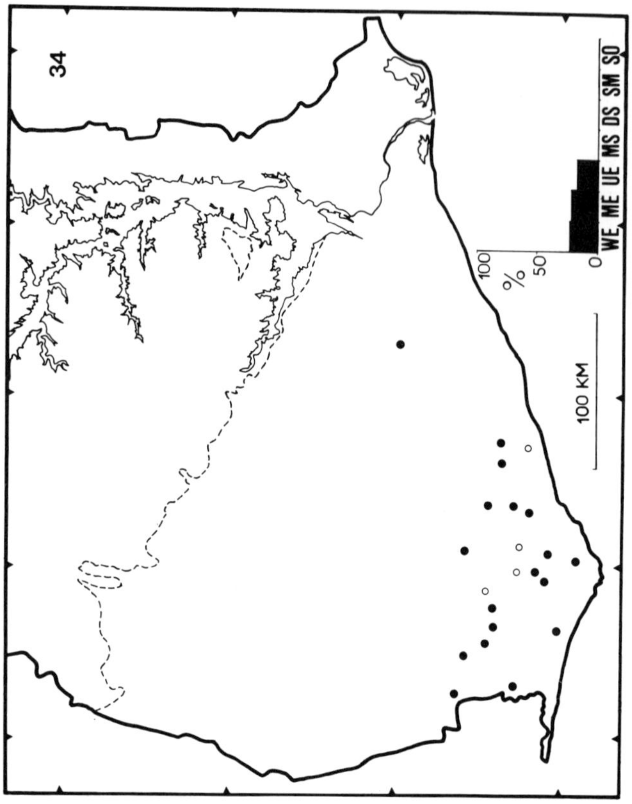
34
100 KM
%
100
50
0
WE ME UE MS DS SM SO

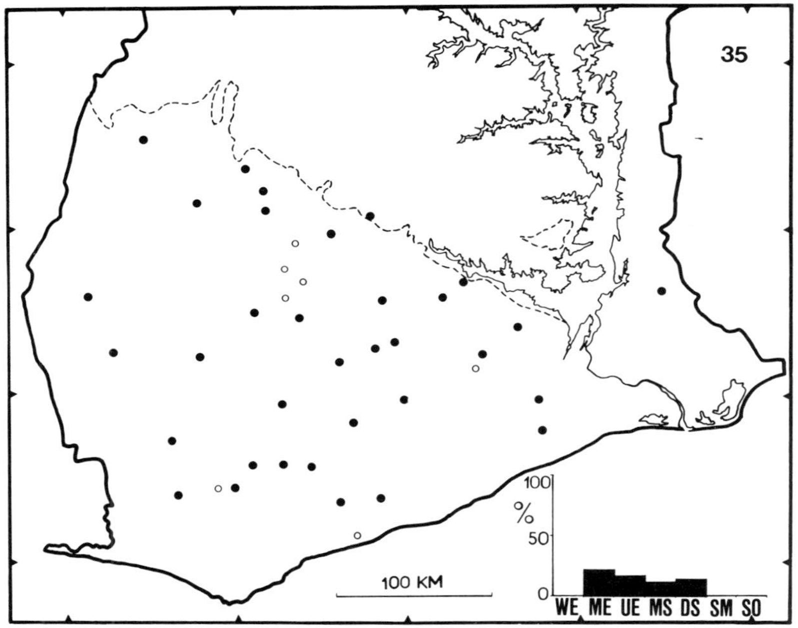

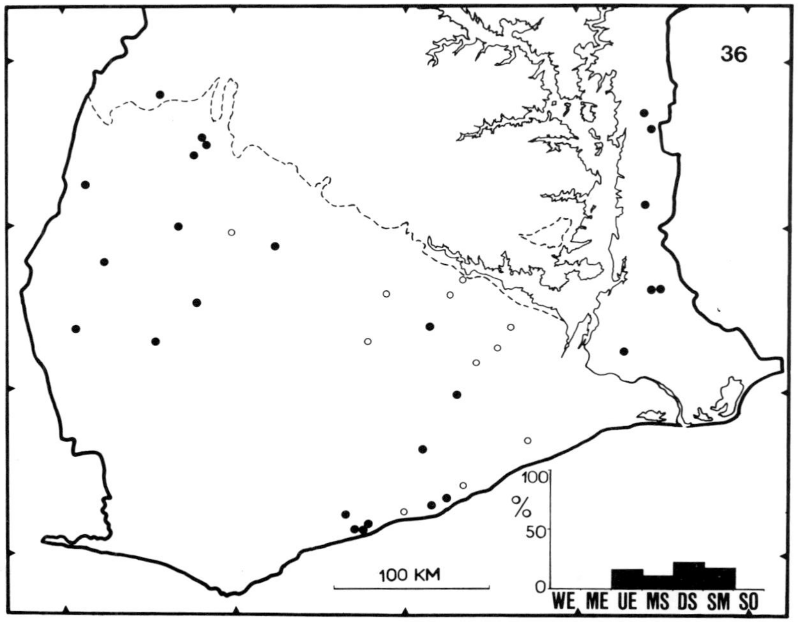

31 ALCHORNEA FLORIBUNDA Muell. Arg., FWTA 1:403. Euphorbiaceae. Akan: gyamanini.

Large shrub or small tree; lvs mesophyll. Fl dioecious, small, inconspicuous, probably wind-pollinated, 2–5. Fr small explosive capsule; seeds medium. Gregarious in understorey of forest. Liberia to Zaire and Uganda; (Guineo-Congolian wide).

32 ALLANBLACKIA PARVIFLORA A. Chev. (18). (*A. floribunda* of FWTA 1:291, not of Oliv.). Guttiferae. Akan: sonkyi.

Medium-sized tree; lvs mesophyll, with clear latex. Rauh's model. Fl monoecious, pink, large, scented, bowl, 4–8. Fr long berry; seeds very large, in gelatinous pulp. Significantly associated with soils low in calcium, potassium, magnesium, base saturation and pH (over half records on soil with pH 3.8–4.1, none >5.3). S. Leone to Ghana; (Upper Guinea endemic). *A. floribunda* Oliv. *sensu stricto* does not occur west of Dahomey.

33 ALLOPHYLUS AFRICANUS P. Beauv. forma *chrysothrix* Radlk., FWTA 1:713. Sapindaceae.

Scrambling shrub; lvs trifoliolate, leaflets notophyll, long stiff hairs on shoots. Fl monoecious, white, scented, small, densely clustered, 5–6. Fr 1-seeded berry, thin flesh, seed medium. Occurs mainly in rather disturbed forest, and roadsides. Guinée to Nigeria. Species is an ecological and chorological transgressor.

34 ALSODEIOPSIS STAUDTII Engl., FWTA 1:638. Icacinaceae.

Pygmy tree; lvs microphyll, slightly pubescent. Troll's model (59). Fl small, white, inconspicuous. Fr a 1-seeded drupe with thin flesh; seed medium. Somewhat gregarious, in dense shade. Iv. Coast to Zaire; (Guineo-Congolian wide).

35 ALSTONIA BOONEI De Wild., FWTA 2:68. Apocynaceae (35). Akan: sinuro, nyamedua.

Large tree with narrow buttresses; lvs mesophyll, deciduous, white latex. Prévost's model (59). Fl yellow-white, small, tube, 10–3. Fr slender follicle, seeds medium, plumose, wind-dispersed, 3–4. Seedlings not found in deep shade; common in secondary forest. Tolerates swampy soil but also in well-drained sites. Senegal to Zaire, Ethiopia and Uganda; (Guineo-Congolian wide). Light timber; trade name: alstonia (40).

36 AMORPHOPHALLUS JOHNSONII N. E. Br., FWTA 3:118. Araceae. (Type from Kwahu).

Medium herb, stem consisting of corm, with solitary leaf to 1.5 m high divided into notophyll segments. Spathe purplish, inflorescence with unpleasant smell, pollinated by flies and dung-beetles, trap. Fl small, monoecious, male yellow, female olive, clustered below purple spadix, 3–6. Fr 1-seeded berry; seed medium. Regenerates in light shade. Mali to Nigeria; (Guinea wide). Plants in Moist Semi-deciduous forest have inflorescences with long peduncle; in dry forest, inflorescence is sessile on ground.

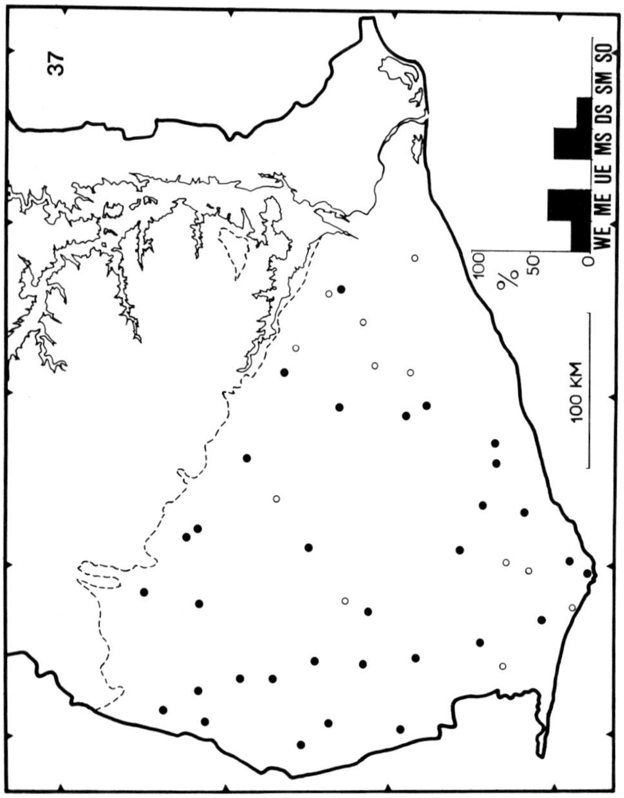
37
100
%
50
0
WE ME UE MS DS SM SO
100 KM

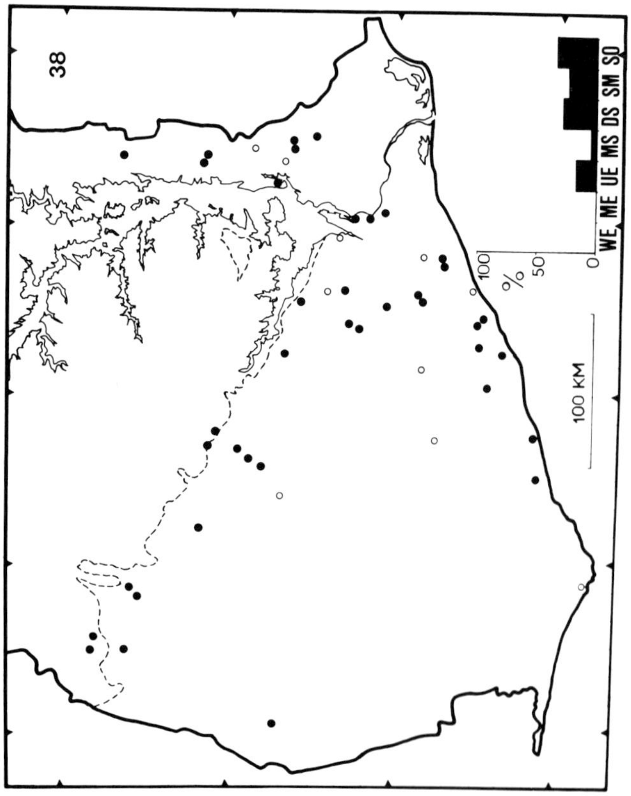
38
100
%
50
0
WE ME UE MS DS SM SO
100 KM

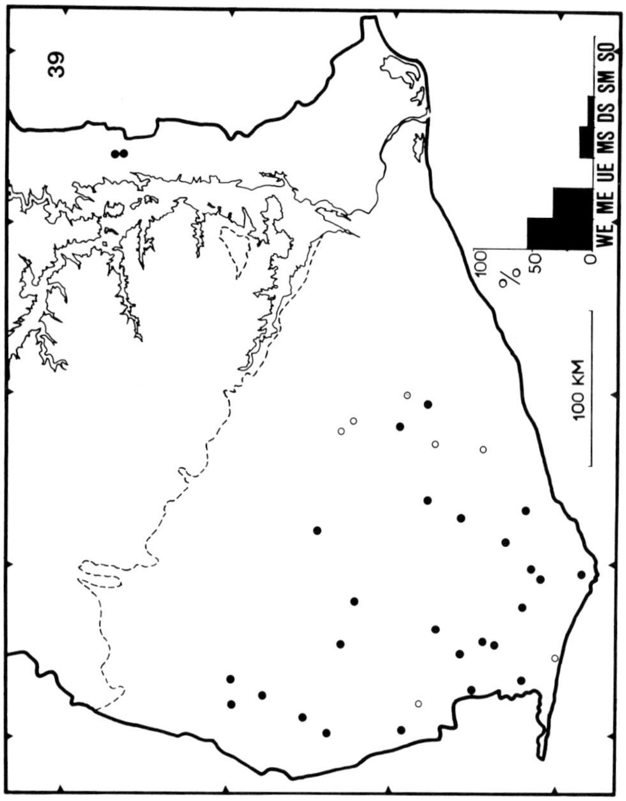
39
100
%
50
0
WE ME UE MS DS SM SO
100 KM

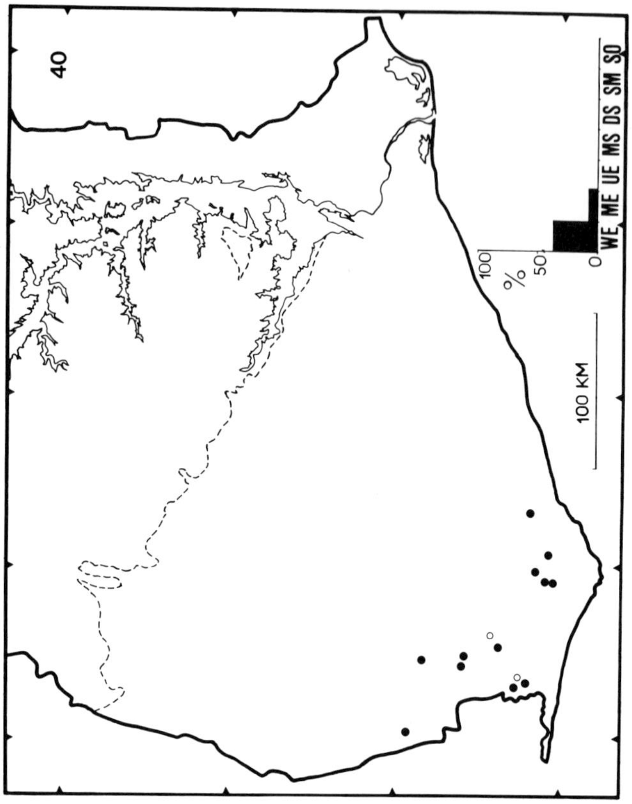
40
100
%
50
0
WE ME UE MS DS SM SO
100 KM

37 AMPHIMAS PTEROCARPOIDES Harms, FWTA 1:448. Caesalpiniaceae. Akan: yaya.
Large tree, red latex in bark; lvs pinnate, leaflets notophyll, deciduous. Massart's model. Fl yellow, small to medium brush inflorescence, 9–10. Fr flat, thin, l-seeded pod, wind-dispersed, 2–4; seed large to very large. Regenerates in shade. Absence from plots in Upland Evergreen forest may be fortuitous, as it occurs in Atewa Range F. R. Guinée to Zaire; (Guineo-Congolian wide). Wood hard, unclassified.

38 ANCHOMANES DIFFORMIS (Bl.) Engl., FWTA 3:121. (Incl. *A. welwitschii* Rendle, FWTA 3:121). Araceae.
Medium herb, similar in morphology to *Amorphophallus johnsonii*, but corm much larger and petiole thorny. Spathe yellowish to purplish, always long-pedunculate. Fl small, cream, monoecious, completely covering the spadix, 2–5. Pollination by flies. Fr l-seeded berry; seed medium. Mainly in forest with light canopy, and secondary forest. In Dry Semi-deciduous forest especially characteristic of 'fire-zone' forest. Extends to woodland in the savanna zone. Sierra Leone to Cameroun; (ecological and chorological transgressor). The colour differences used in FWTA to distinguish *A. welwitschii* are not sufficiently constant to provide reliable characters at specific level.

39 ANCISTROPHYLLUM OPACUM (Mann & Wendl.) Drude, FWTA 3:167. Palmae.
Large woody climber with spiny leaf-sheaths; lvs pinnate, leaflets macrophyll, distal part of leaf rhachis bearing pinnae modified as recurved hooks, to aid climbing. Tomlinson's model. Fl yellowish, medium, bowl, 11–2. Fr red, l-seeded berry covered with scales; seed medium to large. Seedlings in deep shade, but elongating in gaps. Significantly associated with base-poor soils. Ghana to Cameroun; (Guinea wide). Stems split for canes.

40 ANDROSIPHONIA ADENOSTEGIA Stapf, FWTA 1:201. Passifloraceae.
Small tree; lvs mesophyll, toothed. Cook's model (59). Fl greenish, dish, medium. Fr orange, several-seeded berry, 12–1; seeds medium. Regenerates under forest shade. S. Leone to Ghana; (Upper Guinea endemic).

41 ANGYLOCALYX OLIGOPHYLLUS (Bak.) Bak. f., FWTA 1:510. Papilionaceae.
Pygmy tree; lvs pinnate, leaflets notophyll. Corner's model (59). Fl greenish-white with purple spots, cauliflorous, flag, not clearly seasonal. Fr orange, elongated berry; seeds very large. Gregarious under deep forest shade; usually overlooked because it is seldom seen in flower or fruit. Plot records very significantly correlated with clay soils of low exchangeable base status. Excellent indicator of lowland evergreen forest in Ghana, but widespread in drier forest in W. Nigeria. Liberia to Zaire; (Guineo-Congolian wide). Has been regarded (124) as member of Caesalpiniaceae, but probably better kept in Papilionaceae.

42 ANINGERIA ALTISSIMA (A. Chev.) Aubrév. & Pellegr., FWTA 2:24. Sapotaceae. Akan: samfena.
Large tree with buttresses, white latex present; lvs notophyll. Fl creamy-white, small bowl, 12–1. Fr red, l-seeded berry, 3–5; seed large. Seedlings (26) under forest shade. Guinée to Zaire and E. Africa; (Chorological transgressor). Timber suitable for constructional work.

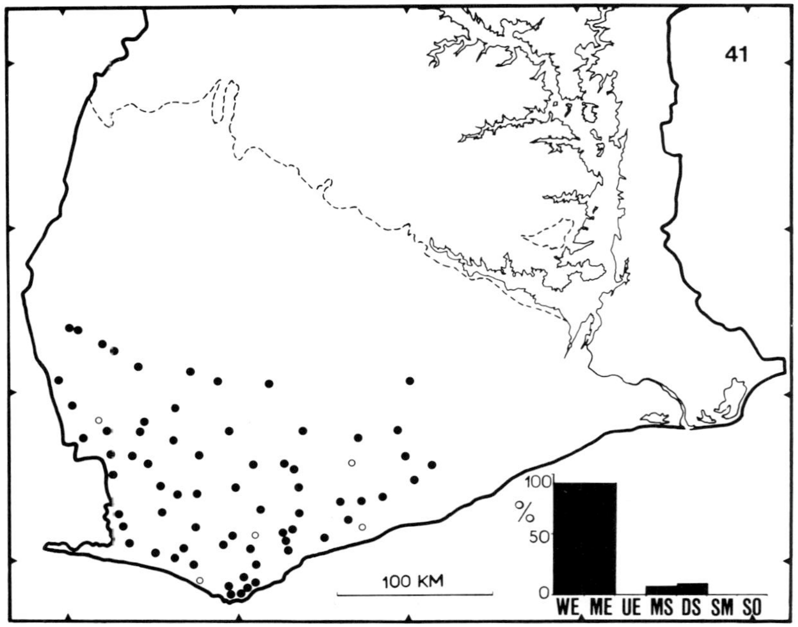

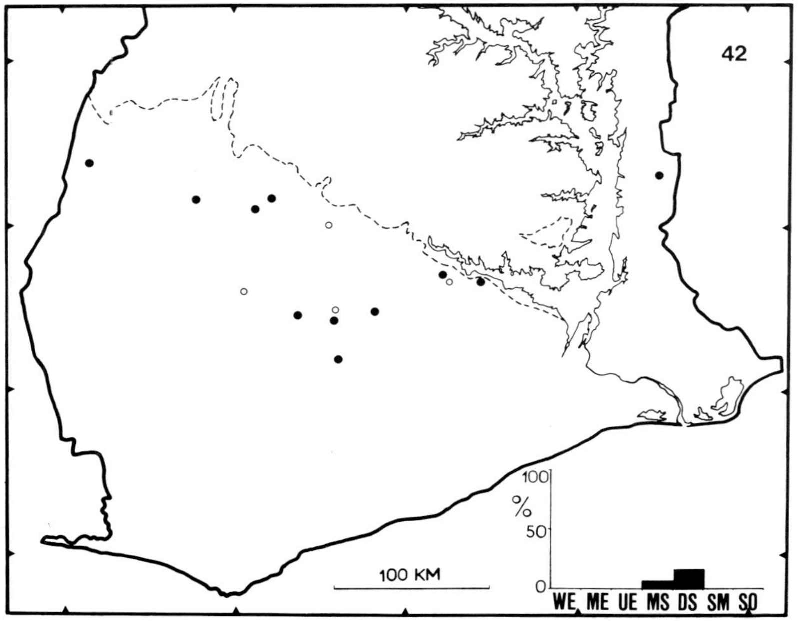

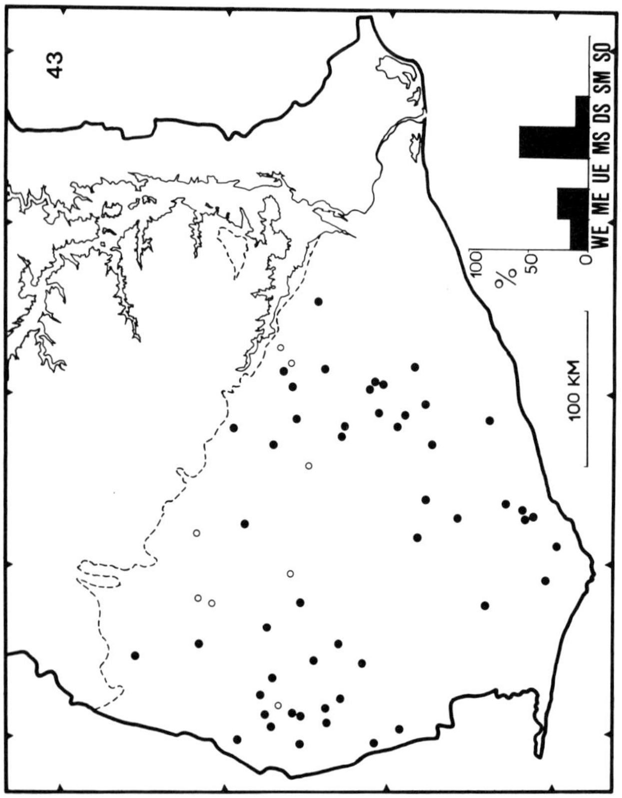
43
100 KM
100
%
50
0
WE ME UE MS DS SM SO

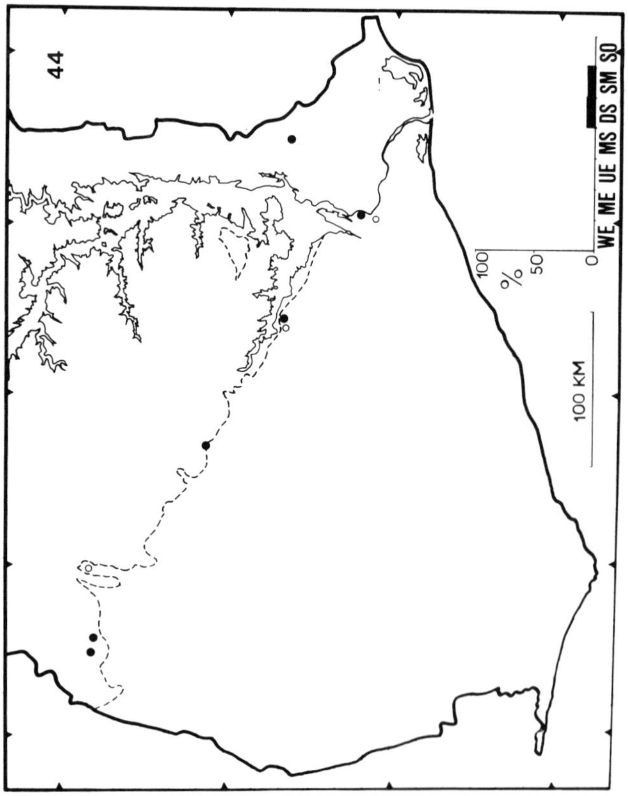
44
100 KM
100
%
50
0
WE ME UE MS DS SM SO

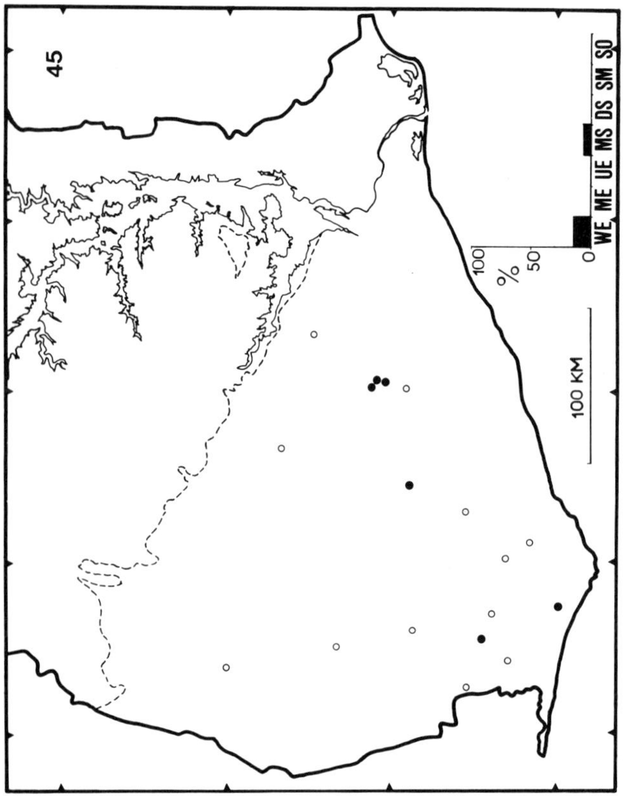
45
100 KM
100
%
50
0
WE ME UE MS DS SM SO

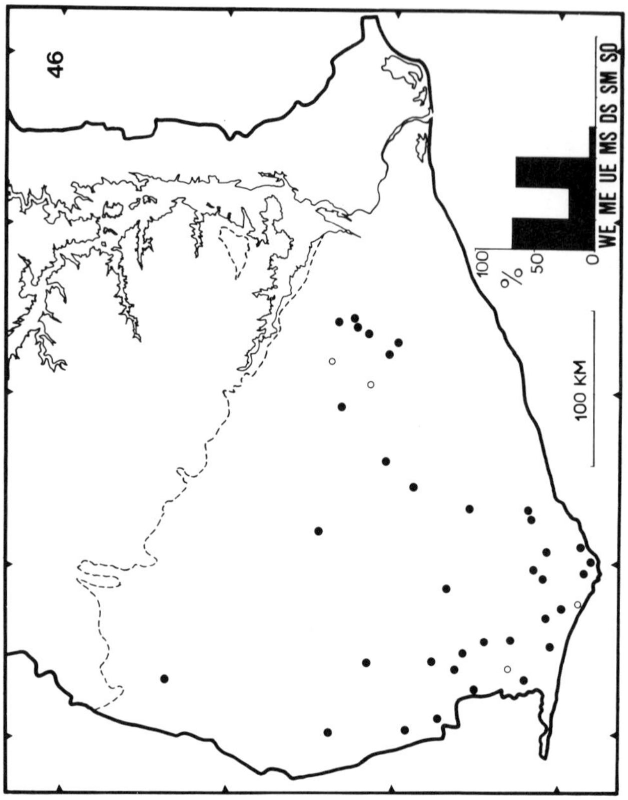
46
100 KM
100
%
50
0
WE ME UE MS DS SM SO

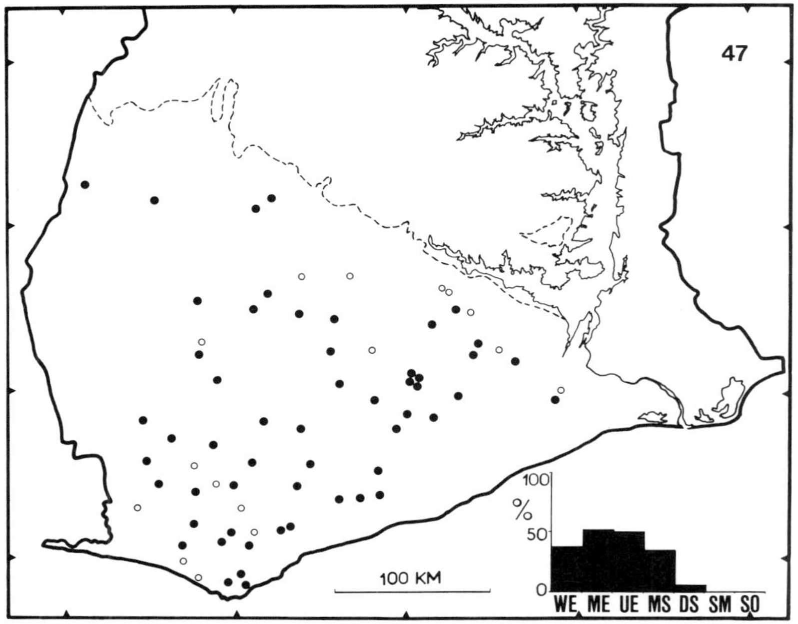

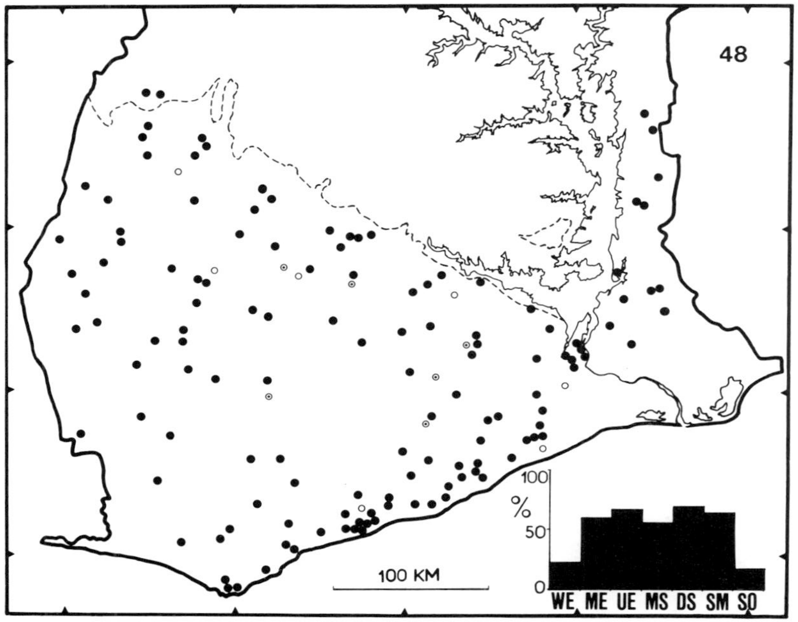

43 ANINGERIA ROBUSTA (A. Chev.) Aubrév. & Pellegr., FWTA 2:24. Sapotaceae. Akan: samfenanini.

Large tree with buttresses, white latex; lvs pubescent, notophyll. Leaves are distinguished from those of *A. altissima* by greater hairiness, but the species are otherwise very similar. Fl creamy-white, small, bowl, 2–3. Fr red, l-seeded berry, 3–4; seed large. Seedlings (26) under forest shade. Guinea-Bissau to Central African Republic; (Guineo-Congolian wide). Timber unclassified, but now being exploited.

44 ANOGEISSUS LEIOCARPUS (DC.) Guill. & Perr., FWTA 1:280. Combretaceae. Ewe: tsetse; Akan: kane.

Medium-sized tree; lvs microphyll; leaf-fall markedly seasonal but tree seldom completely leafless. Roux's model. Fl greenish-yellow, small, brush inflorescence, 7–11. Fr small samara, wind-dispersed, 8–1; seed medium. Principally a tree of river banks and old villages in savanna, but also found in closed forest near the savanna boundary. Regenerates in gaps. Somewhat fire-resistant, but usually absent from regularly-burnt savanna. Senegal to Cameroun and Ethiopia; (ecological and chorological transgressor). Timber resistant to termites; trade name: kane. Graceful avenue-tree (66).

45 ANOPYXIS KLAINEANA (Pierre) Engl., FWTA 1:286. Rhizophoraceae. Akan: kokote.

Large tree; lvs notophyll, seedling leaves hairy. Attims' model (59). Fl white, small, brush inflorescence, 7–9. Fr capsule with small winged seeds, wind-dispersed. Seedlings in open or slight shade; vegetative characters of Rubiaceae. S. Leone to Zaire; (Guineo-Congolian wide). Class 3 timber; trade name: kokoti (40).

46 ANTHONOTHA FRAGRANS (Bak. f.) Exell & Hillcoat, FWTA 1:473. Caesalpiniaceae. Akan: totoro.

Medium-sized tree; lvs pinnate, leaflets mesophyll, leaves brown-hairy below. Fl greenish-yellow, medium, fragrant, flag. Fr a 2-seeded pod which opens after falling; no obvious dispersal mechanism; seed very large. Regenerates in forest shade. Almost restricted to evergreen forest, and plot records associated with unsaturated soils of very low exchangeable base content, and low pH. S. Leone to Zaire; (Guineo-Congolian wide). Trunk straight and cylindrical, but readily attacked by borers and not used as timber (113).

47 ANTHONOTHA MACROPHYLLA P. Beauv., FWTA 1:473. Caesalpiniaceae. Akan: tutuabo.

Medium-sized rather crooked tree; lvs pinnate, leaflets mesophyll, leaves silvery-hairy below. Troll's model. Fl yellow-white, medium, flag, not clearly seasonal. Fr a several-seeded pod which splits after falling; no obvious dispersal mechanism; seeds very large. Regenerates in forest shade. Within the Moist Semi-deciduous forest much more abundant in the south-east than in the north-west subtype. Guinée to Angola; (Guineo-Congolian wide).

48 ANTIARIS TOXICARIA Leschenault, subsp. *welwitschii* (Engl.) C. C. Berg, var. *welwitschii* and var. *africana* (Engl.) C. C. Berg (22, 23), (*Antiaris welwitschii* Engl. and *A. africana* Engl., FWTA 1:612–613). Moraceae. Akan: kyenkyen.

Large deciduous trees, sometimes buttressed, lvs notophyll. Roux's model (59). Var. *welwitschii* when mature has smooth, glabrous leaves, while those of var. *africana* are rough and hairy; seedling and sapling leaves are, however, hairy in both varieties and cannot be certainly identified. Only herbarium records are distinguished on the map; var. *africana* ○; var. *welwitschii* ⊙. Fl dioecious, small, greenish-white, female inconspicuous, male in flattened head, wind-pollinated, 10–1. Fr red 1-seeded berry, 12–4; seed large. Regenerates in dense shade. Var. *africana* extends into fringing forest in savanna as tree 10–15 m high. Var. *welwitschii* from S. Leone to Angola and Tanzania; var. *africana* from Senegal to Zaire and Uganda. (*A. toxicaria* is a chorological transgressor). Formerly used for bark cloth (66). Class 2 timber, trade name: antiaris (40).

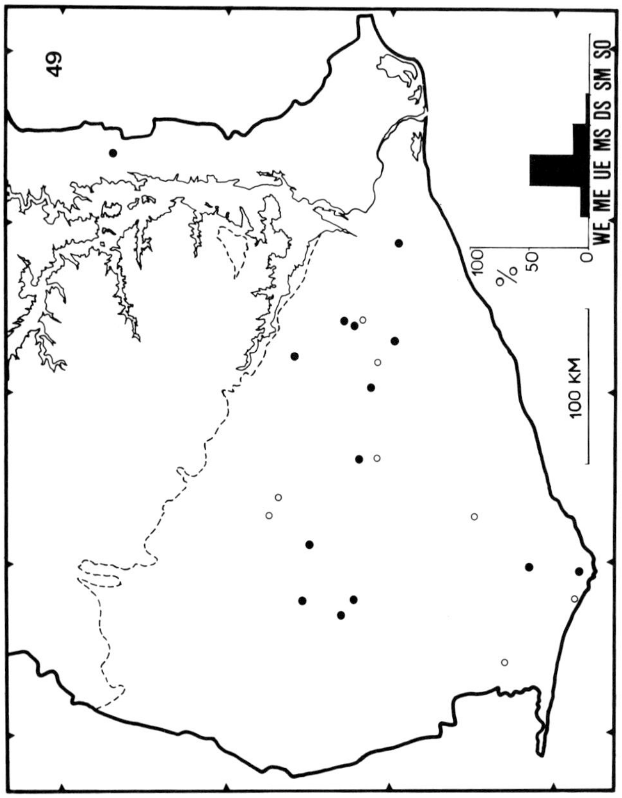
49
100 KM
100
%
50
0
WE ME UE MS DS SM SO

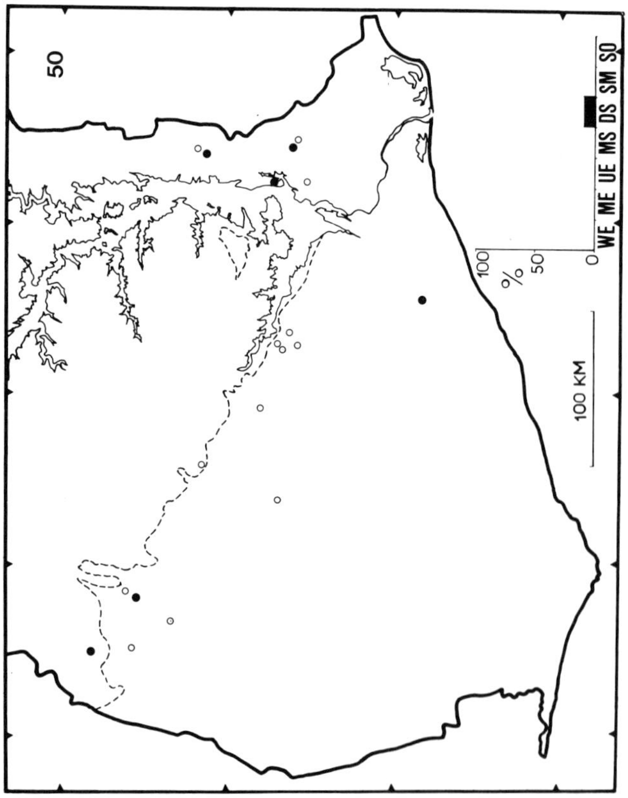
50
100 KM
100
%
50
0
WE ME UE MS DS SM SO

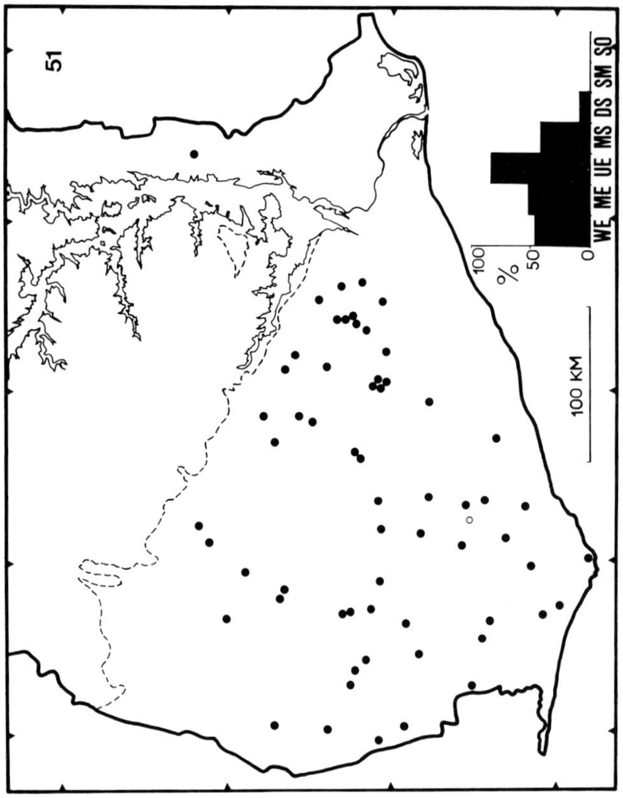
51
100 KM
100
%
50
0
WE ME UE MS DS SM SO

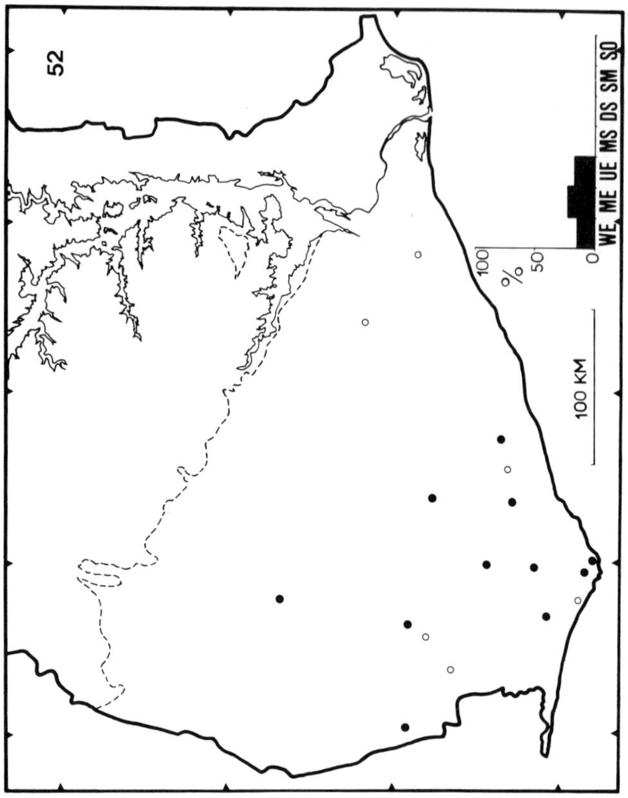
52
100 KM
100
%
50
0
WE ME UE MS DS SM SO

49 ANTIDESMA LACINIATUM Muell. Arg. var. *membranaceum* Muell. Arg., FWTA 1:374. Euphorbiaceae.
Small crooked tree; lvs mesophyll, hairy (var. *laciniatum* is less hairy). Fl dioecious, small, pale red, crowded in brush inflorescences, not clearly seasonal. Fr small, drupaceous; seed medium. Abundant and gregarious in understorey of Upland Evergreen forest. Guinée to Zaire; (Guineo-Congolian wide).

50 ANTIDESMA MEMBRANACEUM Muell. Arg., FWTA 1:375. Euphorbiaceae. Akan: numanumagyama.
Small tree with fluted bole; lvs notophyll. Fl dioecious, small, greenish, crowded in brush inflorescences, 3–5. Fr small, drupaceous; seed medium. Virtually confined to forest near the savanna boundary. Guinea-Bissau to Angola and Mozambique; (Chorological transgressor).

51 APHANOSTYLIS MANNII (Stapf) Pierre. FWTA 2:59. Apocynaceae.
Large woody climber; lvs microphyll, white latex present. Fl white, dish with tube, medium-sized, 3–6. Fr orange berry; seeds medium to large. Regeneration abundant in forest shade. Associated with soils of very low exchangeable base status, and of low pH (mode c.4.3). Guinea Bissau to Zaire; (Guineo-Congolian wide).

52 APORRHIZA UROPHYLLA Gilg, FWTA 1:721. Sapindaceae.
Medium-sized tree; lvs pinnate, leaflets notophyll, with basal marginal domatia. Fl probably monoecious, white, small, clustered, 8–10. Fr capsule; seeds medium, black with yellow aril, 9–11. Regenerates in shade. Closed forest understorey, mainly in evergreen forest. S. Leone to Cameroun; (Guinea wide).

53 APTANDRA ZENKERI Engl., FWTA 1:649. Olacaceae. Akan: ayemtudua.
Small tree; lvs notophyll. Troll's model (59). Fl dioecious small, greenish-white, funnel-shaped, calyx greatly enlarged in fruit, waxy-pink, not clearly seasonal. Fr black drupe; seeds very large. Regenerates in forest shade. Ghana to Angola; (Guineo-Congolian wide).

54 ARGOMUELLERA MACROPHYLLA Pax, FWTA 1:405. Euphorbiaceae. Brong: mprepre.
Small tree or shrub, branching from near base; lvs macrophyll, toothed; twigs and leaves somewhat hairy. Fl monoecious, small, inconspicuous, 12–3. Fr small explosive capsule; seeds medium, 12–3. Gregarious in the understorey; an indicator for the north-west subtype of Moist Semideciduous forest. S. Leone to Angola, Sudan and Mozambique; (Chorological transgressor).

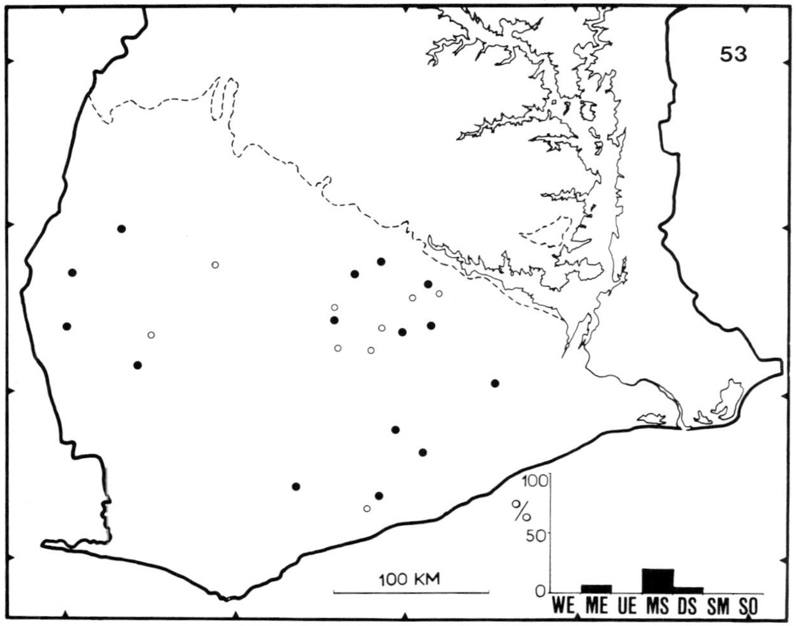

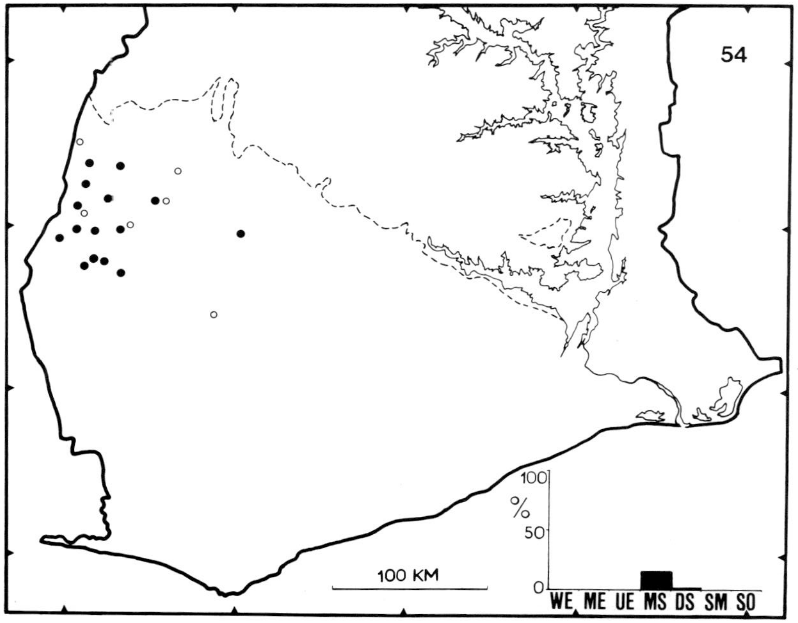

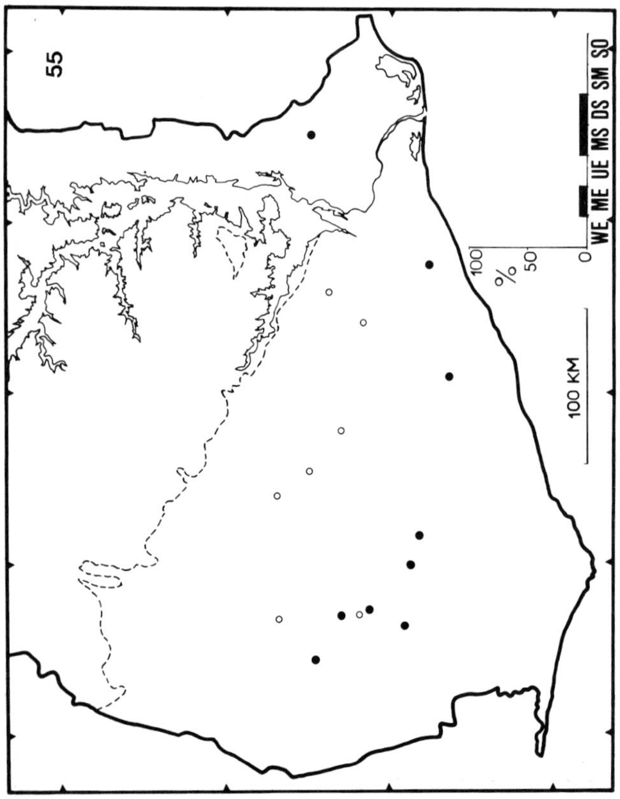
55
100 KM
%
100
50
0
WE ME UE MS DS SM SO

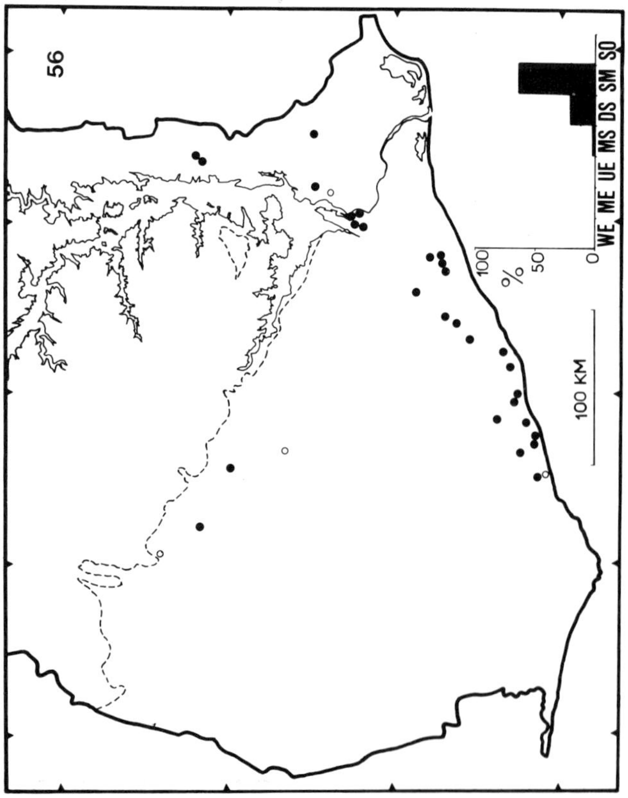
56
100 KM
%
100
50
0
WE ME UE MS DS SM SO

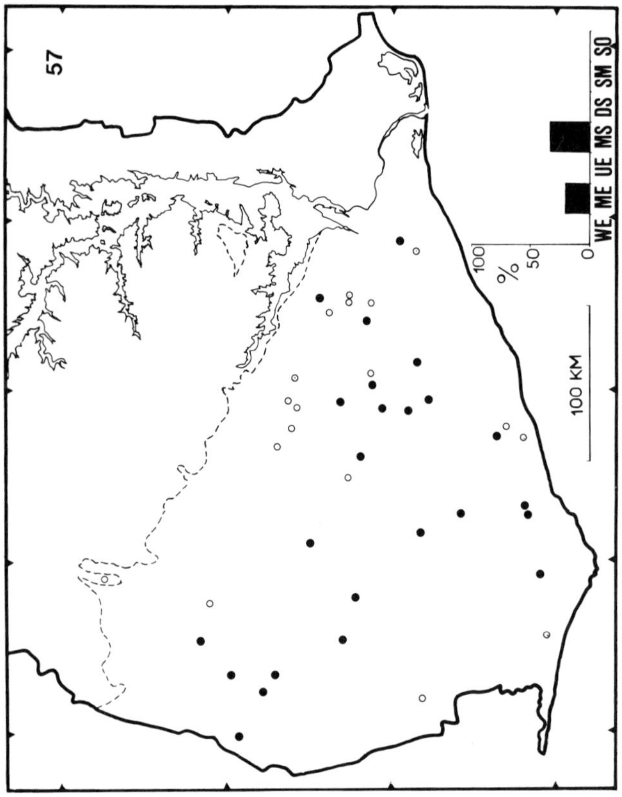
57
100 KM
%
100
50
0
WE ME UE MS DS SM SO

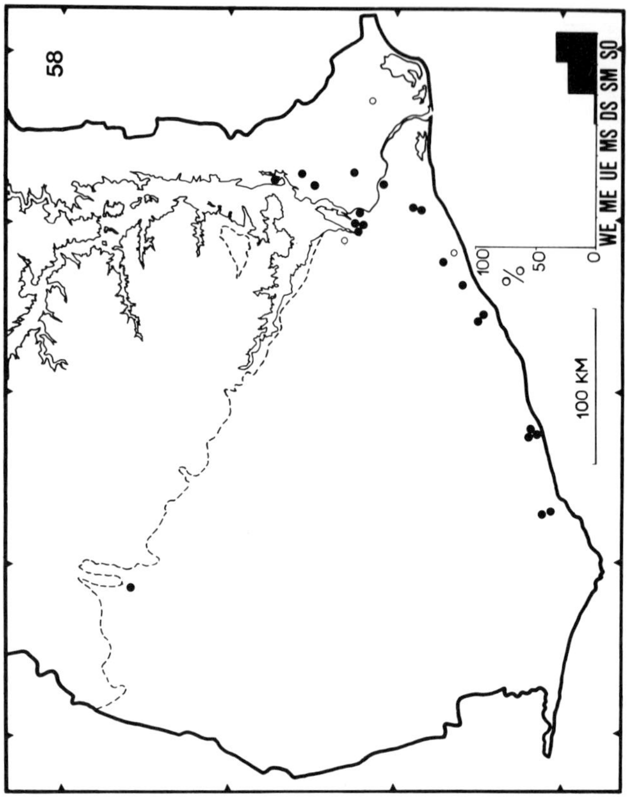
58
100 KM
%
100
50
0
WE ME UE MS DS SM SO

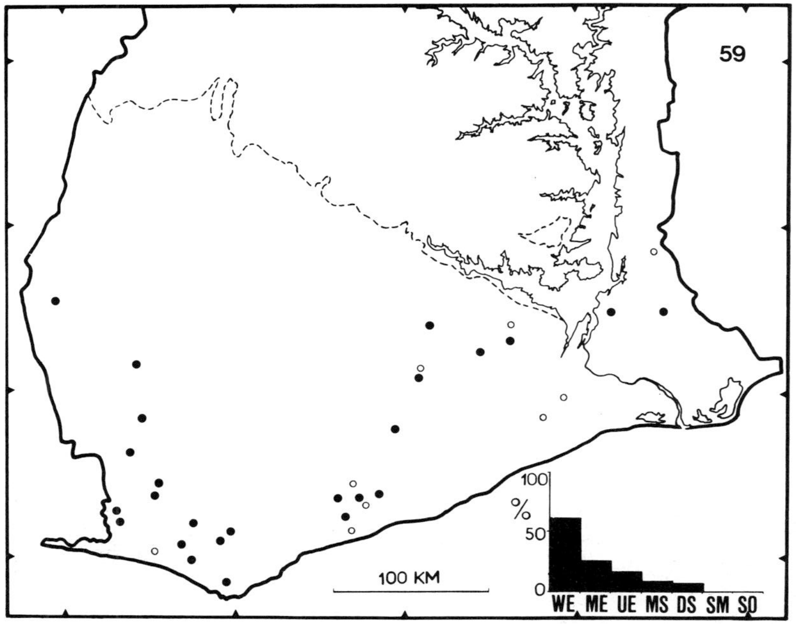

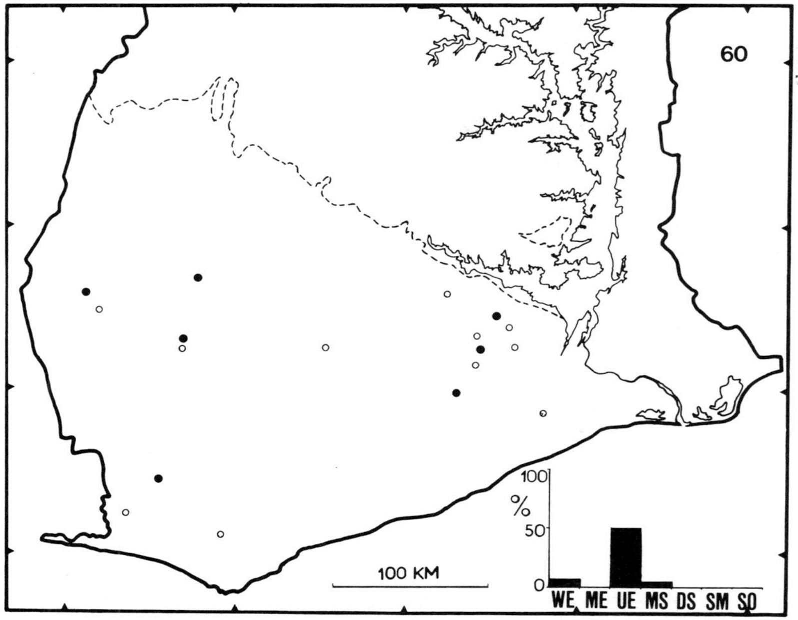

55 ARTABOTRYS STENOPETALUS Engl. & Diels, FWTA 1:41. Annonaceae. Akan: akoo-ano.
Medium-sized woody climber; lvs notophyll, twigs and leaves pubescent, peduncle hook-like, aiding climbing. Roux's model. Fl yellow, medium, funnel, mostly 1–4. Fr red berry with thin flesh; seeds large. Regenerates in shade. Ghana to Zaire; (Guineo-Congolian wide).

56 ARTABOTRYS VELUTINUS Sc. Elliot, FWTA 1:40. Annonaceae. Akan: akoo-ano.
Medium-sized woody climber; lvs notophyll; twigs and leaves hairy (more so than in *A. stenopetalus* to which it is closely related); peduncle hook-like. Roux's model. Fl yellow, medium, funnel, not clearly seasonal. Fr red berry with thin flesh; seeds large. Regenerates in shade. In drier forest than *A. stenopetalus*. Guinée to Nigeria; (Guinea wide).

57 ARTHROPTERIS PALISOTII (Desv.) Alston, FWTA FF:52. (*A. palisoti* in FWTA). Davalliaceae.
Fern, root-climber on stems of understorey trees in closed forest; leaves pinnate, leaflets microphyll. Propagated by dust-like spores. Iv. Coast to Cameroun; (Guinea wide).

58 ASPARAGUS WARNECKEI (Engl.) Hutch., FWTA 3:93. Liliaceae.
Slender woody climber, scale leaves modified as sharp recurved hooks; cladodes microphyll. McClure's model. Fl pale yellow, strongly scented, bowl, small, 1–2. Fr small berry, seeds medium, 1–2. Restricted to dry forest. Guinée to Togo; (Upper Guinea endemic).

59 ASPLENIUM AFRICANUM Desv., FWTA FF:55. (Incl. *A. currori* Hook., FWTA FF:55). Aspleniaceae.
Epiphytic rosette fern on trunks and low branches in forest understorey, macrophyll. Propagated by dust-like spores. Guinée to Angola; (Guineo-Congolian wide).

60 ASPLENIUM BARTERI Hook., FWTA FF:56. Aspleniaceae.
Epiphytic rosette fern on trunks and low branches. Leaves pinnate, leaflets toothed, nanophyll. Propagated by dust-like spores, and by proliferation from near tip of leaf. Guinée to Zaire; (Guineo-Congolian wide).

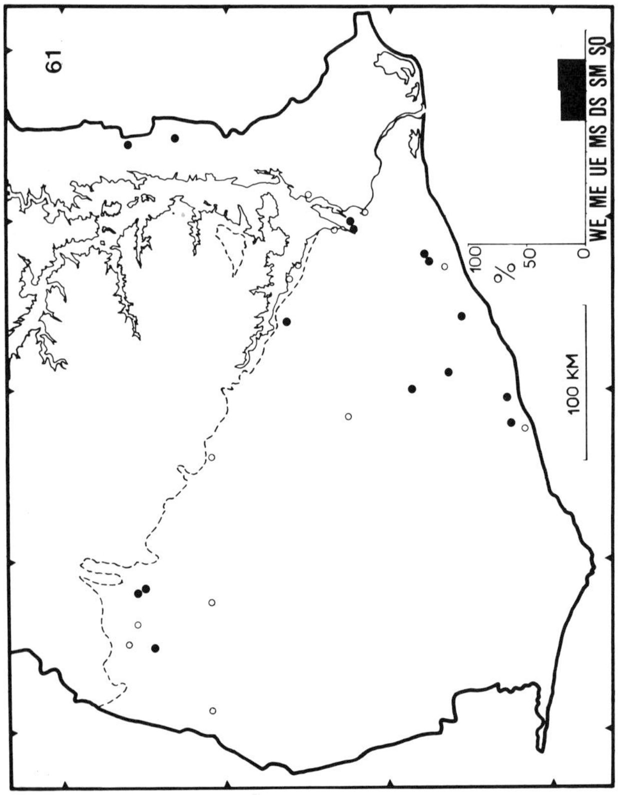
61
100 KM
100
%
50
0
WE ME UE MS DS SM SO

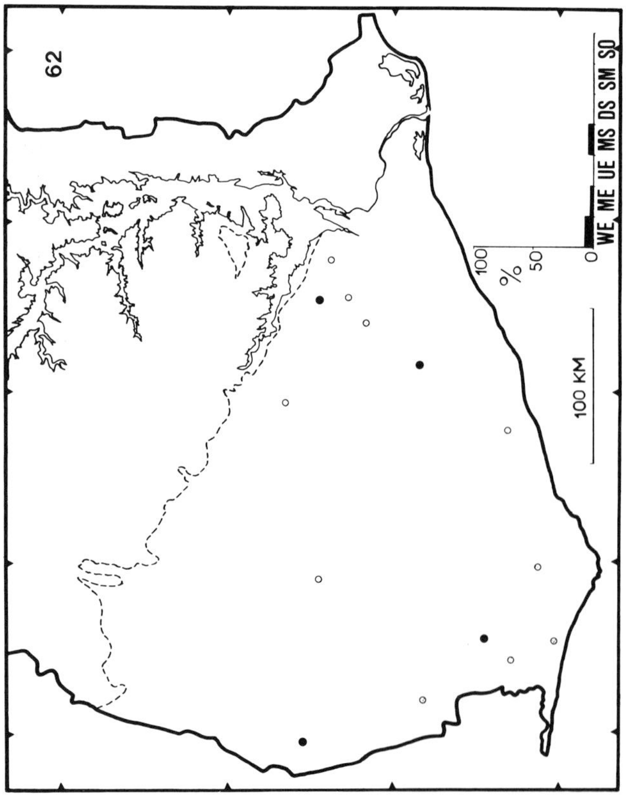
62
100 KM
100
%
50
0
WE ME UE MS DS SM SO

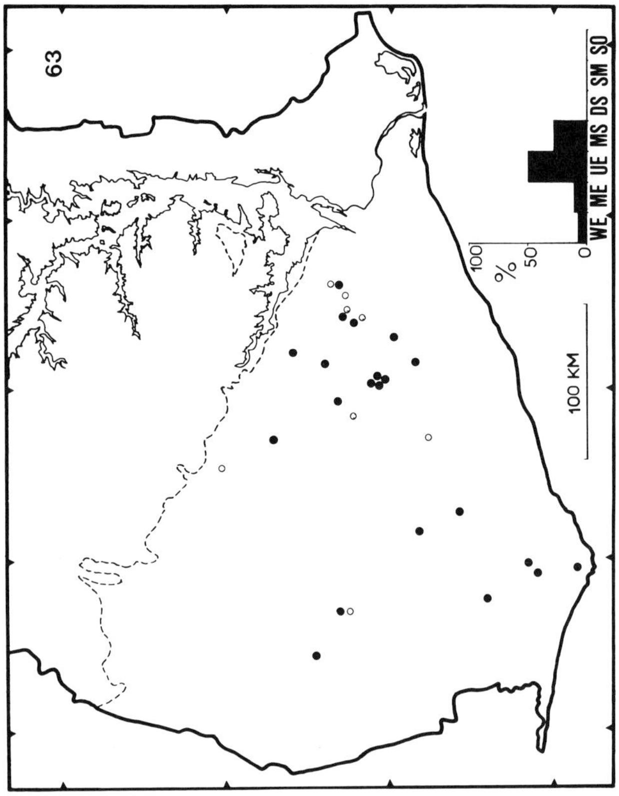
63
100 KM
100
%
50
0
WE ME UE MS DS SM SO

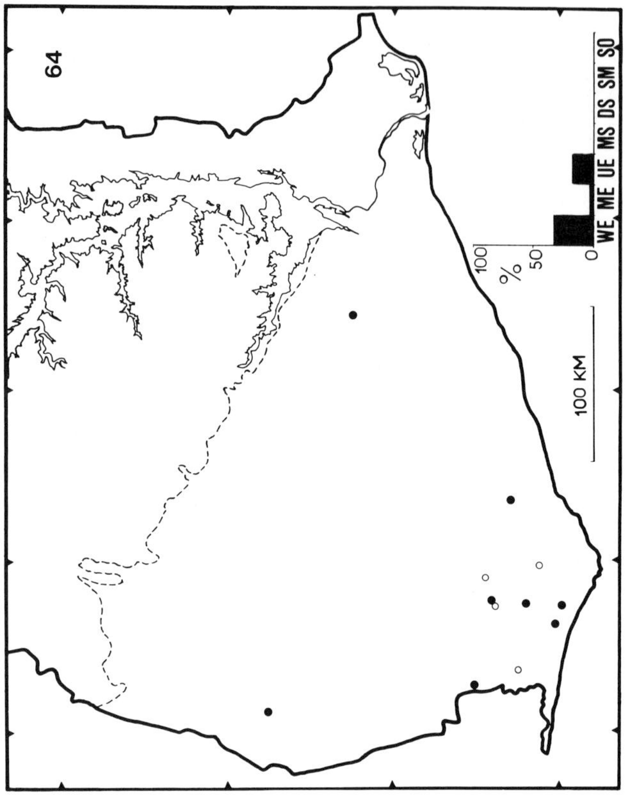
64
100 KM
100
%
50
0
WE ME UE MS DS SM SO

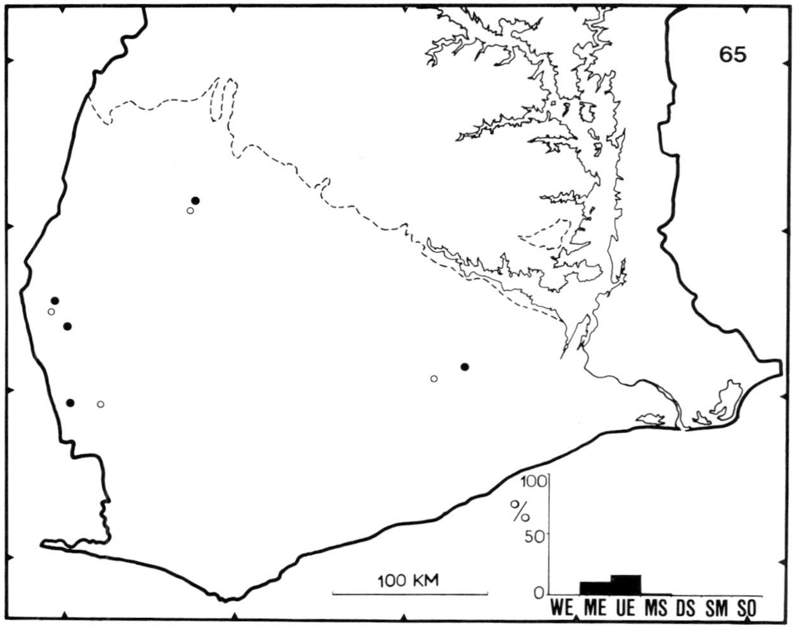

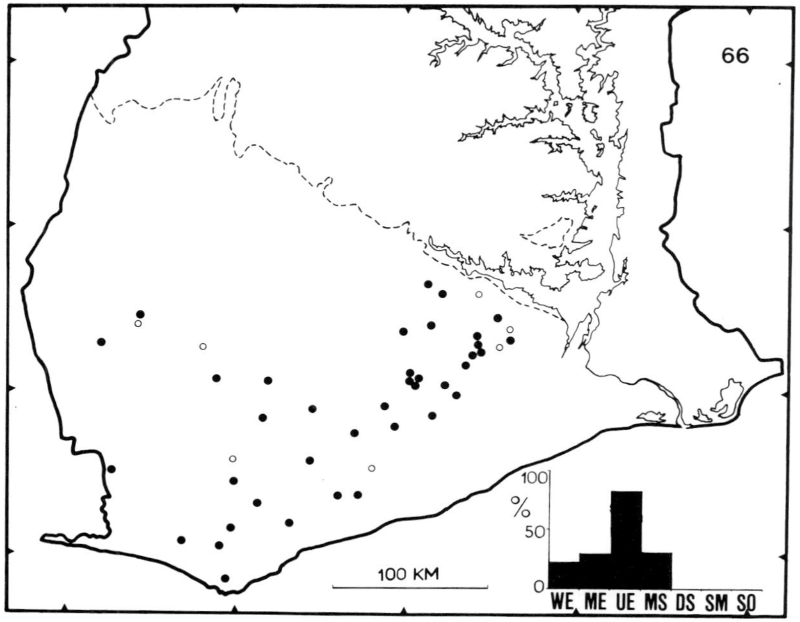

61 ASPLENIUM EMARGINATUM P. Beauv., FWTA FF:56. Aspleniaceae.
Ground fern on floor of dry forest; lvs pinnate, leaflets mesophyll. Propagated by dust-like spores. Guinée to Angola and Uganda; (Guineo-Congolian wide).

62 ASPLENIUM VARIABILE Hook., FWTA FF:56. (Incl. var. *paucijugum* (Ballard) Alston). Aspleniaceae.
Fern of rocky ground on forest floor, and sometimes epiphytic on trunk bases. Lvs simple or pinnate, lvs and leaflets notophyll to mesophyll. Propagated by dust-like spores. Guinée to Zaire; (Guineo-Congolian wide).

63 ATAENIDIA CONFERTA (Benth.) Milne-Redh., FWTA 3:89. Marantaceae.
Perennial ground herb, rosette; lvs pubescent below, macrophyll. Fl pinkish, medium, funnel with tube, almost covered by reddish bracts, not clearly seasonal. Fr dry, indehiscent; seeds medium. Closed forest, and small gaps. Iv. Coast to Zaire and Uganda; (Guineo-Congolian wide).

64 ATROXIMA LIBERICA Stapf, FWTA 1:109. Polygalaceae. (32).
Small to medium woody climber, or shrub, in forest understorey; lvs mesophyll. Mangenot's model (59). Fl cream, medium, flag, 11–2. Fr orange-red berry, 12–3; seeds medium. Liberia to Angola; (Guineo-Congolian wide).

65 AUBREVILLEA PLATYCARPA Pellegr., FWTA 1:492. Mimosaceae.
Large buttressed tree with wide crown; lvs bipinnate, leaflets nanophyll, deciduous while fruiting. Troll's model. Fl yellow, small, brush inflorescence 12–1. Fr thin indehiscent pod (Fig. 6.1b), with one end twisted, wind-dispersed; seeds medium, 2–5. Regenerates in closed forest. Guinée to Zaire; (Guineo-Congolian wide).

66 AULACOCALYX JASMINIFLORA Hook. f., FWTA 2:145. Rubiaceae. Akan: ntweson.
Medium-sized tree; lvs notophyll. Fagerlind's model. Fl white, large, dish with tube, 12–1. Fr berry, seeds medium, 1–4. Regenerates under forest canopy; gregarious in moist forest. S. Leone to Angola; (Guineo-Congolian wide).

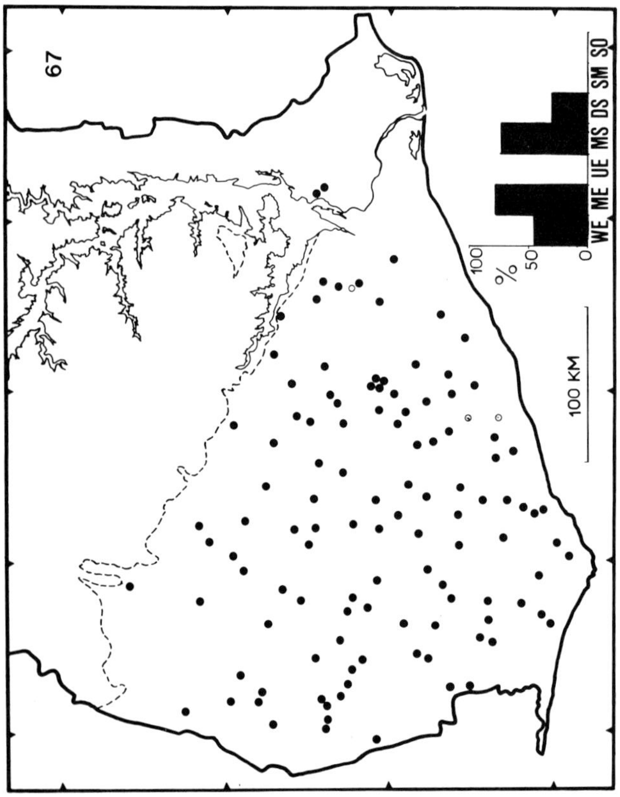
67
100 KM
%
100
50
0
WE ME UE MS DS SM SO

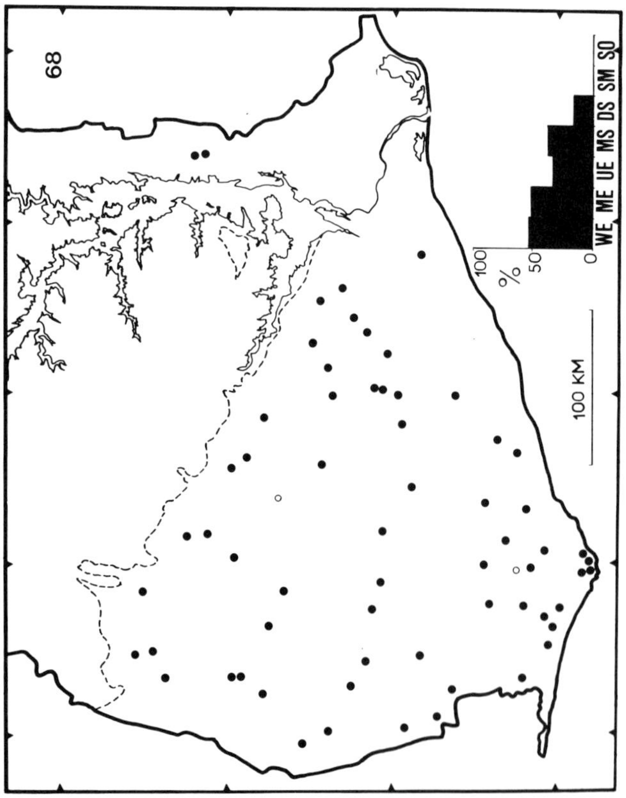
68
100 KM
%
100
50
0
WE ME UE MS DS SM SO

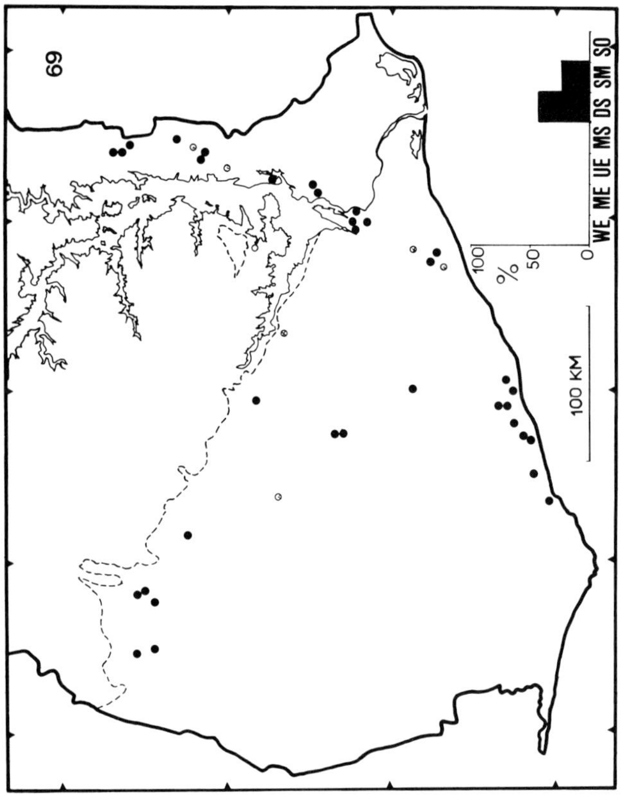
69
100 KM
%
100
50
0
WE ME UE MS DS SM SO

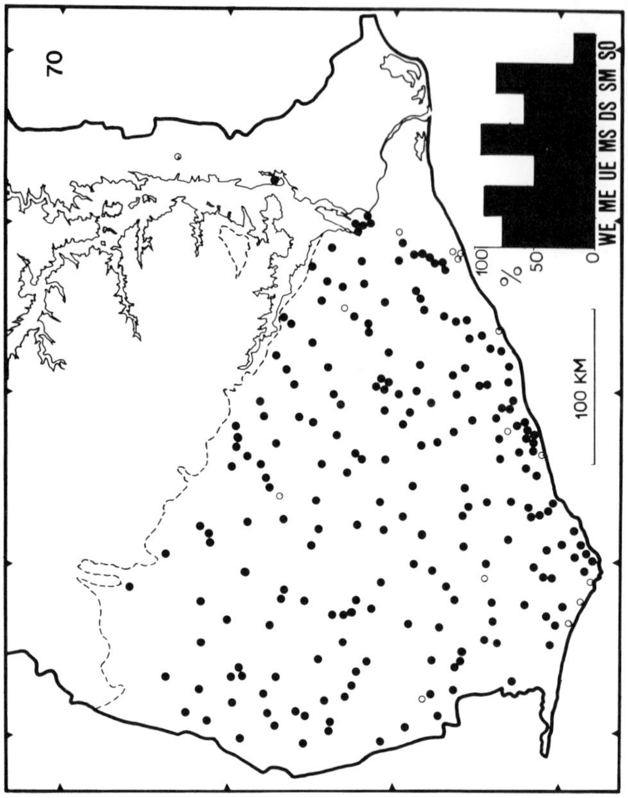
70
100 KM
%
100
50
0
WE ME UE MS DS SM SO

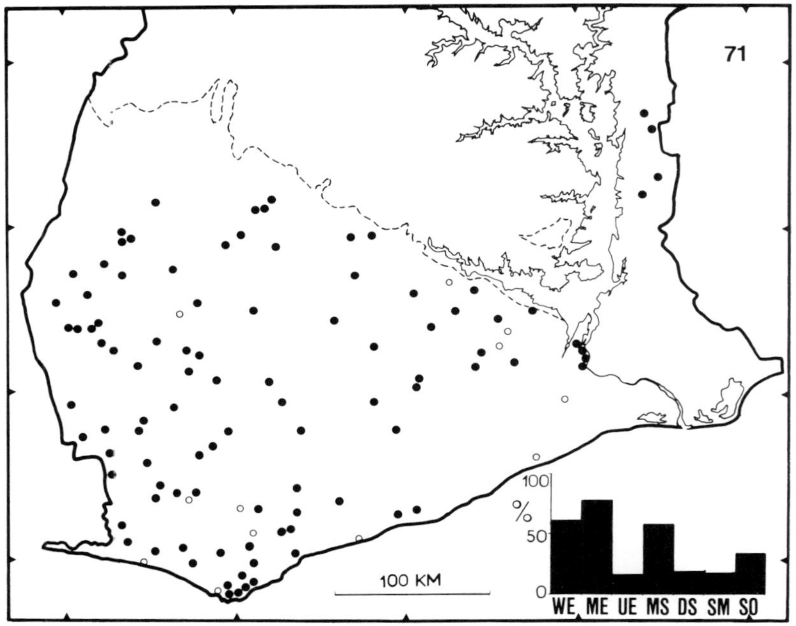

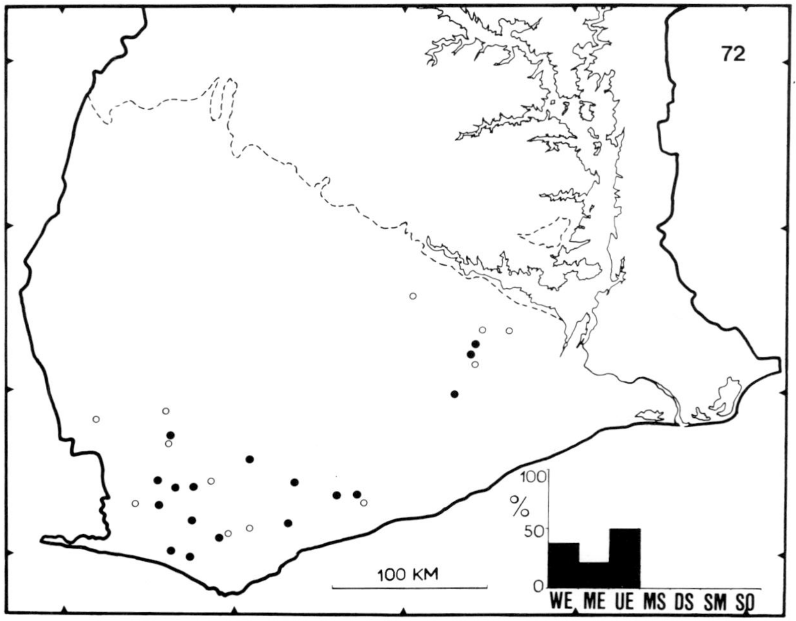

67 BAISSEA BREVILOBA Stapf, FWTA 2:79. Apocynaceae.
Small woody climber in forest understorey, containing white latex, stems pubescent, lvs microphyll. Cook's model. Fl white, medium, dish with tube, 4–6. Fr follicle, medium seeds with parachute of hairs. Abundant regeneration under forest canopy. Guinée to Nigeria; (Guinea wide).

68 BAISSEA MULTIFLORA A. DC., FWTA 2:79. Apocynaceae.
Small woody climber in forest understorey, containing white latex, stems pubescent, lvs notophyll. Roux's model. Fl white, medium, dish with tube, 12–4. Fr follicle, seeds large with parachute of hairs. Regenerates under forest canopy; also recorded from fringing forest in savanna zone. Very significantly associated with base-poor soils; half the survey records came from soils with pH 4.2–4.9. Senegal to Angola; (ecological and chorological transgressor).

69 BAISSEA ZYGODIOIDES (K. Schum.) Stapf, FWTA 2:79. Apocynaceae.
Small woody climber in forest understorey and thickets, containing white latex; lvs microphyll. Cook's model. Closely related to *B. breviloba,* but differing in blunter leaves. Fl white, medium, dish with tube, 3–7. Fr follicle, seeds large with parachute of hairs. Occurs in drier forest than does *B. breviloba.* Guinée to Ghana; (Upper Guinea endemic).

70 BAPHIA NITIDA Lodd., FWTA 1:512. Papilionaceae. Akan: odwen.
Small tree or shrub in forest understorey, or in thicket, lvs notophyll to mesophyll. Troll's model (58). Fl white, medium, flag, more or less throughout the year. Fr explosive pod, seeds medium. Abundant throughout the forest zone; more widely distributed in Ghana than any other tree. S. Leone to Cameroun; (Guinea wide).

71 BAPHIA PUBESCENS Hook. f., FWTA 1:513. Papilionaceae. Akan: odwen-kobiri.
Small to medium tree in forest understorey, lvs notophyll, twigs pubescent. Troll's model. Fl white, medium, flag, not seasonal. Fr explosive pod, seeds medium. Regenerates under forest canopy. Very significantly associated with acid soils; more than half survey records from soils with pH 4.2–5.3. Liberia to Zaire; (Guineo-Congolian wide).

72 BEILSCHMIEDIA MANNII (Meisn.) Benth. & Hook. f., FWTA 1:57. Lauraceae. Akan: tweanka.
Medium tree in lower canopy of evergreen forest, strongly aromatic; lvs mesophyll. Aubréville's model (59). Fl greenish-white, small, inconspicuous, not distinctly seasonal. Fr red drupe, seed very large. Regenerates under forest canopy. Very significantly associated with base-poor soils; only 1 survey record in soil with TEB > 4 m-equiv. per 100 g. Guinée to Zaire; (Guineo-Congolian wide).

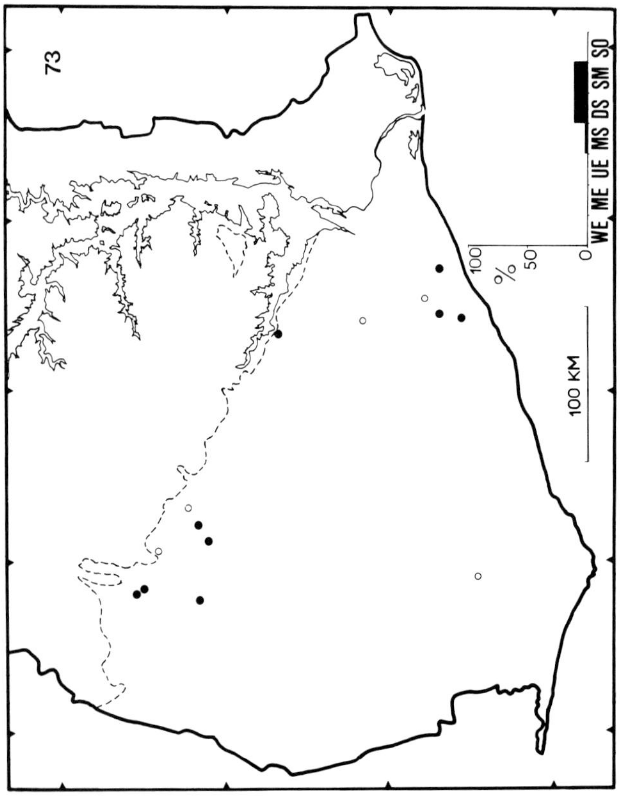
73
100 KM
%
100
50
0
WE ME UE MS DS SM SO

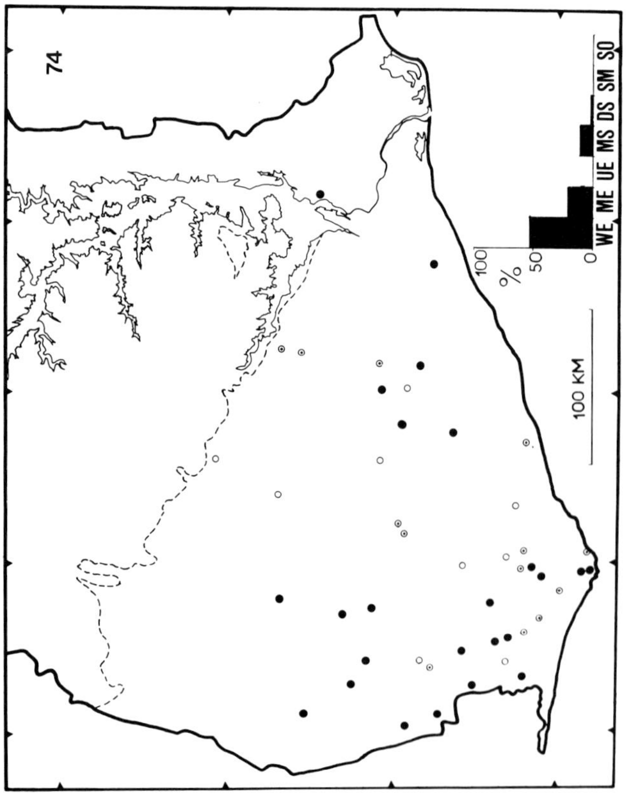
74
100 KM
%
100
50
0
WE ME UE MS DS SM SO

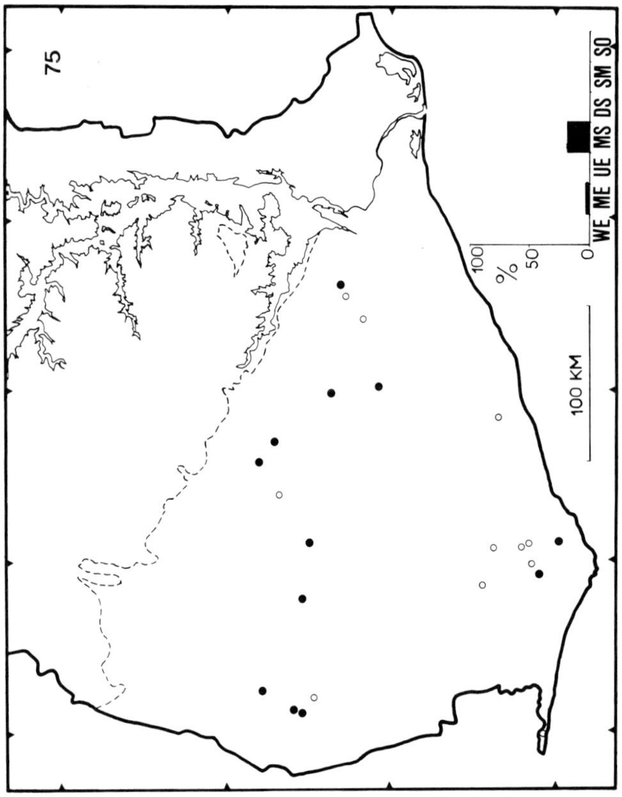
75
100 KM
%
100
50
0
WE ME UE MS DS SM SO

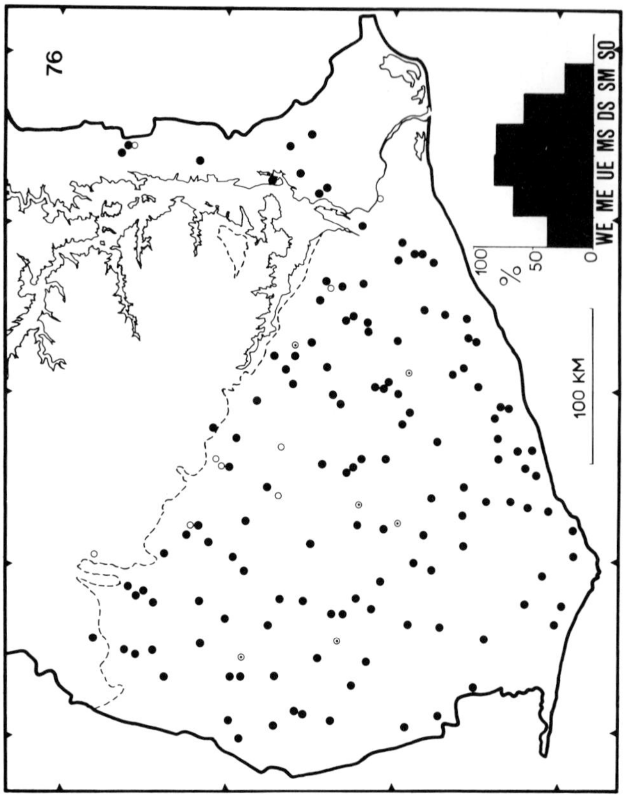
76
100 KM
%
100
50
0
WE ME UE MS DS SM SO

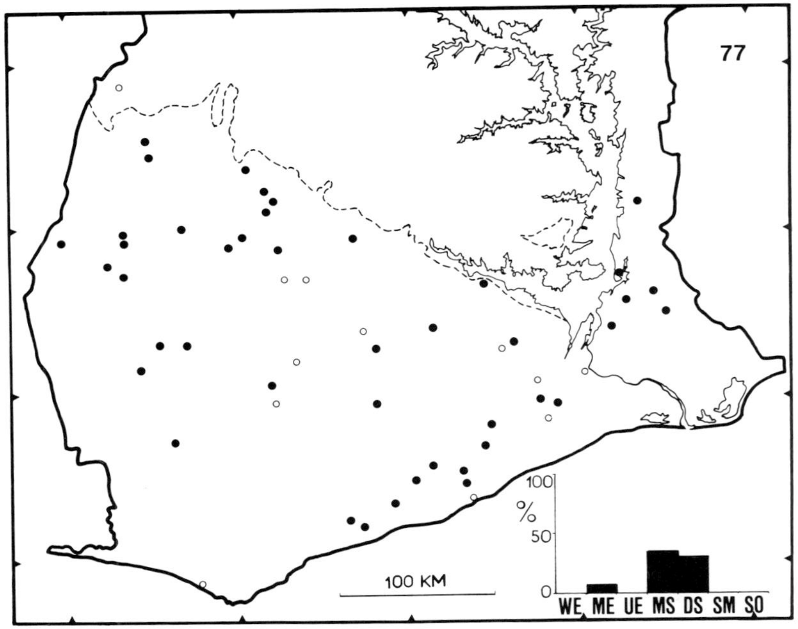

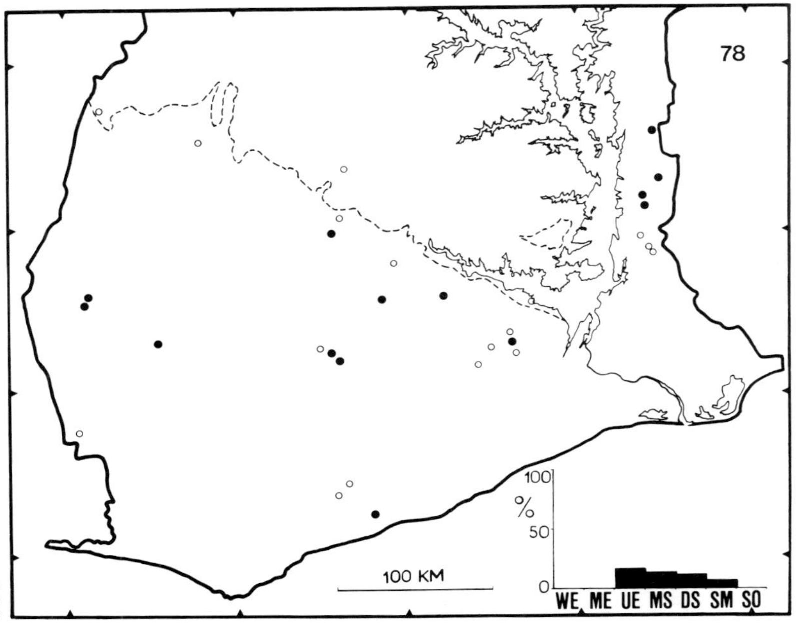

73 BEQUAERTIODENDRON OBLANCEOLATUM (S. Moore) Heine & J. H. Hemsley, FWTA 2:25. Sapotaceae. Akan: nnanfuro.
Medium-sized tree, containing white latex; lvs macrophyll, undersurface of leaf hidden by short white pubescence. Aubréville's model. Fl whitish, small, inconspicuous. Fr 1-seeded red berry, seed medium. Gregarious in dry forest. S. Leone to Ghana, also in Uganda; (Upper Guinea near-endemic).

74 BERLINIA CONFUSA Hoyle, FWTA 1:470, and BERLINIA TOMENTELLA Keay, FWTA 1:470. Caesalpiniaceae. Akan: kwatafompaboa.
Medium-sized trees, lvs pinnate, leaflets notophyll (*B. confusa*) to mesophyll (*B. tomentella*). Troll's model. Seedlings and saplings of the two species cannot always be identified with certainty; only herbarium records are distinguished on the map: *B. confusa* ○; *B. tomentella* ⊙. Fl white, large, flag (Fig. 9.1h), not obviously seasonal. Fr large explosive pod, seeds very large. *B. confusa:* S. Leone to Gabon, (Guinea wide); *B. tomentella:* S. Leone to Ghana, (Upper Guinea endemic).

75 BERSAMA ABYSSINICA Fres. subsp. *paullinioides* (Planch.) Verdc., FWTA 1:726. Melianthaceae. Akan: esonodua.
Small to medium-sized tree; lvs pinnate, leaflets notophyll. Rauh's model (58). Fl greenish-white, fragrant, medium, bowl, not clearly seasonal. Fr red capsule, seeds large, shiny red with yellow aril. Widespread in tropical Africa; (ecological and chorological transgressor).

76 BLIGHIA SAPIDA Konig, FWTA 1:722, and BLIGHIA WELWITSCHII (Hiern) Radlk., FWTA 1:722. Sapindaceae. Akan: akye (*B. sapida*) akyekobiri (*B. welwitschii*).
Large trees; lvs pinnate, leaflets notophyll to mesophyll, more (*B. sapida*) or less (*B. welwitschii*) pubescent. Rauh's model. Because seedlings and saplings cannot always be identified, only herbarium records are mapped separately; *B. sapida* ○, *B. welwitschii* ⊙. Fl monoecious, whitish, small, bowl, not clearly seasonal. Fr red fleshy capsule, seeds very large (larger in *B. sapida*), black, with cream to yellowish aril which is edible in *B. sapida*. Regenerates in shade. *B. welwitschii* is confined to high forest, but *B. sapida* is widely planted as a shade tree in villages both in the forest and savanna zones. *B. sapida:* Senegal to Cameroun (Guinea wide) and cult. elsewhere in tropics; *B. welwitschii:* S. Leone to Angola and Uganda; (Guineo-Congolian wide).

77 BLIGHIA UNIJUGATA Bak., FWTA 1:723. Sapindaceae. Akan: akyebiri.
Medium-sized tree; lvs pinnate, leaflets mesophyll. Fl monoecious, whitish, fragrant, small, bowl, 12–3. Fr red fleshy capsule, seeds large, black with yellow aril, 1–3. Regenerates in shade. S. Leone to Angola and Kenya; (Guineo-Congolian wide).

78 BOLBITIS ACROSTICHOIDES (Afzel. ex Sw.) Ching, FWTA FF:68. Lomariopsidaceae.
Fern of forest floor, with preference for rocky slopes where it may be gregarious; also in cocoa farms. Lvs pinnate, leaflets toothed, mesophyll. Propagated by dust-like spores, and proliferation from terminal leaflets. Guinée to Angola and Madagascar; (Chorological transgressor).

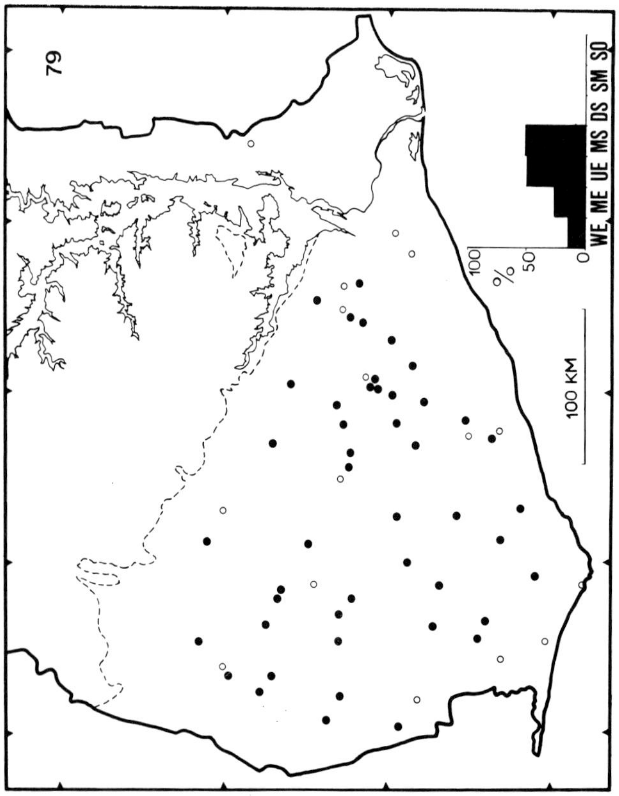
79
100 KM
%
100
50
0
WE ME UE MS DS SM SO

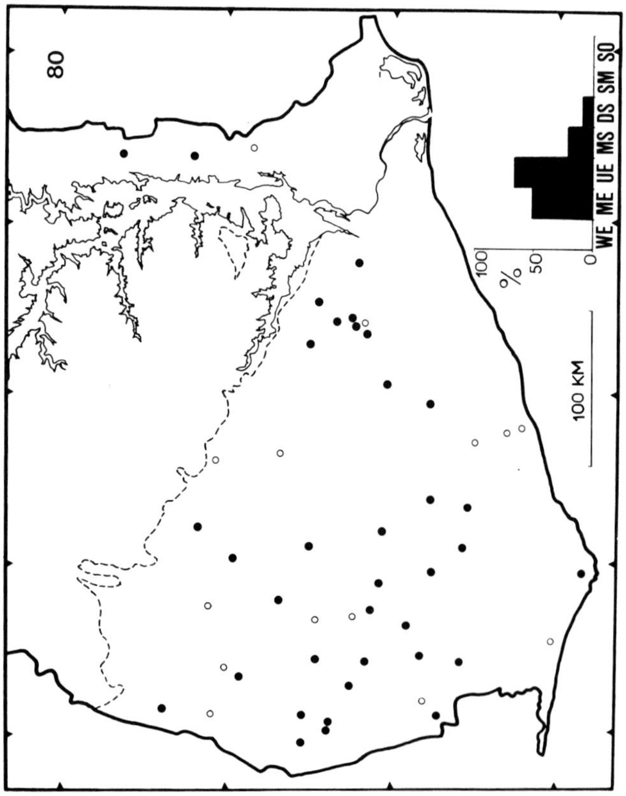
80
100 KM
%
100
50
0
WE ME UE MS DS SM SO

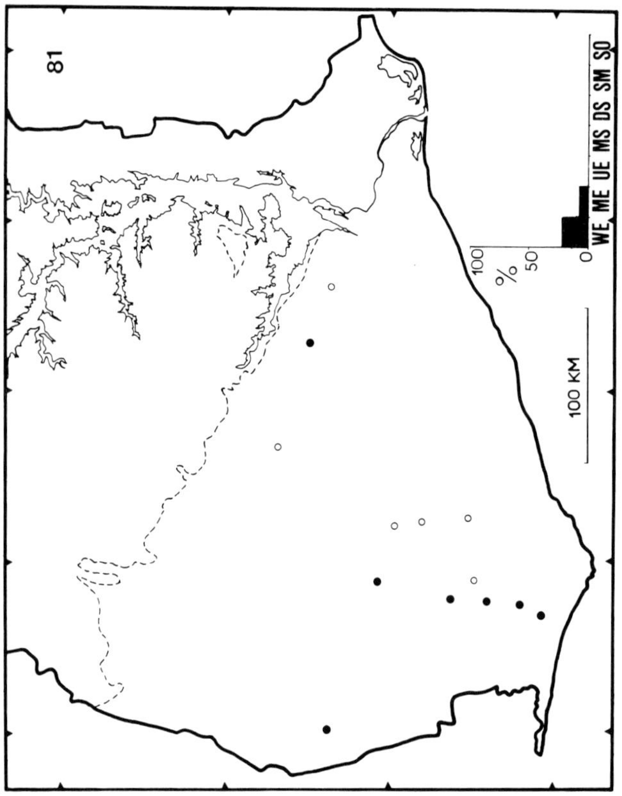
81
100 KM
%
100
50
0
WE ME UE MS DS SM SO

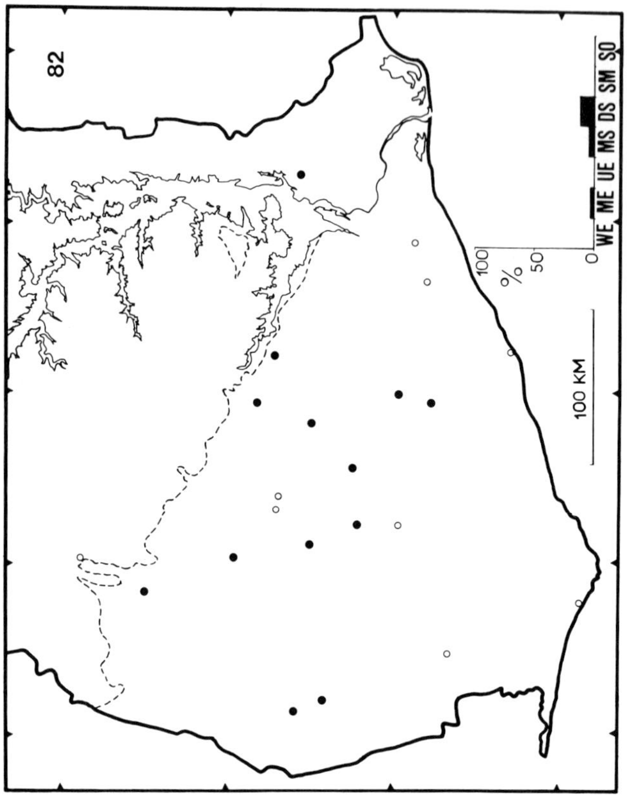
82
100 KM
%
100
50
0
WE ME UE MS DS SM SO

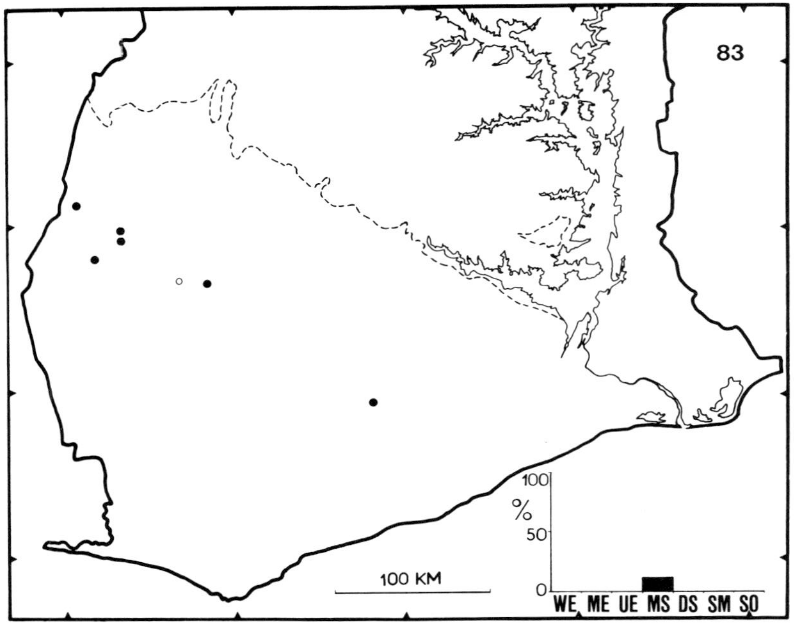

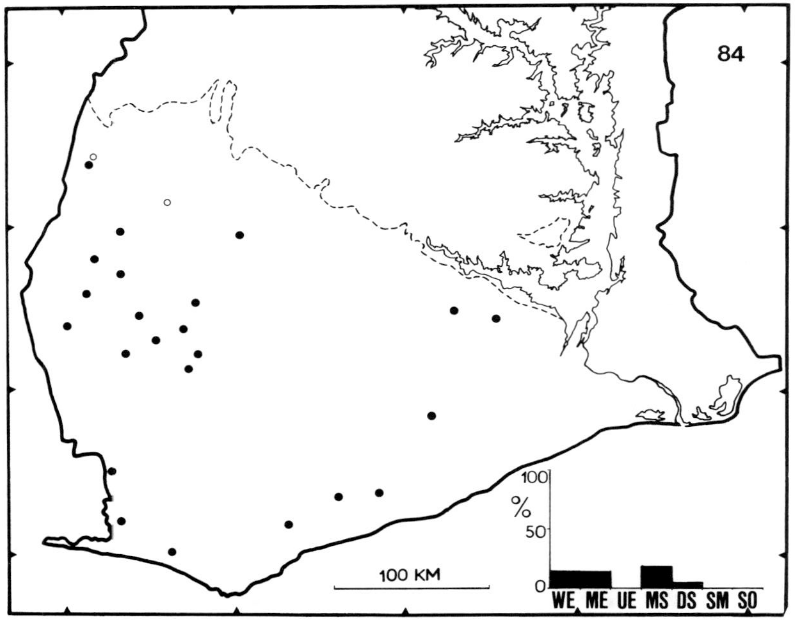

79 BOLBITIS AURICULATA (Lam.) Alston, FWTA FF:68. Lomariopsidaceae.
Fern of forest floor. Lvs pinnate, leaflets mesophyll. Propagated by dust-like spores; creeping rhizome. Guinée to Zaire and Uganda; (Guineo-Congolian wide).

80 BOLBITIS GEMMIFERA (Hieron.) C. Chr., FWTA FF:68. Lomariopsidaceae.
Fern of forest floor. Lvs pinnate, leaflets mesophyll. Propagated by dust-like spores, and proliferation from terminal leaflets. Very significantly associated with base-poor soils of low pH. Guinée to Angola and Tanzania; (Guineo-Congolian wide).

81 BOMBAX BREVICUSPE Sprague, FWTA 1:335. Bombacaceae. Akan: onyinakoben. (Type from Subriso).
Large tree; lvs digitate, leaflets microphyll, deciduous. Massart's model. Fl white, large, funnel, 11–1. Fr dry capsule; seeds medium, wind-dispersed in cottony fluff, 2–5. S. Leone to Cameroun; (Guinea wide).

82 BOMBAX BUONOPOZENSE P. Beauv., FWTA 1:334. Bombacaceae. Akan: akonkodie.
Large tree with buttresses; trunk and branches thorny. Lvs digitate, leaflets notophyll, deciduous. Massart's model. Fl red, large, bell, 11–2; pollinated by sunbirds. Fr dry capsule; seeds medium, wind-dispersed in cottony fluff, 2–4. Regenerates in small gaps. S. Leone to Gabon; (Guinea wide).

83 BOWRINGIA MILDBRAEDII Harms, FWTA 1:511. Papilionaceae.
Medium-sized woody climber, lvs notophyll. Fl white, medium, flag. Fr pod; seeds large, red. Regenerates in shade. Fl and fr not yet found in Ghana. Ghana to Angola; (Guineo-Congolian wide).

84 BUCHHOLZIA CORIACEA Engl., FWTA 1:93. Capparaceae. Akan: esonobese.
Medium-sized tree, lvs mesophyll. Fl cream, medium, bowl. Fr woody berry; seeds very large, dispersed by elephants, and germinating in dung. Bark and seeds with pungent, spicy smell. Guinée to Gabon; (Guinea wide).

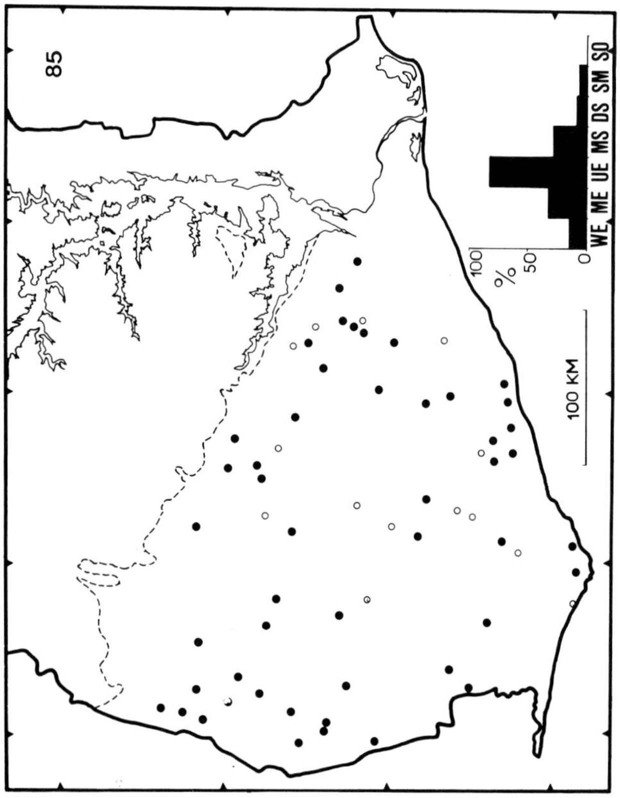
85
100 KM
100
%
50
0
WE ME UE MS DS SM SO

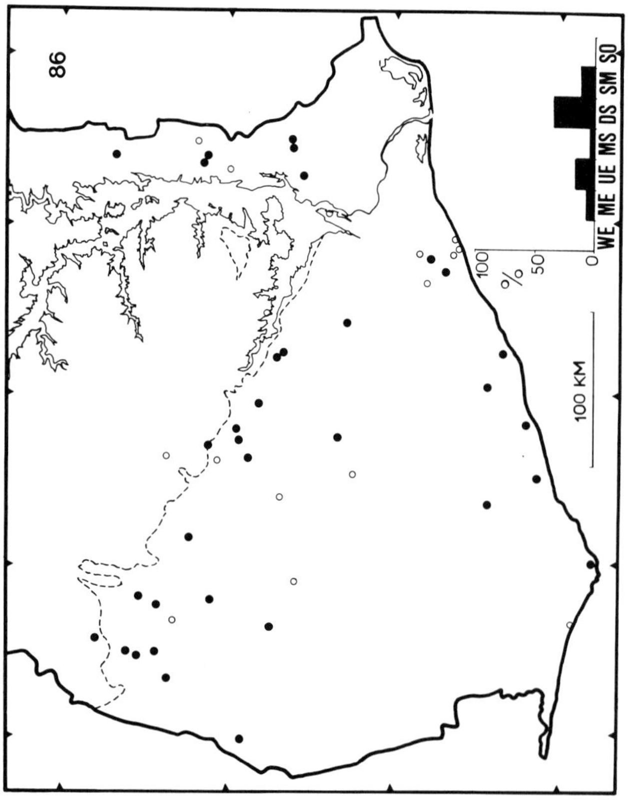
86
100 KM
100
%
50
0
WE ME UE MS DS SM SO

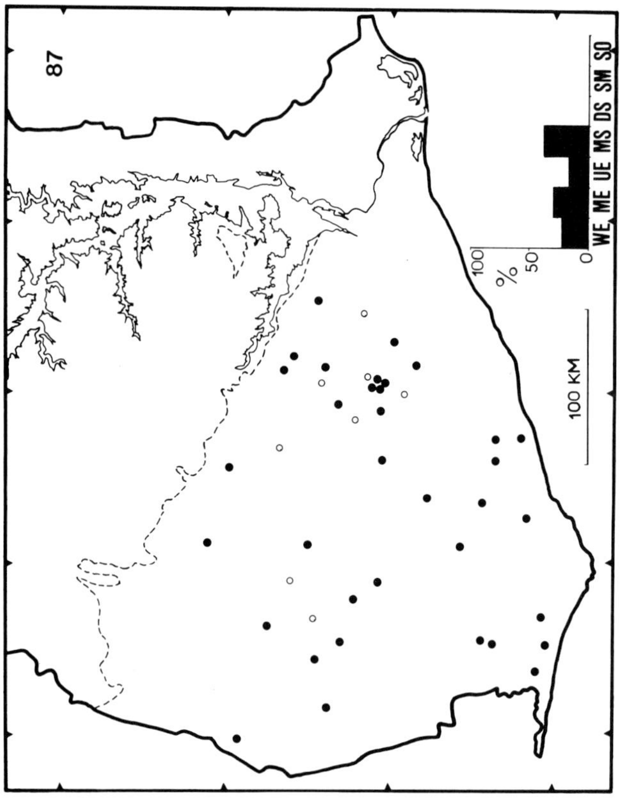
87
100 KM
100
%
50
0
WE ME UE MS DS SM SO

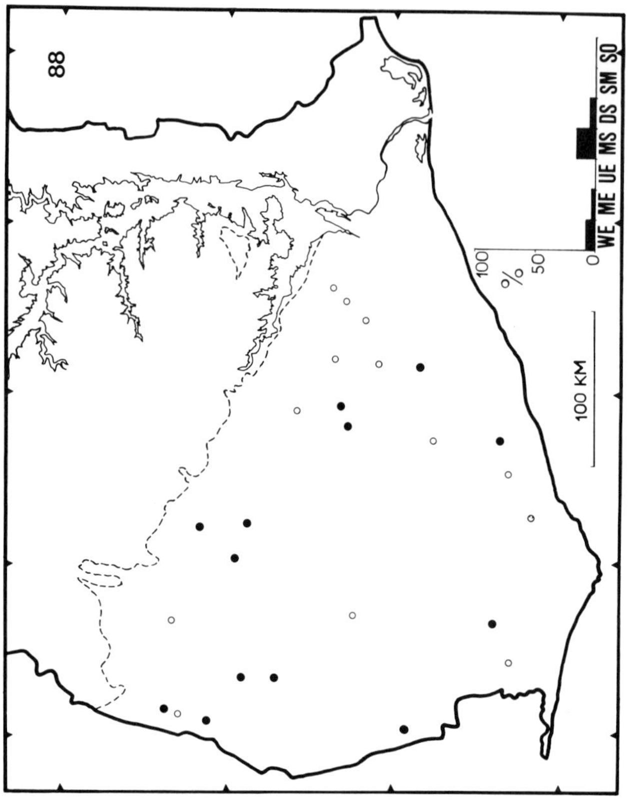
88
100 KM
100
%
50
0
WE ME UE MS DS SM SO

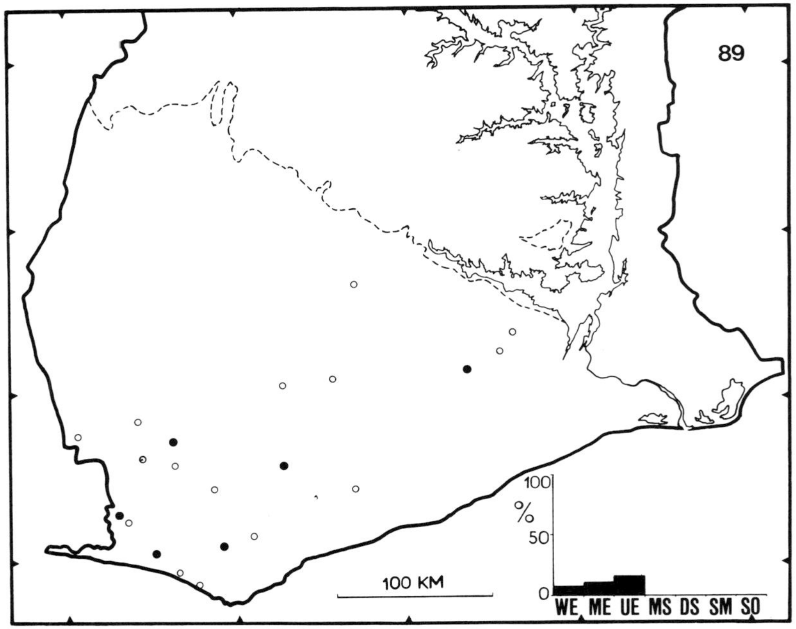

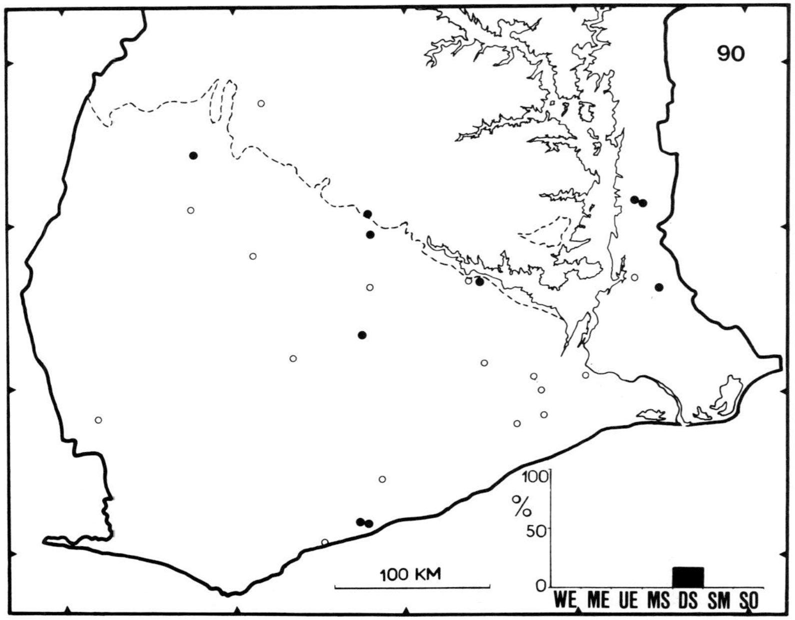

85 BUSSEA OCCIDENTALIS Hutch., FWTA 1:483. Caesalpiniaceae. Akan: koto-prepre. (Type from Dunkwa).
Medium to large tree; lvs bipinnate, leaflets notophyll. Fl yellow, large, bowl, not distinctly seasonal. Fr woody explosive pod; seeds very large, eaten by colobus monkeys. Regenerates in shade. S. Leone to Ghana; (Upper Guinea endemic).

86 BYRSOCARPUS COCCINEUS Schum. & Thonn., FWTA 1:741. Connaraceae. Akan: awendade. (Type from near Accra).
Large woody climber in forest, or shrub in thicket; xylem without anomalies. Lvs pinnate, leaflets nanophyll to microphyll. Fl white, very fragrant, bowl, heterostyled (15), pollinated by bees, flowering abundantly and gregariously a few days after heavy rain; flowers accompanied by young leaves. Fr orange, fleshy, 1-seeded follicle; seed medium, black with yellow aril. Occurs in fringing forest throughout the savanna zone. Widespread in Tropical Africa; (ecological and chorological transgressor).

87 CALAMUS DEERRATUS Mann & Wendl., FWTA 3:166. Palmae. Akan: mfia (used also for other rattans).
Medium to large woody climbing palm, with very thorny leaf-sheaths, leaves, and inflorescence. Lvs pinnate, leaflets macrophyll. Tomlinson's model (59). Fl dioecious, yellowish, small, clustered on spikes, not distinctly seasonal. Fr 1-seeded berry, covered with scales; seed medium, surrounded by thin flesh. Especially abundant in swamp forest, but regenerating under non-swampy closed forest. Senegal to Angola and Uganda; (Guineo-Congolian wide).

88 CALLICHILIA SUBSESSILIS (Benth.) Stapf, FWTA 2:64. Apocynaceae. (20).
Pygmy tree in forest understorey; lvs mesophyll, anisophyllous, containing white latex. Leeuwenberg's model. Fl white, large, dish with tube, not distinctly seasonal. Fr orange berry; seeds medium. Regenerating under closed canopy. S. Leone to Ghana; (Upper Guinea endemic).

89 CALONCOBA ECHINATA (Oliv.) Gilg, FWTA 1:188. Flacourtiaceae.
Small tree or shrub, lvs mesophyll. Fl white, medium, bowl, not distinctly seasonal. Fr orange spiny berry; seeds medium. Under forest canopy, or in gaps; occurs in Moist Semi-deciduous forest, but more commonly in evergreen forest. Guinée to Ghana; (Upper Guinea endemic).

90 CALONCOBA GILGIANA (Sprague) Gilg. FWTA 1:189. Flacourtiaceae. Akan: awiewu.
Small tree, flowering while deciduous. Lvs mesophyll. Fl white, large, bowl, 12–3, very decorative. Fr smooth, yellow berry; seeds medium. Regenerates under forest shade. S. Leone to Nigeria; (Guinea wide).

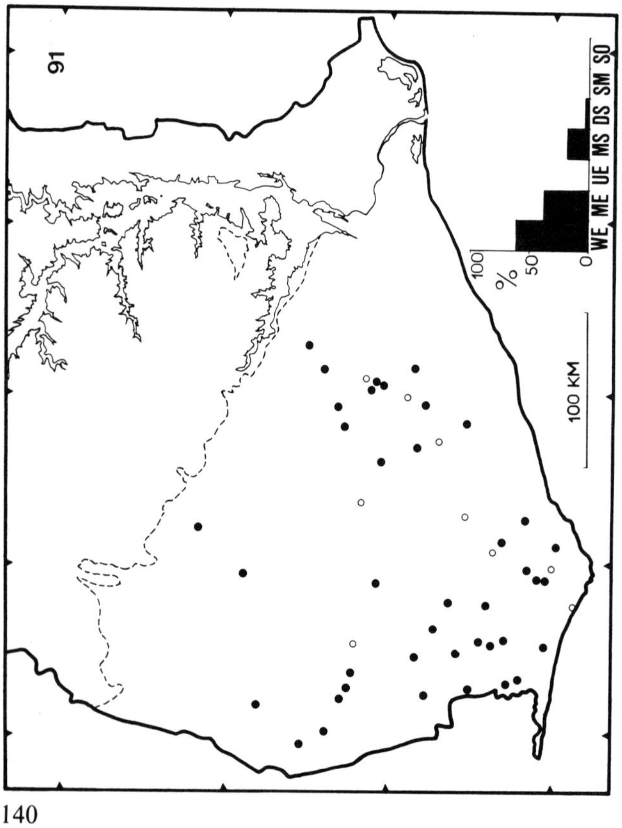
91
100 KM
%
100
50
0
WE ME UE MS DS SM SO

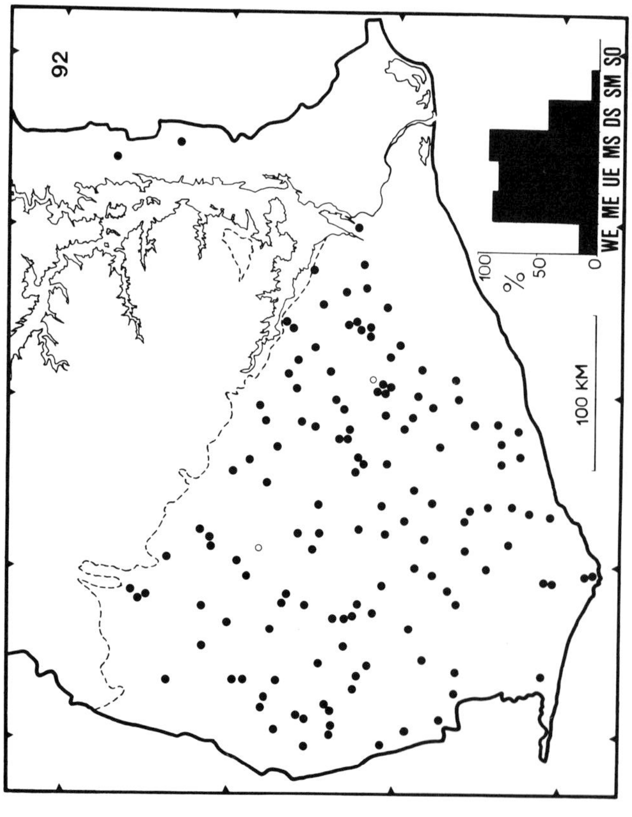
92
100 KM
%
100
50
0
WE ME UE MS DS SM SO

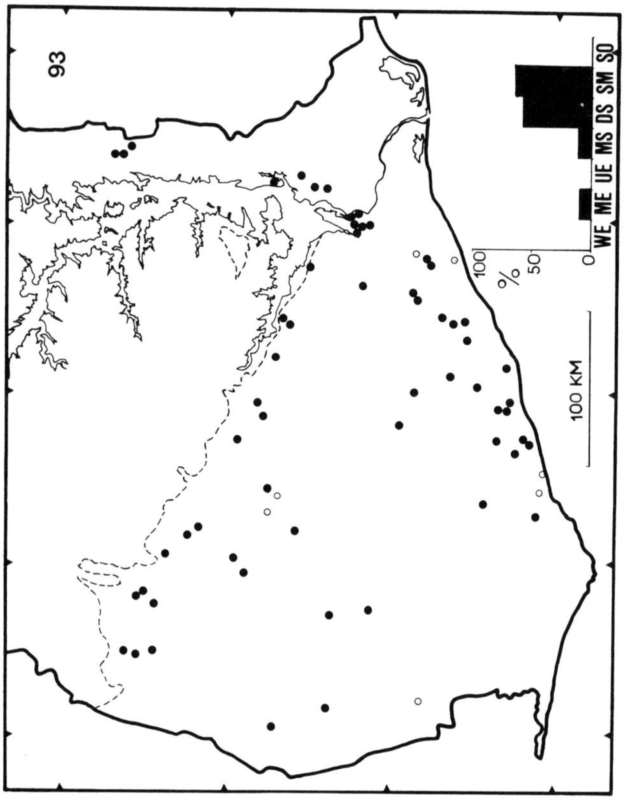
93
100 KM
%
100
50
0
WE ME UE MS DS SM SO

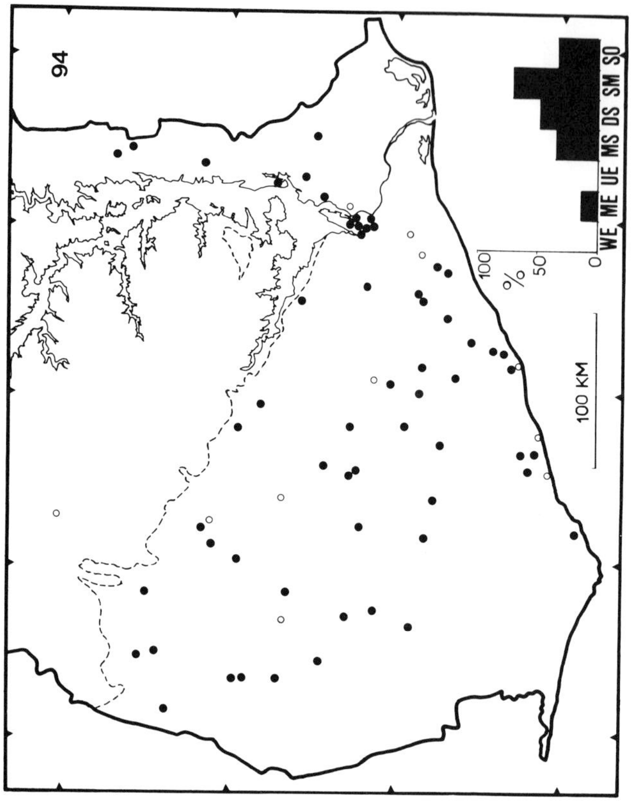
94
100 KM
%
100
50
0
WE ME UE MS DS SM SO

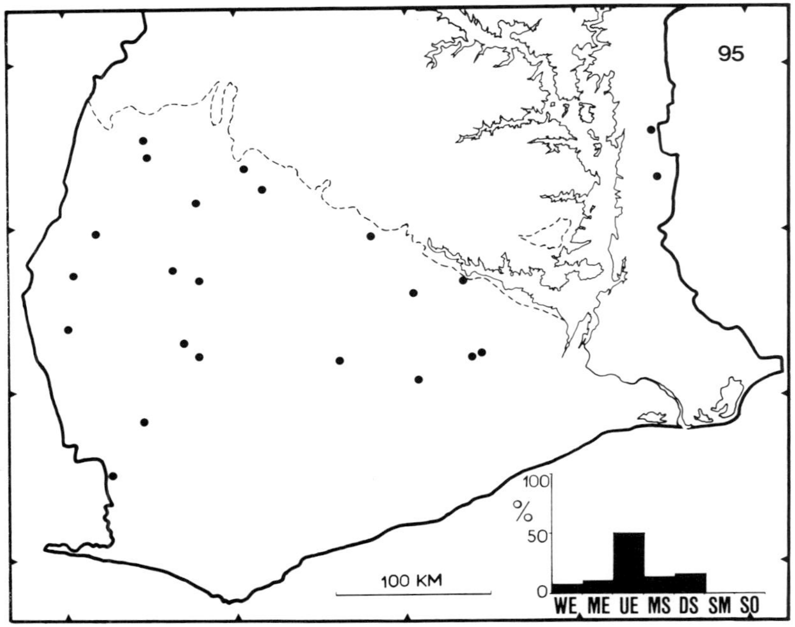

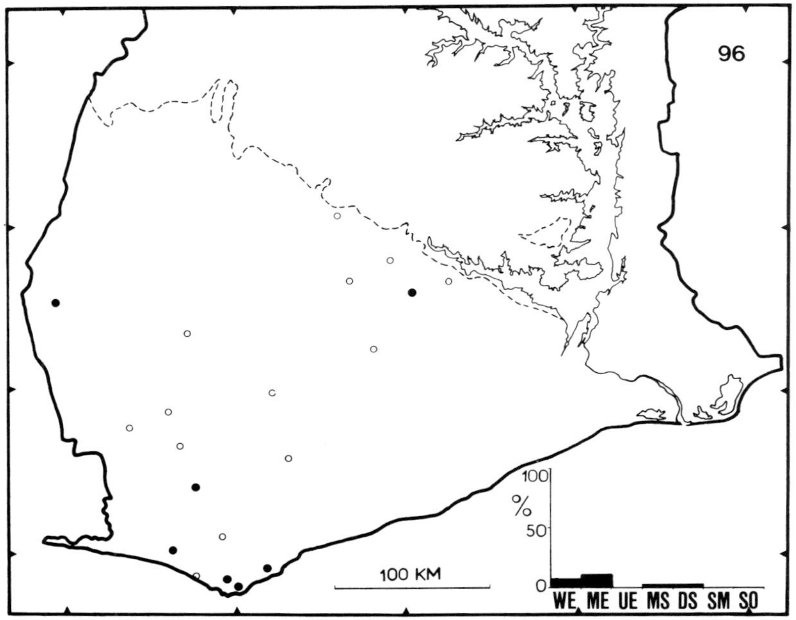

91 CALPOCALYX BREVIBRACTEATUS Harms, FWTA 1:488. Mimosaceae. Akan: atrotre.

Medium tree, shortly buttressed. Lvs bipinnate, leaflets notophyll, sometimes briefly deciduous. Fl small, brownish-orange, brush, clustered in spikes, 9–11. Fr woody explosive pod; seeds large, 12–2. Regenerating freely under forest canopy. Survey records very significantly correlated with base-poor soils and low pH. S. Leone to Nigeria; (Guinea wide).

92 CALYCOBOLUS AFRICANUS (G. Don) Heine, FWTA 2:338. Convolvulaceae. Akan: mutuo.

Large woody climber, xylem dissected into separate strands which are star-shaped in section. Lvs mesophyll. Roux's model. Fl white, medium, funnel, 1–3. Fr wind-dispersed with aid of two greatly-enlarged sepals, 1–4; seeds medium. Profuse regeneration under forest shade. S. Leone to Zaire; (Guineo-Congolian wide).

93 CALYCOBOLUS HEUDELOTII (Bak. ex. Oliv.) Heine, FWTA 2:337. Convolvulaceae.

Large woody climber, xylem as in *C. africanus;* lvs notophyll. Roux's model. Fl white, medium to large, funnel, 12–4, flowering abundantly and gregariously. Fr (Fig. 6.1i) as in *C. africanus;* seeds medium, 1–4. Regenerates profusely under forest shade. Guinea-Bissau to Cameroun; (Guinea wide).

94 CAMPYLOSTEMON ANGOLENSE Oliv., FWTA 1:626. Celastraceae.

Large woody climber, xylem shallowly grooved on periphery; gutta threads present. Lvs notophyll, toothed. Roux's model. Fl yellowish, small, dish, not clearly seasonal. Fr capsule; seeds medium, winged. Regenerates under forest canopy. S. Leone to Angola and Tanzania; (Guineo-Congolian wide).

95 CAMPYLOSTEMON MUCRONATUM (Exell) J. B. Hall (50), (*Hippocratea mucronata* Exell, FWTA 1:629). Celastraceae.

Large woody climber; xylem shallowly grooved on periphery; gutta threads present. Lvs notophyll, toothed. Roux's model. Fl yellowish, small, dish, 4–6. Fr capsule; seeds medium, winged. Regenerates under forest canopy. Ivory Coast to Zaire and Uganda; (Guineo-Congolian wide).

96 CANARIUM SCHWEINFURTHII Engl., FWTA 1:697. Burseraceae. Akan: bediwonua.

Large tree with deeply grooved bark and fragrant slash, exuding clear resin. Lvs pinnate, leaflets mesophyll, hairy when young, briefly deciduous. Rauh's model (59). Fl creamy-white, medium, bowl, 3–5. Fr purple drupe, 6–9; pyrene containing three very large seeds, formerly used as a source of edible oil (56). Regenerates only in gaps and along roadsides. Senegal to Angola, Ethiopia and Tanzania; (Guineo-Congolian wide). Class 3 timber; trade name: canarium (40).

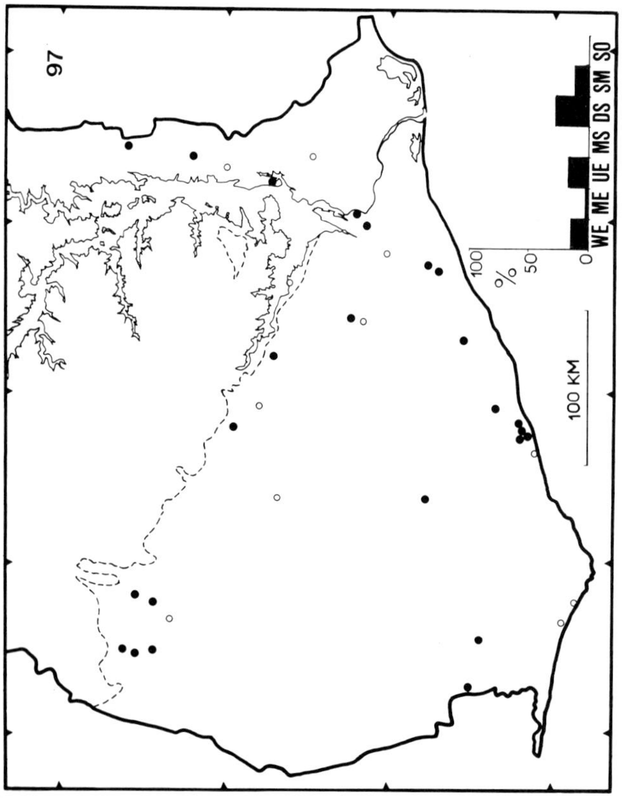
97
100 KM
%
100
50
0
WE ME UE MS DS SM SO

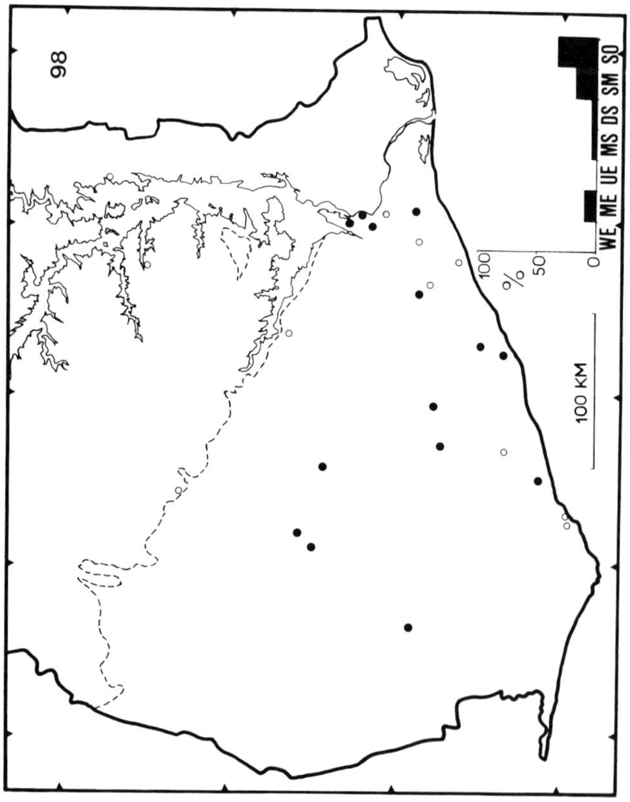
98
100 KM
%
100
50
0
WE ME UE MS DS SM SO

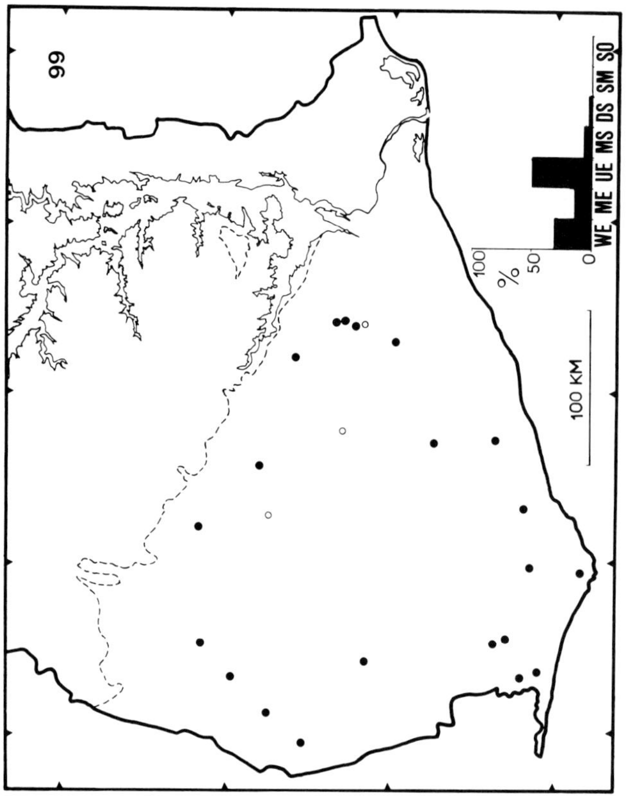
99
100 KM
%
100
50
0
WE ME UE MS DS SM SO

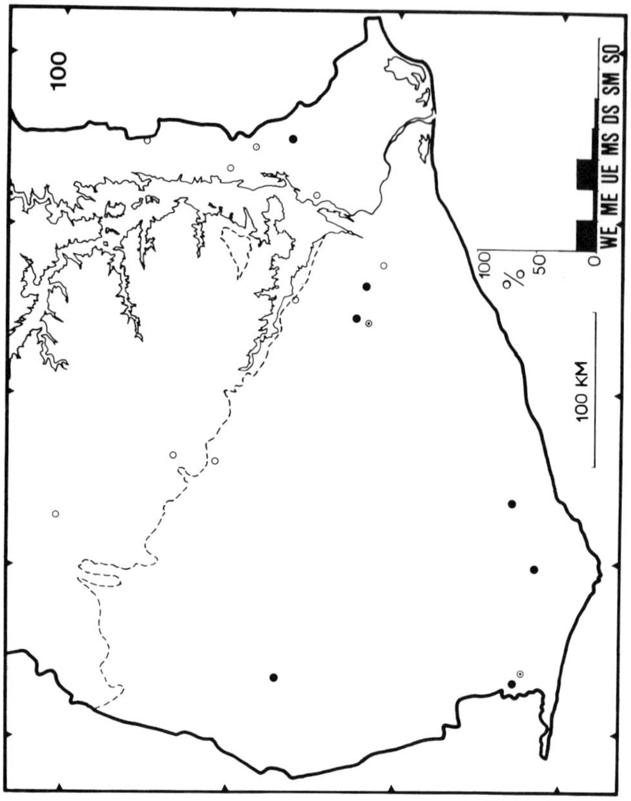
100
100 KM
%
100
50
0
WE ME UE MS DS SM SO

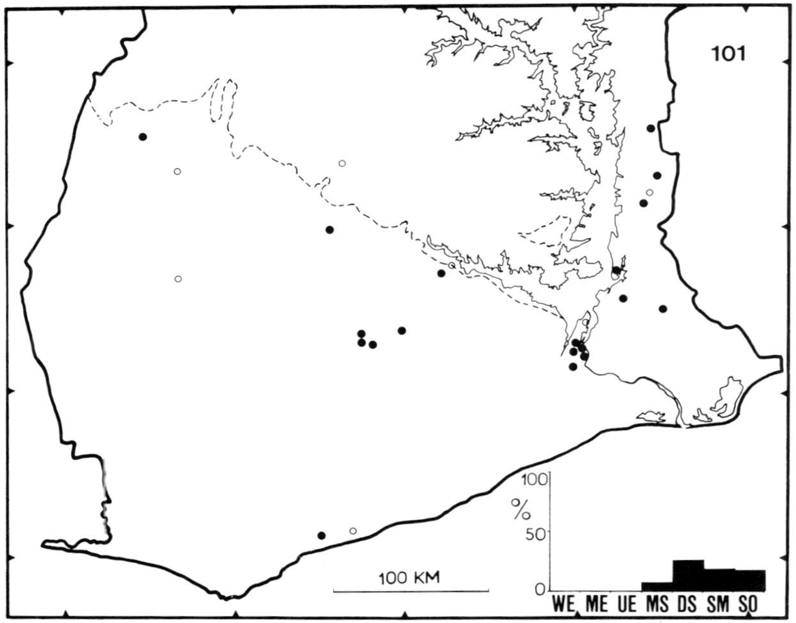

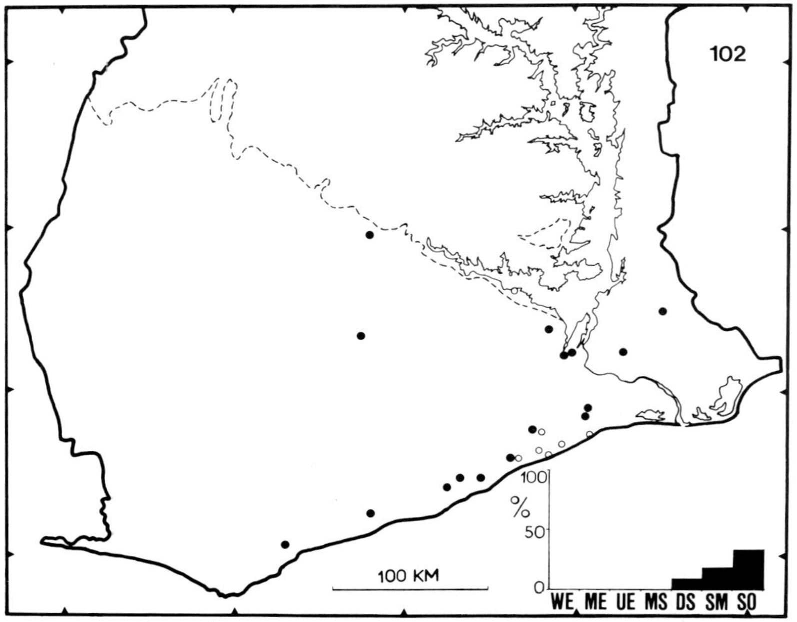

97 CANTHIUM HISPIDUM Benth., FWTA 2:182. Rubiaceae.
Small woody climber; lvs microphyll, hairy; recurved branchlets assist climbing. Cook's model. Fl white, small, bell, not distinctly seasonal. Fr berry, seeds medium. Occurs as a shrub in thickets and secondary forest as well as under forest canopy. S. Leone to Gabon; (Guinea wide).

98 CANTHIUM HORIZONTALE (Schumach.) Hiern, FWTA 2:182. Rubiaceae. (Type probably from near Accra).
Small woody climber with recurved branchlets; lvs microphyll in dry forest to notophyll in moist forest. Cook's model. Fl greenish-white, medium, bell, not distinctly seasonal. Fr black berry, seeds medium. Occurs also as a shrub in thicket; found in fringing forest in the savanna zone. Senegal to Nigeria; (Guinea wide).

99 CANTHIUM SETOSUM Hiern. FWTA 2:182. Rubiaceae.
Medium to large woody climber with recurved branchlets, lvs notophyll, hairy. Cook's model. Fl white, small, bell, not distinctly seasonal. Fr black berry, seeds medium. Regenerates under forest canopy, but usually found in small gaps. S. Leone to gabon; (Guinea wide).

100 CANTHIUM VENOSUM (Oliv.) Hiern. FWTA 2:184. Rubiaceae.
Woody climber with recurved branchlets, pubescent. Cook's model. Two rather distinct forms are represented in Ghana: (i) small to medium climber or shrub, lvs notophyll, tertiary veins slightly prominent below, occurring in forest near the savanna boundary, and in fringing forest within the savanna zone: herbarium records mapped ○; (ii) large climber, lvs microphyll, tertiary veins impressed below, occurring in evergreen forest: herbarium records mapped ⊙. Fl whitish, small, bell, not distinctly seasonal. Fr black berry, seeds medium. Senegal to Angola and Zambia; (ecological and chorological transgressor).

101 CANTHIUM VULGARE (K. Schum.) Bullock, FWTA 2:184. Rubiaceae. Akan: ogyapamnini.
Small tree, branching low (Enti FH 6852, a large tree, cited in FWTA as *C. vulgare*, is actually *C. arnoldianum* (De Wild. & Th. Dur.) Hepper). Lvs notophyll, pubescent in saplings. Roux's model. Fl cream, fragrant, small, bell, 4–6. Fr black berry, seeds medium. Regenerates in shade. Widespread in Tropical Africa; (chorological transgressor).

102 CAPPARIS ERYTHROCARPOS Isert, FWTA 1:89. Capparaceae. (Type from near Accra).
Medium-sized climber, with sharp recurved stipular prickles; lvs microphyll. Roux's model. Fl white, large, bowl, not clearly seasonal. Fr scarlet berry; seeds medium, surrounded by sweet juicy flesh. Regenerates on the forest floor; abundant as a scrambling shrub in thicket clumps on the Accra Plains. Guinée to Angola and Uganda; (Guineo-Congolian wide).

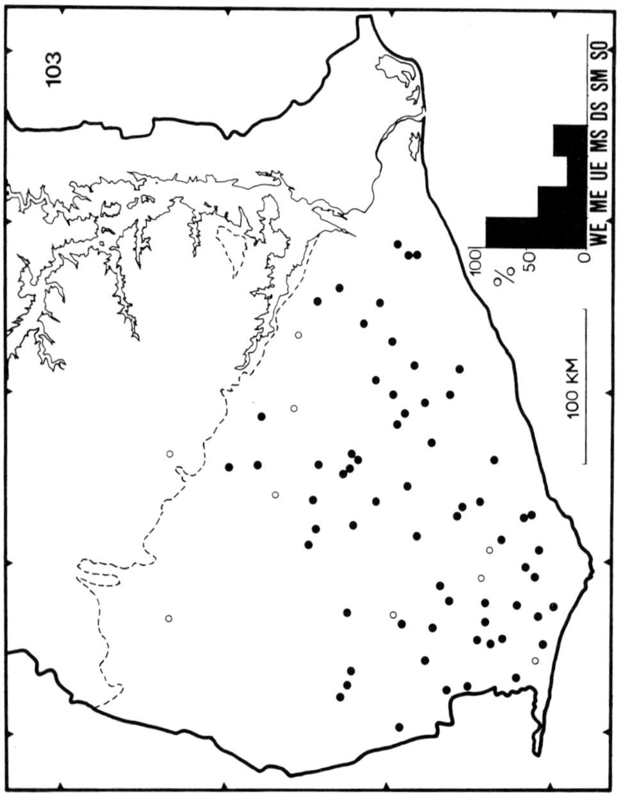
103
100 KM
100
50
0
%
WE ME UE MS DS SM SO

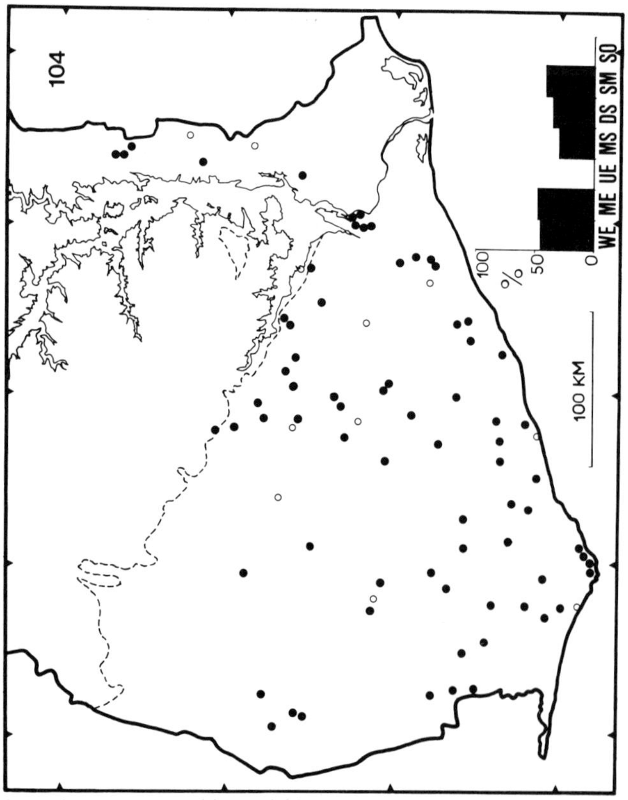
104
100 KM
100
50
0
%
WE ME UE MS DS SM SO

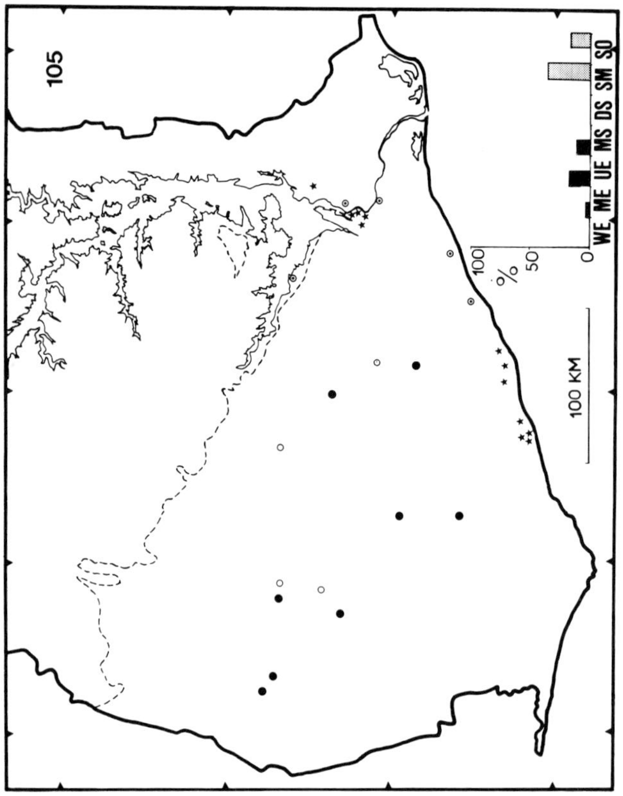
105
100 KM
100
50
0
%
WE ME UE MS DS SM SO

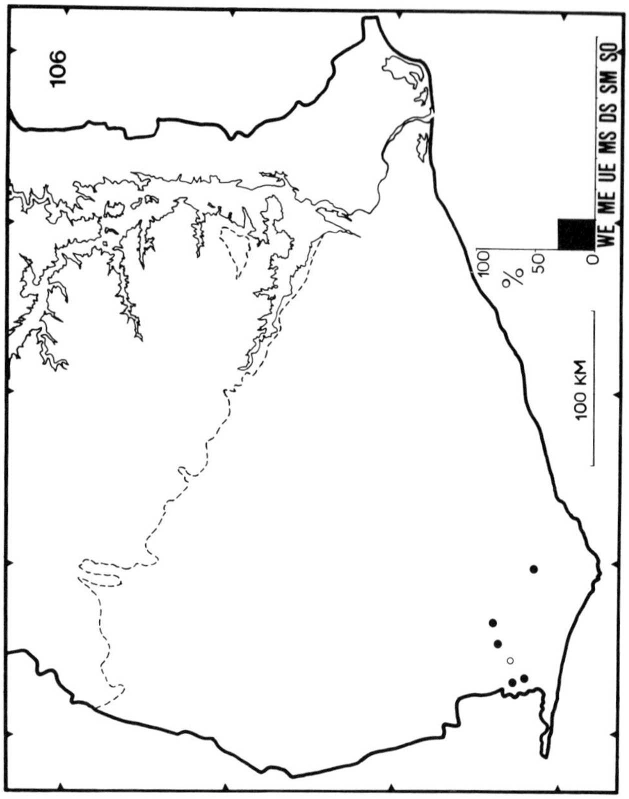
106
100 KM
100
50
0
%
WE ME UE MS DS SM SO

103 CARAPA PROCERA DC., FWTA 1:702. Meliaceae. Akan: kwakuobese.
Medium-sized tree; lvs pinnate, leaflets mesophyll, brilliant red when newly flushed. Scarrone's model. Fl pinkish-white, small, bowl, monoecious, not clearly seasonal. Fr woody capsule, seeds very large. Regenerates in shade. Tolerant of swampy conditions, but not at all confined to swamps. Extends into fringing forest in the southern part of the savanna zone. Senegal to Angola; also in Tropical America; (chorological transgressor).

104 CARPOLOBIA LUTEA G. Don, FWTA 1:109. Polygalaceae. (32). Akan: ofewa.
Small tree or shrub, lvs notophyll. Troll's model (59). Fl white, medium, flag, not clearly seasonal. Fr orange berry, seeds medium. Regenerates in forest shade. Guinée to Nigeria; (Guinea wide).

105 CASSIPOUREA CONGOENSIS R. Br. ex DC., FWTA 1:283. (Incl. *C. sp. B* of FWTA 1:283). Rhizophoraceae. Akan: kokotenua.
Small trees. Roux's model. Two distinct forms occur in Ghana:
(1) Leaves obtuse to subacuminate, subentire or bluntly toothed, hairs on leaf underside restricted to midrib, dry forest – survey records ★; herbarium records ⊙; histogram bars dotted.
(2) Leaves acuminate, larger than (1), sharply toothed, hairy on underside of leaf lamina, moist forest – survey records ●; herbarium records ○; histogram bars solid black; (*Cassipourea sp. B,* FWTA 1:283).
Lvs notophyll (1) or mesophyll (2). Fl brown-orange, small, bowl, 3–6. Fr capsule, seeds medium, with orange-red arils. Form (1) occurs in fringing forest in the savanna zone. Senegal to Zaire and Malawi; (chorological transgressor).

106 CASSIPOUREA HIOTOU Aubrév. & Pellegr. (8). Rhizophoraceae.
Medium-sized tree; lvs pubescent, mesophyll. Roux's model. Fl greenish-white, medium, bowl, Fr not known. Recorded only from Wet Evergreen forest. Iv. Coast and Ghana; (Upper Guinea endemic).

107 CASTANOLA PARADOXA (Gilg) Schellenb. ex Hutch. & Dalz., FWTA 1:746. Connaraceae.
Large woody climber; lvs trifoliolate, leaflets notophyll. Fl white, fragrant, small, bowl, in dense panicles, not clearly seasonal. Fr red velvety follicle; seed large, black, with yellow aril. S. Leone to Zaire; (Guineo-Congolian wide).

108 CEIBA PENTANDRA (Linn.) Gaertn. FWTA 1:335. Bombacaceae. Akan: onyina.
Large tree with wide-spreading crown, tall buttresses (Fig. 5.3), large thorns on trunk and branches. Lvs digitate, leaflets notophyll, deciduous. Massart's model (59). Fl creamy-white, large, bowl, at dusk liberating pungent smell and abundant nectar, 11–2 while leafless; pollinated by bats (16). Fr capsule; seeds medium, wind-dispersed in cottony fluff, 1–3. Regenerates only in gaps and along roadsides. Timber not classified, but now being cut (40). Throughout Tropical Africa, and also in Trop. America (together with other members of the genus) and Tropical Asia; (chorological transgressor); widely planted for kapok.

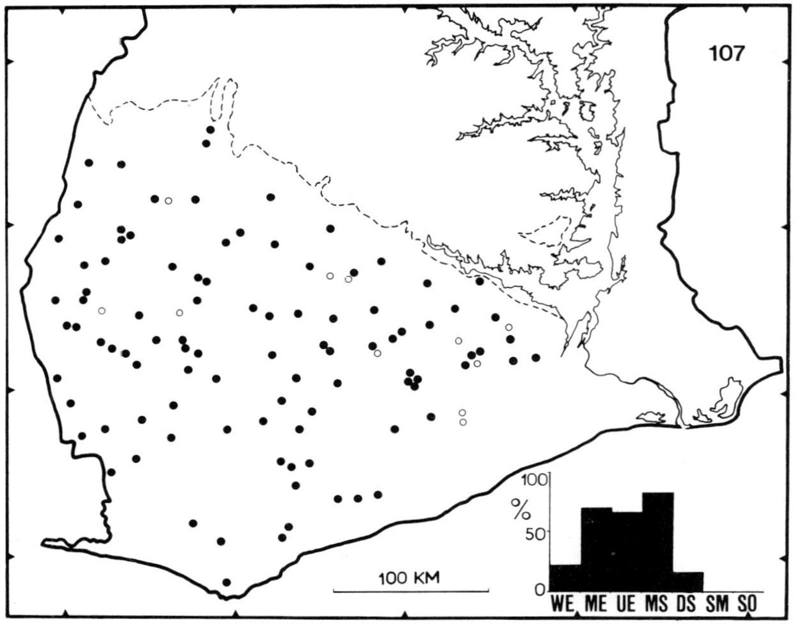

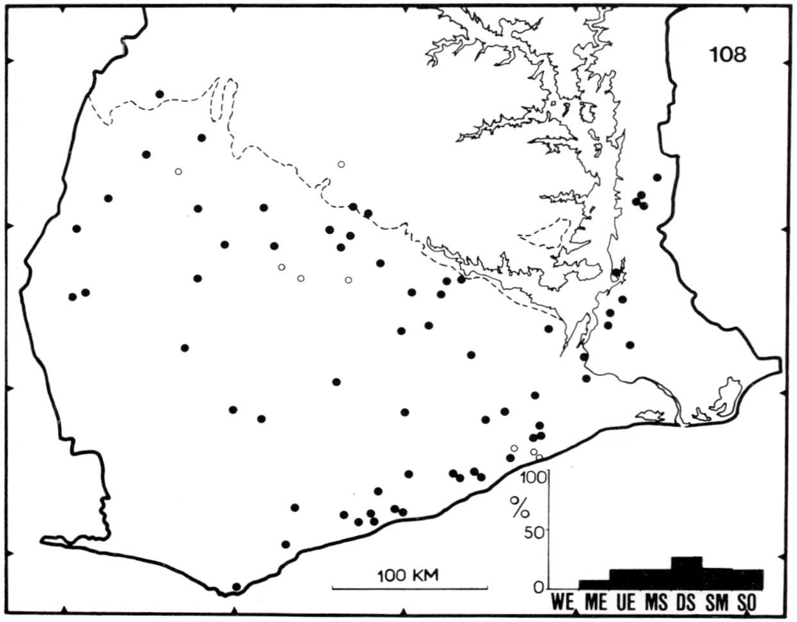

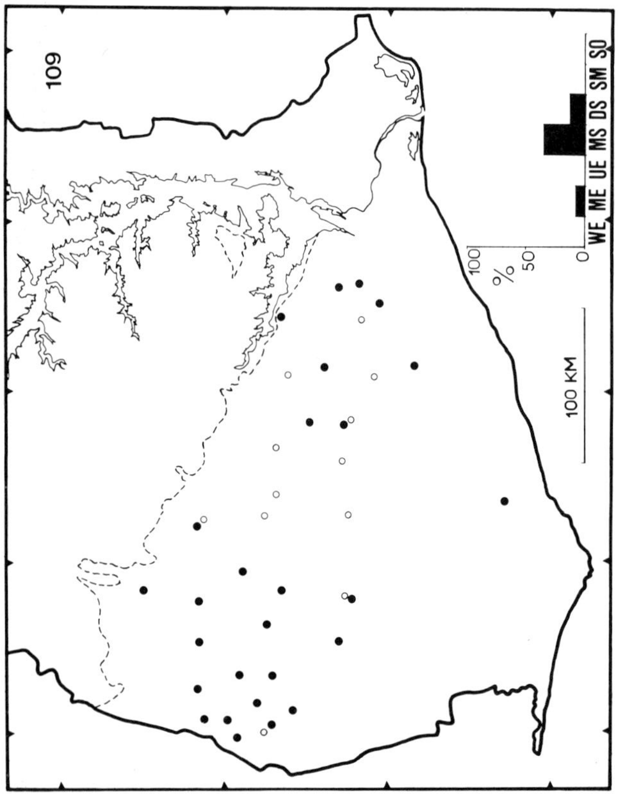
109
100 KM
%
100
50
0
WE ME UE MS DS SM SO

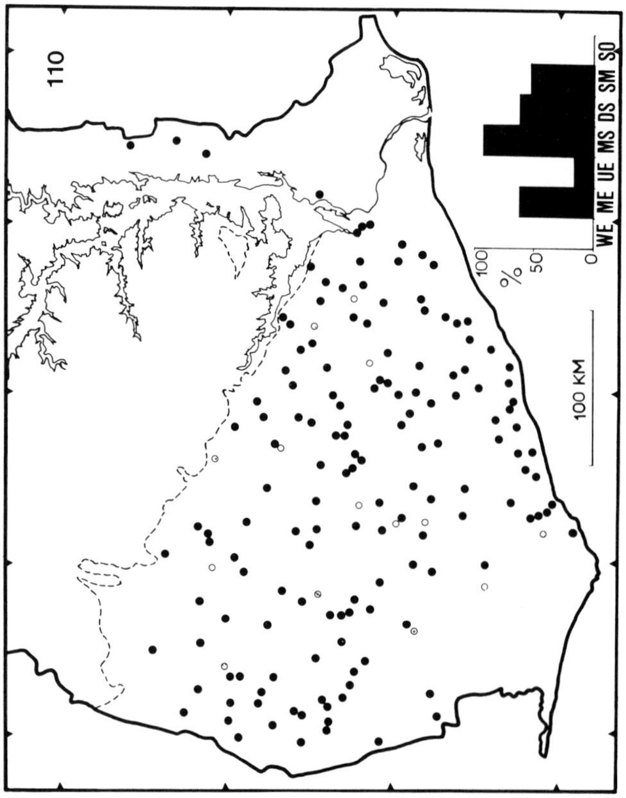
110
100 KM
%
100
50
0
WE ME UE MS DS SM SO

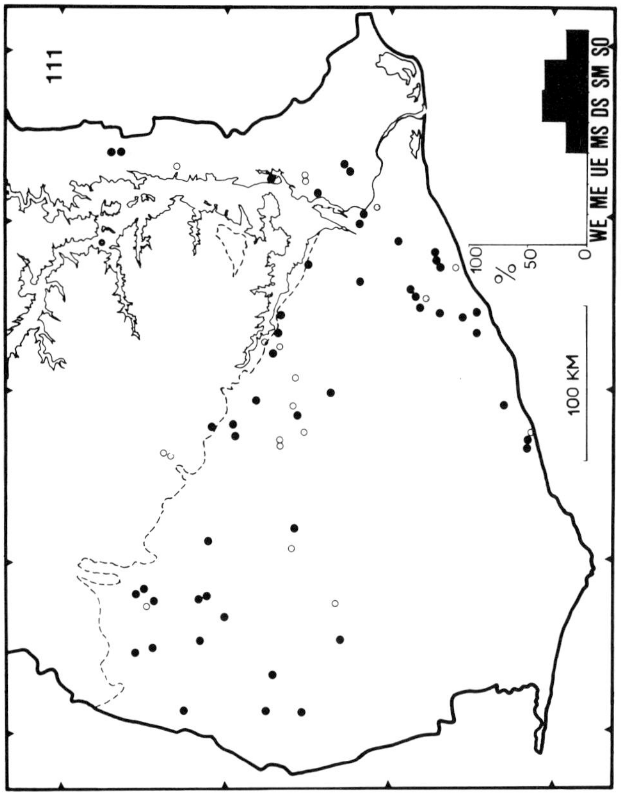
111
100 KM
%
100
50
0
WE ME UE MS DS SM SO

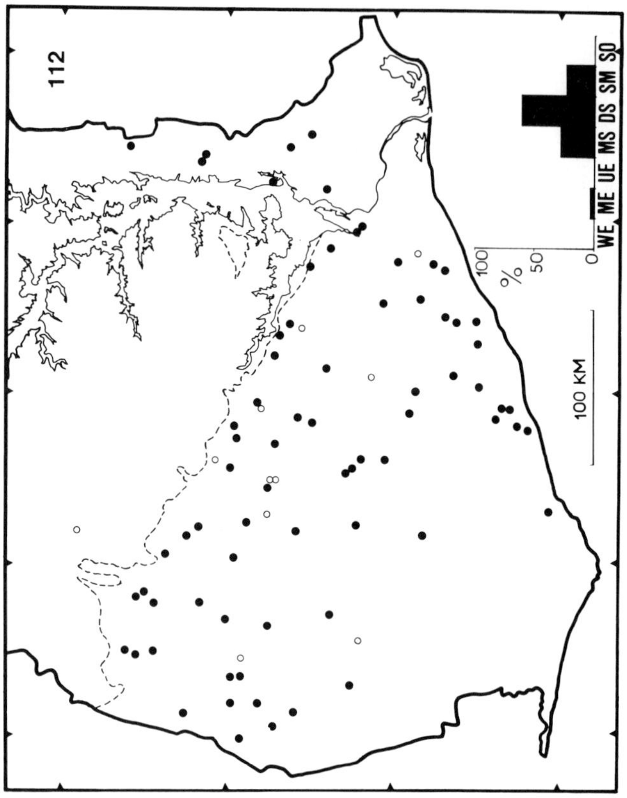
112
100 KM
%
100
50
0
WE ME UE MS DS SM SO

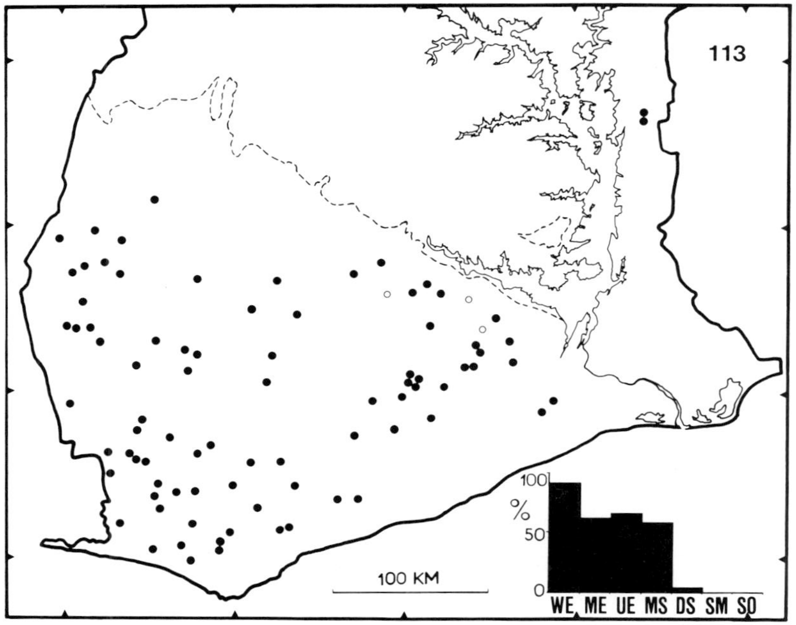

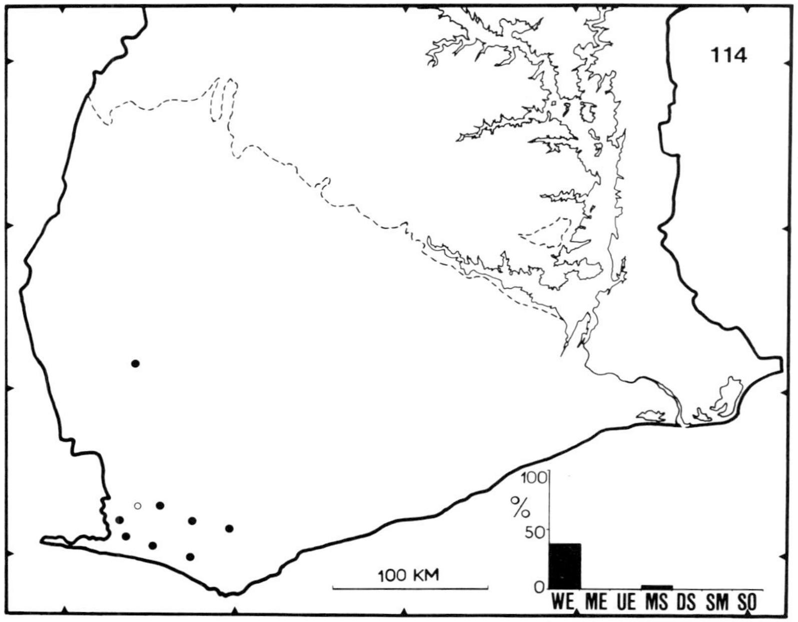

109 CELTIS ADOLFI-FRIDERICI Engl., FWTA 1:592. Ulmaceae. Akan: esakosua.
Large tree with buttresses, lvs notophyll. Fl white, small, polygamous, in large inflorescences, not clearly seasonal. Fr drupe, seed large. Regenerates in forest shade. Iv. Coast to Zaire and Uganda; (Guineo-Congolian wide). Class 3 timber.

110 CELTIS MILDBRAEDII Engl., FWTA 1:592. Ulmaceae. Akan: esa.
Large tree with buttresses; lvs microphyll in canopy, larger in understorey (Fig. 5.8a). Troll's model. Fl greenish, small, polygamous, inconspicuous, not clearly seasonal. Fr red drupe; seed medium. Regenerates freely under forest shade. Iv. Coast to Angola and Tanzania; (Guineo-Congolian wide). Class 3 timber (40).

111 CELTIS WIGHTII Planch. (91), (*C. brownii* Rendle, FWTA 1:592). Ulmaceae. Akan: esafufuo.
Medium tree, branching low; lvs notophyll, toothed. Troll's model. Fl greenish, small, polygamous, inconspicuous, not clearly seasonal. Fr red drupe; seed medium. Regenerates in shade. Survey records show very significant negative correlation with acid soil; 80% of records were from soil with pH > 6. Widespread in Tropical Africa; if included in *C. phillipensis* Blanco (78) then also in Tropical Asia; (chorological transgressor).

112 CELTIS ZENKERI Engl., FWTA 1:592. Ulmaceae. Akan: esakoko.
Large tree with buttresses; lvs notophyll, sometimes briefly deciduous. Troll's model (59). Fl green, small, polygamous, inconspicuous, not clearly seasonal. Fr red drupe; seed medium. Survey records show very significant correlations with the following soil characteristics: low clay content, high base content, pH > 6. Guinée to Angola and Tanzania; (Guineo-Congolian wide). Class 3 timber.

113 CERCESTIS AFZELII Schott, FWTA 3:126. Araceae.
Root climber on the trunks of understorey trees; lvs 3-lobed, mesophyll. Fl white, small, monoecious, crowded in spike, not clearly seasonal. Fr red drupe; seed medium. Senegal to Nigeria; (Guinea wide).

114 CERCESTIS IVORENSIS A. Chev., FWTA 3:127. Araceae.
Herb of forest floor, mesophyll. Fl small, monoecious, crowded in spike. Ripe fr not known. Liberia to Ghana; (Upper Guinea endemic).

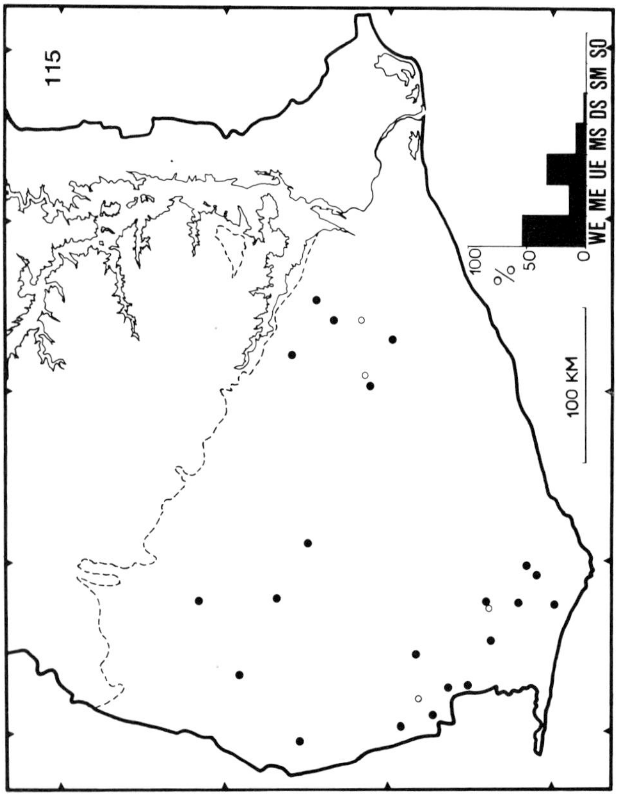
115
100 KM
%
100
50
0
WE ME UE MS DS SM SO

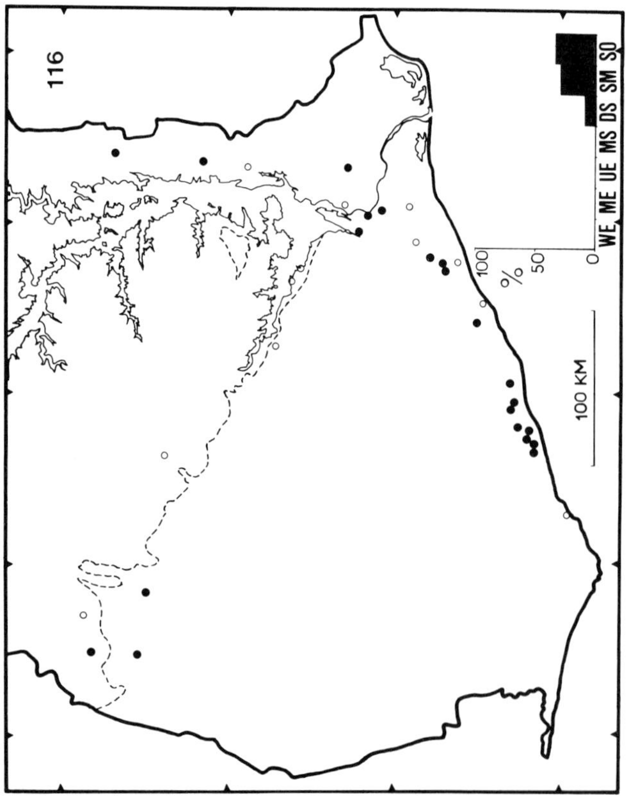
116
100 KM
%
100
50
0
WE ME UE MS DS SM SO

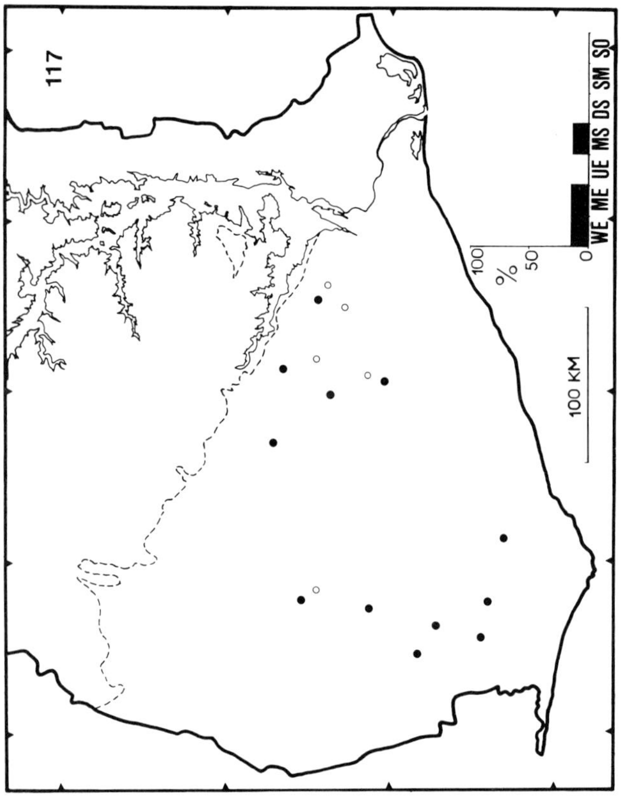
117
100 KM
%
100
50
0
WE ME UE MS DS SM SO

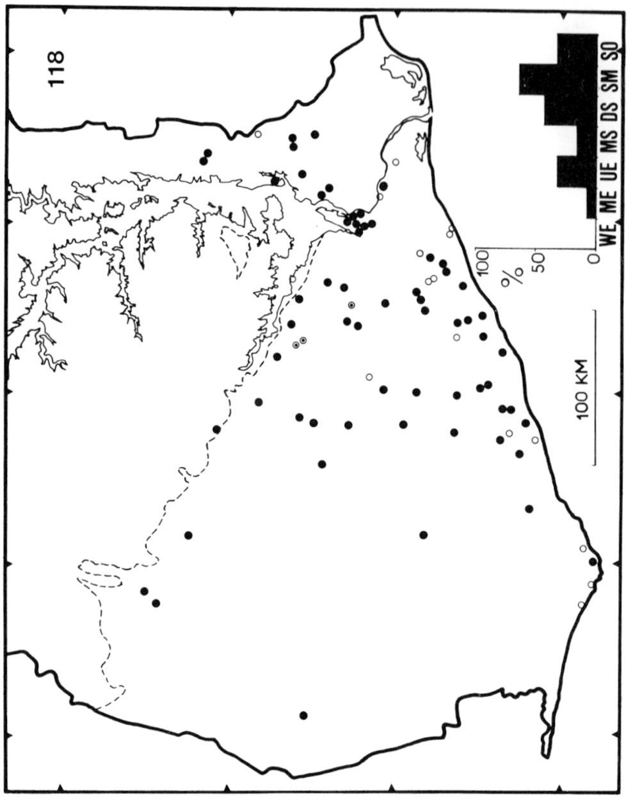
118
100 KM
%
100
50
0
WE ME UE MS DS SM SO

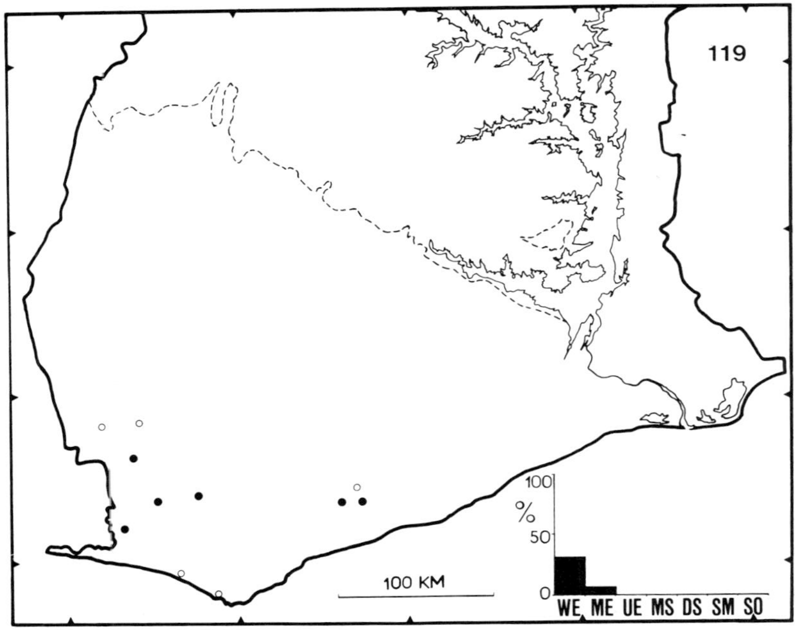

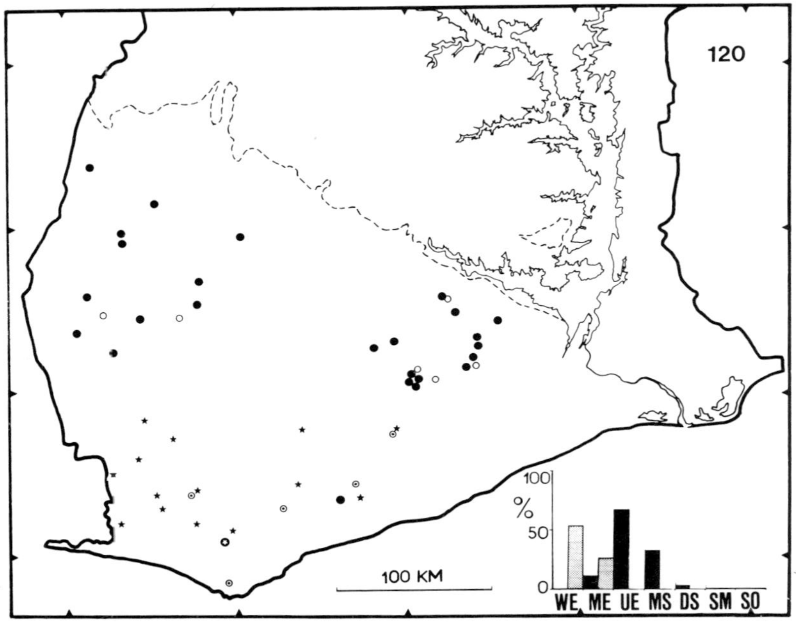

115 CERCESTIS STIGMATICUS N. E. Br., FWTA 3:126. Araceae.
Stoloniferous root climber on tree trunks in forest understorey; lvs mesophyll. Fl cream, small, monoecious, crowded in spike, not clearly seasonal. Fr red drupe; seed medium. Guinée to Nigeria; (Guinea wide).

116 CHAETACME ARISTATA Planch., FWTA 1:593. Ulmaceae. Akan: esonotokwakofuo.
Small tree or shrub, with low arching branches, spiny, lvs toothed, scabrid, microphyll. Troll's model. Fl small, yellowish, monoecious or dioecious, male flowers clustered, female subsolitary, inconspicuous, not clearly seasonal. Fr buff-coloured drupe; seed medium. Regenerates under shade. Iv. Coast to southern and eastern Africa and Madagascar; (chorological transgressor).

117 CHASSALIA CORALLIFERA (A. Chev. ex De Wild.) Hepper, FWTA 2:192. Rubiaceae.
Pygmy tree 30 cm high, lvs notophyll. Chamberlain's model. Fl cream, small, bowl with tube, not clearly seasonal. Fr black berry, on coral pink fleshy pedicel; seeds medium. Grows in dense shade. S. Leone to Ghana; (Upper Guinea endemic).

118 CHASSALIA KOLLY (Schumach.) Hepper, FWTA 2:192. Rubiaceae. (Type from near Accra).
Straggling pygmy tree, lvs notophyll. Leeuwenberg's model. Fl whitish or purplish, heterostyled (14), medium, dish with tube, not seasonal; inflorescence axes white or purple. Fr purplish berry; seeds medium. Variable species: herbarium records of typical form with glabrous inflorescence shown ○; form with pubescent inflorescence and rather thinner subtruncate stipules shown ⊙. Guinée to Cameroun; (Guinea wide).

119 CHASSALIA SUBHERBACEA (Hiern) Hepper, FWTA 2:192. Rubiaceae.
Pygmy tree less than 30 cm high, lvs notophyll. Chamberlain's model. Fl white, medium, bowl with tube, not seasonal. Fr blue-black, on fleshy white pedicel; seed medium. Grows in deep shade. S. Leone to Cameroun; (Guinea wide).

120 CHAZALIELLA SCIADEPHORA (Hiern) Petit & Verdc. (111), (*Psychotria sciadephora* Hiern, FWTA 2:201). Rubiaceae.
Pygmy tree or shrub, lvs notophyll to mesophyll. Leeuwenberg's model. Fl white, small, bell; not clearly seasonal. Fr red berry; seeds medium. Regenerates in shade. Two fairly distinct forms are present in Ghana:

(1) Leaves and twigs glabrous, inflorescence axes glabrous or with a few short hairs, flowering inflorescence 3–8 cm diam., plant 1–2 m high: plot records ●, herbarium records ○, histogram bars solid black.

(2) Twigs, underside of midrib, petioles, inflorescence axes shortly but densely puberulous, flowering inflorescence 1.5–2 cm diam., plant usually less than 1 m high; plot records ★, herbarium records ⊙, histogram bars dotted, (⊛ for occurrence of both forms on the same plot). Note the differing distributions in Ghana. Guinée to Cameroun; (Guinea wide).

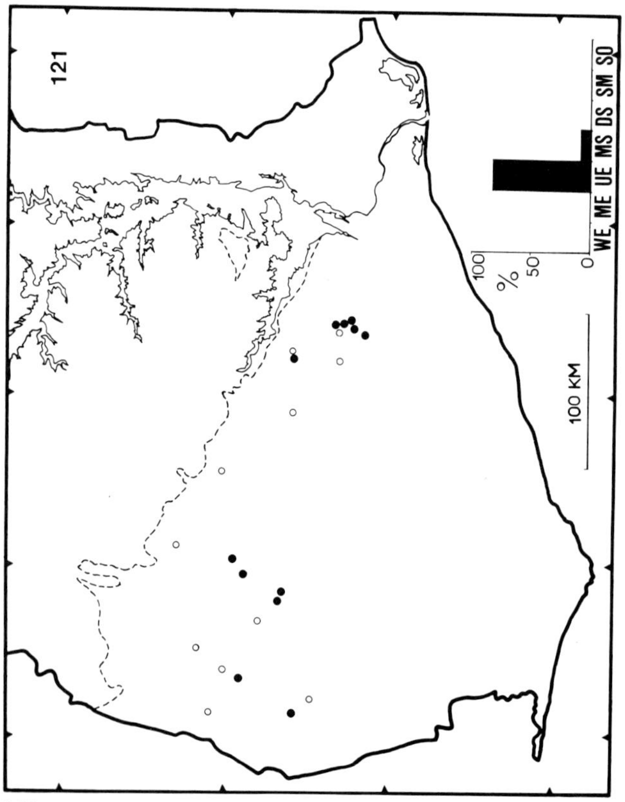
121
100 KM
%
100
50
0
WE ME UE MS DS SM SO

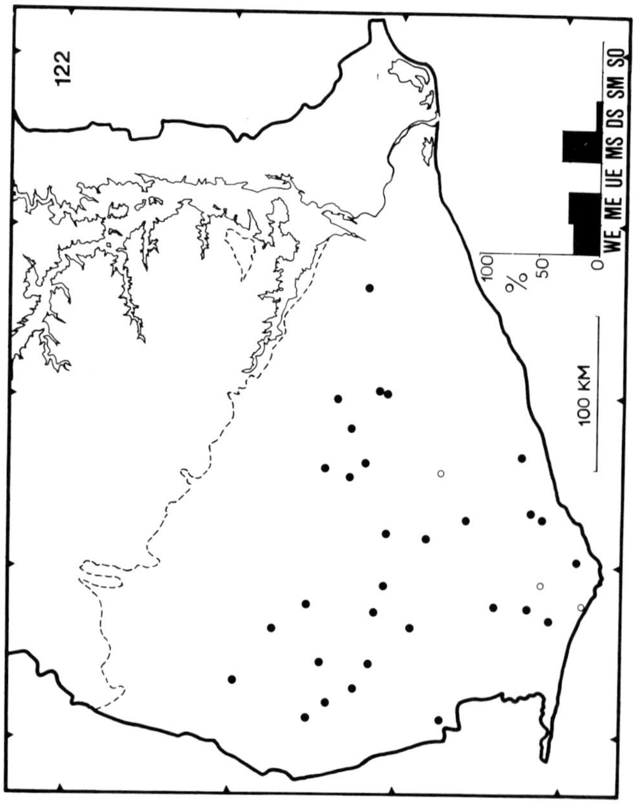
122
100 KM
%
100
50
0
WE ME UE MS DS SM SO

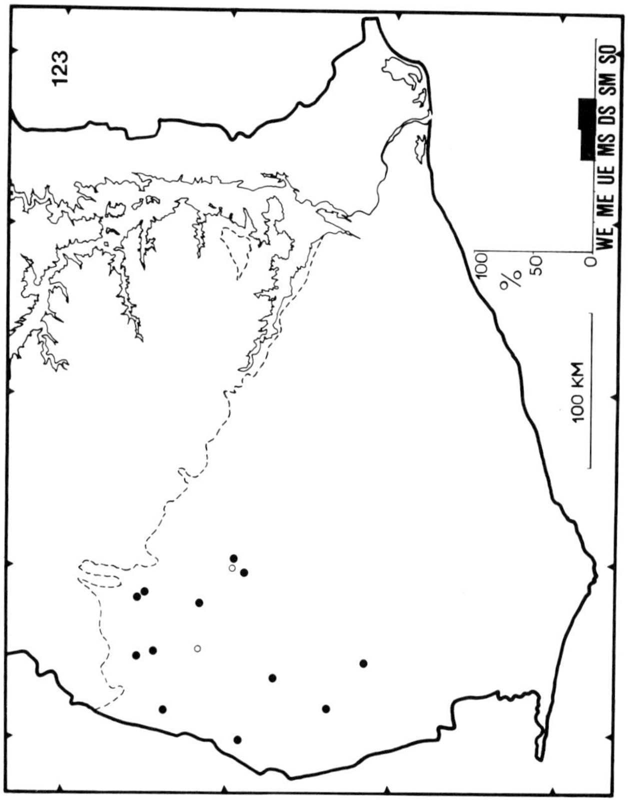
123
100 KM
%
100
50
0
WE ME UE MS DS SM SO

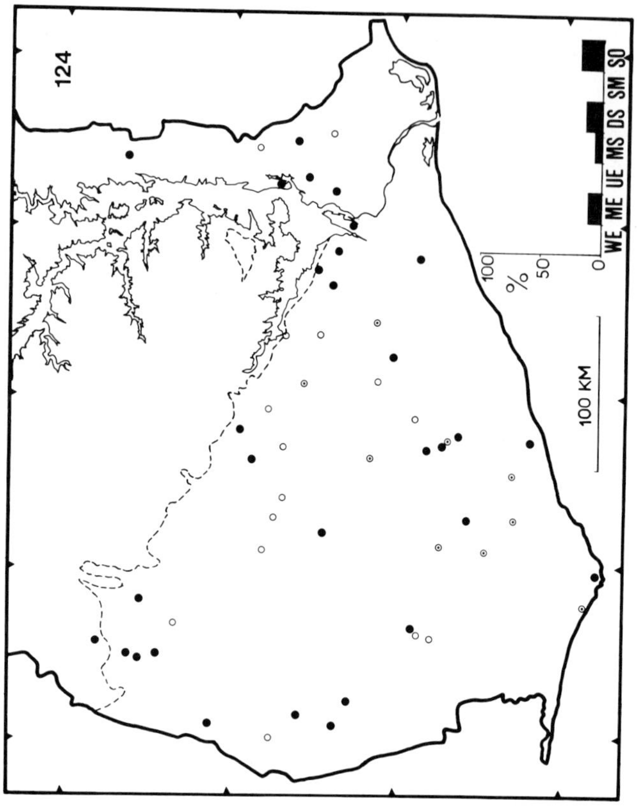
124
100 KM
%
100
50
0
WE ME UE MS DS SM SO

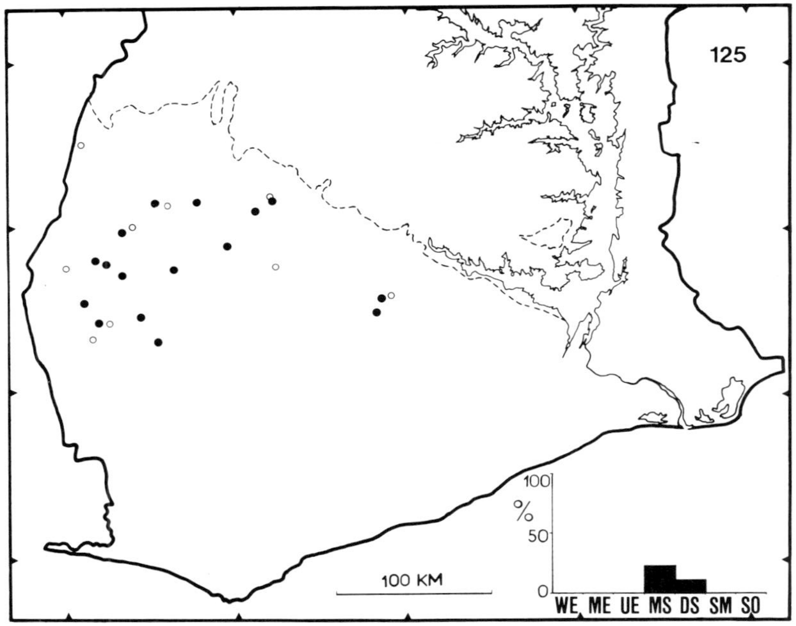

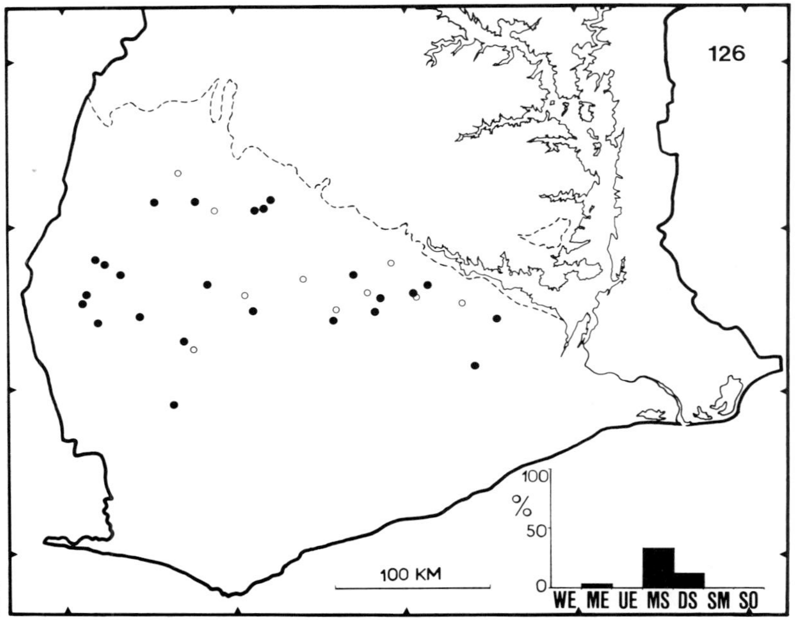

121 CHIDLOWIA SANGUINEA Hoyle, FWTA 1:456. Caesalpiniaceae. Akan: ababima. (Type from Ofin Headwaters F.R.).
Medium-sized tree, sometimes low-branching; lvs pinnate, leaflets notophyll. Troll's model. Fl crimson, medium, bowl, 1–3. Fr large explosive woody pod 3–12; seeds large. Regenerates in shade. S. Leone to Ghana, and Gabon; (Upper Guinea near endemic).

122 CHLAMYDOCARYA MACROCARPA A. Chev. ex Hutch. & Dalz., FWTA 1:642. Icacinaceae.
Slender, woody, twining climber, lvs mesophyll. Fl small, crowded in spikes, dioecious. Fr drupe, covered by greatly enlarged beak-like, bristly orange corolla, 12–2; seed large. Regenerates in forest shade. Iv. Coast and Ghana; (Upper Guinea endemic).

123 CHLAMYDOCARYA THOMSONIANA Baill., FWTA 1:643. Icacinaceae.
Slender, woody, twining climber; lvs hairy, mesophyll. Fl small, crowded in spikes, dioecious. Fr drupe, covered by greatly enlarged, beak-like, bristly, orange corolla, 3–6; seed medium. Regenerates in forest shade. S. Leone to Zaire; (Guineo-Congolian wide).

124 CHLOROPHORA EXCELSA (Welw.) Benth., FWTA 1:595, and CHLOROPHORA REGIA A. Chev., FWTA 1:595. Moraceae. Akan: odum.
Large trees; lvs mesophyll (*C. excelsa*), notophyll (*C. regia*); briefly deciduous; leaf underside minutely puberulous in *C. excelsa* only (illustrated in 113). Roux's model. Cream latex present. Fl small, dioecious, green (female) yellow (male), crowded in dense spikes, wind pollinated, 1–3. Infructescence pulpy, containing many small nuts. Regeneration only in gaps and along roadsides; mostly distorted or killed by the gall bug *Phytolyma* (121). Seedlings and saplings cannot always be identified with certainty, (though sapling leaves of *C. excelsa* are usually hairier), so the species were not distinguished in the survey; herbarium records marked ○ for *C. excelsa* and ⊙ for *C. regia*. Distribution: Iv. Coast to Angola and Mozambique for *C. excelsa;* (chorological transgressor); Senegal to Ghana (Upper Guinea endemic) for *C. regia*. Class 1 timber; trade name: 'iroko' (40). Distribution in Ghana as calculated from timber enumeration records shown in Fig. 11.1.

125 CHRYSOPHYLLUM GIGANTEUM A. Chev., FWTA 2:28. Sapotaceae. Akan: kumfana.
Medium-sized tree with buttresses; lvs mesophyll, densely covered on undersurface by persistent golden or silvery appressed hair. Roux's model. White latex present. Slash with strong, rather unpleasant smell. Fl small, white, inconspicuous, not clearly seasonal. Fr yellowish berry, seeds very large. Seedlings (26) under forest canopy. *C. giganteum* is often confused with *C. albidum,* but is distinguished by its racemose inflorescence, pellucid-dotted leaves, and greenish, scarcely edible fruit. In the survey *C. albidum* was found only in a few sacred groves near villages; it seems not to be native in Ghana, but to occur only as a tree grown for its popular orange-coloured edible fruits. Distribution of *C. giganteum:* S. Leone to Nigeria; (Guinea wide).

126 CHRYSOPHYLLUM PERPULCHRUM Mildbr. ex Hutch. & Dalz., FWTA 2:28. Sapotaceae. Akan: atabene.
Medium to large tree; lvs brown-hairy below, mesophyll. Roux's model. White latex present. Fl small, brownish, inconspicuous, not clearly seasonal. Fr red-brown berry, seeds large. Seedlings (26) under forest canopy. Liberia to Zaire and East Africa; (Guineo-Congolian wide).

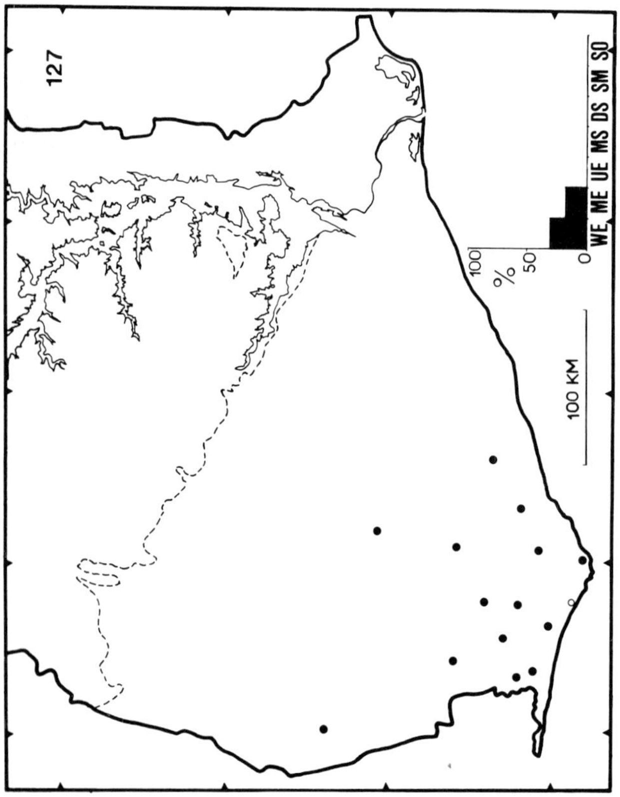
127
100 KM
100
%
50
0
WE ME UE MS DS SM SO

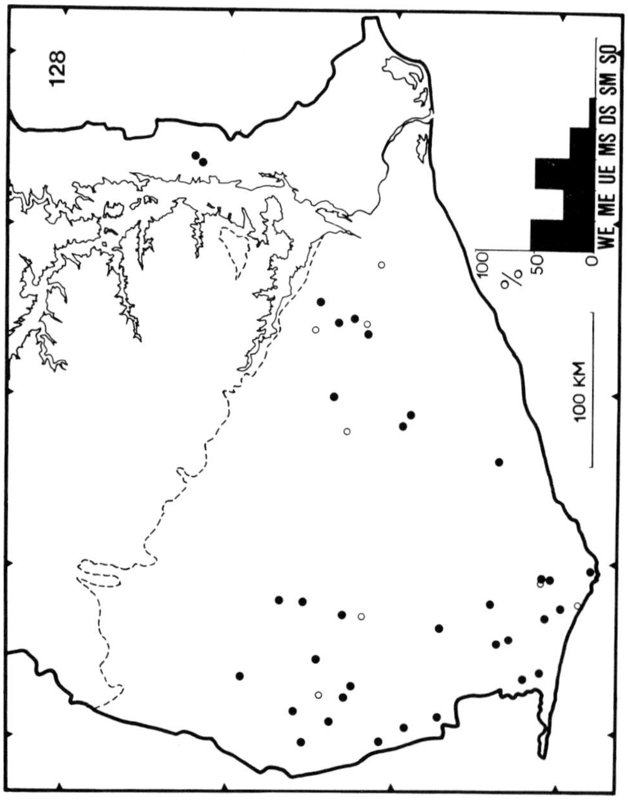
128
100 KM
100
%
50
0
WE ME UE MS DS SM SO

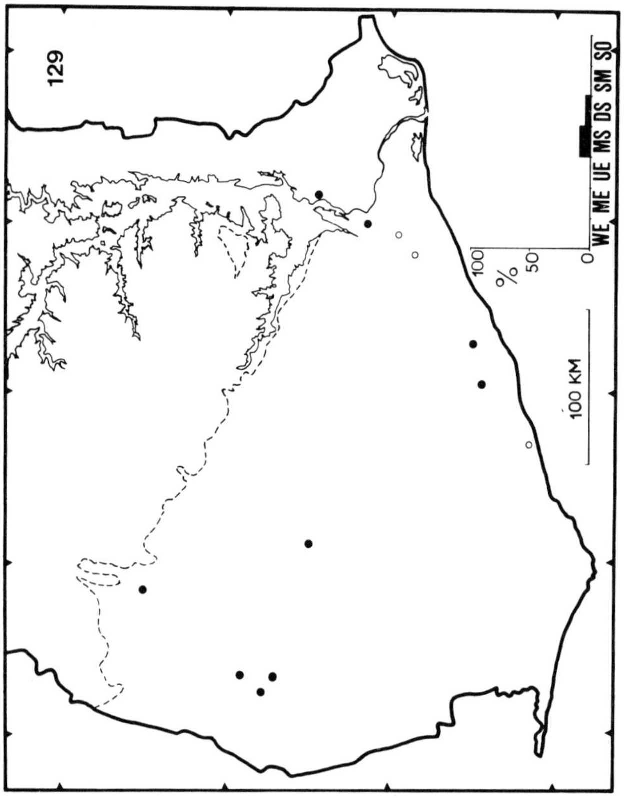
129
100 KM
100
%
50
0
WE ME UE MS DS SM SO

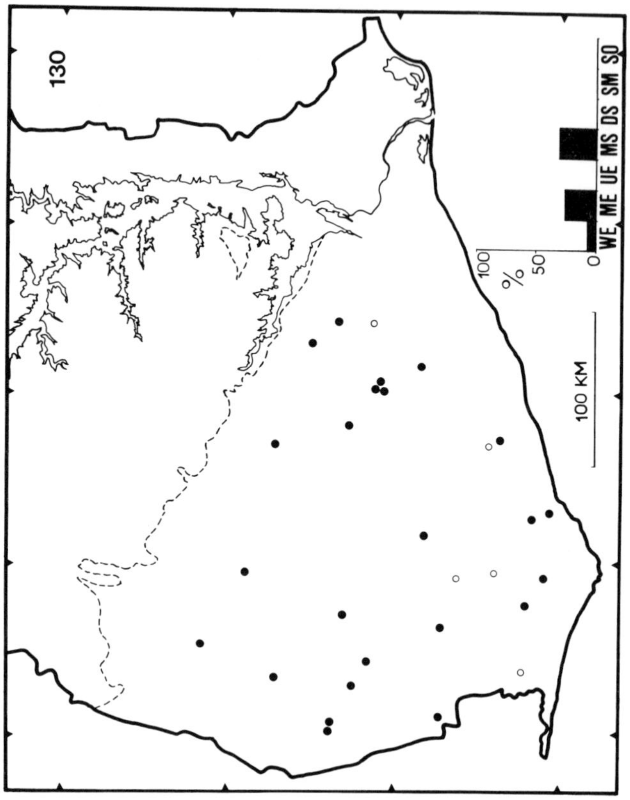
130
100 KM
100
%
50
0
WE ME UE MS DS SM SO

127 CHRYSOPHYLLUM PRUNIFORME Pierre ex Engl., FWTA 2:26. Sapotaceae. Akan: duatadwe.

Medium to large tree, lvs notophyll. Roux's model. White latex present. Fl small, greenish-yellow, inconspicuous, not clearly seasonal. Fr yellow berry; seeds very large. Seedlings (26) in forest shade. S. Leone to Zaire; (Guineo-Congolian wide).

128 CHRYSOPHYLLUM SUBNUDUM Bak., FWTA 2:27. Sapotaceae. Akan: kankabe.

Medium, somewhat buttressed tree; lvs notophyll, undersurface densely covered by appressed silvery hairs when young, later glabrescent. Roux's model (59). White latex present. Fl whitish, small, inconspicuous, not clearly seasonal. Fr reddish berry; seeds large. Seedlings (26) in shade. S. Leone to Zaire; (Guineo-Congolian wide).

129 CHRYSOPHYLLUM WELWITSCHII Engl., FWTA 2:27. Sapotaceae.
Slender woody climber, or shrub in thickets; lvs notophyll. Roux's model. White latex present. Fl whitish, small, inconspicuous, not clearly seasonal. Fr orange berry; seeds large. Seedlings (26) in shade. S. Leone to Angola; (Guineo-Congolian wide).

130 CHYTRANTHUS CARNEUS Radlk. (45), (*C. villiger* Radlk., FWTA 1:717). Sapindaceae. Akan: onibunanua.

Small unbranched tree; lvs pinnate, leaflets macrophyll. Corner's model. Fl whitish, medium, bowl, monoecious, cauliflorous from base of trunk, 9–12. Fr red-brown berry; seeds very large. Regenerates in forest shade. S. Leone to Angola; (Guineo-Congolian wide).

131 CHYTRANTHUS CAULIFLORUS (Hutch. & Dalz.) Wickens (45), (*Laccodiscus cauliflorus* Hutch. & Dalz. in FWTA 1:721). Sapindaceae. (Type from Amaneakrom near Enchi).

Unbranched small tree; lvs pinnate, leaflets mesophyll, very hairy. Fl medium, bowl, probably polygamous, cauliflorous, 7–11. Corner's model (59). Fr yellow berry, 12–1; seeds very large. Regenerates in shade. Plot records very highly correlated with acid, base-poor soils. Liberia to Ghana, and Gabon; (Upper Guinea near-endemic).

132 CHYTRANTHUS MACROBOTRYS (Gilg) Exell & Mendonça, FWTA 1:717. Sapindaceae. Akan: esuinia.

Small tree; lvs pinnate, leaflets mesophyll. Fl medium, whitish, bowl, monoecious, cauliflorous, 12–1. Fr orange berry, 2–3; seeds very large. Regenerates in shade. Ghana to Zaire; (Guineo-Congolian wide).

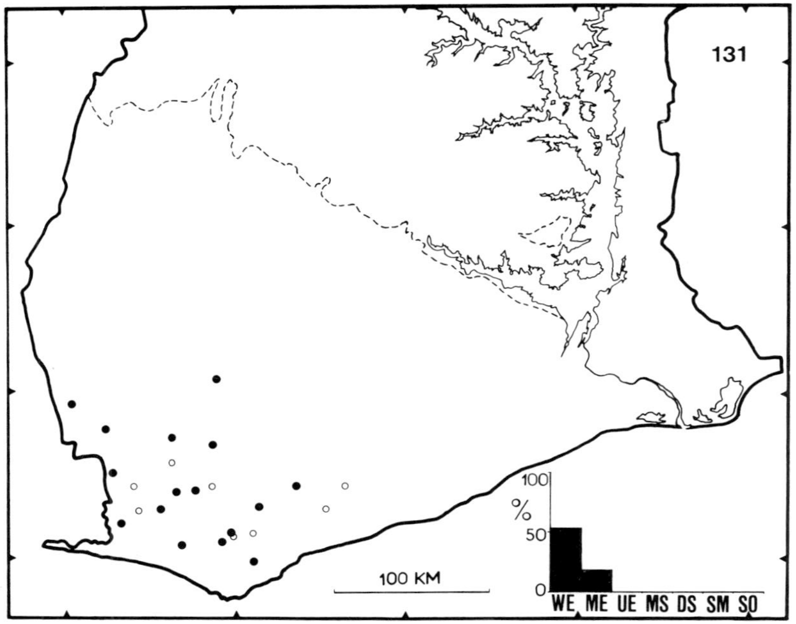

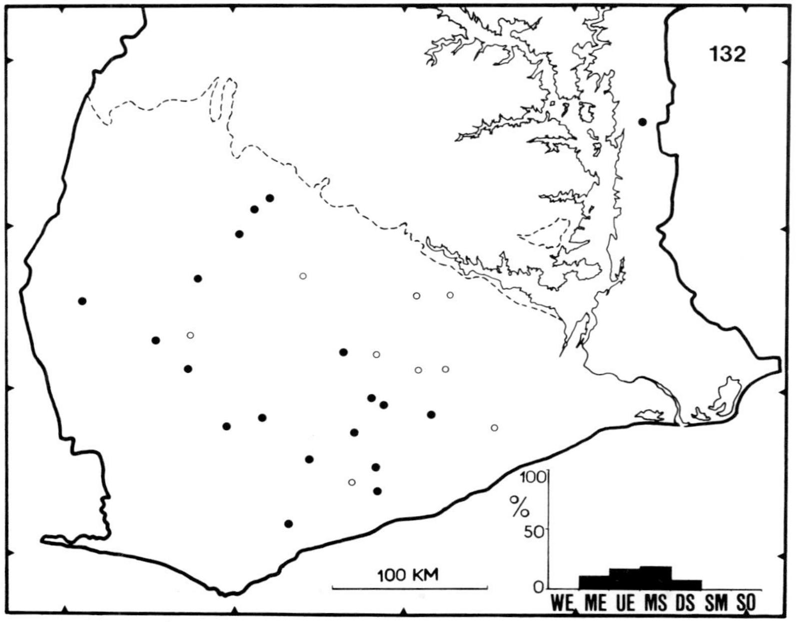

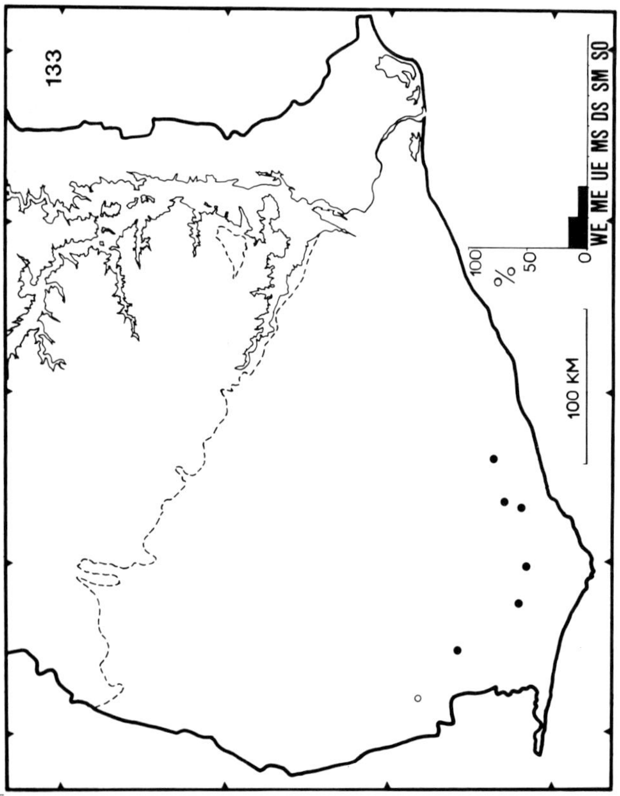
133
100 KM
%
100
50
0
WE ME UE MS DS SM SO

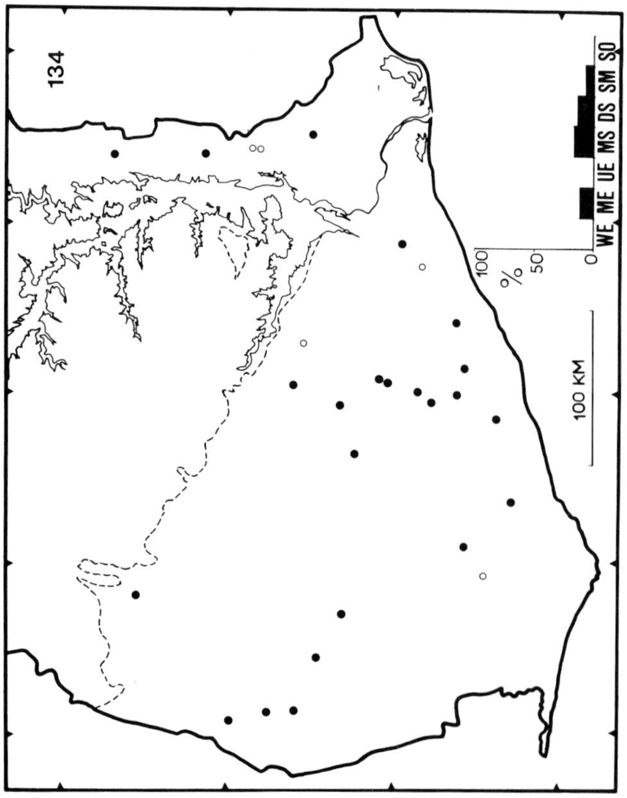
134
100 KM
%
100
50
0
WE ME UE MS DS SM SO

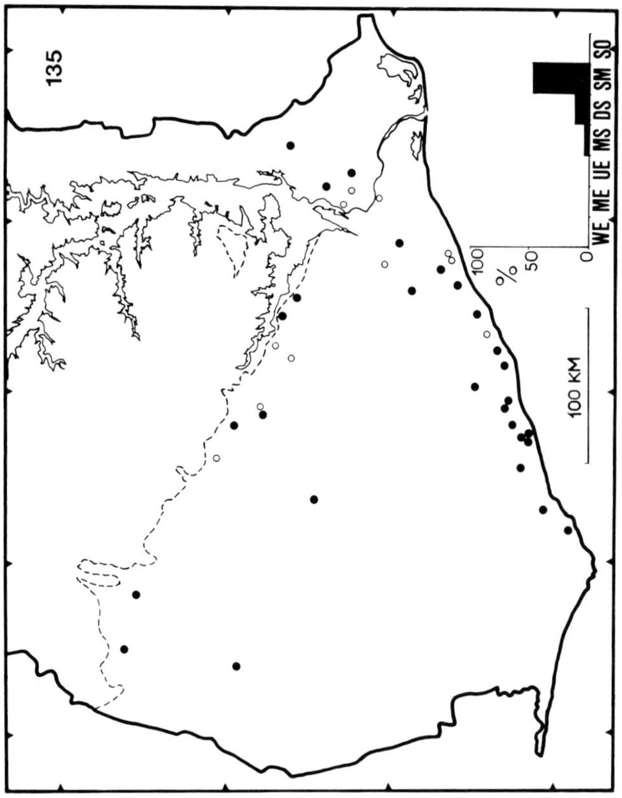
135
100 KM
%
100
50
0
WE ME UE MS DS SM SO

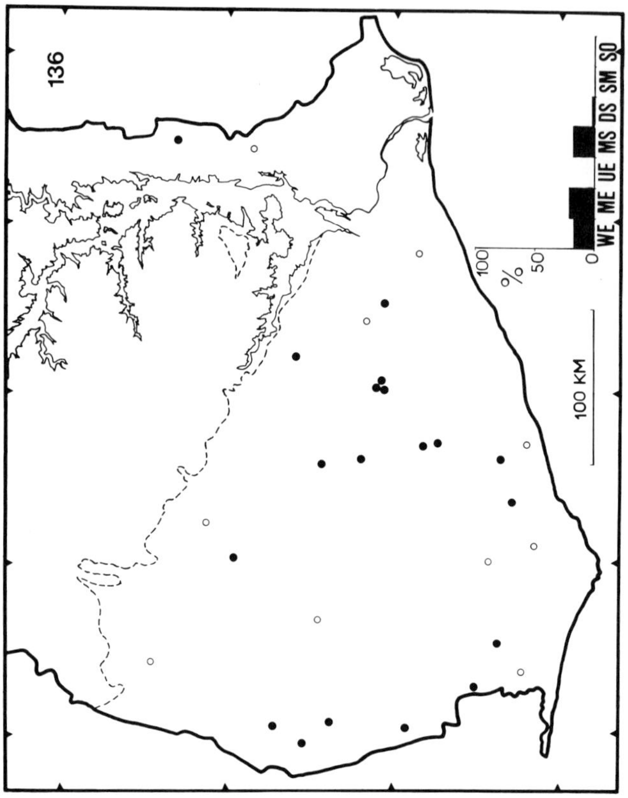
136
100 KM
%
100
50
0
WE ME UE MS DS SM SO

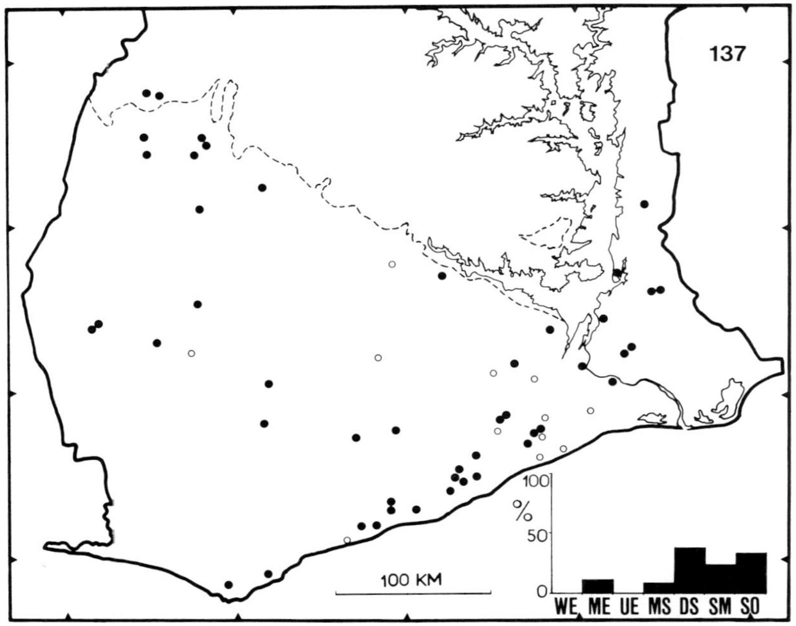

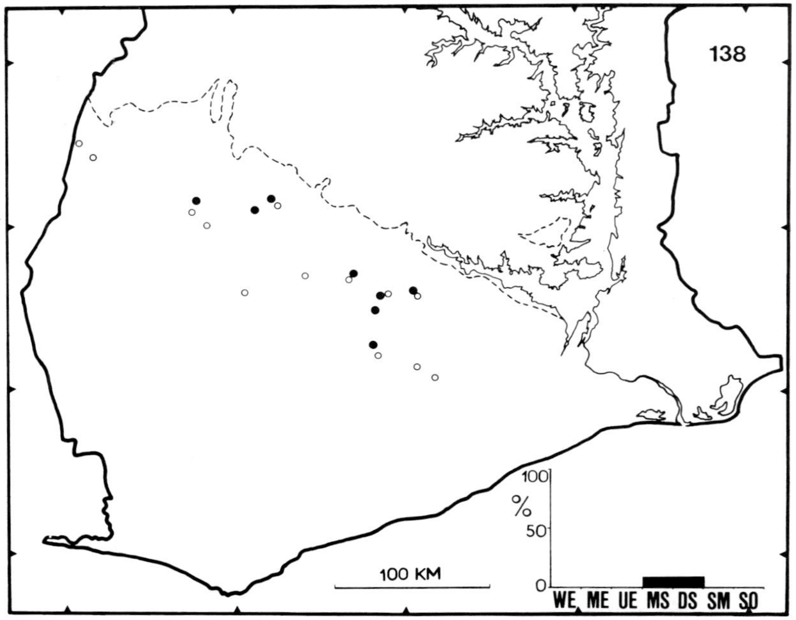

133 CHYTRANTHUS VERECUNDUS N. Hallé & Aké Assi (64). Sapindaceae.
Pygmy tree; lvs pinnate, leaflets mesophyll. Corner's model. Fl cream, small, probably polygamous, collected January. Fr berry, seeds very large. Iv. Coast and Ghana; (Upper Guinea endemic).

134 CISSUS ARALIOIDES (Welw. ex. Bak.) Planch., FWTA 1:679. Vitaceae (Ampelidaceae in FWTA).
Medium-sized tendril climber, stem mostly fleshy but woody at the base; lvs digitate, leaflets notophyll, slightly toothed. Fl small, greenish, dish, 9–11. Fr red one-seeded berry, 12–3; seed large. Regenerates vegetatively from the individual internodes into which the stem may break up. Occurs in fringing forest in the savanna zone. Widely distributed in Tropical Africa; (ecological and chorological transgressor).

135 CISSUS PETIOLATA Hook f., FWTA 1:678. Vitaceae (Ampelidaceae in FWTA).
Medium-sized tendril climber, stem fleshy at first, becoming softly woody and developing thin corky wings; lvs lobed, toothed, mesophyll. Fl small, greenish, dish, not clearly seasonal. Fr reddish 1-seeded berry; seed medium. Regenerates in shade. Guinée to Angola and E. Africa; (chorological transgressor).

136 CISSUS PRODUCTA Afzel., FWTA 1:678. Vitaceae (Ampelidaceae in FWTA).
Medium-sized woody tendril climber; lvs toothed, notophyll. Fl greenish, small, dish, not clearly seasonal. Fr black 1-seeded berry; seed medium to large. Regenerates in shade. S. Leone to Angola and E. Africa; (Guineo-Congolian wide).

137 CLAUSENA ANISATA (Willd.) Hook. f. ex Benth., FWTA 1:686. Rutaceae. Akan: samanankaa. (Type probably from near Accra).
Small tree; lvs pinnate, leaflets microphyll, glandular-punctate, with strong aniseed smell. Fl whitish, small, bowl, not clearly seasonal. Fr blue-black drupe; seed medium. Regenerates under shade. Widespread in Tropical Africa; (chorological transgressor).

138 CLEIDION GABONICUM Baill., FWTA 1:406. Euphorbiaceae. Akan: mpawu.
Small tree, lvs mesophyll. Fl small, monoecious, male in spikes, greenish, female solitary with long red styles; not clearly seasonal. Fr explosive capsule; seeds medium. Gregarious in forest understorey; regenerates in shade. Ghana, and Cameroun to Gabon; (Guinea wide).

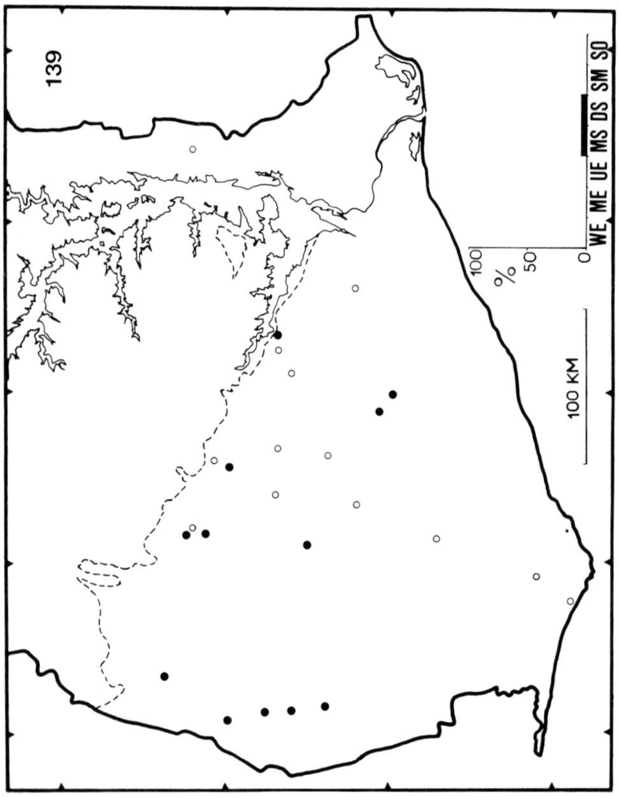
139
100 KM
100
%
50
0
WE ME UE MS DS SM SO

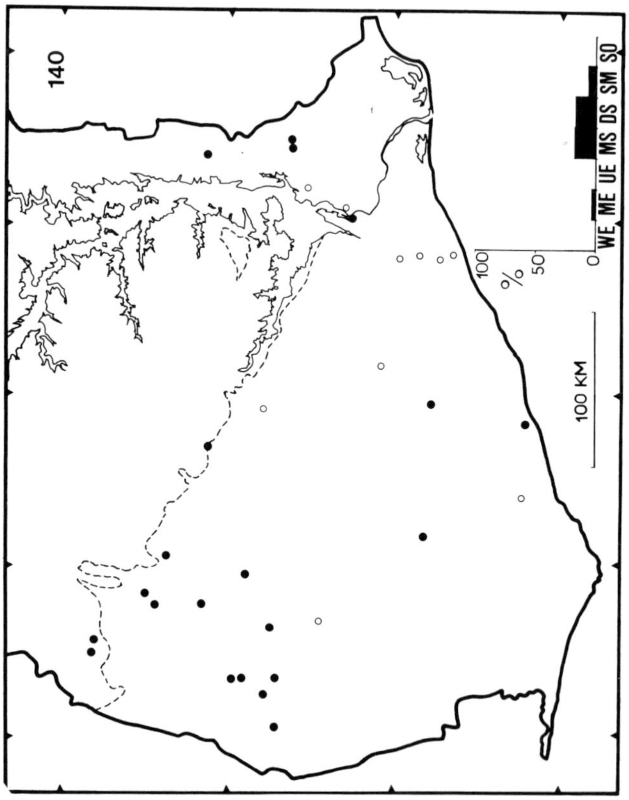
140
100 KM
100
%
50
0
WE ME UE MS DS SM SO

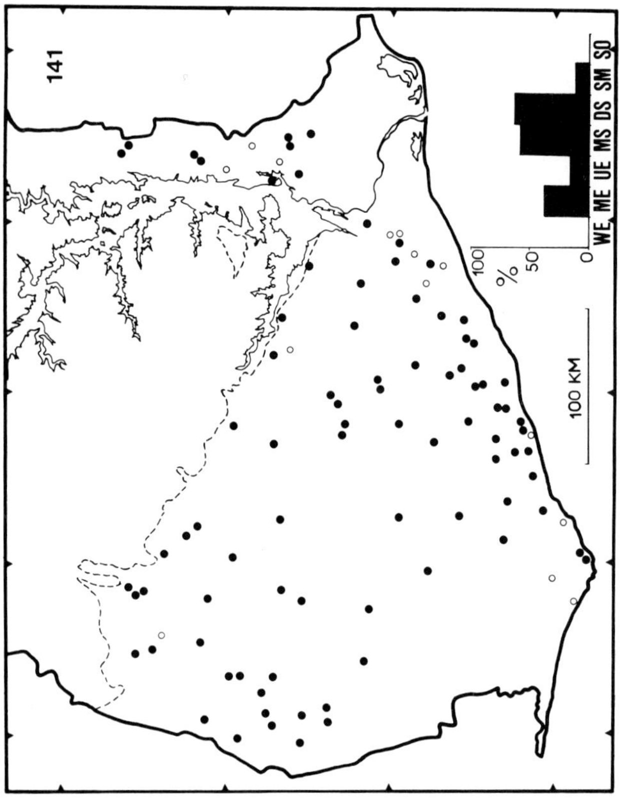
141
100 KM
100
%
50
0
WE ME UE MS DS SM SO

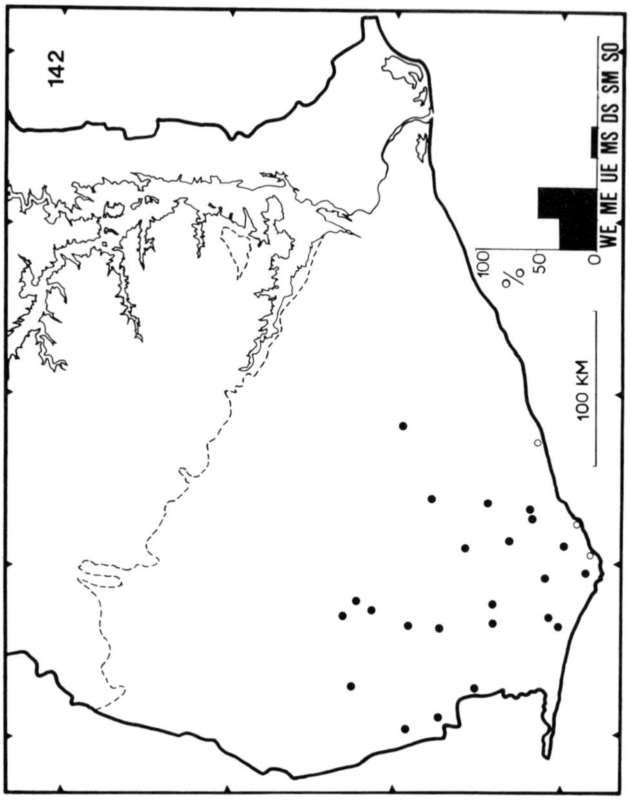
142
100 KM
100
%
50
0
WE ME UE MS DS SM SO

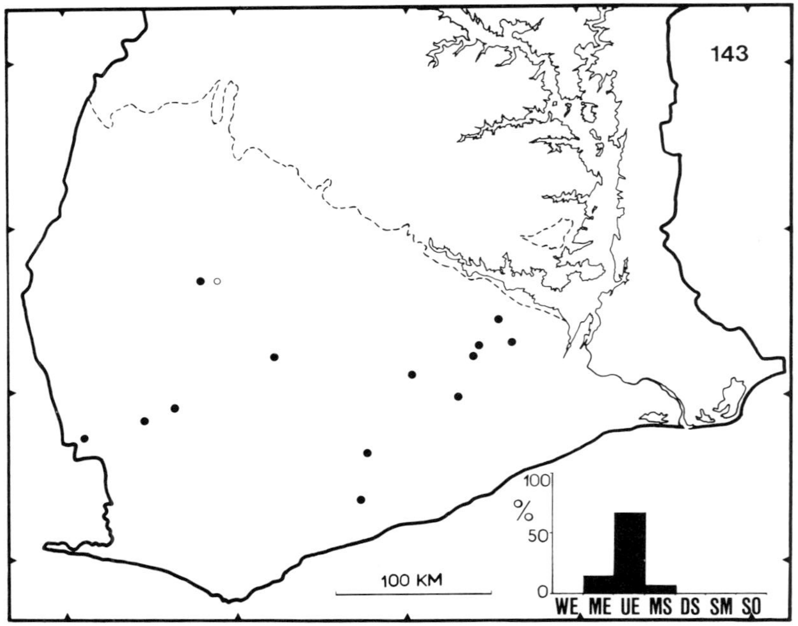

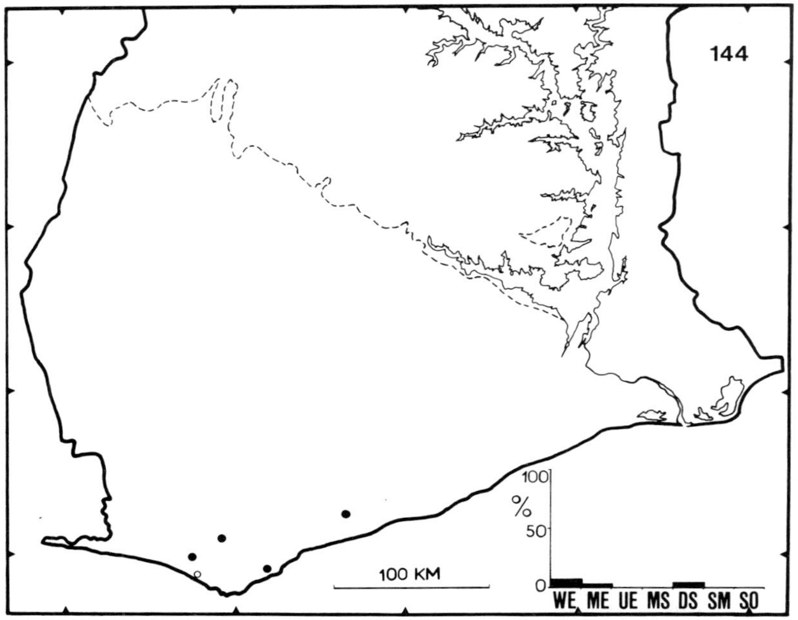

139 CLEISTOPHOLIS PATENS (Benth.) Engl. & Diels, FWTA 1:38. Annonaceae. Akan: ngonenkyene.
Medium-size tree, slash strongly aromatic; lvs mesophyll. Troll's model (59). Fl greenish-yellow, medium, dish, not clearly seasonal. Fr woody, 1–2-seeded, no obvious adaptation to dispersal; seeds large. Regenerates only in gaps. Tolerant of swampy soils but not confined to them. S. Leone to Zaire; (Guineo-Congolian wide).

140 CLERODENDRUM CAPITATUM (Willd.) Schum. & Thonn., FWTA 2:443. Verbenaceae. (Type probably from Akwapim).
Slender woody climber, or shrub in thickets; petioles forming woody hooks; lvs notophyll, usually hairy (though less so in moister forest: see var. *cephalanthum* (Oliv.) H. Huber in FWTA). Fl white, large, bowl with long tube, not clearly seasonal. Fr black berry, seeds medium. Also in fringing forest in savanna zone. Widespread in Tropical Africa; (ecological and chorological transgressor).

141 CNESTIS FERRUGINEA DC., FWTA 1:743. Connaraceae. Akan: apoose.
Pygmy or small tree or sometimes climbing; lvs pinnate, leaflets microphyll, pubescent. Fl white, small, dish, not seasonal. Fr red, fleshy, 1-seeded follicle; seed large, black, with yellow aril. Regenerates in shade, but most abundant in thickets. Senegal to Angola; (Guineo-Congolian wide).

142 CNESTIS LONGIFLORA Schellenb., FWTA 1:743. (Incl. *C. dinklagei* Schellenb., FWTA 1:743). Connaraceae.
Slender woody climber with tendriliform branches; lvs pinnate, leaflets microphyll, hairiness varying from glabrous (as in type of *C. dinklagei*) to pubescent (as in typical *C. longiflora*). Fl white, small, dish, cauliflorous, not clearly seasonal. Fr red, fleshy, 1-seeded, covered with pungent bristles; seed medium. Liberia to Nigeria; (Guinea wide).

143 CNESTIS RACEMOSA G. Don, FWTA 1:743. Connaraceae.
Slender woody climber; lvs pinnate, leaflets microphyll. Fl white, small, dish (not yet found in Ghana). Fr pink-red fleshy 1-seeded follicle; seed medium. S. Leone to Ghana; (Upper Guinea endemic).

144 COFFEA AFZELII Hiern, FWTA 2:156. Rubiaceae.
Slender woody climber, with deflexed branchlets; lvs microphyll. Cook's model. Fl white, large, dish with tube, fragrant. Fr berry, seeds medium. S. Leone to Zaire; (Guineo-Congolian wide).

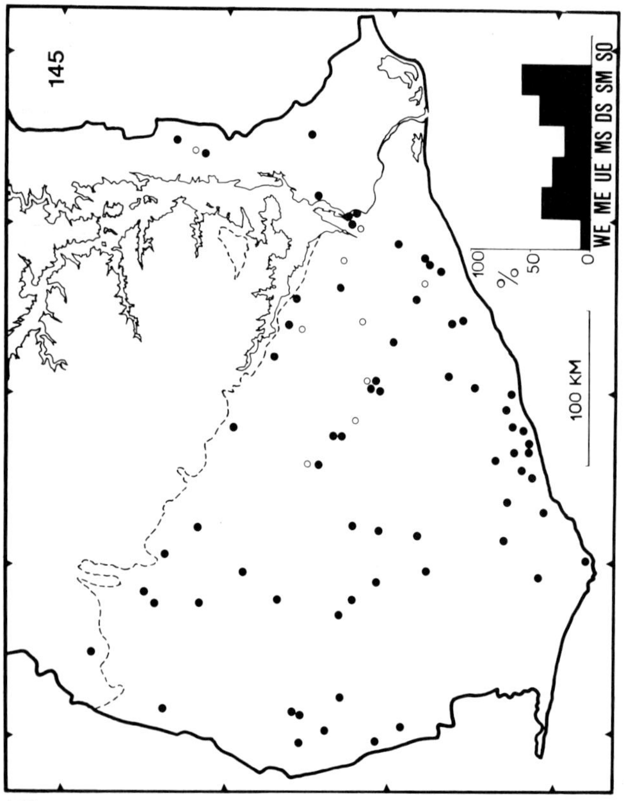
145
100 KM
%
100
50
0
WE ME UE MS DS SM SO

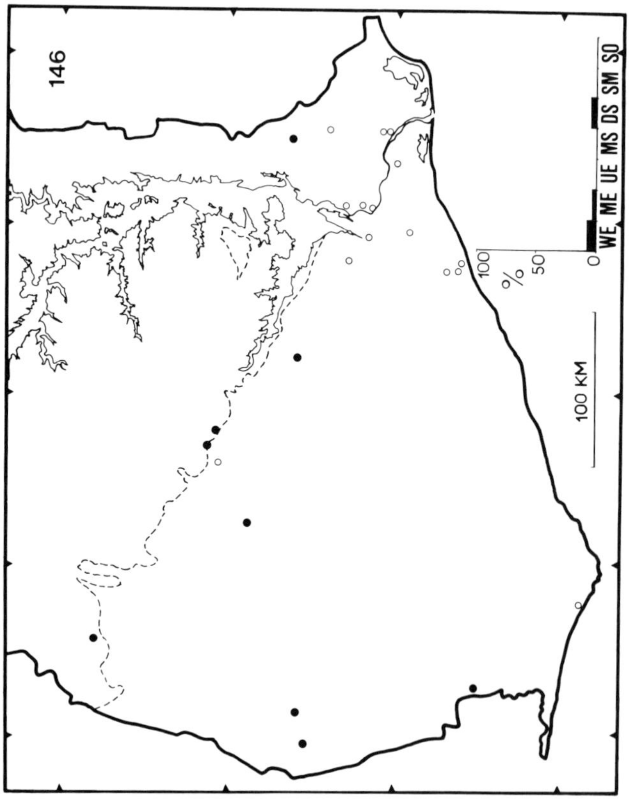
146
100 KM
%
100
50
0
WE ME UE MS DS SM SO

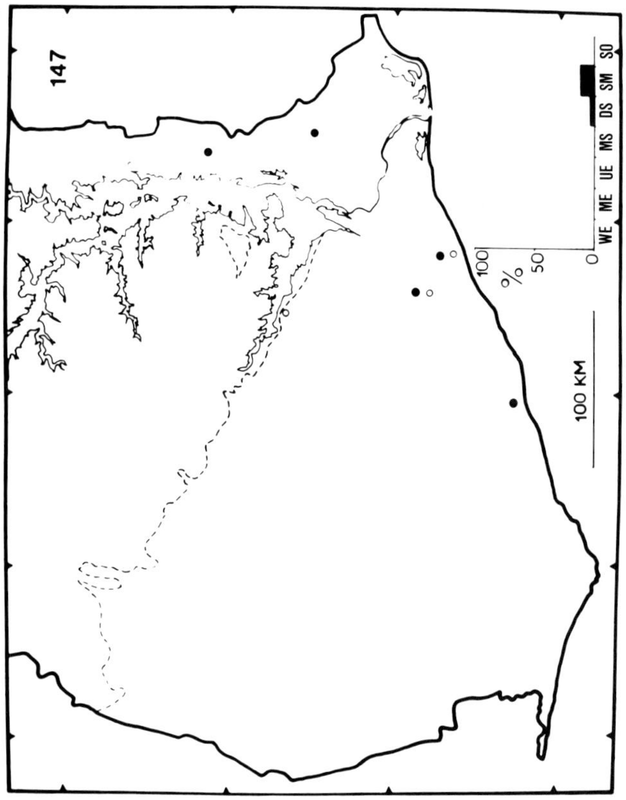
147
100 KM
%
100
50
0
WE ME UE MS DS SM SO

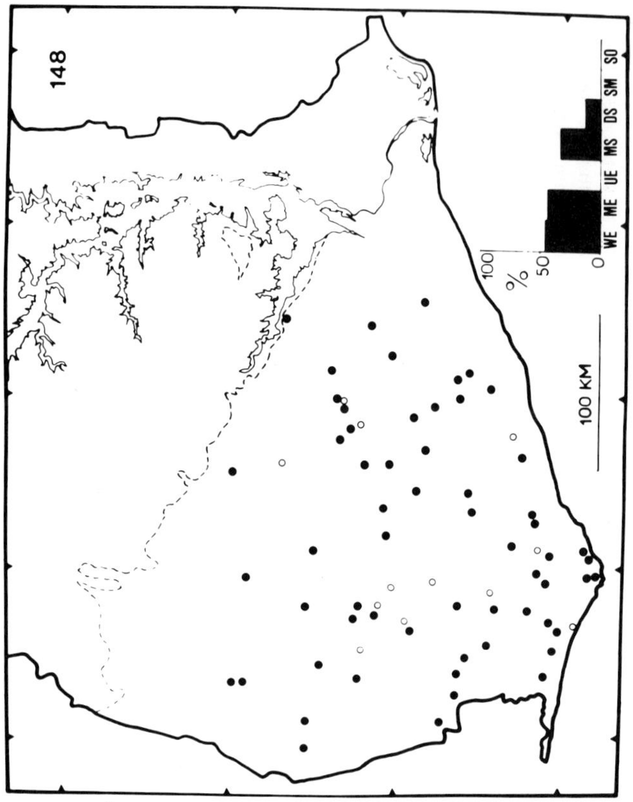
148
100 KM
%
100
50
0
WE ME UE MS DS SM SO

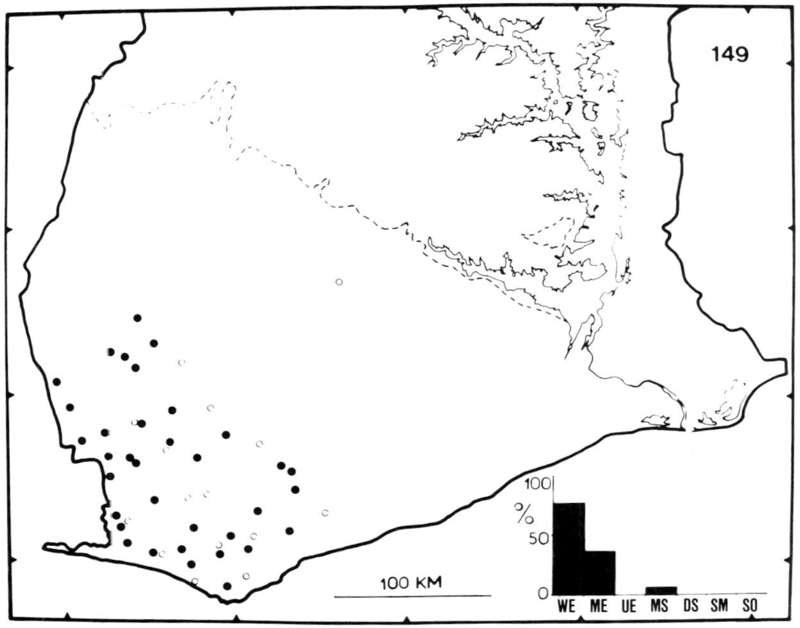

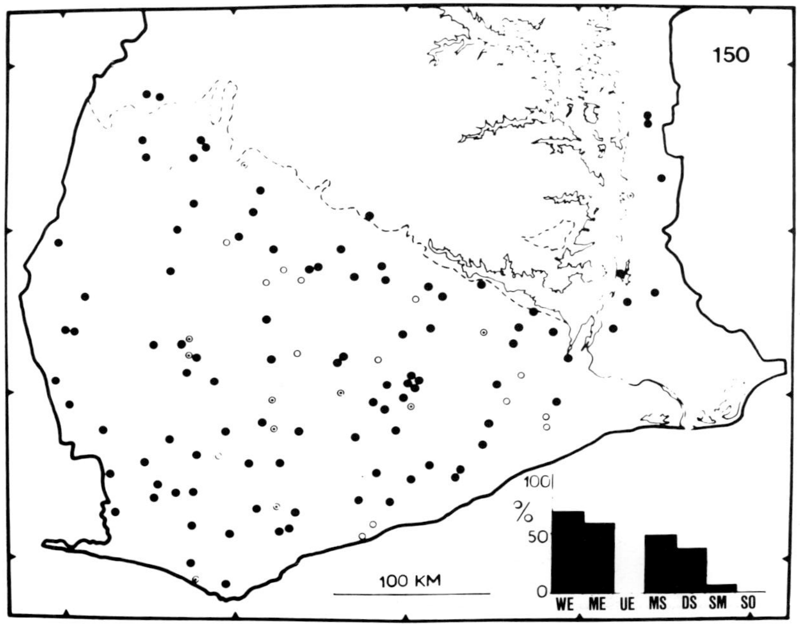

145 COFFEA EBRACTEOLATA (Hiern) Brenan, FWTA 2:157. Rubiaceae.
Pygmy tree, lvs microphyll. Roux's model. Fl white, large, dish with tube, not clearly seasonal. Fr black berry; seeds medium. Guinée to Cameroun; (Guinea wide).

146 COFFEA RUPESTRIS Hiern, FWTA 2:156. Rubiaceae.
Pygmy tree, lvs microphyll, deciduous. Fl white, large, dish with long tube, fragrant, not clearly seasonal. Fr red-orange berry; seeds medium. Guinée to Nigeria; (Guinea wide). The typical form is characteristic of the forest-savanna boundary, and of fringing forest in the savanna zone; the form from within the forest zone is taller and has larger, much less hairy leaves – it is probably worthy of subspecific rank.

147 COFFEA TOGOENSIS A. Chev., FWTA 2:156. Rubiaceae.
Pygmy to small tree, lvs notophyll. Roux's model. Fl white, medium, dish with tube, sweet-scented, not clearly seasonal. Fr black berry; seeds medium. Ghana to Togo; (Upper Guinea endemic). (Of the several species of *Coffea* which are cultivated in Ghana, *C. canephora* Pierre ex Froehner is grown most widely; a specimen seen in Mpameso F.R. could have been indigenous. *C. togoensis* is not cultivated).

148 COLA CARICIFOLIA (G. Don) K. Schum., (*C. caricaefolia* in FWTA 1:328). Sterculiaceae. Akan: osonkurobia.
Medium-sized tree; lvs lobed, mesophyll, hairy. Rauh's model (not Corner, 59). Fl red, medium, bell, monoecious, cauliflorous, 8–11. Fr orange-red fleshy follicle, 9–11; seeds large. Regenerates under shade. S. Leone to Ghana; (Upper Guinea endemic).

149 COLA CHLAMYDANTHA K. Schum., FWTA 1:328. Sterculiaceae. Akan: tananfre.
Medium tree; lvs digitate, leaflets macrophyll. Corner's model. Fl deep red, large, bowl, monoecious, cauliflorous, 10–11. Fr red fleshy follicle, 10–2; seeds scarlet, very large. Regenerates under shade. Plot records very strongly correlated with acid, base-poor soils. S. Leone to Central African Republic; (Guinea wide).

150 COLA GIGANTEA A. Chev. var. *glabrescens* Brenan & Keay, FWTA 1:330, and COLA LATERITIA K. Schum. var. *maclaudii* (A. Chev.) Brenan & Keay, FWTA 1:330. Sterculiaceae. Akan: watapuo.
Large (*C. gigantea*) or medium (*C. lateritia*) trees with buttresses; lvs lobed, mesophyll to macrophyll; *C. gigantea* is deciduous. Rauh's model. Fl yellow changing to pink, medium, bell, monoecious, 8–10 (*C. lateritia*), 9–1 (*C. gigantea*). Fr red fleshy follicle (*C. gigantea*) or berry (*C. lateritia*); seeds large, surrounded by juicy aril. Regenerate in shade. Seedlings and saplings are effectively indistinguishable and survey records are therefore not mapped separately; herbarium records: *C. gigantea* ○; *C. lateritia* ⊙. Distribution: *C. gigantea* var. *glabrescens* Iv. Coast to Nigeria; *C. lateritia* var. *maclaudii* Guinée to Nigeria. (Both species are Guinea wides).

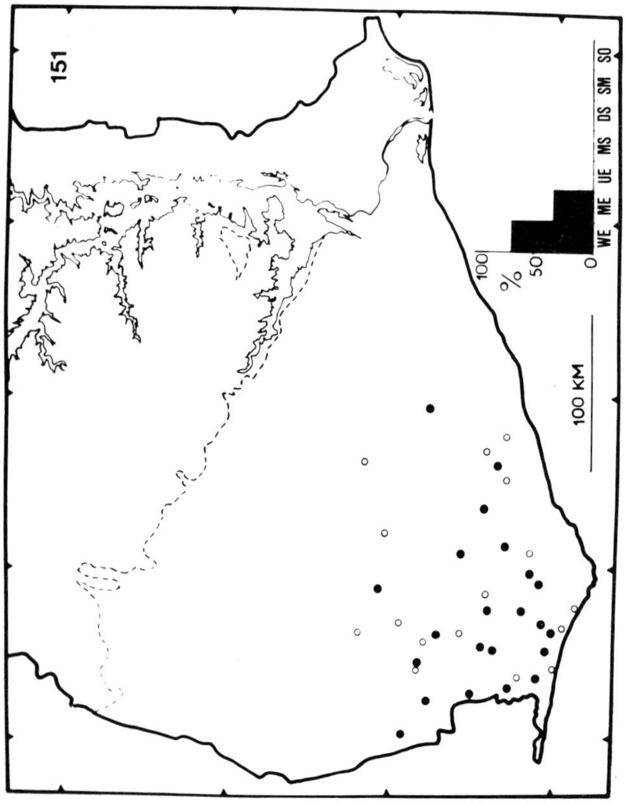
151
100 KM
100
%
50
0
WE ME UE MS DS SM SO

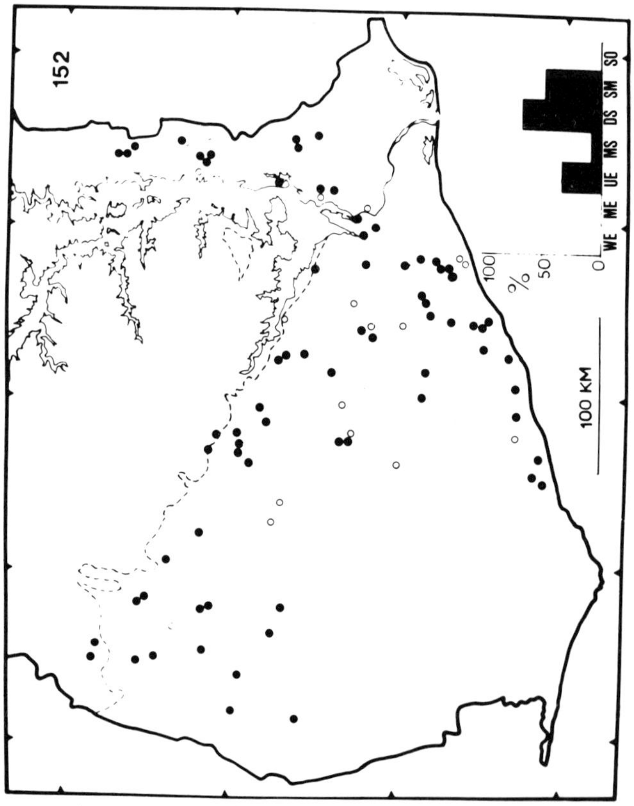
152
100 KM
100
%
50
0
WE ME UE MS DS SM SO

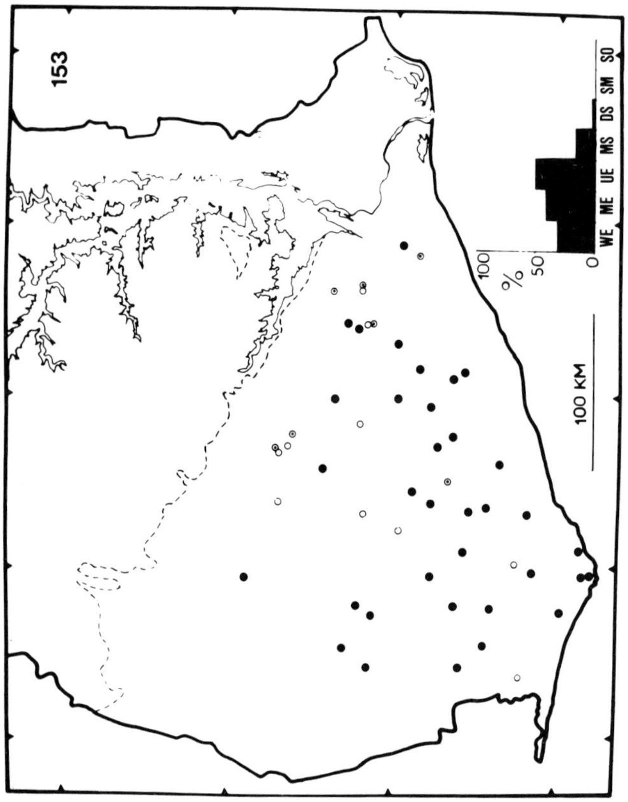
153
100 KM
100
%
50
0
WE ME UE MS DS SM SO

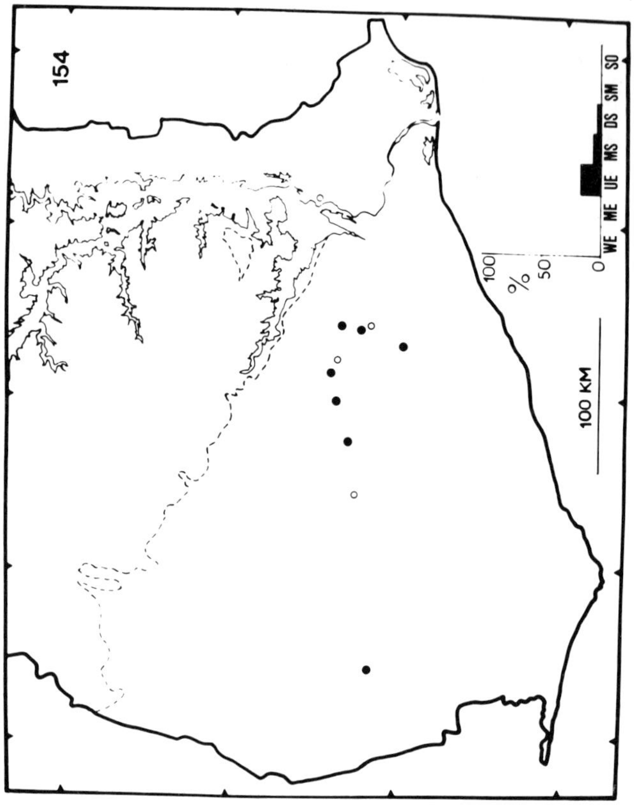
154
100 KM
100
%
50
0
WE ME UE MS DS SM SO

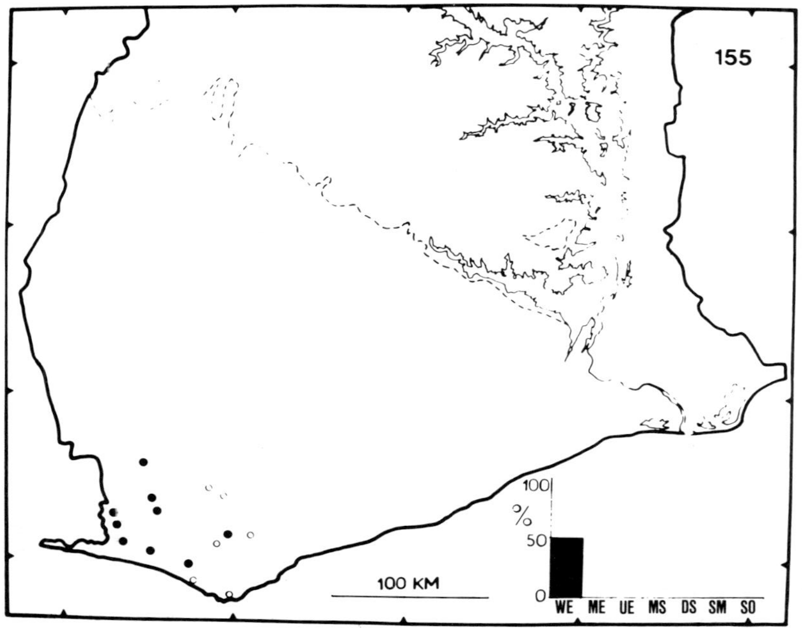

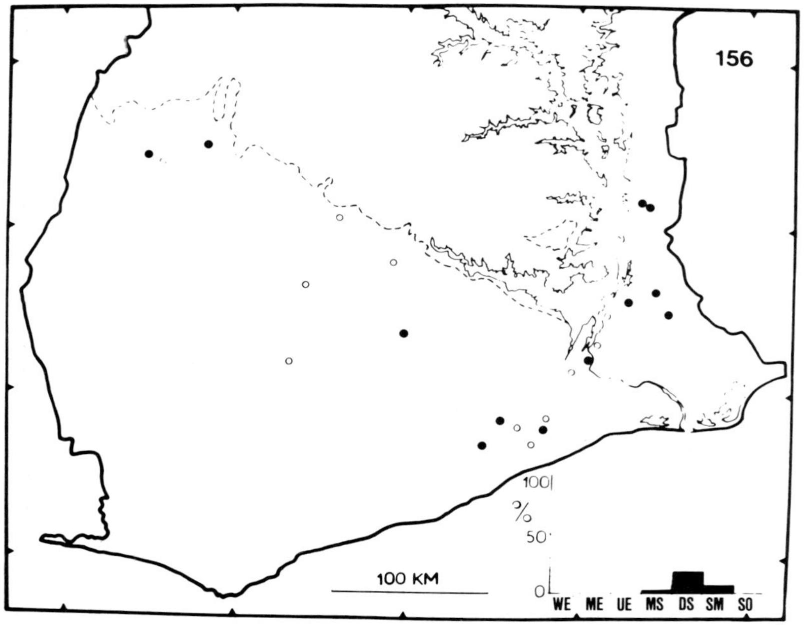

151 COLA HETEROPHYLLA (P. Beauv.) Schott & Endl., FWTA 1:328. Sterculiaceae.
Pygmy tree; lvs mesophyll, sometimes lobed. Fl pale yellow, medium, bowl, monoecious, not clearly seasonal. Fr red berry; seeds large. Regenerates in shade. Plot records very significantly correlated with acid, base-poor soils: never found in soil with T.E.B. > 4 m-equiv. per 100 g. S. Leone to Nigeria and Gabon; (Guinea wide).

152 COLA MILLENII K. Schum., FWTA 1:328. Sterculiaceae. Akan: anansedodowa.
Medium tree; lvs lobed, mesophyll. Rauh's model. Fl reddish, medium, bowl, monoecious, sometimes cauliflorous, 9–1. Fr orange-red berry; seeds large, surrounded by juicy aril. Regenerates in shade. The survey records from Upland Evergreen forest seem rather anomalous; fruit resembles that of *C. millenii* but flowers are yet to be collected. Iv. Coast to Nigeria; (Guinea wide).

153 COLA NITIDA (Vent.) Schott & Endl., FWTA 1:329, and COLA VERTICILLATA (Thonn.) Stapf ex A. Chev., FWTA 1:330. Sterculiaceae. Akan: bese. (Type of *C. verticillata* from Akwapim).
Medium trees, lvs mesophyll. Rauh's model (59). Fl yellow and purple, medium, bowl, monoecious, not clearly seasonal. Fr greenish fleshy follicle; seeds very large. Regeneration in shade. Seeds of *C. nitida* (kola) are eaten, and are an important market commodity. Seeds of *C. verticillata* have a similar appearance, but are very mucilaginous and not generally eaten. Only herbarium records are distinguished on the map: *C. nitida* ○; *C. verticillata* ⊙. The survey records are almost all of spontaneous plants, but it was not usually possible to identify them as one species or the other. Distribution: *C. nitida* Guinée to Ghana, (Upper Guinea endemic); *C. verticillata* Ghana to Cameroun, (Guinea wide).

154 COLA RETICULATA A. Chev., FWTA 1:329. Sterculiaceae.
Small tree, lvs notophyll to mesophyll. Fl pale yellow, medium, bell (Fig. 9.1d), monoecious, not clearly seasonal. Fr red berry; seeds large. Regenerates in shade. Guinée to Ghana; (Upper Guinea endemic). Vegetatively very similar to *C. boxiana* Brenan & Keay, *C. flavo-velutina* K. Schum. and *Octolobus spectabilis* Welw., but distinguishable by characters of the terminal bud.

155 COLA UMBRATILIS Brenan & Keay, FWTA 1:326. Sterculiaceae. Akan: tananfrebere. (Type from Prestea).
Small tree; lvs digitate, leaflets macrophyll. Corner's model. Fl red, medium, bowl, monoecious, cauliflorous. Fr red berry; seeds very large. Regenerates in shade. Endemic to Ghana; (Upper Guinea endemic).

156 COMBRETUM MUCRONATUM Schumach. (65), (*C. smeathmannii* G. Don, FWTA 1:272). Combretaceae. (Type from Akwapim).
Medium-sized woody climber, or scrambling shrub in thicket; lvs notophyll, pellucid dotted. Roux's model. Fl white, small, fragrant, brush inflorescence, 12–4. Fr dry, thin-winged; seed medium. Regenerates in shade. Gambia to Zaire and Uganda; (Guineo-Congolian wide).

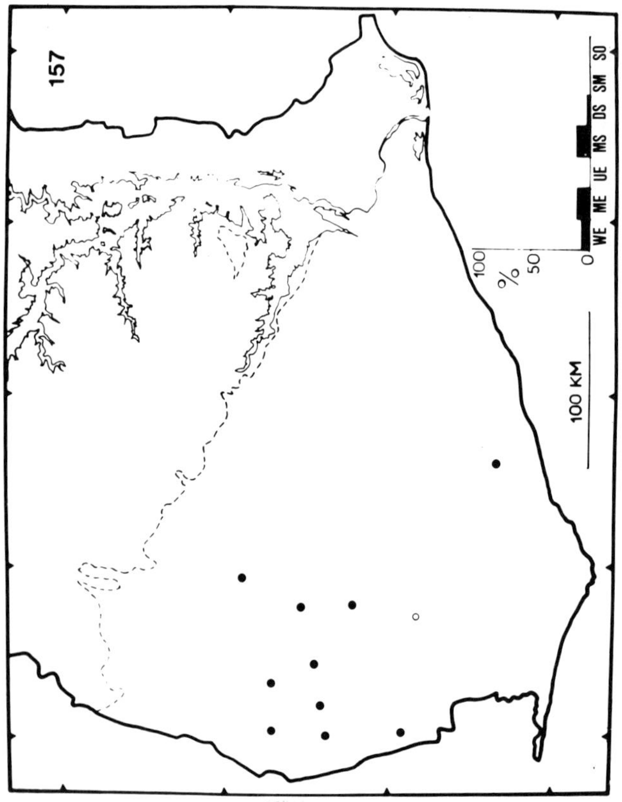

157
100 KM
%
100
50
0
WE ME UE MS DS SM SO

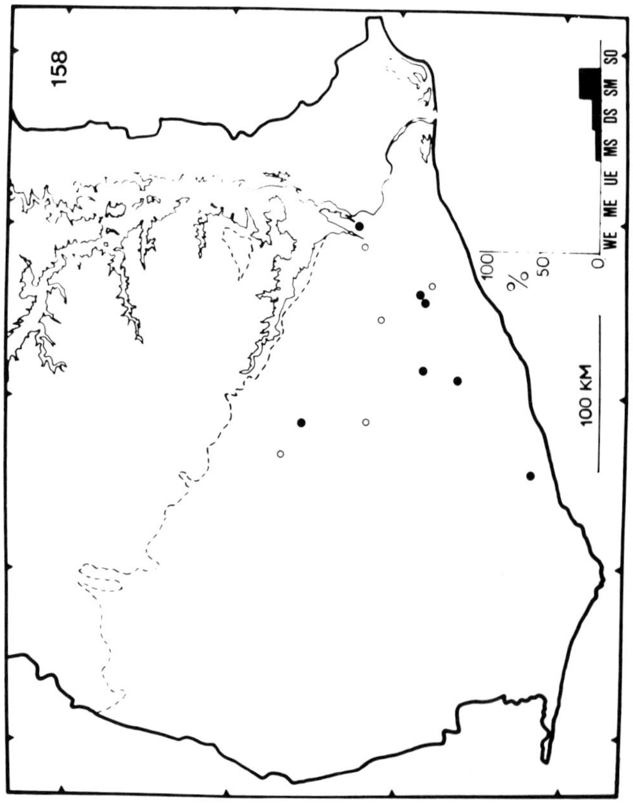

158
100 KM
%
100
50
0
WE ME UE MS DS SM SO

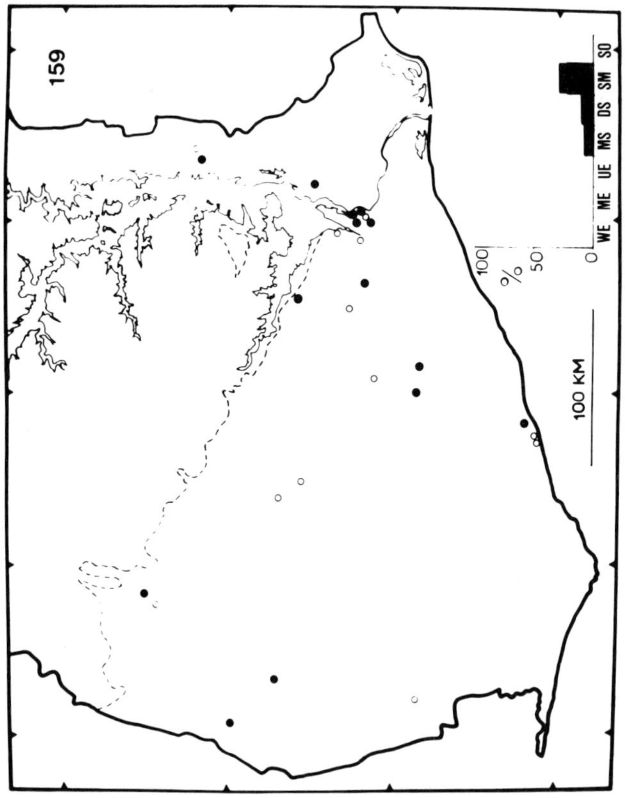

159
100 KM
%
100
50
0
WE ME UE MS DS SM SO

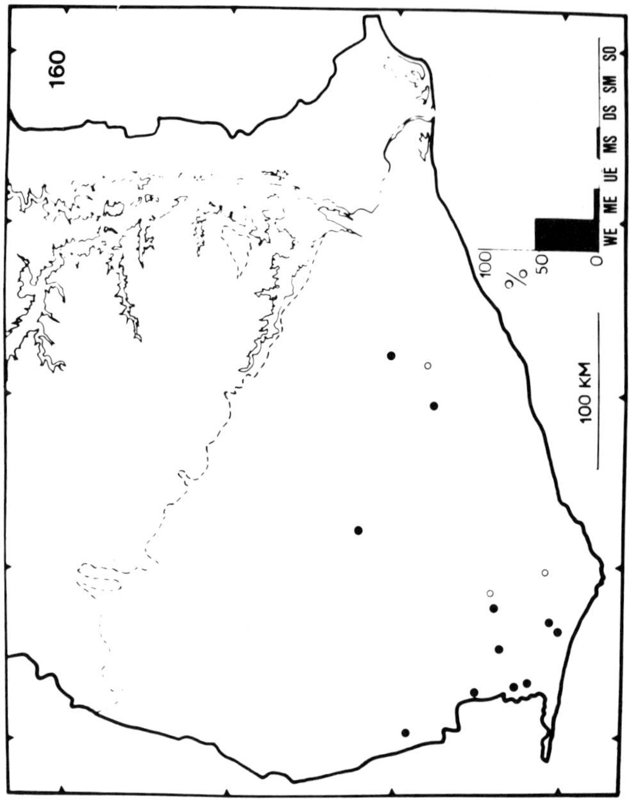

160
100 KM
%
100
50
0
WE ME UE MS DS SM SO

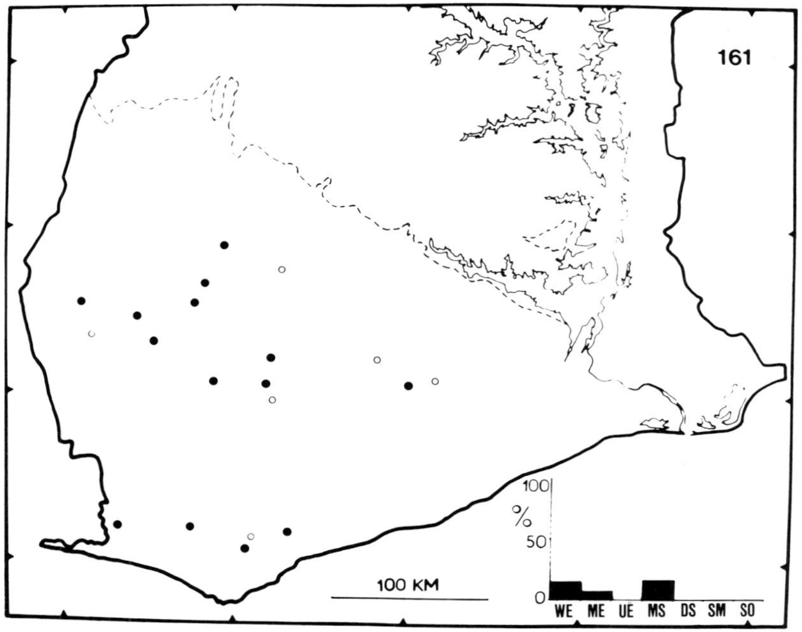

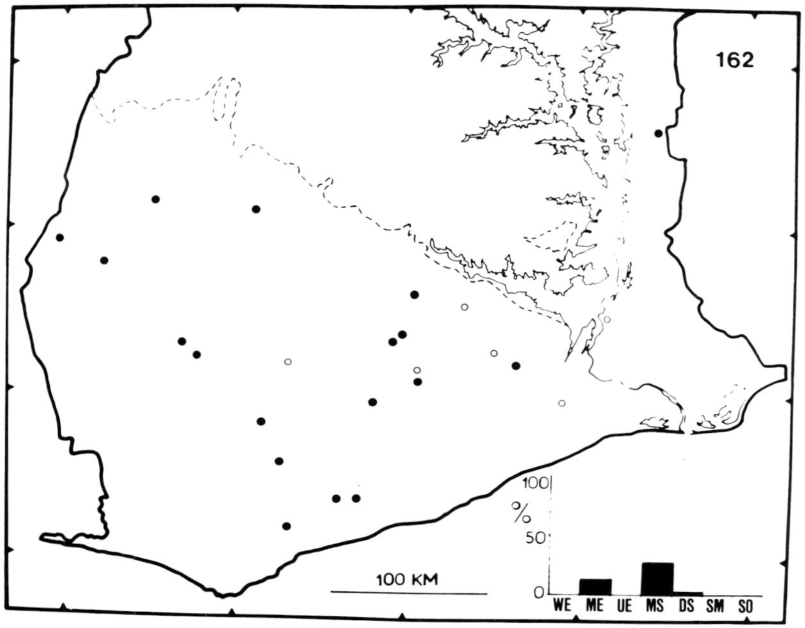

157 COMBRETUM OYEMENSE Exell, FWTA 1:272. Combretaceae.
Medium-sized woody climber; xylem cylindrical, normal; lvs notophyll, with red scales below. Roux's model. Fl whitish, small, brush. Fr dry, thin-winged; seed medium. Regenerates in shade. S. Leone, Ghana, Cameroun and Gabon; (Guinea wide). About 20 species of *Combretum* were recorded during the survey, almost all in the vegetative condition. Of these only 12 could be satisfactorily named – 2 of the 12 are not included in the FWTA account. *Combretum* is probably the least well-known genus of West African forest plants.

158 COMMELINA MACROSPERMA J. K. Morton, FWTA 3:49. Commelinaceae. (Type from Kibi).
Decumbent perennial herb; lvs microphyll. Fl white, medium, bowl, not clearly seasonal. Fr dry capsule; seeds medium. Regenerates in shade. S. Leone to Nigeria; (Guinea wide).

159 COMMELINIDIUM GABUNENSE (Hack.) Stapf, FWTA 3:436. Gramineae.
Decumbent perennial broad-leaved grass, rooting from the nodes; microphyll. Spikelets green, medium, not seasonal. Fr small; no obvious dispersal mechanism. Under forest shade, and in small gaps. Liberia to Angola and Uganda; (Guineo-Congolian wide).

160 CONNARUS AFRICANUS Lam., FWTA 1:748. Connaraceae.
Large woody climber; xylem without anomalies, but red exudate in bark. Lvs trifoliolate, leaflets notophyll. Fl cream-white, medium, bowl, heterostyled; not clearly seasonal. Fr red dry 1-seeded follicle; seed very large, black with orange aril. Senegal to Nigeria; (Guinea wide).

161 COPAIFERA SALIKOUNDA Heckel, FWTA 1:457. Caesalpiniaceae. Akan: entedua.
Large tree; slash yielding fragrant resin; lvs pinnate, leaflets microphyll. Fl whitish, small, crowded in spikes. Fr woody pod; seed very large, black with red aril. Regenerates in shade. Guinée to Ghana; (Upper Guinea endemic).

162 CORYMBORKIS CORYMBIS Thou. (95), (*C. corymbosa* Thou., FWTA 3:211). Orchidaceae.
Perennial erect herb of forest floor; lvs mesophyll to macrophyll. Tomlinson's model. Fl white, large, bowl with long tube, not clearly seasonal. Fr capsule; seeds minute, wind-dispersed. Widely distributed in Tropical Africa, and extending to Madagascar; (chorological transgressor).

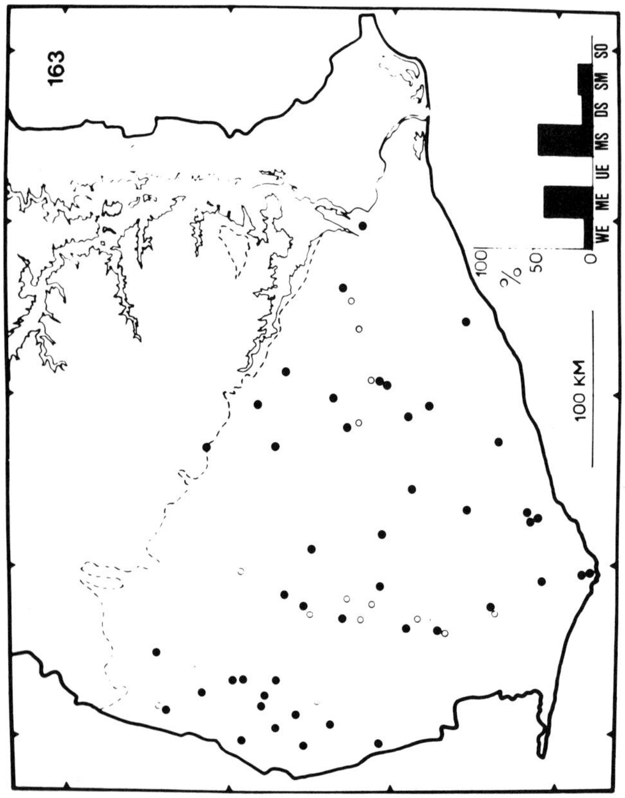
163
100 KM
%
100
50
0
WE ME UE MS DS SM SO

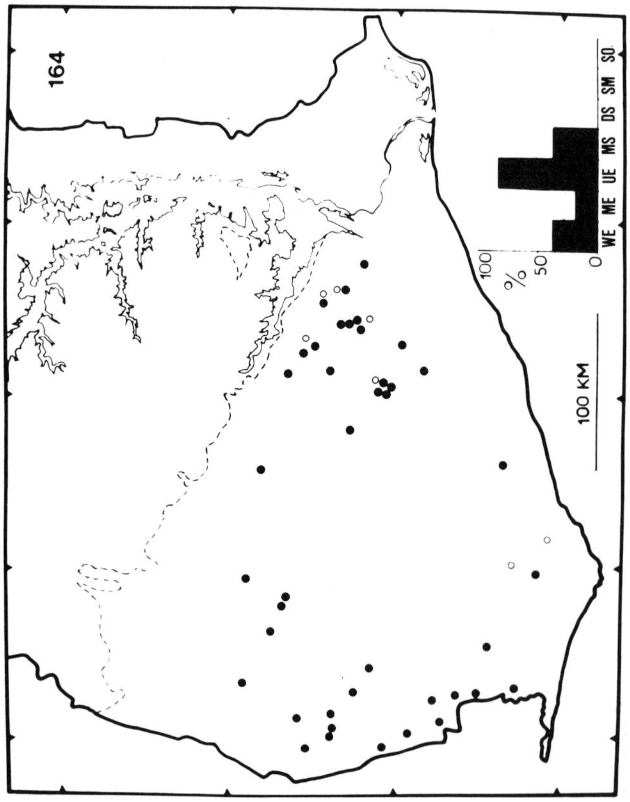
164
100 KM
%
100
50
0
WE ME UE MS DS SM SO

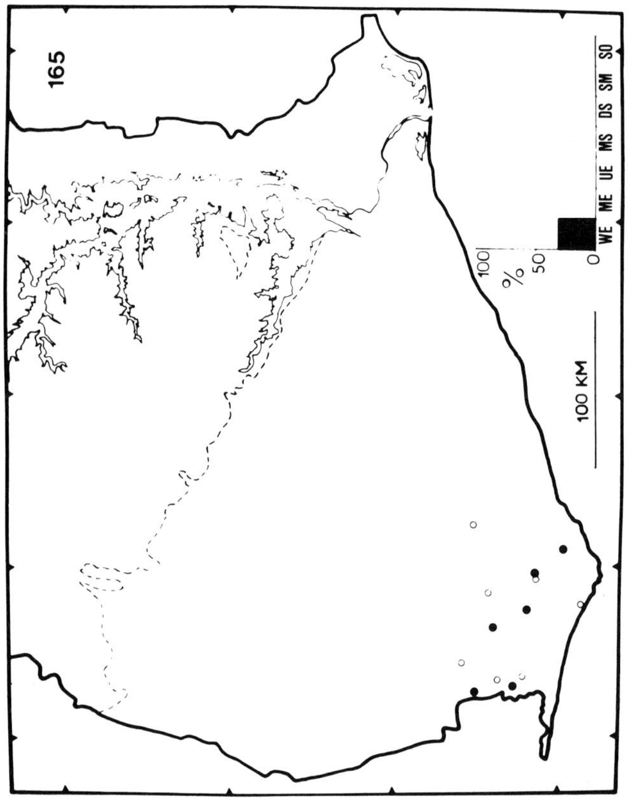
165
100 KM
%
100
50
0
WE ME UE MS DS SM SO

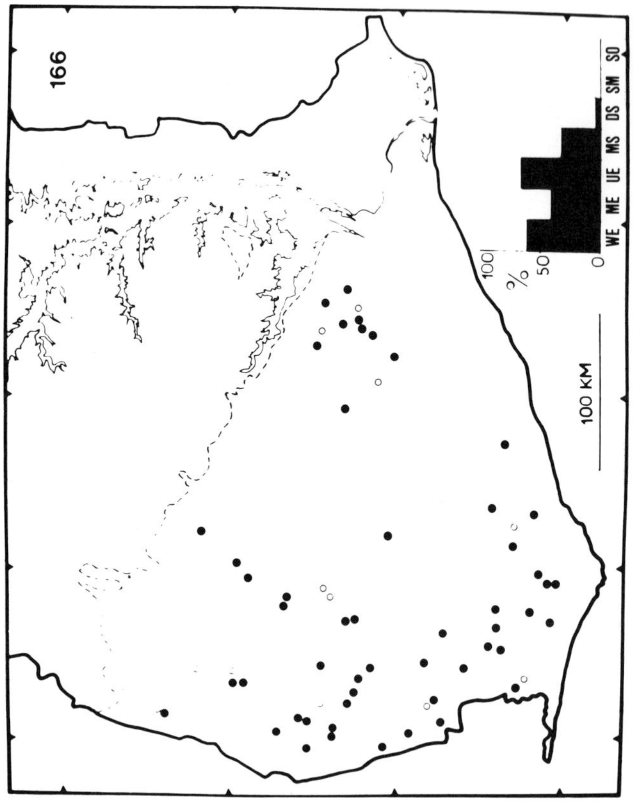
166
100 KM
%
100
50
0
WE ME UE MS DS SM SO

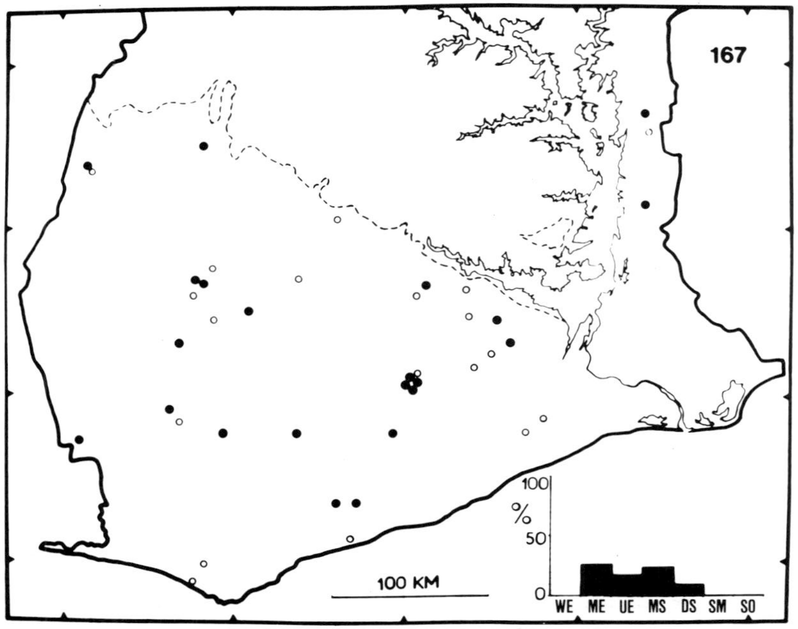

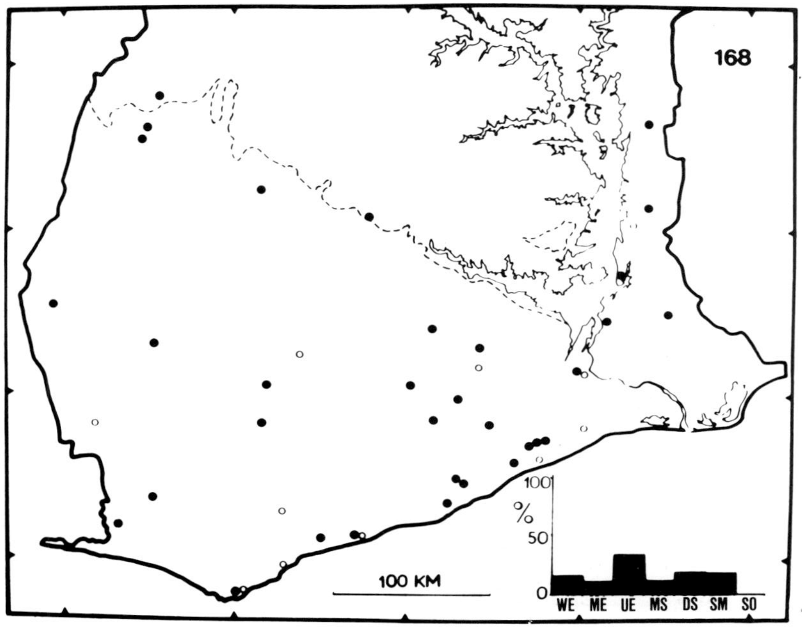

163 CORYNANTHE PACHYCERAS K. Schum., FWTA 2:111. Rubiaceae. Akan: pamprama.
Medium-sized tree, trunk fluted and often with epicormic branches near the base; lvs mesophyll. Fl white, sweetly scented, small, bowl, crowded in panicles, 9–12. Fr reddish capsule; seeds small with two narrow tail-like wings. Regenerates in shade. Gregarious (Fig. 3.6). S. Leone to Zaire; (Guineo-Congolian wide).

164 COSTUS ENGLERANUS K. Schum., FWTA 3:78. Zingiberaceae.
Creeping rhizomatous herb on forest floor (Fig. 5.13b); lvs succulent, notophyll. Fl white with yellow throat, large, funnel with tube. Fr opening irregularly, with no obvious dispersal mechanism; seeds medium. Spreads vegetatively in forest shade, forming large patches. 2/3 of plot records had soil of pH 4.2–4.9, and T.E.B. < 4 m-equiv./100 g. Iv. Coast to Cameroun; (Guinea wide).

165 COULA EDULIS Baill., FWTA 1:645. Olacaceae. Akan: bodwue.
Medium-sized tree; lvs mesophyll. Roux's model (59). Fl yellow-green, small, bowl, clustered on spikes, 1–2. Fr woody nut; seed large, edible, much eaten by rodents. Regenerates in shade. Good indicator of Wet Evergreen forest. S. Leone to Zaire; (Guinea wide).

166 CRATERISPERMUM CAUDATUM Hutch., FWTA 2:188. Rubiaceae. Akan: duade.
Small tree; lvs notophyll. Massart's model (59). Fl white, medium, dish with tube, not clearly seasonal. Fr black berry; seeds medium. Regenerates in shade. Plot records very highly correlated with acid, base-poor soils, of high clay content. S. Leone to Nigeria; (Guinea wide).

167 CRATERISPERMUM CERINANTHUM Hiern, FWTA 2:188. Rubiaceae. Akan: Afra-ni-Afei.
Medium-sized tree; lvs mesophyll. Massart's model. Fl white, medium, dish with tube, not clearly seasonal. Fr greenish berry (possibly black when ripe); seeds medium. Regenerates in shade. 85% of plot records have soil with pH 4.2–4.9. Iv. Coast to Zaire; (Guineo-Congolian wide).

168 CREMASPORA TRIFLORA (Thonn.) K. Schum., FWTA 2:148. Rubiaceae. (Type from Akwapim).
Medium to large woody climber with deeply grooved stem, or shrub in thicket; lvs microphyll and bluntly subacuminate in dry forest to notophyll and acutely acuminate in moist and wet forest. Cook's model. Fl white, fragrant, medium, dish with tube, not clearly seasonal. Fr dark red berry; seeds medium. Regenerates in shade. Widespread in Tropical Africa; (chorological transgressor).

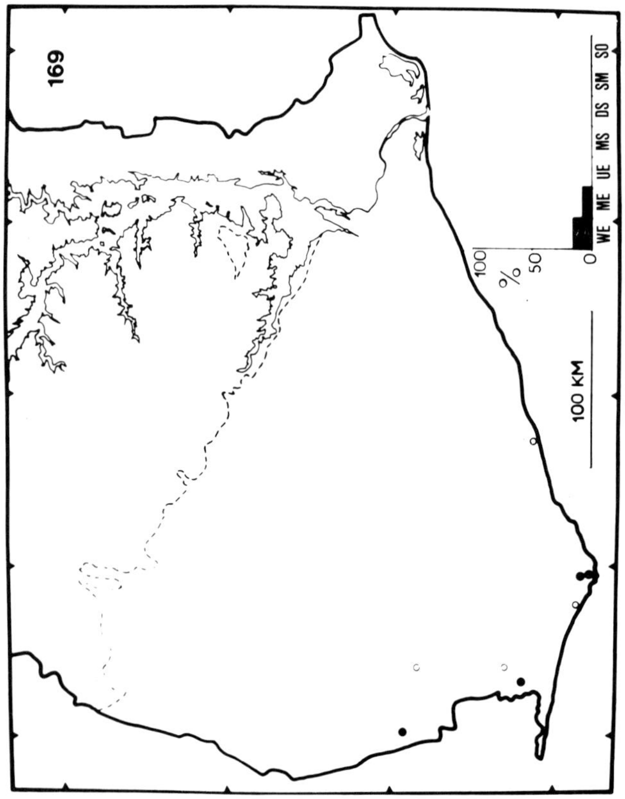
169
100 KM
%
100
50
0
WE ME UE MS DS SM SO

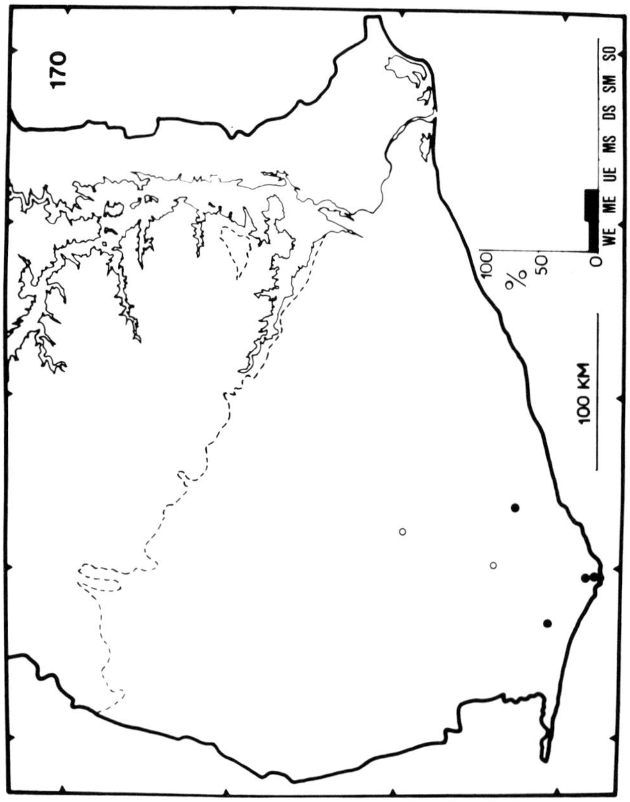
170
100 KM
%
100
50
0
WE ME UE MS DS SM SO

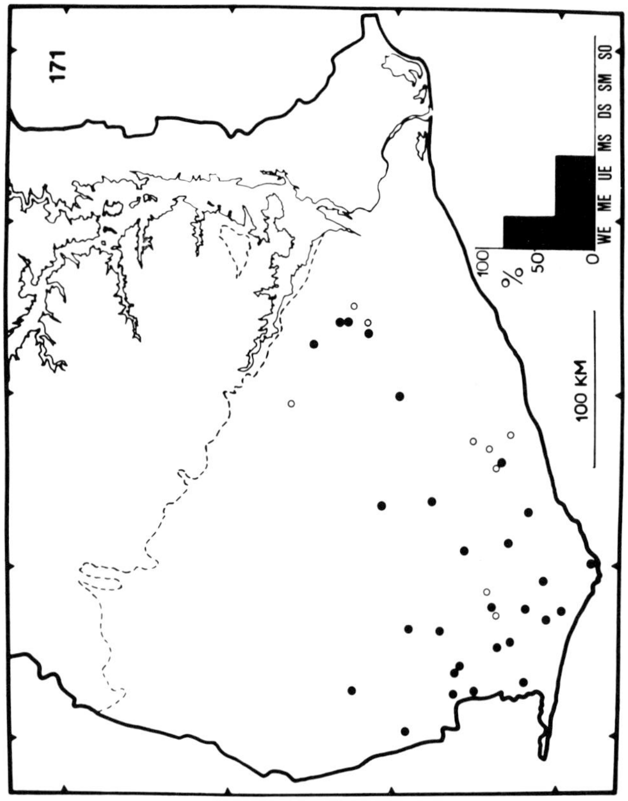
171
100 KM
%
100
50
0
WE ME UE MS DS SM SO

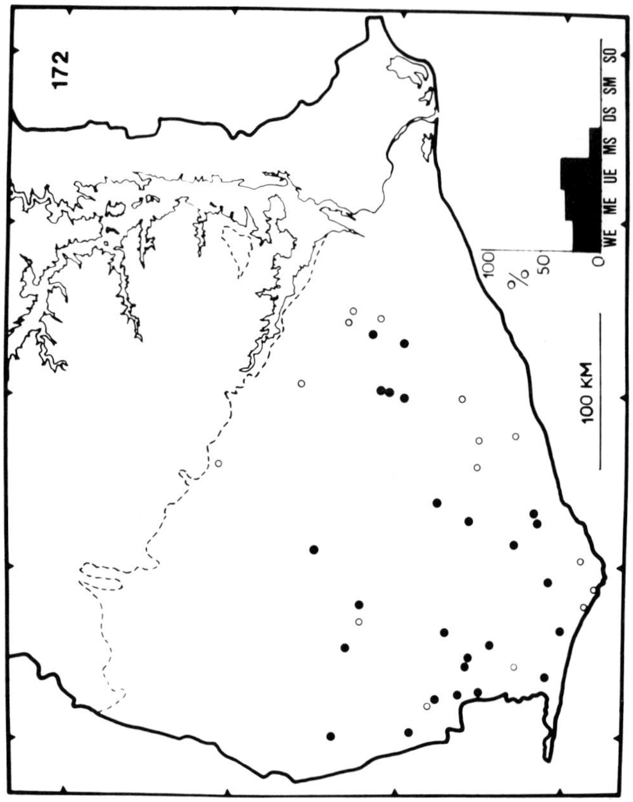
172
100 KM
%
100
50
0
WE ME UE MS DS SM SO

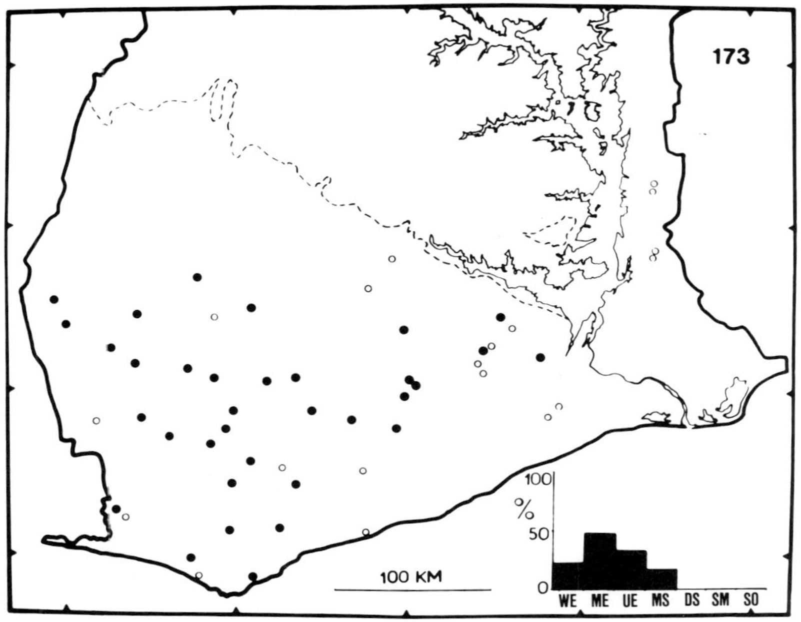

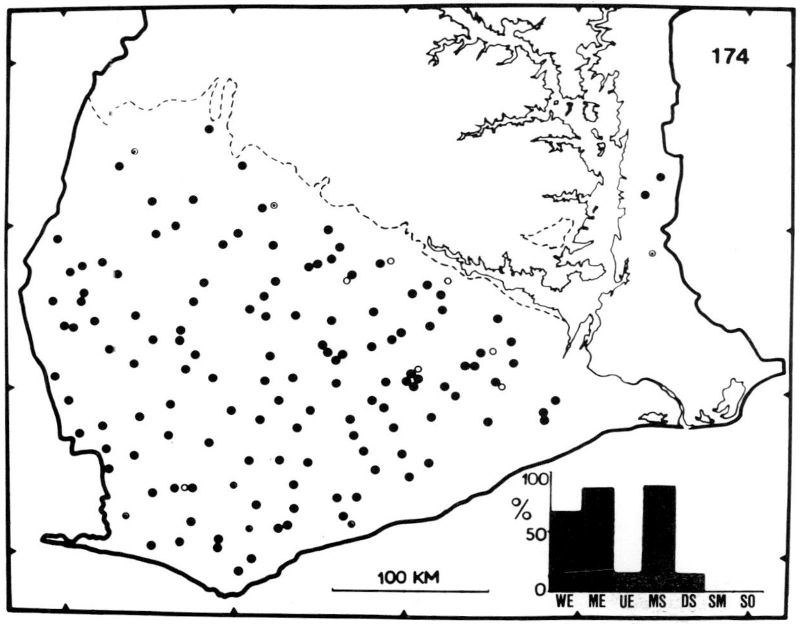

169 CROTONOGYNE MANNIANA Muell. Arg., FWTA 1:400. Euphorbiaceae.
Small tree; lvs mesophyll. Fl cream, dioecious, bowl, male small, female medium, not clearly seasonal. Fr explosive capsule; seeds medium. Regenerates in shade. Gregarious. Ghana and Nigeria; (Guinea wide).

170 CRUDIA GABONENSIS Pierre ex Harms, FWTA 1:467. Caesalpiniaceae. Akan: samantaanini.
Medium tree with buttresses; lvs pinnate, leaflets notophyll. Fl red, small, dish. Fr large pod, seeds very large. Regenerates in shade. Gregarious. Iv. Coast to Ghana and Gabon; (Guinea wide). Wood very hard (113), not classified.

171 CTENITIS JENSENIAE (C. Chr.) Tard., FWTA FF:71. Aspidiaceae.
Fern of forest floor. Lvs bipinnate, leaflets microphyll to notophyll. Propagated by dust-like spores. Plot records very highly correlated with acid, base-poor soils. Guinée to Zaire; (Guineo-Congolian wide).

172 CTENITIS PILOSISSIMA (J. Sm.) Alston, FWTA FF:71. Aspidiaceae.
Fern of forest floor. Lvs tripinnate, leaflets nanophyll. Propagated by dust-like spores. S. Leone to Cameroun; (Guinea wide).

173 CTENITIS PROTENSA (Afzel. ex Sw.) Ching, FWTA FF:71. Aspidiaceae.
Fern of forest floor. Lvs tripinnate; leaflets nanophyll to microphyll. Propagated by dust-like spores, and slender creeping rhizome. Guinée to Cameroun; (Guinea wide).

174 CULCASIA ANGOLENSIS Welw. ex Schott, FWTA 3:124, and CULCASIA STRIOLATA Engl., FWTA 3:124. Araceae.
C. angolensis is a large herbaceous root climber which may climb into the canopy, while *C. striolata* is an erect perennial ground herb with stilt roots. Plot records are mapped together because young plants of the two species are virtually indistinguishable. Herbarium records are distinguished: *C. angolensis* ○; *C. striolata* ⊙. Lvs: *C. angolensis* macrophyll, *C. striolata* mesophyll. Fl small, whitish, monoecious, crowded in spike, not clearly seasonal. Fr reddish berry; seeds medium. Regenerate in shade. Guinée to Angola (*C. angolensis*); (Guineo-Congolian wide); Guinée to Cameroun (*C. striolata*); (Guinea wide).

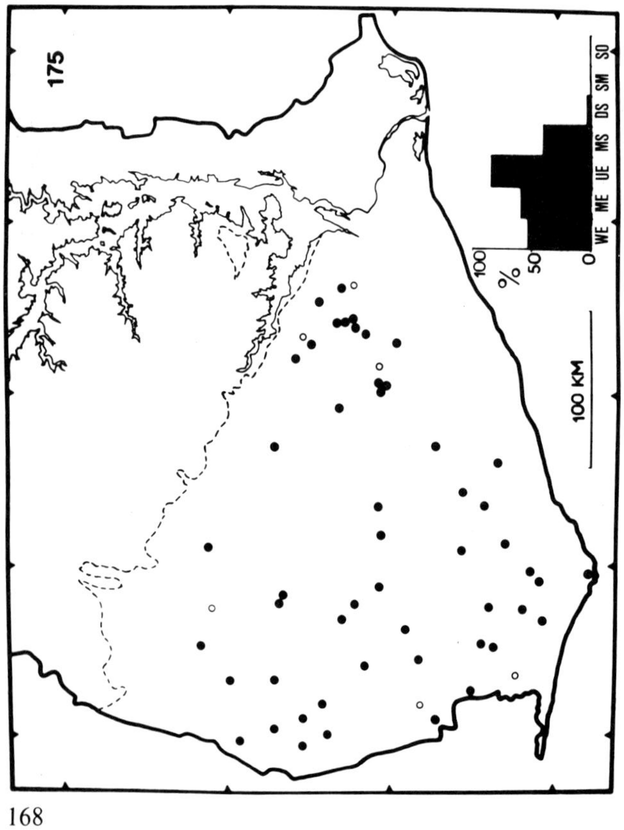
175
100 KM
100
%
50
0
WE ME UE MS DS SM SO

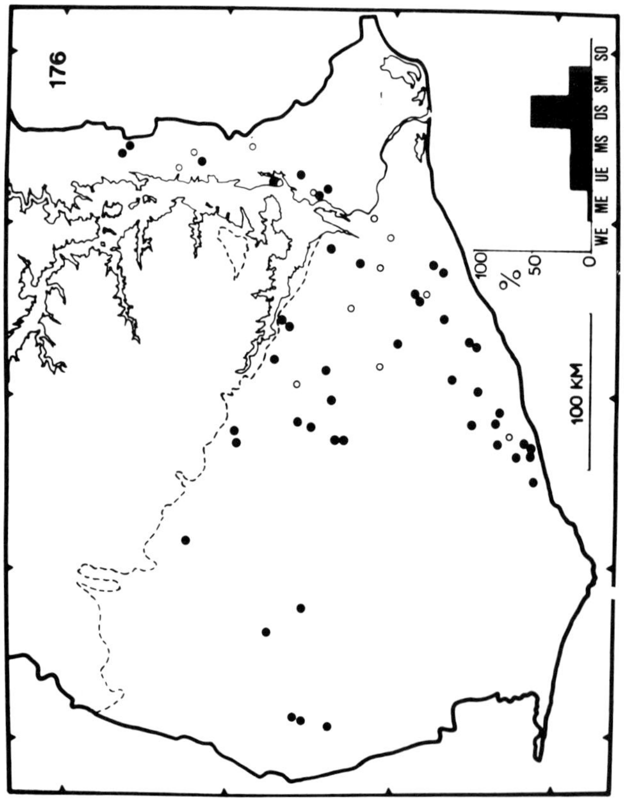
176
100 KM
100
%
50
0
WE ME UE MS DS SM SO

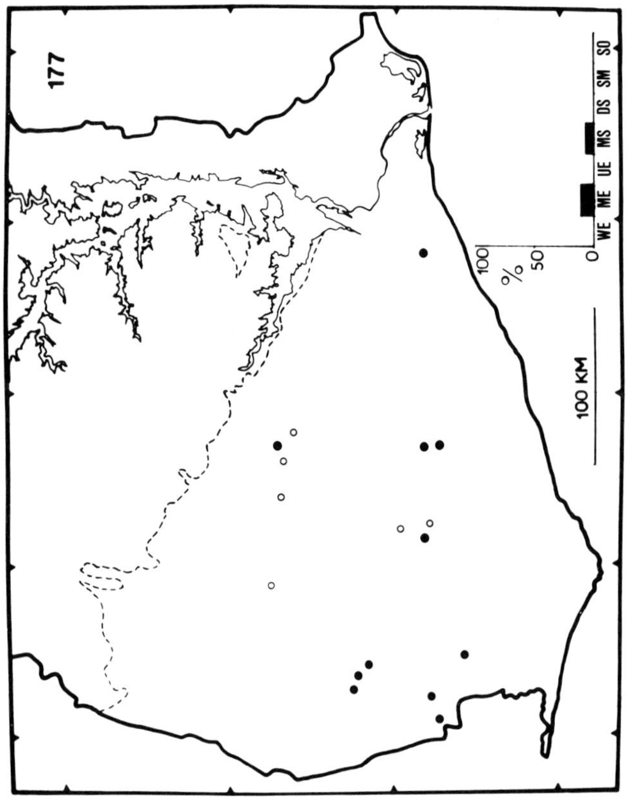
177
100 KM
100
%
50
0
WE ME UE MS DS SM SO

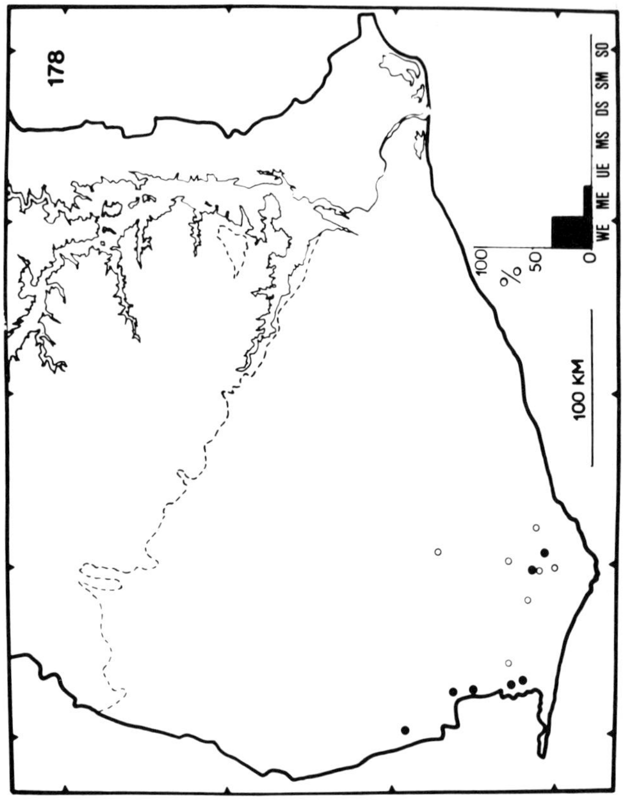
178
100 KM
100
%
50
0
WE ME UE MS DS SM SO

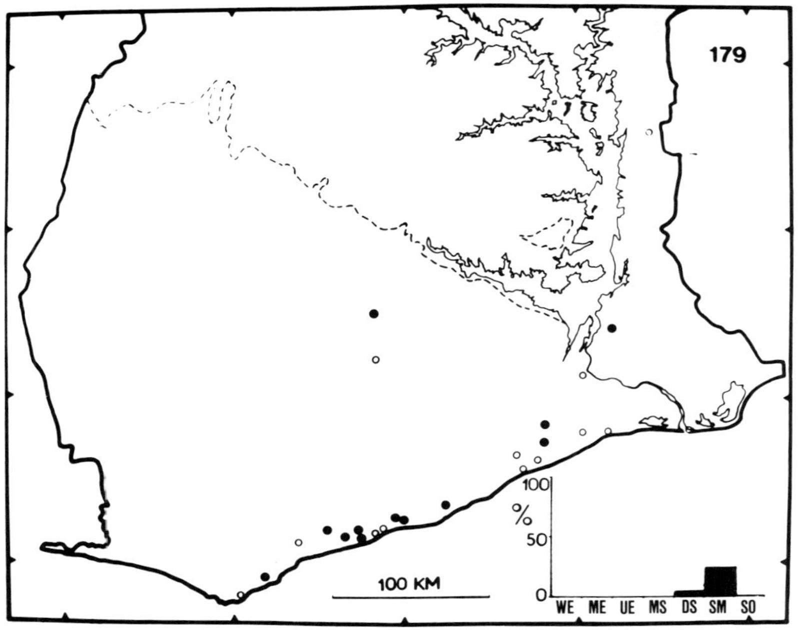

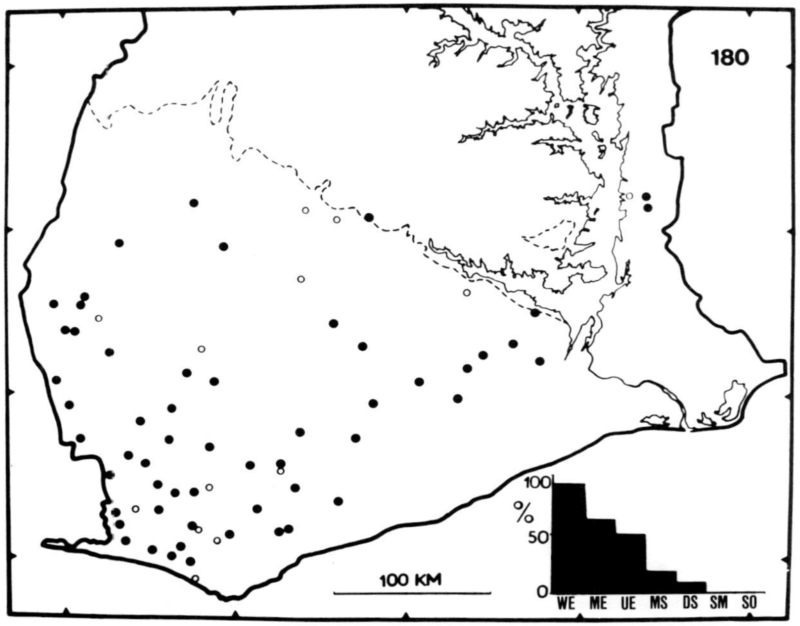

175 CULCASIA PARVIFLORA N. E. Br., FWTA 3:124. Araceae.
Small root climber on tree trunks in forest understorey; lvs notophyll. Fl small, whitish, monoecious, crowded in spike, not clearly seasonal. Fr red berry; seeds medium. S. Leone to Cameroun; (Guinea wide).

176 CULCASIA SCANDENS P. Beauv., FWTA 3:124. Araceae. (Type from Shama).
Root climber on tree trunks in forest understorey; lvs notophyll. Fl small, whitish, monoecious, crowded in spike, not clearly seasonal. Fr red berry; seeds medium. Liberia to Cameroun; (Guinea wide).

177 CYLICODISCUS GABUNENSIS Harms, FWTA 1:489. Mimosaceae. Akan: denya.
Large tree with spreading crown (Fig. 5.2); lvs bipinnate, leaflets microphyll, deciduous. Fl yellowish, small, densely crowded in spikes, not clearly seasonal. Fr long follicle seeds very large, flattened and winged, dispersed by wind. Regenerates in shade. Iv. Coast to Gabon; (Guinea wide). Class 3 timber, very hard and heavy (40).

178 CYNOMETRA ANANTA Hutch. & Dalz., FWTA 1:458. Caesalpiniaceae. Akan: ananta. (Type from Ankobra River basin).
Large tree with thin buttresses; lvs pinnate, leaflets microphyll, brilliant red when newly flushed. Troll's model. Fl white, small, bowl, densely crowded, 10–11. Fr explosive pod, 12–2; seeds very large. Regenerates in shade. Liberia to Ghana; (Upper Guinea endemic). Class 3 timber (40).

179 CYNOMETRA MEGALOPHYLLA Harms, FWTA 1:458. Caesalpiniaceae. Akan: ananta-akoa.
Medium-sized tree, often branching from near base; lvs pinnate, leaflets microphyll, pale at first when flushing. Troll's model. Fl white, small, bowl, densely crowded, 12–2. Fr corky nut, adapted to water dispersal; seeds very large. Restricted to riverbanks in the moister parts of the forest zone, but in dry forest also found away from streams. Regenerates in shade. S. Leone to Nigeria; (Guinea wide).

180 DACRYODES KLAINEANA (Pierre) H. J. Lam, FWTA I:696. Burseraceae. Akan: adwea.
Medium-sized tree; lvs pinnate, leaflets notophyll. Fl pale yellow, small, bowl, polygamous, 10–11. Fr orange drupe with sweet edible flesh, 1–2; seeds large. Regeneration in shade. Plot records very significantly associated with base-poor, acid soils. Occurs in fringing forest in the savanna zone in Mole National Park. S. Leone to Gabon; (Guinea wide).

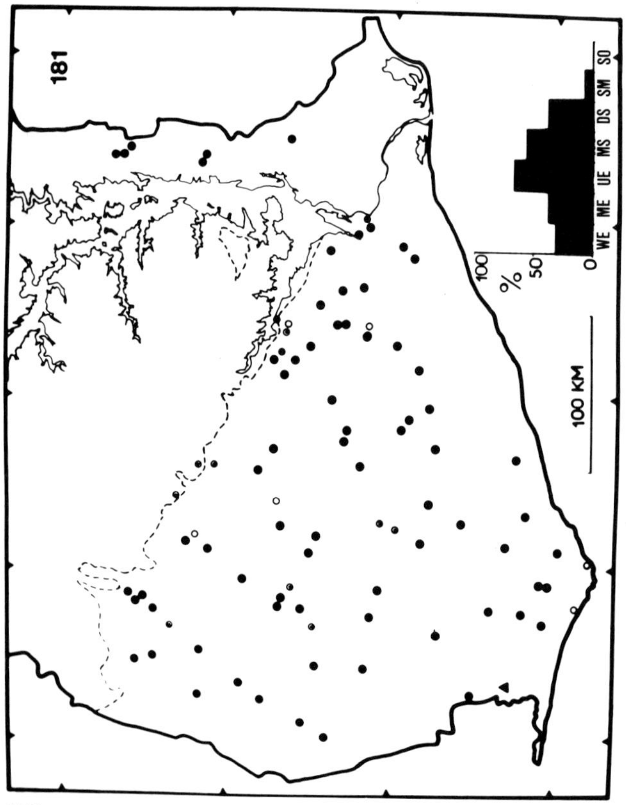
181
100 KM
%
100
50
0
WE ME UE MS DS SM SO

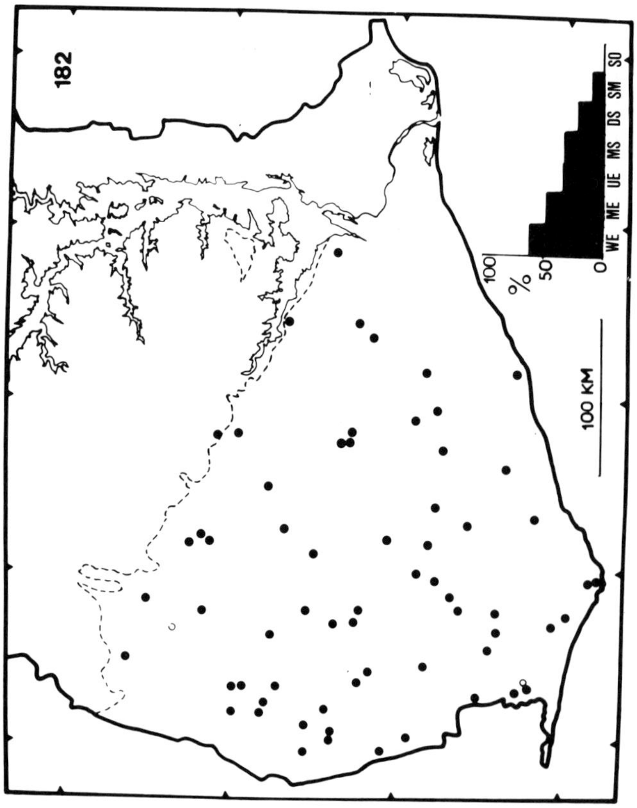
182
100 KM
%
100
50
0
WE ME UE MS DS SM SO

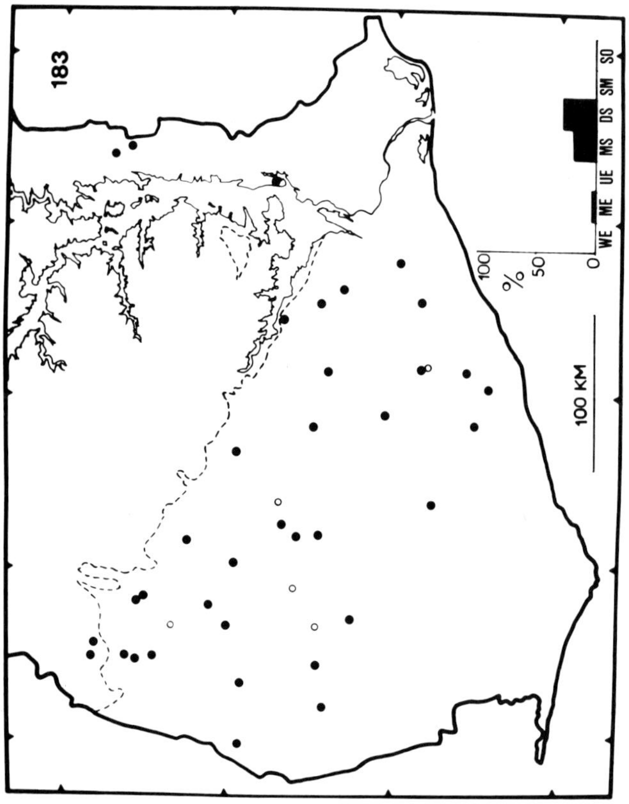
183
100 KM
%
100
50
0
WE ME UE MS DS SM SO

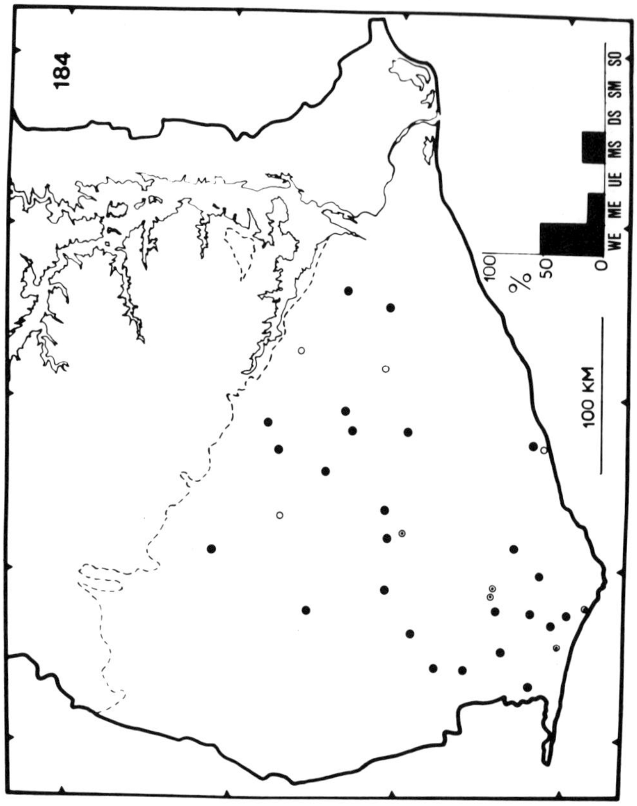
184
100 KM
%
100
50
0
WE ME UE MS DS SM SO

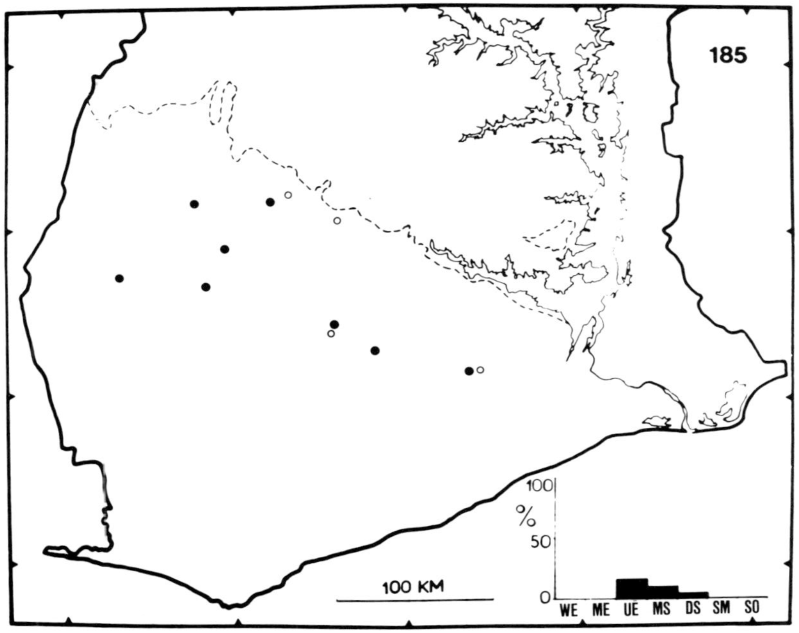

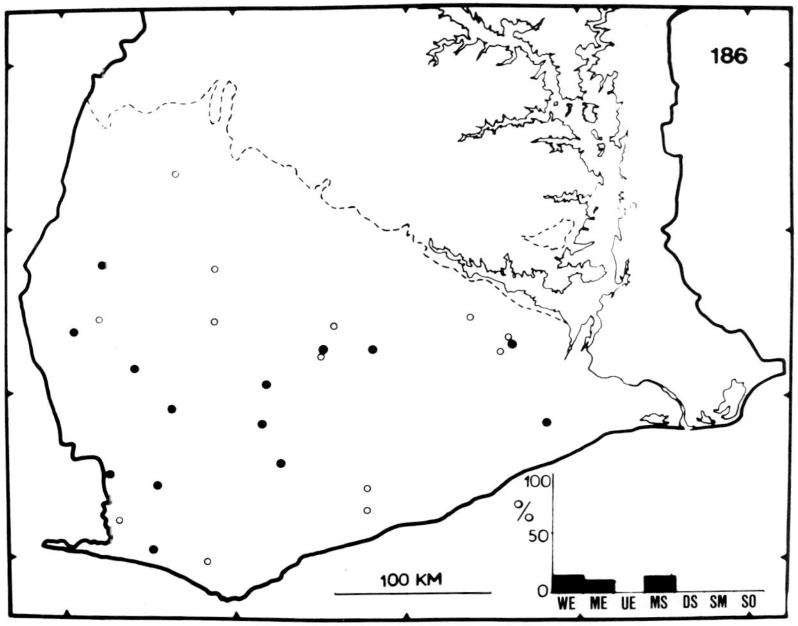

181 DALBERGIA AFZELIANA G. Don, FWTA 1:516, DALBERGIA HOSTILIS Benth., FWTA 1:516, and DALBERGIA sp. Papilionaceae.

Large woody climbers, or scrambling shrubs in thicket; wood normal; some red latex in phloem of *D. afzeliana,* long spines on stems of *D. hostilis.* Lvs pinnate; leaflets nanophyll (*D. afzeliana* and *D. hostilis*), microphyll (*D.* sp.). Troll's model. Seedlings and young saplings cannot be certainly identified, hence plot records are not distinguished. Herbarium records: *D. afzeliana* ○; *D. hostilis* ⊙; *D. sp.* ▲. Fl white, small flag, not clearly seasonal. Fr flat, thin, indehiscent pod, wind dispersed; seeds medium. Regenerate in shade. Distribution: *D. afzeliana* Guinea-Bissau to Angola, (Guineo-Congolian wide); *D. hostilis* Guinea-Bissau to Zaire, (Guineo-Congolian wide); *D.* sp. Ghana and Ivory Coast (Aké Assi 7065), (Upper Guinea endemic).

182 DALBERGIA OBLONGIFOLIA G. Don, FWTA 1:516. Papilionaceae.

Large woody climber, or scrambling shrub in thicket; wood becoming dissected by phloem bands, and stem becoming flattened and ribbon-like. Lvs pinnate, leaflets microphyll. Fl white, small, flag. Fr thin flattened indehiscent pod, wind-dispersed, 9–11; seeds large. Regenerates in shade. S. Leone to Ghana; (Upper Guinea endemic).

183 DALBERGIELLA WELWITSCHII (Bak.) Bak. f., FWTA 1:517. Papilionaceae.

Large woody climber; xylem becoming dissected by phloem bands, phloem containing red latex, stem becoming flattened and ribbon-like. Lvs pinnate, leaflets microphyll. Troll's model. Fl yellowish with red streak, fragrant, medium, flag, 11–12. Fr thin flattened indehiscent pod; seed large. Regenerates in shade. Guinée to Angola; (Guineo-Congolian wide).

184 DANIELLIA OGEA (Harms) Rolfe ex Holl., FWTA 1:463, and DANIELLIA THURIFERA Benn., FWTA 1:463, Caesalpiniaceae. Akan: ehyedua (*D. ogea*), sopi (*D. thurifera*).

Large trees; lvs pinnate, leaflets notophyll, bright red when flushing, glandular-punctate, deciduous. Troll's model. Fl lilac, medium, bowl, 10–1. Fr papery, dehiscent pod, seed remaining attached to a valve, and wind dispersed; seed very large. Regenerates in shade and small gaps. The only reliable character to separate the two species is the hairiness of the inflorescence (113), hence plot records have not been distinguished. Herbarium records: *D. ogea* ○; *D. thurifera* ⊙. Distribution: *D. ogea* Senegal to Gabon, (Guinea wide); *D. thurifera* Guinea Bissau to Ghana, (Upper Guinea endemic). Timber easily worked, exploited (40), but not classified. A fragrant copal resin may be collected from around the base of the trunk.

185 DASYLEPIS BREVIPEDICELLATA Chipp, FWTA 1:186. Flacourtiaceae. Akan: asratoaduanini. (Type from Mim).

Small to medium tree; lvs mesophyll, toothed. Fl white, medium, bowl, 3–4. Fr berry; seeds medium to large. Regenerates in shade. Guinée to Zaire; (Guineo-Congolian wide, 100). *D. assinensis* Chev. has been considered a synonym of *D. brevipedicellata* (100), but as it is quite distinct in Ghana, both in morphology and distribution, it has not been shown on the map.

186 DEINBOLLIA GRANDIFOLIA Hook. f., FWTA 1:715. Sapindaceae. Akan: mmata.

Pygmy to small tree; lvs pinnate, leaflets mesophyll to macrophyll. Fl white, small, polygamous, bowl, 11–2. Fr orange berry with sweet flesh around the large seeds. In evergreen forest occurs as a pygmy tree with rather short cauliflorous inflorescences, but in semi-deciduous forest is taller, with large panicles among the leaves, and with much bigger leaflets. S. Leone to Gabon and Uganda; (Guineo-Congolian wide, 45).

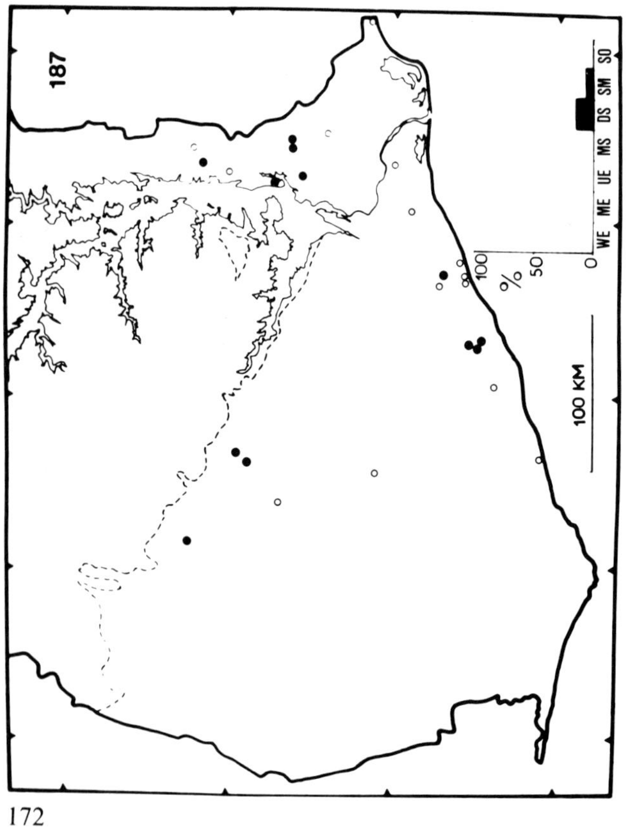
187
100 KM
%
100
50
0
WE ME UE MS DS SM SO

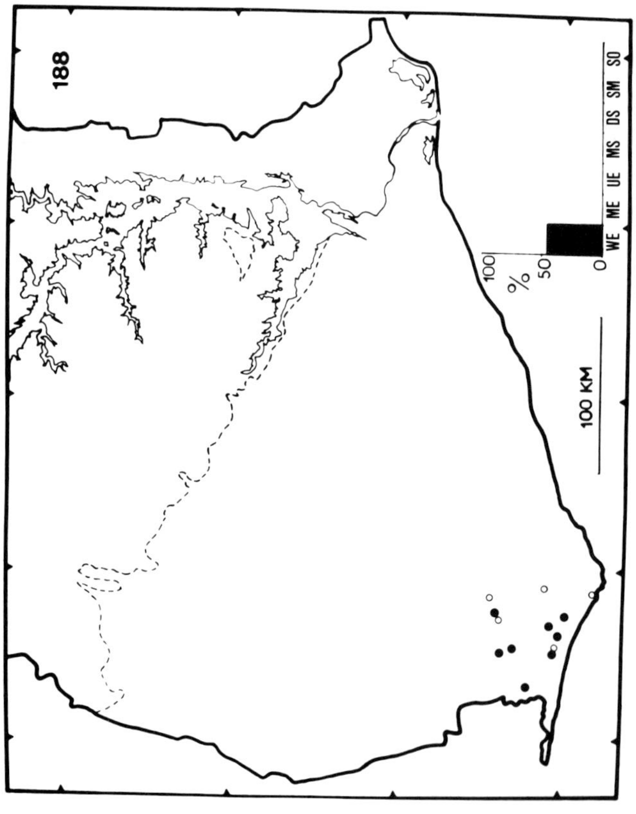
188
100 KM
%
100
50
0
WE ME UE MS DS SM SO

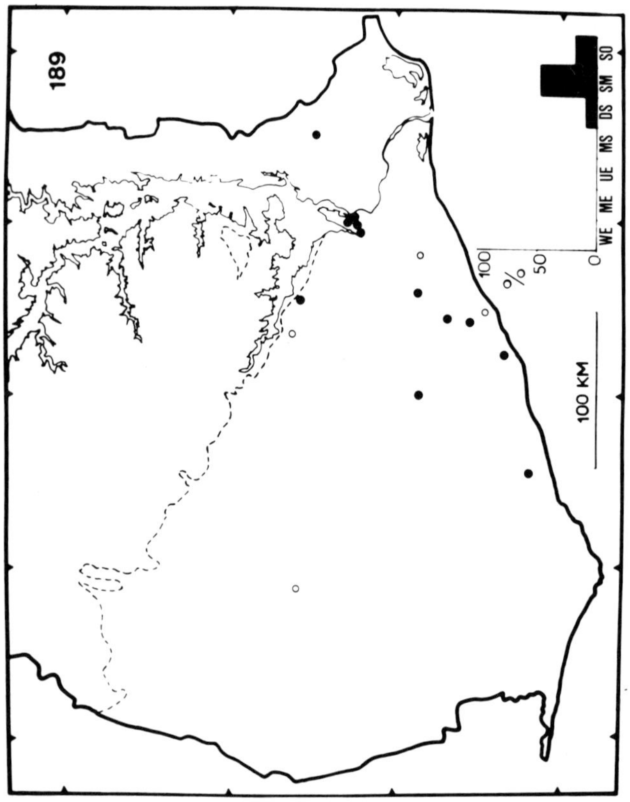
189
100 KM
%
100
50
0
WE ME UE MS DS SM SO

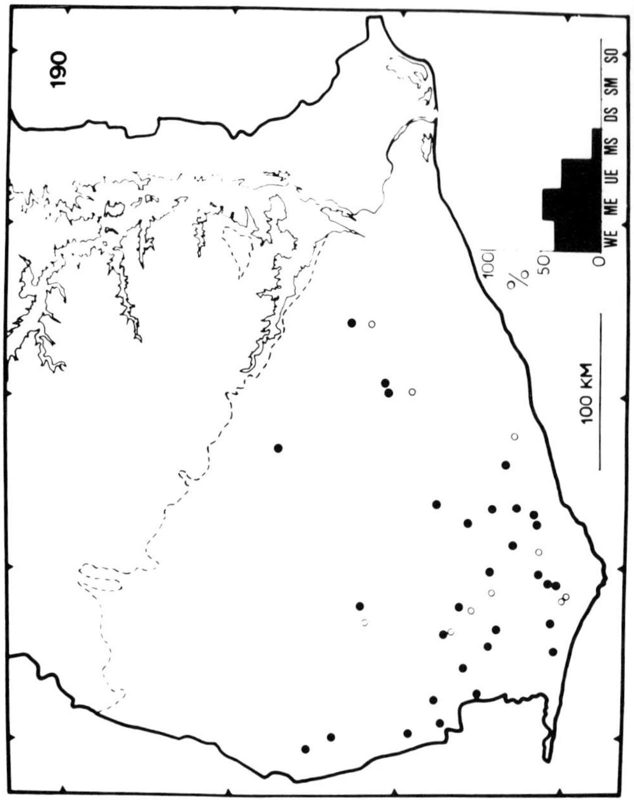
190
100 KM
%
100
50
0
WE ME UE MS DS SM SO

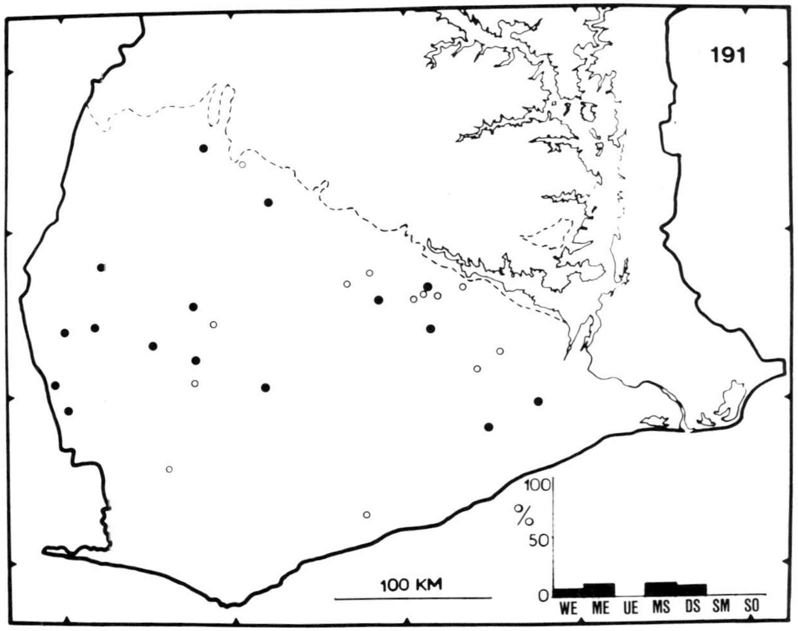

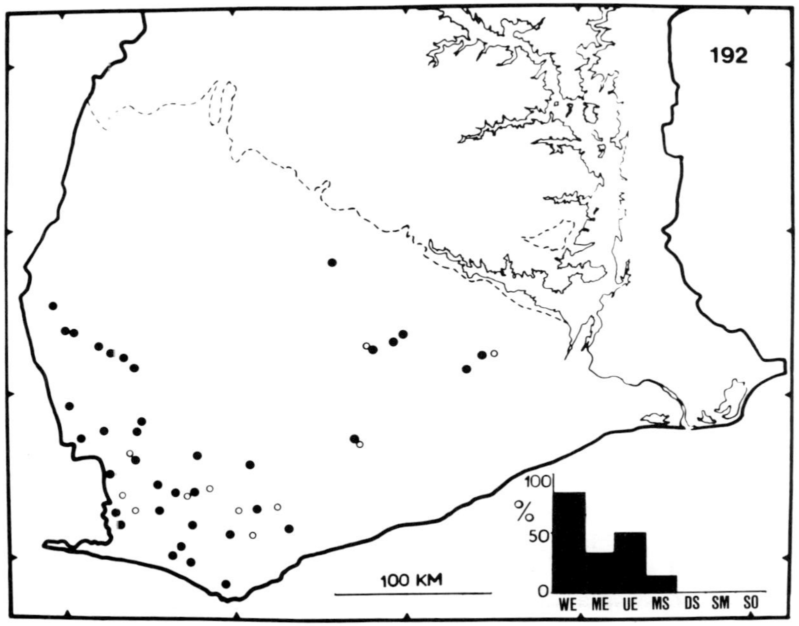

187 DEINBOLLIA PINNATA (Poir.) Schum. & Thonn., FWTA 1:715. Sapindaceae. Akan: woagye-akoa. (Type probably from near Accra).
Small tree; lvs pinnate, leaflets mesophyll, pubescent. Corner's model. Fl cream, polygamous, small, bowl, 1–5. Fr orange berry with sweet flesh around the large seeds. Regenerates in shade. Guinée to Nigeria; (Guinea wide).

188 DELPYDORA GRACILIS A. Chev., FWTA 2:25. Sapotaceae.
Pygmy tree, containing white latex, very hairy; lvs mesophyll. Corner's model (59). Fl white, small to medium, bowl, not clearly seasonal. Fr orange berry, seeds large. Regenerates in shade. Good indicator of Wet Evergreen forest. Liberia to Ghana; (Upper Guinea endemic).

189 DENNETTIA TRIPETALA Bak. f., FWTA 1:51. Annonaceae.
Small tree, slash strongly and distinctively aromatic; lvs notophyll. Roux's model. Fl dark purple with cream centre, medium, bowl, cauliflorous, 12–3. Fr reddish berry; seeds medium to large. Regenerates in shade. Occurs in Moist Semi-deciduous forest only on shallow soil over rock. Iv. Coast to Cameroun; (Guinea wide). In Ivory Coast (3) and Ghana (85) has been misidentified as *Uvariopsis congensis* Robyns & Ghesq.

190 DESPLATSIA CHRYSOCHLAMYS (Mildbr. & Burret) Mildbr. & Burret, FWTA 1:307. Tiliaceae. Akan: esonowisamfie (also other species of *Desplatsia*).
Small tree; lvs mesophyll to macrophyll, toothed; hairy. Massart's model (59). Fl yellowish-white, medium, bowl, 7–10. Fr large orange fibrous berry, covered with bristly hairs; seeds large. Dispersed by elephants, the seeds germinating in the dung. Half the plot records were on soil with pH 4.2–4.5. S. Leone to Zaire and Uganda; (Guineo-Congolian wide).

191 DESPLATSIA SUBERICARPA Bocq., FWTA 1:307. Tiliaceae.
Small tree; lvs mesophyll, toothed. Fl pinkish, medium, bowl, not clearly seasonal. Fr large, yellow, fibrous berry; seeds large, dispersed by elephants. S. Leone and Zaire; (Guineo-Congolian wide).

192 DIALIUM AUBREVILLEI Pellegr., FWTA 1:449. Caesalpiniaceae. Akan: dua-bankye.
Large tree with buttresses, red latex in slash; lvs pinnate, leaflets notophyll. Troll's model. Fl small, whitish, bowl, in large inflorescences, 7–11. Fr 1-seeded black berry with acid edible pulp surrounding the large seed. Regenerates in shade. Plot records very significantly associated with base-poor, acid soils of high clay content. S. Leone to Ghana; (Upper Guinea endemic). Timber is very hard, and not used (113).

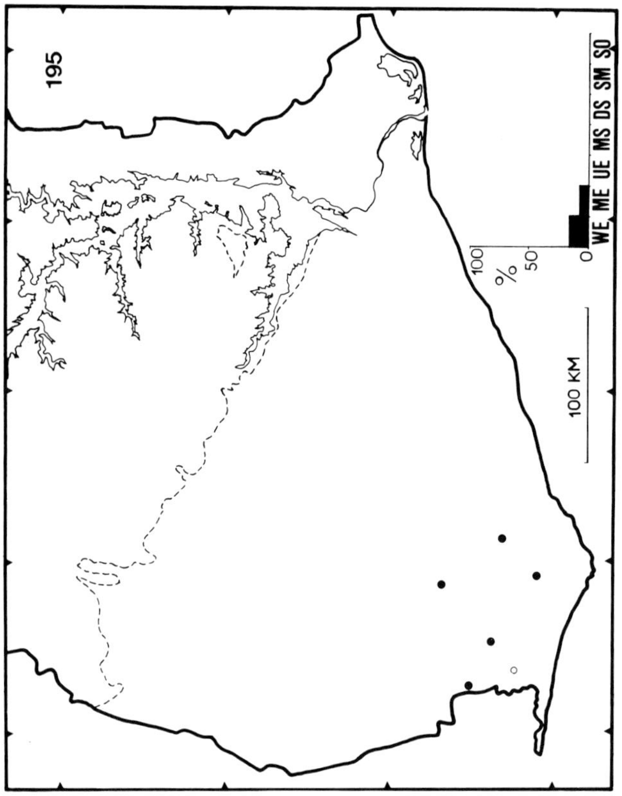
195
100 KM
%
100
50
0
WE ME UE MS DS SM SO

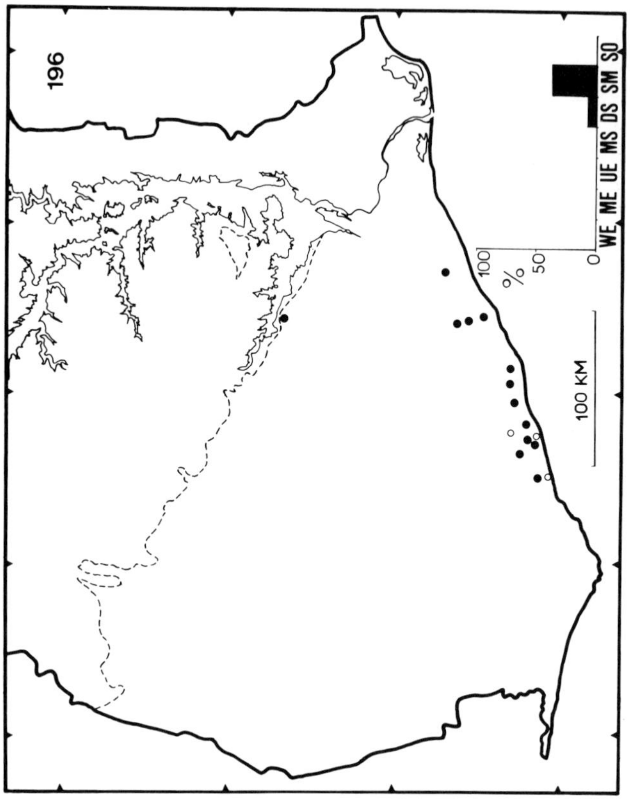
196
100 KM
%
100
50
0
WE ME UE MS DS SM SO

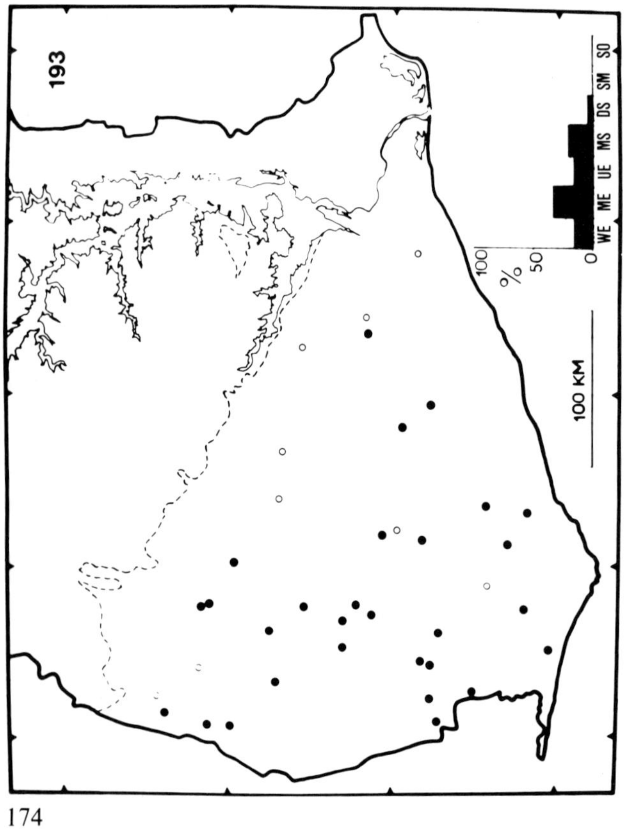
193
100 KM
%
100
50
0
WE ME UE MS DS SM SO

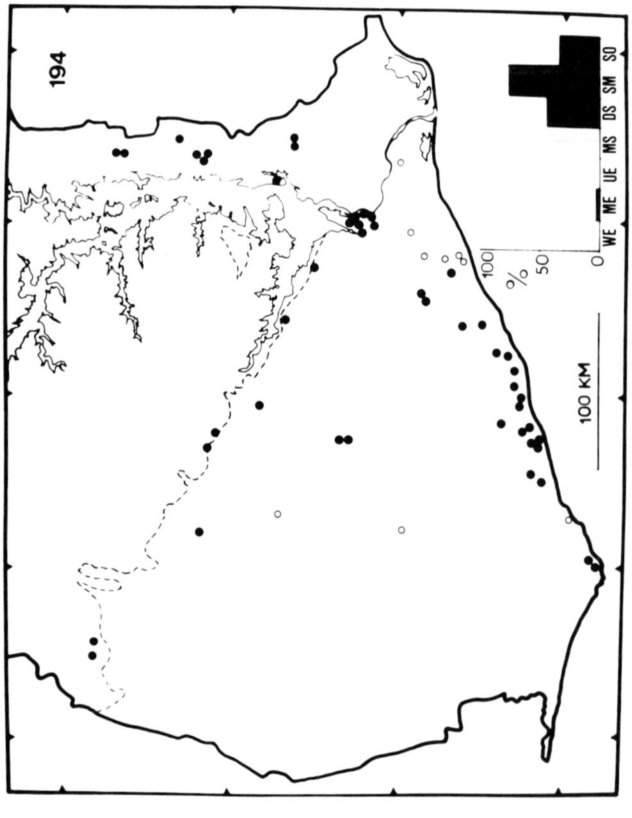
194
100 KM
%
100
50
0
WE ME UE MS DS SM SO

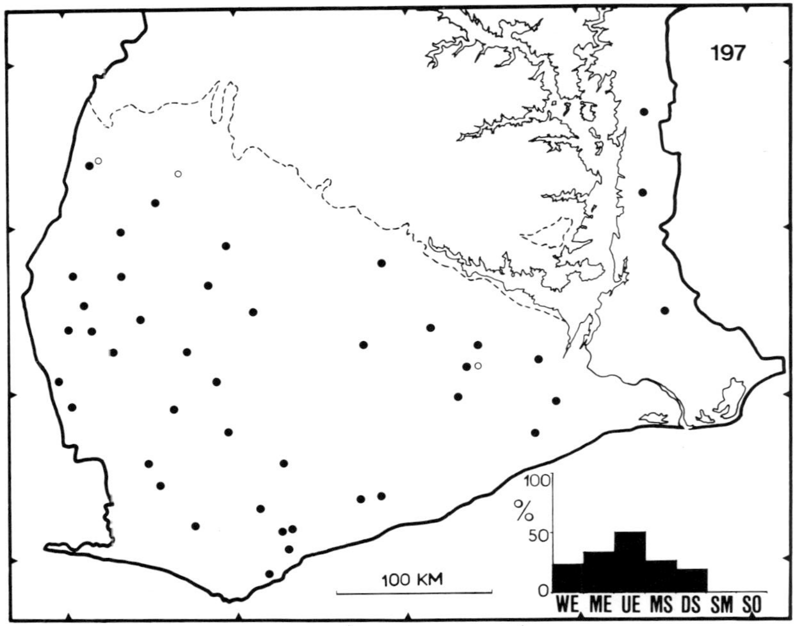

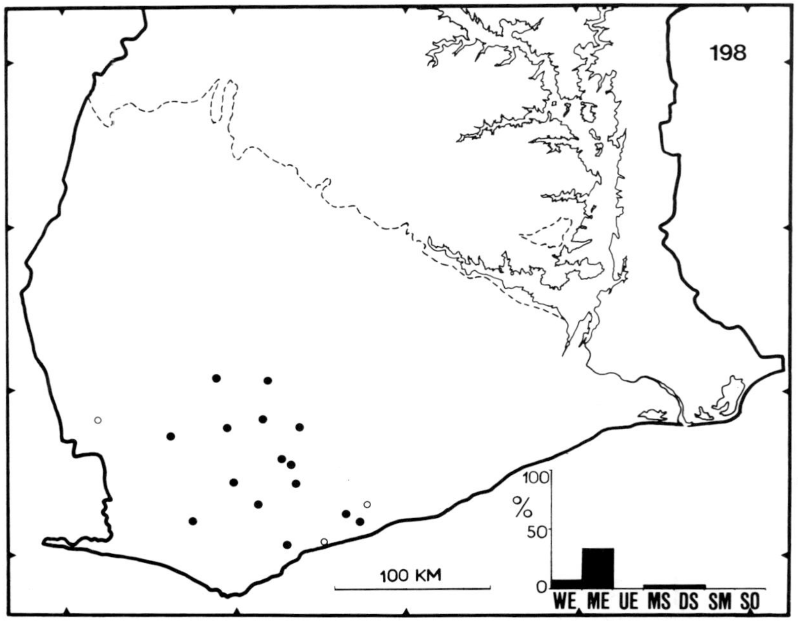

193 DIALIUM DINKLAGEI Harms, FWTA 1:449. Caesalpiniaceae. Akan: dwedweedwe.

Medium-sized tree, red latex in slash; lvs pinnate, leaflets microphyll, deciduous. Troll's model. Fl brownish yellow, small, bowl, in large inflorescences, 3–9. Fr 1-seeded dark brown berry, with edible pulp surrounding the medium-sized seed. Regenerates in shade. Guinée to Zaire; (Guineo-Congolian wide).

194 DIALIUM GUINEENSE Willd., FWTA 1:449. Caesalpiniaceae. Akan: asenaa. Ga: yoye. (Type probably from near Accra).

Small to medium tree, often branching low, red latex in slash; lvs pinnate, leaflets microphyll. Troll's model. Fl brownish-cream, small, bowl, in large inflorescences, 9–11. Fr black 1-seeded berry, with edible pulp surrounding the medium-sized seed. Regenerates in shade. Occurs in fringing forest in the savanna zone. Senegal to Nigeria; (Guinea wide).

195 DICHAPETALUM ANGOLENSE Chod., FWTA 1:436. Dichapetalaceae (Chailletiaceae in FWTA).

Large woody climber, with normal xylem; lvs mesophyll, with extra-floral nectaries on lower surface; twigs and leaves pubescent. Roux's model (59). Fl white, small, bowl; not yet found in Ghana. Fr yellow-orange drupe, not yet found in Ghana; seeds very large. Regenerates in shade. Liberia to Angola and Uganda; (Guineo-Congolian wide, 29).

196 DICHAPETALUM BARTERI Engl., FWTA 1:436. Dichapetalaceae (Chailletiaceae in FWTA). Akan: akusakusa.

Small tree; lvs notophyll to mesophyll, with extra-floral nectaries on lower surface. Roux's model. Fl white, small, bowl, not clearly seasonal. Fr yellow, tuberculate drupe; seed large. Regenerates in shade. Ivory Coast to Nigeria; (Guinea wide, 29).

197 DICHAPETALUM CRASSIFOLIUM Chod. (30). Dichapetalaceae (Chailletiaceae in FWTA).

Large woody climber, xylem deeply dissected by radial phloem intrusion, reddish translucent exudate from cut stem; lvs notophyll, with extra-floral nectaries on lower surface. Roux's model. Fl white, small, bowl, not clearly seasonal. Fr orange drupe; seeds medium to large. Regenerates in shade. Liberia to Angola and Zambia; (Guineo-Congolian wide). The only specimen of this species listed in FWTA ed. 2 (Vigne FH 2485) was identified as *D. toxicarium* (G. Don) Baill.; but *D. crassifolium* is easily recognised in the field by its leaves which are very shiny on both surfaces, and by the densely appressed pubescent petiole.

198 DICHAPETALUM FILICAULE Bretel. (30). (*D. cymulosum* of FWTA, partly, not of (Oliv.) Engl.). Dichapetalaceae (Chailletiaceae in FWTA).

Slender woody climber, or shrub in thicket; lvs microphyll. Fl white, small, bowl. Fr orange drupe; seeds large. Regenerates in shade. Ivory Coast and Ghana; (Upper Guinea endemic).

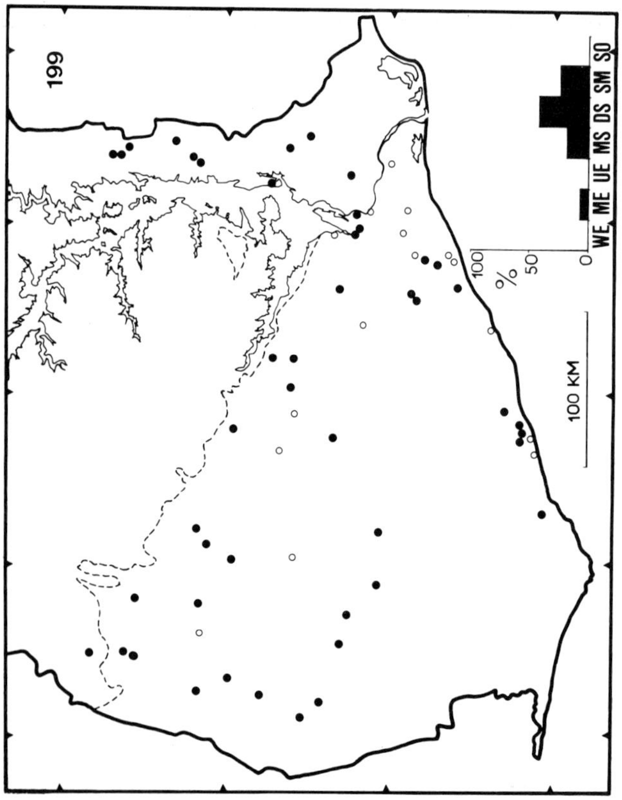
199
100 KM
%
100
50
0
WE ME UE MS DS SM SO

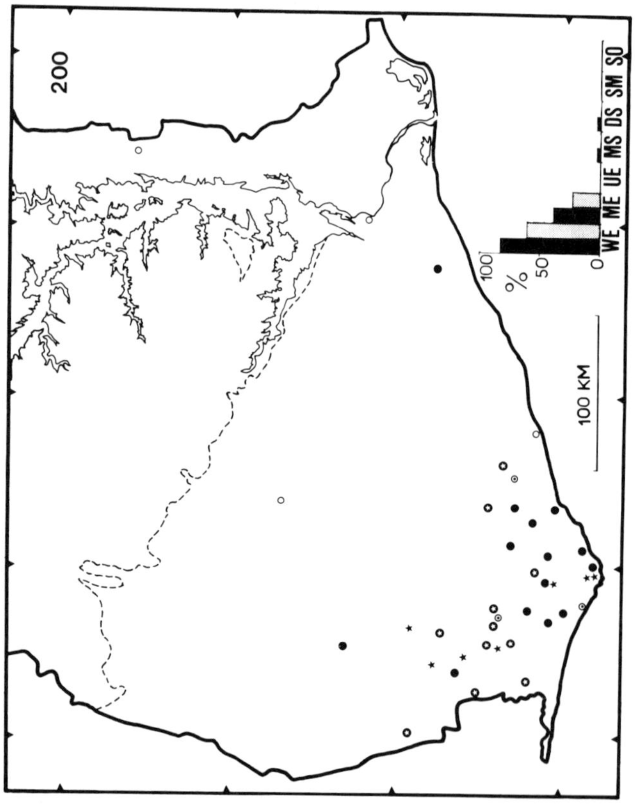
200
100 KM
%
100
50
0
WE ME UE MS DS SM SO

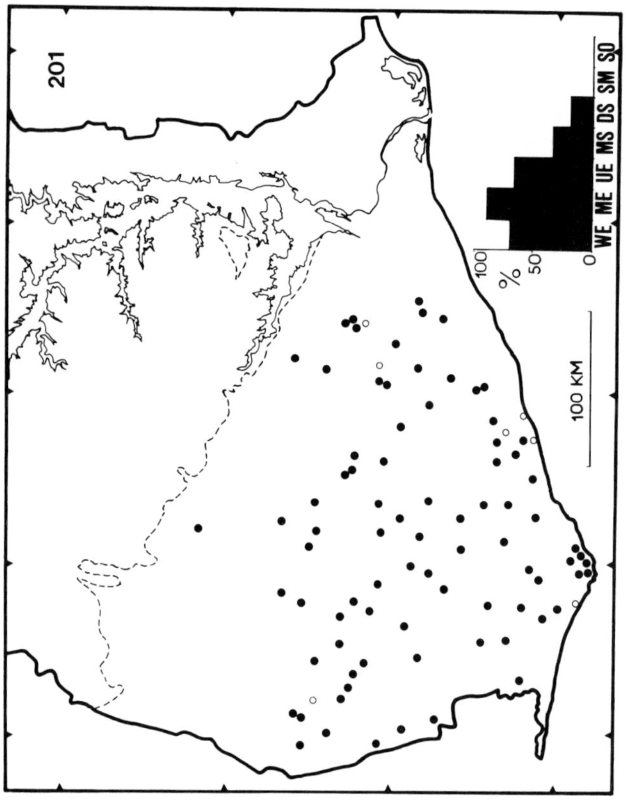
201
100 KM
%
100
50
0
WE ME UE MS DS SM SO

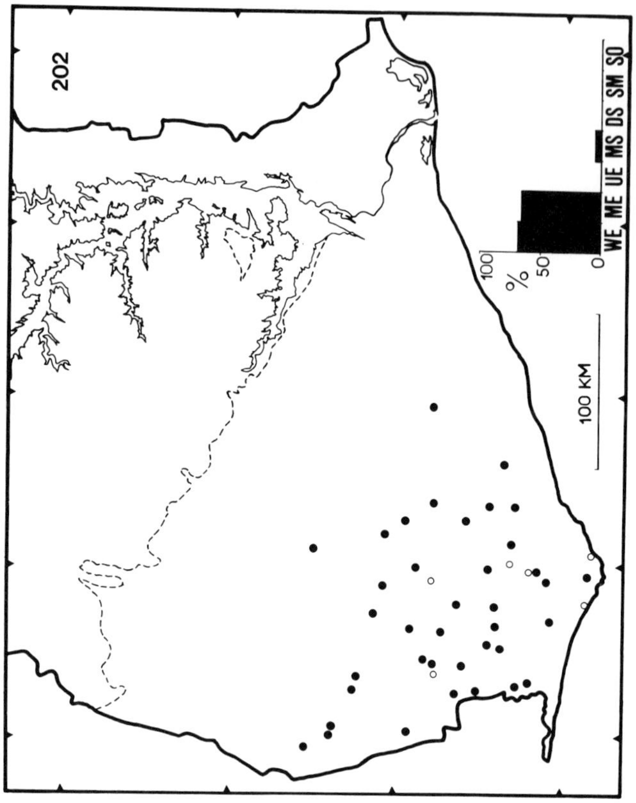
202
100 KM
%
100
50
0
WE ME UE MS DS SM SO

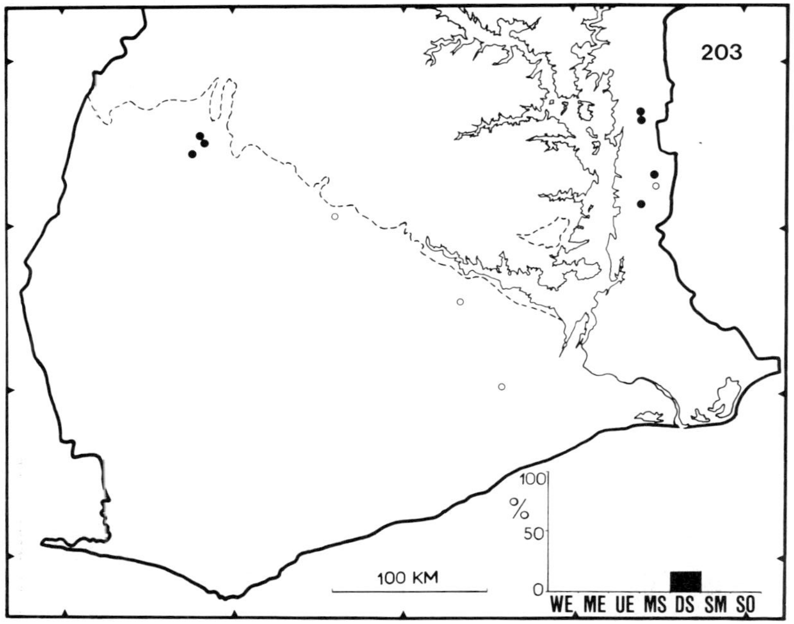

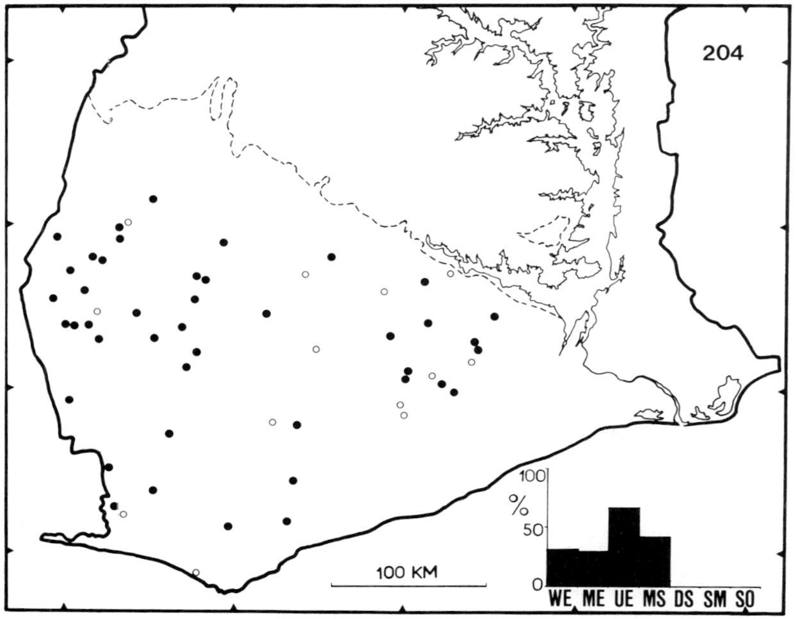

199 DICHAPETALUM GUINEENSE (DC.) Keay, FWTA 1:436. Dichapetalaceae (Chailletiaceae in FWTA). Akan: esonowedie. (Type from Accra Plains).

Small to medium tree, or shrub, when large with fluted trunk and flaking papery bark, and basal branches; lvs microphyll to notophyll, pubescent, with extra-floral nectaries below. Roux's model. Fl white, small, bowl, sweet-scented, often showing profuse gregarious flowering, not clearly seasonal. Fr orange berry; seeds large. Regenerates in shade. S. Leone to Nigeria; (Guinea wide). This species is sometimes (31) included in *D. madagascariense* Poir., which is then considered to be very variable and widespread in Africa; in West Africa, however, *D. guineense* is distinctive and relatively constant.

200 DICHAPETALUM HEUDELOTII (Planch. ex Oliv.) Baill., FWTA 1:438. (Incl. *D. kumasiense* Hoyle, FWTA 1:438, and *D. johnstonii* Engl., FWTA 1:438, (31). Dichapetalaceae (Chailletiaceae in FWTA).

Medium to large woody climbers; old stems deeply furrowed, xylem dissected by radial intrusion of phloem. Extra-floral nectaries on leaf undersurface. Roux's model. Two entities are mapped:

(1) lvs mesophyll, twigs and leaf undersurface persistently and densely velutinous – survey records ★; herbarium records ⊙; histogram bars dotted (such specimens have sometimes been named (31) *D. johnstonii* or *D. varians* Pellegr.).

(2) lvs microphyll to notophyll, sparsely pubescent below on midrib and primary veins only, twigs pubescent but not velutinous – survey records ●; (joint occurrence with (1) ⊛); herbarium records ○; histogram bars solid black. Fl white, small, bowl. Fr orange drupe; seeds large. Regenerate in shade. Guinée to Zaire; (Guineo-Congolian wide). Some of the survey records are *D. heudelotii* var. *ndongense* (Engl.) Bretel. (*D. martineaui* Aubrév. & Pellegr. (31) in FWTA 1:438), distinguished by the spreading, bristly hair on midrib and twigs.

201 DICHAPETALUM PALLIDUM (Oliv.) Engl., FWTA 1:436. Dichapetalaceae (Chailletiaceae in FWTA).

Medium-sized woody climber, or shrub in thicket; lvs notophyll, covered by dense white hair below. Roux's model. Fl small, white, bowl, not clearly seasonal. Fr orange drupe; seeds medium. Regenerates in shade. Survey records very highly correlated with acid, base-poor soils. Guinée to Gabon; (Guinea wide). *D. albidum* A. Chev. ex Pellegr. is sometimes (29) kept distinct because of its bristly fruits; this taxon occurs in Ghana, but can only be identified on the rare occasions when fruits are present.

202 DICHAPETALUM TOXICARIUM (G. Don) Baill., FWTA 1:438. Dichapetalaceae (Chailletiaceae in FWTA).

Medium to large woody climber, old stem with rounded ridges, and xylem deeply dissected by radial intrusion of phloem; or shrub in thicket. Lvs notophyll to mesophyll, with extra-floral nectaries below. Roux's model. Fl small, white, bowl, not clearly seasonal; peduncle often fused with petiole. Fr yellow drupe; seeds very large. Regenerates in shade. Plot records very highly correlated with acid, base-poor soils. Guinée to Ghana; (Upper Guinea endemic). Seeds are rich in toxic fluorine-containing organic substances (58).

203 DICRANOLEPIS GRANDIFLORA Engl., FWTA 1:173. Thymelaeaceae.

Pygmy tree; lvs microphyll, slightly pubescent below. Mangenot's model. Fl white, large, dish with tube, 11–1. Fr orange berry; seeds medium. Regenerates in shade. Ghana to Cameroun; (Guinea wide).

204 DICRANOLEPIS PERSEI Cummins, FWTA 1:173. Thymelaeaceae. (Type from Kumasi).

Pygmy tree; lvs notophyll. Mangenot's model (59). Fl white, large, dish with tube, not clearly seasonal. Fr orange berry; seeds medium to large. Regenerates in shade. Ivory Coast and Ghana; (Upper Guinea endemic).

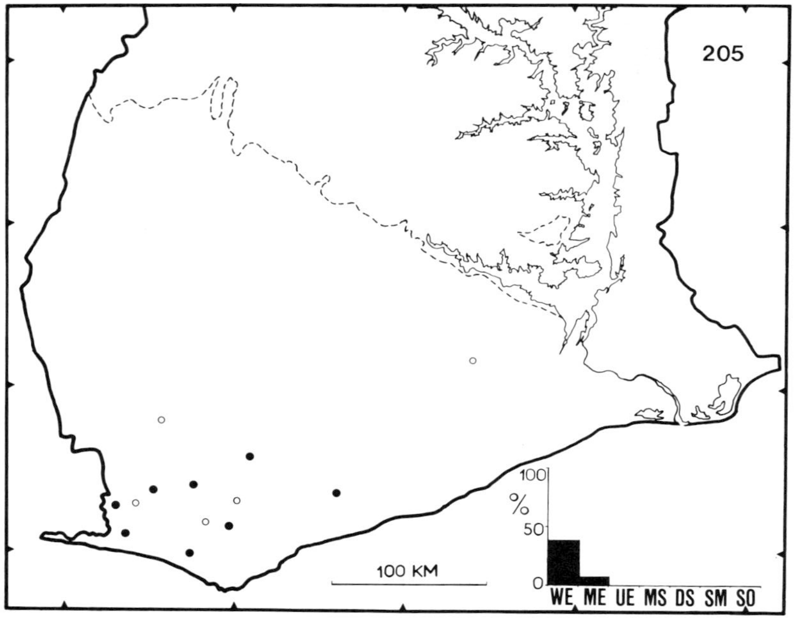
205
100 KM
%
100
50
0
WE ME UE MS DS SM SO

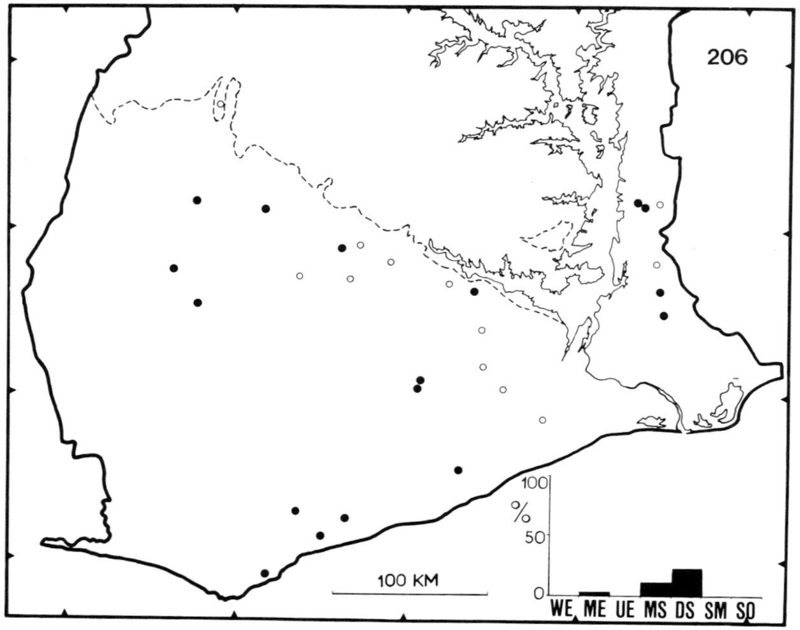
206
100 KM
%
100
50
0
WE ME UE MS DS SM SO

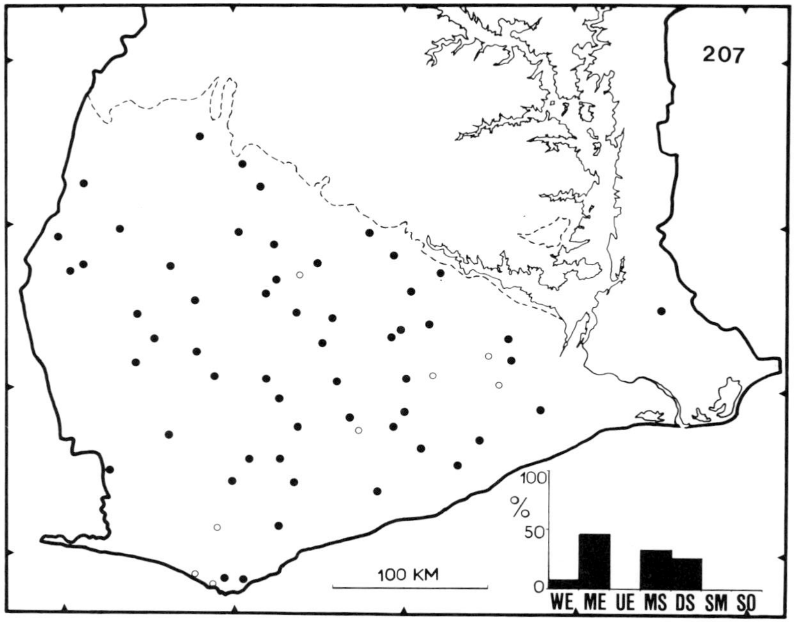
207
100 KM
%
100
50
0
WE ME UE MS DS SM SO

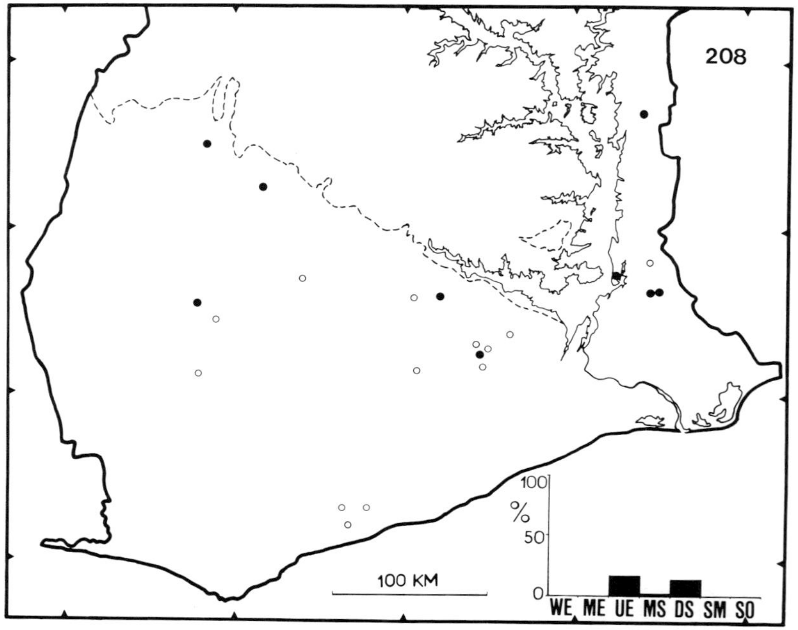
208
100 KM
%
100
50
0
WE ME UE MS DS SM SO

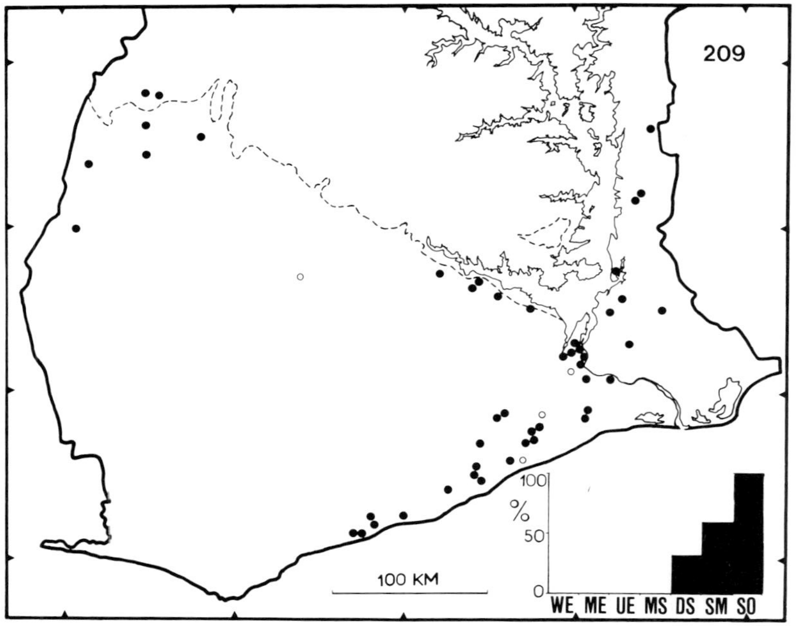

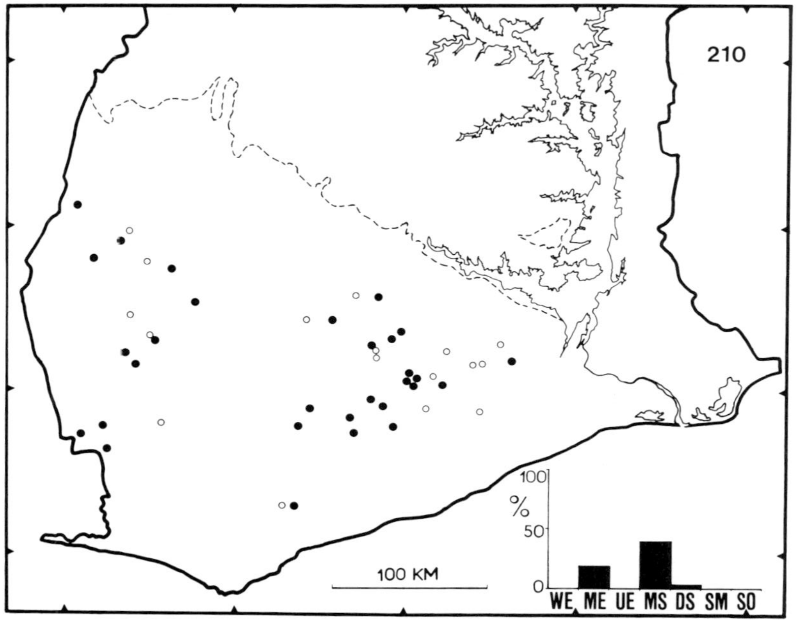

205 DICTYOPHLEBA LEONENSIS (Stapf) Pichon, FWTA 2:59. Apocynaceae.
Large woody climber, bark thick and rough, exuding white latex when cut; lvs mesophyll; lvs and twigs hairy when young. Leeuwenberg's model (59). Fl white, large, dish with tube, 4–8. Fr orange berry, shaped like eggcup containing small egg; seeds large to very large. Regenerates in shade. Guinée to Ghana; (Upper Guinea endemic).

206 DIOSCOREA PRAEHENSILIS Benth., FWTA 3:153. (Incl. *D. liebrechtsiana* De Wild., FWTA 3:154). Dioscoreaceae. Akan: ahabayere.
Medium-sized climber with prickly annual stems and large underground tuber protected by spiny roots; lvs microphyll. McClure's model. Fl whitish, small, bowl, dioecious, 5–8. Fr 3-lobed capsule; seeds medium, winged, wind-dispersed. Regenerates in shade, but grows best in forest margins and gaps. S. Leone to Nigeria; (Guinea wide). Tuber is edible, and the plant is sometimes semi-cultivated (94).

207 DIOSCOREA SMILACIFOLIA De Wild., FWTA 3:153. Dioscoreaceae.
Medium-sized climber with prickly perennial stems and woody underground tuber; lvs notophyll. McClure's model. Fl white, small, bowl, dioecious, 11–2. Fr 3-lobed capsule; seeds medium, winged, wind-dispersed. Regenerates in shade, but grows best in forest margins and gaps. S. Leone to Cameroun; (Guinea wide).

208 DIOSCOREOPHYLLUM CUMMINSII (Stapf) Diels, FWTA 1:73. Menispermaceae. (Type from Kumasi).
Slender softly woody climber; lvs more or less lobed, mesophyll; lvs and twigs hairy. Undescribed model (59). Fl yellowish, small, dish, dioecious, usually cauliflorous, 1–8. Fr red drupe, with extremely sweet pulp; seed medium. Typically occurs in disturbed or secondary forest; the survey records therefore underestimate its abundance. Guinée to Zaire and Sudan; (Guinea-Congolian wide). The sweetness of the pulp is due to a protein, monellin, which (weight for weight) is 3 000 times sweeter than sucrose (34).

209 DIOSPYROS ABYSSINICA (Hiern) F. White, FWTA 2:10. Ebenaceae. Ga: gblitso.
Medium-sized tree; lvs notophyll. Massart's model. Fl cream, small, bowl, dioecious, c.6. Fr black 1-seeded berry, 8–1; seed medium. Regenerates in shade. Widespread in Tropical Africa; (chorological transgressor).

210 DIOSPYROS CANALICULATA De Wild., FWTA 2:14. Ebenaceae (120). Akan: otwabere.
Small tree, branching low; lvs mesophyll. Massart's model (59). Fl white, fragrant, dish with tube, dioecious (male medium, female large), cauliflorous on branches, not clearly seasonal. Fr yellow berry, completely covered by yellow to purplish-brown accrescent calyx; seeds large. Regenerates in shade. Liberia to Angola; (Guinea-Congolian wide).

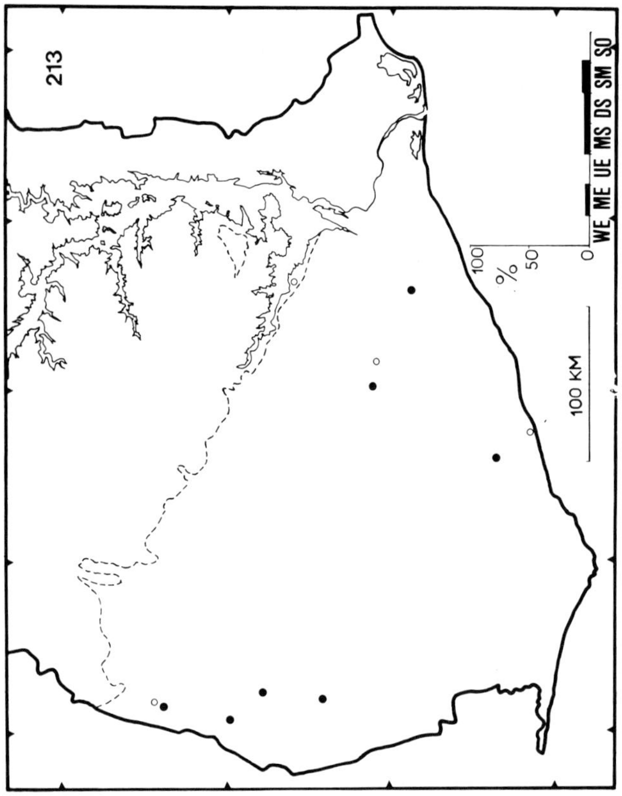
213
100 KM
%
100
50
0
WE ME UE MS DS SM SO

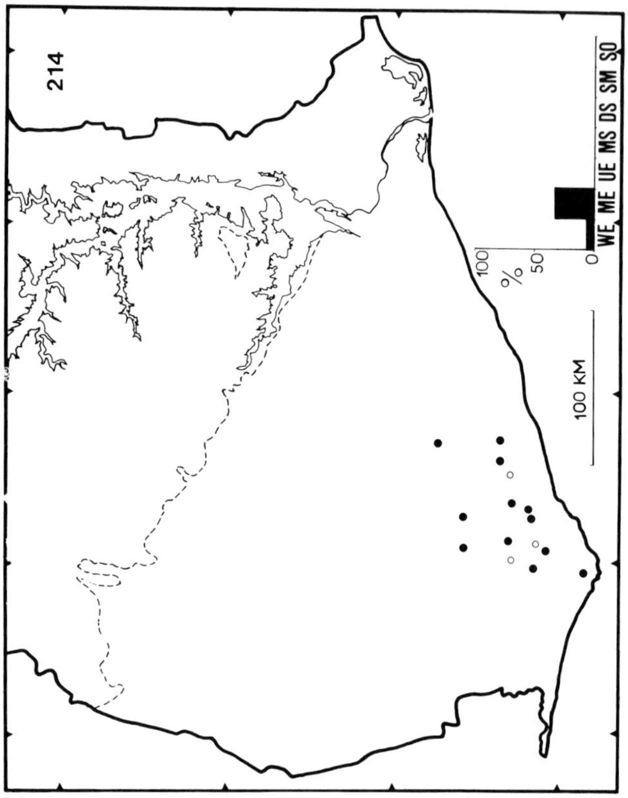
214
100 KM
%
100
50
0
WE ME UE MS DS SM SO

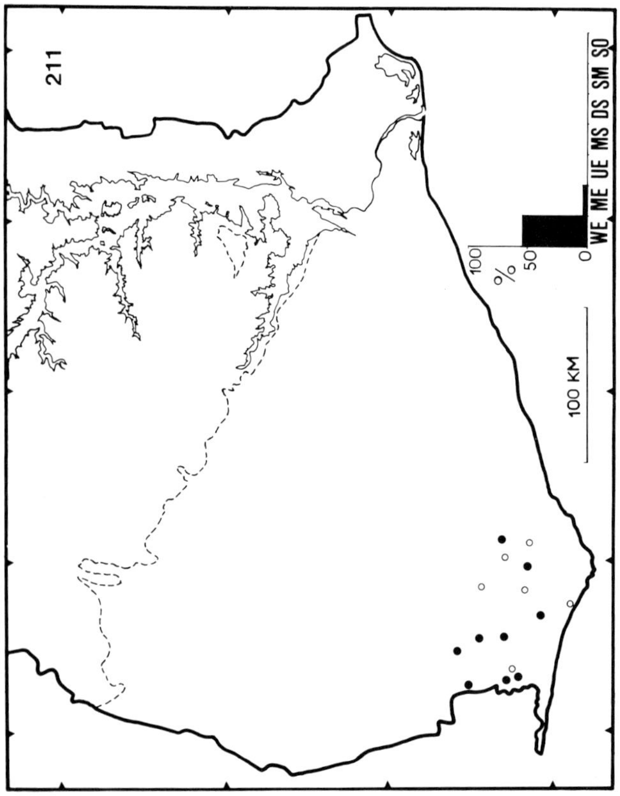
211
100 KM
%
100
50
0
WE ME UE MS DS SM SO

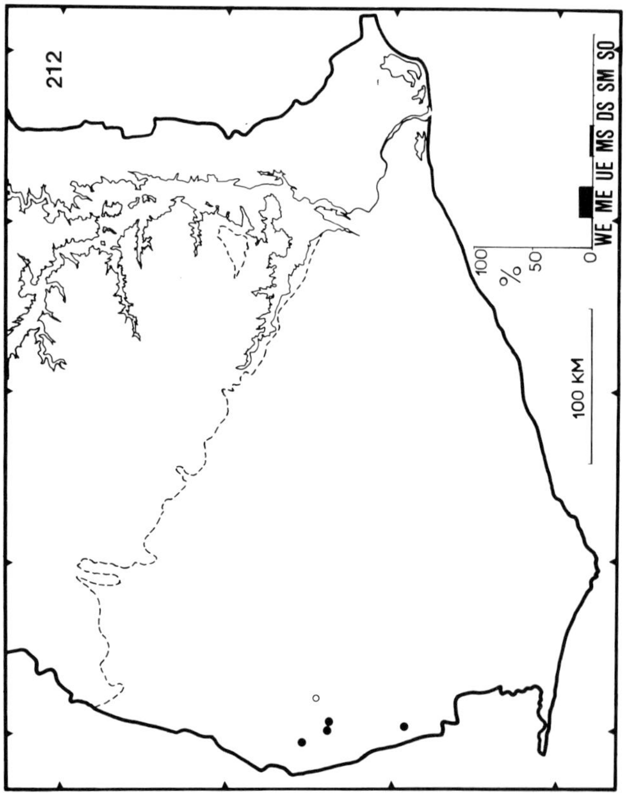
212
100 KM
%
100
50
0
WE ME UE MS DS SM SO

211 DIOSPYROS CHEVALIERI De Wild., FWTA 2:12. Ebenaceae (120).
Pygmy tree, with densely pubescent twigs; lvs mesophyll. Massart's model. Fl white, medium, dish with tube, dioecious, 6–9. Fr bristly orange berry, 11–2; seeds large to very large. Regenerates in shade. S. Leone to Ghana; (Upper Guinea endemic).

212 DIOSPYROS COOPERI (Hutch. & Dalz.) F. White, FWTA 2:10. Ebenaceae (120).
Small to medium-sized tree; lvs microphyll. Massart's model. Fl pale yellow, dish with tube, dioecious, medium. Fr orange berry; seeds large. Regenerates in shade. S. Leone to Ghana; (Upper Guinea endemic).

213 DIOSPYROS FERREA (Willd.) Bakh., FWTA 2:11. Ebenaceae (120).
Small to medium-sized tree, or shrub in maritime thicket; lvs microphyll. Massart's model. Fl white, small to medium, bell, dioecious, not clearly seasonal. Fr 1–2-seeded brownish-orange berry; seeds medium. Regenerates in shade. Widespread in Africa, and unique in the genus, being also widespread in Asia; (chorological transgressor).

214 DIOSPYROS GABUNENSIS Gürke, FWTA 2:12. Ebenaceae (120). Akan: kusibiri.
Medium-sized tree; lvs macrophyll. Massart's model. Fl white, fragrant, medium, dish with tube, dioecious, not clearly seasonal; may be cauliflorous. Fr brownish berry; seeds large. Regenerates in shade. S. Leone to Zaire and Zambia; (Guineo-Congolian wide).

215 DIOSPYROS HEUDELOTII Hiern, FWTA 2:14. Ebenaceae (120). Akan: omenewabere.
Medium-sized tree; lvs microphyll. Massart's model (59). Fl white, small to medium, bell, dioecious, 9–11. Fr orange berry; seeds large. Regenerates in shade. Senegal to Ghana; (Upper Guinea endemic).

216 DIOSPYROS KAMERUNENSIS Gürke, FWTA 2:11. Ebenaceae (120). Akan: omenewa.
Medium-sized tree; lvs notophyll to mesophyll; twigs and leaf underside pubescent. Massart's model. Fl white, medium, dish with tube, dioecious, 7–11. Fr orange berry, 1–3; seeds very large. Regenerates in shade. Plot records very significantly associated with acid, base-poor soils. Liberia to Ghana, and Cameroun to Gabon; (Guinea wide). Timber hard, used for telegraph poles (40).

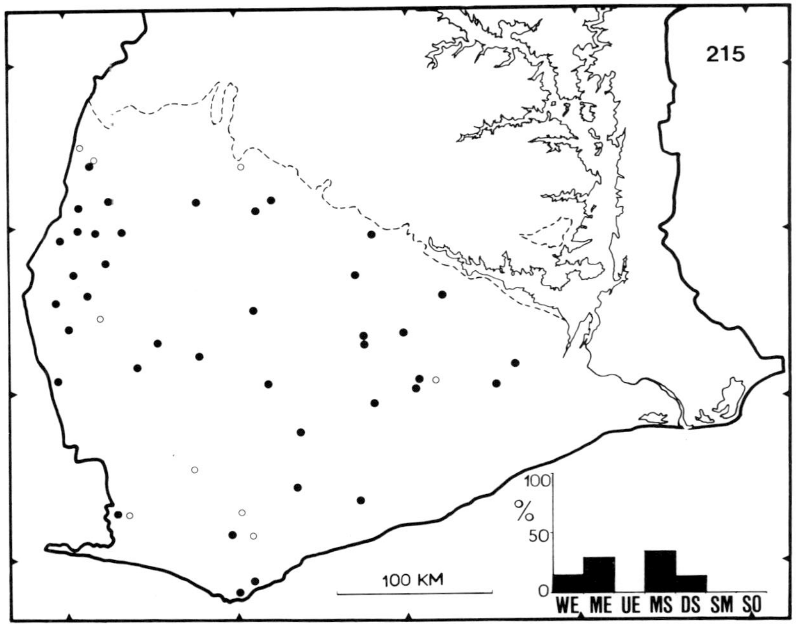

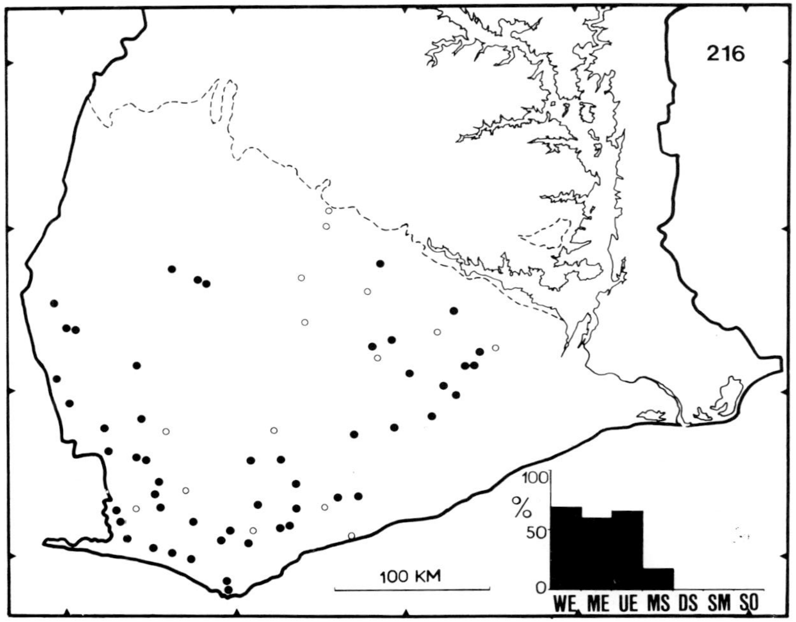

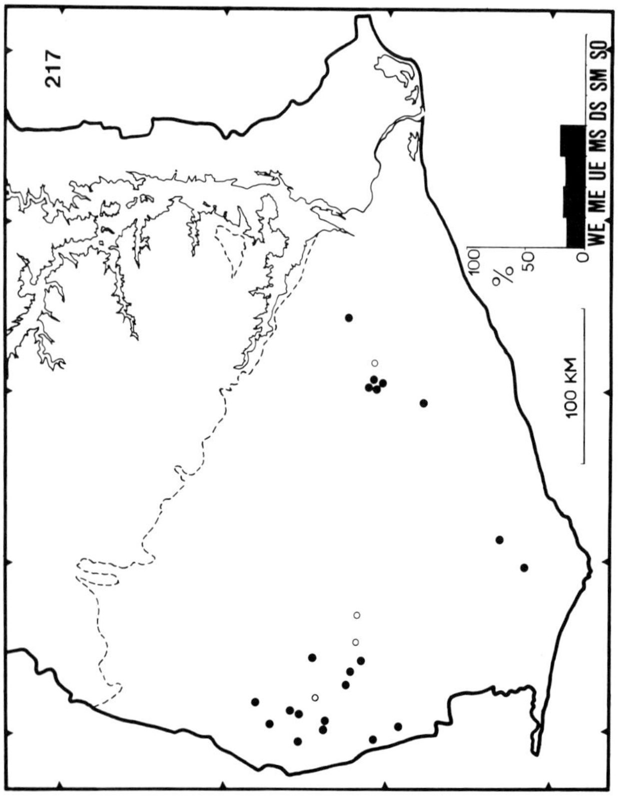
217
100 KM
%
100
50
0
WE ME UE MS DS SM SO

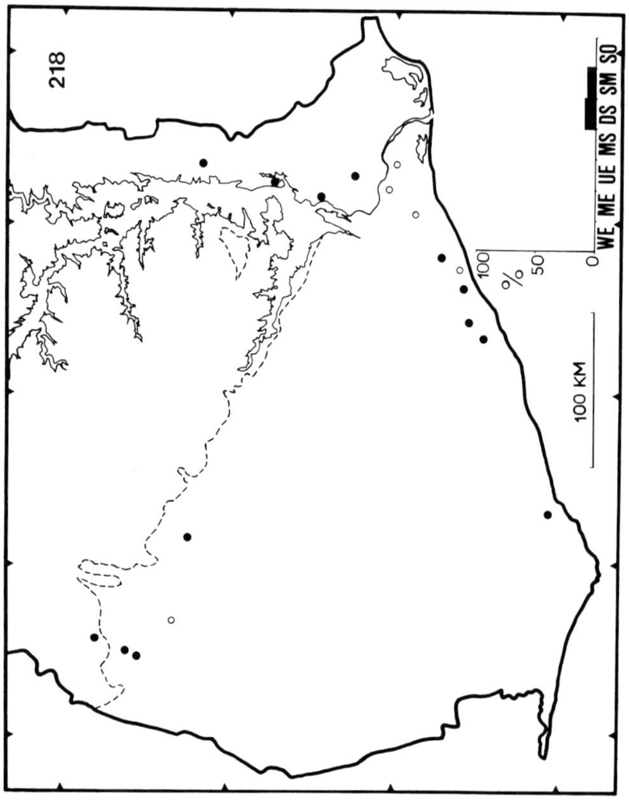
218
100 KM
%
100
50
0
WE ME UE MS DS SM SO

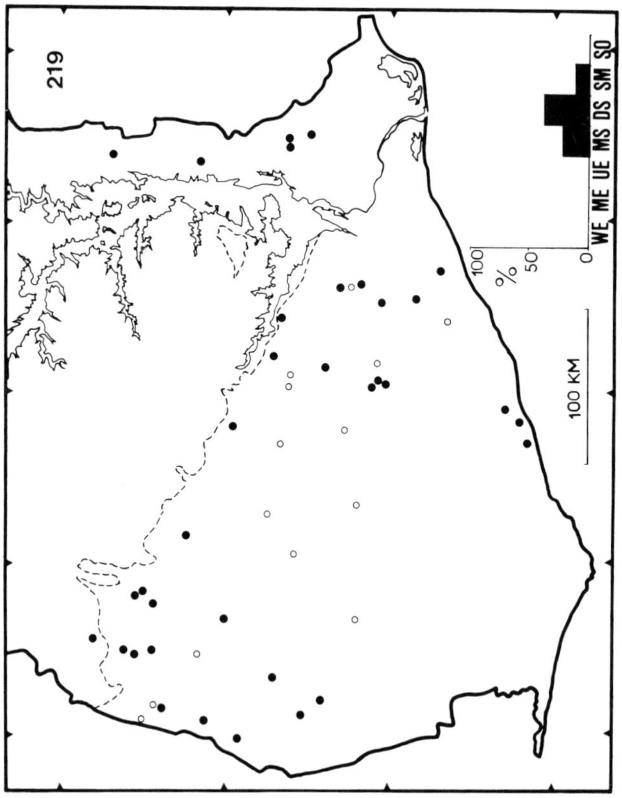
219
100 KM
%
100
50
0
WE ME UE MS DS SM SO

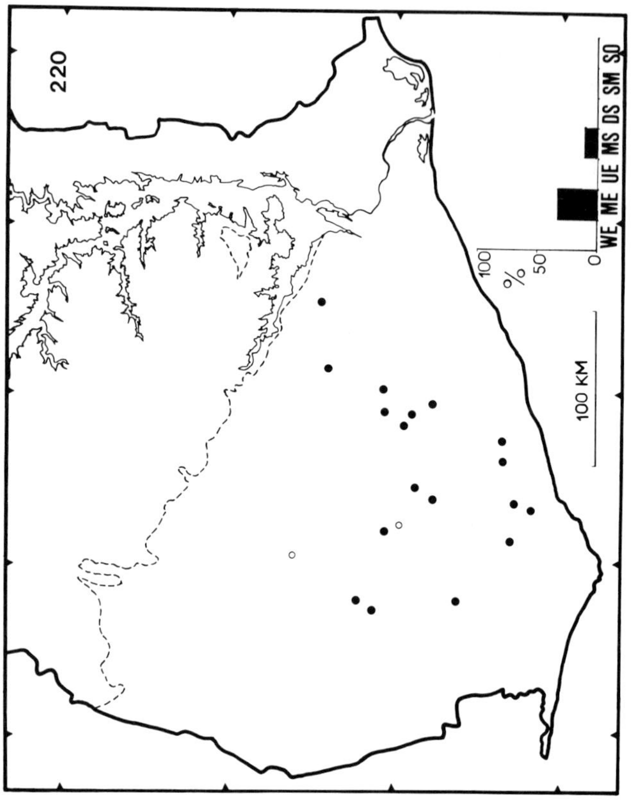
220
100 KM
%
100
50
0
WE ME UE MS DS SM SO

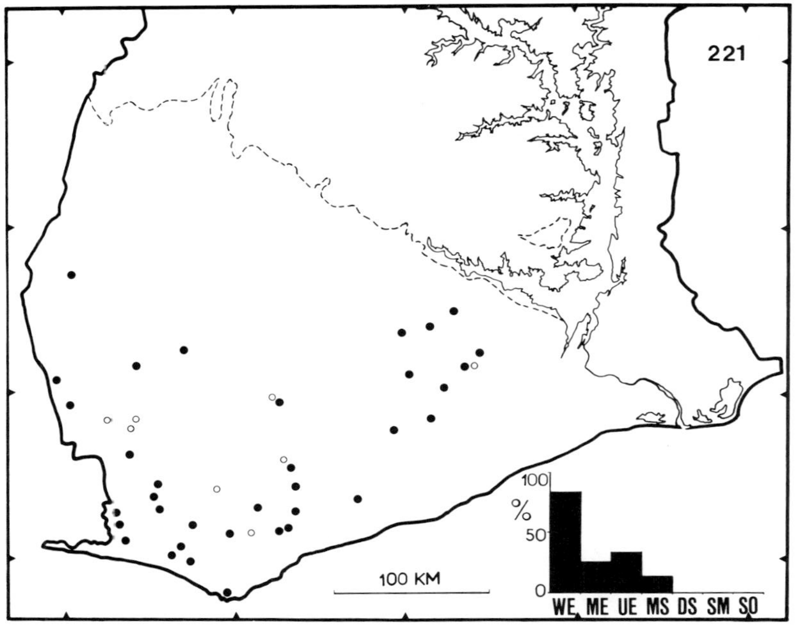

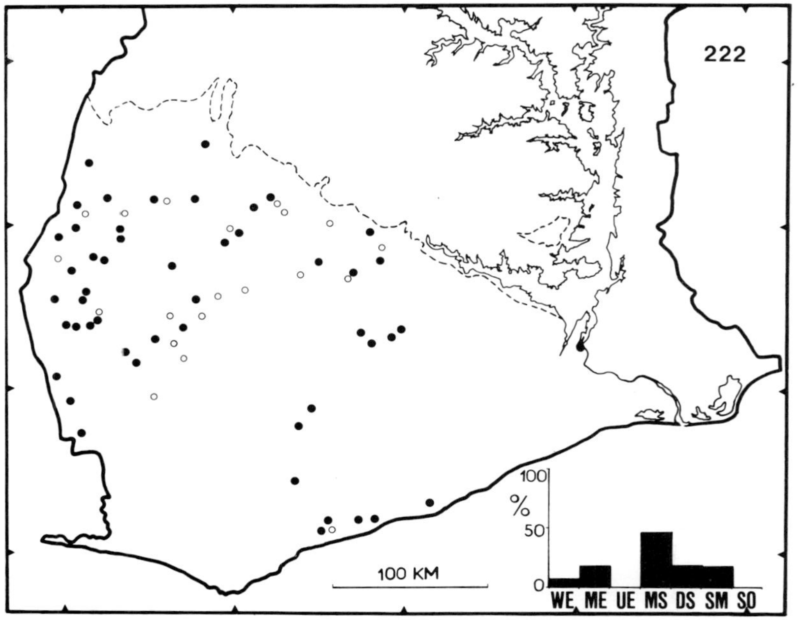

217 DIOSPYROS MANNII Hiern, FWTA 2:11. Ebenaceae (120). Akan: atweafufuo. Medium-sized tree; lvs mesophyll. Massart's model. Fl whitish or brownish, medium, bowl, dioecious, 4–5. Fr bristly orange berry; seeds very large. Regenerates in shade. S. Leone to Angola; (Guineo-Congolian wide).

218 DIOSPYROS MESPILIFORMIS Hochst. ex. A. DC., FWTA 2:12. Ebenaceae. Ga: nokotso.

Medium-sized tree; lvs notophyll; more or less deciduous. Massart's model. Fl white, fragrant, medium, bell, dioecious, not clearly seasonal. Fr yellow berry; seeds large, dispersed by baboons (81). Regenerates in shade. Widespread in savanna, where it tends to occur on termite mounds, in fringing forest, and on rocky slopes. Widespread in Africa; (ecological and chorological transgressor). The heartwood of large trees is blackish, streaked with brown, and is used for carving.

219 DIOSPYROS MONBUTTENSIS Gürke, FWTA 2:14. Ebenaceae (120). Akan: atwerenantin.

Small tree, often branching low, with papery flaking bark, and spines on branches; lvs notophyll to mesophyll. Fl cream, fragrant, medium, dish with tube, 3–6. Fr orange berry, 9–12; seeds large. Regenerates in shade. Iv. Coast to Zaire; (Guineo-Congolian wide).

220 DIOSPYROS PISCATORIA Gürke, FWTA 2:10. Ebenaceae (120). Akan: benkyi. Medium-sized tree; lvs notophyll. Massart's model. Fl medium, bowl, dioecious. Fr red berry; seeds medium. Regenerates in shade. Saplings encountered frequently in the survey, but mature trees are very rare. Guinée to Angola; (Guinea wide).

221 DIOSPYROS SANZA-MINIKA A. Chev., FWTA 2:12. Ebenaceae (120). Nzima: sanzamulike.

Medium-sized tree, with extraordinarily thick, brittle, deeply-fissured outer bark which is much more resistant to decay than the wood. Lvs mesophyll. Massart's model (59). Fl white, medium to large, bell, dioecious, 3–5. Fr yellow berry, 1–4; seeds very large. Regenerates in shade. Plot records very significantly associated with acid, base-poor soils. S. Leone to Ghana and Cameroun to Congo Republic; (Guinea wide). Class 3 timber; trade name: 'kusibiri' (40).

222 DIOSPYROS SOUBREANA F. White, FWTA 2:14. Ebenaceae (120). Akan: otweto.

Shrub or pygmy tree; lvs notophyll. Massart's model. Fl white or cream, bowl, dioecious (male small, female medium), not clearly seasonal. Fr dark red berry; seeds large. Regenerates in shade. Iv. Coast to Nigeria; (Guinea wide).

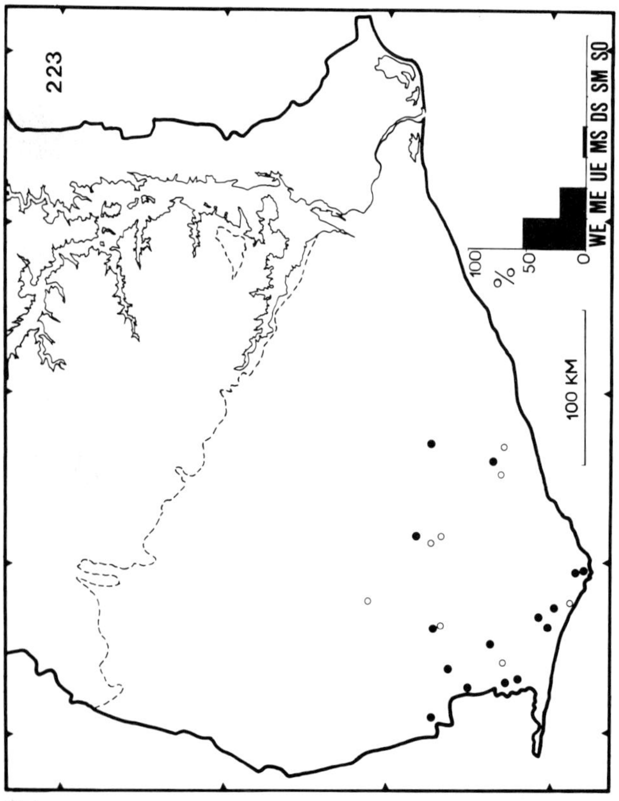
223
100 KM
%
100
50
0
WE ME UE MS DS SM SO

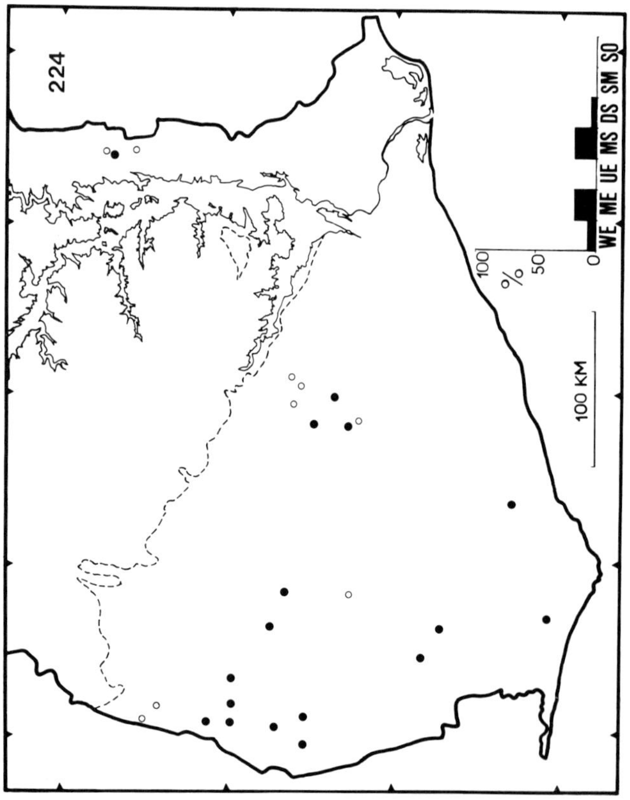
224
100 KM
%
100
50
0
WE ME UE MS DS SM SO

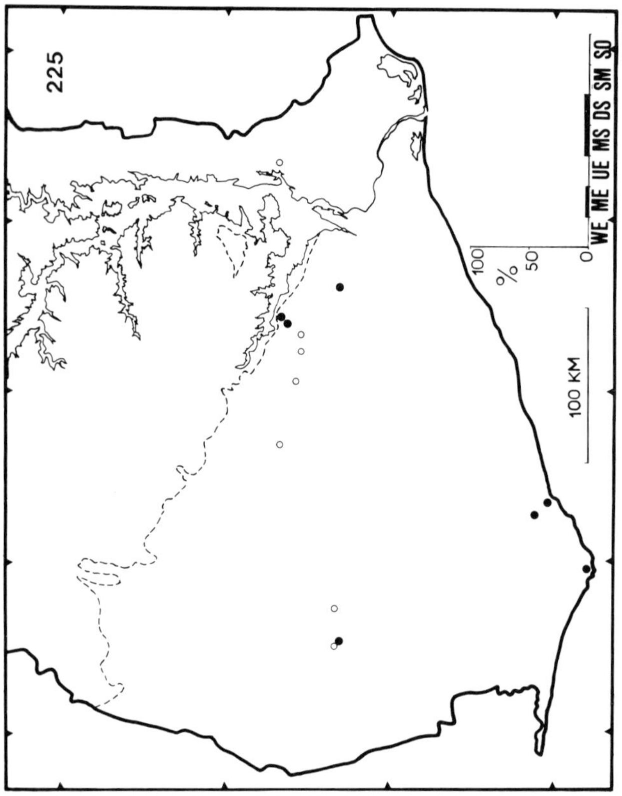
225
100 KM
%
100
50
0
WE ME UE MS DS SM SO

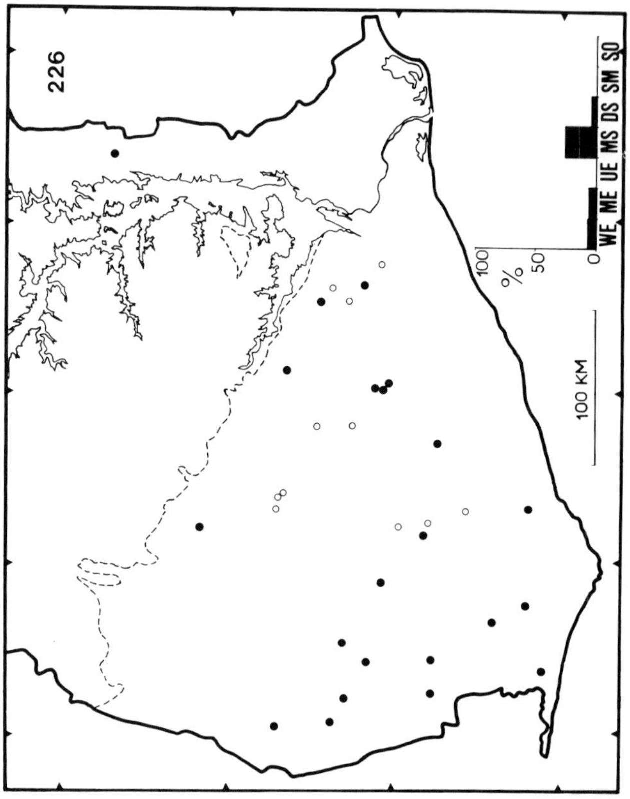
226
100 KM
%
100
50
0
WE ME UE MS DS SM SO

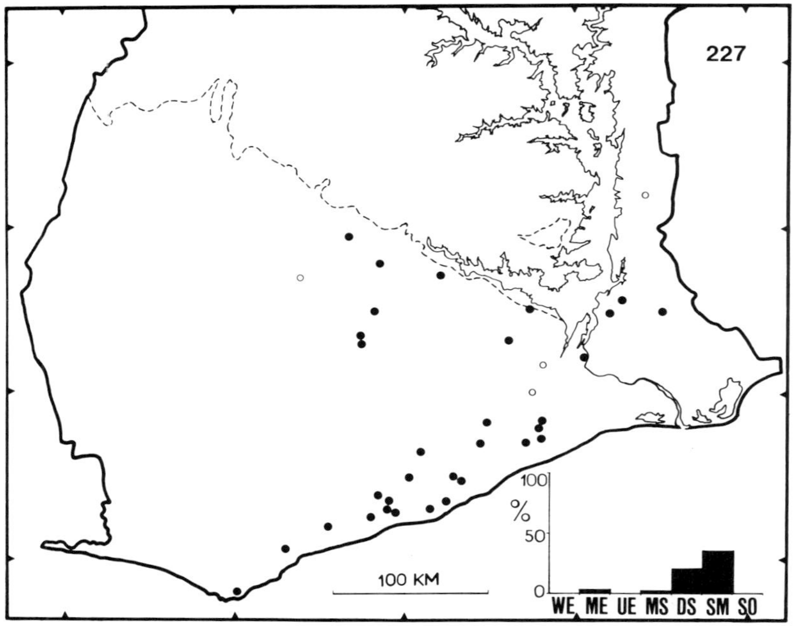

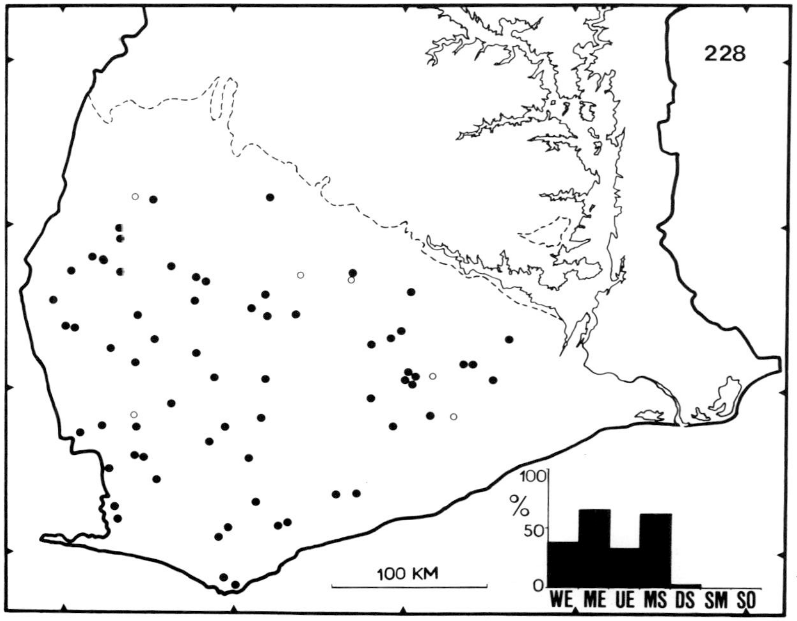

223 DIOSPYROS VIGNEI F. White, FWTA 2:15. Ebenaceae (120).
Pygmy to small tree; lvs notophyll; twigs and underside of midrib pubescent. Massart's model. Fl medium, bell, dioecious (male axillary and cauliflorous, female cauliflorous), not clearly seasonal. Fr orange berry; seeds very large. Regenerates in shade. Liberia to Ghana. (Upper Guinea endemic).

224 DIOSPYROS VIRIDICANS Hiern, FWTA 2:10. Ebenaceae (120). Akan: atwea.
Medium-sized tree; lvs notophyll. Massart's model. Fl greenish-white, small, bowl, dioecious, not clearly seasonal. Fr black berry; seeds large. Regenerates in shade. S. Leone to Angola; (Guinea wide).

225 DIPHASIA ANGOLENSIS (Hiern) Verdoorn (77), (*D. klaineana* Pierre, FWTA 1:689). Rutaceae. Akan: amuduro.
Small to medium-sized tree; lvs trifoliolate, leaflets notophyll, with translucent glands; leaves and twigs densely pubescent. Fl white, small, bowl; hermaphrodite in Gabon and Angola, dioecious in Iv. Coast and Ghana; mainly 2–5. Fr orange 2-seeded berry, mainly 7–8; seeds medium. Regenerates in shade. Guinée to Ghana, and Gabon to Angola; (Guinea wide).

226 DISTEMONANATHUS BENTHAMIANUS Baill., FWTA 1:449. Caesalpiniaceae. Akan: bonsamdua.
Large tree with red bark; lvs pinnate, leaflets microphyll; deciduous. Troll's model. Fl white and brown, dish, medium, not clearly seasonal. Fr papery, indehiscent pod, wind-dispersed; seeds medium. Regenerates mainly in gaps and along roadsides. The young shoots are covered with small hooked hairs. S. Leone to Gabon; (Guinea wide). Class 3 timber; trade name: ayan (40).

227 DRACAENA ARBOREA (Willd.) Link, FWTA 3:156. (Incl. *D. smithii* Bak. ex Hook. f., FWTA 3:156). Agavaceae. Akan: ntomme.
Medium-sized tree, often branching near the ground; lvs megaphyll. Leeuwenberg's model. Fl white, narrowly funnel-shaped, medium to large, not clearly seasonal. Fr orange berry; seeds medium. Regenerates in shade. Guinée to Cameroun; (Guinea wide). Often planted to mark boundaries, and at shrines.

228 DRACAENA CAMEROONIANA Bak., FWTA 3:157. Agavaceae.
Pygmy tree or shrub; lvs mesophyll. Chamberlain's model. Fl white, medium to large, narrowly funnel-shaped, 12–1. Fr orange berry, 1–5; seeds medium. Regenerates in shade. More than half of plot records on soil with pH 4.2–4.9. Guinée to Angola and Zambia; (Guineo-Congolian wide).

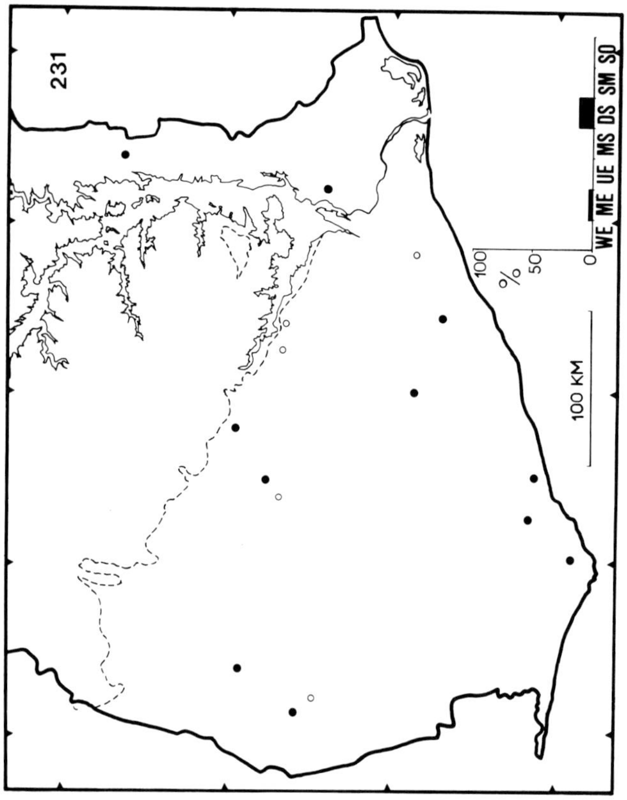
231
100 KM
%
100
50
0
WE ME UE MS DS SM SO

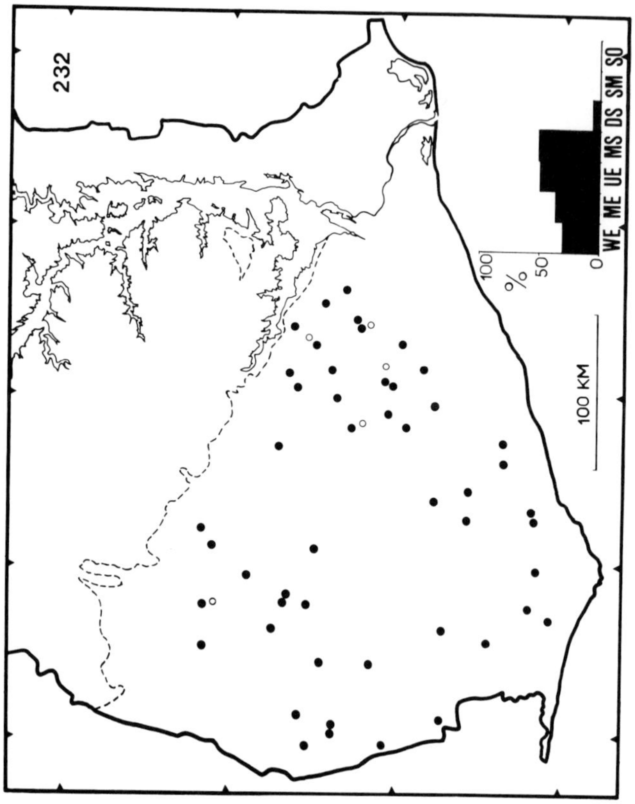
232
100 KM
%
100
50
0
WE ME UE MS DS SM SO

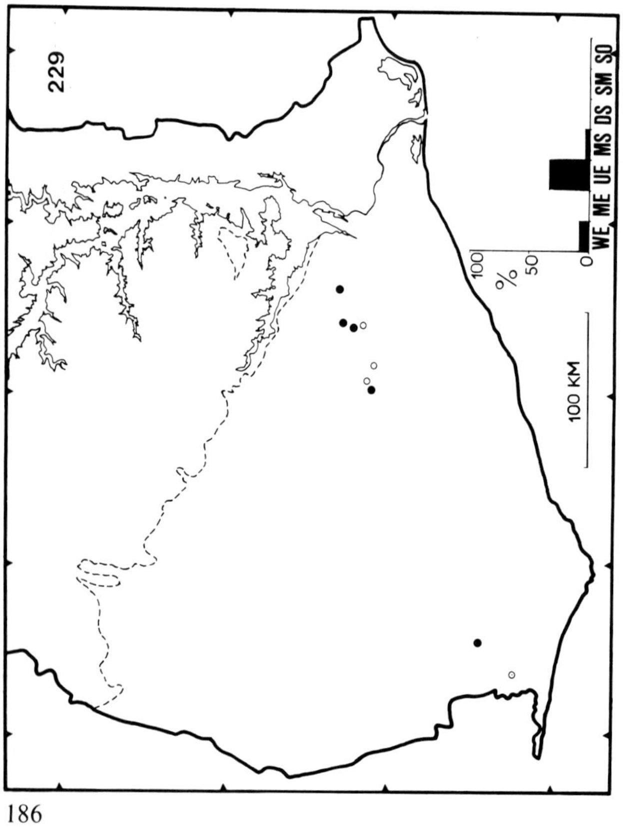
229
100 KM
%
100
50
0
WE ME UE MS DS SM SO

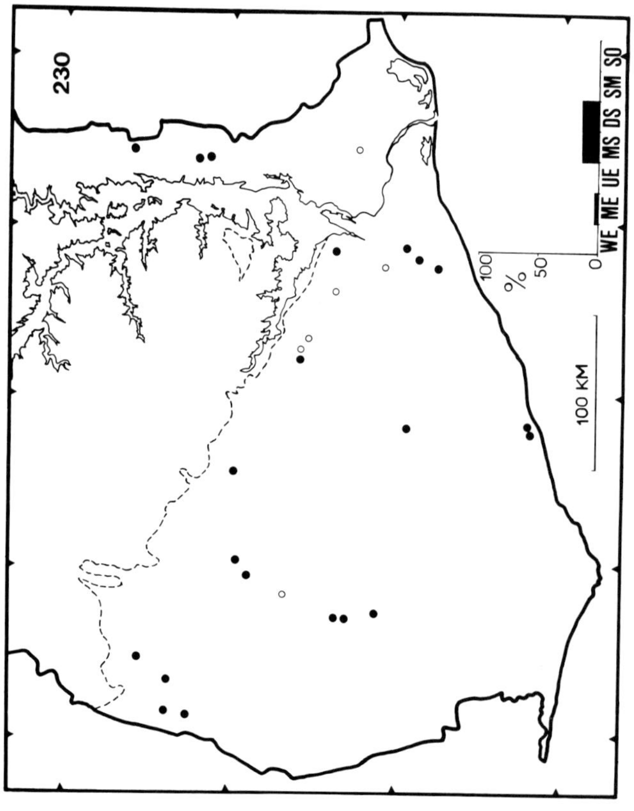
230
100 KM
%
100
50
0
WE ME UE MS DS SM SO

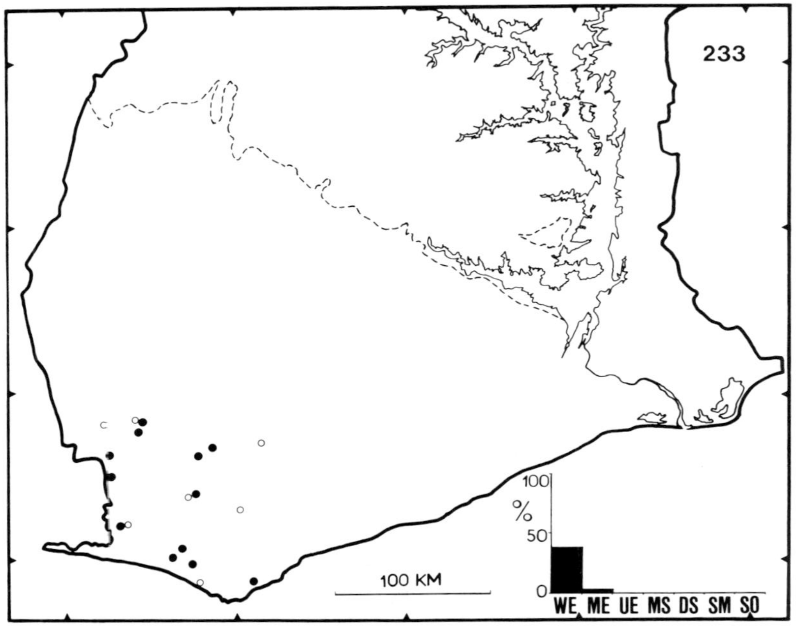

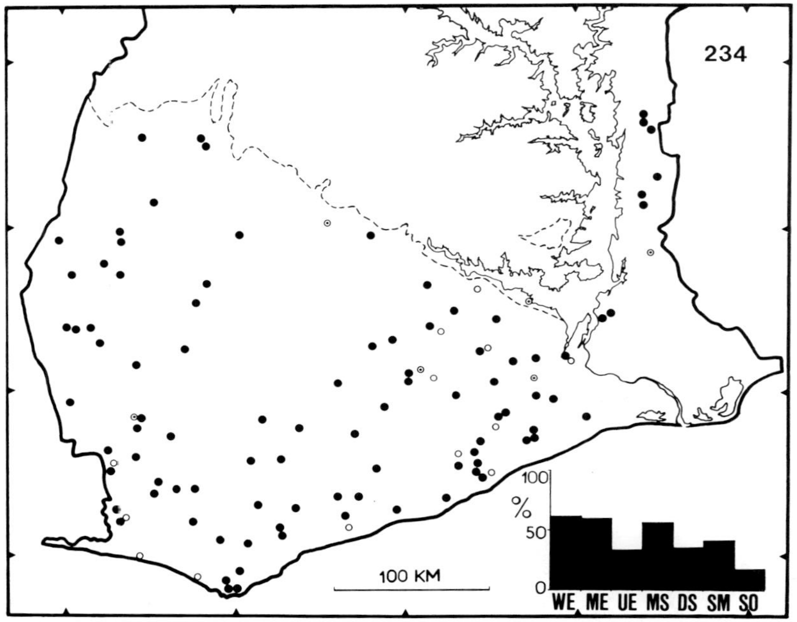

229 DRACAENA ELLIOTII Bak., FWTA 3:156. Agavaceae.
Unbranched pygmy tree; lvs mesophyll. Chamberlain's model. Fl medium, narrowly funnel-shaped. Fr orange berry; seeds medium. Regenerates in shade. Guinée to Ghana; (Upper Guinea endemic).

230 DRACAENA FRAGRANS (Linn.) Ker-Gawl., FWTA 3:157. Agavaceae.
Pygmy tree or shrub; lvs mesophyll to macrophyll. Chamberlain's model. Fl pinkish-white, medium to large, narrowly funnel-shaped, mainly 1–3. Fr orange berry; seeds medium. Regenerates in shade. Gregarious on forest floor. Ghana to Nigeria and East Africa; (Guineo-Congolian wide). The name *D. fragrans* is applied here as in FWTA, but this usage is almost certainly incorrect.

231 DRACAENA PERROTTETII Bak., FWTA 3:157. Agavaceae. Akan: kesene.
Medium-sized tree, in habit much resembling dicotyledonous trees; lvs mesophyll. Scarrone's model. Fl white, fragrant, large, narrowly funnel-shaped, not clearly seasonal. Fr red berry; and infructescence axes also red; seeds large. Regenerates in shade. Senegal to Nigeria; (Guinea wide).

232 DRACAENA PHRYNIOIDES Hook., FWTA 3:156. Agavaceae.
Perennial acaulescent herb; lvs mesophyll. Fl medium, narrowly funnel-shaped, not clearly seasonal. Fr red berry; seeds medium. Regenerates in shade. Liberia to Cameroun; (Guinea wide).

233 DRACAENA SCOPARIA A. Chev. ex Hutch., FWTA 3:157.
Shrub; lvs mesophyll. Leeuwenberg's model. Fl white, medium, narrowly funnel-shaped, 12–2. Fr orange berry; seeds medium. Regenerates in shade. Iv. Coast and Ghana; (Upper Guinea endemic).

234 DRACAENA SURCULOSA Lindl. var. *surculosa,* FWTA 3:157, and var. *capitata* Hepper, FWTA 3:159. Agavaceae.
Shrubs to 1 m (var *surculosa*) or up to 5 m (var. *capitata*); lvs notophyll (var. *capitata*) or notophyll to mesophyll (var. *surculosa*). McClure's model. Fl white, medium to large, dish with tube, not clearly seasonal. Fr orange or red berry; seeds medium. Regenerate in shade. Var. *capitata* may form fleshy, subterranean tubers in dry localities. The varieties are not always distinguishable in the field so survey records are not mapped separately. Herbarium records: var. *surculosa* ⊙. var. *capitata* ○. Distribution (both varieties): S. Leone to Cameroun; (Guinea wide).

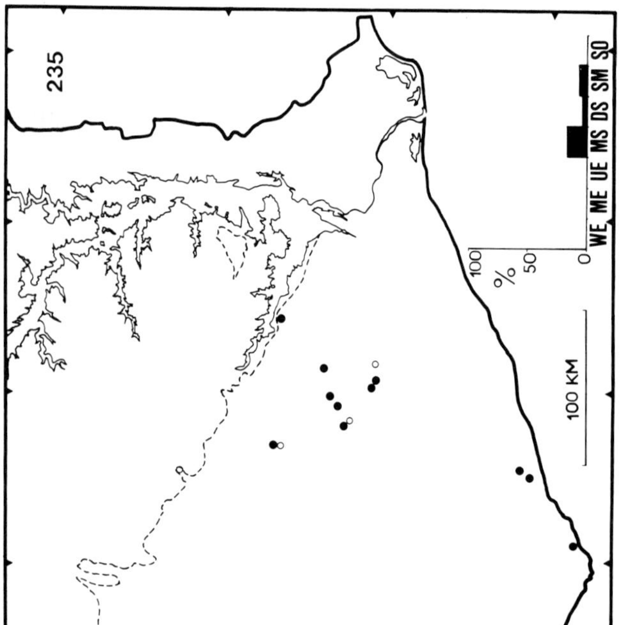
235
100 KM
%
100
50
0
WE ME UE MS DS SM SO

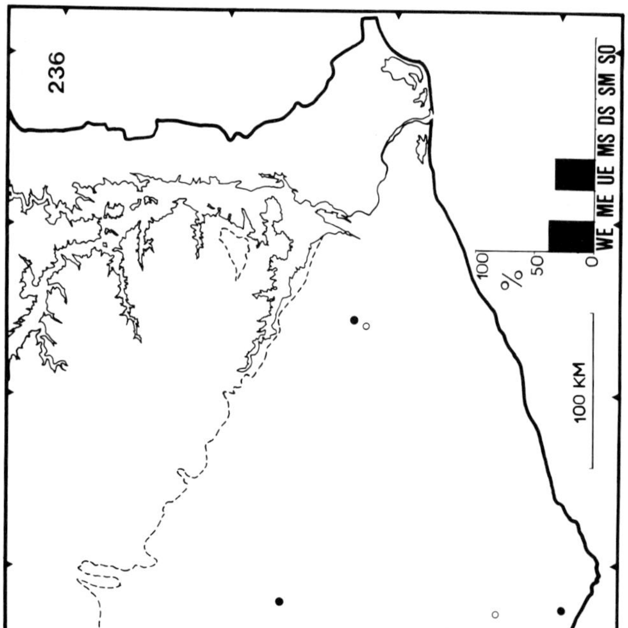
236
100 KM
%
100
50
0
WE ME UE MS DS SM SO

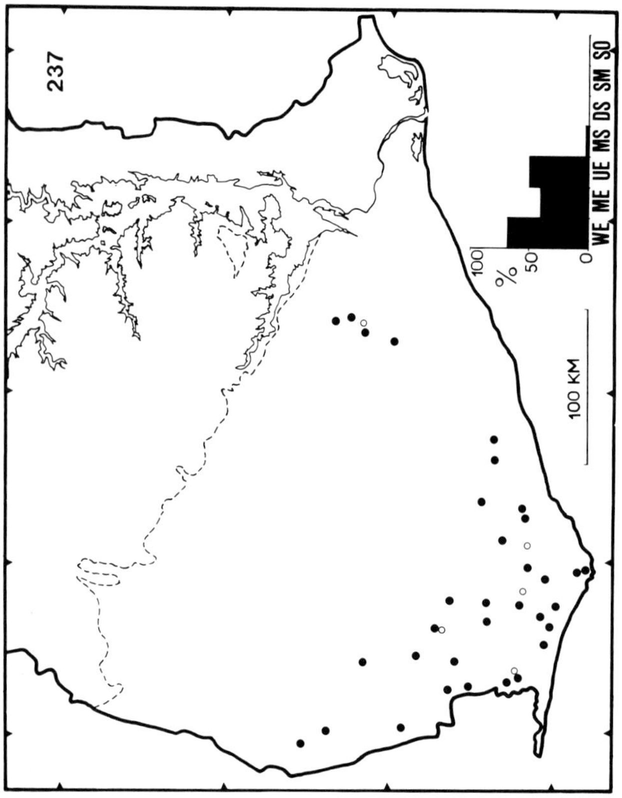
237
100 KM
%
100
50
0
WE ME UE MS DS SM SO

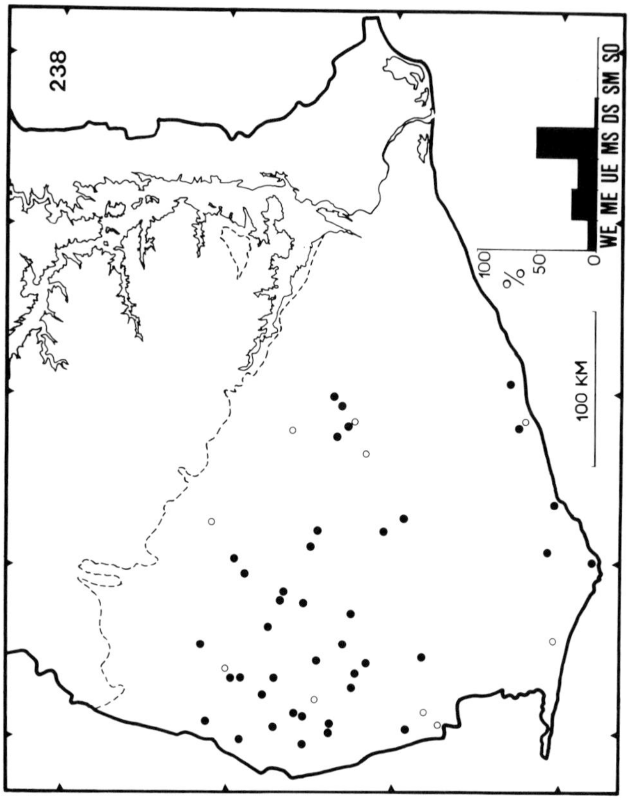
238
100 KM
%
100
50
0
WE ME UE MS DS SM SO

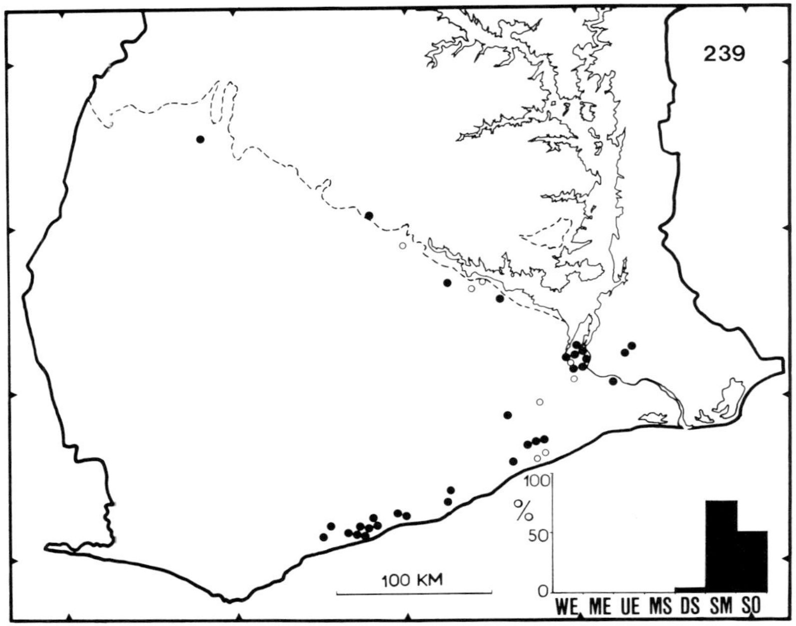

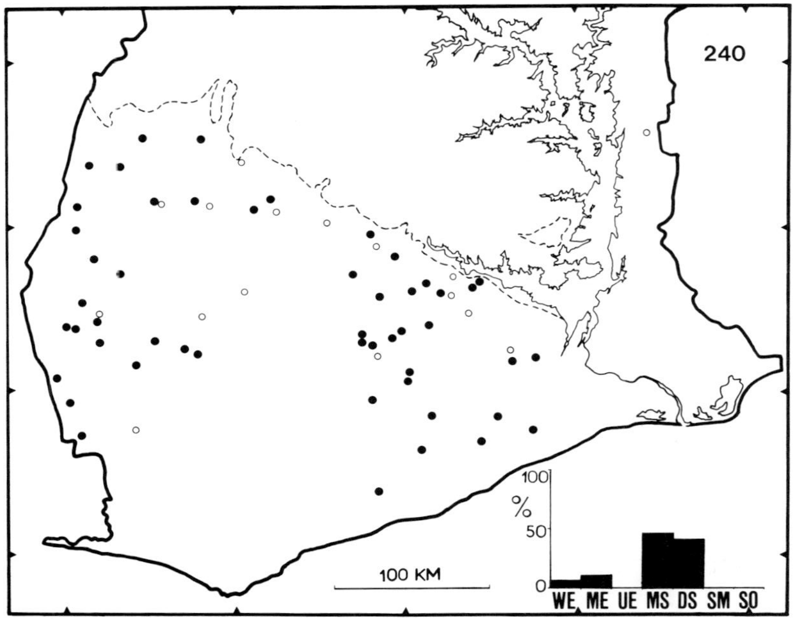

235 DRYPETES AFRAMENSIS Hutch., FWTA 1:381. Euphorbiaceae. (Type from 'Afram Plain', probably just south of R. Afram).
Small to medium-sized tree; lvs mesophyll. Roux's model. Fl white, small, bowl, dioecious, cauliflorous on branches and among leaves, 1–3. Fr orange 2-seeded berry; seeds large. Regenerates in shade. Often gregarious. Ghana to Nigeria; (Guinea wide).

236 DRYPETES AUBREVILLEI Léandri, FWTA 1:381. Euphorbiaceae. Akan: duamoko.
Medium-sized tree; lvs microphyll, slightly toothed. Roux's model. Fl small, bowl, dioecious. Fr yellow 2-seeded berry; seeds medium. Regenerates in shade. S. Leone to Ghana (Upper Guinea endemic). Bark has a peppery taste.

237 DRYPETES AYLMERI Hutch. & Dalz., FWTA 1:381. Euphorbiaceae. Akan: opahafufuo.
Medium-sized tree; lvs mesophyll. Roux's model (59). Fl yellow, small, bowl, dioecious, cauliflorous on branches, 9–10. Fr 1-seeded berry; seed large. Regenerates in shade. Often abundant in evergreen forest. S. Leone to Ghana; (Upper Guinea endemic). Distinctive mustard smell from freshly-cut slash.

238 DRYPETES CHEVALIERI Beille, FWTA 1:382. Euphorbiaceae. Akan: katrika.
Small tree; twigs pubescent; lvs notophyll, toothed. Roux's model (59). Fl cream, small, bowl, dioecious, 9–1. Fr orange 2-seeded berry; seeds large. Regenerates in shade. Gregarious. Liberia to Nigeria; (Guinea wide).

239 DRYPETES FLORIBUNDA (Muell. Arg.) Hutch., FWTA 1:381. Euphorbiaceae. Ga: tetso. Akan: bedibesa.
Small tree, or shrub in thicket; lvs microphyll to notophyll, toothed. Roux's model. Fl cream, small, bowl, dioecious, cauliflorous on branches and trunk, 2–5. Fr yellow berry; seeds medium. Regenerates in shade. Extends into forest outliers on rocky scarps in the savanna zone. Senegal to Zaire; (Guineo-Congolian wide).

240 DRYPETES GILGIANA (Pax) Pax & K. Hoffm., FWTA 1:382. Euphorbiaceae. Akan: katrikanini.
Small tree; lvs notophyll, slightly toothed. Roux's model. Fl yellowish, small, bowl, dioecious, 9–11. Fr orange berry, 10–4; seeds large. Regenerates in shade. Guinea Bissau to Cameroun; (Guinea wide). Fruits edible and sometimes sold in the market.

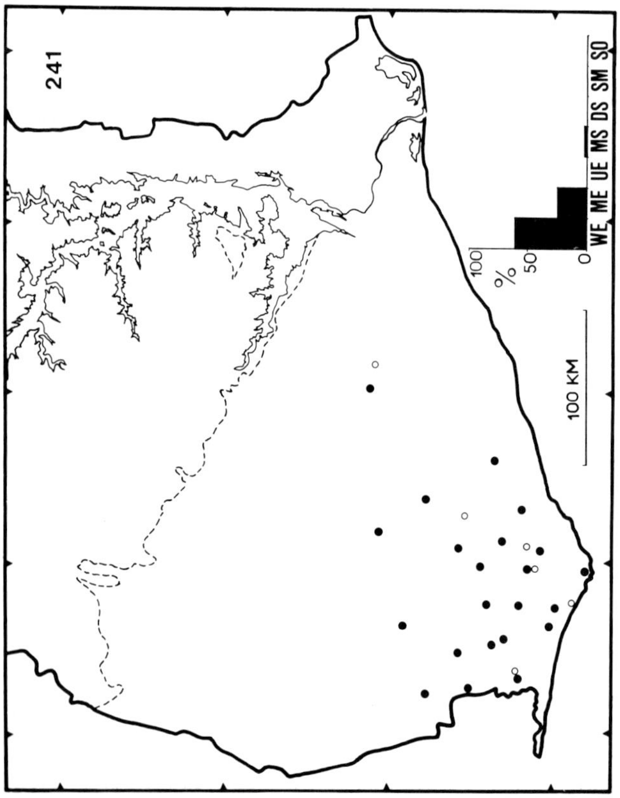
241
100 KM
100
%
50
0
WE ME UE MS DS SM SO

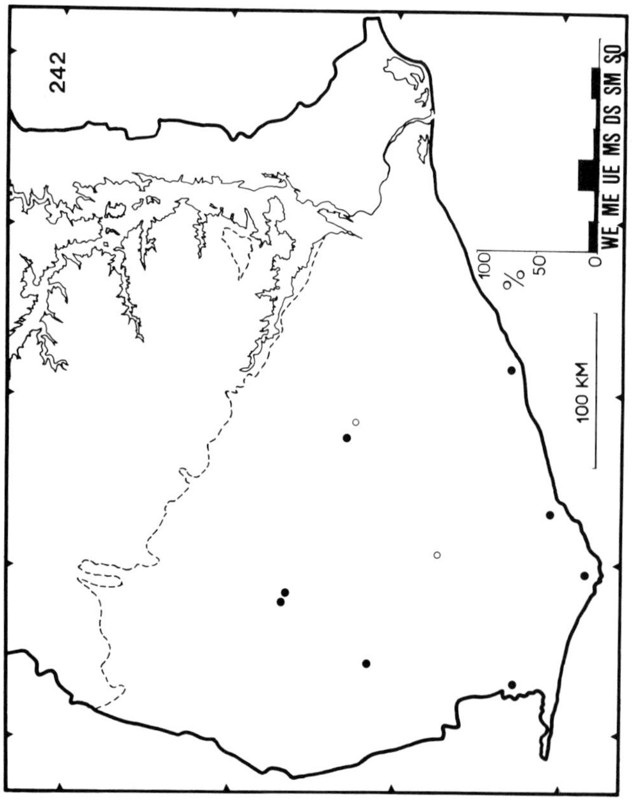
242
100 KM
100
%
50
0
WE ME UE MS DS SM SO

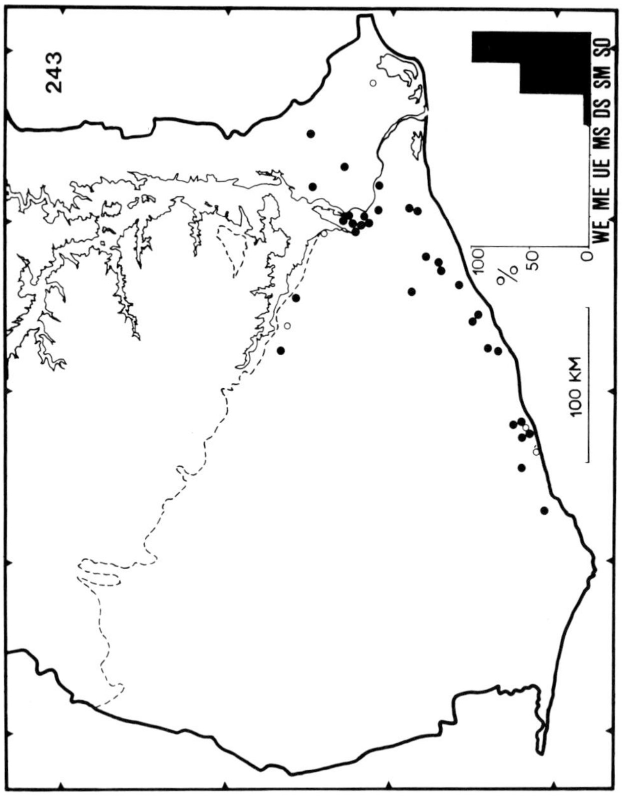
243
100 KM
100
%
50
0
WE ME UE MS DS SM SO

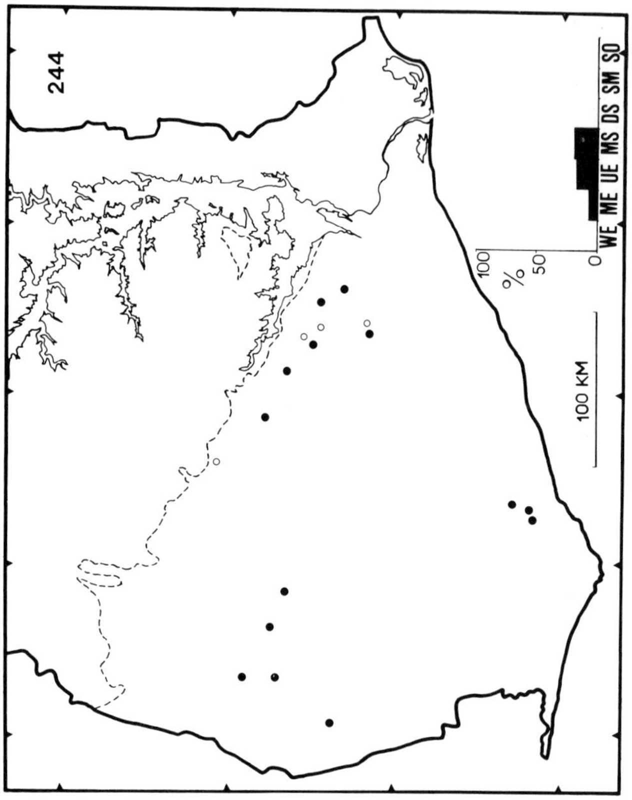
244
100 KM
100
%
50
0
WE ME UE MS DS SM SO

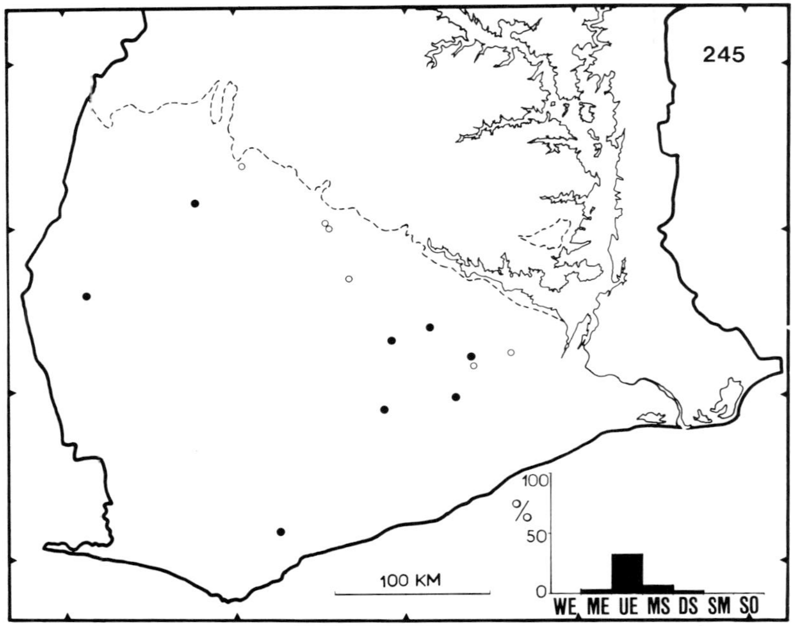

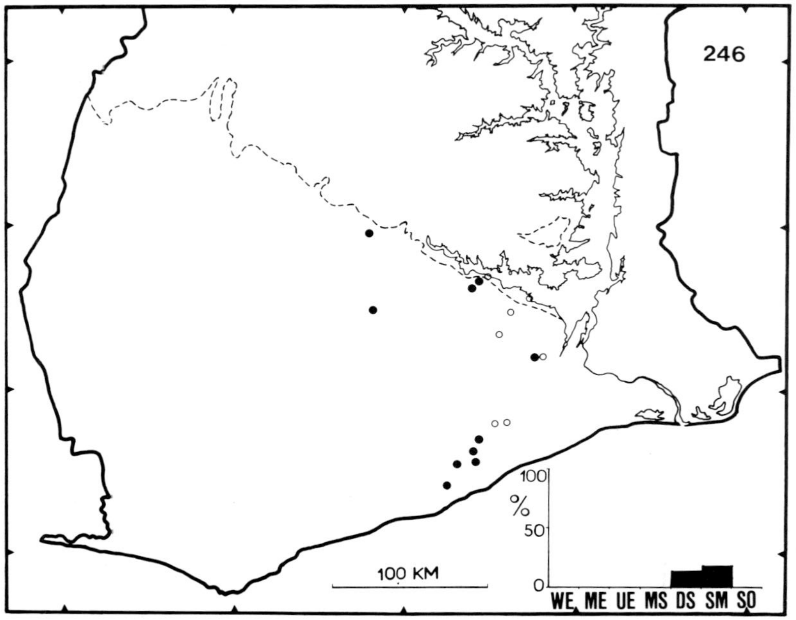

241 DRYPETES IVORENSIS Hutch. & Dalz., FWTA 1:381. Euphorbiaceae.
Small slender tree; lvs mesophyll, toothed. Roux's model. Fl white, small, bowl, dioecious, cauliflorous on main trunk, not clearly seasonal. Fr orange-red berry; seeds large. Regenerates in shade. Plot records very highly correlated with acid, base-poor soils. Liberia to Ghana; (Upper Guinea endemic).

242 DRYPETES LEONENSIS Pax, FWTA 1:381. Euphorbiaceae.
Medium-sized tree, slash not usually strong-smelling; twigs pubescent; lvs notophyll. Roux's model. Fl small, bowl, dioecious. Fr yellow 2-seeded berry; seeds large. Regenerates in shade. A rare species of sporadic occurrence in most forest types. Guinée to Cameroun; (Guinea wide).

243 DRYPETES PARVIFOLIA (Muell. Arg.) Pax & K. Hoffm. FWTA 1:382. Euphorbiaceae. Ga: osadentso. Akan: katrikabere.
Small tree, often with strong epicormic shoots from near the base; lvs microphyll, slightly toothed. Roux's model. Fl white, small, bowl, dioecious, mostly 10–11. Fr orange-red berry, mostly 5; seeds medium. Regenerates in shade. Strongly gregarious, often forming pure stands in the understorey of dry forest (Fig. 8.12). Extends to fringing forest in the southern part of the savanna zone. S. Leone to Nigeria; (Guinea wide).

244 DRYPETES PELLEGRINII Léandri, (*D. pellegrini,* FWTA 1:381). Euphorbiaceae. Akan: opahakokoo.
Medium-sized tree, slash with strong mustard smell; lvs mesophyll, toothed. Roux's model. Fl white, small, bowl, dioecious, 2–3. Fr yellowish-brown, 2-seeded berry, 3–5; seeds large. Regenerates in shade. Iv. Coast and Ghana; (Upper Guinea endemic).

245 DRYPETES PRINCIPUM (Muell. Arg.) Hutch., FWTA 1:381. Euphorbiaceae. Akan: opaha.
Medium-sized tree; slash almost scentless; lvs mesophyll, toothed. Roux's model. Fl cream, small, bowl, dioecious, among leaves and on twigs below leaves, mainly 10–12. Fr yellowish, 2-seeded berry, mainly 11–3; seeds large. Regenerates in shade. Guinée to Cameroun; (Guinea wide).

246 DRYPETES SINGROBOENSIS Aké Assi (2), (*D.* sp., FWTA 1:382). Euphorbiaceae. Akan: opaha-akoa.
Small tree, branching low; lvs notophyll to mesophyll, with persistent foliaceous stipules. Roux's model. Fl white, small, bowl, dioecious, not clearly seasonal. Fr pale yellow berry; seeds large. Regenerates in shade. Ivory Coast and Ghana; (Upper Guinea endemic).

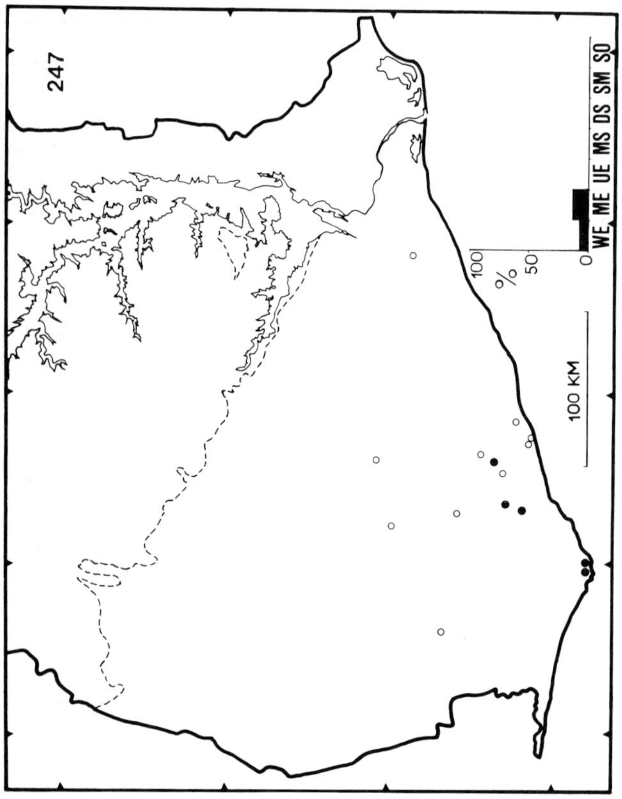
247
100 KM
%
100
50
0
WE ME UE MS DS SM SO

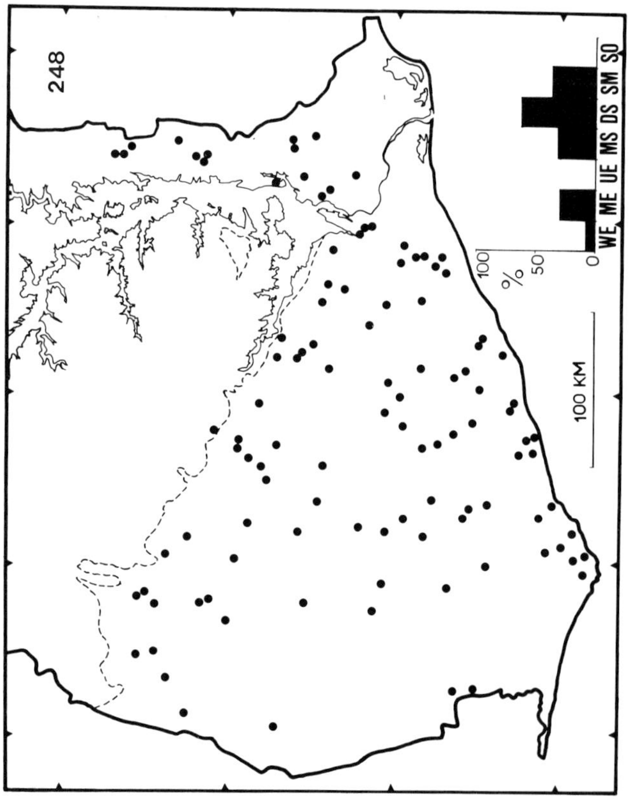
248
100 KM
%
100
50
0
WE ME UE MS DS SM SO

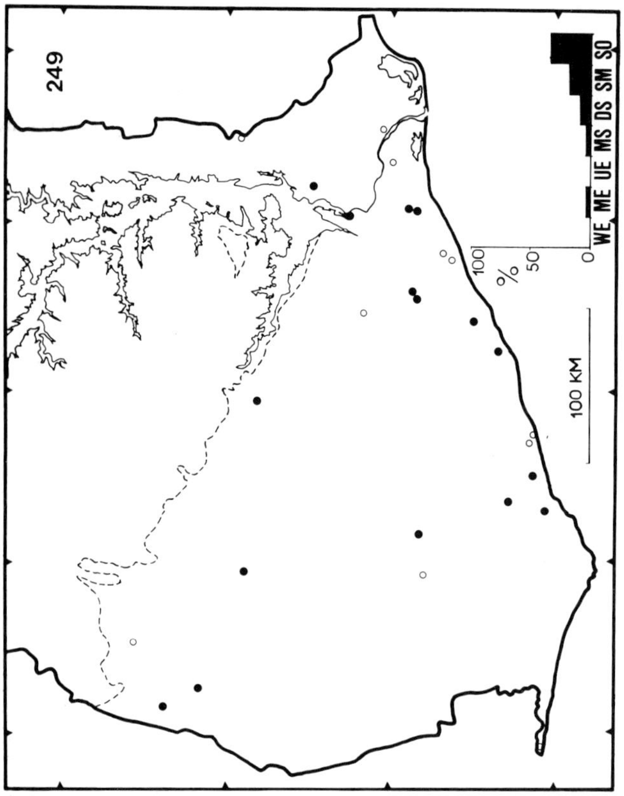
249
100 KM
%
100
50
0
WE ME UE MS DS SM SO

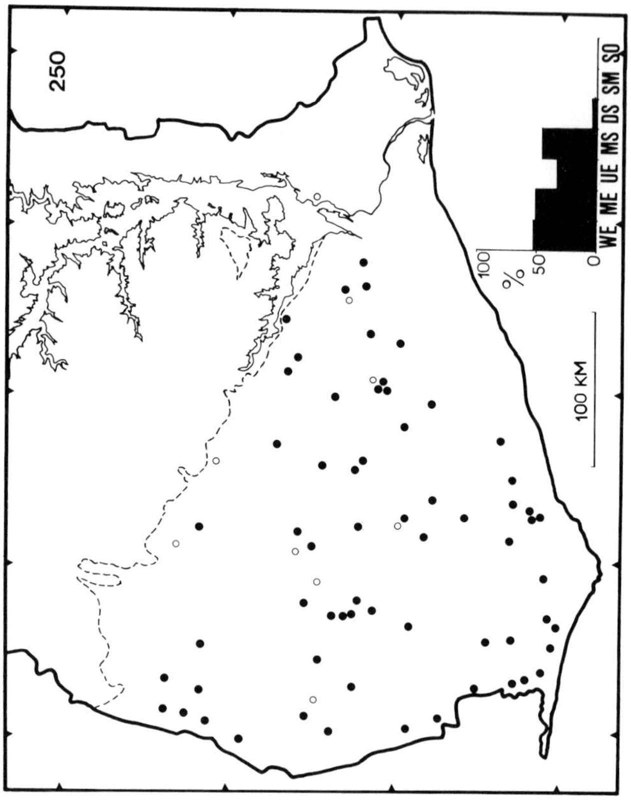
250
100 KM
%
100
50
0
WE ME UE MS DS SM SO

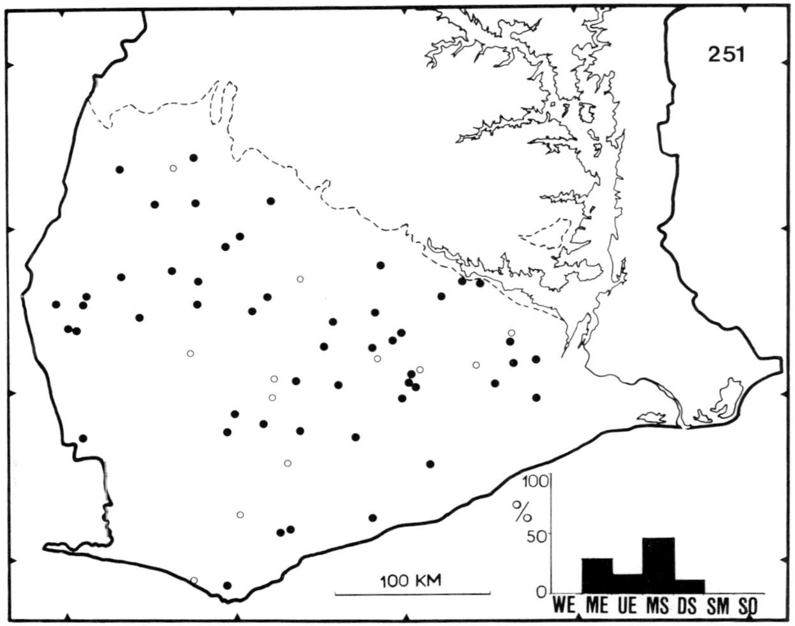

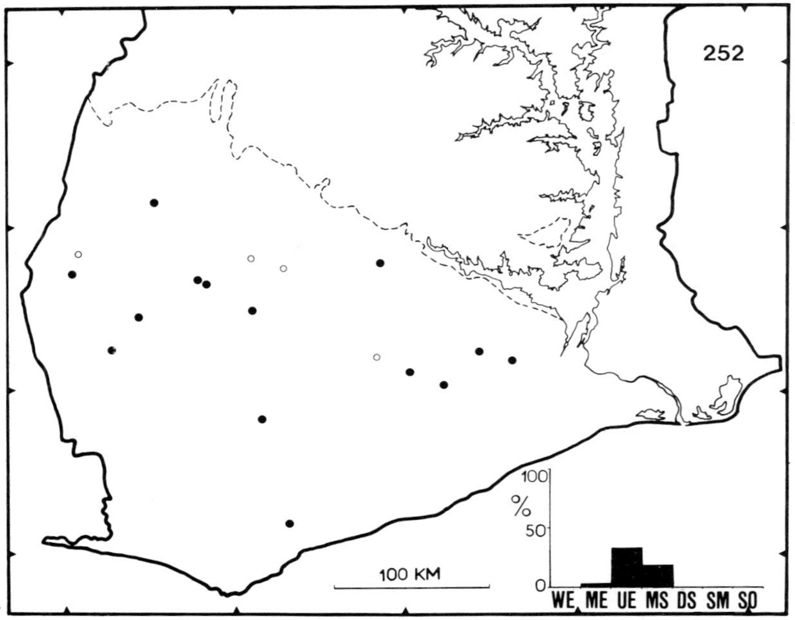

247 EHRETIA TRACHYPHYLLA C. H. Wright, FWTA 2:318. Boraginaceae. Akan: okyini. (Type from Aburi).
Medium-sized tree, branching low and often with epicormic shoots; lvs mesophyll, with characteristic appressed pubescence on underside. Fl white, small, bell, 6–8. Fr drupe; seeds medium. Regenerates in shade. Iv. Coast and Ghana; (Upper Guinea endemic).

248 ELAEIS GUINEENSIS Jacq., FWTA 3:161. Palmae. Akan: abe. English: oilpalm.
Medium-sized palm tree; lvs pinnate, leaflets macrophyll, basal leaflets modified as spines. Corner's model (59). Fl monoecious, not seasonal, clustered in unisexual inflorescences; male small, yellowish, bowl, strongly scented, attractive to stingless bees; female medium, almost embedded in spadix. Fr reddish-purple drupe with oily flesh; seed large to very large. Seedlings and young plants widespread on forest floor; mature trees especially abundant in forest near the savanna boundary, the 'fire-zone' subtype of Dry Semi-deciduous type (55). Widespread in Tropical Africa; (chorological transgressor). Cultivated for the oil in its fruits throughout the tropics.

249 ELAEOPHORBIA GRANDIFOLIA (Haw.) Croizat, FWTA 1:423. (Incl. *E. drupifera* (Thonn.) Stapf, FWTA 1:423). Euphorbiaceae. Akan: akani. Ga: tunyotso.
Small to medium tree, containing abundant caustic white latex; lvs notophyll to mesophyll, succulent. Rauh's model (59). Cyathia yellowish, small, bowl, not clearly seasonal; flowers monoecious. Fr yellow drupe; seed medium. Regenerates in shade. Typical of thicket clumps on the Accra Plains; leaves smaller in this situation than on trees in forest, but all intermediates occur, and it does not seem possible to follow FWTA in recognising two species of *Elaeophorbia* in West Africa. Guinée to Central African Republic; (Guinea wide).

250 ENANTIA POLYCARPA (DC.) Engl. & Diels, FWTA 1:51. Annonaceae. Akan: duasika.
Medium-sized tree with unique bright yellow slash; lvs mesophyll. Roux's model. Fl large, greenish-yellow, bowl, not clearly seasonal. Fr cluster of black berries on red stalks; seed medium. Regenerates in shade. S. Leone to Nigeria; (Guinea wide).

251 ENTANDROPHRAGMA ANGOLENSE (Welw.) DC., FWTA 1:700. Meliaceae. Akan: edinam.
Large tree with high buttresses and spreading surface roots; lvs pinnate, leaflets notophyll, deciduous. Rauh's model (59). Fl small, greenish, bowl, monoecious, mainly 12–2, appearing with the young leaves. Fr woody capsule, ripening 9–11; seeds large, winged, wind-dispersed. Germination occurs in shade, but saplings establish best in small gaps. Guinée to Angola; (Guineo-Congolian wide). Class 1 timber; trade name: gedu nohor (40).

252 ENTANDROPHRAGMA CANDOLLEI Harms, FWTA 1:700. Meliaceae. Akan: penkwa-akoa.
Large tree with slight buttresses; lvs pinnate, leaflets notophyll, deciduous. Rauh's model. Fl cream, small, bowl, monoecious, 9–12. Fr woody capsule; seeds large, winged, wind-dispersed. Regenerates in shade. Guinée to Zaire; (Guineo-Congolian wide). Class 2 timber; trade name: omu (40).

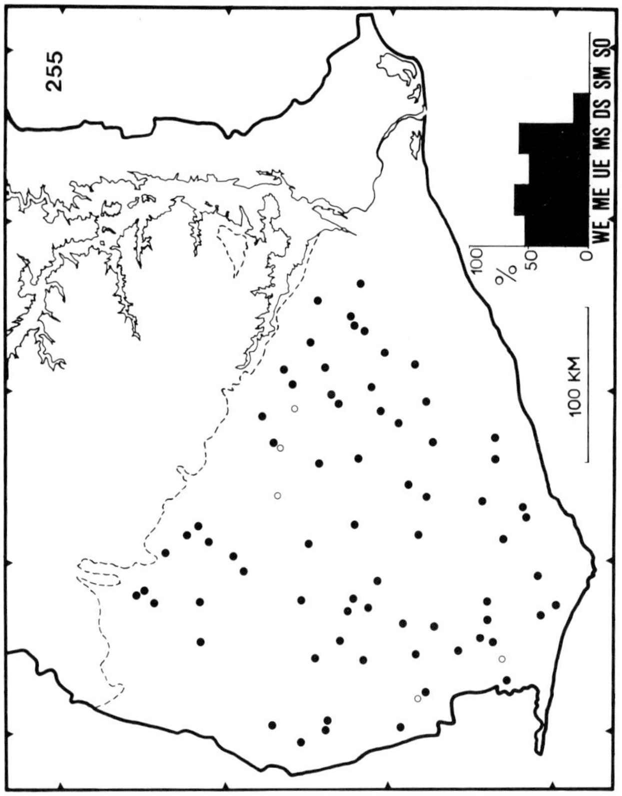
255
100 KM
100
%
50
0
WE ME UE MS DS SM SO

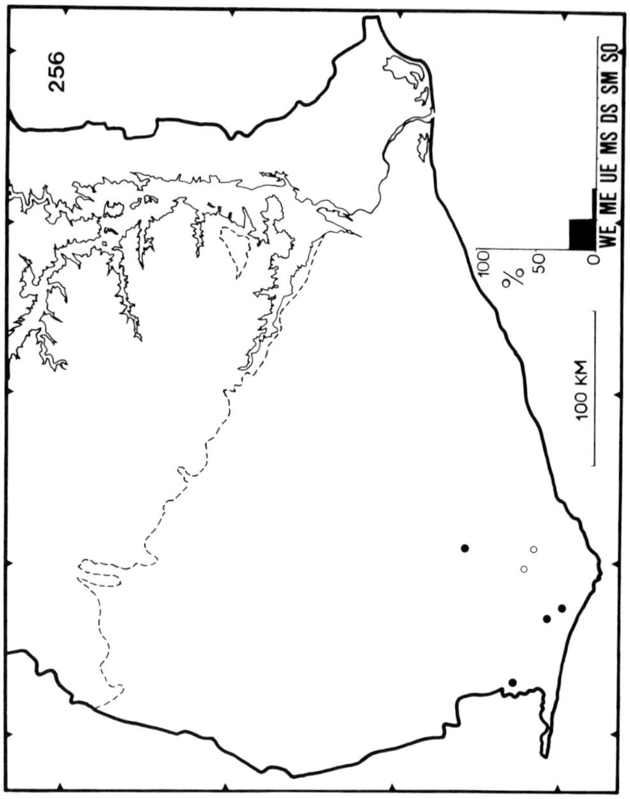
256
100 KM
100
%
50
0
WE ME UE MS DS SM SO

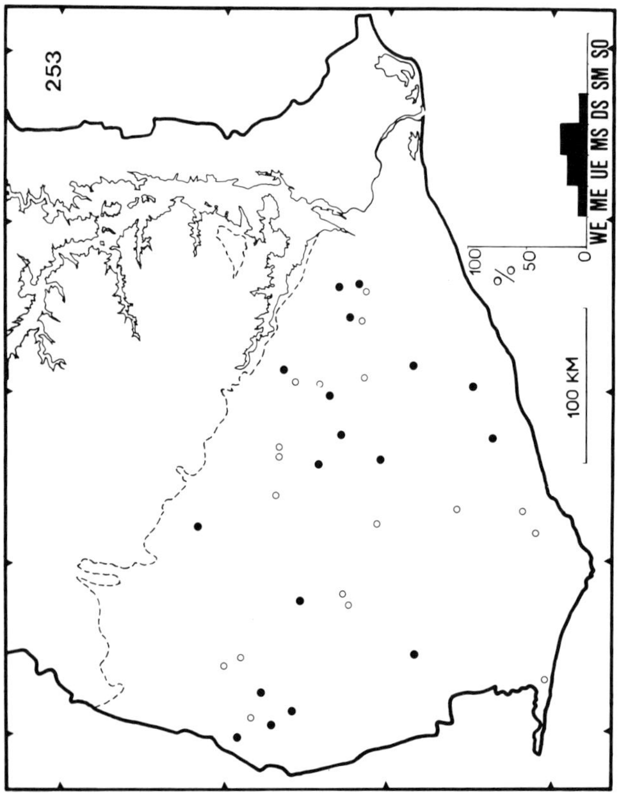
253
100 KM
100
%
50
0
WE ME UE MS DS SM SO

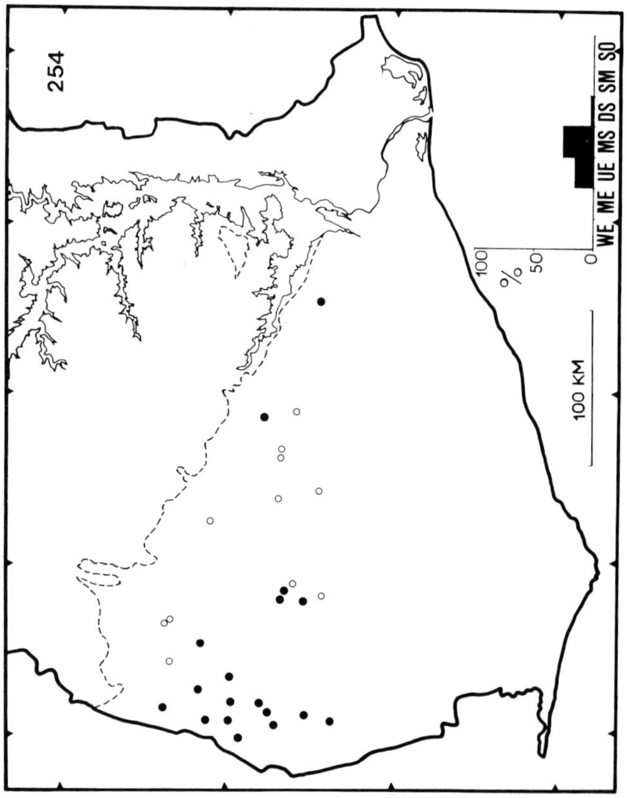
254
100 KM
100
%
50
0
WE ME UE MS DS SM SO

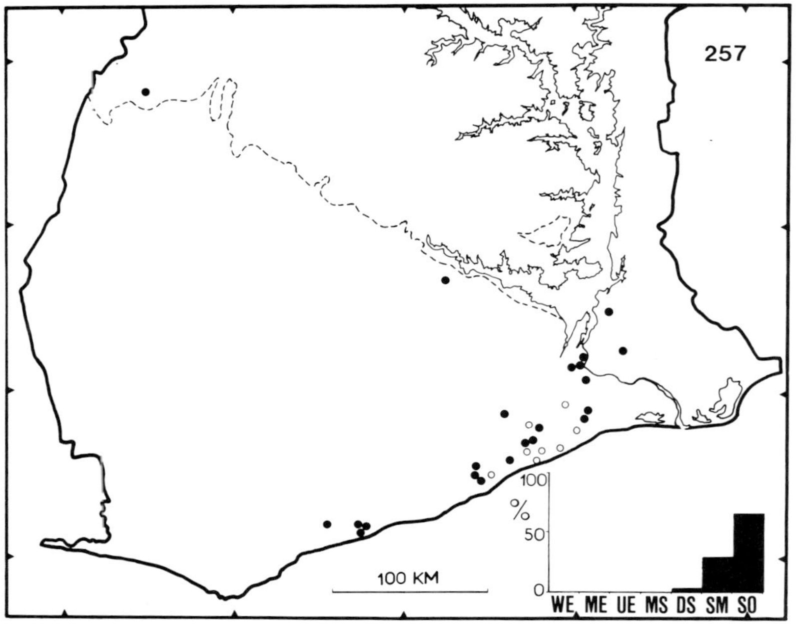

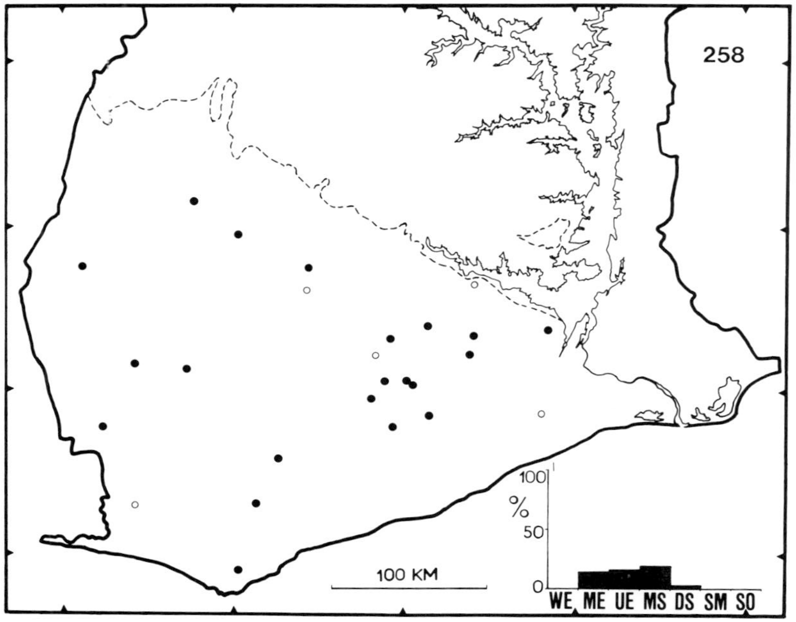

253 ENTANDROPHRAGMA CYLINDRICUM (Sprague) Sprague, FWTA 1:701. Meliaceae. Akan: penkwa.

Large tree with small buttresses; slash scented; lvs pinnate, leaflets notophyll, deciduous. Rauh's model. Fl small, dish, yellowish-white, monoecious, 10–3. Fr woody capsule, 5–8; seeds large, winged, wind-dispersed. Regenerates in shade. S. Leone to Zaire and Uganda; (Guineo-Congolian wide). Class 1 timber; trade name: sapele (40).

254 ENTANDROPHRAGMA UTILE (Dawe & Sprague) Sprague, FWTA 1:700. Meliaceae. Akan: efoobrodedwo.

Large tree with large buttresses and deeply fissured bark; lvs pinnate, leaflets notophyll, deciduous. Rauh's model (59). Fl white, medium, bowl, monoecious, 12–3 with young leaves. Fr woody capsule, taking a year to develop and ripening with the flowers; seeds large, winged, wind-dispersed. Seeds germinate in shade, but development to sapling requires thinning of the canopy. S. Leone to Angola and Uganda; (Guineo-Congolian wide). Class 1 timber; trade name: utile (40).

255 EREMOSPATHA MACROCARPA (Mann & Wendl.) Wendl., FWTA 3:168. Palmae. Akan: mfia (rattans in general).

Large climbing palm, with smooth leaf-sheaths; lvs pinnate, rhachis prickly on margin, leaflets macrophyll, spiny on margin, terminal leaflets reduced to strong deflexed spines. Tomlinson's model. Fl yellowish, medium, funnel, not clearly seasonal. Fr brownish 1-seeded berry; seed large. Regenerates in shade, but mature plants commonest in gaps and swamp forest. Plot records very highly correlated with acid soils. S. Leone to Cameroun; (Guinea wide).

256 ERIOCOELUM RACEMOSUM Bak., FWTA 1:724. Sapindaceae. Akan: onibonakokoo.

Small to medium tree, branching low; lvs pinnate, leaflets notophyll; twigs and leaves hairy. Fl white, small, dish, unisexual and probably dioecious, not clearly seasonal. Fr woody, yellow, indehiscent; seeds large. Regenerates in shade. S. Leone to Ghana; (Upper Guinea endemic).

257 ERYTHROXYLUM EMARGINATUM Thonn., FWTA 1:356. Erythroxylaceae. Ga: taamitso. (Type from near Accra).

Pygmy tree or shrub; lvs microphyll. Fl white, small, bowl, not clearly seasonal. Fr red drupe; seed medium. Regenerates in shade. Extends to forest patches on rocky scarps in the savanna zone. Widespread in Tropical Africa; (chorological transgressor).

258 EUADENIA EMINENS Hook. f., FWTA 1:93. Capparaceae (Capparidaceae in FWTA). Akan: dinsinkoro.

Unbranched pygmy tree; lvs trifoliolate, leaflets mesophyll. Chamberlain's model (59). Fl large, cream, flag, 1–4. Fr orange fleshy pod; seeds large. Regenerates in shade. S. Leone to Ghana; (Upper Guinea endemic).

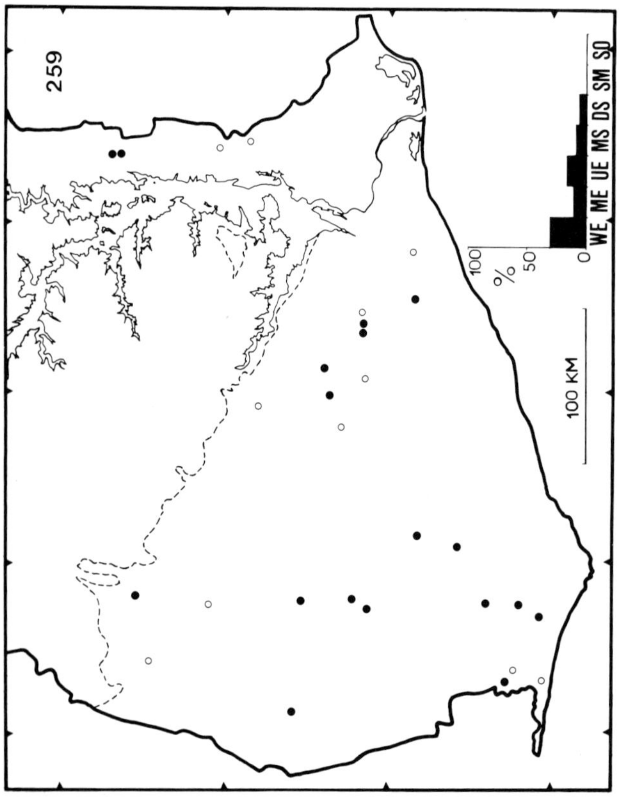
259
100 KM
%
100
50
0
WE ME UE MS DS SM SO

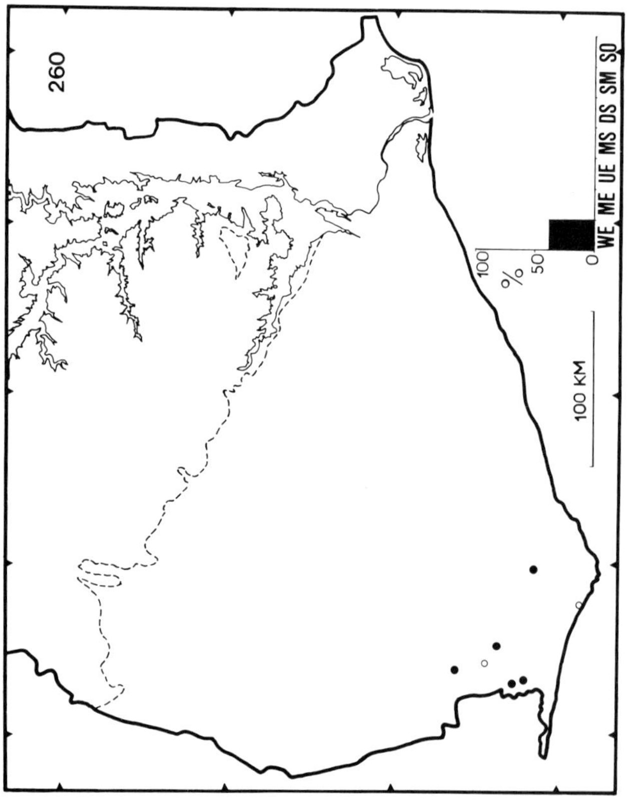
260
100 KM
%
100
50
0
WE ME UE MS DS SM SO

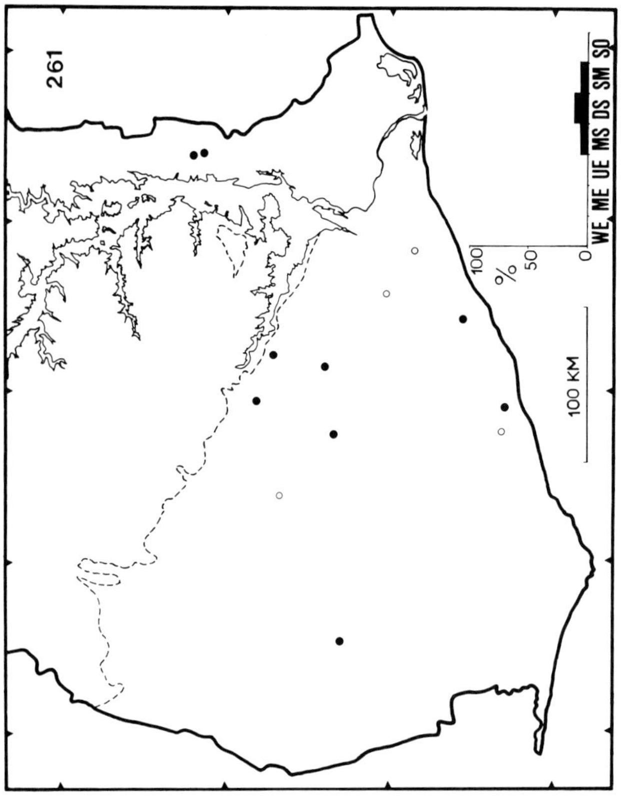
261
100 KM
%
100
50
0
WE ME UE MS DS SM SO

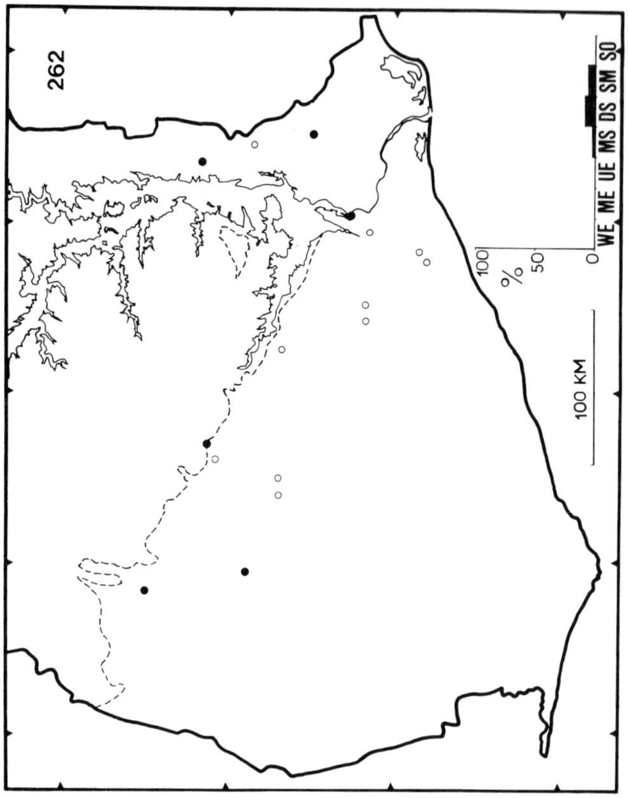
262
100 KM
%
100
50
0
WE ME UE MS DS SM SO

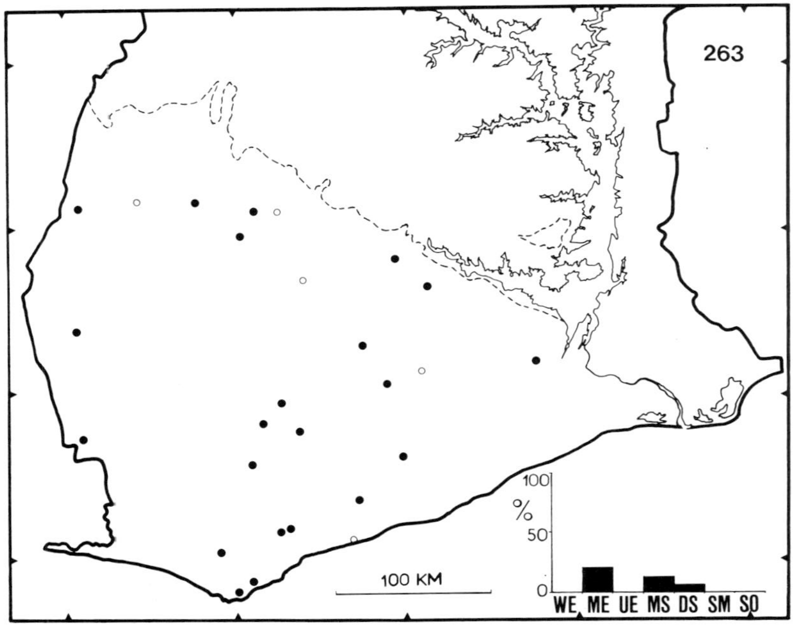

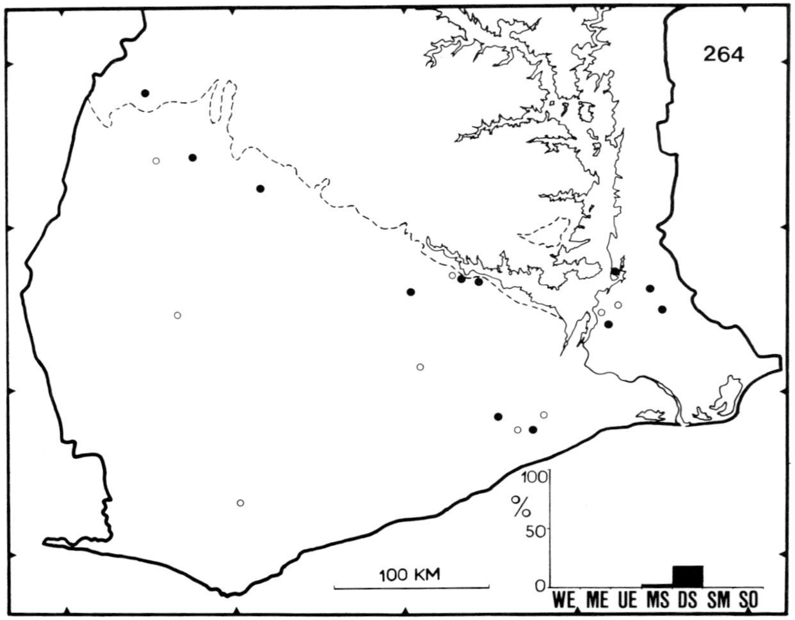

259 EUCLINIA LONGIFLORA Salisb., FWTA 2:121. Rubiaceae.
Pygmy tree; lvs mesophyll. Fagerlind's model (59). Fl large, cream, funnel with long tube, 11–4. Fr yellowish berry, seeds medium to large. Regenerates in shade. Guinea Bissau to Angola and Uganda; (Guineo-Congolian wide).

260 EUGENIA CALOPHYLLOIDES DC., FWTA 1:238. Myrtaceae.
Small tree; lvs mesophyll, pellucid glandular-punctate; twigs and underside of midrib pubescent. Fl white, small, densely clustered, bowl, mostly 11–4. Fr drupe, seed medium. Regenerates in shade. Guinée to Ghana; (Upper Guinea endemic).

261 EUGENIA OBANENSIS Bak. f., FWTA 1:238. Myrtaceae.
Pygmy tree or shrub; lvs microphyll to notophyll, pellucid glandular-punctate; young twigs densely pubescent. Fl white, small, densely-clustered, bowl, 10–3. Fr reddish drupe, seed medium. Regenerates in shade. Ghana to Cameroun; (Guinea wide). It is doubtful whether the Ghanaian material is really *E. obanensis* (type from wet forest in SE Nigeria.)

262 EULOPHIA GUINEENSIS Lindl., FWTA 3:246. Orchidaceae.
Perennial ground herb with tuberous rhizome; lvs macrophyll. Tomlinson's model. Fl pale purple, large gullet, 7–12. Fr capsule; seeds dust-like. Regenerates in shade. Gambia to Angola and Tanzania; (Guineo-Congolian wide).

263 FARQUHARIA ELLIPTICA Stapf, FWTA 2:74. Apocynaceae.
Large woody climber containing clear latex; lvs notophyll. Fl white with yellow centre, large, dish with tube, 1–5. Fr long woody follicle; seeds medium, with plume of long hairs, wind-dispersed. Regenerates in shade. Iv. Coast to Congo Republic; (Guinea wide).

264 FICUS EXASPERATA Vahl, FWTA 1:605. Moraceae. Akan: nyankyerene.
Small to medium-sized tree, containing brownish latex; lvs notophyll, very scabrid (used as sandpaper); deciduous. Figs greenish-yellow, medium, trap, mostly 11–3 while leafless. Regenerates in gaps and farms, so the survey underestimates its abundance. Widespread in Tropical Africa; (chorological transgressor).

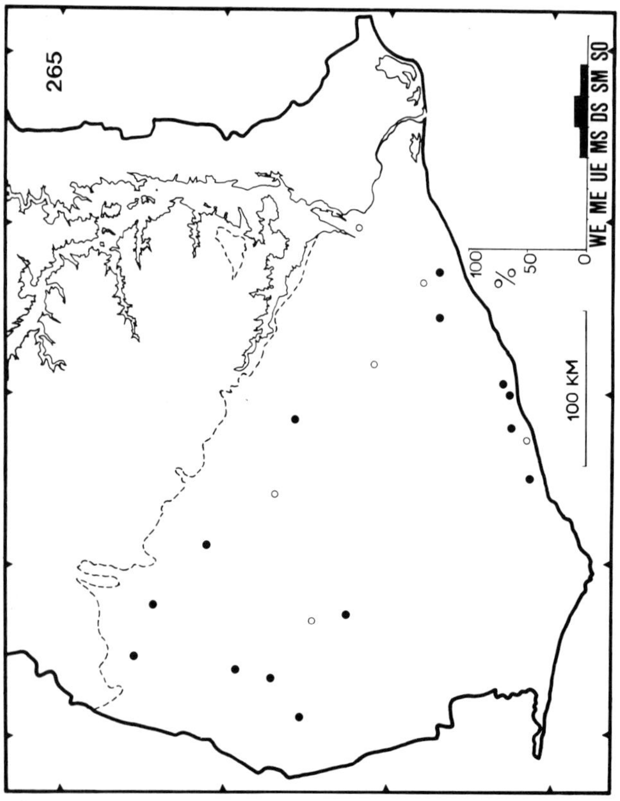
265
100 KM
100
%
50
0
WE ME UE MS DS SM SO

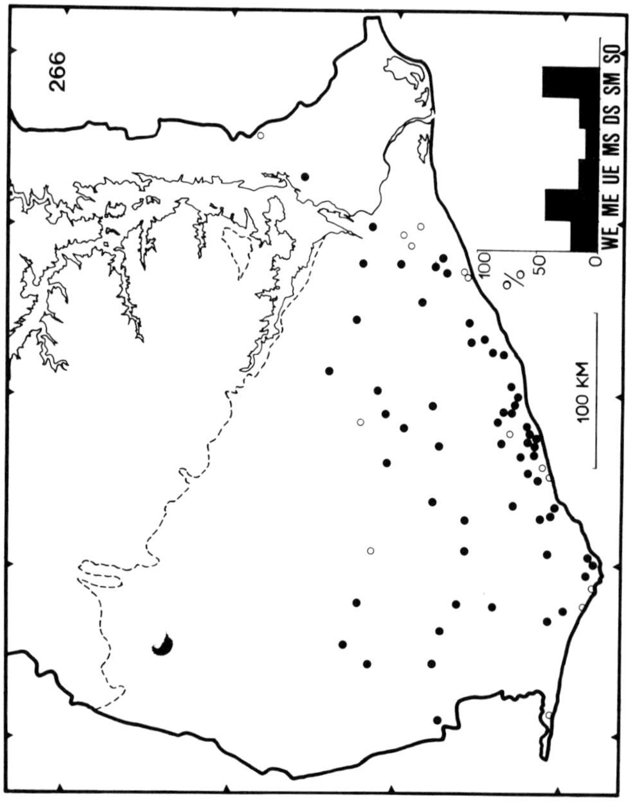
266
100 KM
100
%
50
0
WE ME UE MS DS SM SO

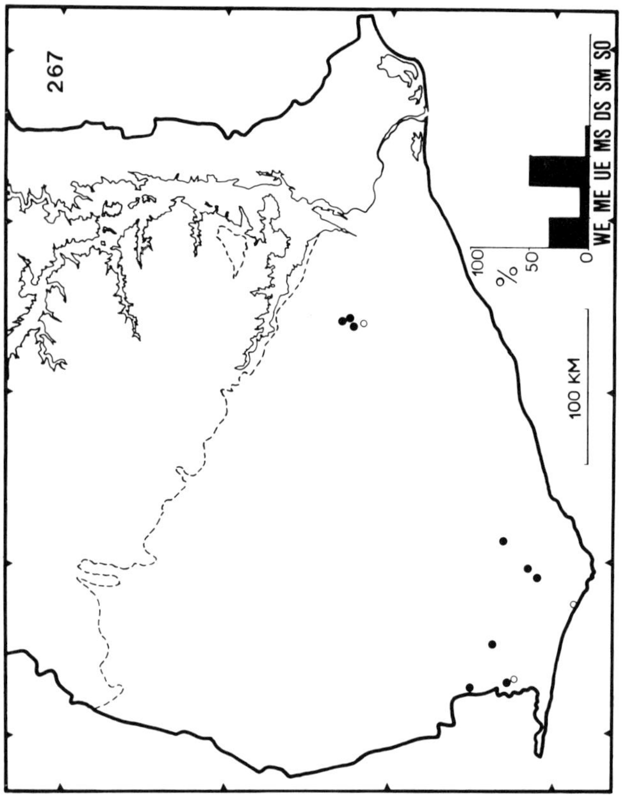
267
100 KM
100
%
50
0
WE ME UE MS DS SM SO

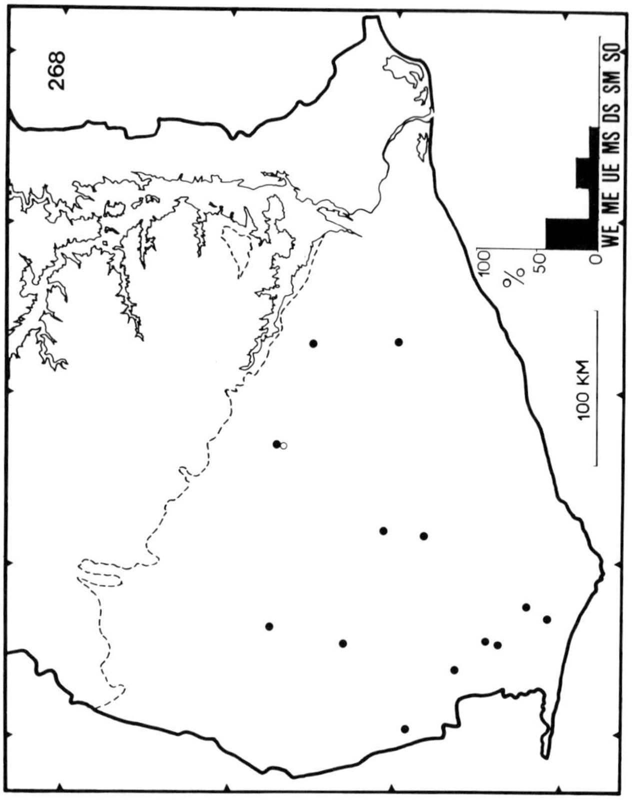
268
100 KM
100
%
50
0
WE ME UE MS DS SM SO

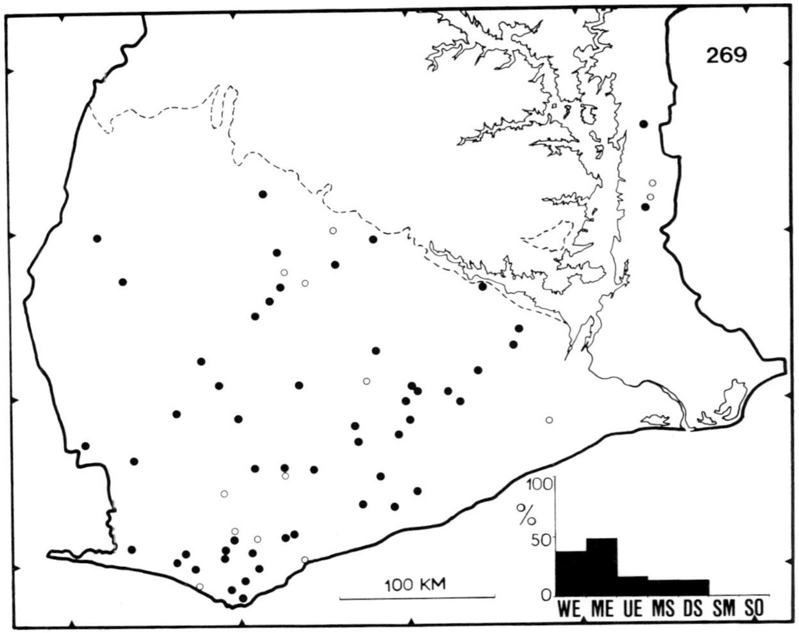

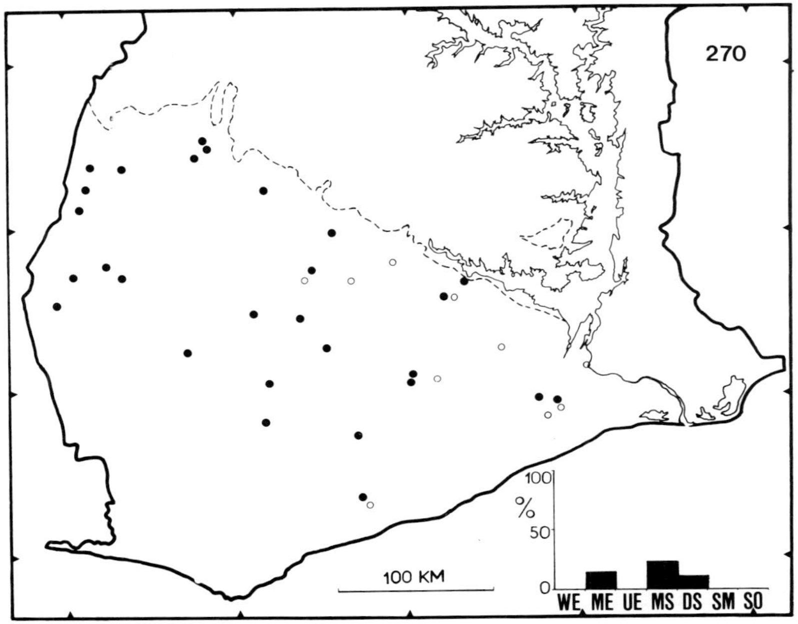

265 FICUS SAGITTIFOLIA Warb. ex Mildbr. & Burret, FWTA 1:607. Moraceae.
Epiphytic shrub, containing white latex; lvs macrophyll. Figs large, yellowish, trap, not clearly seasonal. Regenerates on low branches within the canopy. Guinea Bissau to Nigeria; (Guinea wide).

266 FLAGELLARIA GUINEENSIS Schumach., FWTA 3:51. Flagellariaceae. (Type probably from near Accra).
Slender climber with perennial herbaceous stems; lvs mesophyll, grass-like, tip forming a coiled tendril. Schoute's model. Fl yellowish, small, bowl, not seasonal. Fr red drupe; seed medium. Regenerates in shade, but commonly seen in secondary thickets and farms. Widespread from Iv. Coast across Africa; (chorological transgressor).

267 FRIESODIELSIA ENGHIANA (Diels) Verdc. (109). Annonaceae.
Large woody climber, wood outline shallowly lobed in transverse section; lvs notophyll to mesophyll, pale blue-grey below; twigs and underside of midrib densely hairy. Fl reddish-brown, medium to large, bowl. Fr berry; seed large. Regenerates in shade. Plot records very significantly correlated with base-poor soils. S. Leone to Zaire; (Guineo-Congolian wide).

268 FRIESODIELSIA VELUTINA (Sprague & Hutch.) Van Steenis (109), (*Oxymitra velutina* Sprague & Hutch. in FWTA 1:45). Annonaceae.
Large woody climber; lvs mesophyll; twigs and midrib densely hairy. Fl medium, bowl, not yet collected in Ghana. Fr berry, not yet collected in Ghana. Regenerates in shade. S. Leone to Ghana; (Upper Guinea endemic).

269 FUNTUMIA AFRICANA (Benth.) Stapf, FWTA 2:74. Apocynaceae. Akan: okae.
Medium-sized tree, containing sticky white latex; lvs mesophyll. Prévost's model. Fl white, medium to large, dish with tube, mostly 1–3. Fr woody follicle, 11–4; seeds large, with plume of long hairs (Fig. 6.1a), wind-dispersed. Seeds germinate in light shade, but only become established in gaps and old farmland. Guinea Bissau to Angola and Mozambique; (chorological transgressor).

270 FUNTUMIA ELASTICA (Preuss) Stapf, FWTA 2:74. Apocynaceae. Akan: fruntum.
Medium-sized tree, containing cream-coloured latex becoming rubbery as it dries; lvs mesophyll. Prévost's model (59). Fl white, medium, dish with tube, 1–4. Fr woody follicle, 1–4; seeds large with plume of long hairs, wind-dispersed. Seeds germinate in shade. Guinée to Zaire and Uganda; (Guineo-Congolian wide). Formerly important as a source of commercial rubber.

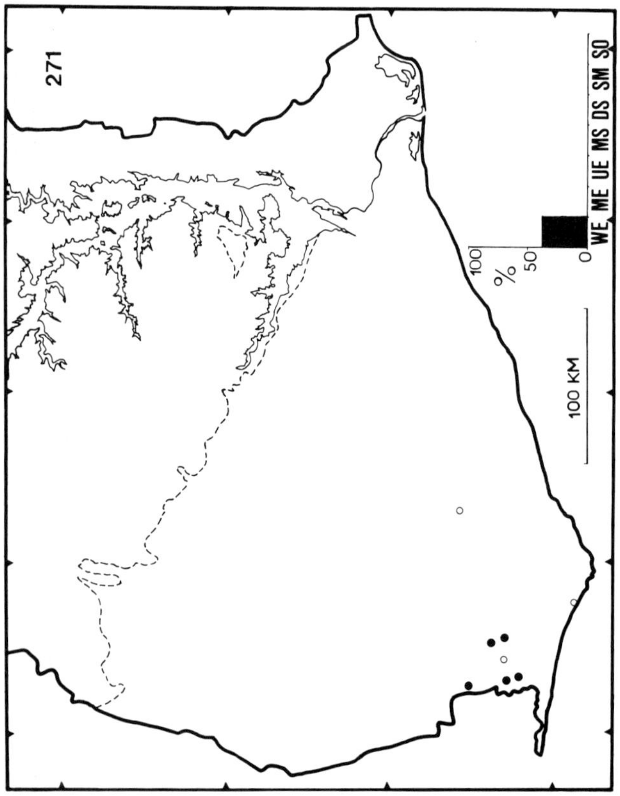
271
100 KM
100
%
50
0
WE ME UE MS DS SM SO

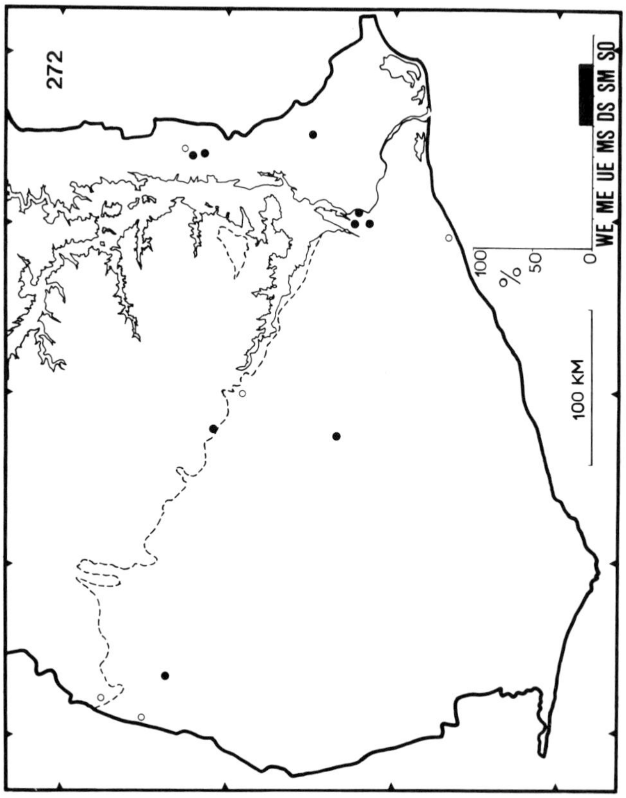
272
100 KM
100
%
50
0
WE ME UE MS DS SM SO

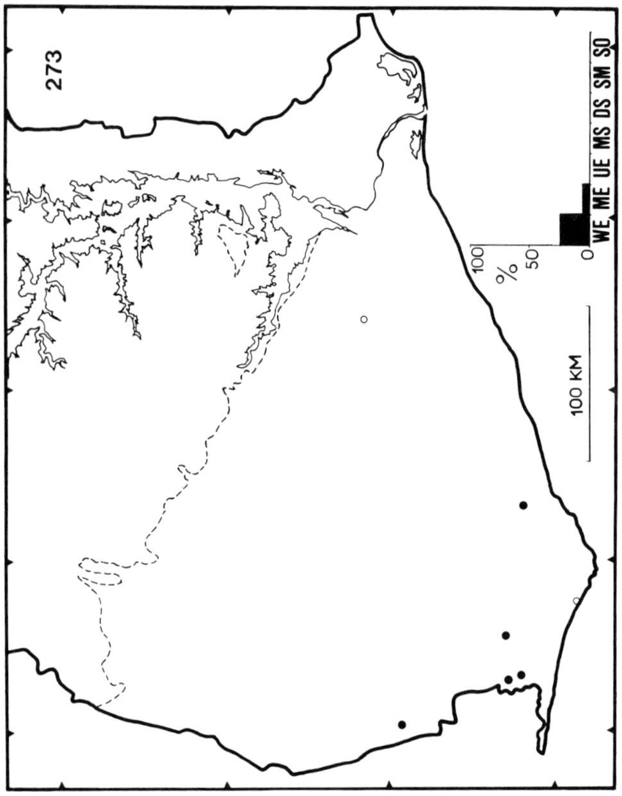
273
100 KM
100
%
50
0
WE ME UE MS DS SM SO

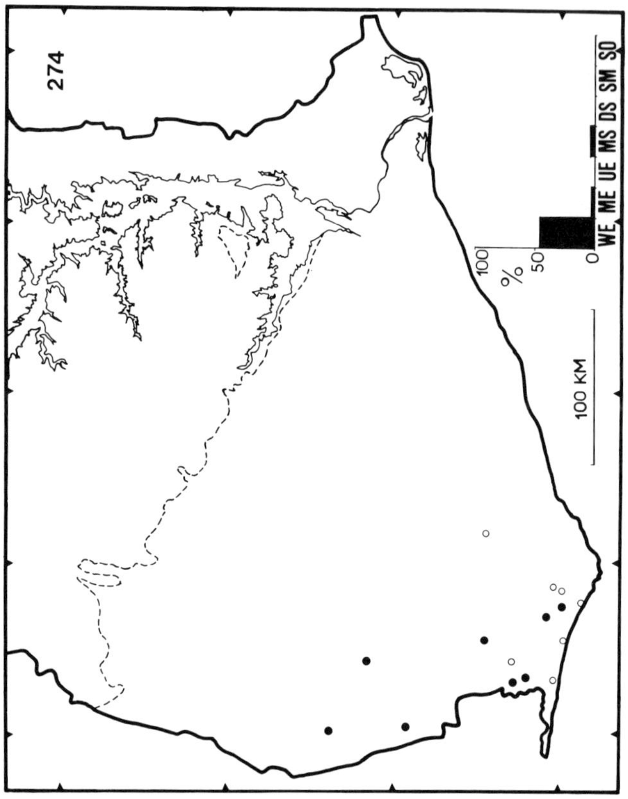
274
100 KM
100
%
50
0
WE ME UE MS DS SM SO

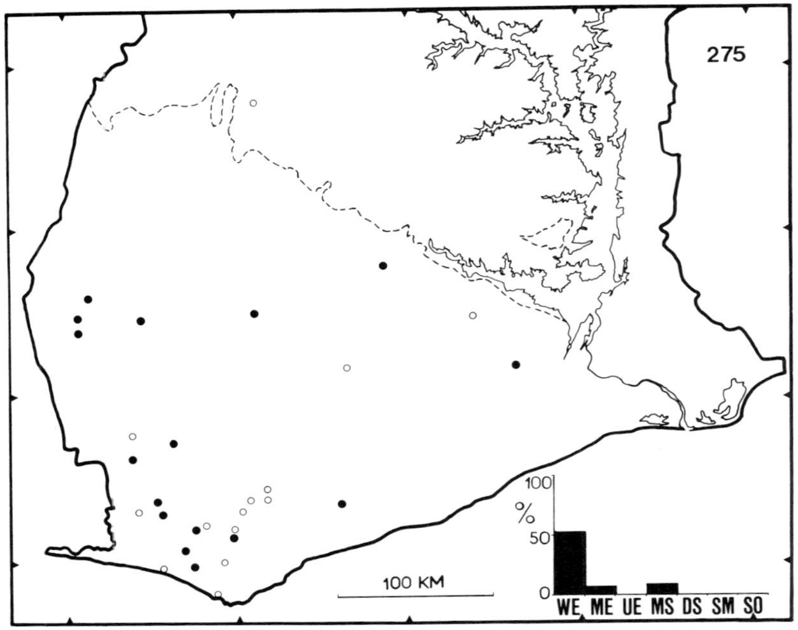

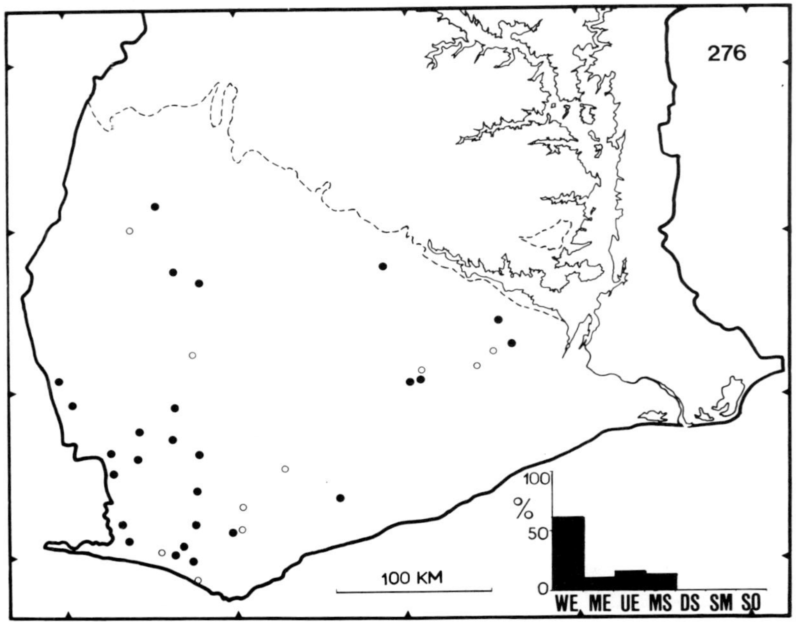

271 GAERTNERA COOPERI Hutch. & M. B. Moss, FWTA 2:191. Rubiaceae.
Pygmy tree; lvs notophyll to mesophyll. Fl white, small, bell, mainly 12–4. Fr black 2-seeded berry; seeds medium. Regenerates in shade. Liberia to Ghana; (Upper Guinea endemic).

272 GARCINIA AFZELII Engl., FWTA 1:295. Guttiferae. Akan: nsoko.
Small tree; lvs notophyll with conspicuous resin canals in fresh and dried leaves. Fl greenish-yellow, medium, bowl, monoecious, 11–4. Fr brownish-yellow berry; seeds large. Regenerates in shade. Guinée to Nigeria; (Guinea wide). Extends to forest patches on rocky scarps in savanna zone. Extensively used as chewing stick (1).

273 GARCINIA EPUNCTATA Stapf, FWTA 1:295. Guttiferae. Akan: nsokonua.
Small to medium tree; slash with clear latex; lvs notophyll with resin canals conspicuous only when leaves are dried. Fl greenish-yellow, medium, bowl, monoecious. Fr berry; seeds large. Regenerates in shade. S. Leone to Angola; (Guineo-Congolian wide, 18, 19). Extensively used as chewing stick (1), but little fertile material has been collected in Ghana.

274 GARCINIA GNETOIDES Hutch. & Dalz., FWTA 1:294. Guttiferae.
Medium-sized tree, with cream latex in slash; lvs mesophyll. Attims' model (59). Fl white, medium, bowl, monoecious. Fr orange berry; seeds large. Regenerates in shade. Liberia to Nigeria; (Guinea wide).

275 GARCINIA SMEATHMANNII (Planch. & Triana) Oliv. (19), (*G. polyantha* Oliv., FWTA 1:294). Guttiferae. Akan: bohwe.
Medium-sized tree, slash with yellow latex; lvs mesophyll, with resin canals faintly visible below. Fl white, small, bowl, in clusters among leaves or on branches below leaves, 8–1. Fr berry; seeds large. Regenerates in shade. Sometimes gregarious in seasonal swamp forest. Guinée to Angola and Zambia; (Guineo-Congolian wide).

276 GEOPHILA AFZELII Hiern, FWTA 2:206. Rubiaceae.
Stoloniferous perennial herb; lvs microphyll, stems and petioles hairy. Fl white or yellow, medium, dish with tube, not clearly seasonal. Fr red berry; seeds medium. Regenerates in shade. Guinée to Congo Republic; (Guinea wide).

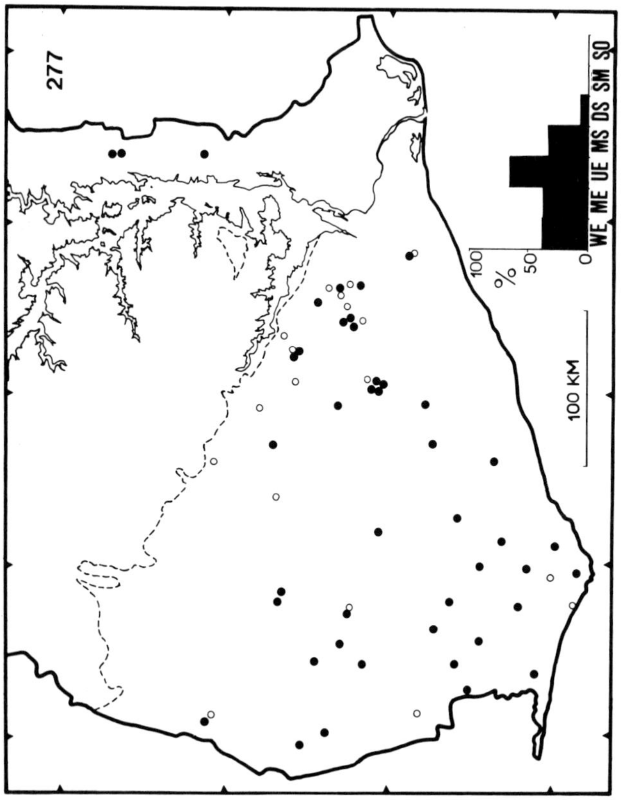
277
100 KM
%
100
50
0
WE ME UE MS DS SM SO

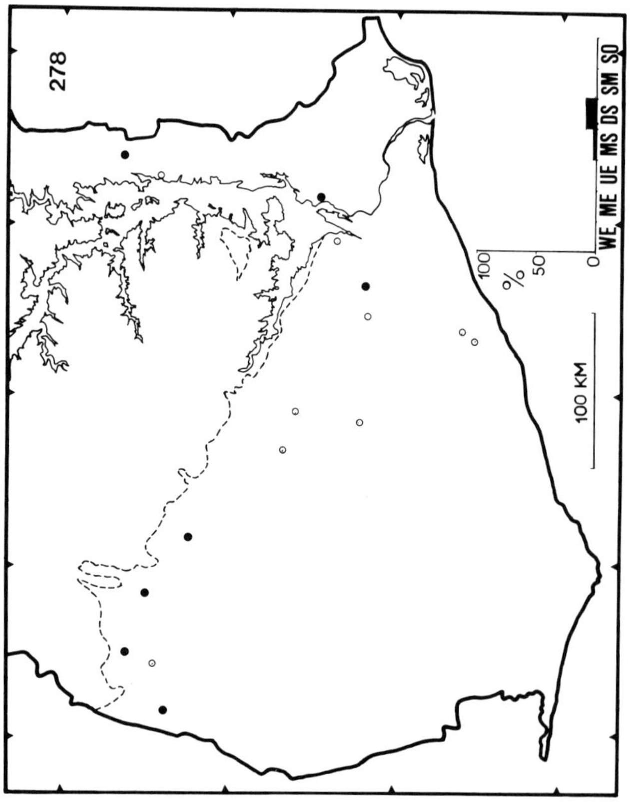
278
100 KM
%
100
50
0
WE ME UE MS DS SM SO

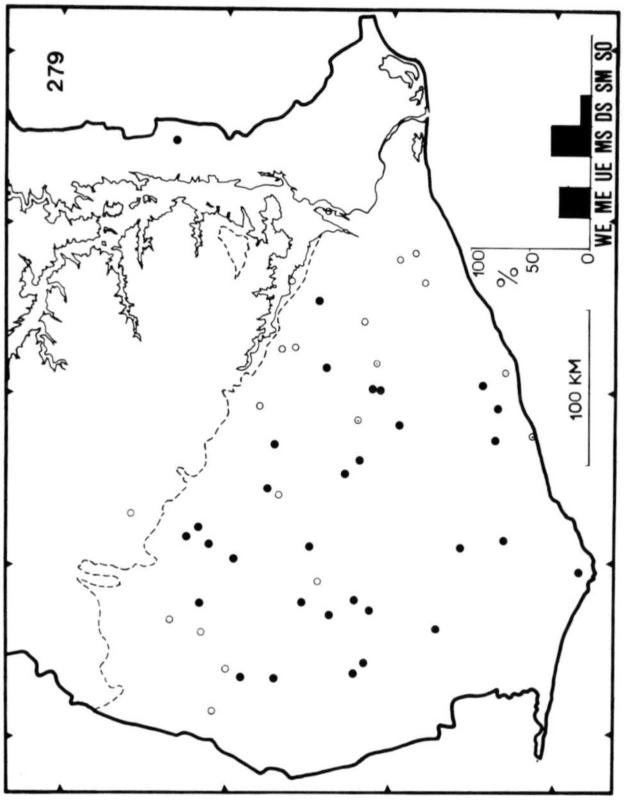
279
100 KM
%
100
50
0
WE ME UE MS DS SM SO

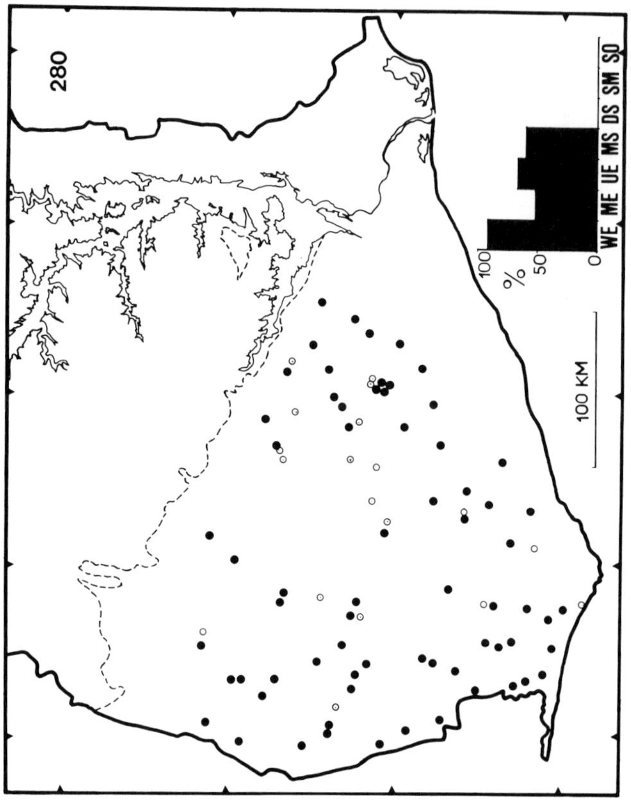
280
100 KM
%
100
50
0
WE ME UE MS DS SM SO

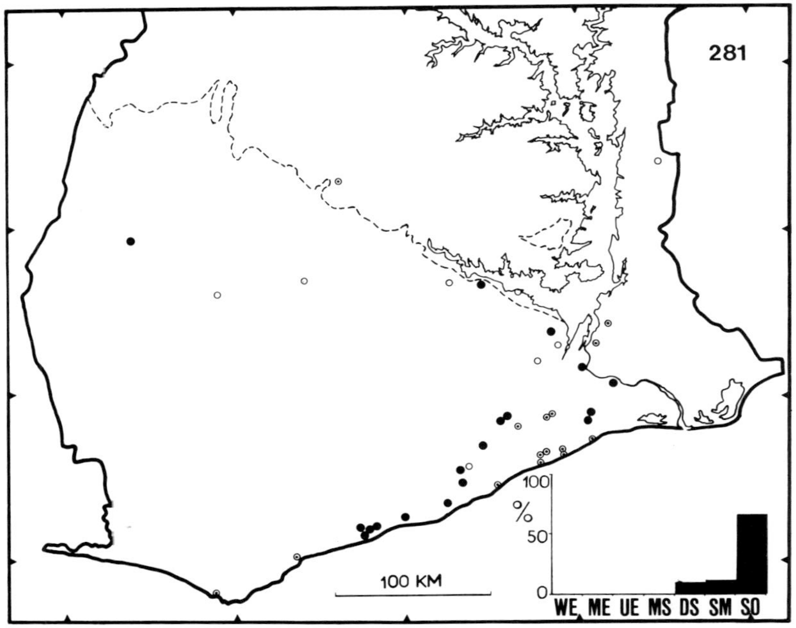

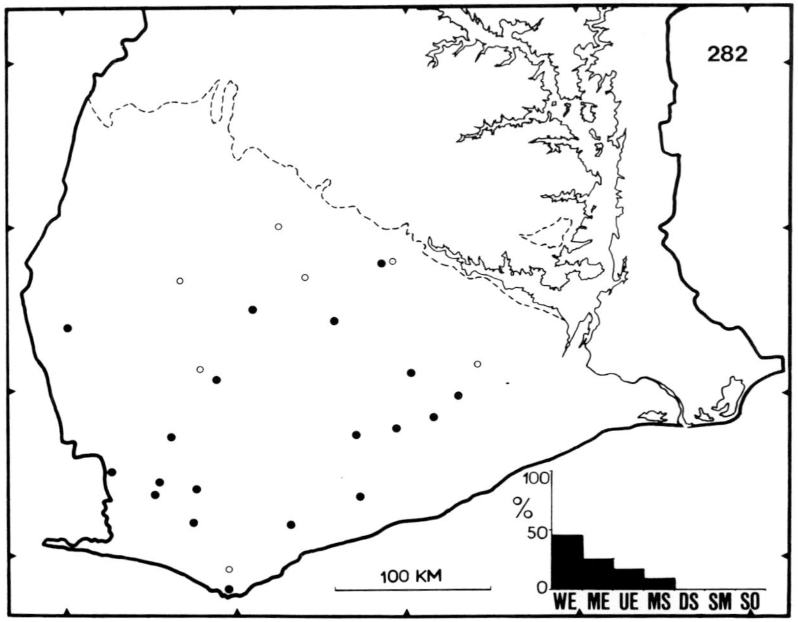

277 GEOPHILA OBVALLATA (Schumach.) F. Didr. subsp. *obvallata*, FWTA 2:206. Rubiaceae. (Type from Akwapim).
Stoloniferous perennial herb; lvs microphyll; petioles and stems hairy. Fl white, medium, dish with tube, not clearly seasonal. Fr dark blue berry; seeds medium. Regenerates in shade. Plot records very significantly correlated with base-poor, rather acid soils. Guinea Bissau to Congo Republic; (Guinea wide).

278 GEOPHILA REPENS (Linn.) I. M. Johnston, FWTA 2:205. Rubiaceae.
Stoloniferous perennial herb; lvs microphyll; petioles and stems hairy. Fl white, medium, dish with tube, not clearly seasonal. Fr red berry; seeds medium. Regenerates in shade, but seen mainly on paths through the forest. Widespread in the tropics; (chorological transgressor).

279 GLYPHAEA BREVIS (Spreng.) Monachino, FWTA 1:308. Tiliaceae. Akan: foto.
Small tree or shrub; lvs mesophyll. Fl yellow, medium, dish, not clearly seasonal. Fr woody, indehiscent; no obvious dispersal mechanism. Seeds medium. Regenerates in shade, but also occurs in secondary forest. Widespread in Tropical Africa; (chorological transgressor).

280 GREENWAYODENDRON OLIVERI (Engl.) Verdc. (108), (*Polyalthia oliveri* Engl., FWTA 1:43). Annonaceae. Akan: duabiri.
Small to medium tree; slash strongly scented; lvs notophyll to mesophyll. Fl white, medium, bowl, 1–5. Fr black berry; seeds medium. Regenerates in shade. Often gregarious in forest understorey. Plot records very significantly correlated with acid, base-poor soils. S. Leone to Gabon; (Guinea wide). Pollen evidence (114) has confirmed the distinctness of *Greenwayodendron* from *Polyalthia*.

281 GREWIA CARPINIFOLIA Juss., FWTA 1:305. Var. *carpinifolia* and var. *rowlandii* (K. Schum.) Burret (54). Tiliaceae. (Type of var. *carpinifolia* probably from Shama).
Large woody climbers, with lobed xylem, or shrubs in thicket; lvs microphyll and scabrid (var. *carpinifolia*), to notophyll and smooth (var. *rowlandii*), deciduous. Roux's model. Fl yellow, medium, dish, not seasonal (var. *carpinifolia*) or 3–6 (var. *rowlandii*). Fr drupaceous, pyrenes 2–4, medium. Regenerates in shade. Ghana to Dahomey and São Tomé (var. *carpinifolia*); S. Leone to Cameroun (var. *rowlandii*). The varieties were not distinguished in the survey, but herbarium records are mapped: var. *carpinifolia* ⊙; var. *rowlandii* ○. Species is a Guinea wide.

282 GREWIA HOOKERANA Exell & Mendonça, FWTA 1:303. Tiliaceae.
Large woody climber, or shrub in thicket; xylem cylindrical; lvs notophyll; twigs and leaf underside pubescent. Fl pinkish-white, small, dish, 1–5. Fr reddish drupe; seeds medium, Regenerates in shade. Plot records very significantly associated with acid soils. S. Leone to Nigeria; (Guinea wide).

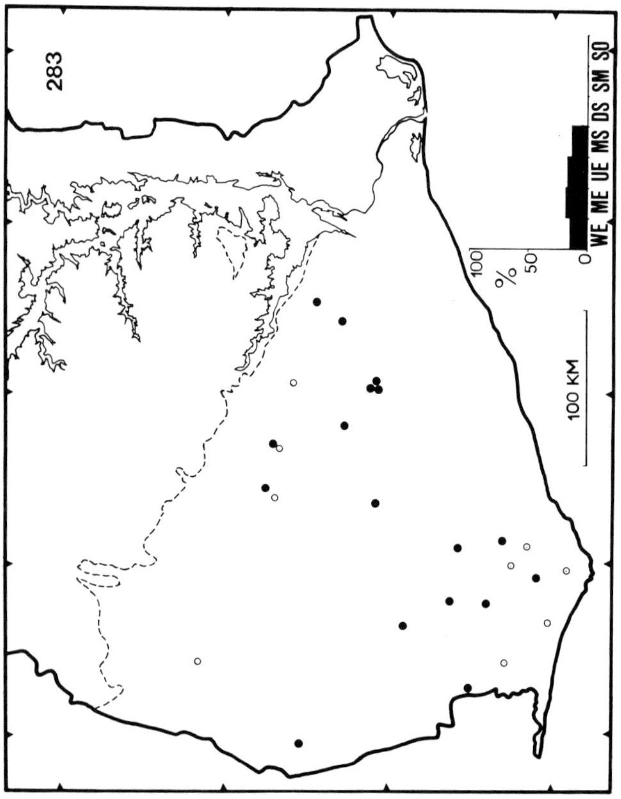
283
100
%
50
0
WE ME UE MS DS SM SO
100 KM

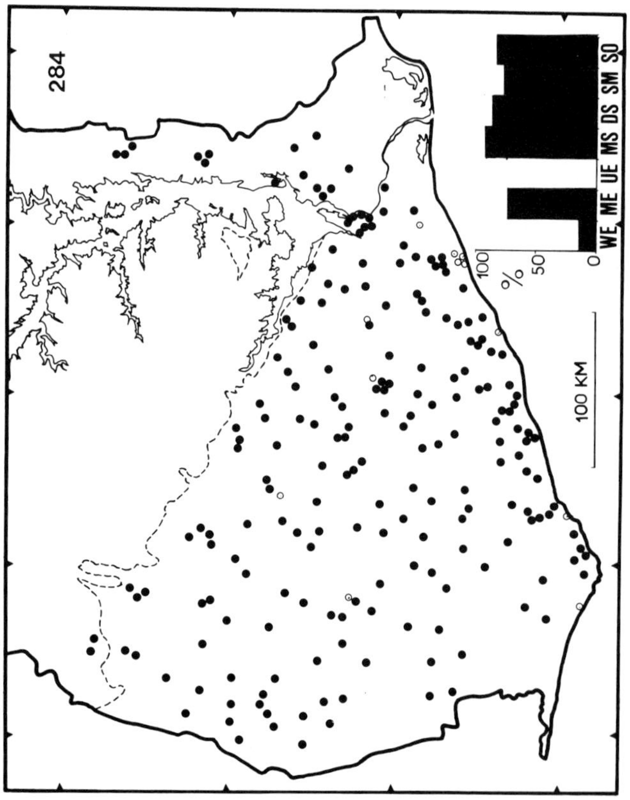
284
100
%
50
0
WE ME UE MS DS SM SO
100 KM

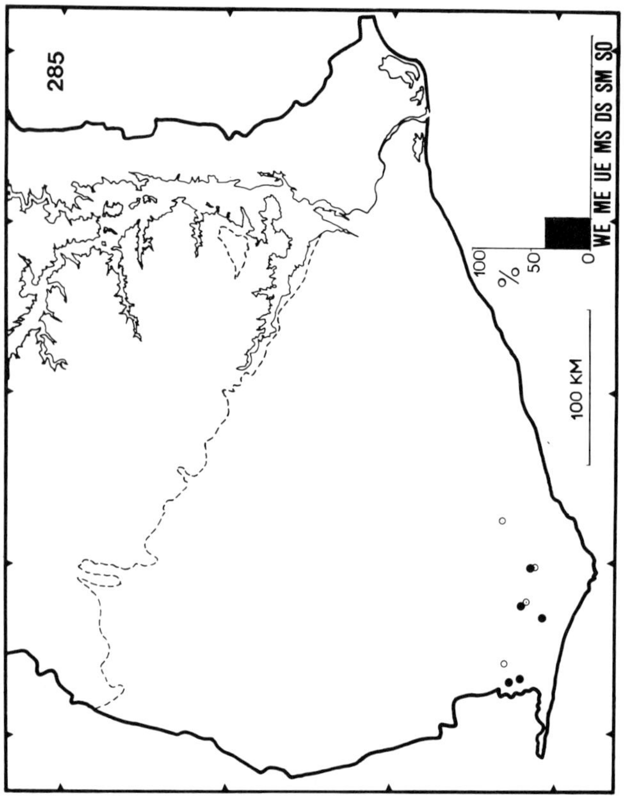
285
100
%
50
0
WE ME UE MS DS SM SO
100 KM

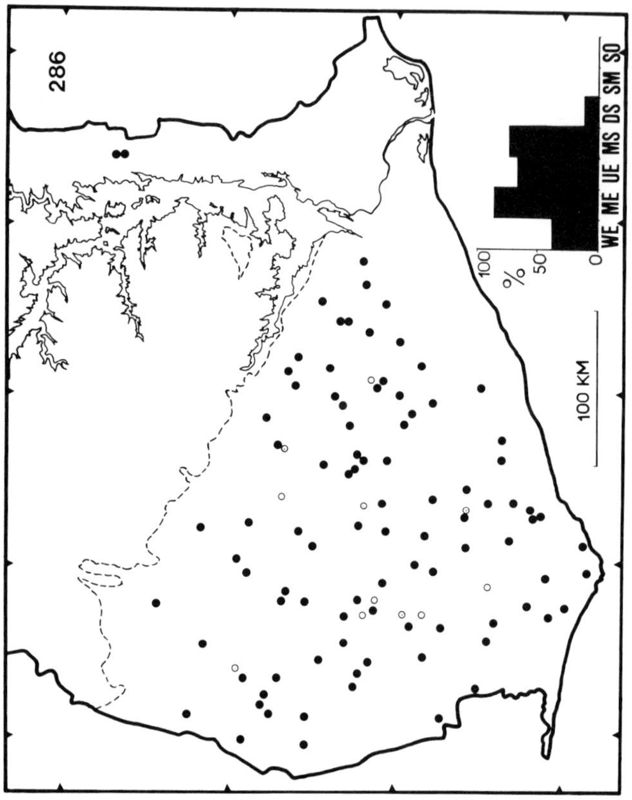
286
100
%
50
0
WE ME UE MS DS SM SO
100 KM

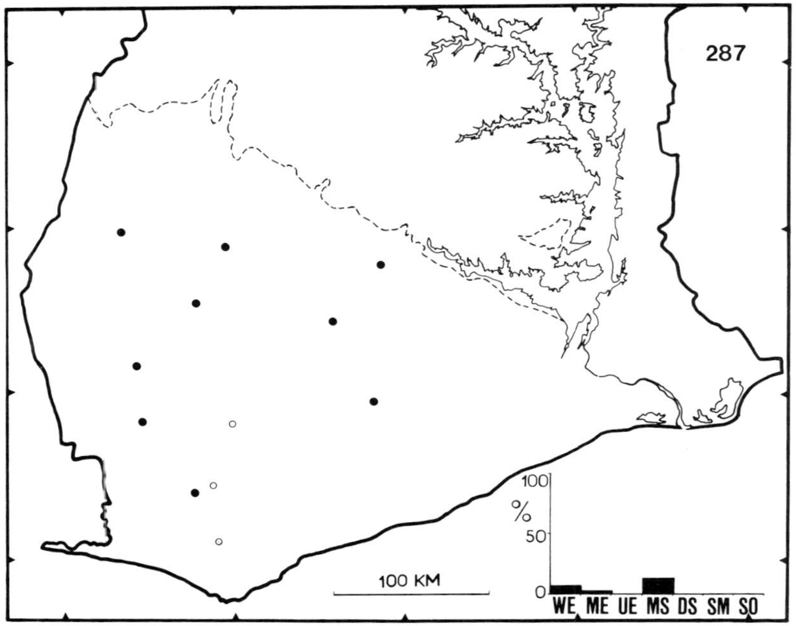

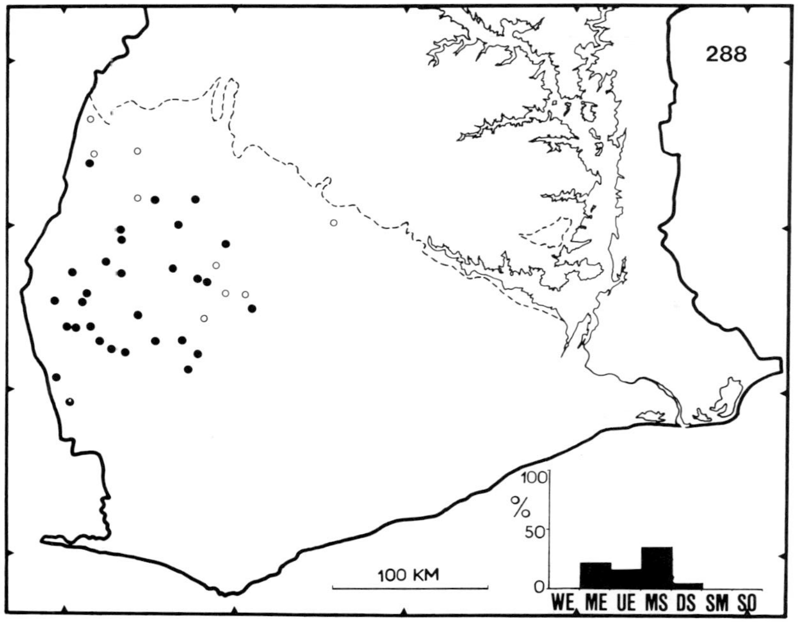

283 GREWIA MALACOCARPA Mast., FWTA 1:303. Tiliaceae.
Medium-sized woody climber, or shrub in thicket; lvs notophyll, densely whitish-tomentellous below. Fl orange, small, not clearly seasonal. Fr drupe; seeds medium. Regenerates in shade. S. Leone to Cameroun; (Guinea wide).

284 GRIFFONIA SIMPLICIFOLIA (Vahl ex DC.) Baill. Caesalpiniaceae. (Type from near Accra or Akwapim). Akan: kegya.
Large woody climber, or shrub in thicket; wood cylindrical; lvs microphyll. Fl greenish, medium, funnel, mostly 8–10. Fr inflated explosive pod, 10–12; seeds large. Regenerates abundantly in shade; more widely and commonly distributed in Ghanaian forest than any other plant, but very significantly less common in the most acid, base-poor, and clayey soils. Liberia to Nigeria, and one record from Gabon; (Guinea wide, 10). All parts of the plant, and especially the seeds, are very rich in 5-hydroxytryptophan (21, 39).

285 GUADUELLA MACROSTACHYS (K. Schum.) Pilger, FWTA 3:360. Gramineae.
Perennial herb (dwarf bamboo); lvs mesophyll. Tomlinson's model. Spikelet green, large, not clearly seasonal. Fr small; no obvious dispersal mechanism. Regenerates in shade. Ghana to Cameroun; (Guinea wide).

286 GUAREA CEDRATA (A. Chev.) Pellegr., FWTA 1:706. Meliaceae. Akan: kwabohoro.
Large tree with buttresses; slash scented; lvs pinnate, leaflets mesophyll. Rauh's model. Fl greenish-yellow, medium, dish, dioecious, mainly 6–8. Fr yellowish capsule, mainly 6–9; seeds large to very large, covered by red arillode. Regenerates profusely in shade; far more common and widespread in Ghanaian forest than other large Meliaceae. Plot records very significantly correlated with acid soils. S. Leone to Zaire and Uganda; (Guineo-Congolian wide). Class 2 timber; trade name: scented guarea (40).

287 GUAREA THOMPSONII Sprague & Hutch., FWTA 1:707. Meliaceae. Akan: kwadwuma.
Large tree with buttresses; slash containing white latex; lvs pinnate, leaflets mesophyll. Rauh's model. Fl yellow, medium, dish, dioecious. Fr. orange capsule; seeds large to very large, with red arillode. Regenerates in shade. Liberia to Zaire; (Guineo-Congolian wide). Class 2 timber; trade name; black guarea (71).

288 GUIBOURTIA EHIE (A. Chev.) J. Léonard, FWTA 1:465. Caesalpiniaceae. Akan: anokye-hyedua.
Large tree with buttresses; lvs bifoliolate, leaflets microphyll, deciduous. Troll's model. Fl white, small, bowl, crowded in large inflorescences, 9–10. Fr papery, indehiscent, 1-seeded pod, 1–2 with young leaves; seed large. Regenerates in shade. Ivory Coast to Nigeria and Gabon; (Guinea wide). Class 2 timber; trade name; shedua.

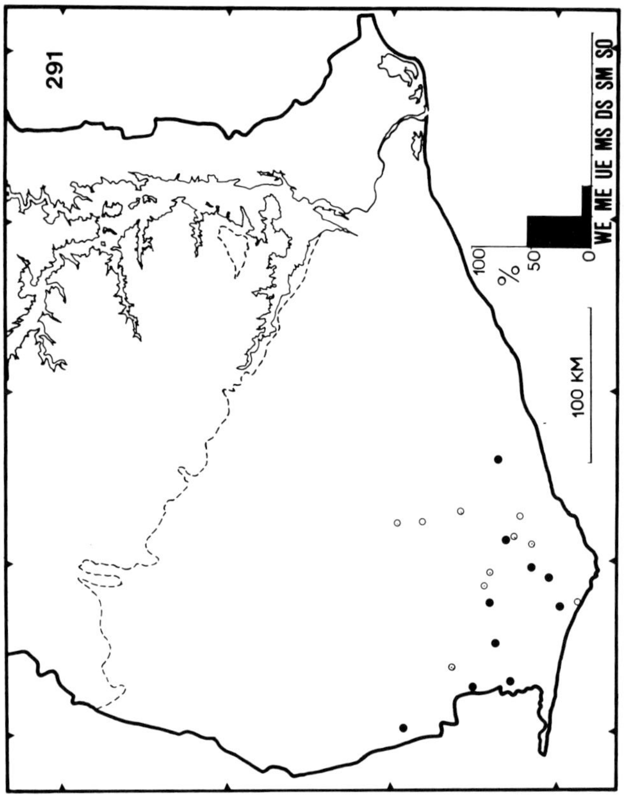
291
100 KM
100
%
50
0
WE ME UE MS DS SM SO

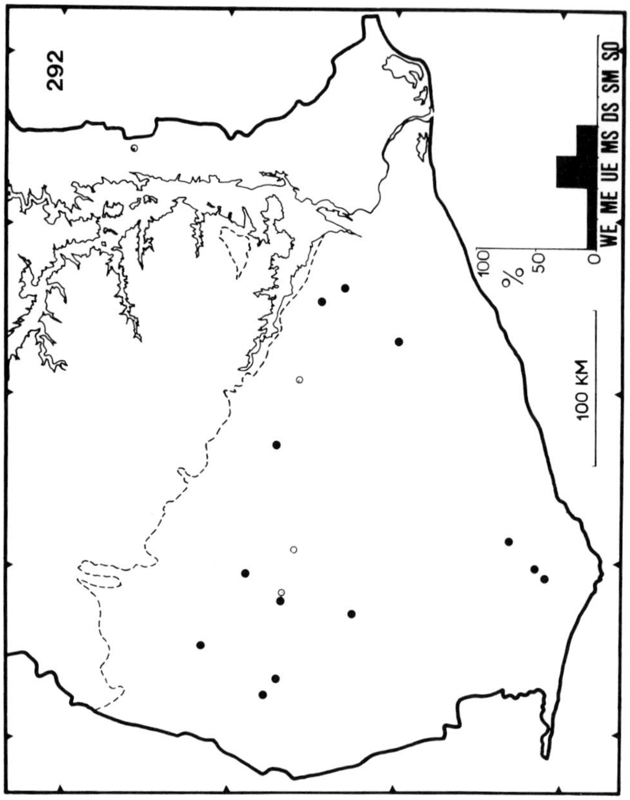
292
100 KM
100
%
50
0
WE ME UE MS DS SM SO

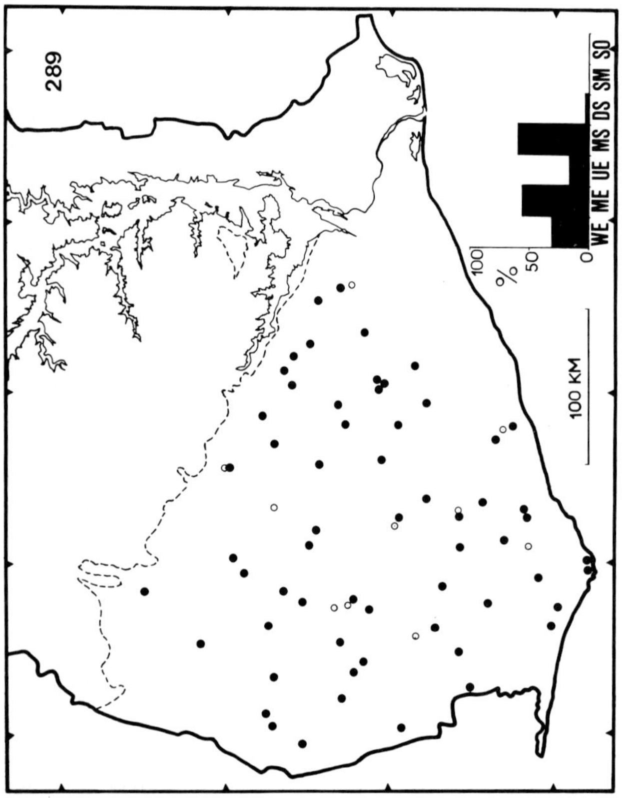
289
100 KM
100
%
50
0
WE ME UE MS DS SM SO

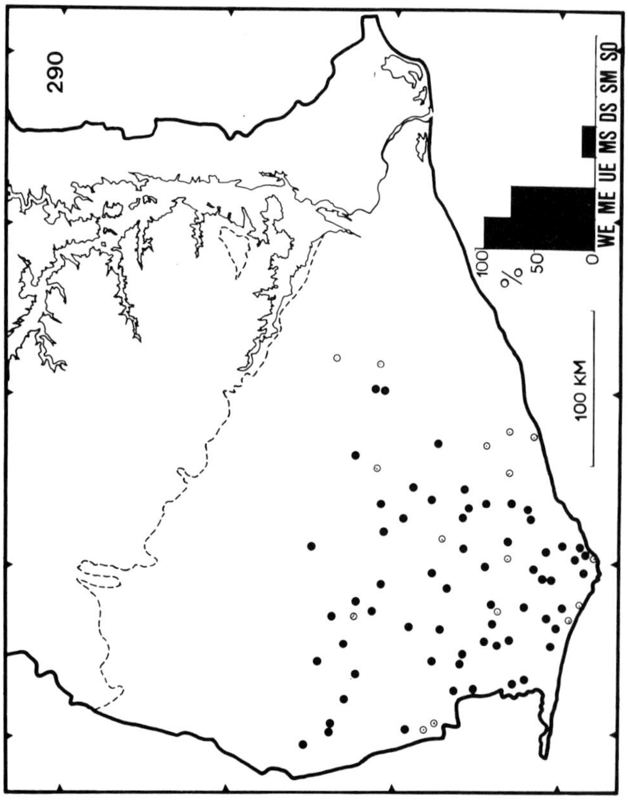
290
100 KM
100
%
50
0
WE ME UE MS DS SM SO

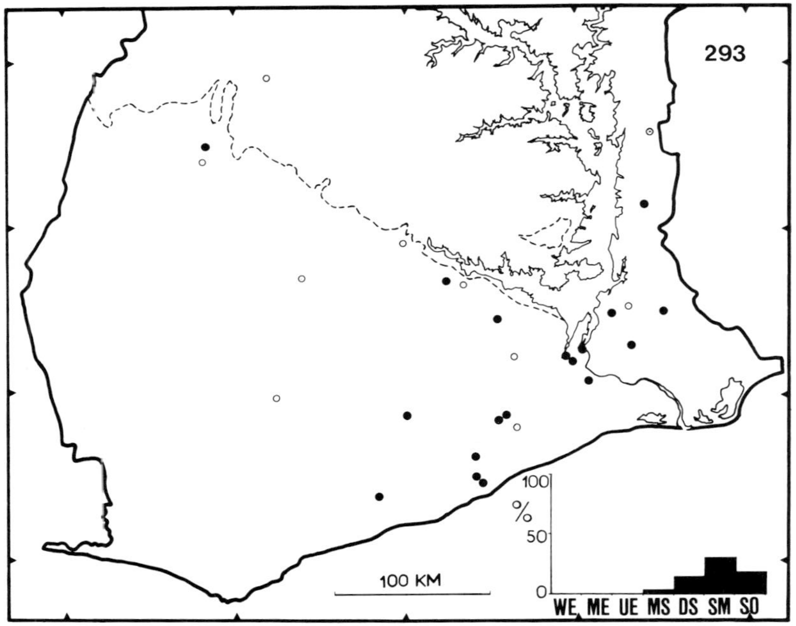

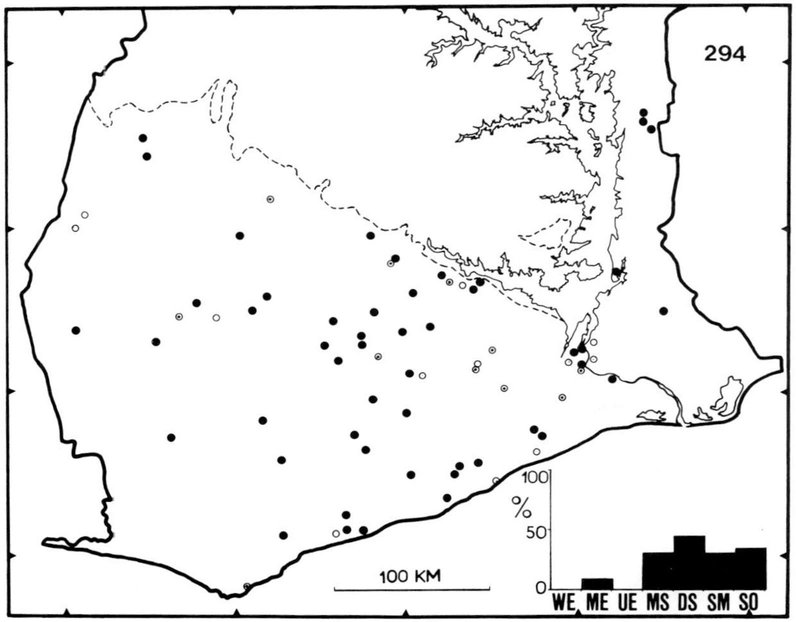

289 HANNOA KLAINEANA Pierre & Engl., FWTA 1:691. Simaroubaceae. Akan: fotie.
Large tree; lvs pinnate, leaflets notophyll, briefly deciduous. Fl white, small, dish, polygamous, 8–11. Fr purple-black drupe; seed large. Regenerates in shade. Plot records very significantly associated with acid soils. Guinée to Angola; (Guineo-Congolian wide).

290 HEISTERIA PARVIFOLIA Sm., FWTA 1:645. Olacaceae. Akan: sikakyia.
Small tree or shrub; lvs notophyll to mesophyll. Troll's model (59). Fl greenish-white, small, inconspicuous, not clearly seasonal. Fr greyish drupe, smaller than the enlarged red calyx; seed medium. Regenerates in shade. Plot records very significantly correlated with acid, base-poor soils. Guinea Bissau to Angola; (Guineo-Congolian wide).

291 HERITIERA UTILIS (Sprague) Sprague (113), (*Tarrietia utilis* (Sprague) Sprague, FWTA 1:332). Sterculiaceae. Akan: nyankom.
Large tree with buttresses, and sometimes stilt roots; leaves digitate or unifoliolate; leaflets mesophyll, densely covered below by small golden scales. Rauh's model. Fl small to medium, white, bowl, monoecious, 9–11. Fr a samara, wind-dispersed, 1–3; seed very large. Regenerates in shade. S. Leone to Ghana, and Gabon; (Upper Guinea near-endemic). Class 1 timber; trade name: niangon (40).

292 HEXALOBUS CRISPIFLORUS A. Rich., FWTA 1:47. Annonaceae. Akan: duabaha.
Medium-sized tree, with fluted trunk; lvs notophyll, pubescent below. Fl cream, large, fragrant, dish, not clearly seasonal. Fr dark brown, mucilaginous-fleshed berry; seeds very large. Regenerates in shade. Guinée to Zaire; (Guineo-Congolian wide).

293 HILDEGARDIA BARTERI (Mast.) Kosterm. Sterculiaceae. Ga: bronyatso. Akan: akyerekyerewa.
Medium-sized tree with buttresses; lvs mesophyll, deciduous. Rauh's model (59). Fl red, medium to large, narrow funnel, monoecious, 11–2, produced while the tree is leafless, attractive to sunbirds; flowering and leaf abscission are inhibited by continuous illumination (53). Fr inflated red 1-seeded pod (Fig. 6.1c); seed large, undergoing secondary dormancy when dry (38). Gregarious on rocky hills. Iv. Coast to Nigeria; (Guinea wide).

294 HIPPOCRATEA AFRICANA (Willd.) Loes. ex Engl., FWTA 1:628, and HIPPOCRATEA ROWLANDII Loes., FWTA 1:628. Celastraceae. (Type of *H. africana* probably from Aflao).
Large woody climbers; xylem in transverse section with radial streaks of included phloem, outline more or less circular in *H. africana* but rather deeply lobed in *H. rowlandii* (62). Lvs microphyll (or sometimes notophyll in *H. africana*), slightly toothed. Roux's model. Fl medium, orange, dish *(H. rowlandii)*, medium, orange or green, bowl *(H. africana)*, not clearly seasonal. Fr flat capsule; seed large, winged (Fig. 6.1j), wind-dispersed. Regenerates in shade. The species were not distinguished in the survey. Herbarium records: *H. africana* ○; *H. rowlandii* ⊙. Distribution: *H. africana* is widespread in Tropical Africa and Asia, (chorological transgressor); *H. rowlandii* Guinée to Cameroun, (Guinea wide).

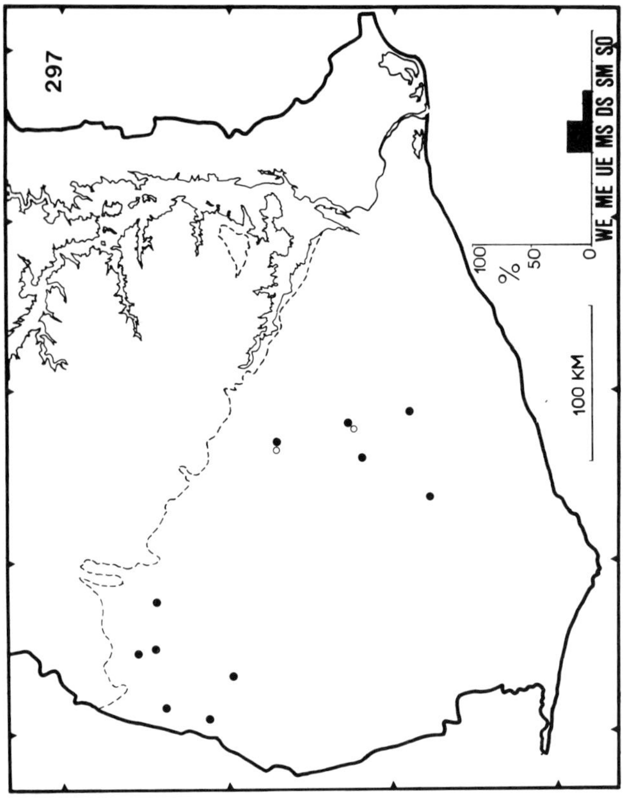
297
100 KM
%
100
50
0
WE ME UE MS DS SM SO

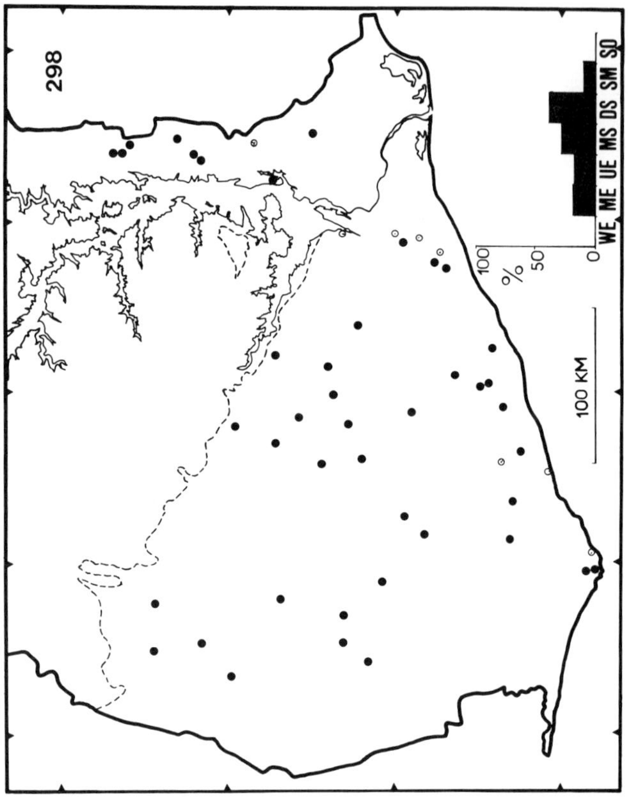
298
100 KM
%
100
50
0
WE ME UE MS DS SM SO

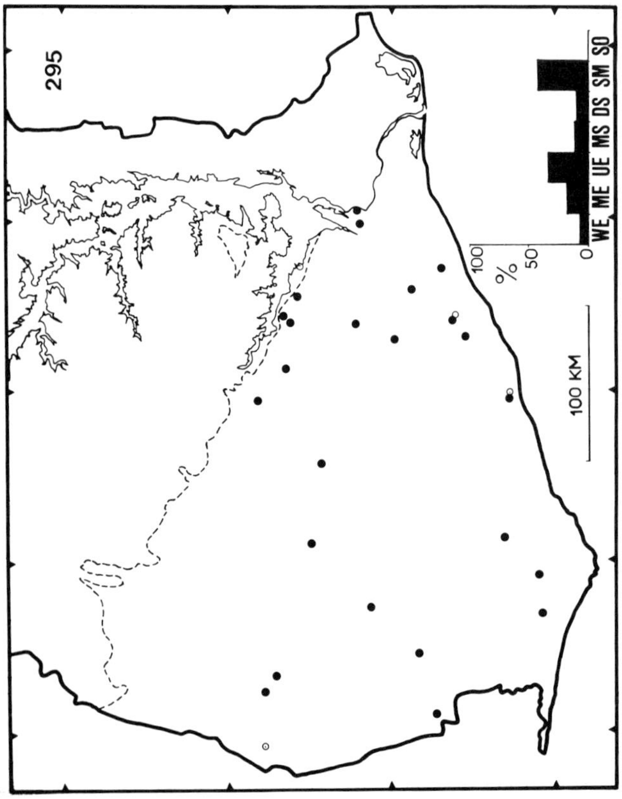
295
100 KM
%
100
50
0
WE ME UE MS DS SM SO

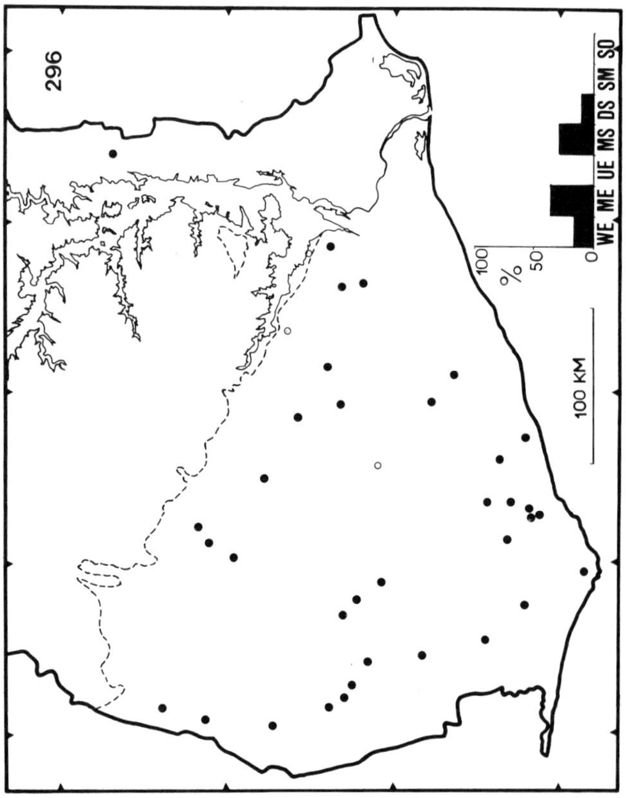
296
100 KM
%
100
50
0
WE ME UE MS DS SM SO

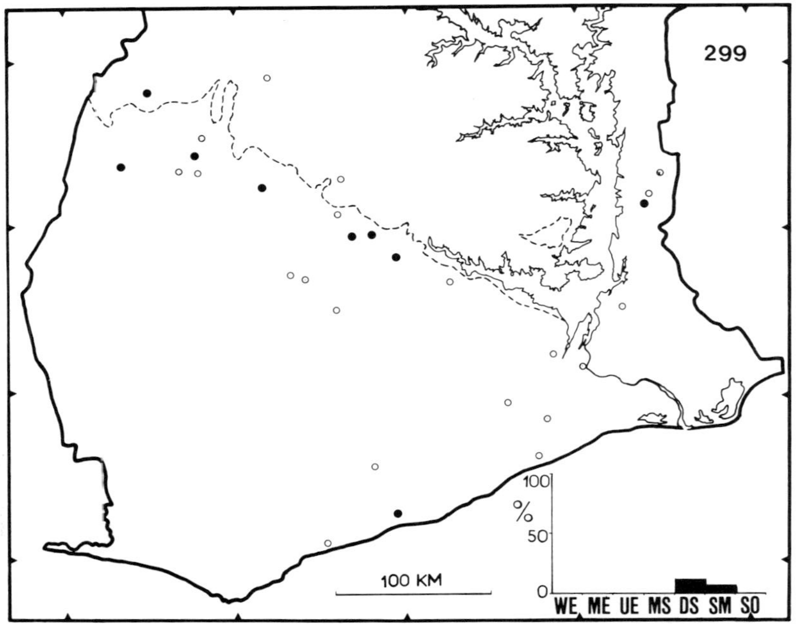

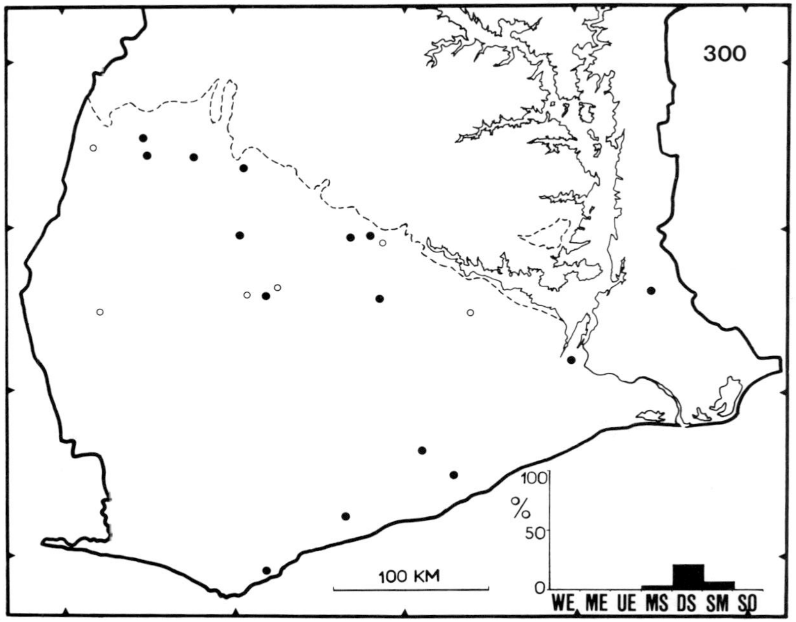

295 HIPPOCRATEA DEWILDEMANIANA (N. Hallé) J. B. Hall (50). Celastraceae.
Large woody climber; xylem cylindrical, normal (62); lvs notophyll, toothed. Roux's model. Fl small, pale green, dish. Fr flat capsule, not yet seen in Ghana; seeds large, winged, wind-dispersed. Regenerates in shade. Iv. Coast to Angola and Uganda; (Guineo-Congolian wide).

296 HIPPOCRATEA MACROPHYLLA Vahl, FWTA 1:629. Celastraceae.
Large woody climber; xylem normal; lvs mesophyll. Roux's model (59). Fl white, medium, dish, 12–5. Fr flat capsule; seeds very large, wingless, no obvious dispersal mechanism. Regenerates in shade. Guinée to Angola; (Guineo-Congolian wide).

297 HIPPOCRATEA PLUMBEA Blakelock & Wilczek (62). Celastraceae.
Large woody climber; lvs notophyll, toothed. Roux's model. Fl greenish, small, dish. Fr flat capsule; seeds large, winged, wind-dispersed. Flowers and fruit not yet known from Ghana, but the leaves are highly distinctive (62). Iv. Coast to Ghana, and Zaire to Uganda; (Guineo-Congolian wide).

298 HIPPOCRATEA WELWITSCHII Oliv., FWTA 1:628. Celastraceae.
Large woody climber, or scrambling shrub in thicket; xylem cylindrical, normal; lvs microphyll. Roux's model. Fl greenish-yellow, small, dish, not clearly seasonal. Fr flat capsule; seeds large, winged, wind-dispersed. Regenerates in shade. Guinée to Angola and Tanzania; (Guineo-Congolian wide).

299 HOLARRHENA FLORIBUNDA (G. Don) Dur. & Schinz., FWTA 2:68. Akan: sese.
Small to medium-sized tree, containing white latex; lvs notophyll. Roux's model. Fl white, medium, dish with tube, mainly 2–4, flowering gregariously. Fr slender follicle; seeds medium, with plume of hairs, wind-dispersed. Regenerates mainly in gaps and farmland. Guinea Bissau to Congo Republic; (Guinea wide). Used for carving traditional stools (40).

300 HOLOPTELEA GRANDIS (Hutch.) Mildbr., FWTA 1:593. Ulmaceae. Akan: nakwa.
Large tree with buttresses, slash smelling of tincture of iodine; lvs notophyll, deciduous. Troll's model (59). Fl small, green, inconspicuous, polygamous, 1–2 just before the new leaves. Fr samara, wind-dispersed; seed medium. Regenerates in moderate shade. Iv. Coast to Zaire and Uganda; (Guineo-Congolian wide). Class 3 timber.

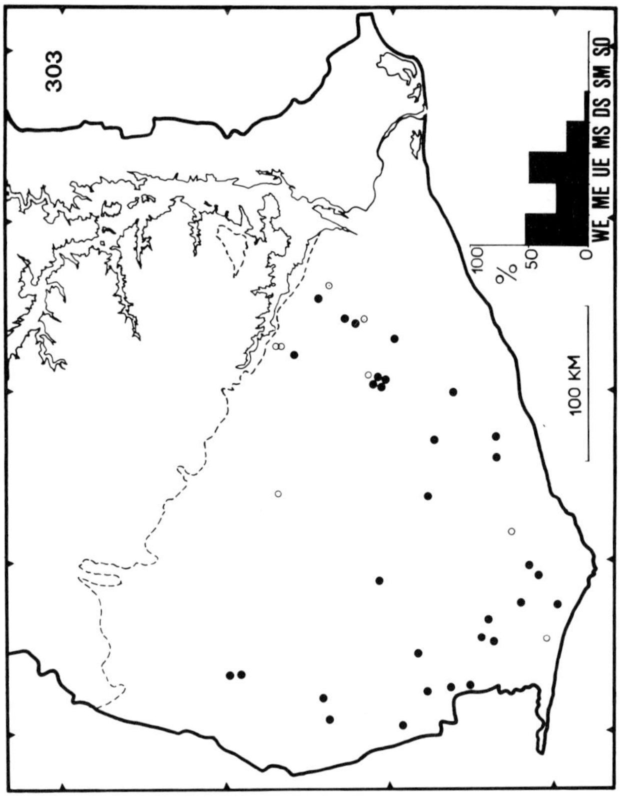
303
100
%
50
0
100 KM
WE ME UE MS DS SM SO

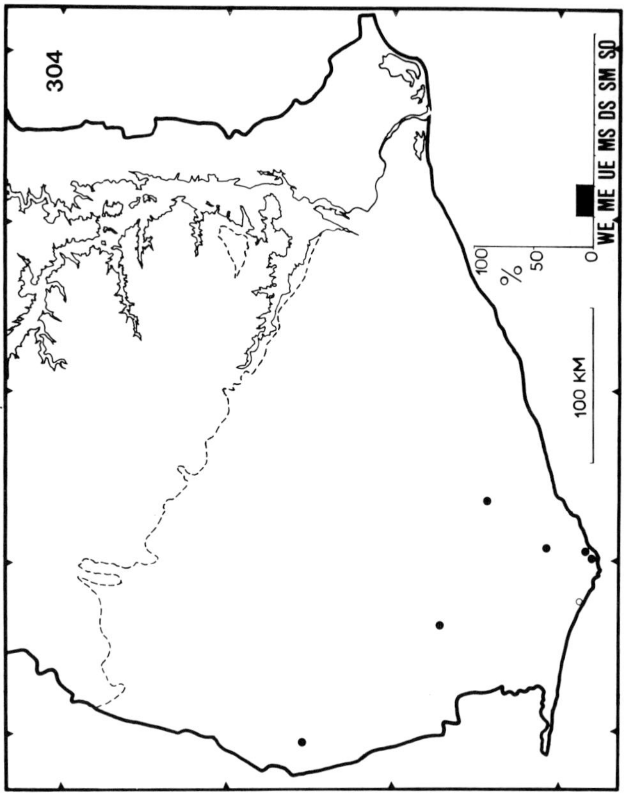
304
100
%
50
0
100 KM
WE ME UE MS DS SM SO

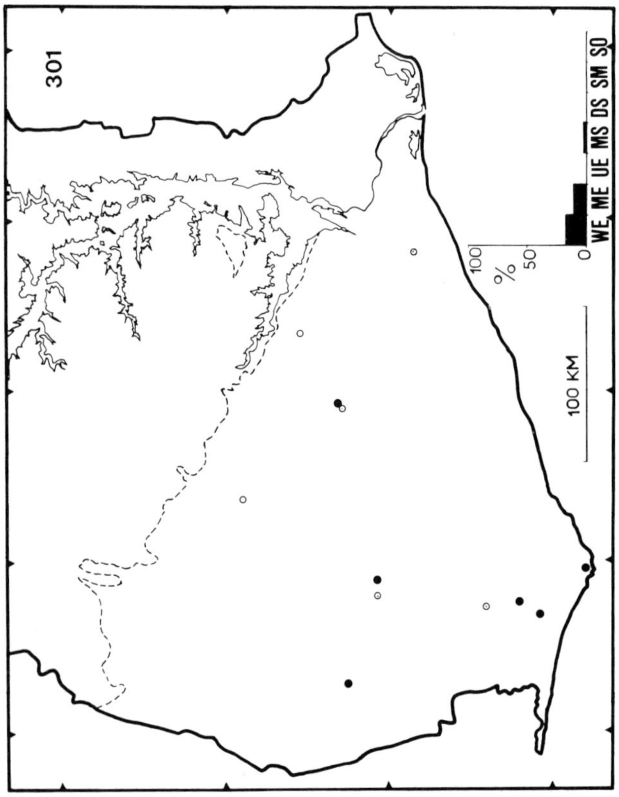
301
100
%
50
0
100 KM
WE ME UE MS DS SM SO

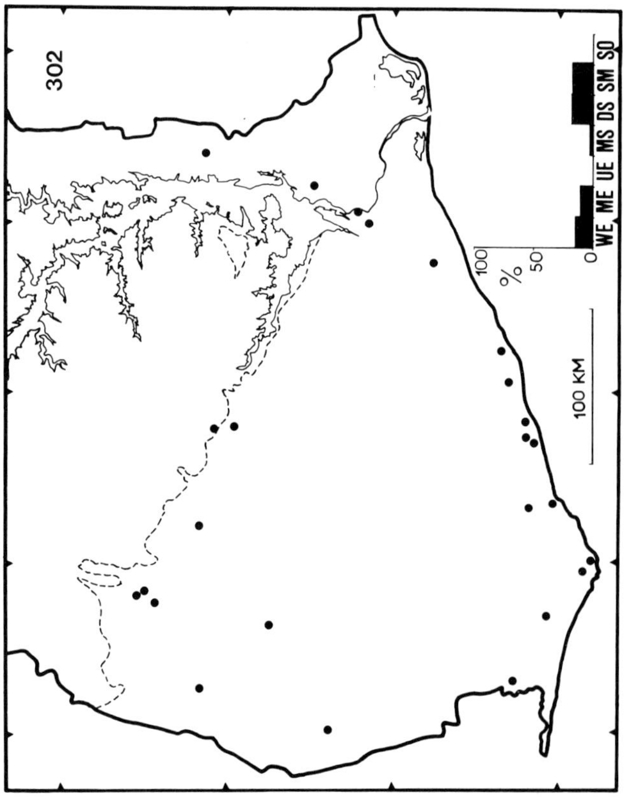
302
100
%
50
0
100 KM
WE ME UE MS DS SM SO

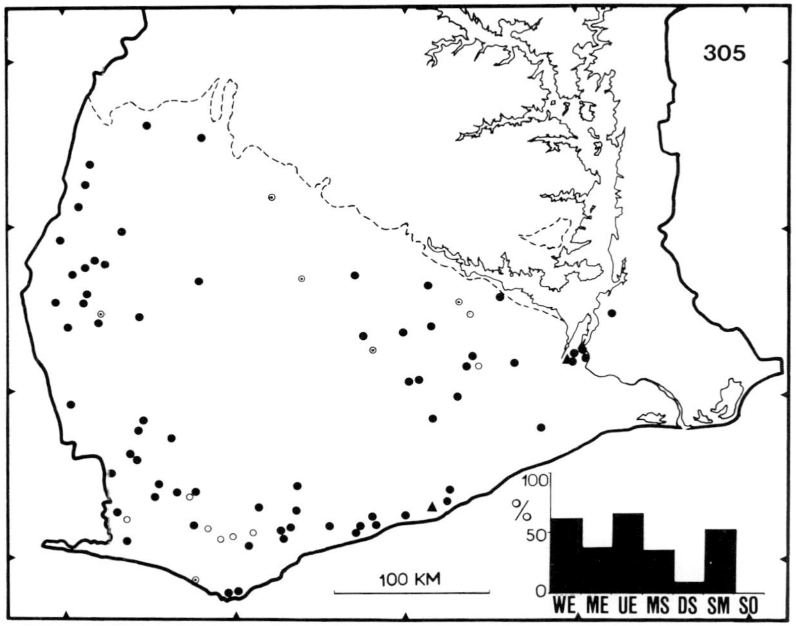

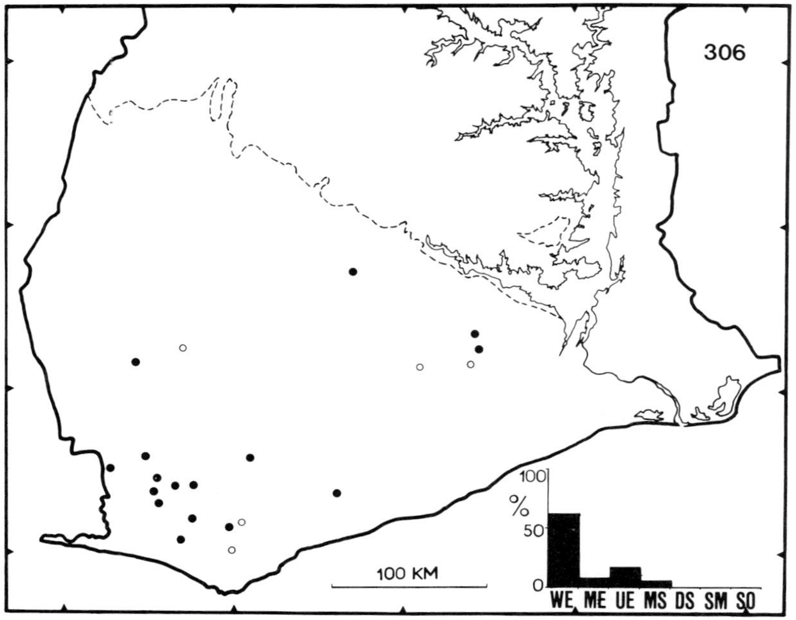

301 HOMALIUM STIPULACEUM Welw. ex Mast. (102), (*H. neurophyllum* Hoyle, FWTA 1:195). Flacourtiaceae (Samydaceae in FWTA).
Medium-sized tree with short buttresses; lvs mesophyll, toothed; twigs and underside of midrib pubescent. Roux's model. Fl cream, small, dish, not clearly seasonal. Fr dry, wind-dispersed with the aid of the greatly accrescent petals; seeds small. Regenerates in shade. Liberia to Angola; (Guineo-Congolian wide).

302 HUGONIA PLANCHONII Hook. f., FWTA 1:359. Linaceae.
Large woody climber, or scrambling shrub in thicket, with woody tendrils; xylem cylindrical, normal. Lvs notophyll, toothed. Fl bright yellow, large, bowl, opening gregariously, mostly 3–6. Fr orange, drupaceous, pyrene several-seeded (12); seeds medium. Regenerates in shade; sapling leaves macrophyll. Guinea Bissau to Zaire; (Guineo-Congolian wide).

303 HUGONIA PLATYSEPALA Welw. ex Oliv., FWTA 1:359. Linaceae.
Large woody climber, or scrambling shrub in thicket, with woody tendrils; xylem cylindrical, normal. Lvs notophyll, slightly toothed. Fl bright yellow, large, bowl, mostly 4–5. Fr orange, drupaceous, pyrene several-seeded (12); seeds medium. Regenerates in shade; sapling leaves macrophyll (Fig. 5.8b). Plot records very significantly correlated with acid, base-poor soils. Guinée to Angola and Uganda; (Guineo-Congolian wide).

304 HUGONIA RUFIPILIS A. Chev. ex Hutch. & Dalz., FWTA 1:359. Linaceae.
Woody climber or scrambling shrub, with woody tendrils; lvs notophyll to mesophyll, toothed; leaf undersurface and twigs hairy. Fl yellow, medium, bowl. Fr unknown. Regenerates in shade; sapling leaves macrophyll. S. Leone to Ghana, and Zaire; (Upper Guinea near-endemic).

305 HUNTERIA EBURNEA Pichon, FWTA 2:62, HUNTERIA UMBELLATA (K. Schum.) Hallier f., FWTA 2:62, and HUNTERIA GHANENSIS Hall & Leeuwenberg (51). Apocynaceae. Akan: kanwene-akoa. (Type of *H. ghanensis* from Apam).
Small to medium trees; scanty white latex in slash and twigs; lvs notophyll (*H. ghanensis* only) or mesophyll. Fl yellowish (*H. ghanensis* only) or white, medium, dish with tube; not clearly seasonal. Fr yellow-orange berry, seeds medium (*H. ghanensis* only) or large. Regenerates in shade. The species were not distinguished in the survey. Herbarium records are mapped as follows: *H. eburnea* ○; *H. umbellata* ⊙; *H. ghanensis* ▲. *H. eburnea* Distribution: Guinée to Nigeria, (Guinea wide); *H. umbellata* Ghana to Central African Republic, (Guinea wide); *H. ghanensis* Iv. Coast and Ghana, (Upper Guinea endemic). Timber tough and fine-grained, used for carving (40).

306 HYMENOCOLEUS HIRSUTUS (Benth.) Robbrecht (96), (*Geophila hirsuta* Benth., FWTA 2:205). Rubiaceae.
Perennial stoloniferous herb; lvs microphyll; stems and leaves hairy. Fl white, medium, dish with tube, not clearly seasonal. Fr orange 2-seeded berry; seeds medium. Regenerates in shade. Plot records very significantly correlated with acid, base-poor soils. Guinée to Zaire and Tanzania; (Guineo-Congolian wide).

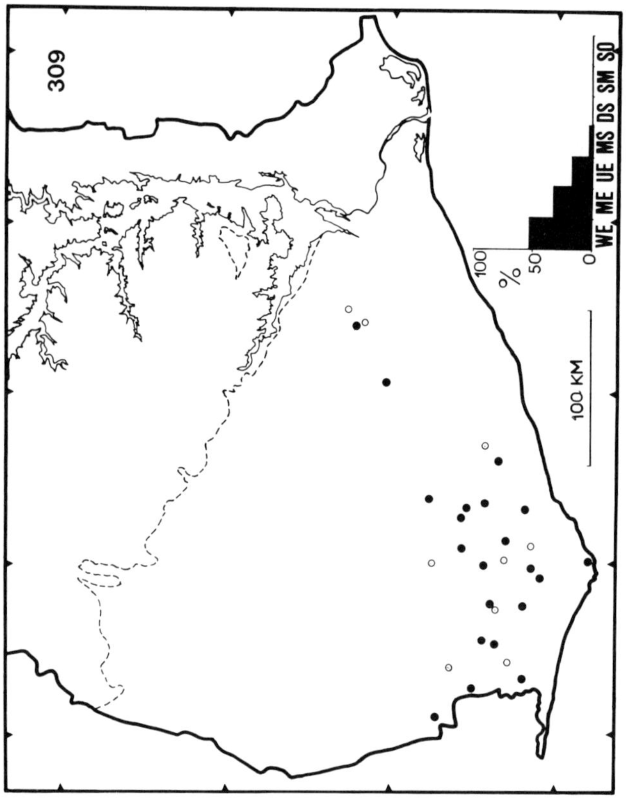
309
100 KM
%
100
50
0
WE ME UE MS DS SM SO

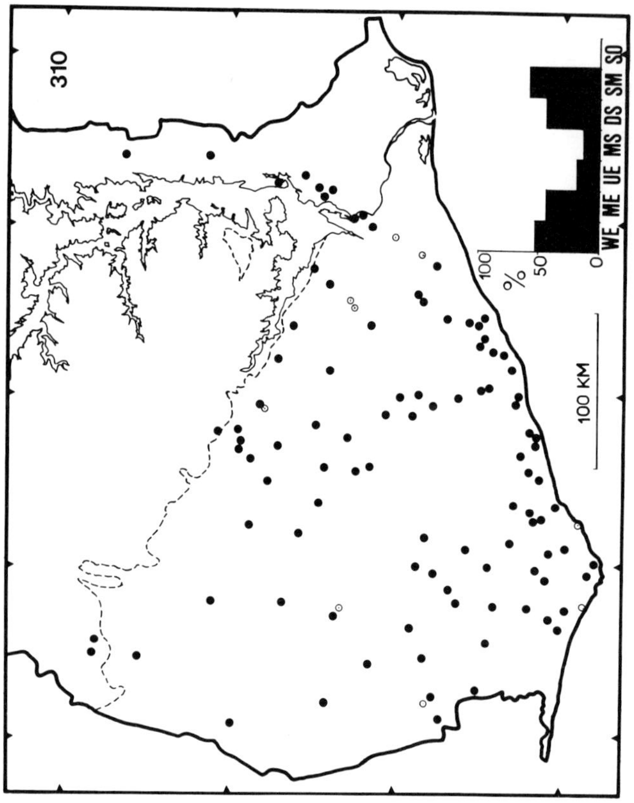
310
100 KM
%
100
50
0
WE ME UE MS DS SM SO

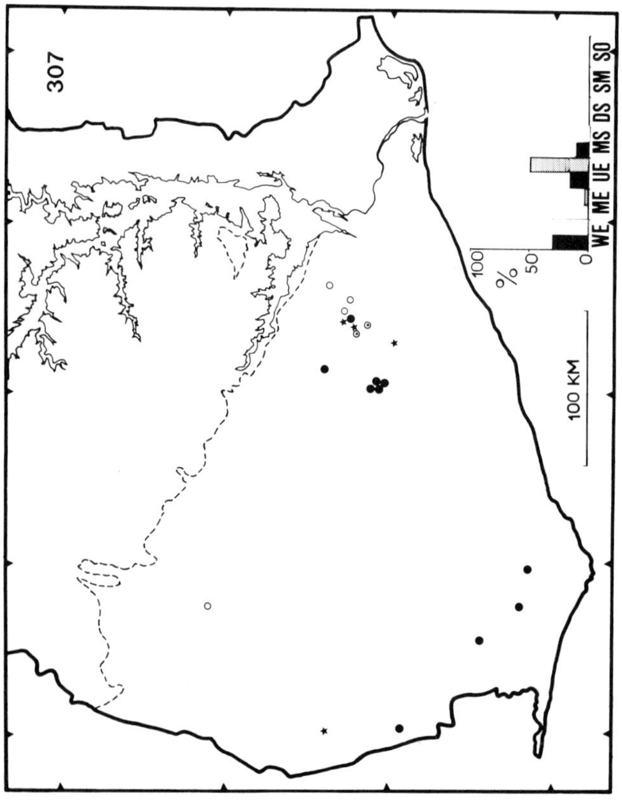
307
100 KM
%
100
50
0
WE ME UE MS DS SM SO

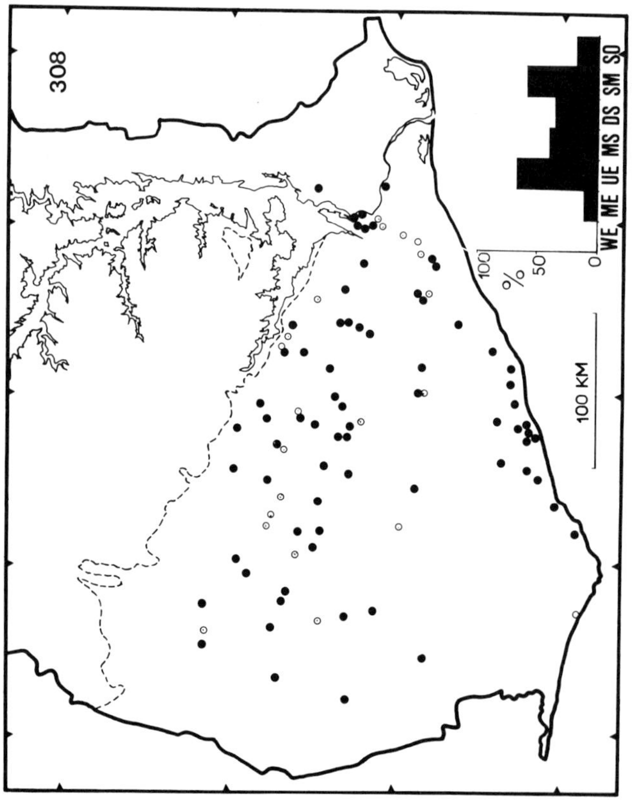
308
100 KM
%
100
50
0
WE ME UE MS DS SM SO

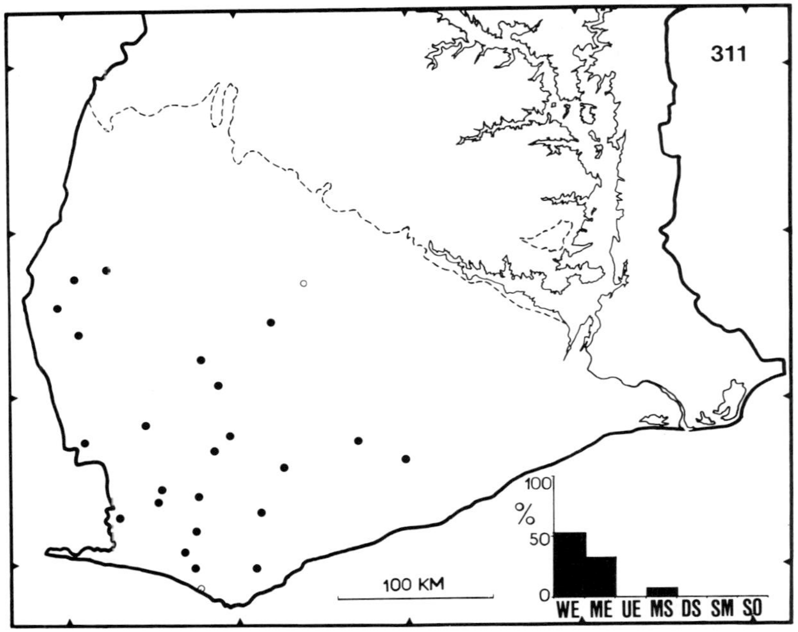

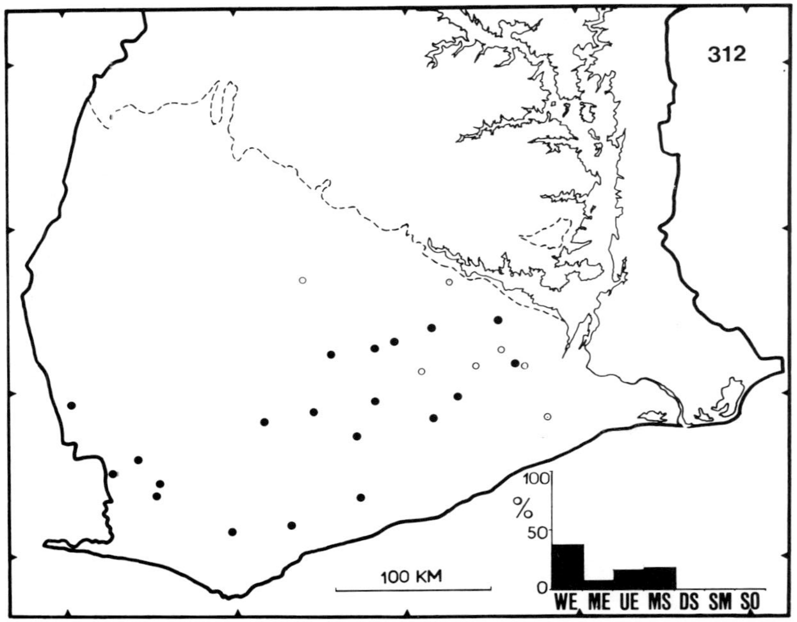

307 HYMENOCOLEUS NEURODICTYON (K. Schum.) Robbrecht (96), (*Geophila neurodictyon* (K. Schum.) Hepper, FWTA 2:206), var. *neurodictyon* and var. *orientalis* (Verdc.) Robbrecht (96, 97). Rubiaceae.

Perennial stoloniferous herb; lvs microphyll; stems and petioles hairy, also underside of veins in var. *neurodictyon*. Fl white, medium, funnel; not clearly seasonal. Fr orange 2-seeded berry; seeds medium. Regenerates in shade, but flourishes best on paths and where canopy is light. Distribution: var. *neurodictyon* Sierra Leone to Central African Republic; var. *orientalis* Liberia to Zaire, Sudan and Tanzania. Survey records: var. *neurodictyon* ● and solid bars in histogram; var. *orientalis* ★ and dotted bars in histogram. Herbarium records: var. *neurodictyon* ○; var. *orientalis* ⊙. Species is Guineo-Congolian wide.

308 HYMENOSTEGIA AFZELII (Oliv.) Harms, FWTA 1:464. Caesalpiniaceae. Akan: takorowa.

Small to medium-sized tree; lvs pinnate, leaflets microphyll. Troll's model. Fl pinkish-white, medium, dish, mostly 10–2, opening gregariously. Fr explosive pod; seed large. Regenerates in shade. Sometimes grows in pure stands in the drier forest types. Guinée to Cameroun; (Guinea wide).

309 HYPOLYTRUM POECILOLEPIS Nelmes, FWTA 3:336. Cyperaceae.

Perennial tufted herb; lvs macrophyll. Corner's model (59). Spikes small to medium, brown; individual florets monoecious, wind-pollinated. Achenes small, no obvious dispersal mechanism, 11–4. Regenerates in shade. S. Leone to Ghana; (Upper Guinea endemic).

310 HYPSELODELPHYS VIOLACEA (Ridl.) Milne-Redh., FWTA 3:88. Marantaceae.

Scrambling tall perennial herb with woody, bamboo-like stems; lvs mostly notophyll on flowering shoots. McClure's model (59). Fl whitish or bluish, medium, bowl; not clearly seasonal. Fr indehiscent, 3-seeded, no obvious dispersal mechanism. Seeds large. Regenerates in shade, but characteristic of secondary thickets. S. Leone to Angola; (Guineo-Congolian wide). *H. poggeana* (K. Schum.) Milne-Redh. is cited for Ghana in FWTA, but not in Flora of Cameroun (72); available herbarium material does not unequivocally support the occurrence in Ghana of this species. On the other hand the clearly bimodal distribution of our plot records strongly suggests the existence of two taxa with preferences for wet and dry forest respectively; this problem would repay further investigation.

311 ICACINA MANNII Oliv., FWTA 1:641. Icacinaceae.

Medium to large woody climber (or shrub in thicket); xylem divided into concentric rings and blocks by parenchyma and phloem; clear exudate; root large and tuberous (86). Lvs notophyll, with sparse biramous hairs on lower surface. Mangenot's model. Fl cream, small, bowl. Fr drupe; seed large. Regenerates in shade. Plot records very significantly correlated with acid, base-poor soils. S. Leone to Zaire; (Guineo-Congolian wide).

312 ILLIGERA PENTAPHYLLA Welw. FWTA 1:59. Hernandiaceae.

Large woody climber; wood normal, cylindrical; lvs digitate, leaflets notophyll, with tendriliform petiolules. Fl small, bowl, greenish, 6. Fr samara, dispersed by wind; seed large. Regenerates in shade, but flourishes in gaps and on forest margin. Iv. Coast to Angola and Uganda; (Guineo-Congolian wide).

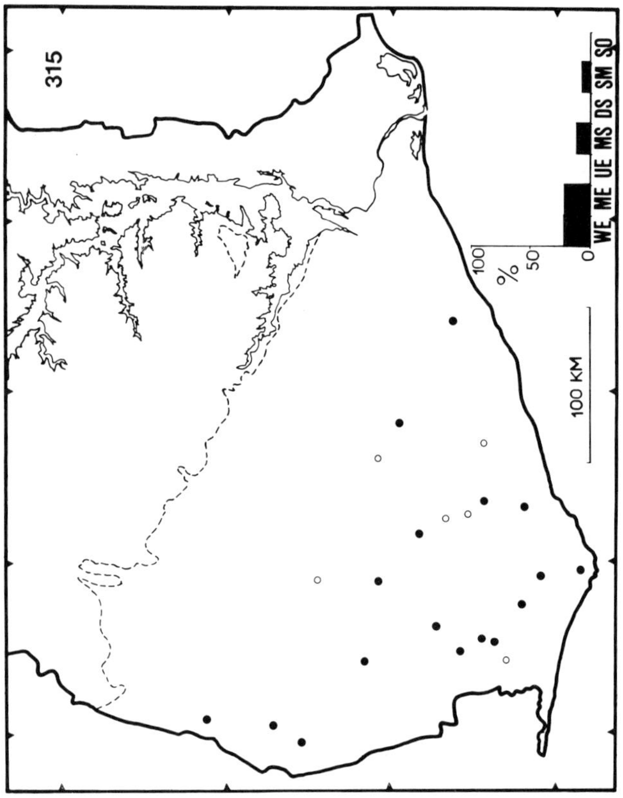
315
100 KM
%
100
50
0
WE ME UE MS DS SM SO

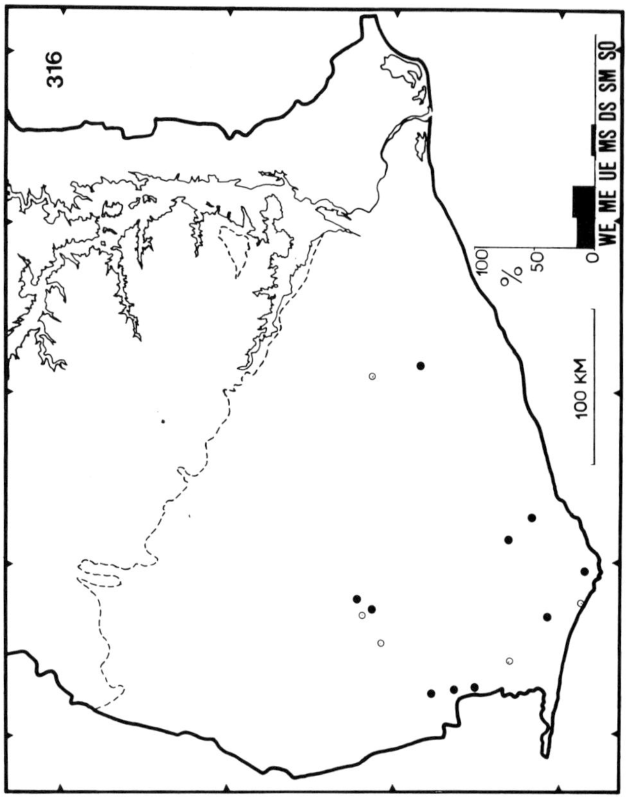
316
100 KM
%
100
50
0
WE ME UE MS DS SM SO

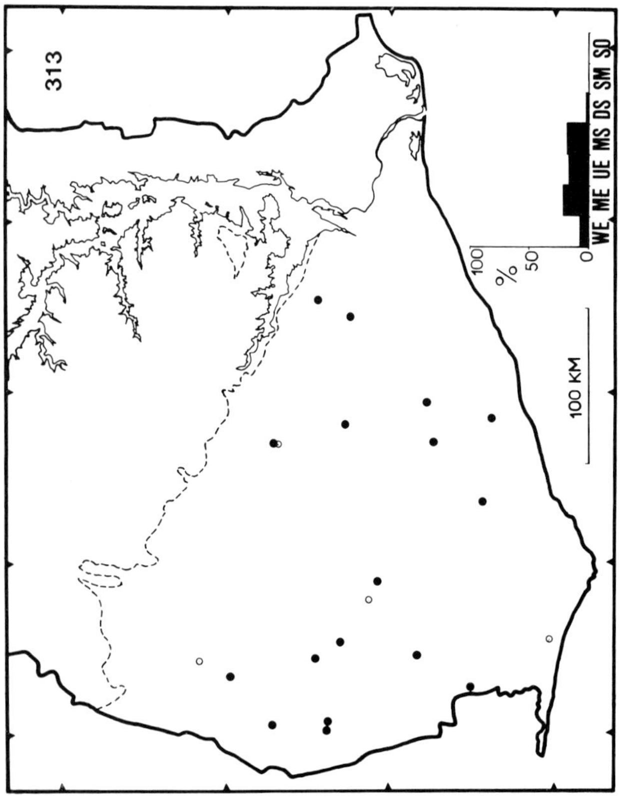
313
100 KM
%
100
50
0
WE ME UE MS DS SM SO

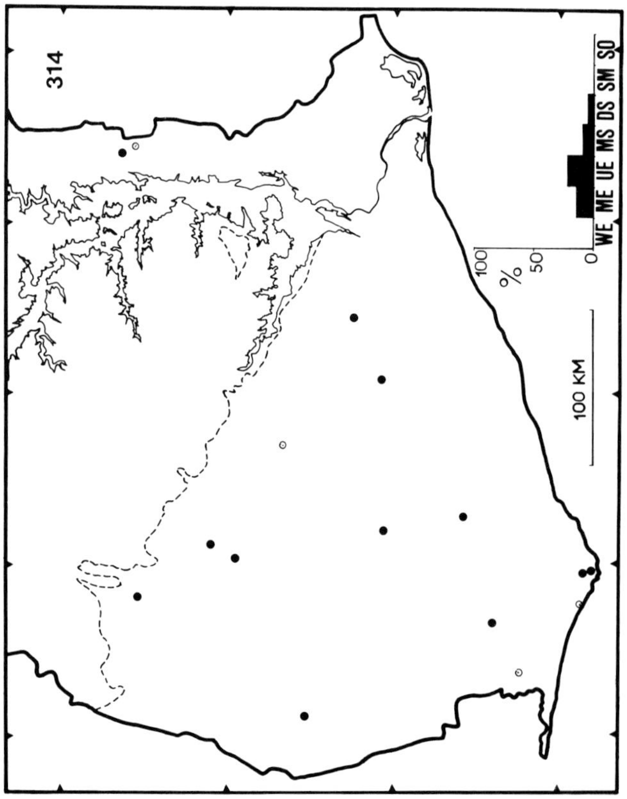
314
100 KM
%
100
50
0
WE ME UE MS DS SM SO

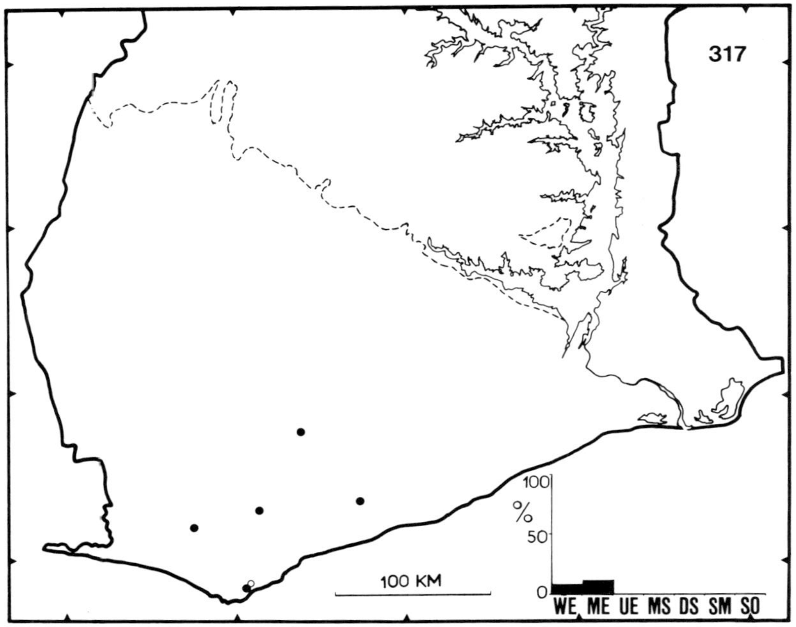

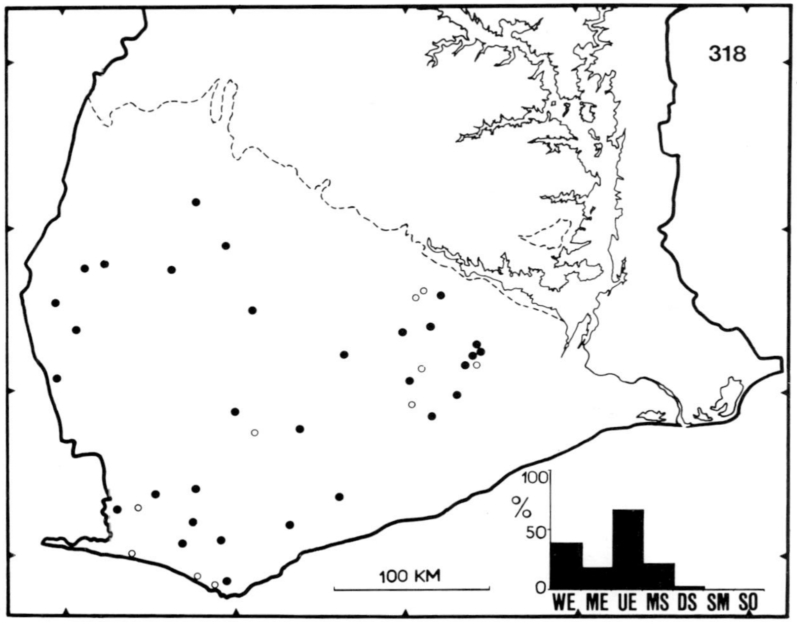

313 IODES LIBERICA Stapf, FWTA 1:643. Icacinaceae.
Medium-sized woody climber with tendrils; xylem becoming divided into two blocks; lvs notophyll. Undescribed model (59). Fl small, yellowish, bowl, dioecious, not clearly seasonal. Fr red drupe; seed medium. Regenerates in shade. Guinée to Ghana, and Zaire; (Upper Guinea near-endemic). Stem pounded and used as sponge. (Akan: akoannguare).

314 IRVINGIA GABONENSIS (Aubry-Lecomte ex O'Rorke) Baill. FWTA 1:693. Irvingiaceae. Akan: abesebuo.
Large tree with buttresses; lvs microphyll. Troll's model (59). Fl yellowish-green, small, bowl, 10–3. Fr yellowish drupe with aromatic flesh, dispersed by elephants (5); seed very large. Regenerates in shade. Senegal to Angola; (Guineo-Congolian wide). A smaller tree with larger, sweeter, less fibrous edible fruits is semi-cultivated, especially in the Volta Region. The tall forest form has been distinguished (87) as var. *excelsa* (Mildbr.) Okafor, and the smaller form as var. *gabonensis;* two such forms are also present in Liberia (113).

315 ISOLONA CAMPANULATA Engl. & Diels, FWTA 1:53. Annonaceae.
Small to medium tree, slash somewhat scented; lvs notophyll. Roux's model. Fl yellowish-green, medium, bowl, not obviously seasonal. Fr yellow berry; seeds large. Regenerates in shade. S. Leone to Gabon; (Guinea wide). Fruit shape and pedicel length seem to be variable, and we cannot accept the opinion (3) that *I. dewevrei* (De Wild. & Th. Dur.) Engl. & Diels can be distinguished in the FWTA area.

316 ISOLONA COOPERI Hutch. & Dalz. ex Cooper & Record, FWTA 1:53. Annonaceae.
Shrub or small tree; slash fragrant; lvs mesophyll. Roux's model. Fl medium, greenish-yellow, bowl, not clearly seasonal. Fr yellow berry; seeds large. Regenerates in shade. Liberia to Ghana; (Upper Guinea endemic).

317 ISOLONA DEIGHTONII Keay, FWTA 1:53. Annonaceae.
Small tree; lvs mesophyll; midrib, underside of lamina and twigs pubescent. Roux's model. Fl green, medium, bowl. Fr yellowish berry; flowers and fruits not yet seen in Ghana. S. Leone to Ghana; (Upper Guinea endemic).

318 IXORA LAXIFLORA Sm., FWTA 2:144. (Incl. *I. nimbana* Schnell, FWTA 2:144).
Shrub or pygmy tree; lvs mesophyll. Fl pink to red, large, dish with tube, not clearly seasonal. Fr orange berry; seeds medium. Regenerates in shade. Guinea Bissau to Ghana; (Upper Guinea endemic). The characters of corolla colour and proportions used in FWTA to distinguish *I. nimbana* are subject to too much variation to be useful taxonomically.

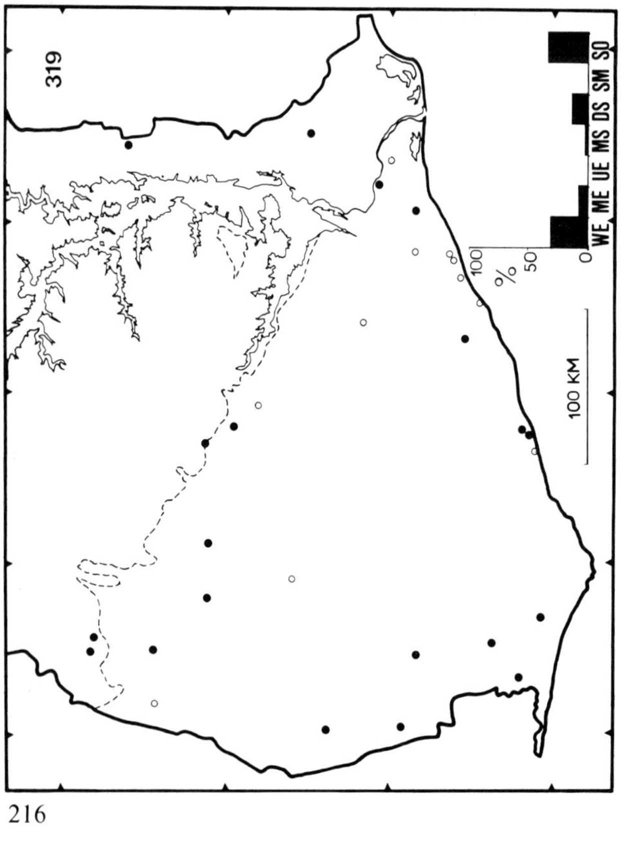
319
100 KM
%
100
50
0
WE ME UE MS DS SM SO

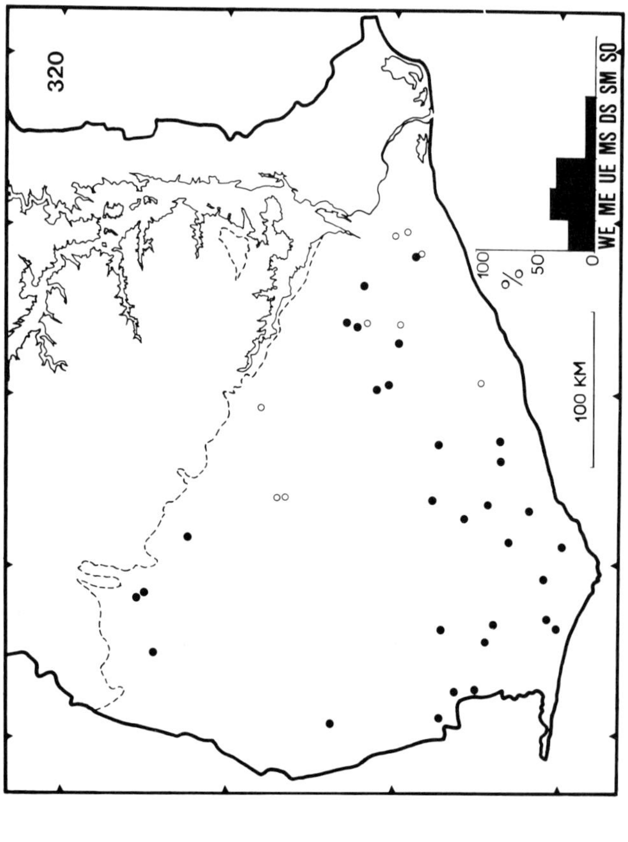
320
100 KM
%
100
50
0
WE ME UE MS DS SM SO

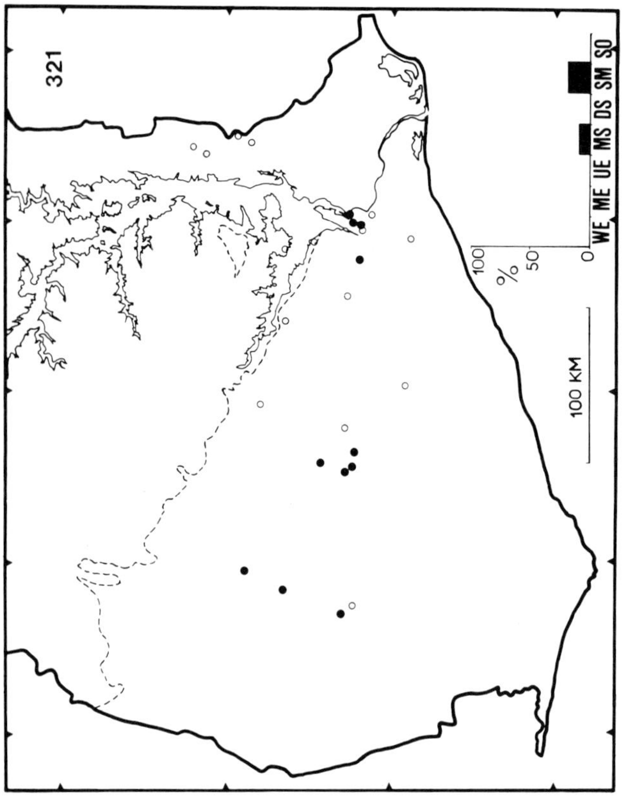
321
100 KM
%
100
50
0
WE ME UE MS DS SM SO

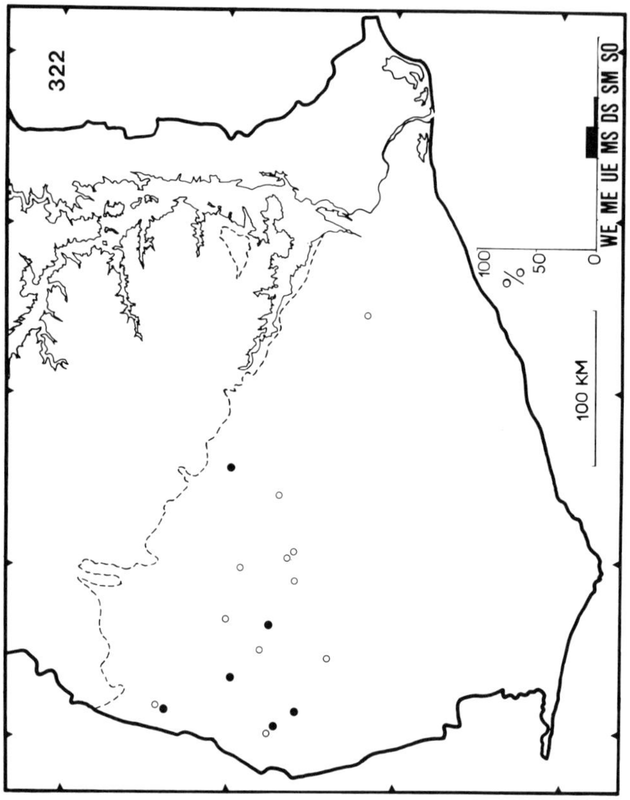
322
100 KM
%
100
50
0
WE ME UE MS DS SM SO

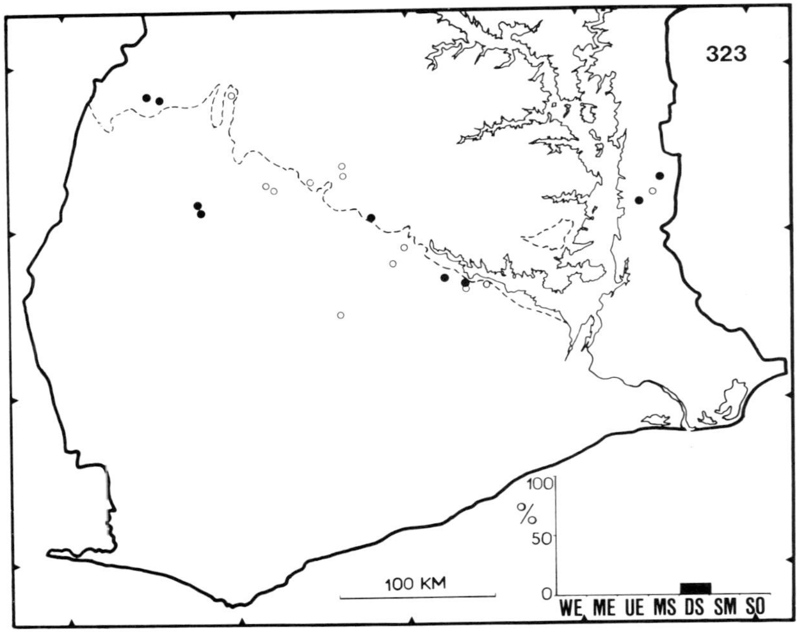

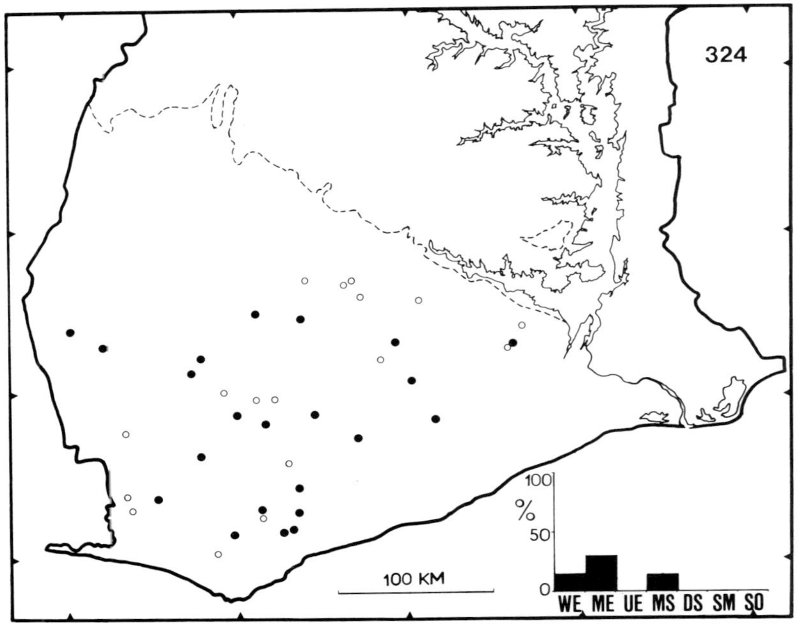

319 JASMINUM PAUCIFLORUM Benth; FWTA 2:50. Oleaceae.
Slender woody climber; lvs microphyll; stems and leaves densely pubescent (dry forest form), or puberulous (wet forest form). Cook's model. Fl white, fragrant, large, dish with tube, mostly 4–5. Fr black berry; seeds medium. Regenerates in shade. Guinea Bissau to Congo Republic and Kenya; (chorological and ecological transgressor).

320 JAUNDEA PINNATA (P. Beauv.) Schellenb., FWTA 1:742. Connaraceae.
Large woody climber, or shrub in thicket; xylem cylindrical, normal; lvs pinnate, leaflets notophyll. Fl white, medium, bowl, mostly 10–1. Fr reddish 1-seeded follicle; seed black with shiny red aril, large. Regenerates in shade. Guinée to Angola and Zambia; (Guineo-Congolian wide).

321 JUSTICIA EXTENSA T. Anders., FWTA 2:428. Acanthaceae.
Soft-wooded shrub; lvs notophyll to mesophyll. Fl pale greenish-yellow with mauve marks, medium, gullet, 10–12. Fr explosive capsule; seeds medium. Regenerates in shade, but flourishes best under light canopy. Guinée to Congo Republic; (Guinea wide).

322 KHAYA ANTHOTHECA (Welw.) C. DC., FWTA 1:699. Meliaceae. Akan: krumben.
Large tree with buttresses; slash scented; lvs pinnate, leaflets notophyll, sometimes briefly deciduous. Rauh's model. Fl small, white, bowl, monoecious, not clearly seasonal. Fr woody capsule; seeds very large, winged, wind-dispersed. Regenerates in shade. S. Leone to Angola and Uganda; (Guineo-Congolian wide). Class 1 timber; trade name: anthotheca (40).

323 KHAYA GRANDIFOLIOLA C. DC., FWTA 1:699. Meliaceae. Akan: kruba.
Large tree with buttresses; slash scented; lvs pinnate, leaflets mesophyll, briefly deciduous. Rauh's model. Fl cream, small, bowl, monoecious, 12–3. Fr woody capsule; seeds very large, winged, wind-dispersed. Extends to fringing forest just north of the savanna boundary. Guinée to Zaire and Uganda; (Guineo-Congolian wide).

324 KHAYA IVORENSIS A. Chev., FWTA 1:699. Meliaceae. Akan: dubini.
Large tree with buttresses; slash scented; lvs pinnate, leaflets notophyll, usually evergreen. Rauh's model (59). Fl white, small, bowl, monoecious, mostly 9–12. Fr woody capsule; seeds very large, winged, wind-dispersed. Regenerates in shade. Iv. Coast to Gabon; (Guinea wide). Class 1 timber; trade name: african mahogany (40). Fig. 11.2 shows distribution in Ghana calculated from timber enumeration records.

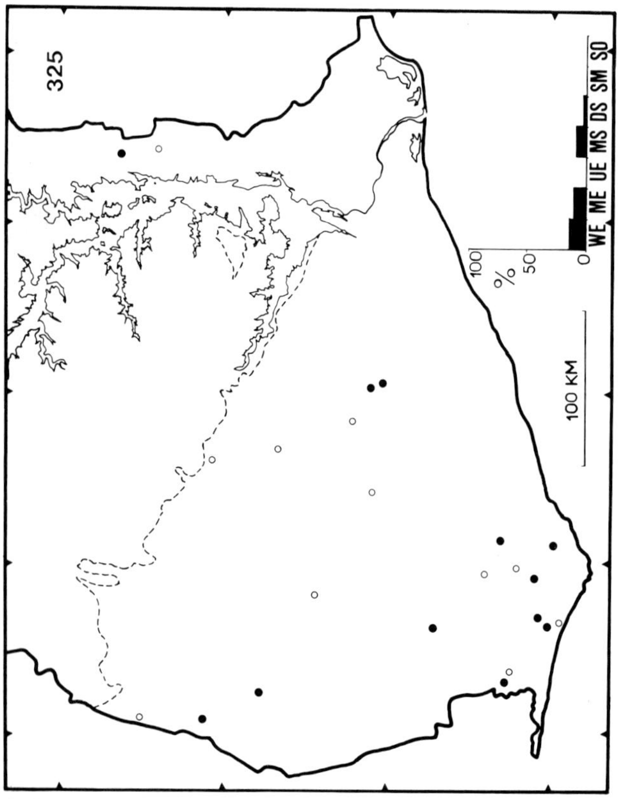
325
100 KM
%
100
50
0
WE ME UE MS DS SM SO

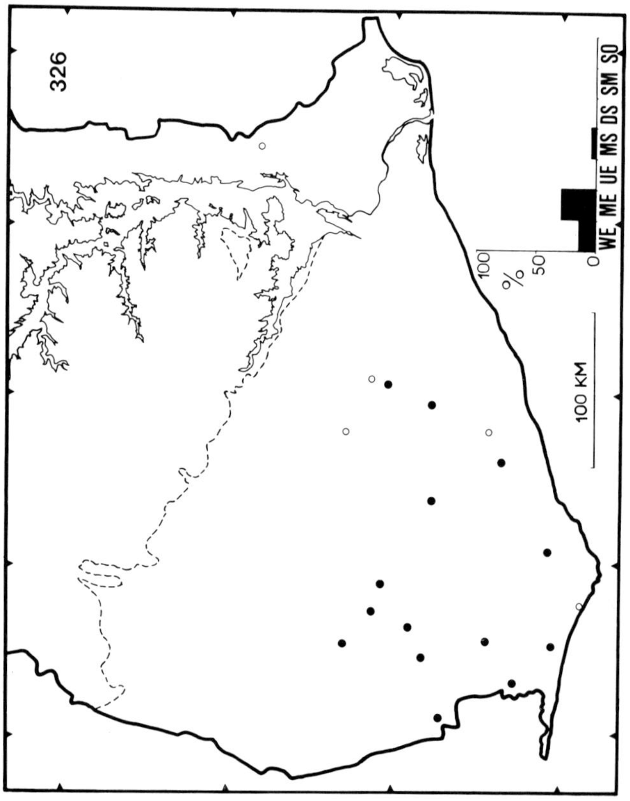
326
100 KM
%
100
50
0
WE ME UE MS DS SM SO

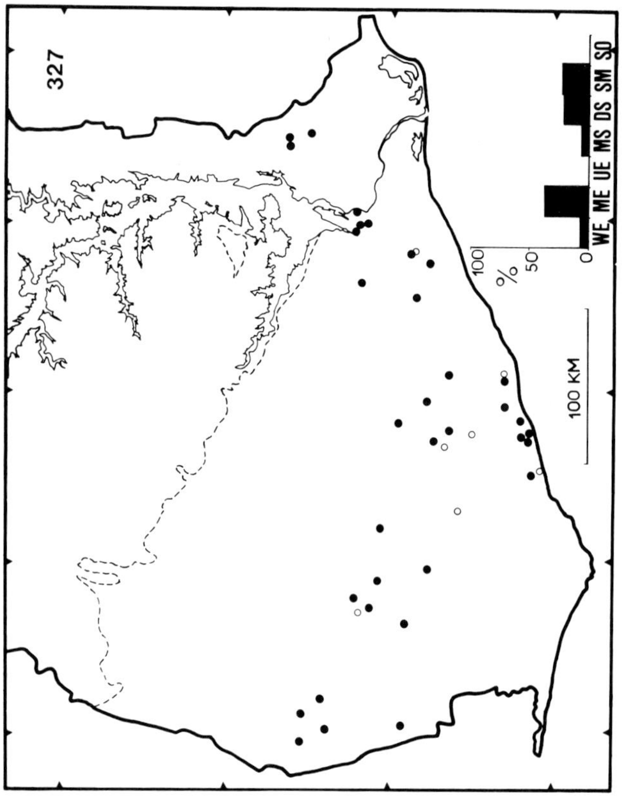
327
100 KM
%
100
50
0
WE ME UE MS DS SM SO

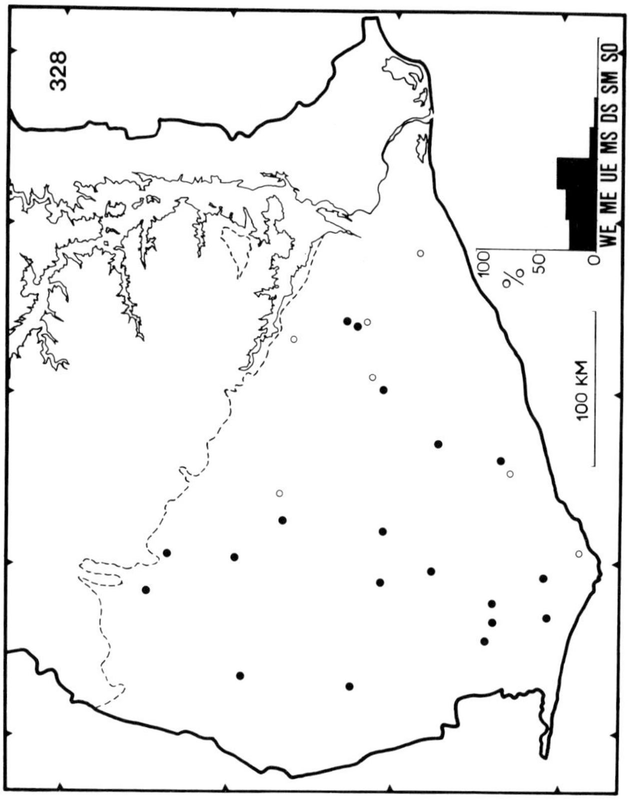
328
100 KM
%
100
50
0
WE ME UE MS DS SM SO

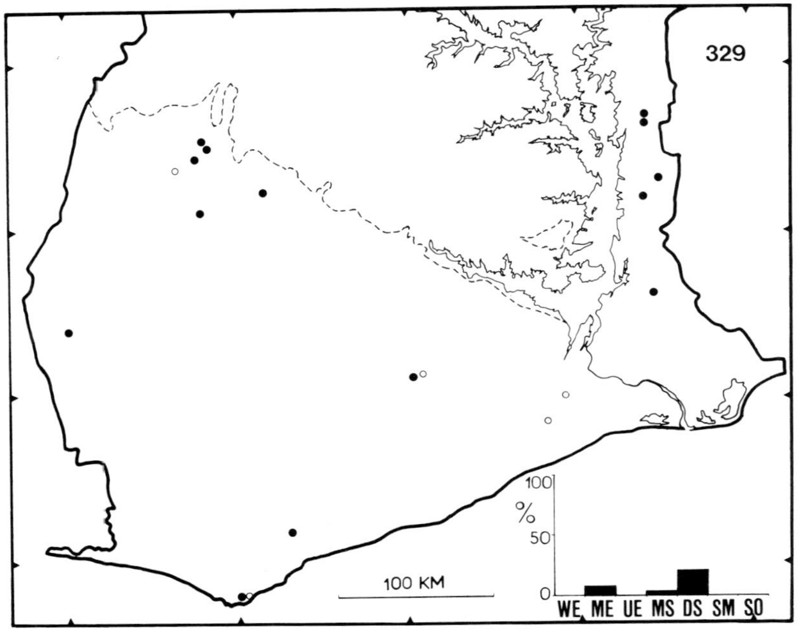

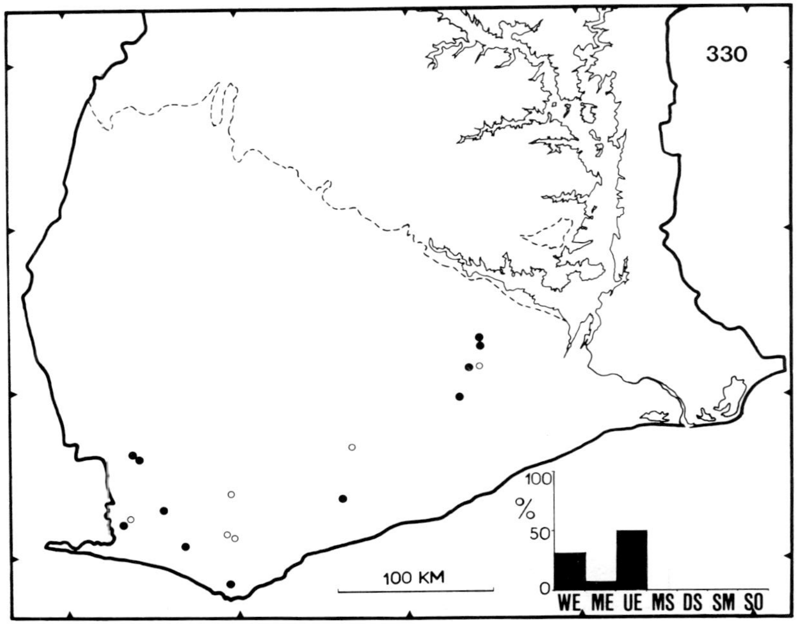

325 KLAINEDOXA GABONENSIS Pierre ex Engl., FWTA 1:693. Irvingiaceae. Akan: kroma.
Large tree with large buttresses; lvs microphyll to notophyll. Fl mauve, small, dish, not clearly seasonal. Fr black, hard drupe with 5 woody pyrenes, dispersed by elephants (5); seeds very large. Regenerates in shade. Guinea Bissau to Zaire and Uganda; (Guineo-Congolian wide). There seems little basis for recognition of varieties in this rather variable species (113).

326 KOLOBOPETALUM LEONENSE Hutch. & Dalz., FWTA 1:72. Menispermaceae.
Slender woody climber; lvs mesophyll. Fl pinkish-yellow, small, bowl, dioecious, cauliflorous. Fr drupe; seed large. Regenerates in shade. S. Leone to Ghana; (Upper Guinea endemic).

327 LANDOLPHIA CALABARICA (Stapf) E. A. Bruce, FWTA 2:55. Apocynaceae.
Large woody climber with tendrils, containing sticky white latex; wood cylindrical, normal; lvs notophyll; leaf undersurface and twigs hairy. Fl white, large, dish with tube, not clearly seasonal. Fr orange berry; seeds large. Regenerates in shade. S. Leone to Nigeria; (Guinea wide).

328 LANDOLPHIA DULCIS (R. Br. ex Sabine) Pichon var. *barteri* (Stapf) Pichon, FWTA 2:57. Apocynaceae.
Large woody climber with tendrils, containing white sticky latex; xylem cylindrical, normal; lvs notophyll, with sessile glands on underside. Leeuwenberg's model (59). Fl pale blue, large, dish with tube, 3–6. Fr orange berry; seeds large. Regenerates in shade. Plot records very significantly associated with base-poor soils. Guinée to Gabon; (Guinea wide).

329 LANDOLPHIA HIRSUTA (Hua) Pichon, FWTA 2:57. Apocynaceae.
Large woody climber with tendrils, containing abundant, white very sticky latex; xylem cylindrical, normal. Lvs notophyll to mesophyll; nerves below and twigs densely spreading-hirsute. Fl yellowish, small to medium, dish with tube. Fr berry, not yet found in Ghana. Regenerates in shade. Senegal to Cameroun; (Guinea wide).

330 LANDOLPHIA MICRANTHA (A. Chev.) Pichon, FWTA 2:55. Apocynaceae.
Medium-sized woody climber with tendrils, containing sticky white latex; xylem cylindrical, normal; lvs notophyll. Fl cream, sweet-scented, small, dish with tube, 12–4. Fr yellow berry; seeds large. Regenerates in shade. Liberia to Ghana; (Upper Guinea endemic). If, as is quite probable, this species proves to be conspecific with *Aphanostylis leptantha* (K. Schum.) Pierre, the latter is the correct name.

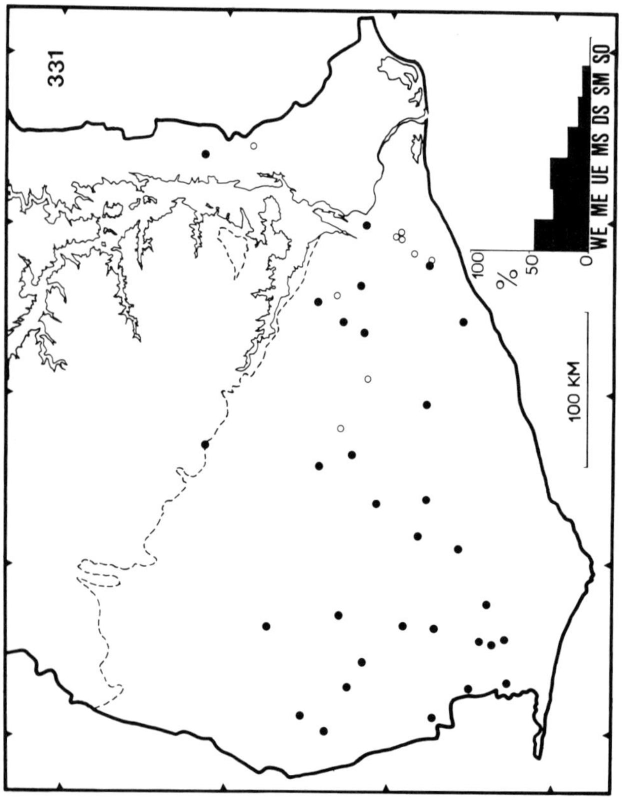
331
100 KM
%
100
50
0
WE ME UE MS DS SM SO

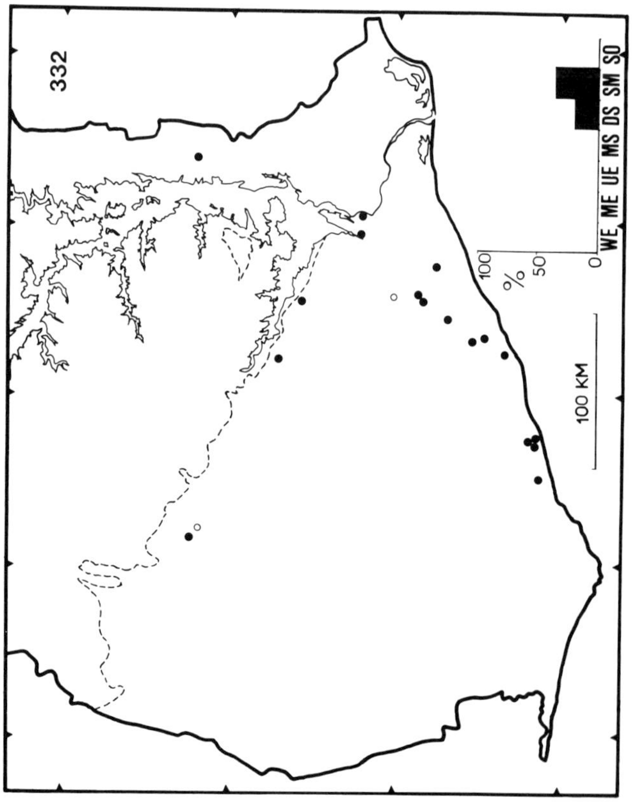
332
100 KM
%
100
50
0
WE ME UE MS DS SM SO

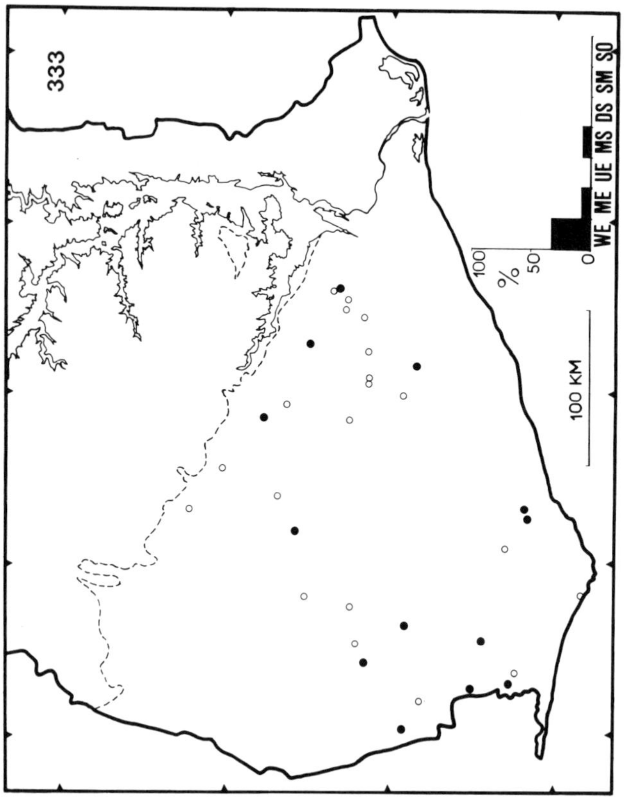
333
100 KM
%
100
50
0
WE ME UE MS DS SM SO

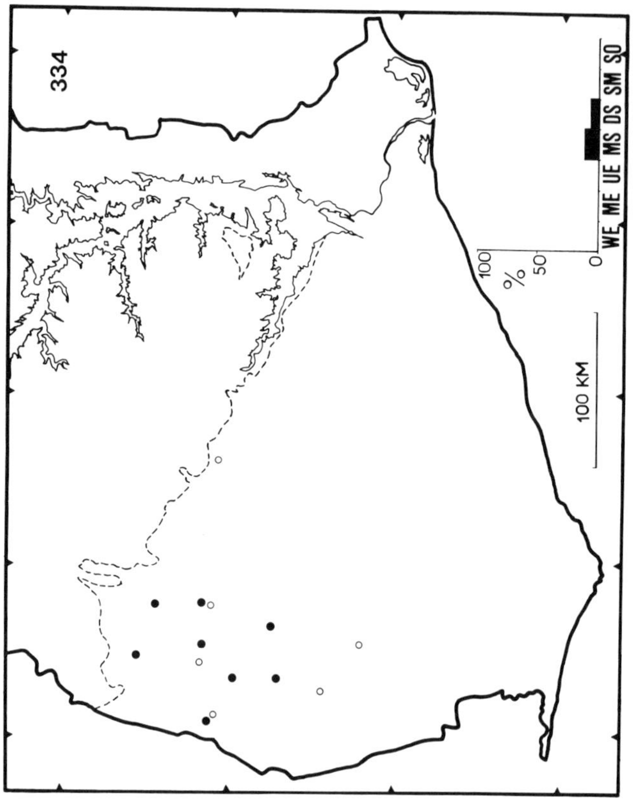
334
100 KM
%
100
50
0
WE ME UE MS DS SM SO

331 LANDOLPHIA OWARIENSIS P. Beauv. var. *owariensis*, FWTA 2:55. Apocynaceae. Akan: abontore.
Large woody climber with tendrils, containing white latex which is easily rolled into rubber strands; xylem cylindrical, normal; lvs notophyll. Fl pinkish-white, medium, dish with tube, not clearly seasonal. Fr orange berry; seeds large. Regenerates in shade. Guinée to Tanzania; (Guineo-Congolian wide). Formerly a source of commercial rubber.

332 LANDOLPHIA TOGOLANA (Hallier f.) Pichon, FWTA 2:55. Apocynaceae.
Medium-sized woody climber with tendrils, containing white latex; xylem cylindrical, normal; lvs microphyll. Fl whitish, small, dish with tube, 2–5. Fr orange berry; seeds large. Regenerates in shade. Iv. Coast to Nigeria; (Guinea wide).

333 LANKESTERIA BREVIOR C.B. Cl., FWTA 2:407. Acanthaceae.
Low shrub; lvs notophyll; twigs pubescent. Fl white with yellow throat, large, dish with tube, mostly 12–3. Fr explosive capsule; seeds medium. Regenerates in shade, but more abundant at forest margins and beside paths. S. Leone to Cameroun; (Guinea wide).

334 LANKESTERIA ELEGANS (P. Beauv.) T. Anders., FWTA 2:407. Acanthaceae.
Shrub; lvs mesophyll; twigs pubescent. Fl orange-red, large, dish with tube, 12–3. Fr explosive capsule; seeds medium. Regenerates in shade. S. Leone to Congo Republic and Uganda; (Guineo-Congolian wide).

335 LANNEA NIGRITANA (Sc. Elliot) Keay var. *nigritana*, FWTA 1:733. Anacardiaceae.
Small to medium tree; lvs pinnate, leaflets microphyll, deciduous. Rauh's model. Fl cream, small, bowl, dioecious, 1–3 while tree is leafless. Fr black drupe; seed medium. Seems to regenerate mainly from rootsuckers. Senegal to Nigeria; (Guinea wide). Specimens with pubescent leaves (var. *pubescens* Keay) have been collected from rocky outcrops in the forest zone. Herbarium records: var. *nigritana* ○; var. *pubescens* ⊙.

336 LANNEA WELWITSCHII (Hiern) Engl., FWTA 1:732. Anacardiaceae. Akan: kumanini.
Medium to large tree; lvs pinnate, leaflets mesophyll, deciduous. Rauh's model. Fl yellow, small, bowl, monoecious, 12–4. Fr black drupe; seed medium. Regenerates in shade. Iv. Coast to Uganda and Angola; (Guineo-Congolian wide).

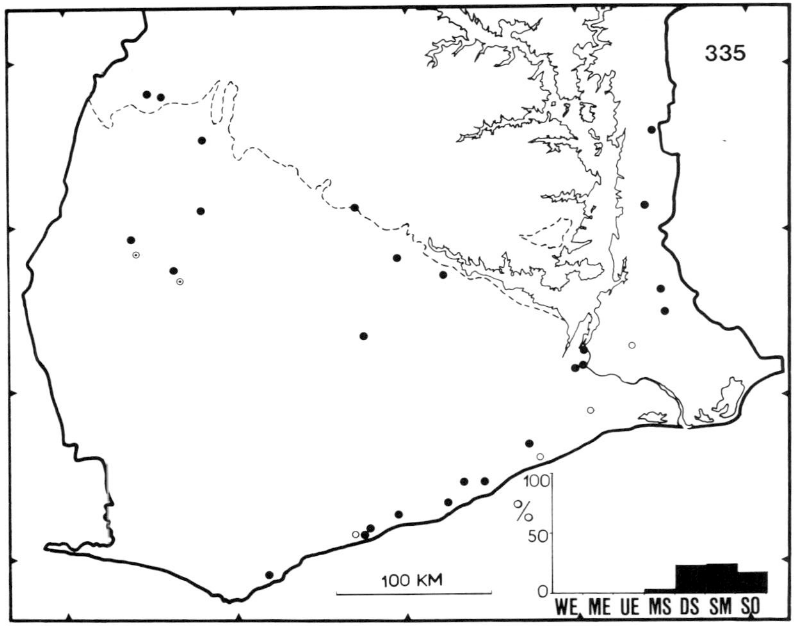

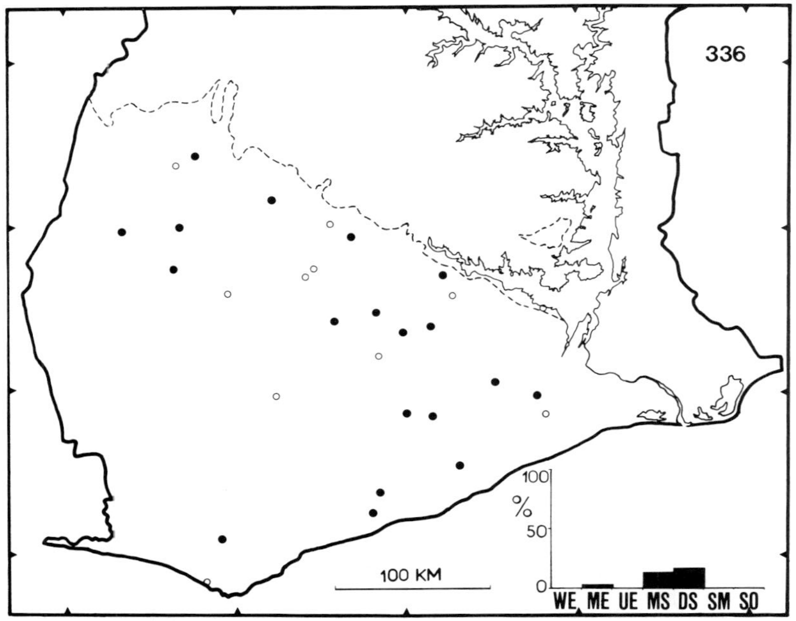

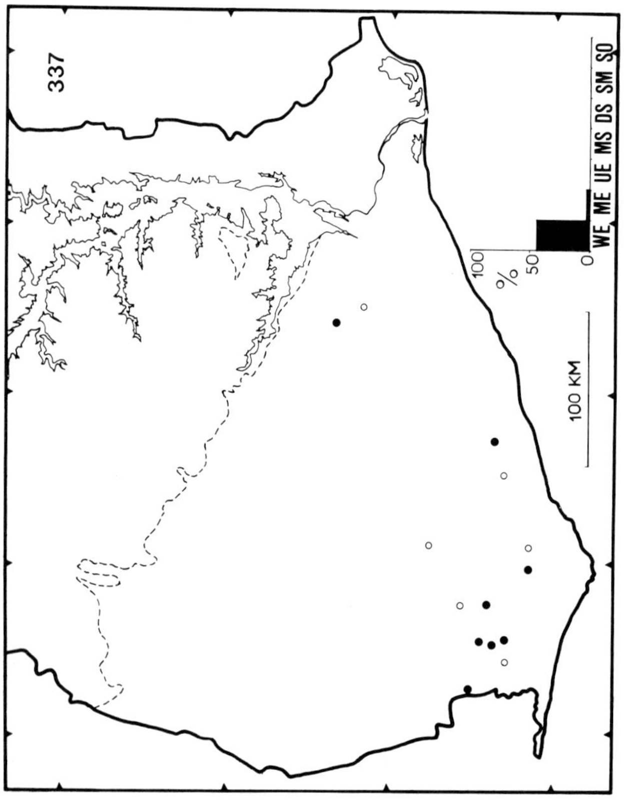
337
100 KM
%
100
50
0
WE ME UE MS DS SM SO

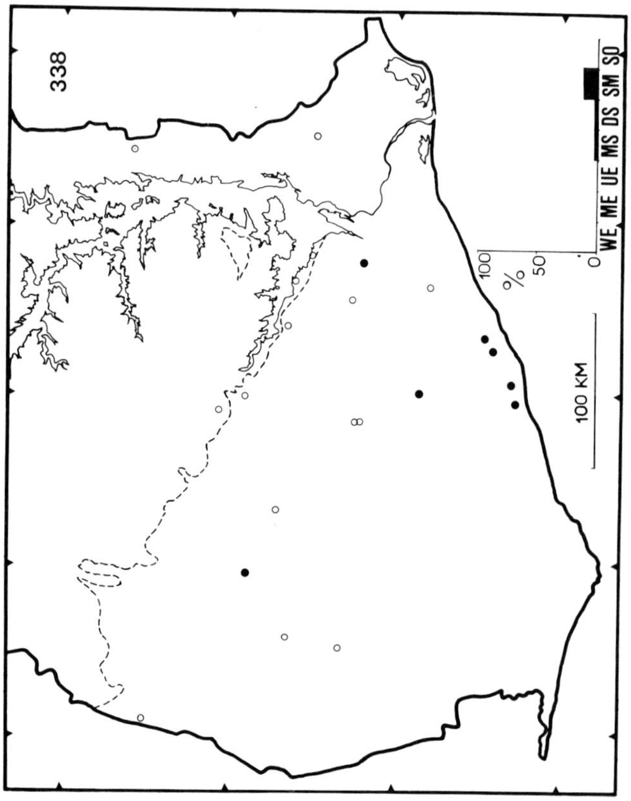
338
100 KM
%
100
50
0
WE ME UE MS DS SM SO

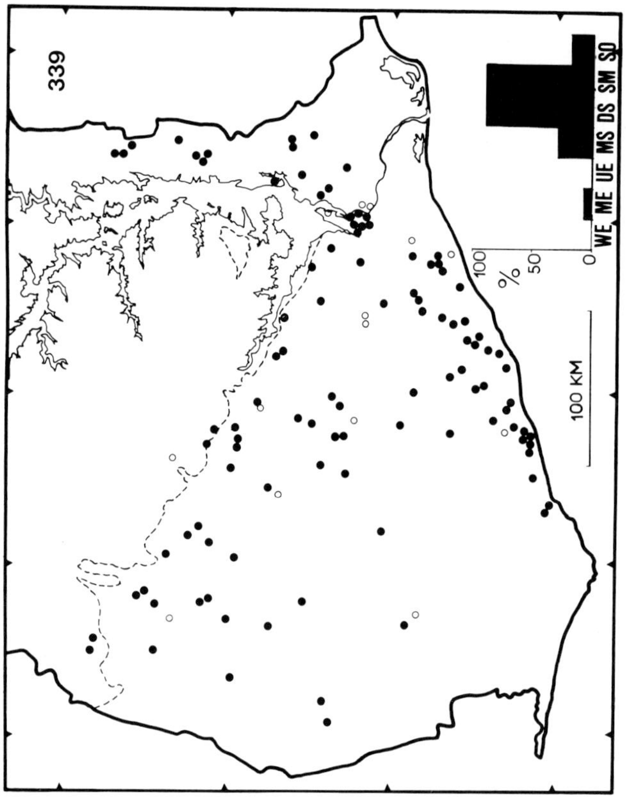
339
100 KM
%
100
50
0
WE ME UE MS DS SM SO

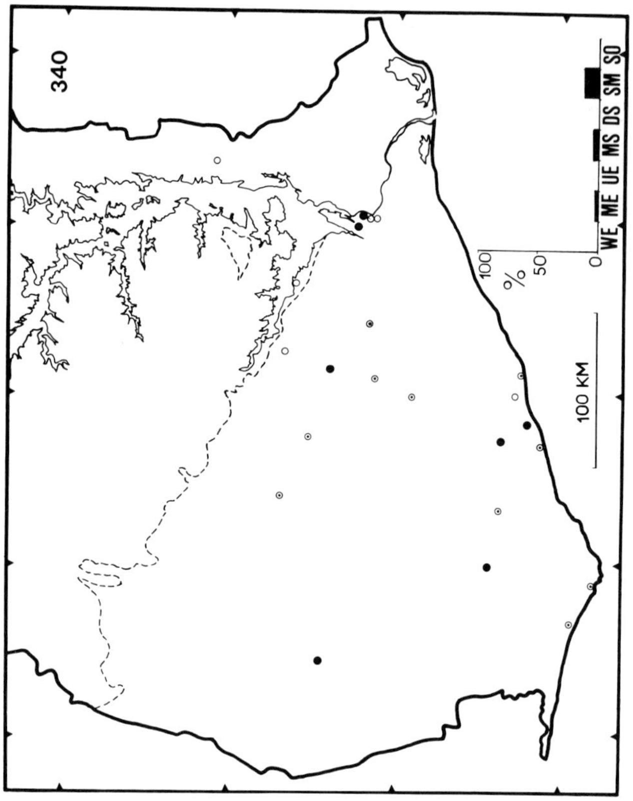
340
100 KM
%
100
50
0
WE ME UE MS DS SM SO

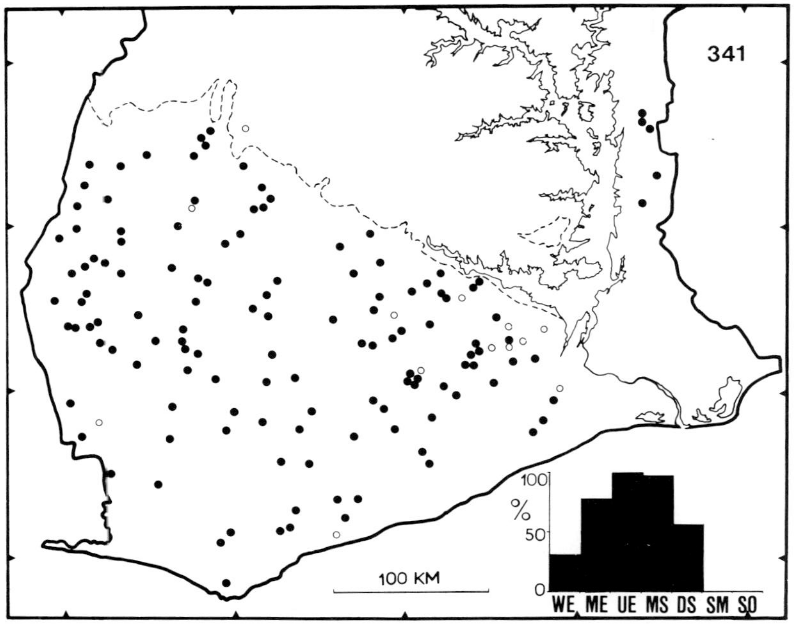

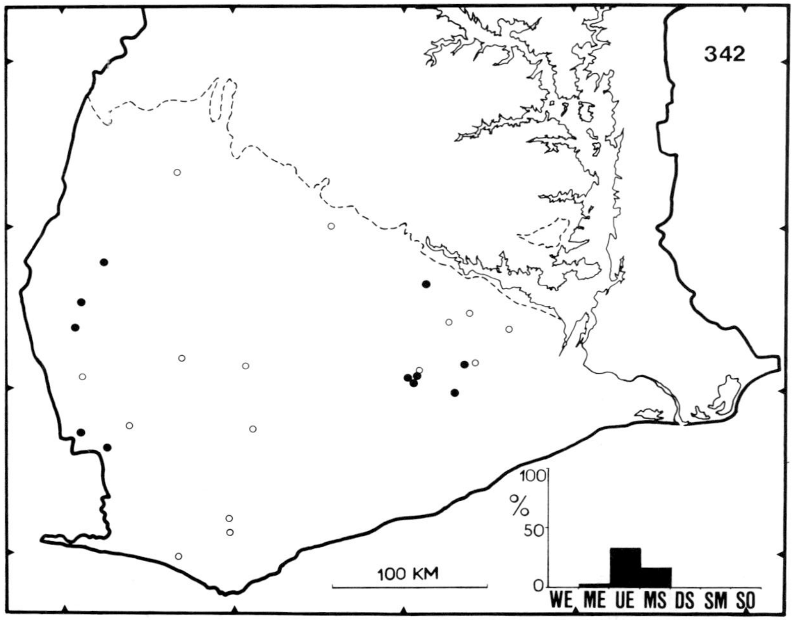

337 LASIANTHUS BATANGENSIS K. Schum., FWTA 2:190. Rubiaceae.
Pygmy tree or shrub; lvs mesophyll; stems, petiole and midrib sparsely hairy. Corner's model. Fl whitish tinged with pink or blue, small, bowl, not clearly seasonal. Fr blue berry; seeds small to medium. Regenerates in shade. S. Leone to Zaire; (Guineo-Congolian wide).

338 LASIODISCUS MANNII Hook. f. var. *chevalieri* (Hutch.) N. Hallé (61), (*L. mildbraedii* Engl., FWTA 1:671). Rhamnaceae. Akan: adafa.
Small tree; lvs notophyll to mesophyll, toothed; twigs, petioles and underside of midrib hairy. Roux's model (59). Fl white, small, dish, in large inflorescences, 10–5. Fr explosive capsule; seeds medium. Regenerates in shade. Sometimes gregarious in the understorey of dry forest. Iv. Coast to Zaire and Tanzania; (Guineo-Congolian wide).

339 LECANIODISCUS CUPANIOIDES Planch. ex Benth., FWTA 1:720. Sapindaceae. Akan: dwindwera.
Small to medium tree, branching low; lvs pinnate, leaflets notophyll; twigs and petiolules densely pubescent. Fl greenish-yellow, small, bowl, probably polygamous, 2–6. Fr yellow 1–2-seeded berry; seeds large. Regenerates in shade. Plot records very significantly associated with soils of higher pH and better base content. Extends into fringing forest and forest on rocky scarps in the savanna zone. Senegal to Angola and Uganda; (Guineo-Congolian wide).

340 LEPTACTINA DENSIFLORA Hook. f. var. *densiflora* and var. *glabra* Hutch. & Dalz., FWTA 2:132. Rubiaceae.
Small woody climber or shrub in thicket; lvs mesophyll, hairy on underside of nerves (var. *densiflora*) or subglabrous (var. *glabra*). Roux's model. Fl white, large, dish with tube, mostly 11–1 and 5–6, opening gregariously. Fr orange berry, seeds small to medium. Regenerates in shade, but more frequent in secondary forest. The varieties were not distinguished during the survey. Herbarium records: var. *densiflora* ○; var. *glabra* ⊙. Distribution: var. *densiflora* S. Leone to Nigeria; var. *glabra* Ghana to Nigeria. Species is a Guinea wide.

341 LEPTASPIS COCHLEATA Thwaites, FWTA 3:362. Gramineae.
Perennial ground herb (Fig. 5.13d); lvs broad, mesophyll. Tomlinson's model. Spikelets small, reddish-brown, monoecious, not clearly seasonal. Fr small to medium, surrounded by persistent lemma which is covered by hooked hairs and so aids dispersal by attachment. Regenerates in shade. Plot records very significantly under-represented in soils of both highest and lowest pH categories. Widespread in Tropical Africa and Asia; (chorological transgressor).

342 LEPTAULUS DAPHNOIDES Benth., FWTA 1:637. Icacinaceae. Akan: afena-akoa.
Small to medium tree; lvs notophyll. Petit's model. Fl white, medium, dish with tube, not clearly seasonal. Fr orange-red drupe; seed medium. Regenerates in shade. Guinée to Zaire and Tanzania; (Guineo-Congolian wide).

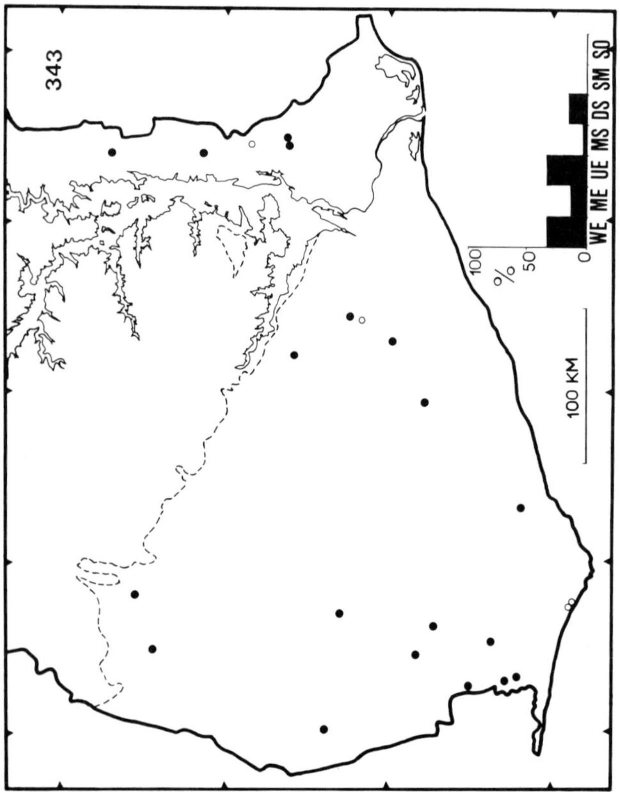
343
100 KM
100
%
50
0
WE ME UE MS DS SM SO

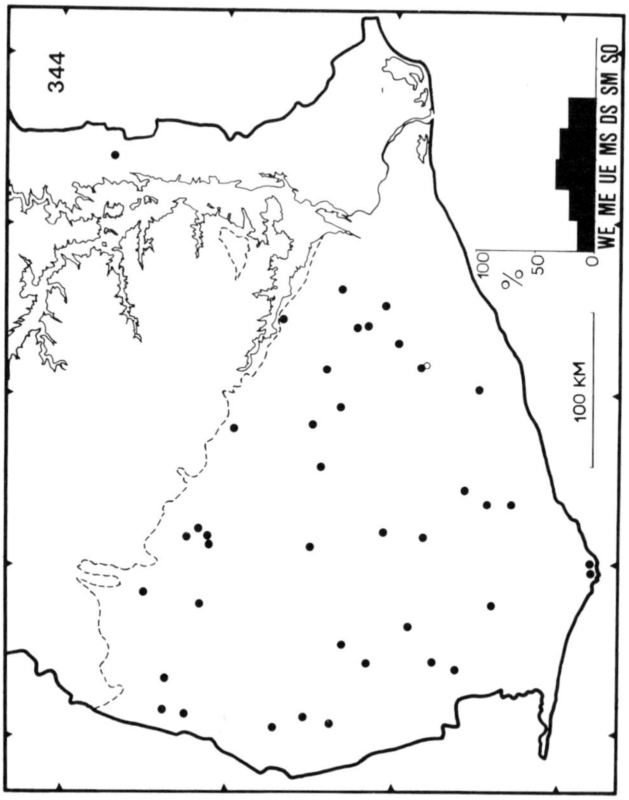
344
100 KM
100
%
50
0
WE ME UE MS DS SM SO

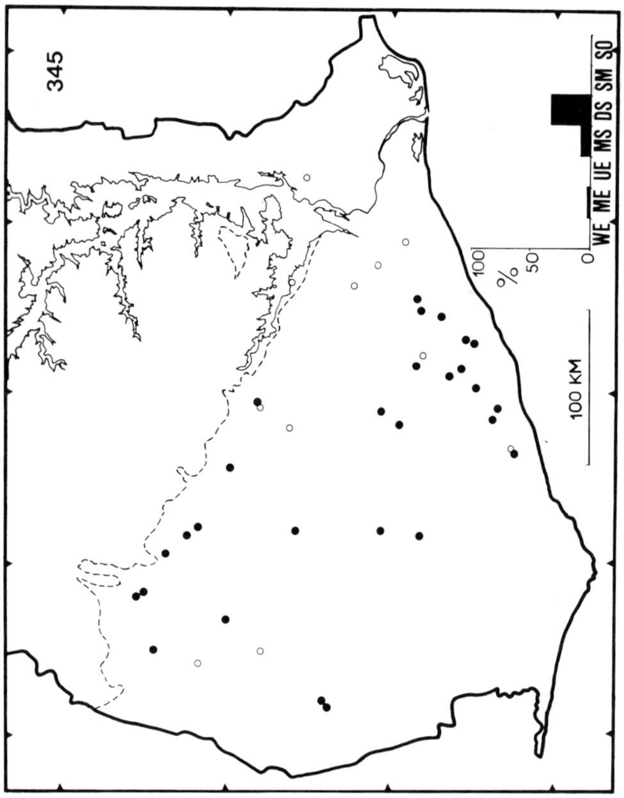
345
100 KM
100
%
50
0
WE ME UE MS DS SM SO

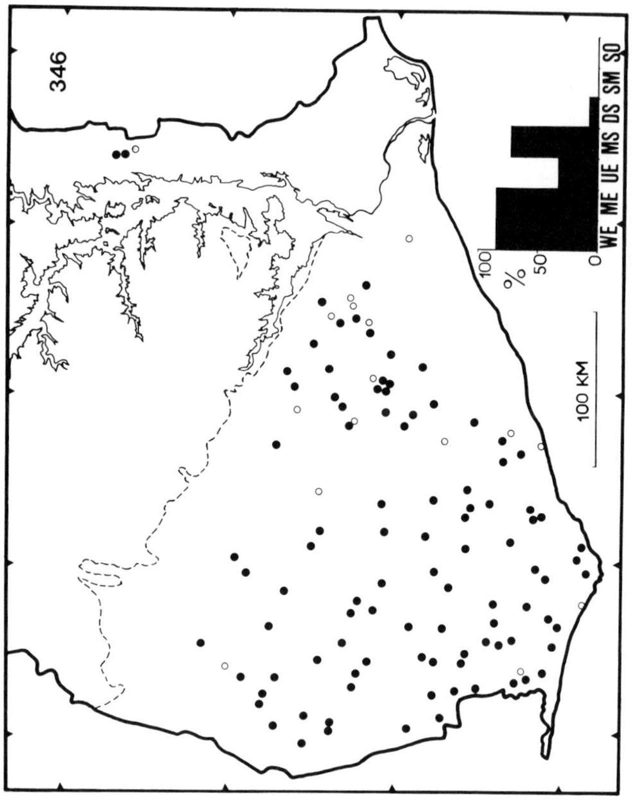
346
100 KM
100
%
50
0
WE ME UE MS DS SM SO

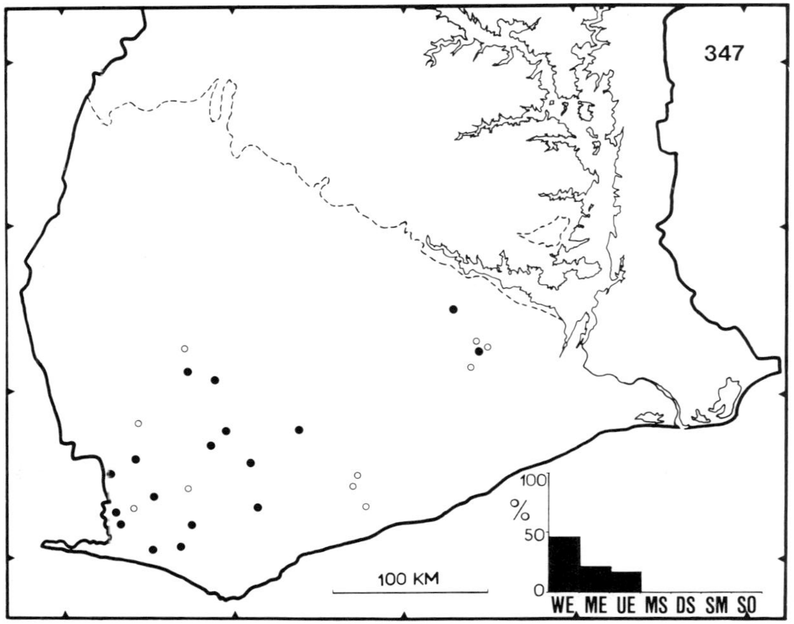

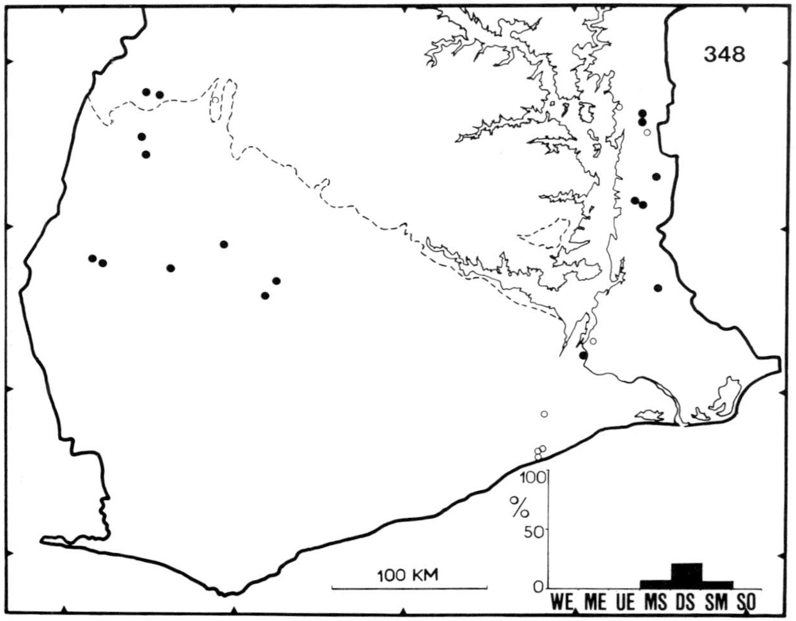

343 LEPTODERRIS BRACHYPTERA (Benth.) Dunn, FWTA 1:521. Papilionaceae. Large woody climber; xylem separated into concentric rings by anomalous phloem; thick red latex exuded by cut phloem. Lvs pinnate, leaflets mesophyll, irregularly shallowly lobed; twigs and underside of leaflets densely pubescent. Troll's model. Fl cream, medium, flag, mostly 2–3. Fr thin, indehiscent, 1-seeded pod; not yet found in Ghana. Regenerates in shade. Senegal to Angola; (Guineo-Congolian wide).

344 LEPTODERRIS FASCICULATA (Benth.) Dunn, FWTA 1:521. Papilionaceae. Large woody climber; xylem normal, cylindrical; thick red latex exuded by cut phloem. Lvs pinnate, leaflets mesophyll; twigs, petiole, rhachis and petiolules densely pubescent. Troll's model. Fl white, medium, flag. Fr indehiscent pod; flowers and fruits not yet found in Ghana. Regenerates in shade. Senegal to Central African Republic; (Guinea wide).

345 LEPTONYCHIA PUBESCENS Keay, FWTA 1:316. Sterculiaceae. Akan: fotonua. Small tree; lvs mesophyll. Troll's model (59). Fl greenish-white, medium, dish, mostly 2–4. Fr capsule; seeds black with red aril. Regenerates in shade. Iv. Coast to Nigeria; (Guinea wide).

346 LOMARIOPSIS GUINEENSIS (Underw.) Alston, FWTA FF:67. Lomariopsidaceae.
Fern climbing by its roots on stems in the understorey; lvs pinnate, leaflets notophyll. Corner's model (59) Propagated by dust-like spores. Plot records very significantly correlated with acid, base-poor soils. Guinée to Angola; (Guineo-Congolian wide).

347 LOMARIOPSIS ROSSII Holtt., FWTA FF:67. Lomariopsidaceae.
Fern climbing by its roots on stems in the understorey; lvs pinnate, leaflets notophyll. Propagated by dust-like spores. Plot records very significantly correlated with acid soils. Guinée to Zaire; (Guineo-Congolian wide).

348 LONCHOCARPUS CYANESCENS (Schum. & Thonn.) Benth., FWTA 1:523. Papilionaceae. Akan: akase. (Type from near Accra).
Large woody climber, or shrub in thicket; xylem divided into concentric rings by anomalous phloem; thick red latex exuded by cut phloem. Lvs pinnate, notophyll, turning blue when crushed. Troll's model. Fl bluish-white, medium, flag, mostly 4–9. Fr flat indehiscent pod; seeds large. Extends to fringing forest in the savanna zone. Guinea Bissau to Cameroun; (Guinea wide). Source of indigo.

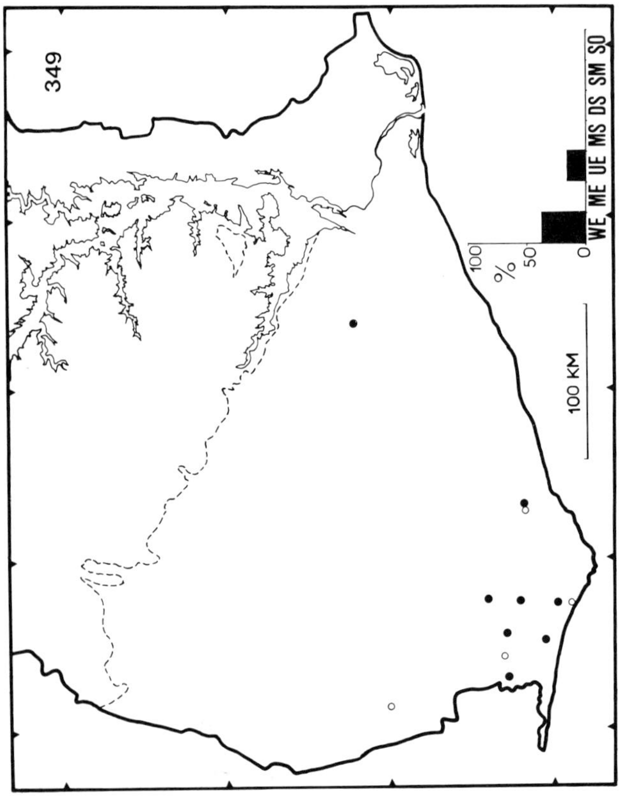
349
100 KM
100
50
0
%
WE ME UE MS DS SM SO

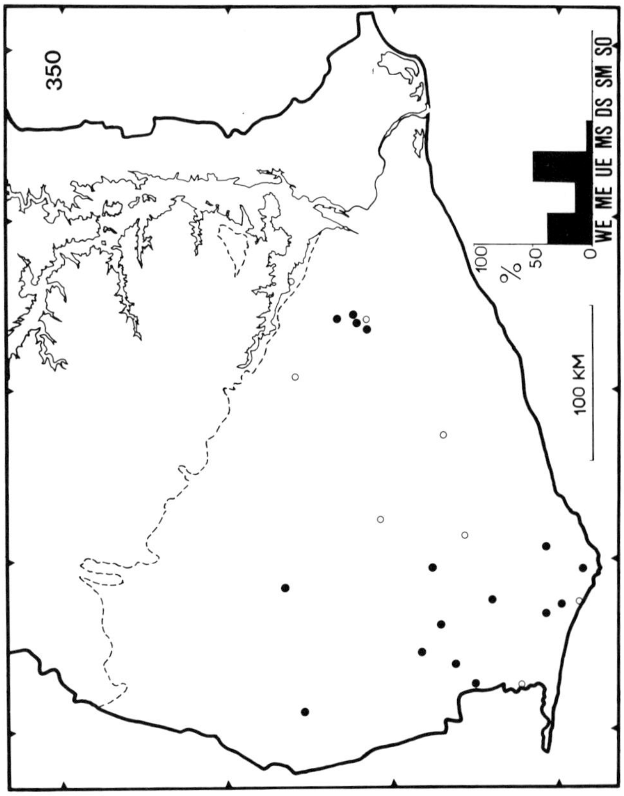
350
100 KM
100
50
0
%
WE ME UE MS DS SM SO

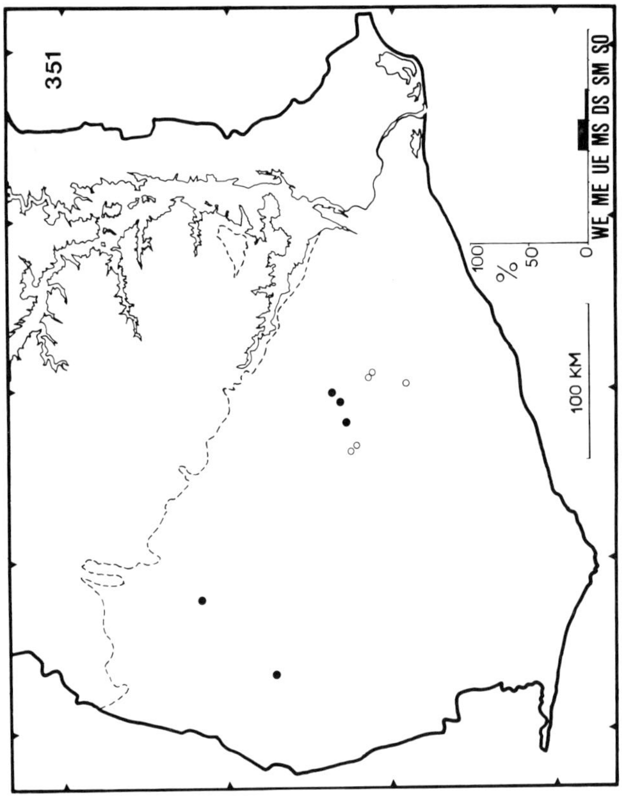
351
100 KM
100
50
0
%
WE ME UE MS DS SM SO

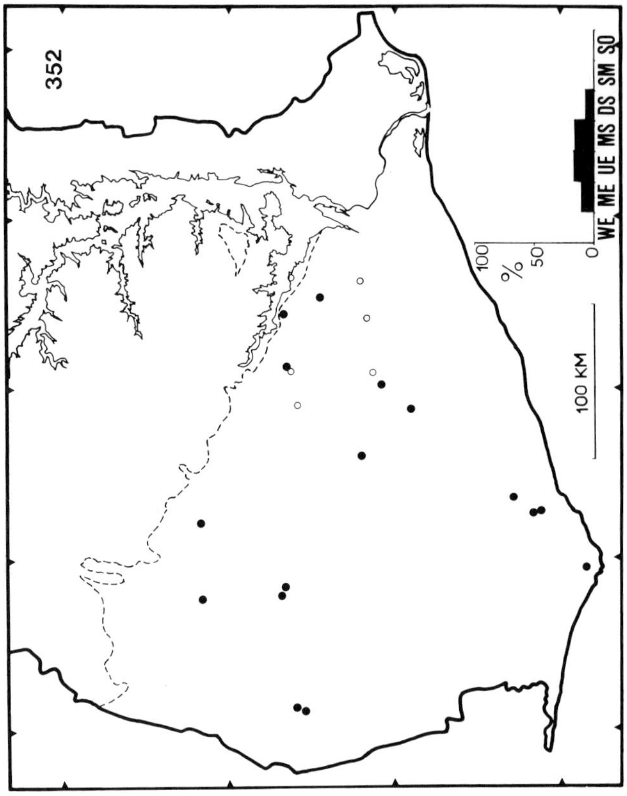
352
100 KM
100
50
0
%
WE ME UE MS DS SM SO

349 LOPHIRA ALATA Banks ex Gaertn. f., FWTA 1:231. Ochnaceae. Akan: kaku.
Large tree; lvs mesophyll, flushing bright red as the old leaves fall. Koriba's model (59). Fl white, fragrant, bowl, medium, 11–12. Fr nut, 2–6, wind-dispersed with aid of enlarged calyx; seed large. Regenerates only in gaps and along roadsides. S. Leone to Zaire; (Guineo-Congolian wide). Class 2 timber; trade name: ekki (40). Distribution in Ghana as calculated from timber enumeration records shown in Fig. 11.3.

350 LOVOA TRICHILIOIDES Harms, FWTA 1:702. Meliaceae. Akan: dubinibiri.
Large tree with buttresses; slash scented; lvs pinnate, leaflets notophyll. Fl white, small, dish, monoecious, 11–1, in large inflorescences. Fr woody capsule, 2–4; seeds large, winged, wind-dispersed. Regenerates in shade. Plot records very significantly associated with acid, base-poor soils. S. Leone to Angola; (Guineo-Congolian wide). Class 1 timber; trade name: african walnut (40).

351 LYCHNODISCUS DANANENSIS Aubrév. & Pellegr., FWTA 1:722. Sapindaceae.
Small to medium tree; lvs pinnate, leaflets mesophyll, toothed; underside of leaflets and twigs pubescent. Fl cream, medium, bowl, probably polygamous, 3–4. Fr capsule, valves reddish inside; seeds large to very large, covered by red aril. Regenerates in shade. S. Leone to Ghana; (Upper Guinea endemic).

352 LYCHNODISCUS RETICULATUS Radlk., FWTA 1:722, Sapindaceae.
Small to medium tree; lvs pinnate, leaflets notophyll to mesophyll, toothed. Fl white, small, bowl, probably polygamous, 1–3. Fr capsule, valves reddish inside; seeds very large, covered by red aril. Regenerates in shade. Ghana to Fernando Po; (Guineo wide).

353 MACROSPHYRA LONGISTYLA (DC.) Hiern, FWTA 2:116. Rubiaceae.
Small woody climber, or shrub in thicket. Lvs notophyll, hairy. Fl cream, large, fragrant, dish with tube, 2–5. Fr berry; seeds medium. Mainly found under broken forest canopy, or in thicket clumps; extends to fringing forest in the savanna zone. Senegal to Cameroun and Uganda; (Guineo-Congolian wide).

354 MAERUA DUCHESNEI (De Wild.) F. White (116), (*Ritchiea duchesnei* (De Wild.) Keay, FWTA 1:92). Capparaceae (Capparidaceae in FWTA).
Small tree, branching low; roots large, soft and tuberous; lvs notophyll. Champagnat's model. Fl large, white or pinkish, bowl, 11–4. Fr green berry; seeds medium. Regenerates in shade. S. Leone to Zaire and Uganda; (Guineo-Congolian wide).

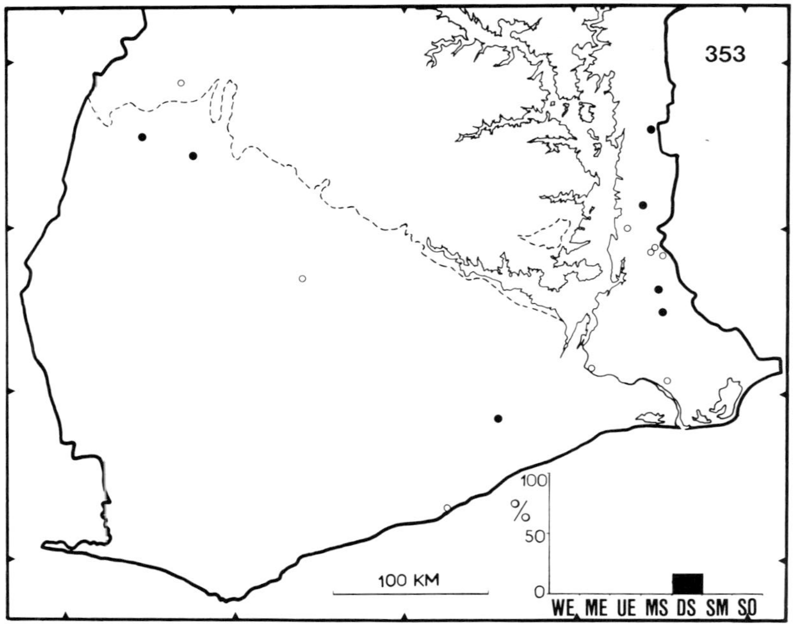

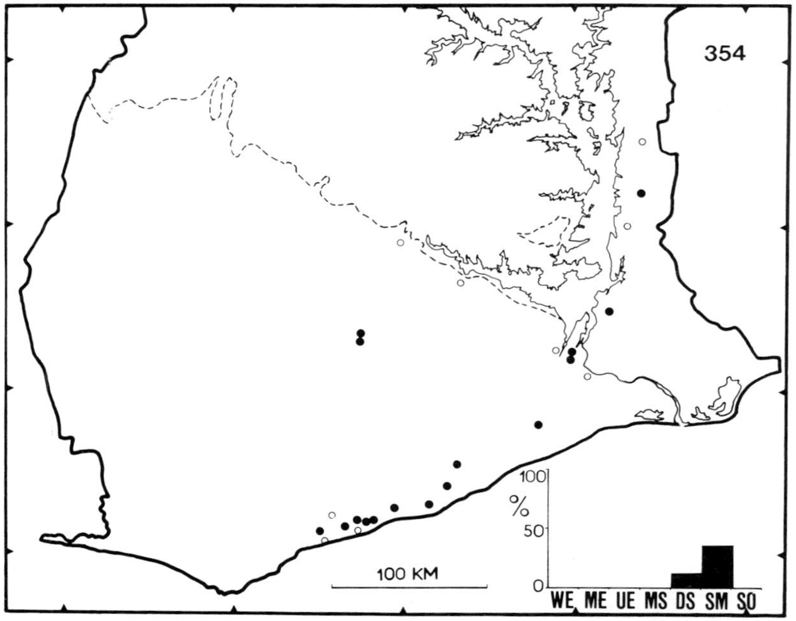

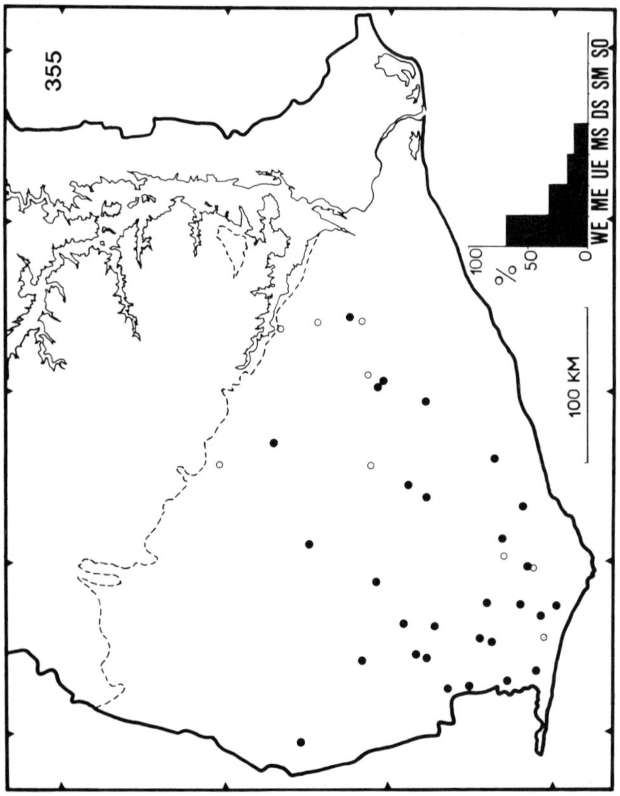
355
100
%
50
0
WE ME UE MS DS SM SO
100 KM

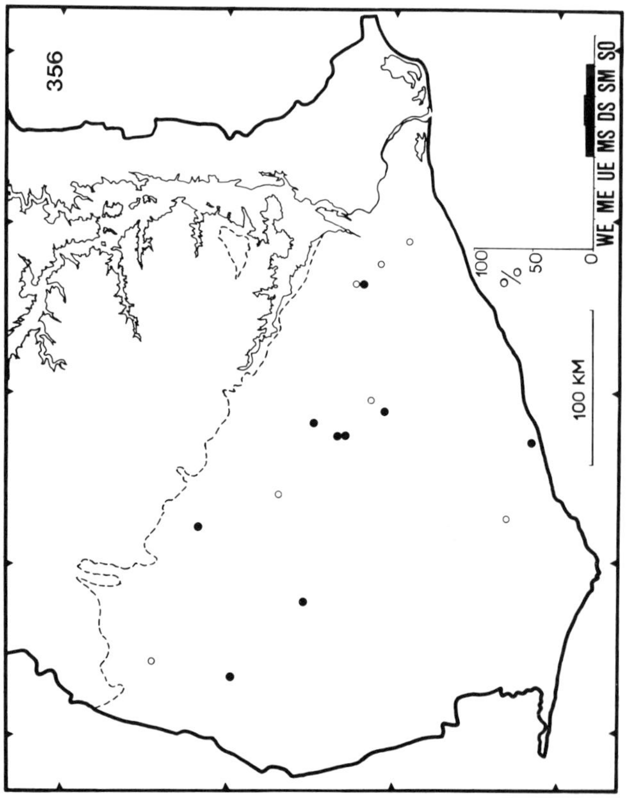
356
100
%
50
0
WE ME UE MS DS SM SO
100 KM

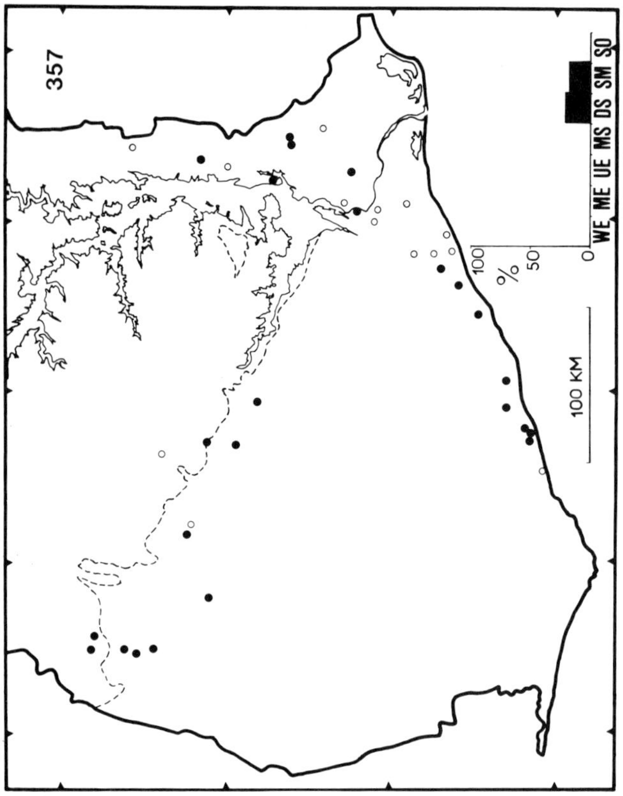
357
100
%
50
0
WE ME UE MS DS SM SO
100 KM

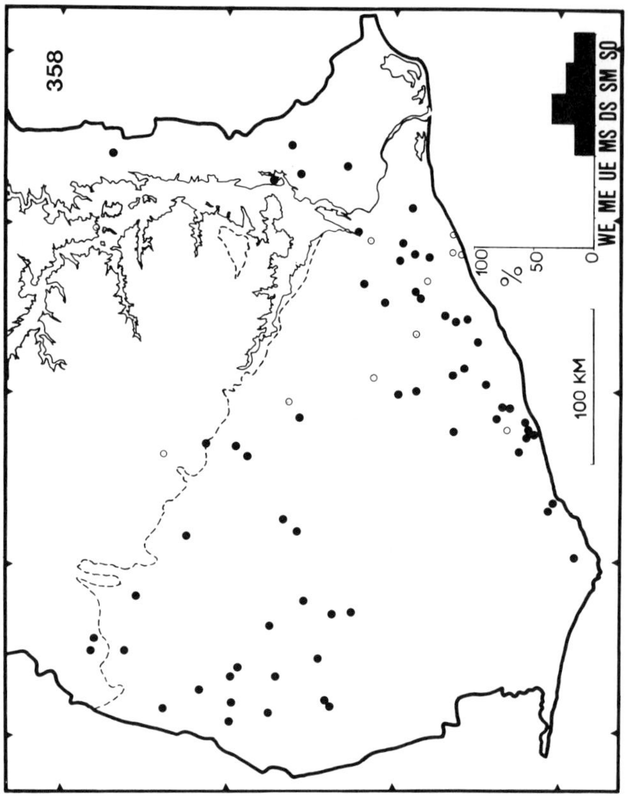
358
100
%
50
0
WE ME UE MS DS SM SO
100 KM

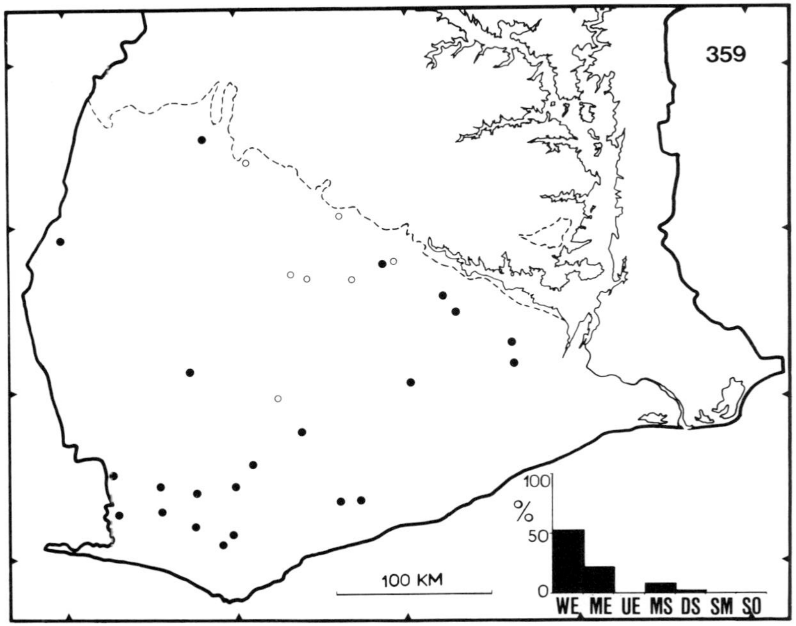

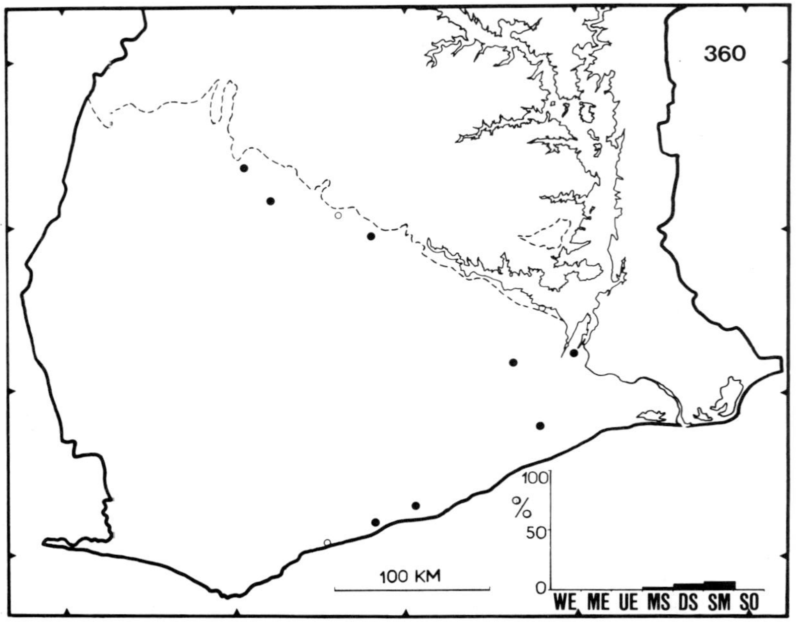

355 MAESOBOTRYA BARTERI (Baill.) Hutch. var. *sparsiflora* (Sc. Elliot) Keay, FWTA 1:374. Euphorbiaceae. Akan: apotrewa.
Small trees; lvs mesophyll. Fl small, whitish, inconspicuous, monoecious, mostly 5–9, male spikes on twigs, and main trunk, female spikes clustered on main trunk. Fr green 1-seeded follicle; seed bright pink, medium. Regenerates in shade. Plot records very significantly correlated with acid, base-poor soils. Guinée to Ghana; (Upper Guinea endemic).

356 MAJIDEA FOSTERI (Sprague) Radlk., FWTA 1:725. Sapindaceae. Akan: ankyewa.
Medium-sized, tree, slash smelling of oil of wintergreen; lvs pinnate; leaflets microphyll. Fl yellowish, small, dish, monoecious, not clearly seasonal. Fr capsule, valves papery and reddish within; seeds blue, hairy, medium. Regenerates in shade. Iv. Coast to Zaire and Uganda; (Guineo-Congolian wide).

357 MALACANTHA ALNIFOLIA (Bak.) Pierre, FWTA 2:24. Sapotaceae.
Small tree, low-branching, with fluted bole; white latex in slash. Lvs mesophyll, deciduous; twigs and underside of nerves pubescent. Troll's model. Fl white, small, bowl, on twigs below leaves, 9–4. Fr red 1-seeded berry, ripe when tree is leafless; seed large. Regenerates in shade. Extends to forest patches on rocky scarps in the savanna zone. Widespread in Tropical Africa; (chorological transgressor).

358 MALLOTUS OPPOSITIFOLIUS (Geisel.) Muell. Arg., FWTA 1:402. Euphorbiaceae. Akan: anyanyanforowa.
Small tree or (more commonly) shrub; lvs notophyll, slightly toothed. Fl (83a) greenish-yellow, small, inconspicuous, dioecious, not clearly seasonal. Fr explosive capsule; seeds medium. Regenerates in shade, but more characteristic of farm regrowth. Widespread in Tropical Africa and Madagascar; (ecological and chorological transgressor).

359 MAMMEA AFRICANA Sabine, FWTA 1:293. Guttiferae. Akan: bompagya.
Large tree, somewhat buttressed; scanty yellow latex exuding from slash; lvs mesophyll. Rauh's model (59). Fl white, large, bowl, not clearly seasonal. Fr yellow berry; seeds very large. Dispersed by elephants. Regenerates in shade. Plot records very significantly correlated with base-poor soils. S. Leone to Zaire; (Guineo-Congolian wide). Class 3 timber; trade name: african apple (40).

360 MANGENOTIA EBURNEA Pichon, FWTA 2:84. Periplocaceae.
Slender woody climber, containing white latex; lvs microphyll. Fl white, medium, dish with tube. Fr dry follicle; seeds medium, wind-dispersed with aid of plume of long hairs. Regenerates in shade. S. Leone to Nigeria; (Guinea wide).

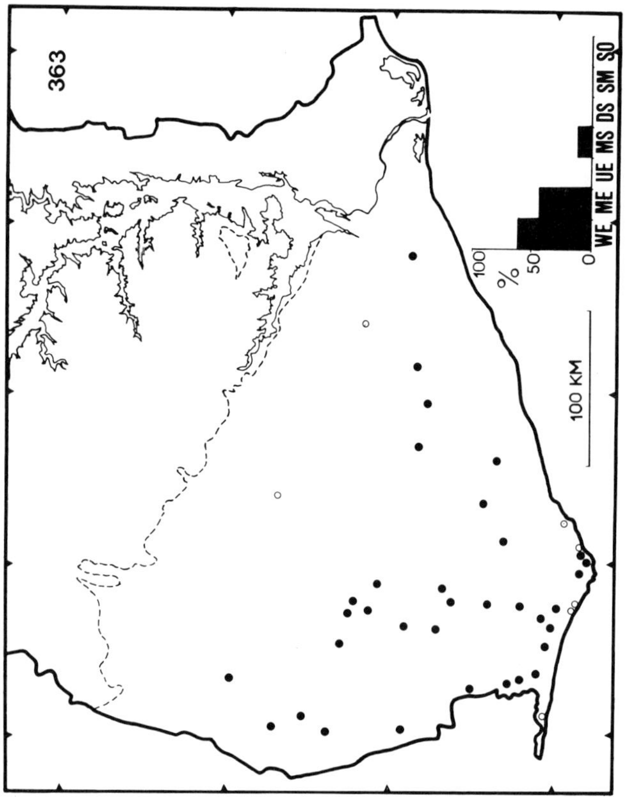
363
100 KM
%
100
50
0
WE ME UE MS DS SM SO

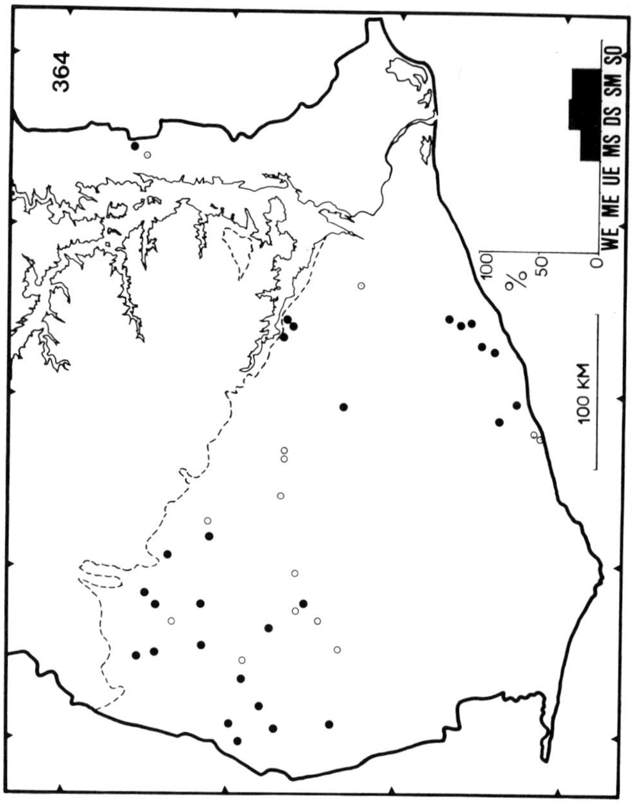
364
100 KM
%
100
50
0
WE ME UE MS DS SM SO

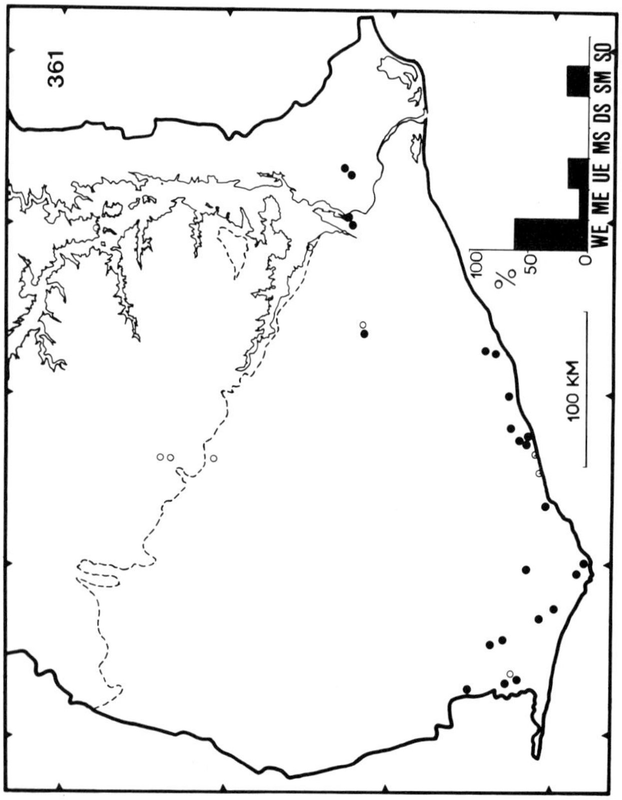
361
100 KM
%
100
50
0
WE ME UE MS DS SM SO

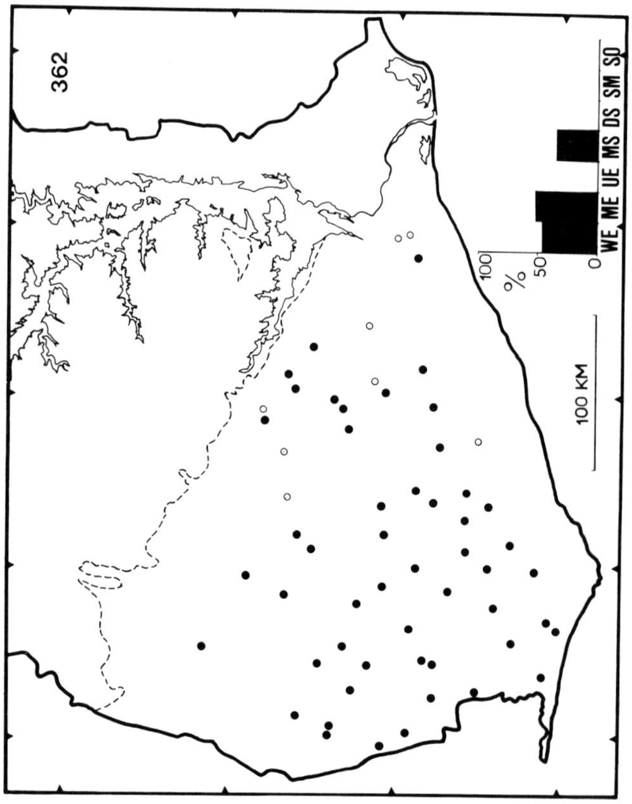
362
100 KM
%
100
50
0
WE ME UE MS DS SM SO

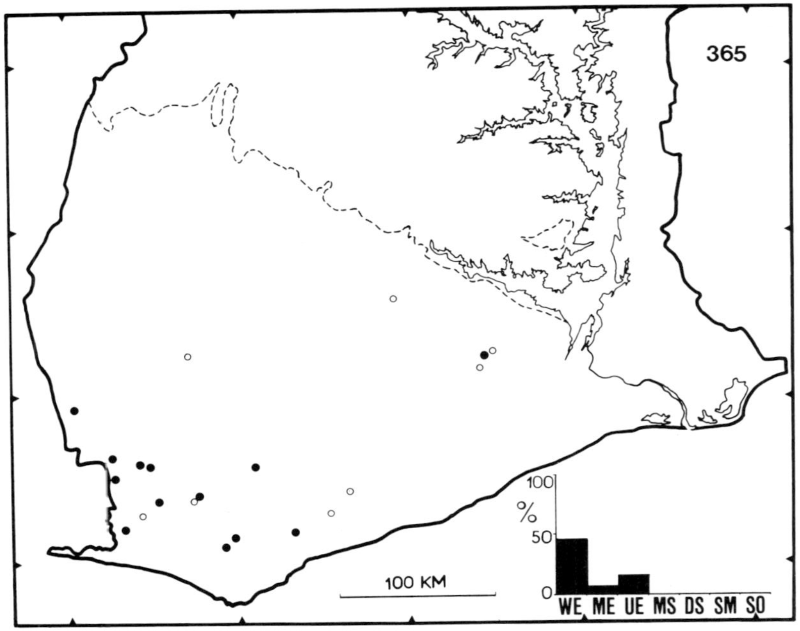

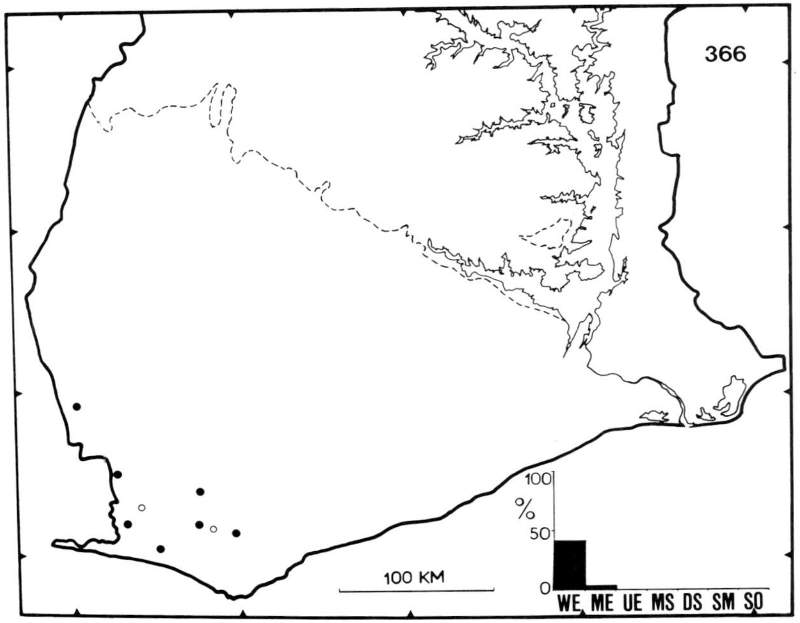

361 MANILKARA OBOVATA (Sabine & G. Don) J. H. Hemsley, FWTA 2:20. (Incl. *M. multinervis* (Bak.) Dubard, FWTA 2:20. Sapotaceae. Akan: berekankum.
Small to medium-sized to large tree, or shrub in thicket; scanty white latex in slash. Lvs microphyll (dry forest near coast) to notophyll or mesophyll; underside covered by more-or-less persistent silky cover of silvery appressed hair. Aubréville's model (59). Fl creamy, fragrant, small, bowl, mostly 10–3. Fr 1-seeded yellow berry; seed large. Regenerates in shade. Guinée to Angola; (ecological and chorological transgressor). Several forms present in Ghana: (a) small tree or shrub with small, obovate leaves, very silvery below; in coastal dry forest; (b) small to medium tree with elliptic to oblong, obtuse, leaves, not very silvery below; in hill forest near savanna boundary, and extending into fringing forest in savanna zone; (c) medium-sized to large tree, leaves similar to (b) but more acuminate in sapling and with fewer, more steeply ascending nerves, in Upland Evergreen forest; (d) medium-sized to large tree, leaves elliptic, very silvery below, numerous nerves spreading at right angles to the midrib, in Wet Evergreen forest. The existence of numerous intermediates has frustrated attempts (9) to distinguish these and other forms as species (11, 113, 117); it should, however, be possible (after further work) to recognise them as varieties.

362 MANNIOPHYTON FULVUM Muell. Arg., FWTA 1:400. Euphorbiaceae.
Large woody climber, or scrambling shrub in thicket; xylem more or less deeply grooved; red latex in slash. Lvs mesophyll, rough-hairy. Fl cream, small, bowl, dioecious, not clearly seasonal. Fr capsule; seeds large. Regenerates in shade. Plot records very significantly correlated with acid, base-poor soils. S. Leone to Angola; (Guineo-Congolian wide).

363 MANOTES LONGIFLORA Bak., FWTA 1:747. Connaraceae.
Small liane or shrub; lvs pinnate, leaflets microphyll; twigs and leaf underside pubescent. Fl pink, medium, bowl, not seasonal. Fr red 1-seeded follicle; seed medium, orange red. Regenerates in shade, but more commonly seen on roadsides and in old farmland. Plot records very significantly correlated with base-poor soils. Iv. Coast to Nigeria.

364 MANSONIA ALTISSIMA (A. Chev.) A. Chev. var. *altissima*, FWTA 1:313. Sterculiaceae. Akan: oprono.
Medium to large tree, with buttresses; lvs mesophyll to macrophyll, densely hairy below, deciduous. Rauh's model. Fl white, fragrant, medium, bowl, 6–8. Fr samara (Fig. 6.1f), dispersed by wind; seed medium to large. Regenerates in shade. Iv. Coast to Nigeria; (Guinea wide). Class 2 timber: trade name: mansonia (40).

365 MAPANIA BALDWINII Nelmes, FWTA 3:335. Cyperaceae.
Perennial caespitose ground herb; lvs macrophyll. Corner's model (59). Spikes medium, clustered, brown; individual florets monoecious, wind-pollinated. Achenes small; no obvious dispersal mechanism, not clearly seasonal. Regenerates in shade. Liberia to Ghana; (Upper Guinea endemic).

366 MARANTHES AUBREVILLEI (Pellegr.) Prance (118), (*Parinari aubrevillei* Pellegr., FWTA 1:428). Chrysobalanaceae (Rosaceae in FWTA).
Medium to large tree; lvs notophyll, loose cottony hair on underside, slightly toothed, two extrafloral nectaries at leaf base. Fl cream, medium, bowl. Fr not known, but probably woody drupe. Regenerates in shade. S. Leone to Ghana, and Cameroun; (Guinea wide, 118).

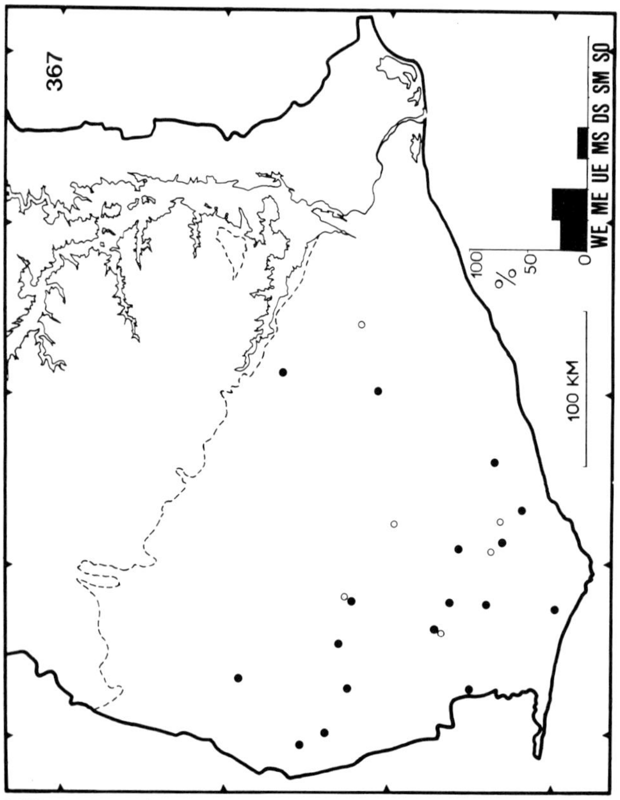
367
100 KM
%
100
50
0
WE ME UE MS DS SM SO

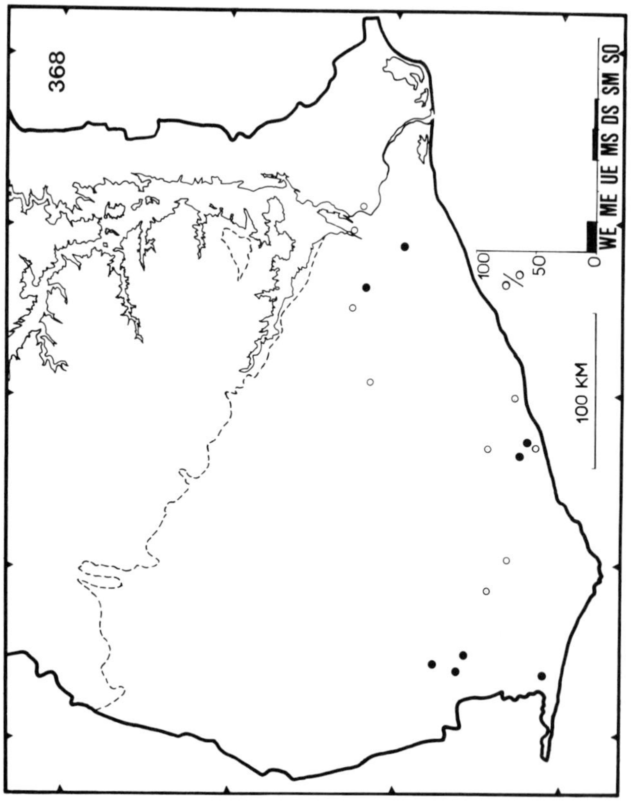
368
100 KM
%
100
50
0
WE ME UE MS DS SM SO

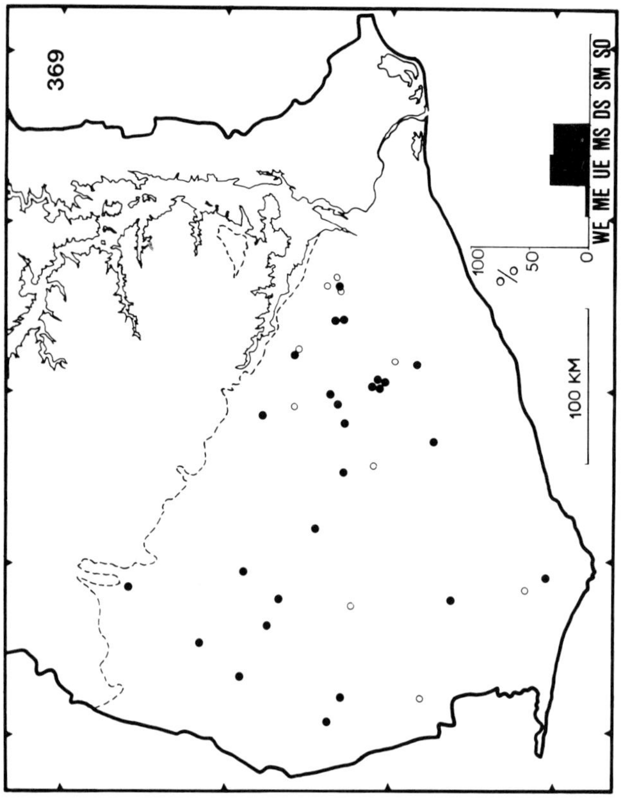
369
100 KM
%
100
50
0
WE ME UE MS DS SM SO

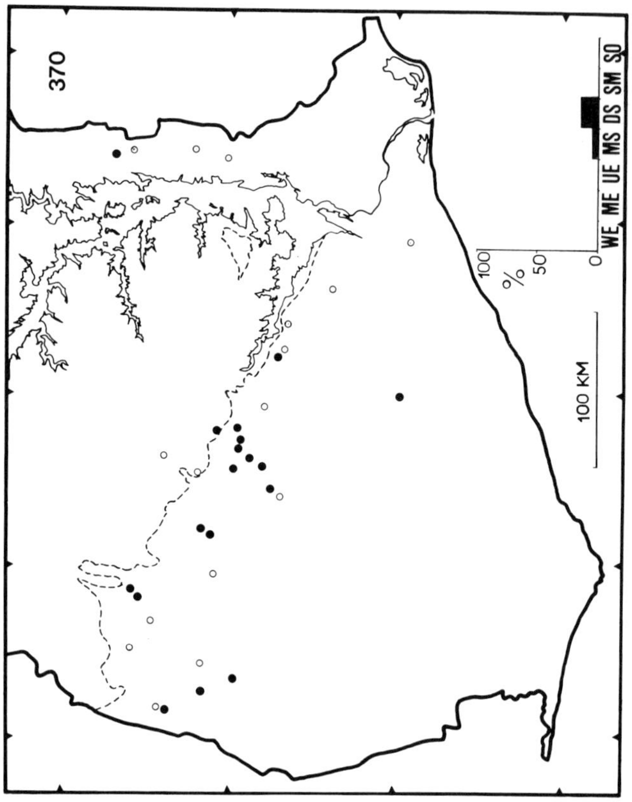
370
100 KM
%
100
50
0
WE ME UE MS DS SM SO

367 MARANTHES GLABRA (Oliv.) Prance (118), (*Parinari glabra* Oliv., FWTA 1:428). Chrysobalanaceae (Rosaceae in FWTA). Akan: afamnini.
Medium to large tree; lvs notophyll, two extrafloral nectaries at leaf base. Fl white, medium, bowl, not clearly seasonal. Fr woody drupe with two pyrenes; seeds large. Regenerates in shade. Plot records very significantly correlated with base-poor soils. S. Leone to Angola; (Guineo-Congolian wide, 118).

368 MARANTHES ROBUSTA (Oliv.) Prance (118), (*Parinari robusta* Oliv., FWTA 1:428). Chrysobalanaceae (Rosaceae in FWTA). Akan: afambere.
Medium-sized tree, sometimes branching low; lvs notophyll, loose cottony hair on underside, two extrafloral nectaries at leaf base. Fl cream, medium, bowl, not clearly seasonal. Fr woody drupe; seed large. Regenerates in shade. Characteristically, though not exclusively, a river bank species. Iv. Coast to Nigeria; (Guinea wide).

369 MARANTOCHLOA CONGENSIS (K. Schum.) Léonard & Mullend., FWTA 3:82. Marantaceae.
Perennial herb; lvs notophyll to mesophyll, glabrous or pubescent below. McClure's model. Fl white, small, bowl, not seasonal. Fr pale capsule; seeds medium, arillate. Regenerates in shade. S. Leone to Zaire; (Guineo-Congolian wide).

370 MARANTOCHLOA CUSPIDATA (Rosc.) Milne-Redh., FWTA 3:85. Marantaceae.
Perennial herb; lvs macrophyll. McClure's model. Fl yellow, medium, bowl, not clearly seasonal. Fr pale capsule; seeds medium, arillate. Regenerates mainly in gaps, and on forest floor after ground fire. Guinée to Ghana; (Upper Guinea endemic).

371 MAREYA MICRANTHA (Benth.) Muell. Arg., FWTA 1:404. Euphorbiaceae. Akan: dubrafo.
Small to medium tree; lvs notophyll to mesophyll, toothed; twigs and underside of midrib pubescent. Fl whitish, small, clustered in spikes, monoecious, not clearly seasonal. Fr red capsule; seeds small. Regenerates in shade. Guinea to Zaire; (Guineo-Congolian wide). Notoriously poisonous; Akan name means 'executioner medicine'.

372 MARKHAMIA LUTEA (Benth.) K. Schum., FWTA 2:387. Bignoniaceae. Akan: obogyaneboobere.
Small to medium tree; lvs pinnate, leaflets notophyll to mesophyll. Scarrone's model. Fl yellow, large, bell, 9–1. Fr dry capsule; seeds medium, winged, wind-dispersed. Regenerates mainly under broken canopy or in gaps. Ghana to Cameroun; (Guinea wide).

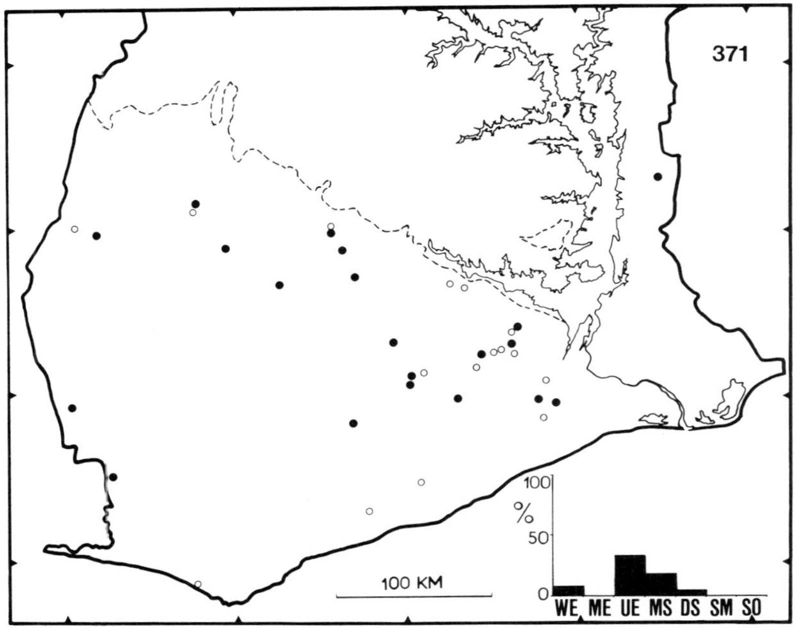

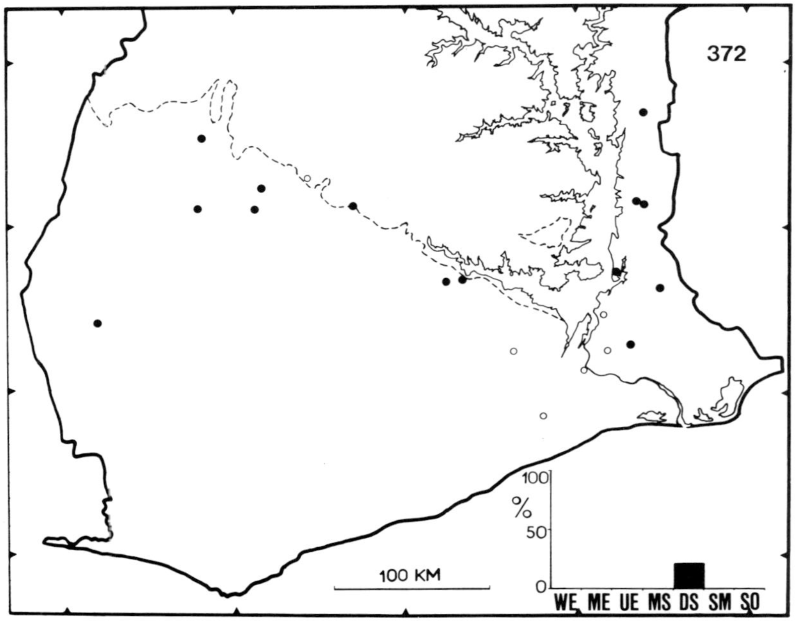

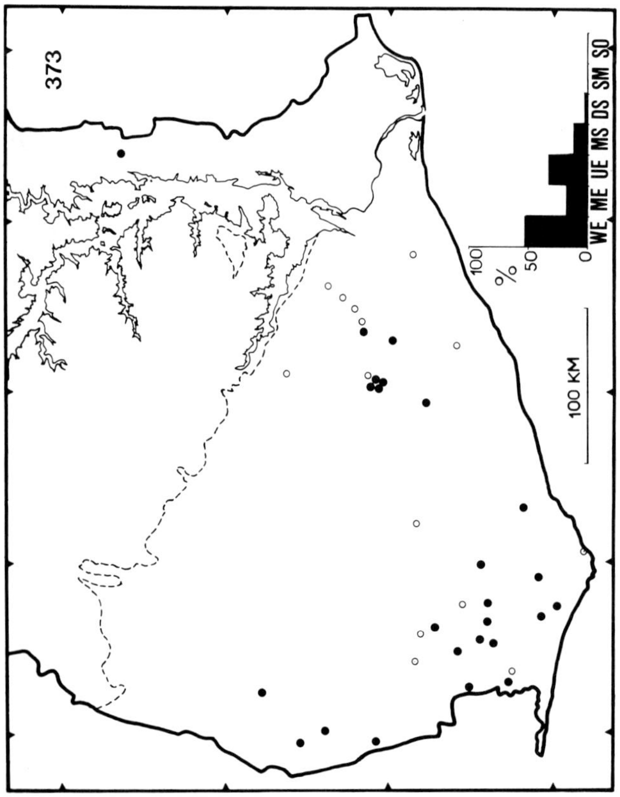
373
100 KM
%
100
50
0
WE ME UE MS DS SM SO

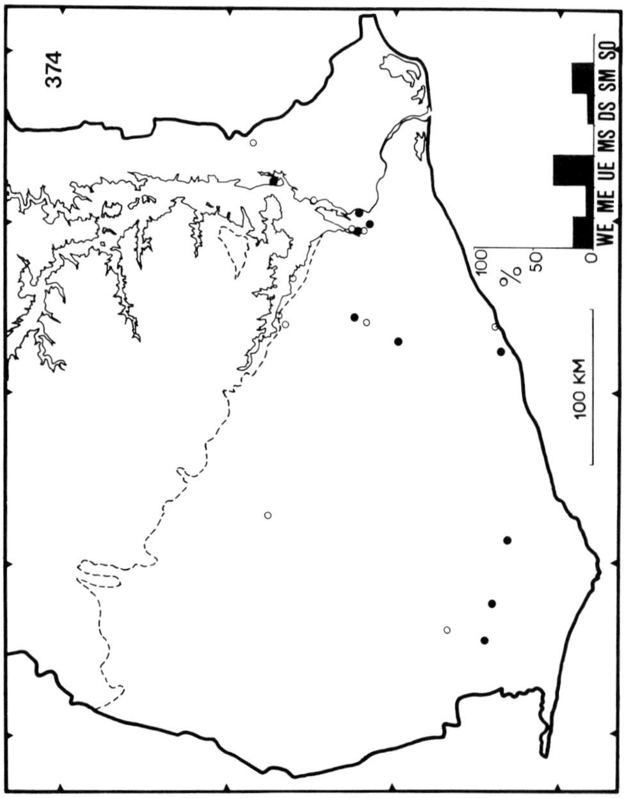
374
100 KM
%
100
50
0
WE ME UE MS DS SM SO

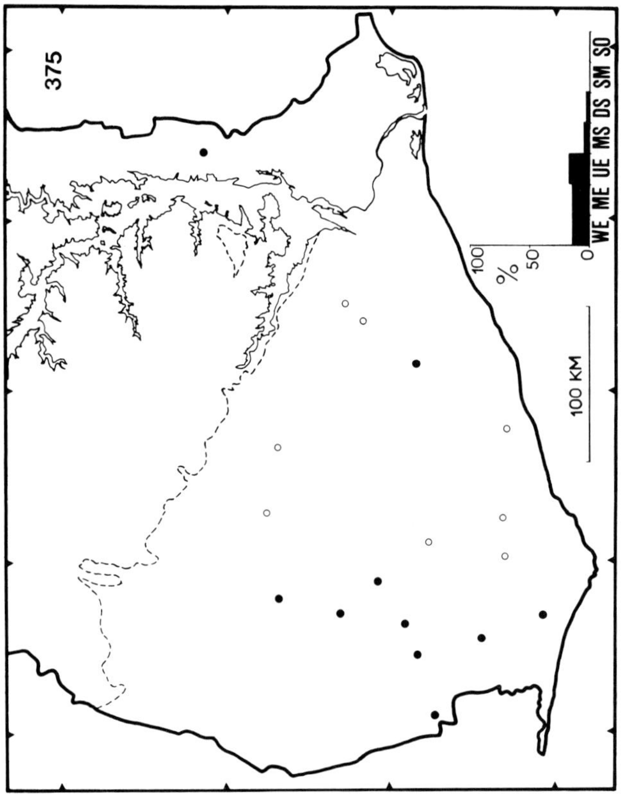
375
100 KM
%
100
50
0
WE ME UE MS DS SM SO

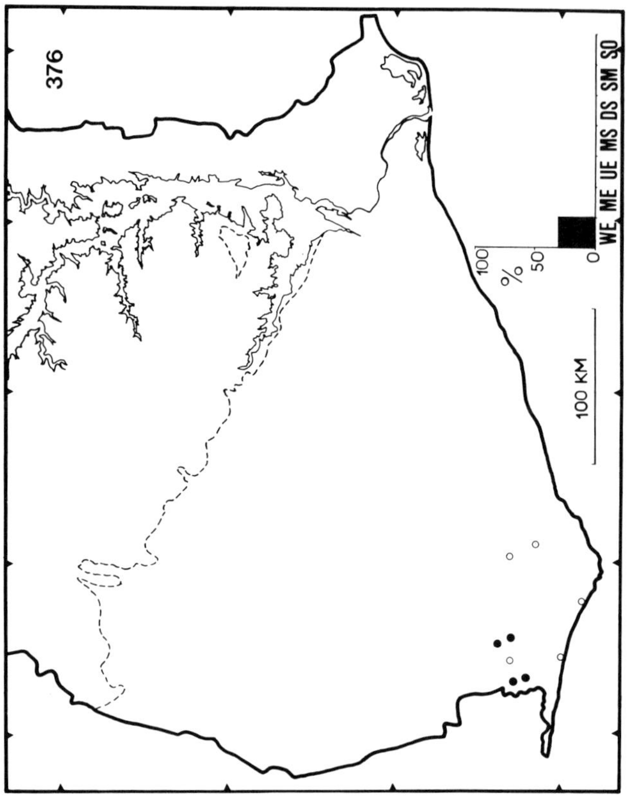
376
100 KM
%
100
50
0
WE ME UE MS DS SM SO

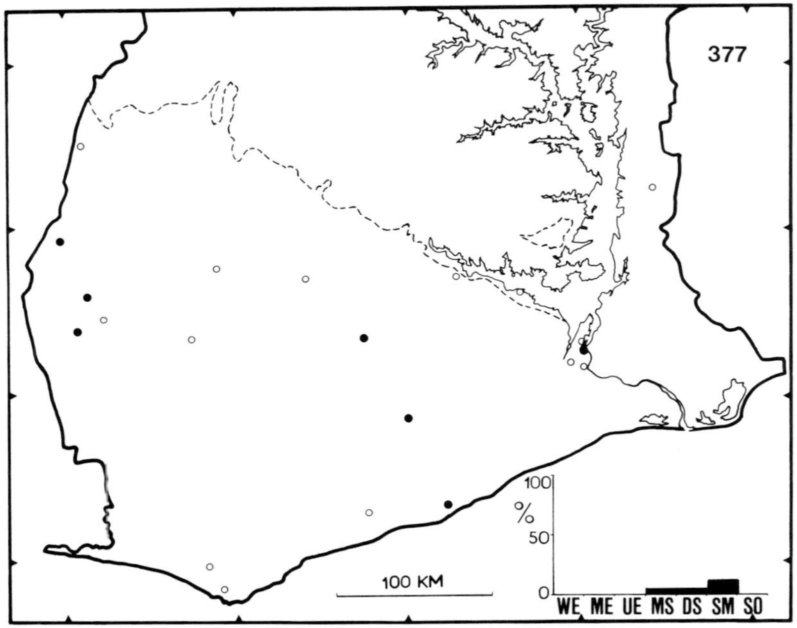

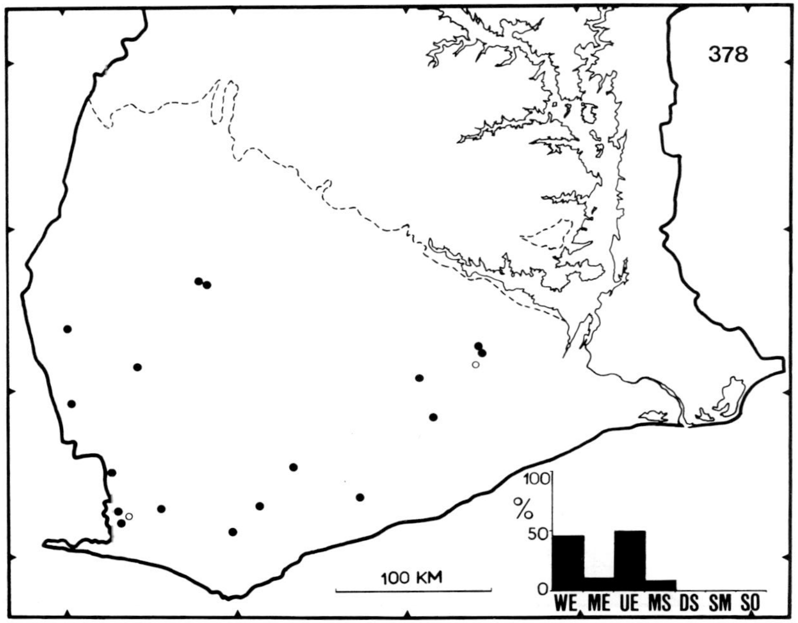

373 MASSULARIA ACUMINATA (G. Don) Bullock ex Hoyle, FWTA 2:114. Rubiaceae. Akan: pobe.
Small slender tree; lvs mesophyll to macrophyll. Petit's model (59). Fl reddish, large, dish with tube, not clearly seasonal. Fr greenish-yellow berry; seeds medium, dispersed by elephants (5). Regenerates in shade. Plot records very significantly associated with acid, base-poor soils. Guinée to Congo Republic; (Guinea wide). A form with narrow, smaller leaves which occurs in SW Ghana may prove to be taxonomically distinct.

374 MEMECYLON AFZELII G. Don, FWTA 1:262. Melastomataceae. Akan: otwe-ani.
Small tree or shrub; lvs microphyll to notophyll. Fl blue, small, dish, 3–4 and 10–12. Fr bluish berry; seeds medium. Regenerates in shade, and forest margins. Guinée to Gabon; (Guinea wide, 67). Plants from evergreen forest have thinner, more acuminate leaves than those from dry forest; the absence of the species from Moist Semi-deciduous forest suggests that the two forms might be varietally distinct.

375 MEMECYLON AYLMERI Hutch. & Dalz., FWTA 1:262 (except Andoh FH 4381 which is *M. afzelii*). Melastomataceae.
Pygmy tree; lvs notophyll. Fl cream to blue, small,; dish, 12–1. Fr blue berry; seeds medium. Regenerates in shade. Guinée to Ghana; (Upper Guinea endemic). Distinguished in Ghana from *M. afzelii* by its visible lateral nerves, shorter pedicels, and larger, more clearly ellipsoid fruit.

376 MEMECYLON BLAKEOIDES G. Don, FWTA 1:263. (Incl. *M. fleuryi* Jac.-Fél., FWTA 1:263). Melastomataceae. Akan: otwesenini.
Medium-sized tree; lvs notophyll to mesophyll. Fl white with blue anthers, small, dish, not clearly seasonal. Fr black berry; seeds medium. Regenerates in shade. Senegal to Gabon; (Guinea wide).

377 MEMECYLON CINNAMOMOIDES G. Don, FWTA 1:263. Melastomataceae.
Small tree or shrub; lvs microphyll to notophyll. Fl blue, small, dish, mostly 12–3. Fr black berry; seeds medium. Characteristic of forest margins, or forest with broken canopy. S. Leone to Ghana; (Upper Guinea endemic).

378 MEMECYLON GUINEENSE Keay, FWTA 1:263. Melastomataceae.
Pygmy to small tree; lvs notophyll. Mangenot's model. Fl white or blue, small, dish. Fr white or purple berry; seeds medium. Regenerates in shade. Liberia to Nigeria; (Guinea wide). Easily recognised by its cordate leaves and narrowly winged twigs.

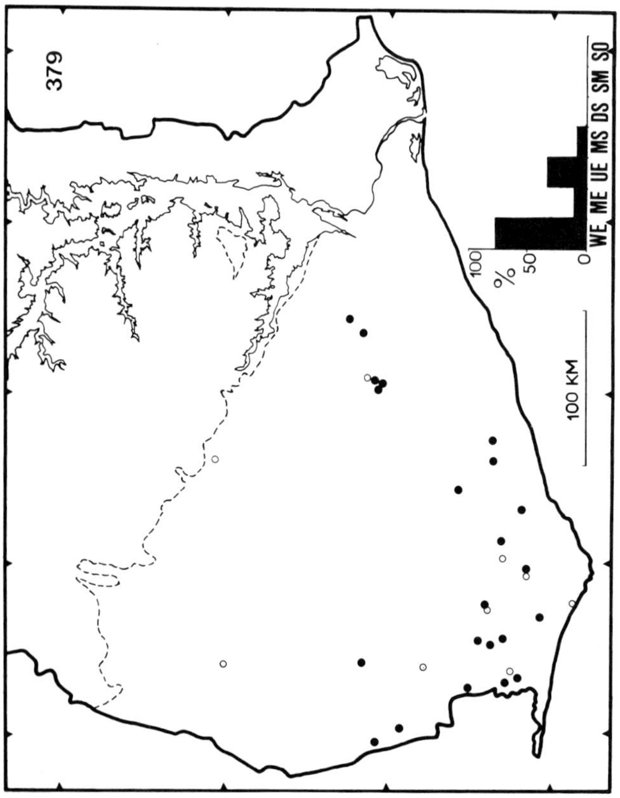
379
100 KM
%
100
50
0
WE ME UE MS DS SM SO

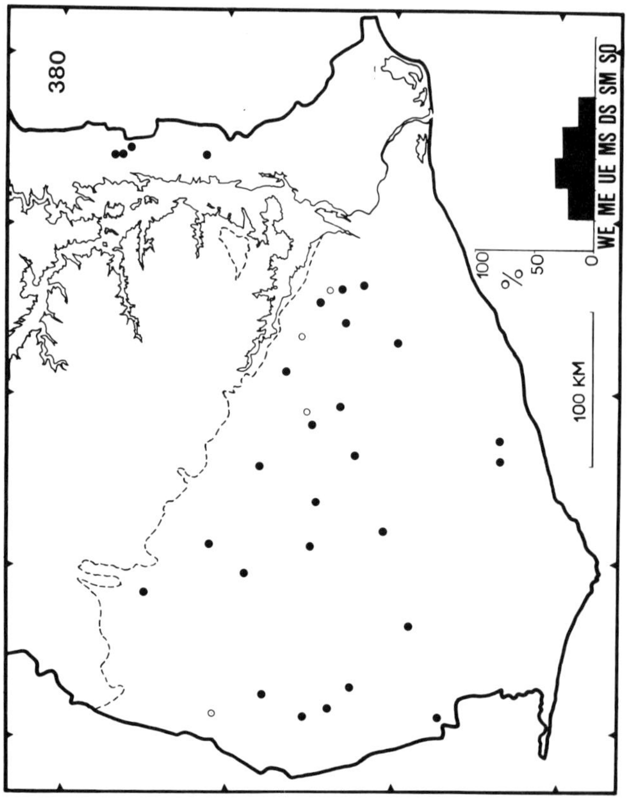
380
100 KM
%
100
50
0
WE ME UE MS DS SM SO

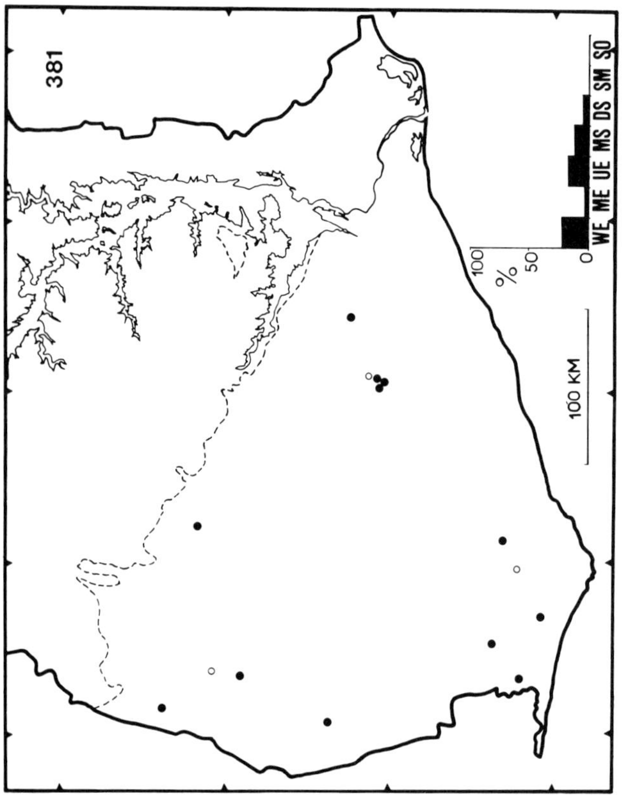
381
100 KM
%
100
50
0
WE ME UE MS DS SM SO

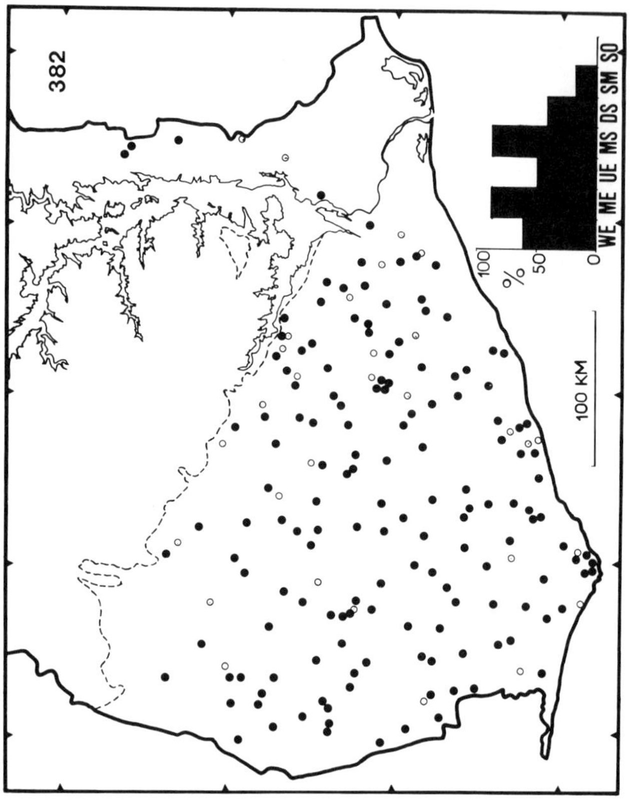
382
100 KM
%
100
50
0
WE ME UE MS DS SM SO

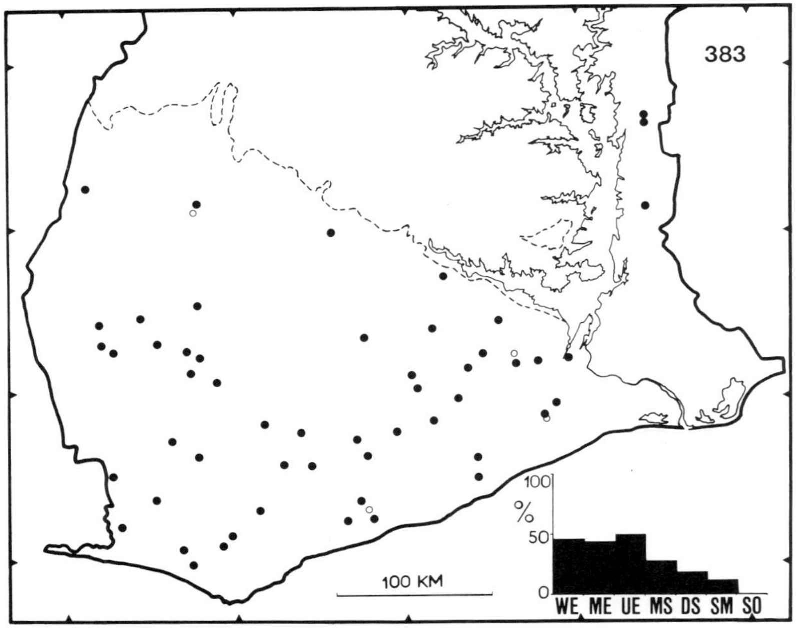

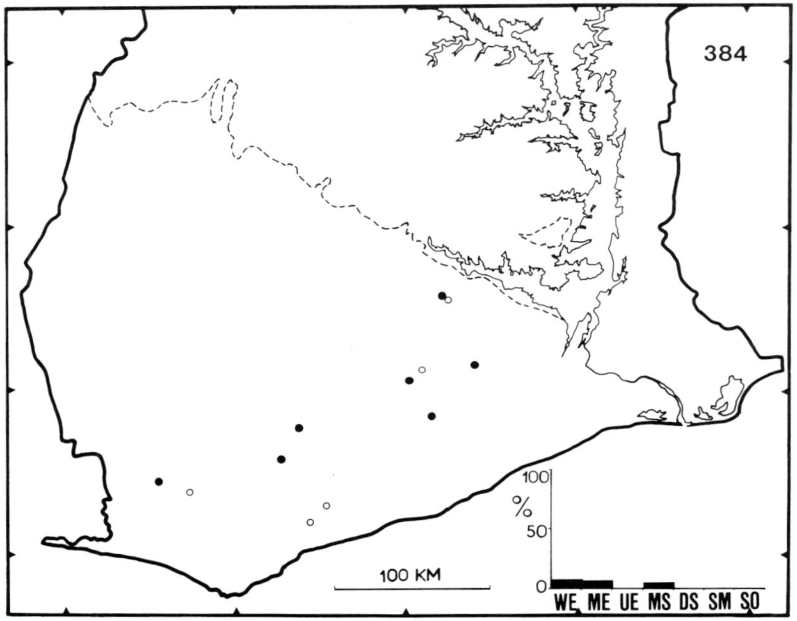

379 MEMECYLON LATERIFLORUM (G. Don) Bremek., FWTA 1:262. Melastomataceae. Akan: otwese.

Medium-sized tree with extraordinarily thin bark; lvs notophyll to mesophyll. Fl pinkish-violet, bowl, small, 3–5. Fr berry; seeds large. Regenerates in shade. Plot records very significantly correlated with acid, base-poor soils. Guinée to Nigeria; (Guinea wide).

380 MEMECYLON MEMBRANIFOLIUM Hook. f., FWTA 1:263. Melastomataceae.

Pygmy tree; lvs notophyll. Fl blue, small, dish, 9–2. Fr blue berry; seeds medium. Regenerates in shade. Iv. Coast to Cameroun; (Guinea wide). Very similar to *M. cinnamomoides*, but leaves thinner.

381 MEMECYLON NORMANDII Jac.-Fél., FWTA 1:262. Melastomataceae.
Medium-sized tree, bark very thin; lvs notophyll. Fl white with blue anthers, small, dish, not clearly seasonal. Fr purple berry; seeds medium. Regenerates in shade. Guinée to Ghana; (Upper Guinea endemic). Similar to *M. lateriflorum*, but leaves more elliptic than oblong, and bark more corky.

382 MICRODESMIS PUBERULA Hook. f. ex Planch., FWTA 1:392. Pandaceae (Euphorbiaceae in FWTA). Akan: ofema.

Small tree; lvs notophyll, toothed. Roux's model (59). Fl yellowish, small, inconspicuous, unisexual (probably dioecious), not seasonal; inflorescences often galled. Fr small, red, 3-celled drupe; seeds medium. Regenerates in shade. Guinée to Angola and Uganda; (Guineo-Congolian wide). Transferred to Pandaceae on grounds of morphology and anatomy (43).

383 MICROSORIUM PUNCTATUM (Linn.) Copel., FWTA FF:49. Polypodiaceae.
Epiphytic caespitose fern (Fig. 5.7); lvs macrophyll. Propagated by dust-like spores; grows on branches in the understorey. Widespread in Tropical Africa and Asia; (chorological transgressor). Ghana's most abundant epiphyte.

384 MIKANIOPSIS TEDLIEI (Oliv. & Hiern) C. D. Adams, FWTA 2:243. Compositae.

Slender woody climber; lvs notophyll, toothed, bright red-purple below. Fl white, medium, tubular, clustered in capitula, 8–9. Fr achene, wind-dispersed with aid of long hairs; seed small to medium. Regenerates on path margins and in small gaps. Ghana and Cameroun; (Guinea wide).

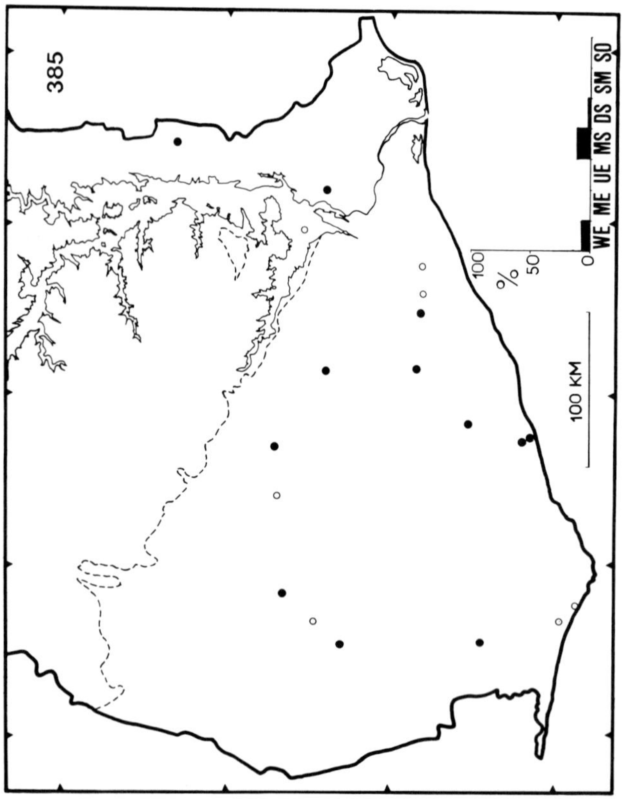
385
100 KM
%
100
50
0
WE ME UE MS DS SM SO

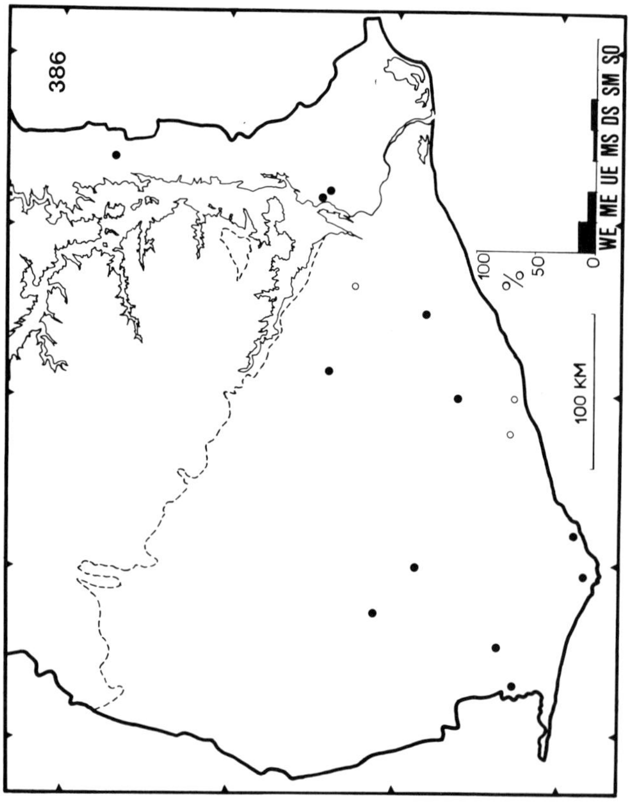
386
100 KM
%
100
50
0
WE ME UE MS DS SM SO

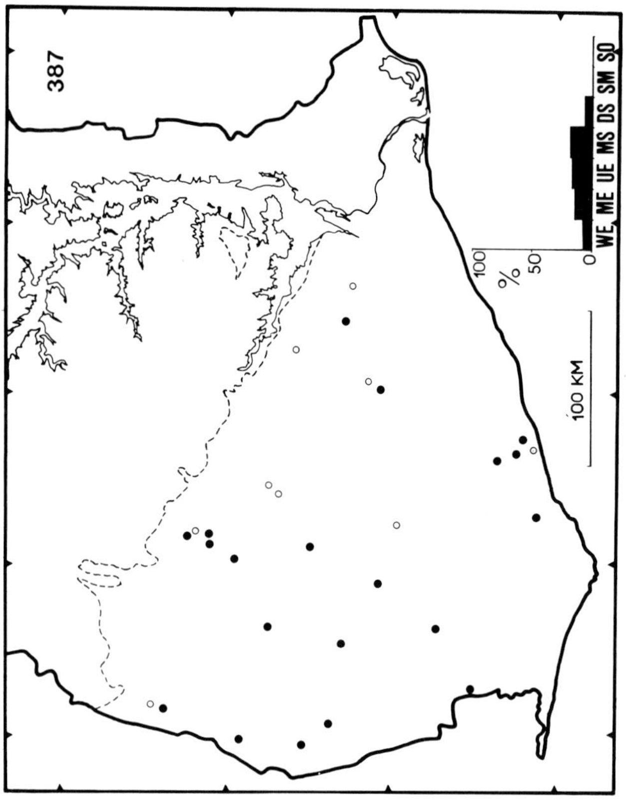
387
100 KM
%
100
50
0
WE ME UE MS DS SM SO

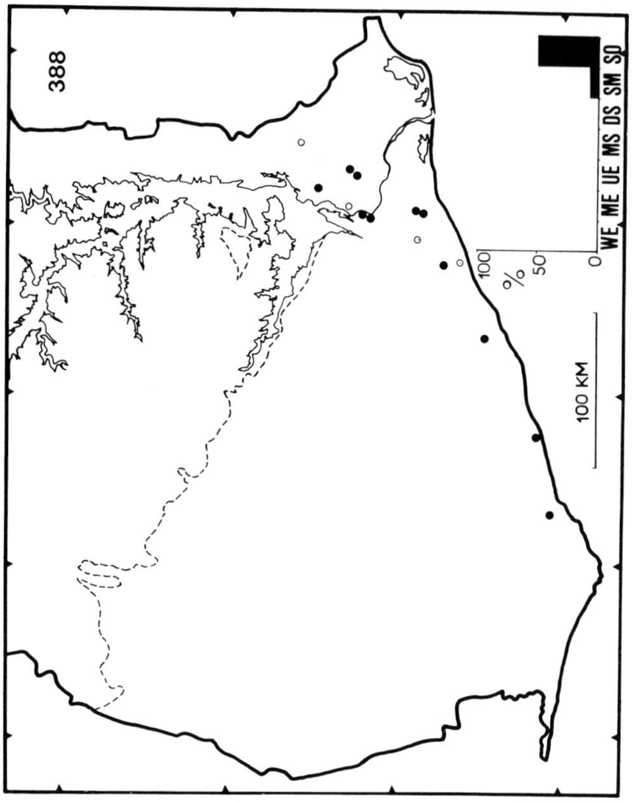
388
100 KM
%
100
50
0
WE ME UE MS DS SM SO

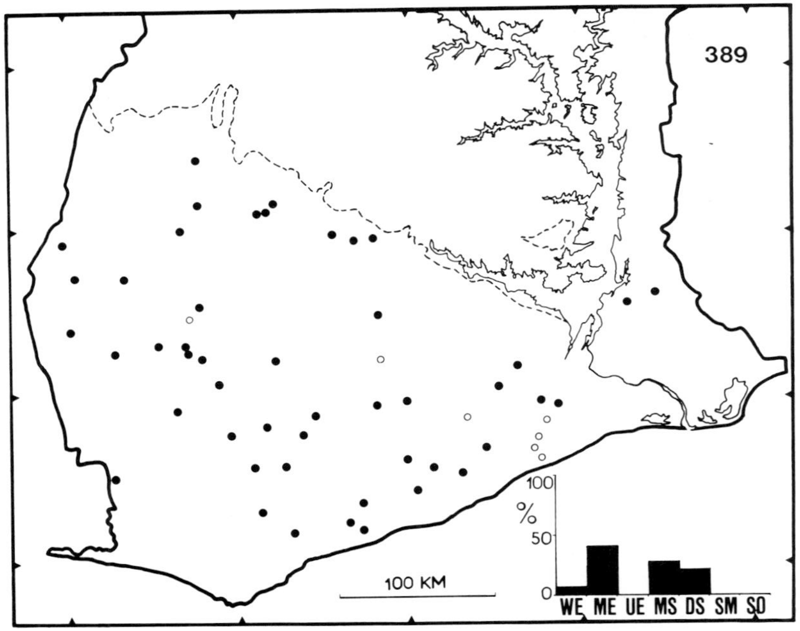

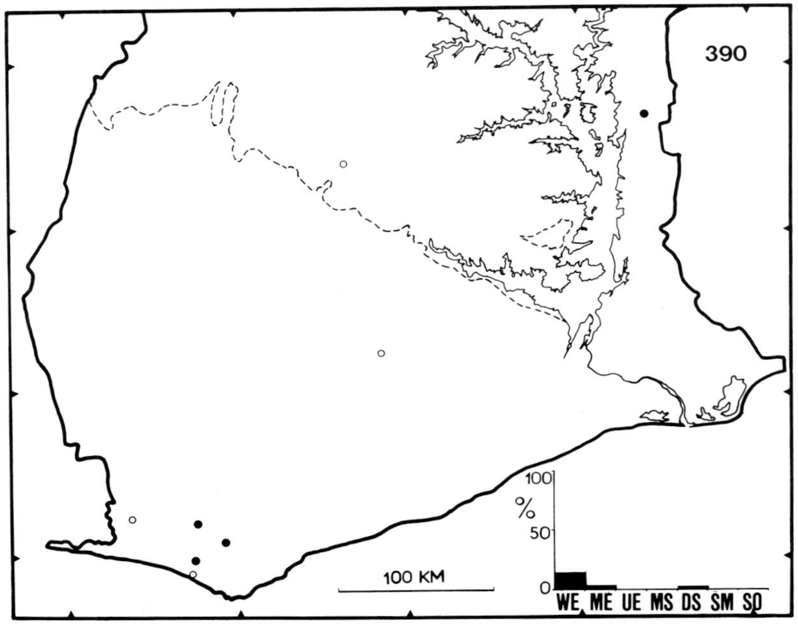

385 MILLETTIA BARTERI (Benth.) Dunn, FWTA 1:526. Papilionaceae.
Large woody climber; bark thick and furrowed, exuding red latex when cut; accessory interrupted ring of xylem formed outside the phloem in old stems. Lvs pinnate; leaflets notophyll. Troll's model. Fl purplish, flag, medium, not clearly seasonal. Fr explosively dehiscent pod, not yet collected in Ghana; seeds large. Senegal to Zaire and Sudan; (Guineo-Congolian wide).

386 MILLETTIA CHRYSOPHYLLA Dunn, FWTA 1:526. Papilionaceae.
Large woody climber, or scrambling shrub in thickets; lvs pinnate, leaflets mesophyll, undersurface covered by appressed, silky hair. Troll's model. Fl white, medium, flag, 7–9. Fr explosively dehiscent pod; seeds large. Regenerates in shade. Guinée to Cameroun; (Guinea wide).

387 MILLETTIA RHODANTHA Baill., FWTA 1:526. Papilionaceae. Akan: tetetoa.
Small to medium tree; red latex exudes from slash. Lvs pinnate, leaflets microphyll, pubescent below. Troll's model. Fl purple, medium, flag, 2–5. Fr explosively dehiscent pod; seeds large. Regenerates mostly in small gaps. Guinée to Ghana (Upper Guinea endemic). *Millettia lane-poolei* Dunn seems to differ only in being less pubescent.

388 MILLETTIA THONNINGII (Schum. & Thonn.) Bak., FWTA 1:527. Papilionaceae. Ga: taatso. (Type from near Ada).
Small to medium tree; lvs pinnate, leaflets microphyll, deciduous, often with characteristic galls. Troll's model. Fl bright blue, medium, flag, 1–3, produced while the tree is leafless. Fr explosively dehiscent pod, taking a year to mature; seeds large, dispersed in harmattan season when humidity is very low. Regenerates mainly from root suckers. Gregarious (Fig. 8.13). Ghana to Nigeria; (Guinea wide).

389 MILLETTIA ZECHIANA Harms, FWTA 1:526. Papilionaceae. Akan: fafraha.
Small tree or shrub; lvs pinnate, leaflets notophyll. Troll's model. Fl purplish, medium, flag, mostly 10–1. Fr explosively dehiscent pod; seeds large. Regenerates in gaps. Guinée to Cameroun; (Guinea wide).

MITRAGYNA CILIATA Aubrév. & Pellegr., FWTA 2:161. Rubiaceae. Akan: subaha.
Medium to large tree; lvs macrophyll. Fl white, small, crowded into heads, not clearly seasonal. Fr dry capsule; seeds small, slightly winged, wind-dispersed. Grows gregariously in freshwater swamps, to which it is adapted by its characteristic 'knee-root' pneumatophores (68). Liberia to Zaire; (Guineo-Congolian wide). Class 2 timber; trade name: abura (40). Not recorded in survey; for distribution in Ghana as calculated from timber enumeration records, see Fig. 11.4. *M. stipulosa* (DC.) O. Ktze. is a closely similar species which has been recorded from swamp forest near the forest-savanna boundary.

390 MONANTHOTAXIS WHYTEI (Stapf) Verdc. (109), (*Popowia whytei* Stapf, FWTA 1:44). Annonaceae.
Large woody climber, or scrambling shrub in thicket; stem and xylem grooved; branchlets tendriliform; lvs notophyll. Fl yellowish, small to medium, bowl, cauliflorous. Fr orange berry; seeds medium. Regenerates in shade. Liberia to Ghana; (Upper Guinea endemic). The other species of *Monanthotaxis* recorded from the Ghana forest zone are rare, and several cannot be satisfactorily identified.

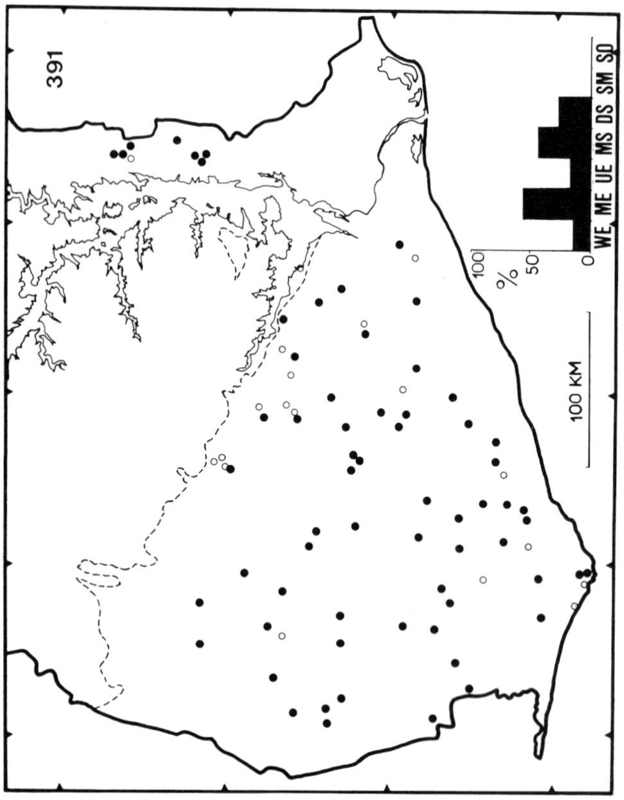
391
100 KM
%
100
50
0
WE ME UE MS DS SM SO

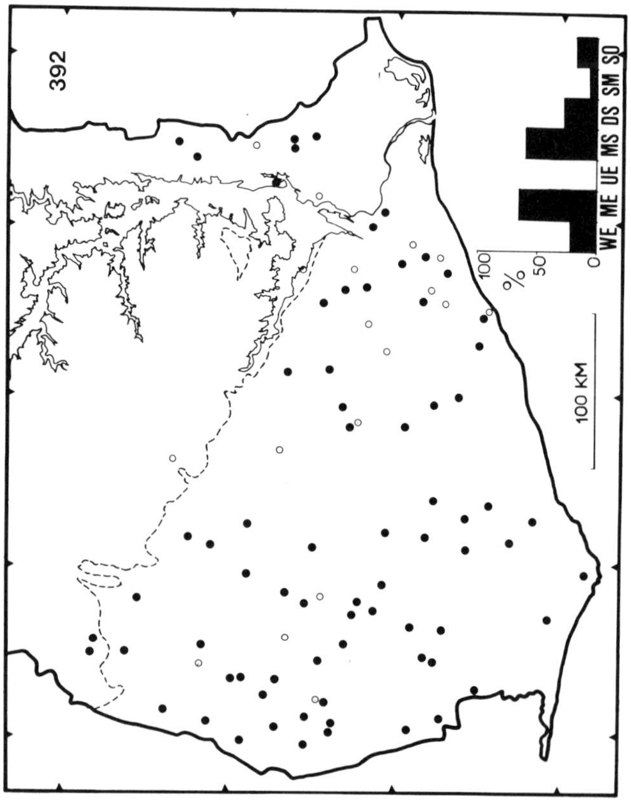
392
100 KM
%
100
50
0
WE ME UE MS DS SM SO

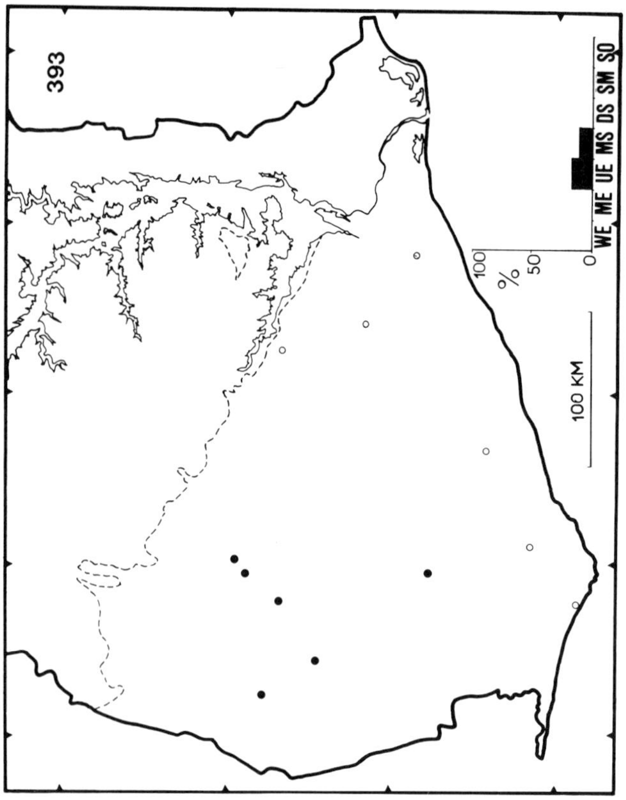
393
100 KM
%
100
50
0
WE ME UE MS DS SM SO

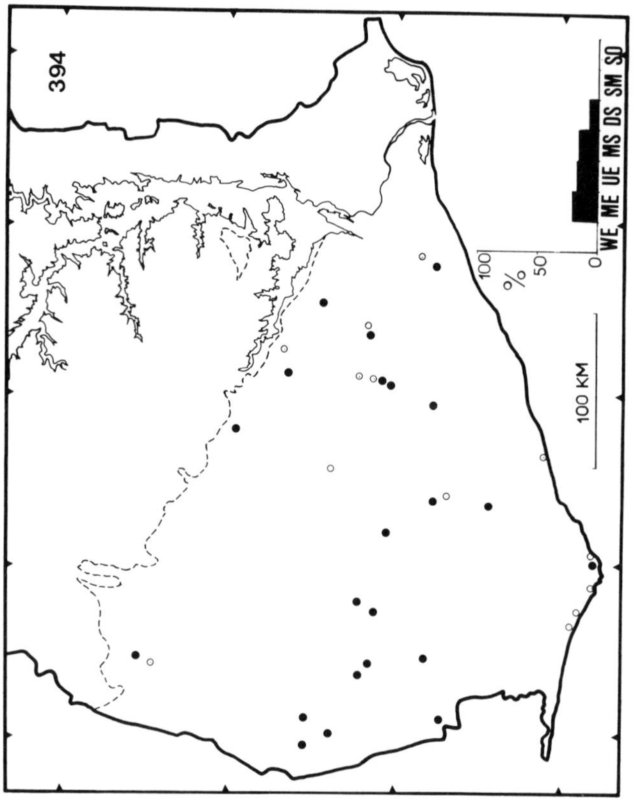
394
100 KM
%
100
50
0
WE ME UE MS DS SM SO

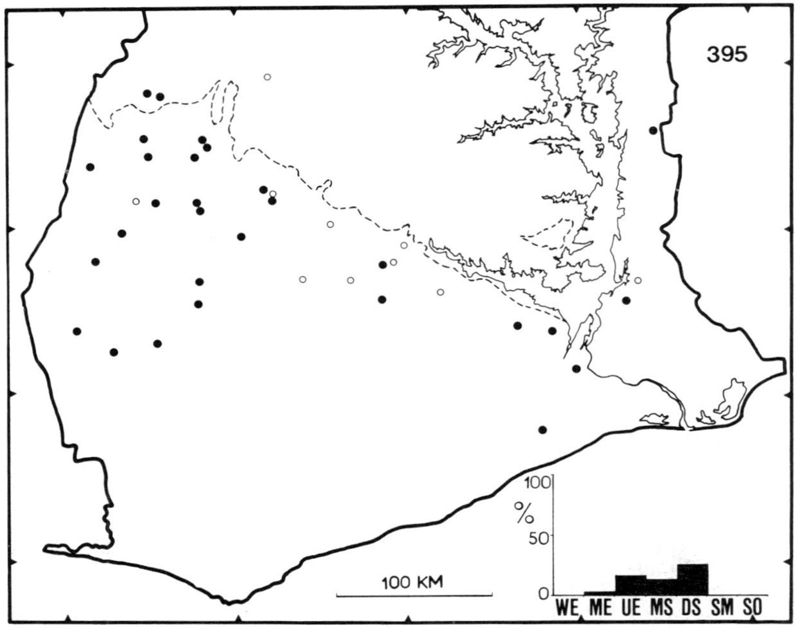

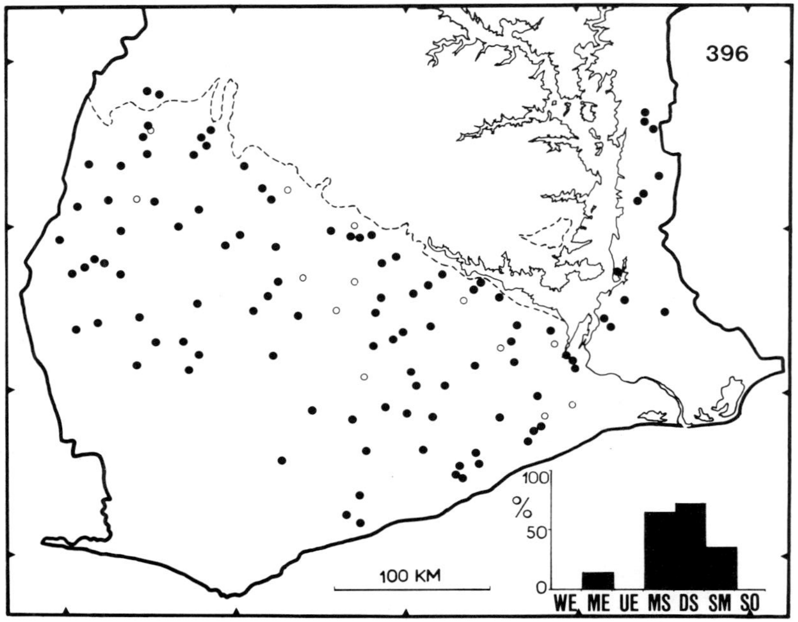

391 MONODORA MYRISTICA (Gaertn.) Dunal, FWTA 1:54. Annonaceae. Akan: wedeaba.

Medium-sized tree; slash aromatic; lvs macrophyll. Troll's model (59). Fl large, yellow and red, bowl with trap, hanging on long stalks, mainly 9–3, appearing with the young leaves. Fr large greenish berry; seeds very large, dispersed by elephants (5). Regenerates in shade. Liberia to Angola and Uganda; (Guineo-Congolian wide). Seeds widely sold in markets as a spice and ingredient of traditional medicines.

392 MONODORA TENUIFOLIA Benth., FWTA 1:54. Annonaceae. Akan: moto-kuradua.

Small tree; slash aromatic; lvs notophyll, deciduous. Troll's model. Fl large, mottled cream and brown, bowl with trap, mostly 12–4, appearing with the young leaves. Fr greenish berry; seeds large. Regenerates in shade. Extends to forest patches on rocky scarps in the savanna zone. Guinée to Cameroun; (Guinea wide).

393 MORINDA LONGIFLORA G. Don, FWTA 2:189. Rubiaceae.

Slender woody climber; lvs notophyll, rather thick. Fl white, fragrant, large, dish with tube, clustered in heads, probably heterostyled, mostly 12–4. Fr yellow berry, developing from the concrescent ovaries of all the flowers in a head; seeds large. Regenerates in shade; sapling leaves are thinner, and mesophyll; flowering and fruiting seen only in gaps and secondary forest. Guinée to Zaire; (Guineo-Congolian wide).

394 MORINDA MORINDOIDES (Bak.) Milne-Redh., FWTA 2:189. Rubiaceae.

Medium-sized woody climber, or scrambling shrub in thicket; lvs notophyll. Fl white, fragrant, large, dish with tube, clustered in heads, heterostyled; 12–4. Fr yellow berry, developing from the concrescent ovaries of all the flowers in a head; seeds large. Regenerates in shade; sapling leaves are thinner and mesophyll to macrophyll; flowering and fruiting seen mainly in gaps and secondary forest. Guinea Bissau to Angola and Sudan; (Guineo-Congolian wide).

395 MORUS MESOZYGIA Stapf, FWTA 1:594. Moraceae. Akan: wonton.

Large tree, containing white latex; lvs notophyll, toothed, deciduous. Fl greenish, small, brush, dioecious, probably wind-pollinated, 1–2 with young leaves. Fr cluster of small greenish drupes, 3–5; seed medium. Regenerates mainly in small gaps. Senegal to Angola, Sudan and Natal (chorological transgressor, 22). Class 3 timber.

396 MOTANDRA GUINEENSIS (Thonn.) A. DC., FWTA 2:80. Apocynaceae. (Type from Akwapim).

Medium-sized woody climber, containing white latex; wood normal. Lvs microphyll to notophyll, with tufts of brown hair in nerve axils on undersurface. Fl white, medium, dish with tube, 3–5. Fr follicle, densely brown-tomentose, 12. Seeds medium to large, wind-dispersed with aid of parachute of silky hairs. Regenerates in shade. Plot samples very significantly associated with base-rich soils of relatively high pH, and with soils of low clay content. Guinée to Angola and Uganda; (Guineo-Congolian wide).

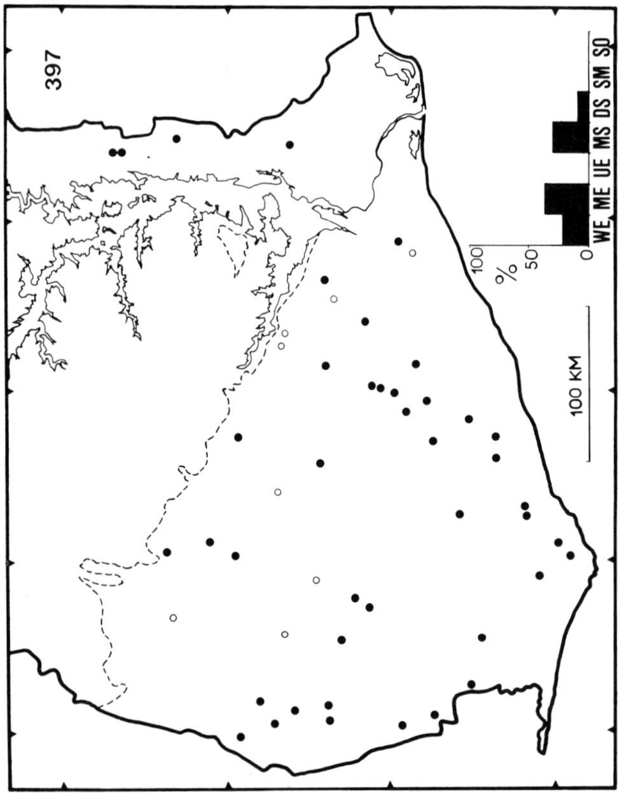
397
100 KM
100
%
50
0
WE ME UE MS DS SM SO

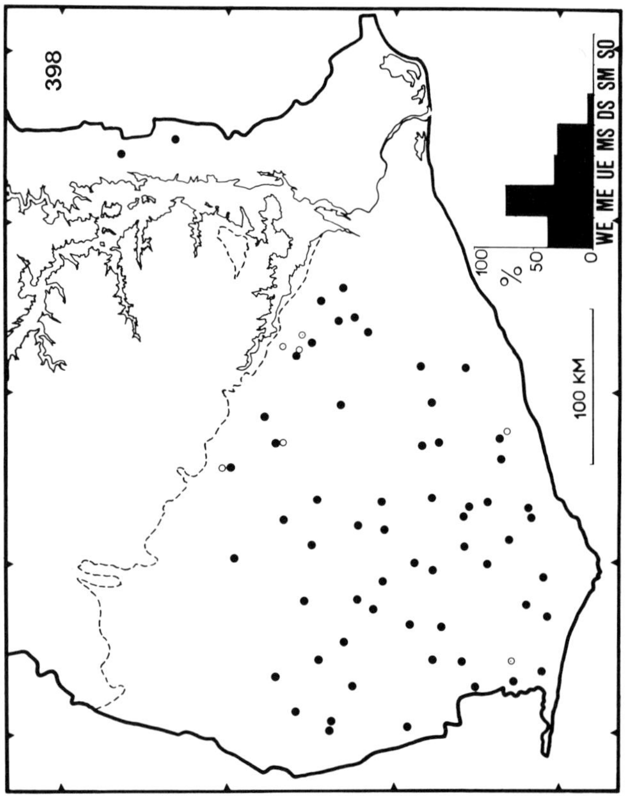
398
100 KM
100
%
50
0
WE ME UE MS DS SM SO

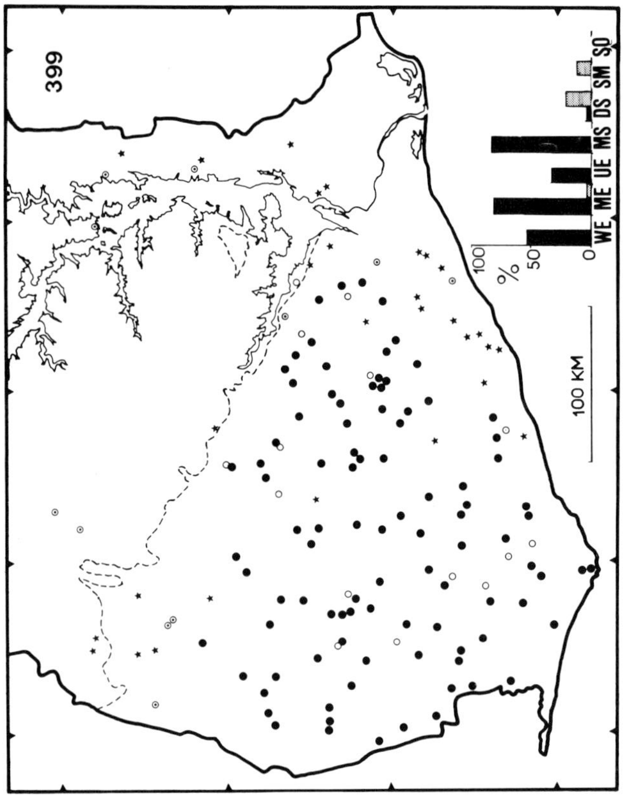
399
100 KM
100
%
50
0
WE ME UE MS DS SM SO

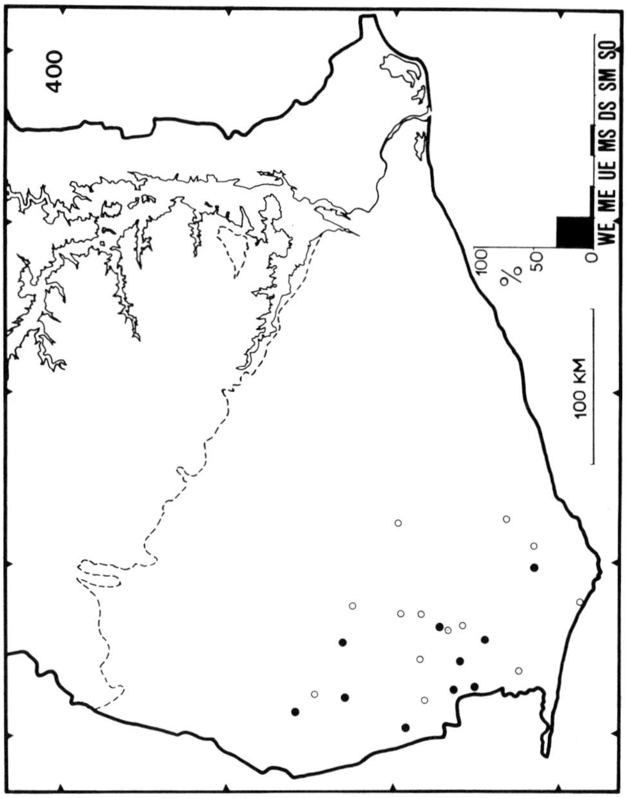
400
100 KM
100
%
50
0
WE ME UE MS DS SM SO

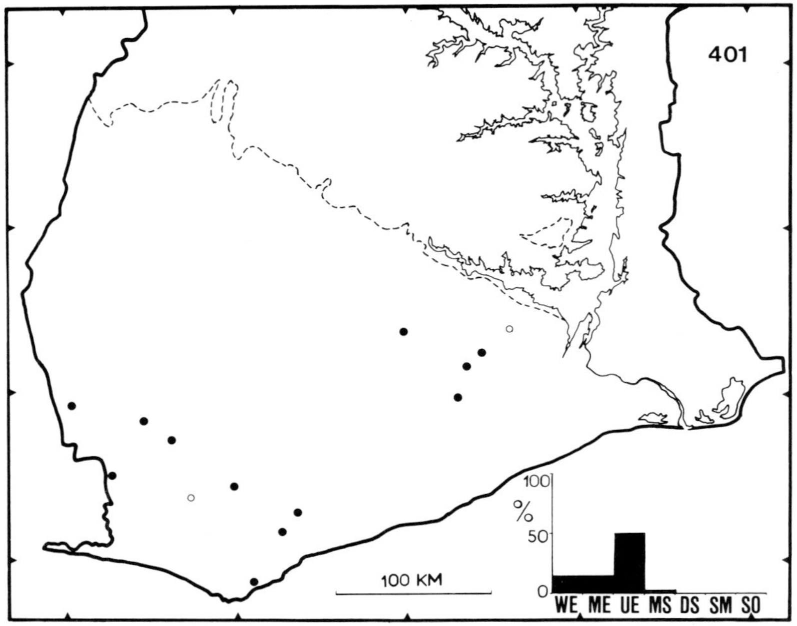

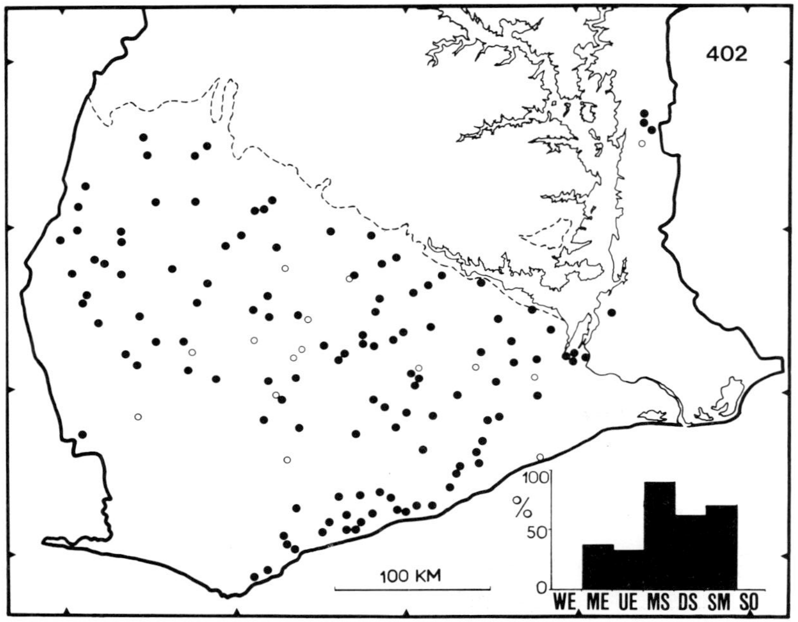

397 MYRIANTHUS ARBOREUS P. Beauv., FWTA 1:614. Moraceae (36). Akan: nyankuma.
Medium-sized tree, sometimes with stilt-roots. Lvs digitate, leaflets macrophyll, toothed, covered below with white, short hairs. Fl small, brush, dioecious, male yellow, female green, 12–4. Fr compound, formed from a cluster of yellow, 1-seeded berries; seeds large. Regenerates in shade. Guinée to Angola and Tanzania; (Guineo-Congolian wide).

398 MYRIANTHUS LIBERICUS Rendle, FWTA 1:616. Moraceae (36). Akan: nyankumanini.
Medium-sized tree, occasionally with stilt roots; lvs mesophyll to macrophyll, toothed, white-pubescent below, sometimes 3-lobed. Fl small, brush, dioecious, yellowish, not clearly seasonal. Fr compound, formed from a cluster of brownish 1-seeded berries; seeds large. Regenerates in shade. Plot samples very significantly associated with base-poor, acid soils. Guinée to Ghana; (Upper Guinea endemic).

399 NAPOLEONAEA VOGELII Hook. & Planch. (80), (*Napoleona vogelii,* FWTA 1:244). Lecythidaceae. Akan: obua.
Small to medium trees. Massart's model (59). Two distinct forms occur in Ghana:
(1) Erect tree, leaves microphyll to notophyll, caudate-acuminate with tip of acumen spathulate, all nerves except the lowest clearly looping onto the nerve above; understorey of moist forest – survey records ●; herbarium records ○; histogram bars solid.
(2) Low-branching tree, leaves notophyll to mesophyll, sub-acuminate, lowest three nerves on each side running up within the margin but not clearly looping; banks of streams in dry forest, and extending to fringing forest in the savanna zone – survey records ★; herbarium records ⊙; histogram bars dotted. Fl cream streaked with purple, large, bowl, 10–2. Fr orange berry with thick pericarp; seeds large, surrounded by edible pulp. Guinée to Angola; (Guineo-Congolian wide).

NAUCLEA DIDERRICHII (De Wild. & Th. Dur.) Merrill, FWTA 2:164. Rubiaceae. Akan: kusia.
Large tree; lvs mesophyll. Fl pale yellow, small to medium, crowded into heads, not clearly seasonal. Fr fleshy compound berry, with very numerous small seeds. Regenerates only in gaps and along roadsides; occurs in high forest as scattered large individuals, and hence only once recorded in the survey. S. Leone to Mozambique; (chorological transgressor). Class 1 timber; trade name: opepe (40). Fig. 11.5 gives distribution in Ghana as calculated from timber enumeration records.

400 NEOSTENANTHERA GABONENSIS (Engl. & Diels) Exell, FWTA 1:42. Annonaceae.
Small tree; lvs notophyll, pale and appressed pubescent below. Fl greenish-yellow, large, bowl, not clearly seasonal. Fr yellow 1-seeded berry; seed large. Regenerates in shade. Liberia to Ghana, Gabon and Angola; (Guinea wide).

401 NEPHTHYTIS AFZELII Schott, FWTA 3:122. Araceae.
Rhizomatous ground-living herb; lvs macrophyll, arising from ground level. Fl small, monoecious, clustered on spike. Fr orange-yellow 1-seeded berry; seed medium. S. Leone to Ghana; (Upper Guinea endemic).

402 NESOGORDONIA PAPAVERIFERA (A. Chev.) R. Capuron, FWTA 1:313. Sterculiaceae. Akan: danta.
Large tree with buttresses; lvs notophyll. Fl medium, white, bowl, not clearly seasonal. Fr woody capsule; seeds medium, winged, wind-dispersed. Regenerates profusely in shade. Sometimes gregarious in Southern Marginal forest. Very significantly less frequent in survey plots with base-poor soils. S. Leone to Central African Republic; (Guinea wide). Class 2 timber; trade name: danta (40).

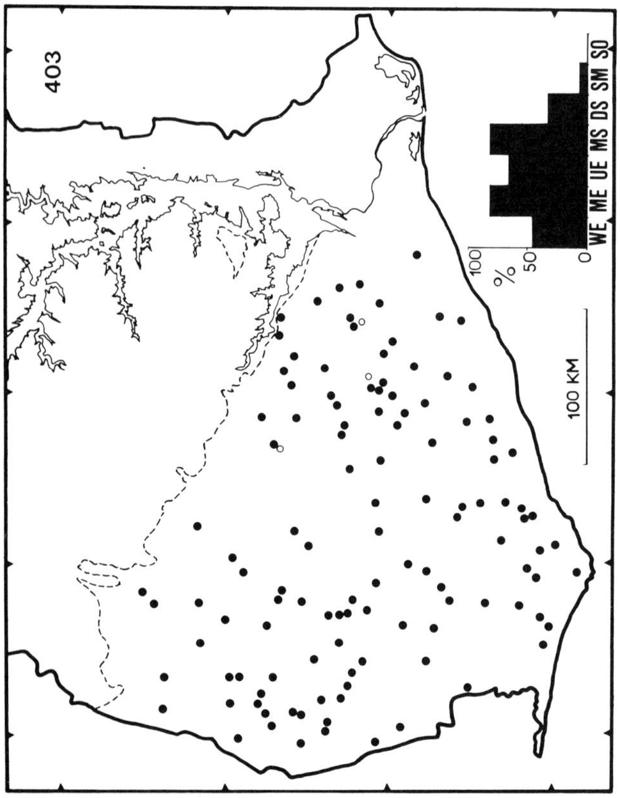
403
100 KM
100
%
50
0
WE ME UE MS DS SM SO

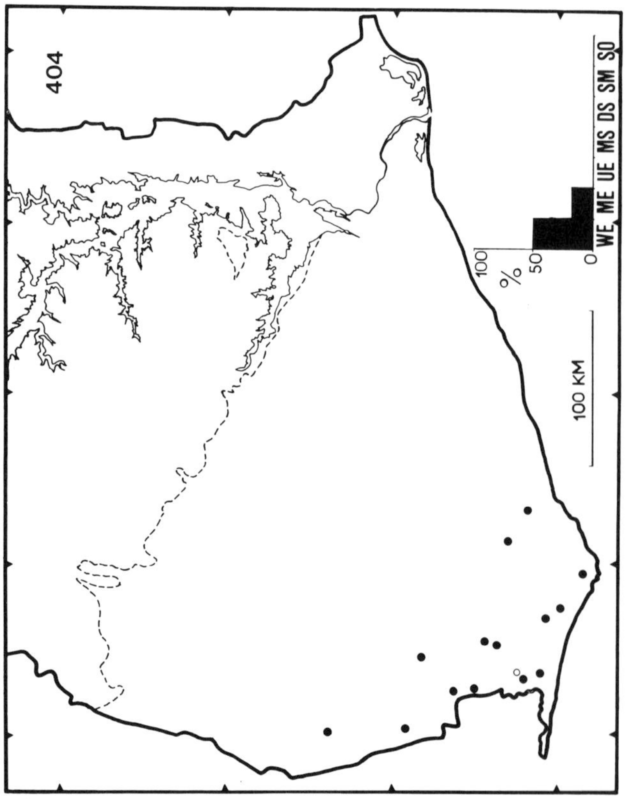
404
100 KM
100
%
50
0
WE ME UE MS DS SM SO

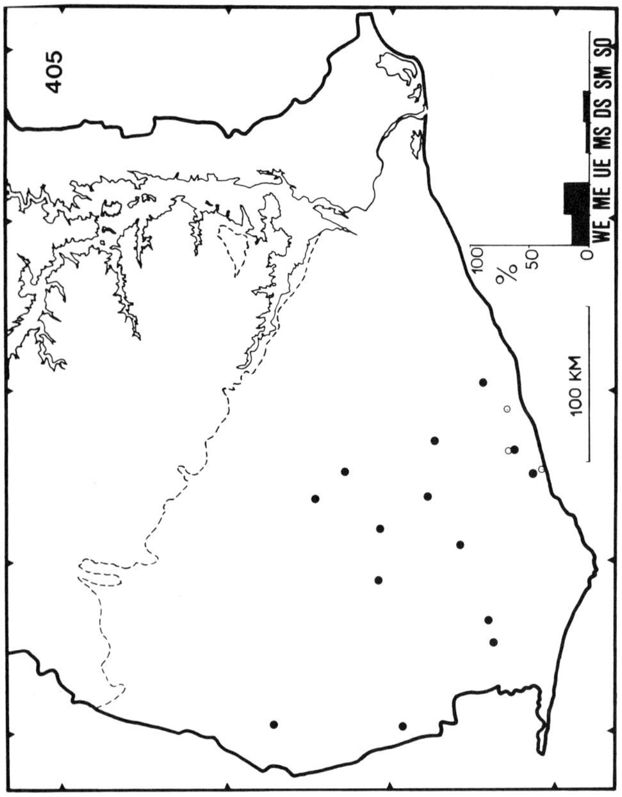
405
100 KM
100
%
50
0
WE ME UE MS DS SM SO

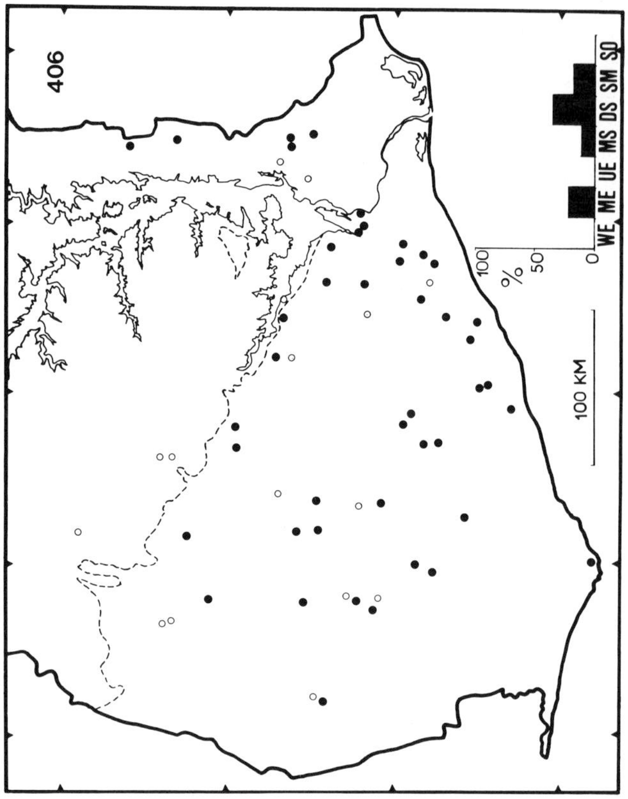
406
100 KM
100
%
50
0
WE ME UE MS DS SM SO

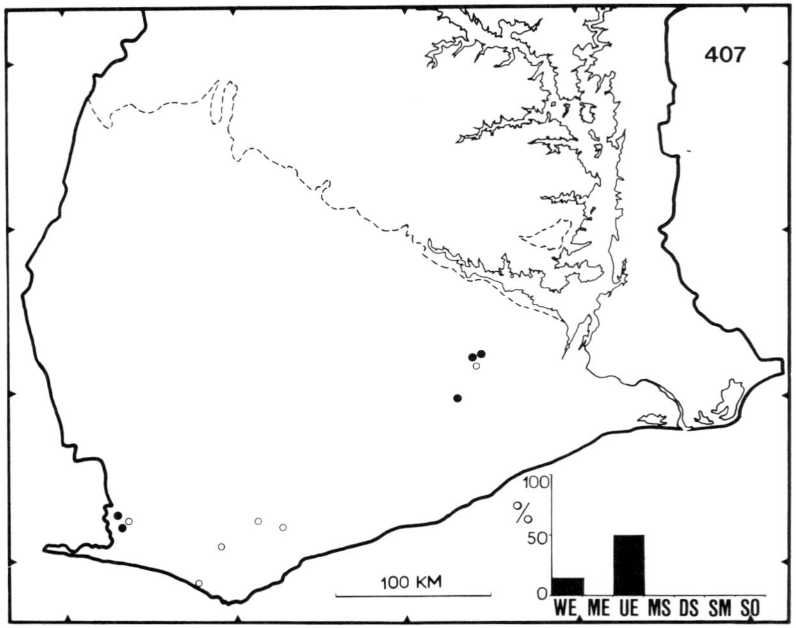

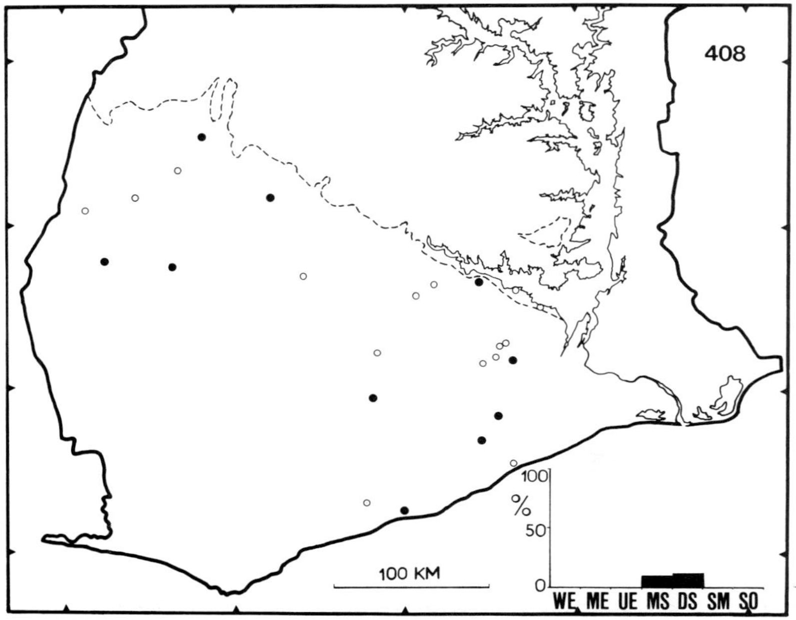

403 NEUROPELTIS ACUMINATA (P. Beauv.) Benth., FWTA 2:338. Convolvulaceae.

Large woody climber; stem slightly grooved or more or less cylindrical, but xylem deeply dissected by narrow radial bands of phloem; lvs notophyll. Fl white, small to medium, bowl, not clearly seasonal. Fr dry, 1-seeded, wind-dispersed with aid of wing-like bract (Fig. 6.1h); seed medium. Regenerates in shade. S. Leone to Angola; (Guineo-Congolian wide).

404 NEUROPELTIS PREVOSTEOIDES Mangenot, FWTA 2:338. Convolvulaceae.
Large woody climber; stem deeply grooved and strongly twisted into complex spirals; xylem as in *N. acuminata;* lvs microphyll to notophyll. Fl white, medium, funnel, not yet collected in Ghana. Fr dry, 1-seeded, wind-dispersed with aid of wing-like bract; seed medium. Regenerates in shade. Liberia to Ghana; (Upper Guinea endemic).

405 NEUROPELTIS VELUTINA Hallier f., FWTA 2:338. Convolvulaceae.
Slender woody climber; twigs and leaves hairy; lvs microphyll to notophyll. Fl white, small to medium, bowl. Fr dry, 1-seeded, wind-dispersed with aid of wing-like bract; seed medium. Flowers and fruits not yet collected in Ghana. Regenerates in shade. S. Leone to Congo Republic; (Guinea wide).

406 NEWBOULDIA LAEVIS (P. Beauv.) Seemann ex Bureau, FWTA 2:388. Bignoniaceae. Akan: sesemasa.

Medium-sized tree; lvs pinnate, leaflets mesophyll, toothed. Scarrone's model. Fl large, purple, bell, mostly 11–2. Fr long capsule, 11–3; seeds large, winged, wind-dispersed. Left standing in farms when other trees are cut down; commonly planted in villages. Senegal to Cameroun; (Guinea wide).

407 NEWTONIA DUPARQUETIANA (Baill.) Keay, FWTA 1:489. Mimosaceae. Akan: adadaba.

Medium to large tree with buttresses. Lvs bipinnate, leaflets microphyll to notophyll. Troll's model. Fl small, greenish-white, clustered in dense spikes, attractive to bees. Fr pod dehiscing along one suture; seeds very large, flattened and narrowly winged, wind-dispersed. Regenerates in shade. S. Leone to Gabon; (Guinea wide).

408 OCHNA STAUDTII Engl. & Gilg (17). (*O. kibbiensis* Hutch. & Dalz., FWTA 1:223). Ochnaceae.

Pygmy to small tree; lvs mesophyll, toothed. Roux's model. Fl yellow-green, medium, bowl, calyx enlarging and turning red in fruit. Fr black drupelet, 12–6; seed medium. Regenerates in shade. Iv. Coast to Zaire; (Guineo-Congolian wide).

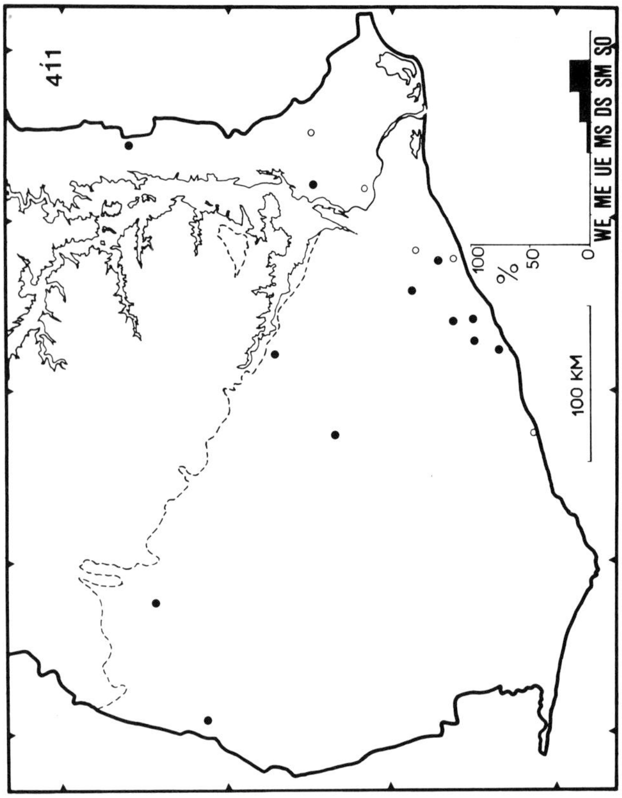
411
100 KM
100
50
0
%
WE ME UE MS DS SM SO

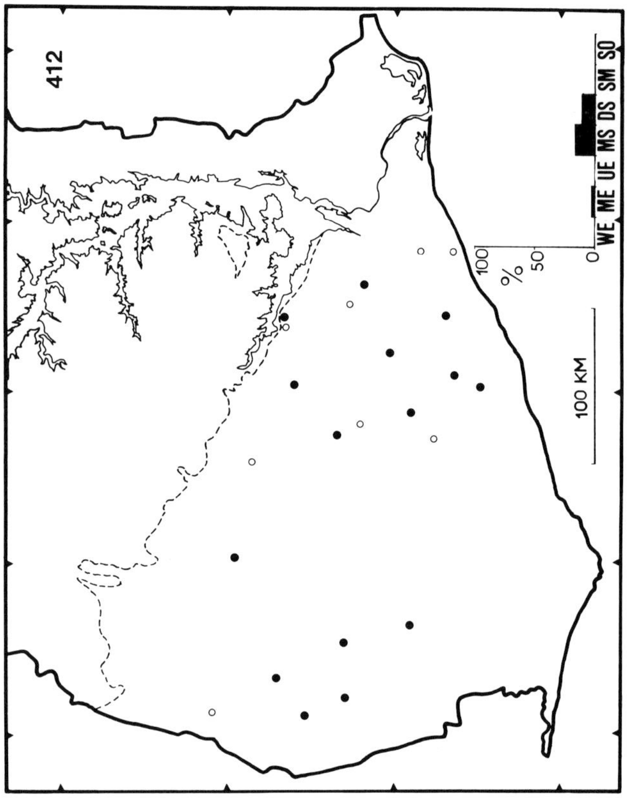
412
100 KM
100
50
0
%
WE ME UE MS DS SM SO

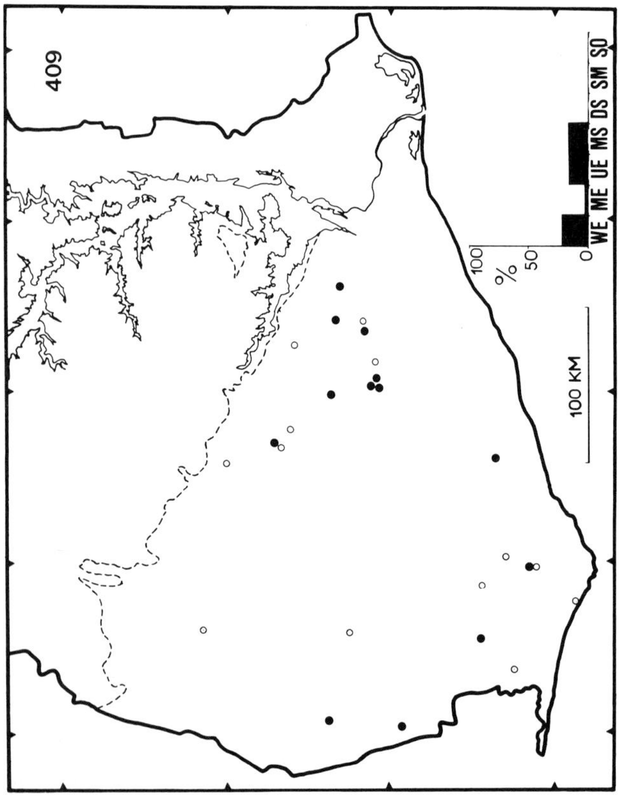
409
100 KM
100
50
0
%
WE ME UE MS DS SM SO

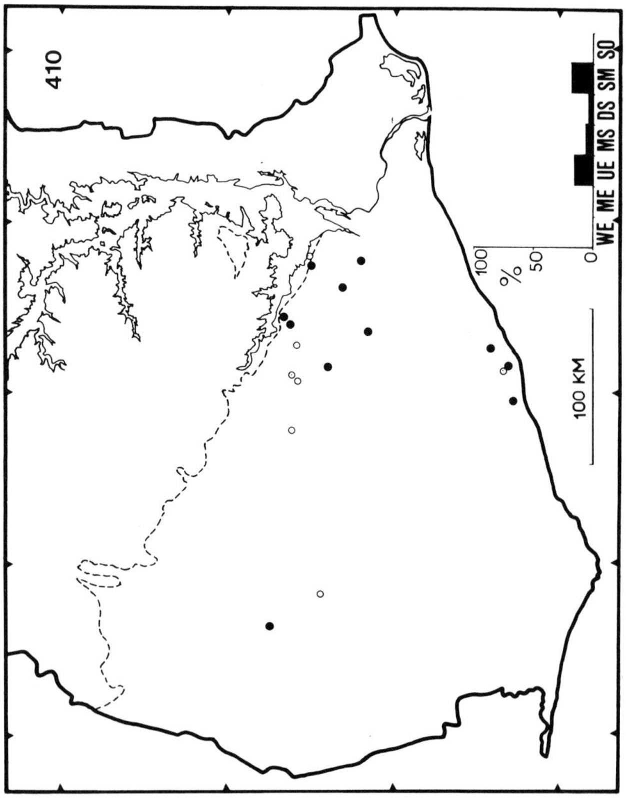
410
100 KM
100
50
0
%
WE ME UE MS DS SM SO

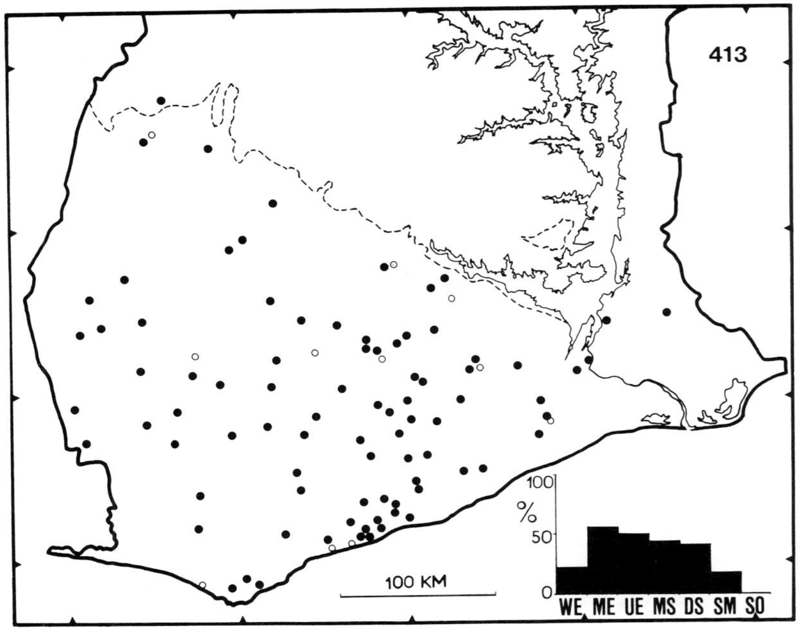

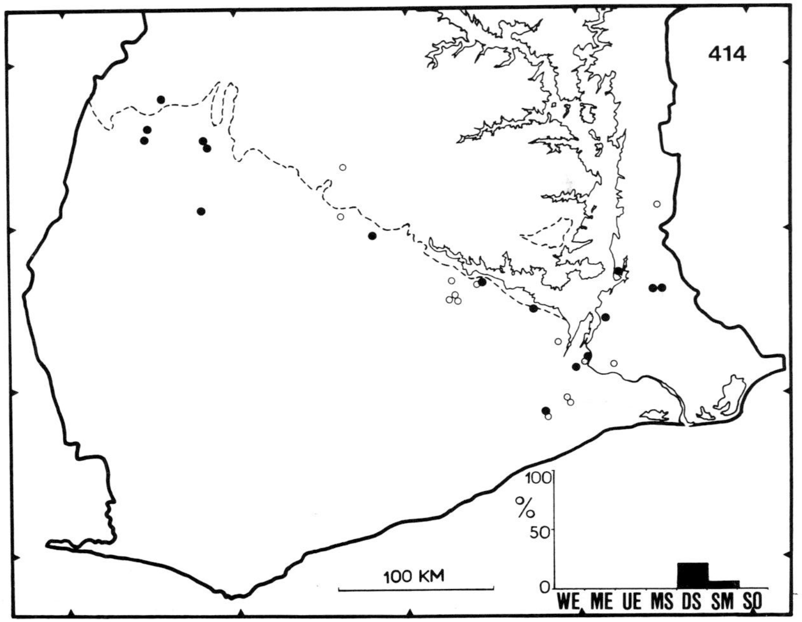

409 OCTOKNEMA BOREALIS Hutch. & Dalz., FWTA 1:656. Olacaceae (73, 123), (Octoknemataceae in FWTA). Akan: wisuboni.
Medium-sized tree; lvs mesophyll, pubescent especially on main nerves below. Roux's model (59). Fl small, yellow-green, clustered in spikes, dioecious, 1–5. Fr red drupe; seed large. Regenerates in shade. Guinée to Ghana; (Upper Guinea endemic).

410 OCTOLOBUS SPECTABILIS Welw. (60), (*O. angustatus* Hutch., FWTA 1:319). Sterculiaceae. Akan: afinafi.
Small tree; lvs notophyll. Fl large, yellow, bell, andromonoecious, 11–3. Fr red berry, 2–6; seeds black, large. Regenerates in shade. S. Leone to Angola; (Guineo-Congolian wide).

411 OECEOCLADES MACULATA (Lindl.) Lindl. (47), (*Eulophidium maculatum* (Lindl.) Pfitz., FWTA 3:243). Orchidaceae.
Perennial ground herb; lvs mesophyll, fleshy, mottled. Tomlinson's model. Fl white with purple streaks, medium, bowl, not clearly seasonal. Fr capsule; seeds dust-like, wind-dispersed. Regenerates in shade. Senegal to Tanzania, and also in Brazil; (chorological transgressor).

412 OECEOCLADES SAUNDERSIANA (Rchb. f.) Garay & Taylor (47), (*Eulophidium saundersianum* (Rchb. f.) Summerh., FWTA 3:243). Orchidaceae.
Perennial ground herb; lvs mesophyll. Tomlinson's model. Fl yellow-green with purple markings, medium, bowl, mostly 12–1. Fr capsule; seeds dust-like, wind-dispersed. Regenerates in shade. S. Leone to Angola and Tanzania; (Guineo-Congolian wide).

413 OLAX GAMBECOLA Baill., FWTA 1:647. Olacaceae.
Pygmy tree; lvs notophyll. Fl small, white, inconspicuous, not clearly seasonal. Fr red drupe; seed medium. Regenerates in shade. Guinée to Angola and Zambia; (Guineo-Congolian wide).

414 OLAX SUBSCORPIOIDEA Oliv., FWTA 1:647. Olacaceae. Akan: ahoohenedua.
Shrub or small tree; cut bark smelling of garlic; lvs microphyll. Fl medium, greenish-white, bell, 11–2. Fr yellow-orange drupe, 2–5; seed medium. Regenerates in shade. Extends to fringing forest in the savanna zone. Senegal to Zaire; (Guineo-Congolian wide).

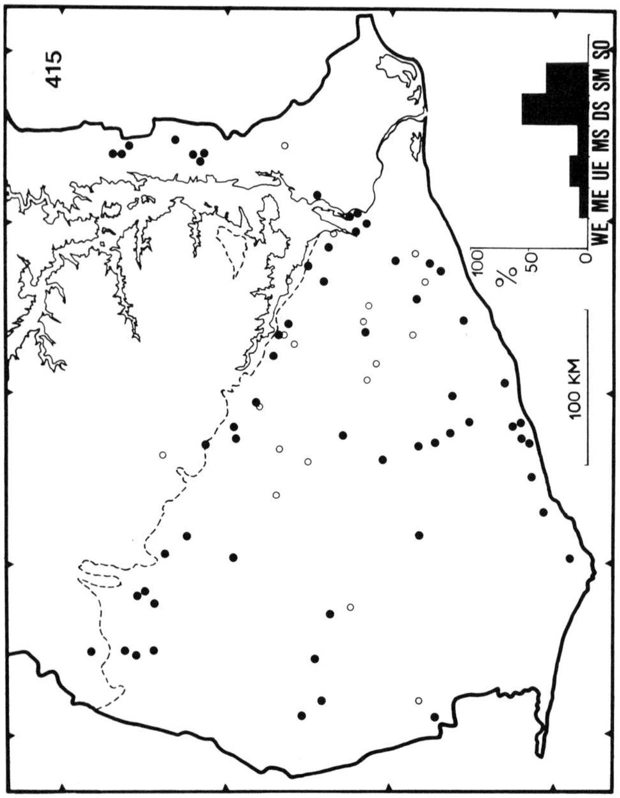
415
100 KM
%
100
50
0
WE ME UE MS DS SM SO

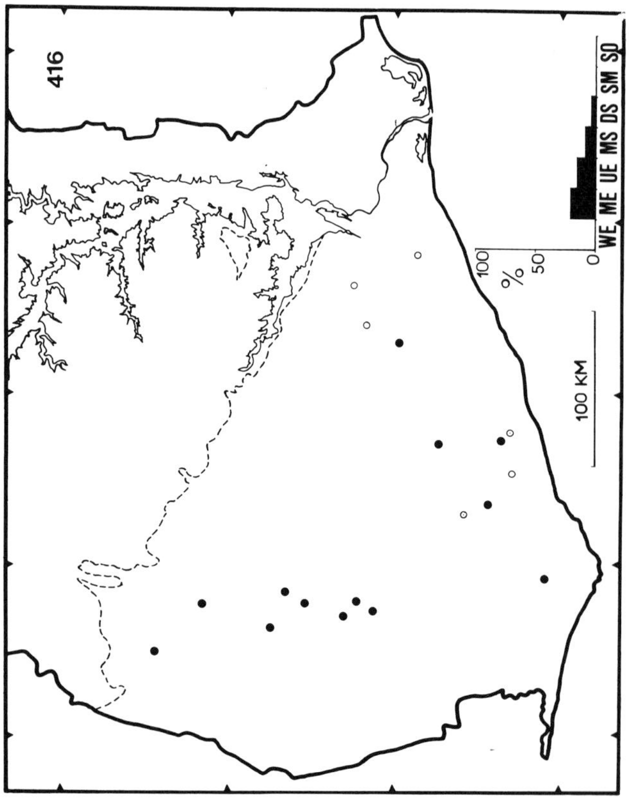
416
100 KM
%
100
50
0
WE ME UE MS DS SM SO

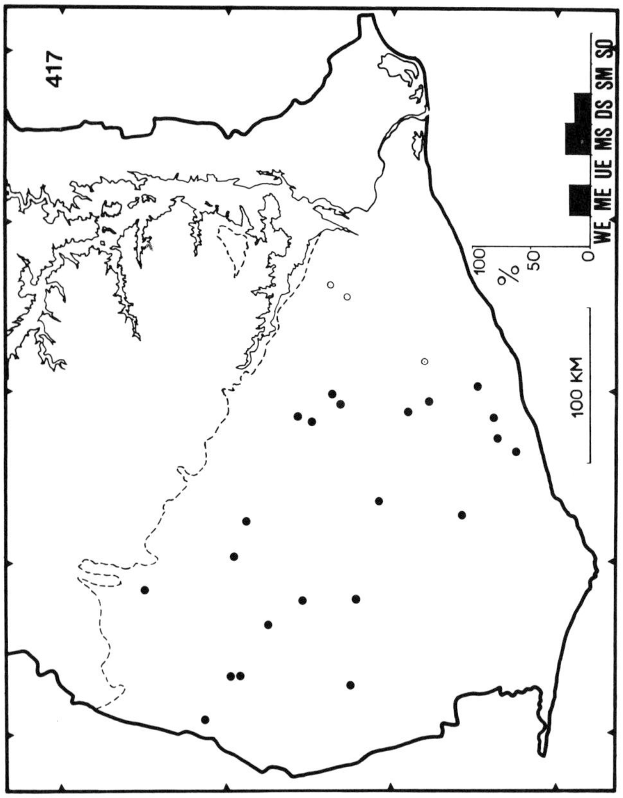
417
100 KM
%
100
50
0
WE ME UE MS DS SM SO

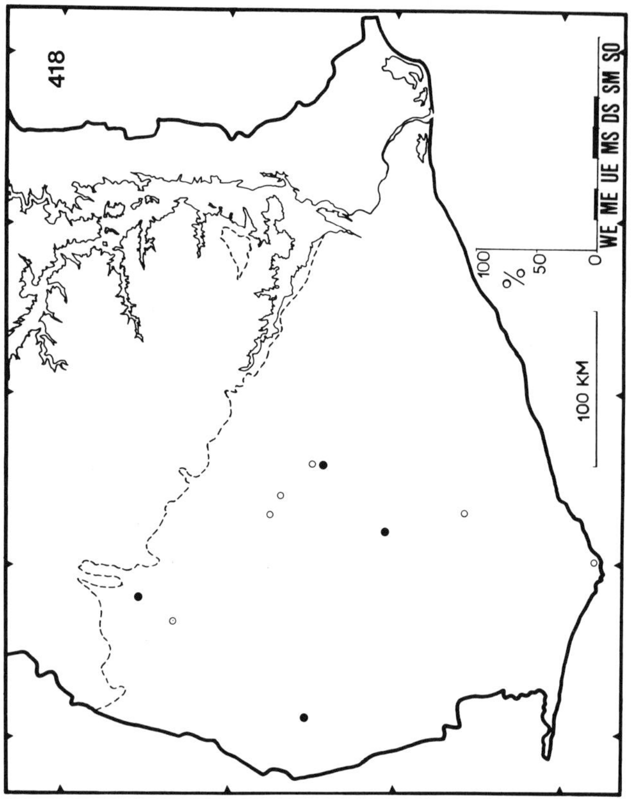
418
100 KM
%
100
50
0
WE ME UE MS DS SM SO

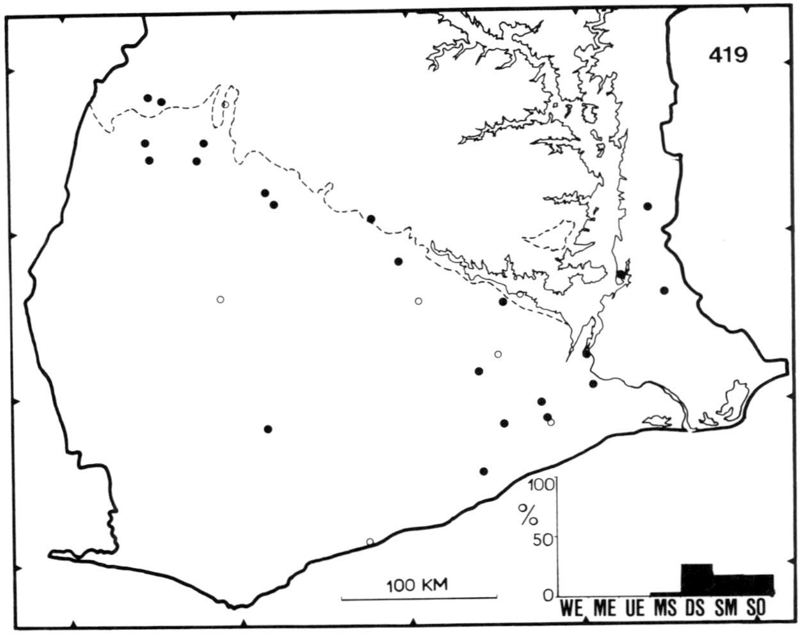

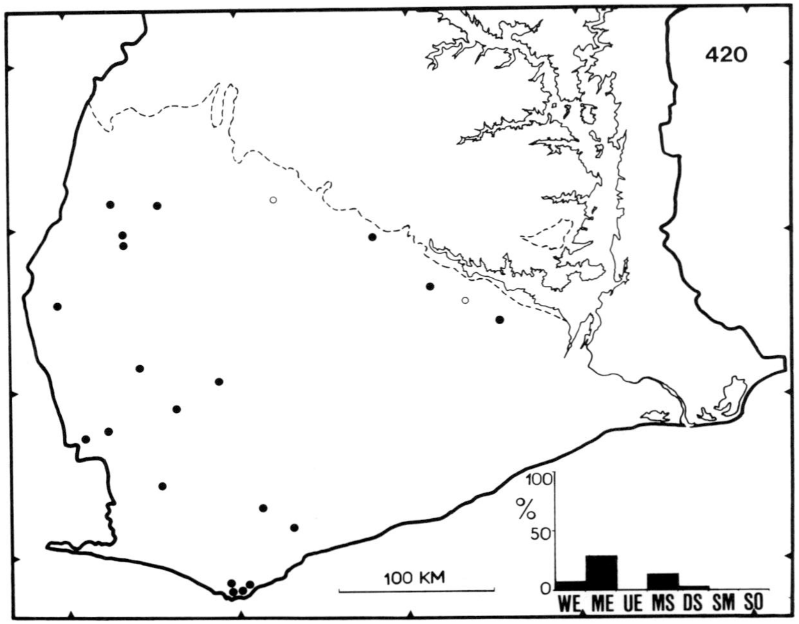

415 OLYRA LATIFOLIA L., FWTA 3:362. Gramineae.
Tall perennial broad-leaved grass; lvs mesophyll. Tomlinson's model. Spikelets monoecious, medium, not clearly seasonal. Seed medium. Most abundant along path margins and under rather open canopy. Throughout tropical Africa and tropical America; (chorological transgressor).

416 ONCINOTIS GLABRATA (Baill.) Stapf ex Hiern, FWTA 2:80. Apocynaceae.
Large woody climber containing white latex; lvs notophyll, with glands on petiole. Fl medium, yellow, clove-scented at night, dish with tube, 3–5. Fr follicle; seeds large, wind-dispersed with aid of tuft of long hairs. Regenerates in shade. Guinée to Angola; (Guineo-Congolian wide).

417 ONCINOTIS PONTYI Dubard (3), (*Alafia lucida* of FWTA 2:73, partly (Baldwin 14023), not of Stapf). Apocynaceae.
Large woody climber containing white latex; stem and xylem sometimes with broad ridges. Lvs notophyll, with glands on petiole. Fl medium, cream, dish with tube. Fr long follicle; seeds large, wind-dispersed with aid of tuft of long hairs. Regenerates in shade. Liberia to Ghana and Zaire to Uganda; (Guineo-Congolian wide).

418 ONGOKEA GORE (Hua) Pierre, FWTA 1:649. Olacaceae. Akan: bodwe.
Large tree; lvs microphyll. Fl small, white, inconspicuous, dioecious, 4–5. Fr yellow drupe, almost covered by the enlarged calyx, 4–6; seed large to very large, edible. S. Leone to Angola; (Guineo-Congolian wide).

419 OPLISMENUS HIRTELLUS (L.) P. Beauv., FWTA 3:437. Gramineae.
Perennial decumbent grass; lvs microphyll. Spikelets small, dispersed by adhesion of sticky awns. Seed small. Most frequent along paths, and where the canopy is rather open. widely distributed in the Old World and New World tropics and subtropics; (ecological and chorological transgressor).

420 OSTRYOCARPUS RIPARIUS Hook. f., FWTA 1:519. Papilionaceae.
Large woody climber; lvs pinnate, leaflets notophyll. Troll's model. Fl medium, yellow-white, flag. Fr 1–2-seeded indehiscent pod; seeds very large. Flowers and fruits not yet collected in Ghana. Regenerates in shade. Although said in floras to be riparian, it is clearly not restricted to river banks but, as in the case of many other large forest climbers, is most readily collected in such situations. Guinea Bissau to Zaire; (Guineo-Congolian wide).

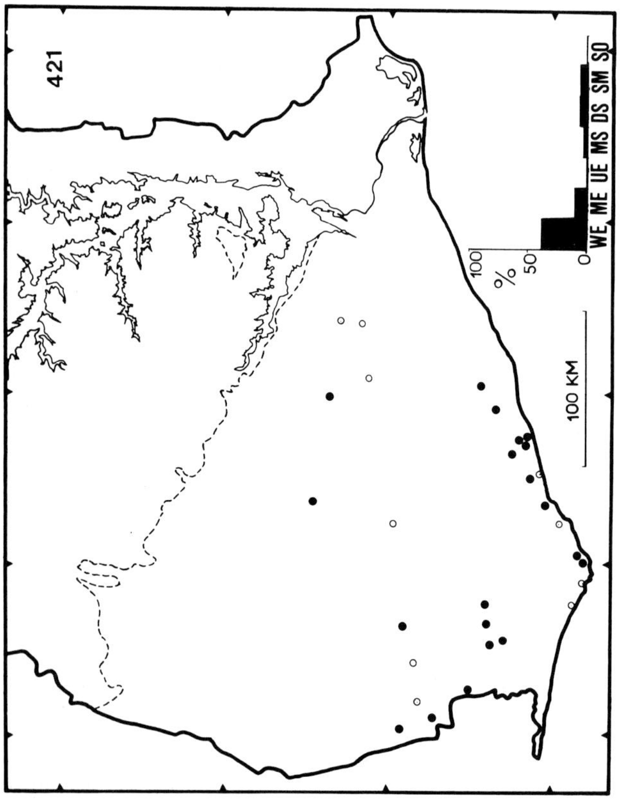
421
100 KM
100
%
50
0
WE ME UE MS DS SM SO

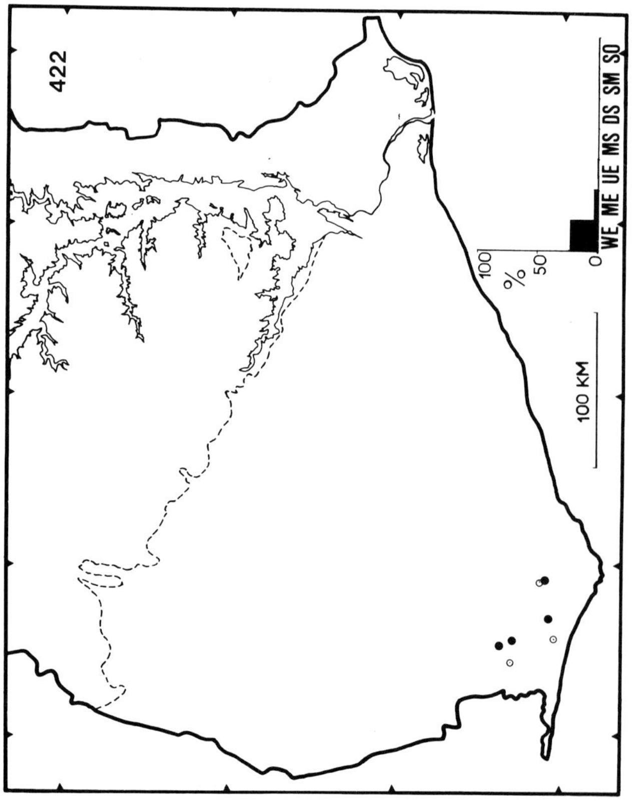
422
100 KM
100
%
50
0
WE ME UE MS DS SM SO

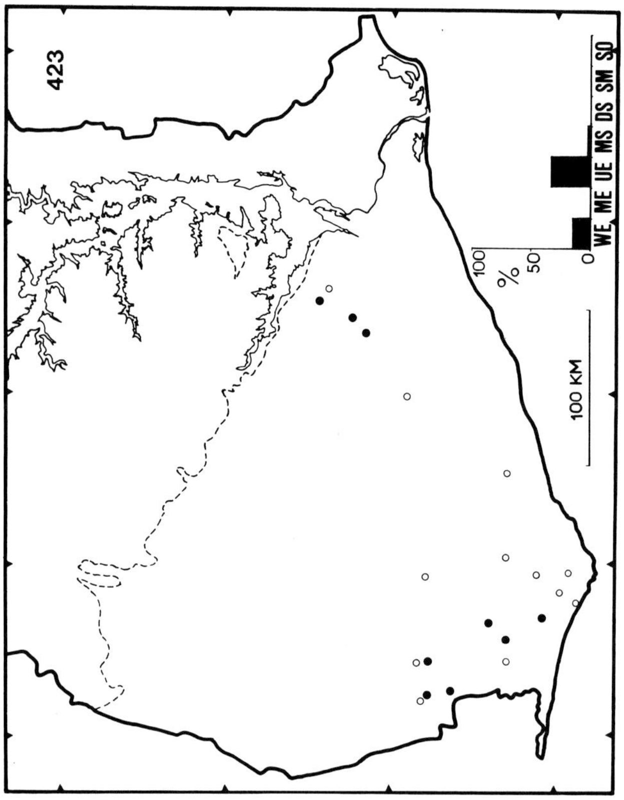
423
100 KM
100
%
50
0
WE ME UE MS DS SM SO

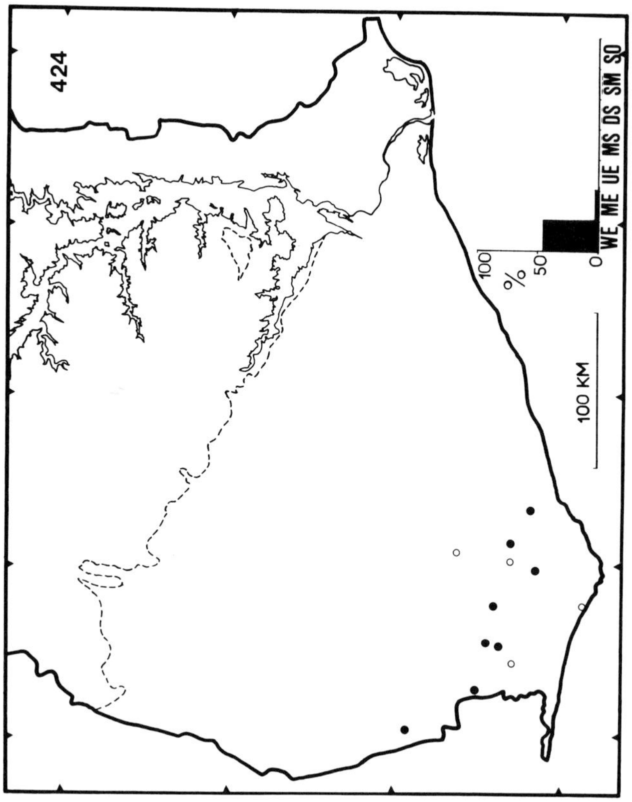
424
100 KM
100
%
50
0
WE ME UE MS DS SM SO

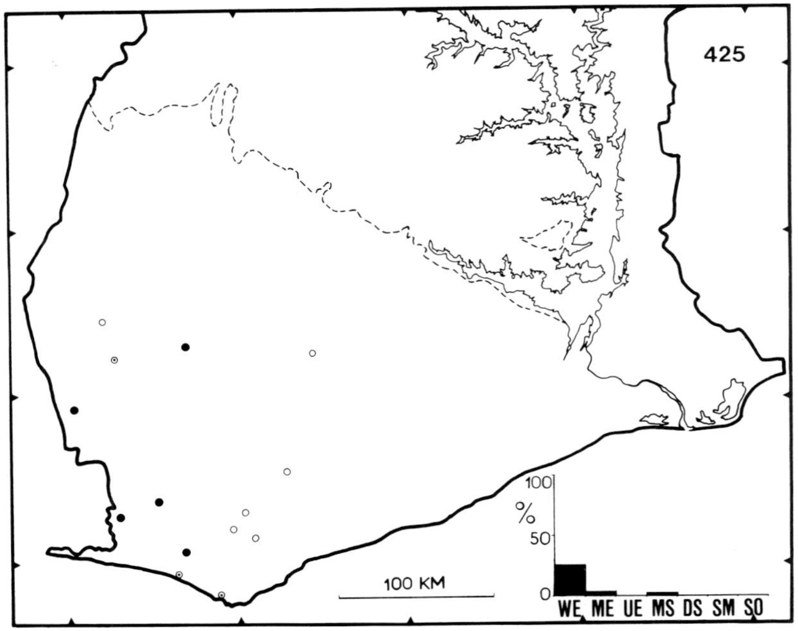

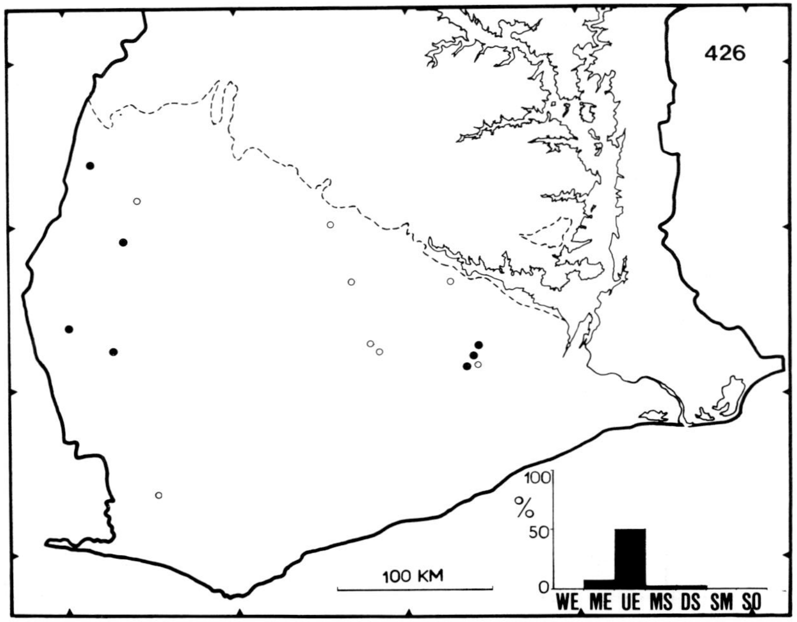

421 OURATEA AFFINIS (Hook. f.) Engl., FWTA 1:229. (Incl. *O. myrioneura* Gilg, FWTA 1:229). Ochnaceae.

Shrub or small tree; lvs notophyll, toothed. Roux's model (59). Fl yellow, medium, bowl, not clearly seasonal; calyx and receptacle red in fruit. Fr black drupelet; seed medium. Regenerates in shade. S. Leone to Cameroun; (Guinea wide).

422 OURATEA AMPLECTENS (Stapf) Engl., FWTA 1:228. Ochnaceae.

Small tree; lvs macrophyll, toothed. Fl yellow, medium, bowl, 11–1; calyx and receptacle red in fruit. Fr black drupelet; seed medium. Regenerates in shade. Liberia to Ghana; (Upper Guinea endemic).

423 OURATEA CALOPHYLLA (Hook. f.) Engl., FWTA 1:229. Ochnaceae. Akan: opunini.

Small tree; lvs mesophyll. Roux's model. Like other species of *Ouratea*, has extraordinarily thin bark. Fl yellow, medium, bowl, 11–3; calyx and receptacle red in fruit. Fr black drupelet; seed medium. Regenerates in shade. S. Leone to Cameroun; (Guinea wide).

424 OURATEA DUPARQUETIANA (Baill.) Gilg, FWTA 1:226.

Unbranched pygmy tree, with terminal cluster of macrophyll, toothed leaves. Corner's model (59). Fl yellow, medium, bowl, not clearly seasonal; calyx and receptacle red in fruit. Fr black drupelet; seed medium. Regenerates in shade. Liberia to Ghana and Gabon; (Guinea wide).

425 OURATEA RETICULATA (P. Beauv.) Engl. var. *reticulata*, FWTA 1:228, and var. *turnerae* (Hook. f.) J. B. Hall (50), (*O. turnerae* (Hook. f.) Hutch. & Dalz., FWTA 1:228). Ochnaceae. Akan: ansedua.

Shrubs or small trees; lvs mesophyll, toothed. Fl yellow, medium, bowl, not clearly seasonal; calyx and receptacle red in fruit. Fr black drupelet; seed medium. Regenerate in shade. Var. *turnerae* differs from var. *reticulata* mainly in its thicker leaves, which turn brown on drying; the varieties were not distinguished during the survey, and plot records are therefore not mapped separately. Herbarium records: var. *reticulata* ○; var. *turnerae* ⊙. Distribution: var. *reticulata* – Guinée to Zaire; var. *turnerae* – Guinée to Zaire. Species is a Guineo-Congolian wide.

426 OURATEA SULCATA (Van Tiegh.) Keay, FWTA 1:229. Ochnaceae.

Pygmy to small tree; lvs notophyll to mesophyll, toothed. Fl yellow, medium, bowl, 2–7; calyx and receptacle red in fruit. Fr black drupelet; seed medium. Regenerates in shade. Iv. Coast to Cameroun; (Guinea wide).

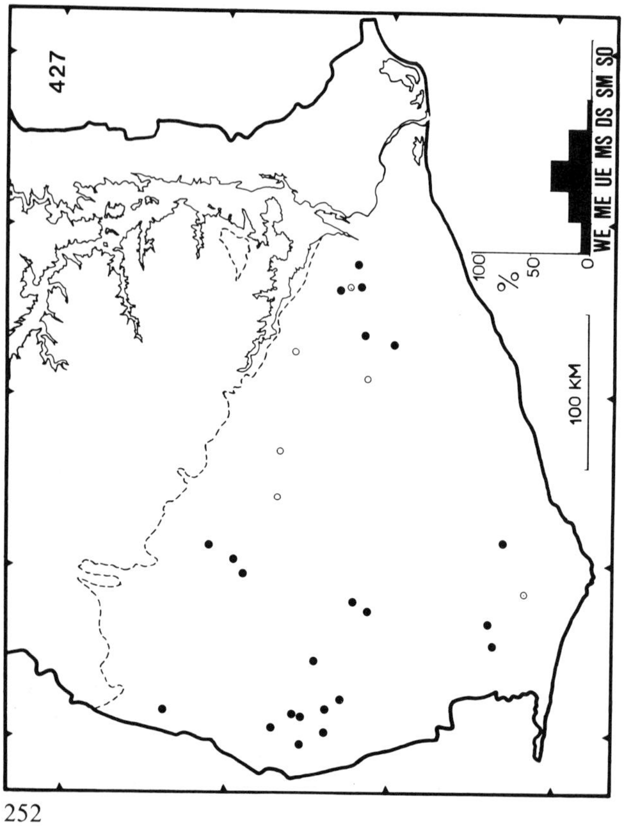
427
100 KM
100
%
50
0
WE ME UE MS DS SM SO

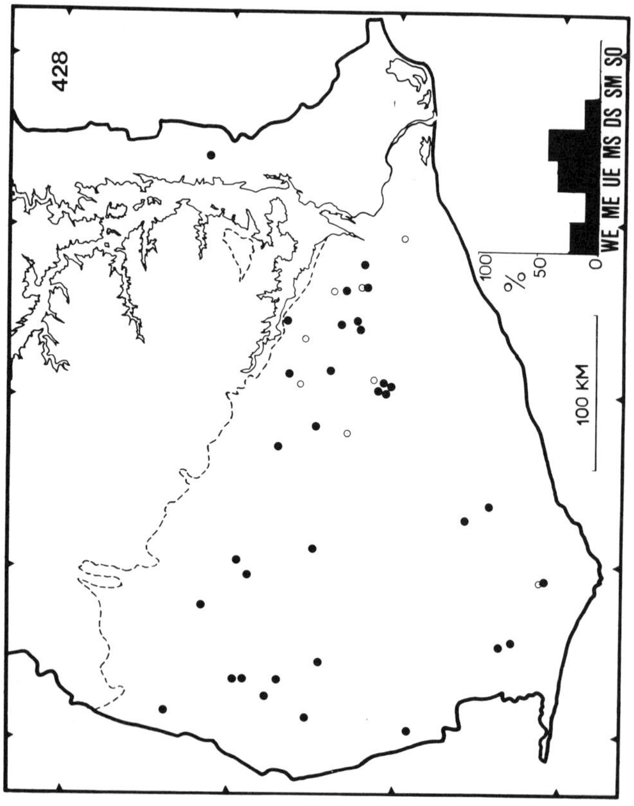
428
100 KM
100
%
50
0
WE ME UE MS DS SM SO

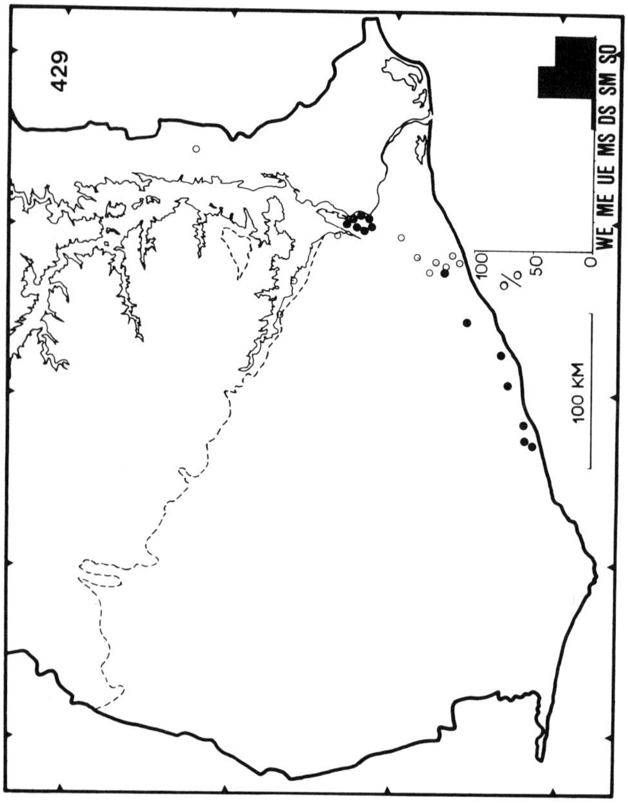
429
100 KM
100
%
50
0
WE ME UE MS DS SM SO

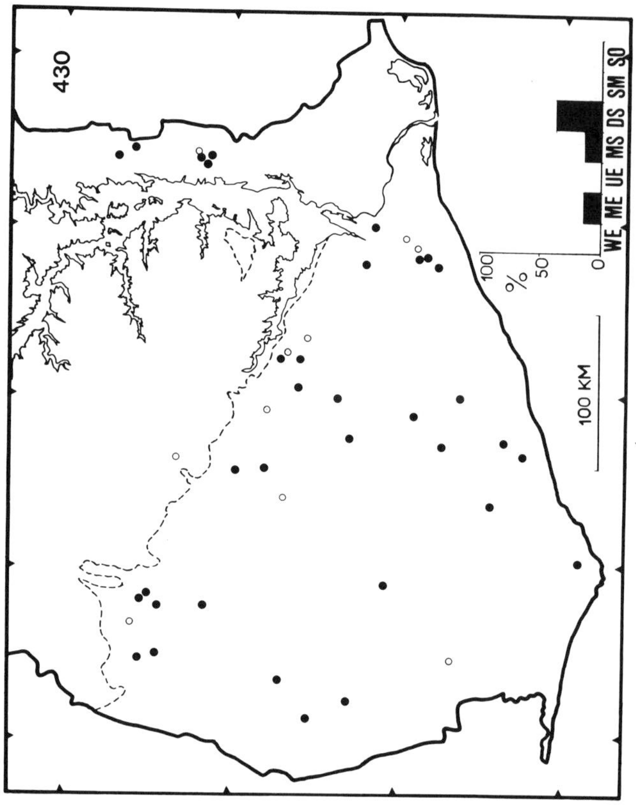
430
100 KM
100
%
50
0
WE ME UE MS DS SM SO

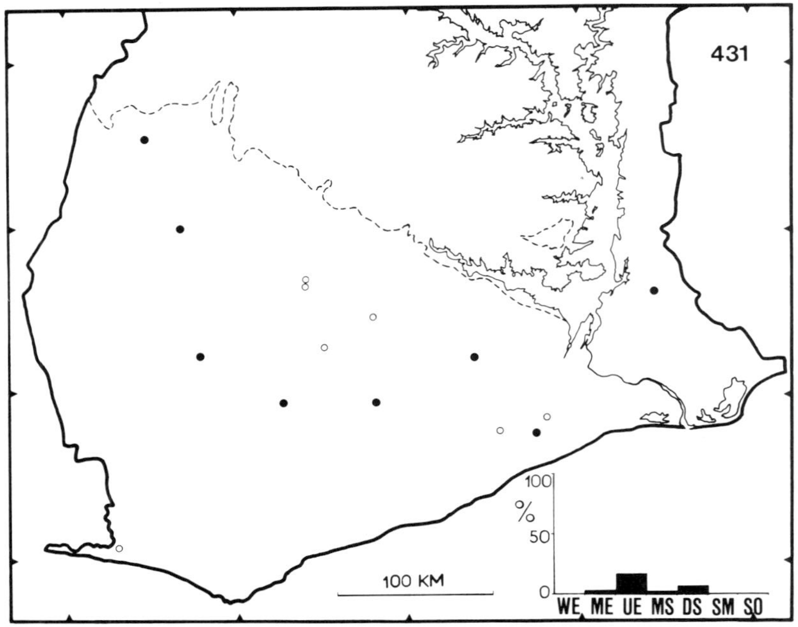

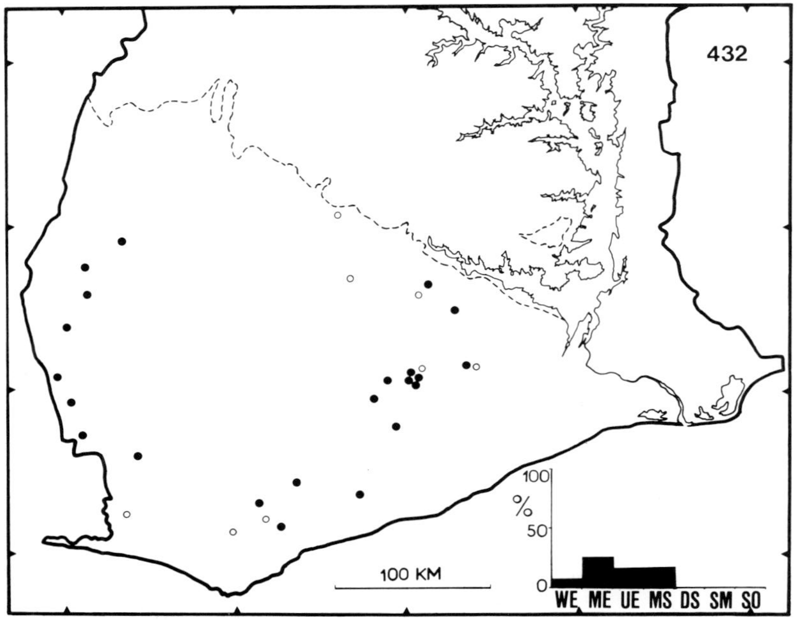

427 OXYANTHUS FORMOSUS Hook. f. ex Planch., FWTA 2:129.
Pygmy or small tree; lvs mesophyll. Petit's model. Fl large, white, dish with tube, 4–6, opening gregariously. Fr woody, orange berry, 8–12; seeds medium. Regenerates in shade. Guinée to Zaire; (Guineo-Congolian wide).

428 OXYANTHUS PALLIDUS Hiern, FWTA 2:129. Rubiaceae.
Pygmy or small tree; lvs mesophyll. Petit's model. Fl large, white, dish with tube, 3–5. Fr yellow berry, c. 8; seeds medium. Regenerates in shade. S. Leone to Angola; (Guineo-Congolian wide).

429 OXYANTHUS RACEMOSUS (Schum. & Thonn.) Keay, FWTA 2:129. Rubiaceae. (Type from Berekuso, Akwapim).
Pygmy or small tree; lvs notophyll to mesophyll, hairy underneath. Petit's model. Fl large, white, dish with tube, 3–7. Fr yellow-orange berry, 8–2; seeds medium. Regenerates in shade. Guinea Bissau to Nigeria and Angola; (Guinea wide).

430 OXYANTHUS SPECIOSUS DC., FWTA 2:129. Rubiaceae. Akan: korantema.
Small to medium-sized tree with fluted bole, branching near the base; lvs mesophyll. Petit's model. Fl large, white, dish with tube, 3–5 and 11–12. Fr orange berry, 7–10 and 1; seeds medium. Regenerates in shade. Widespread in Tropical Africa; (chorological transgressor).

431 OXYANTHUS UNILOCULARIS Hiern, FWTA 2:129. Rubiaceae.
Pygmy or small tree; lvs macrophyll, hairy. Petit's model. Fl large, white, dish with tube, 3–4 and 10–11. Fr green berry; seeds medium. Mainly in forest with thin canopy. Reaches fringing forest in the southern part of the savanna zone. Guinée to Angola and Uganda; (Guineo-Congolian wide).

432 PACHYPODANTHIUM STAUDTII Engl. & Diels, FWTA 1:39. Annonaceae. Akan: kumdwie.
Medium to large tree, slash strongly scented. Lvs mesophyll, pubescent on twigs and underside of midrib, puberulous on underside of lamina. Roux's model (59). Fl cream with purple spots, medium to large, bowl. Fr reddish berry; seeds large. Regenerates in shade. S. Leone to Gabon; (Guinea wide).

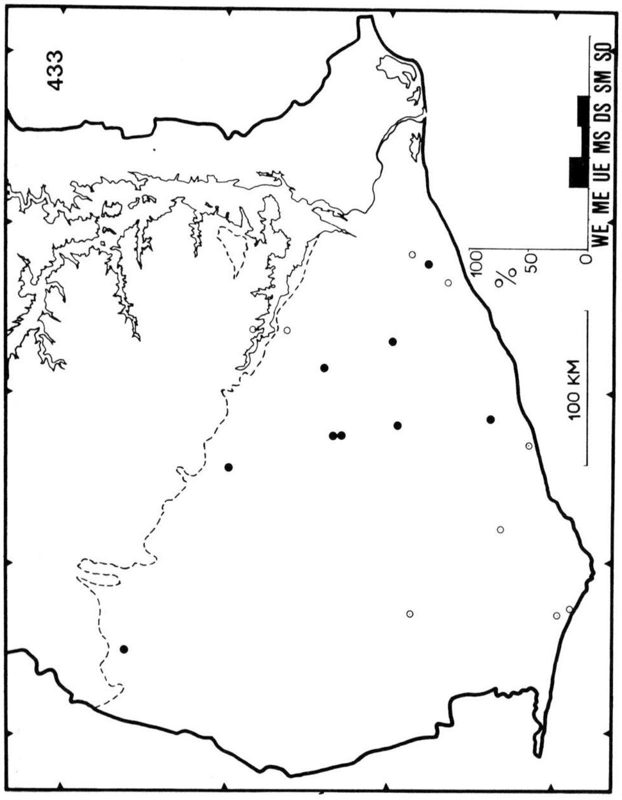
433
100 KM
%
100
50
0
WE ME UE MS DS SM SO

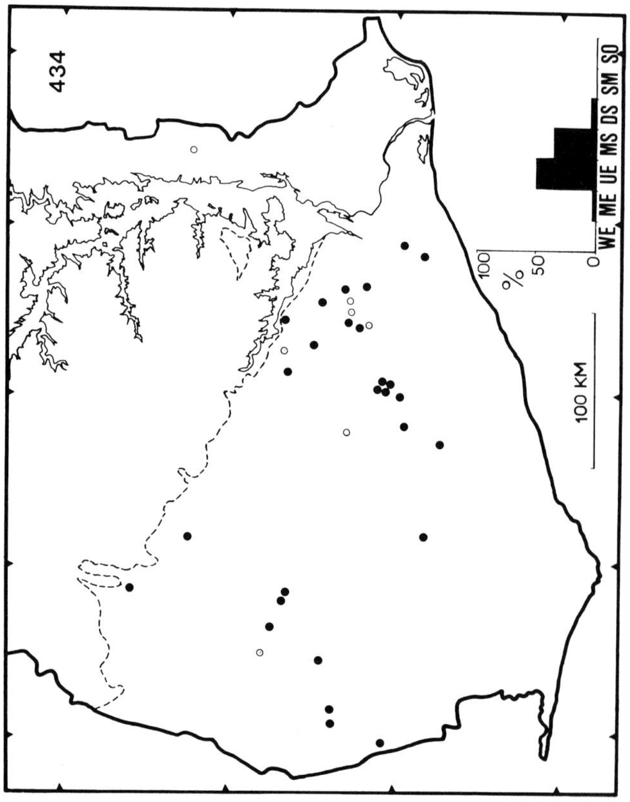
434
100 KM
%
100
50
0
WE ME UE MS DS SM SO

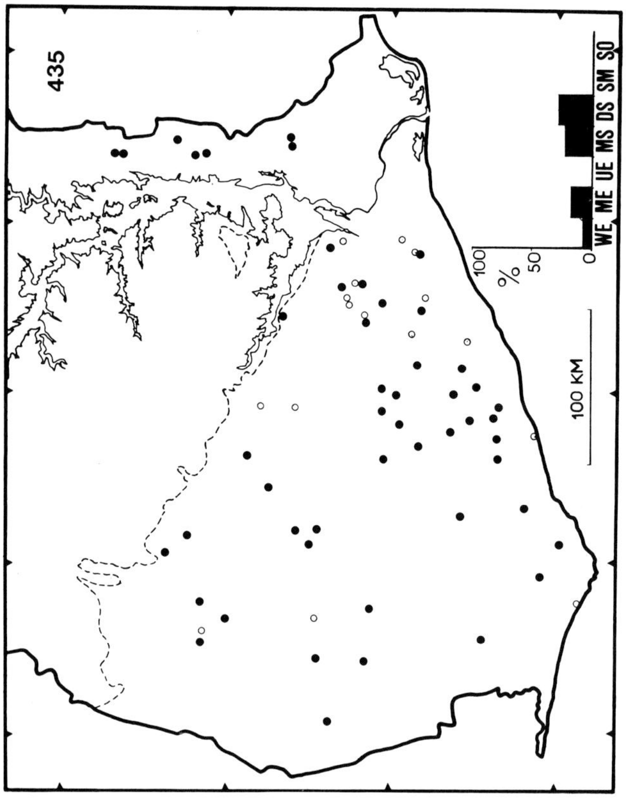
435
100 KM
%
100
50
0
WE ME UE MS DS SM SO

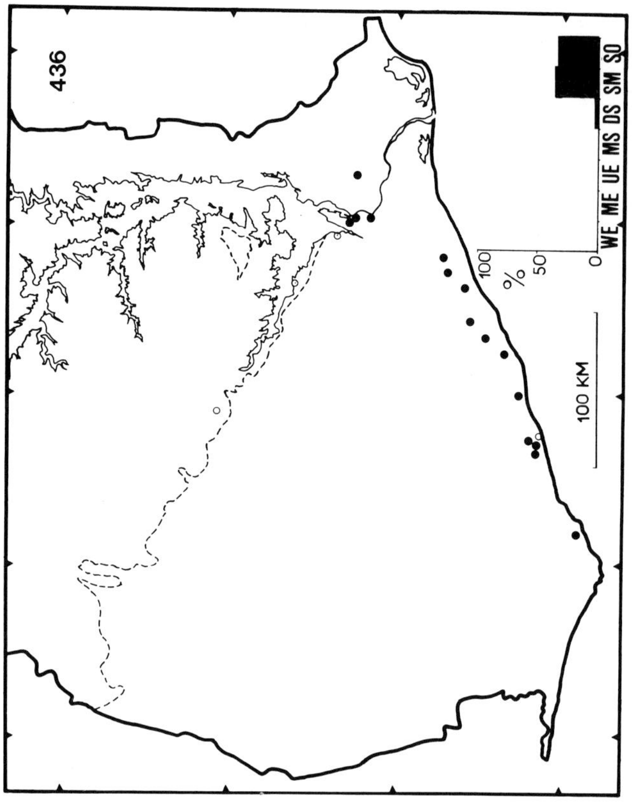
436
100 KM
%
100
50
0
WE ME UE MS DS SM SO

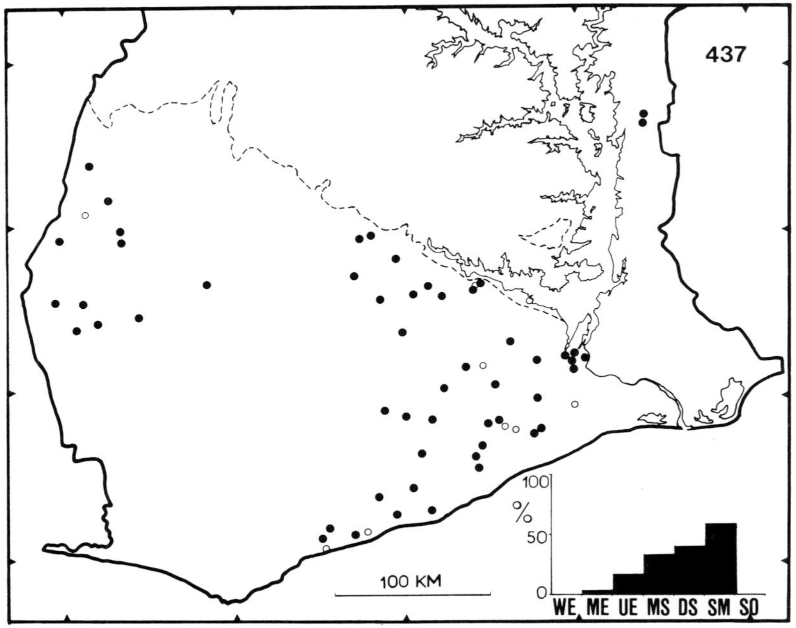

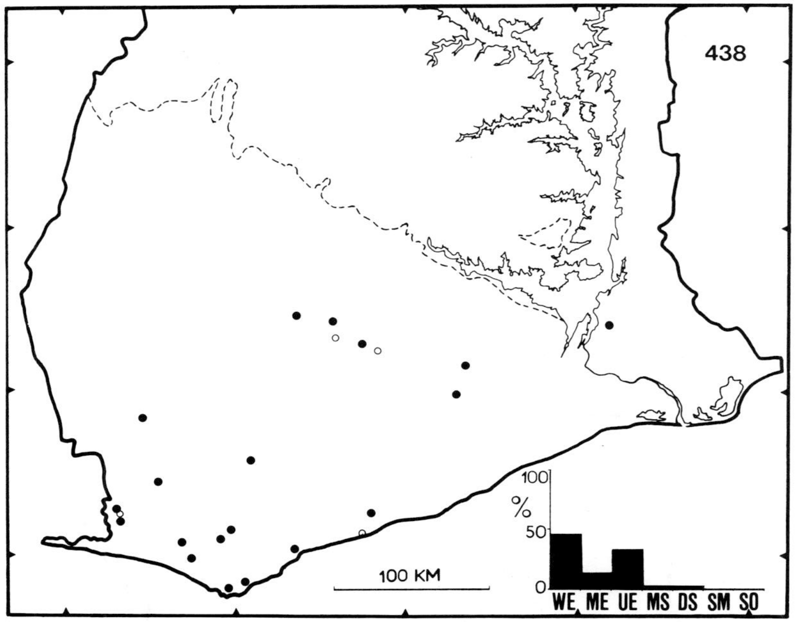

433 PACHYSTELA BREVIPES (Bak.) Baill. ex Engl., FWTA 2:28. Sapotaceae. Akan: aframsua.
Small to medium tree, containing white latex; lvs mesophyll. Fl small, whitish, bowl, mainly on twigs and branches below the leaves, 9–11. Fr orange 1-seeded berry, 11–1; seed large to very large. Mainly, but not exclusively on river banks; extends to fringing forest in the southern part of the savanna zone. Widespread in Tropical Africa; (ecological and chorological transgressor).

434 PALISOTA BARTERI Hook., FWTA 3:35. Commelinaceae.
Rosette-forming perennial herb; lvs macrophyll, hairy on margin and sheath. Fl white, small, bowl, not clearly seasonal. Fr red berry; seeds medium. Regenerates in shade. S. Leone to Congo Republic; (Guinea wide).

435 PALISOTA HIRSUTA (Thunb.) K. Schum., FWTA 3:35. Commelinaceae.
Tall, branching ground herb; lvs mesophyll to macrophyll, hairy on sheath, margin, and underside of midrib. Fl purple, small, bowl, not clearly seasonal. Fr black berry; seeds medium. Occurs mainly in gaps and by paths. Senegal to Zaire; (Guineo-Congolian wide).

436 PANCOVIA BIJUGA Willd., FWTA 1:718. Sapindaceae. (A syntype probably from near Accra).
Small tree; lvs pinnate, leaflets notophyll. Fl cream, small, bowl, monoecious, among leaves and on the branches below leaves, 2–4. Fr orange berry; seeds large. Regenerates in shade. Senegal to Central African Republic; (Guinea wide).

437 PANCOVIA SESSILIFLORA Hutch. & Dalz., FWTA 1:718. Sapindaceae. Akan: iguan ntohwe.
Pygmy tree; lvs pinnate, leaflets mesophyll. Fl cream, small, bowl, monoecious, usually on stem just below leaves, not clearly seasonal. Fr orange berry; seeds large. Regenerates in shade. Iv. Coast to Nigeria; (Guinea wide).

438 PANCOVIA TURBINATA Radlk., FWTA 1:718. (Incl. *P. pedicellaris* Radlk. & Gilg, FWTA 1:718). Sapindaceae.
Small tree or shrub, lvs pinnate, leaflets mesophyll. Fl cream, small, bowl, monoecious, not clearly seasonal, often cauliflorous. Fr yellow-orange berry; seeds large. Regenerates in shade. Guinée to Central African Republic; (Guinea wide).

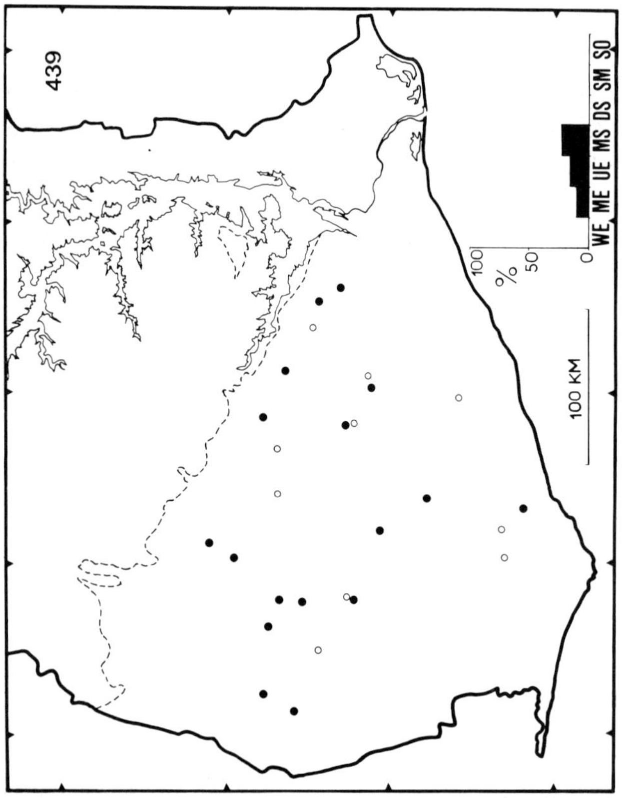
439
100 KM
100
%
50
0
WE ME UE MS DS SM SO

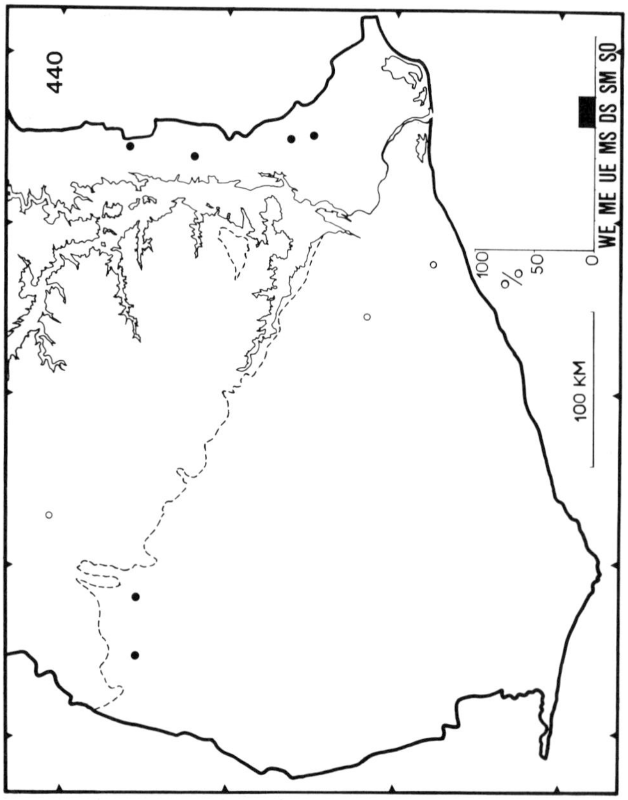
440
100 KM
100
%
50
0
WE ME UE MS DS SM SO

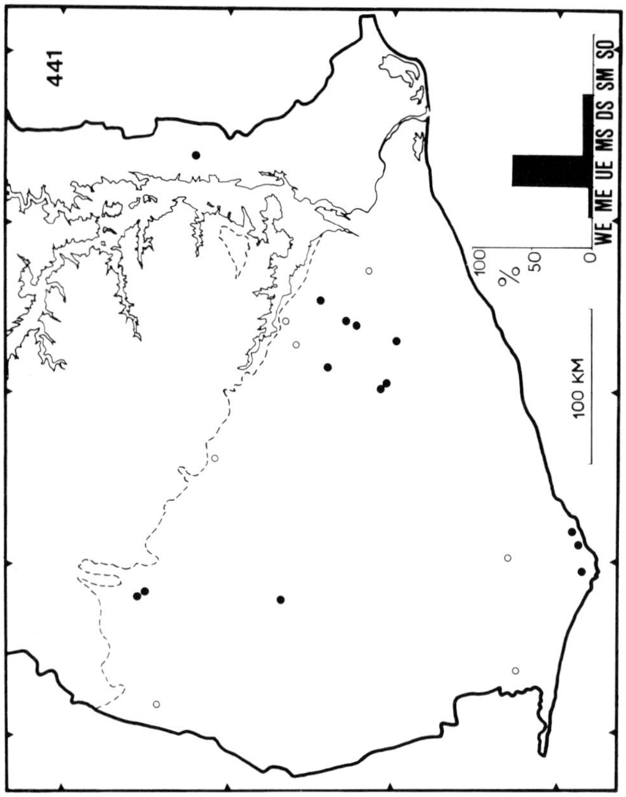
441
100 KM
100
%
50
0
WE ME UE MS DS SM SO

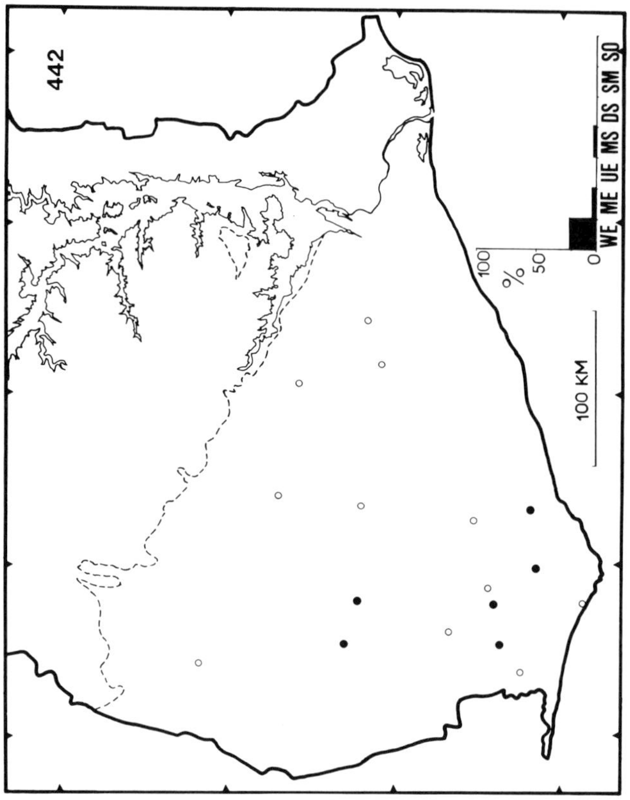
442
100 KM
100
%
50
0
WE ME UE MS DS SM SO

439 PANDA OLEOSA Pierre, FWTA 1:636. Pandaceae. Akan: kokroboba.
Medium-sized tree; lvs mesophyll, toothed. Cook's model (59). Fl small, purplish, dish, dioecious, not clearly seasonal. Fr large greenish drupe; seeds very large, embedded in hard woody endocarp, dispersed by elephants (5). Germination slow. Regenerates in shade. Liberia to Zaire; (Guineo-Congolian wide).

440. PARARISTOLOCHIA GOLDIEANA (Hook. f.) Hutch. & Dalz., FWTA 1:79. Aristolochiaceae.
Medium-sized softly woody climber; lvs mesophyll. Fl enormous (to 45 cm long) purple and white, trap, cauliflorous, 3–5. Fr greenish berry; seeds large. Regenerates in shade. S. Leone to Cameroun; (Guinea wide).

441 PARARISTOLOCHIA MACROCARPA (Duch.) Poncy (93), (*P. flos-avis* (A. Chev.) Hutch & Dalz., FWTA 1:79). Aristolochiaceae.
Medium-sized woody climber; stem becoming flattened. Lvs mesophyll. Fl large, purplish, trap, cauliflorous, 2–5 and 9–11. Fr greenish berry; seeds large. Regenerates in shade. S. Leone to Gabon; (Guinea wide).

442 PARINARI EXCELSA Sabine, FWTA 1:429. Chrysobalanaceae (Rosaceae in FWTA). Akan: afam.
Large tree with buttresses. Lvs microphyll to notophyll, densely white-tomentose below, with glands on petiole. Troll's model (59). Fl white, small, bowl, 1–5. Fr brownish drupe with edible flesh, 10–1; seed very large, dispersed by elephants (5). Widespread in Tropical Africa; (ecological and chorological transgressor). This is a variable species, but it cannot be satisfactorily subdivided (118).

443 PARKIA BICOLOR A. Chev., FWTA 1:487. Mimosaceae. Akan: asoma.
Large tree with buttresses; lvs bipinnate, leaflets leptophyll. Troll's model. Fl small, red, crowded into large spherical clusters, 12–1. Fr fleshy pod, 2–4; seeds large to very large. Regenerates in shade. Guinée to Zaire; (Guineo-Congolian wide). Specimens with much larger leaflets are sometimes found in dry forest near the savanna boundary; these have sometimes been referred to *P. filicoidea* Welw. ex Oliv., but further collecting is needed to decide whether this identification is correct.

444 PAULLINIA PINNATA Linn., FWTA 1:710. Sapindaceae. Akan: toantini.
Medium-sized woody climber; xylem of older stems divided into three strands. Lvs pinnate, leaflets notophyll, toothed. Undescribed model (59). Tendrils present on inflorescences. Fl small, bowl, white, not clearly seasonal. Fr red capsule; seeds black, medium, with white aril. Mainly in forest with thin canopy; extends to fringing forest in the savanna zone. Widespread in Tropical Africa, Madagascar, and Tropical America; (ecological and chorological transgressor).

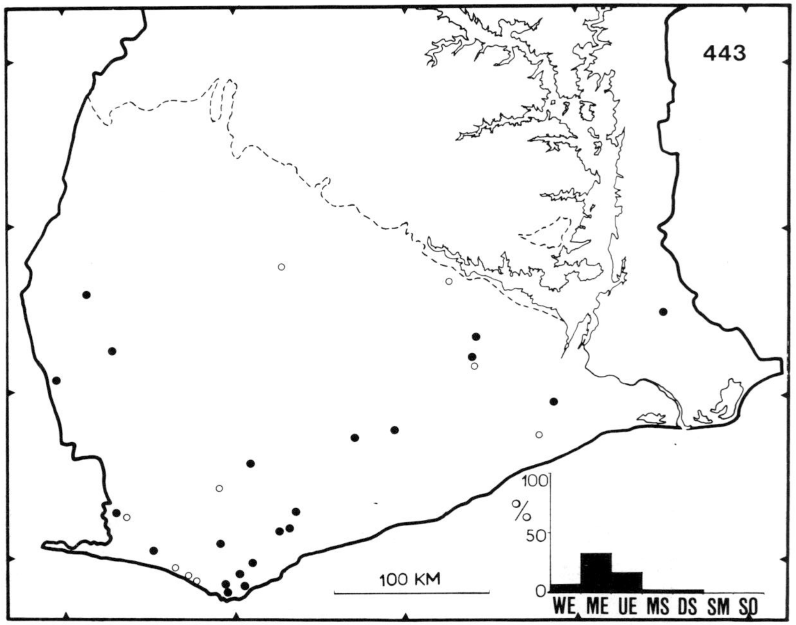

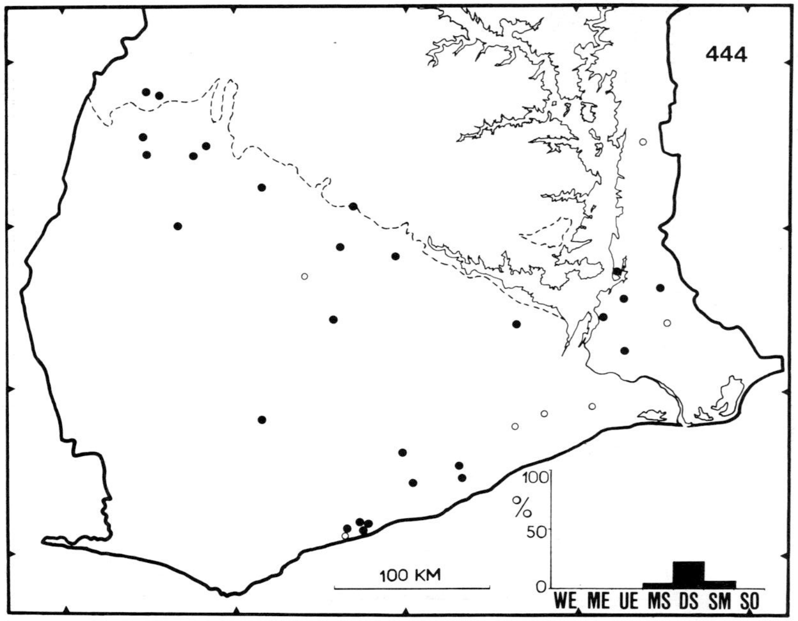

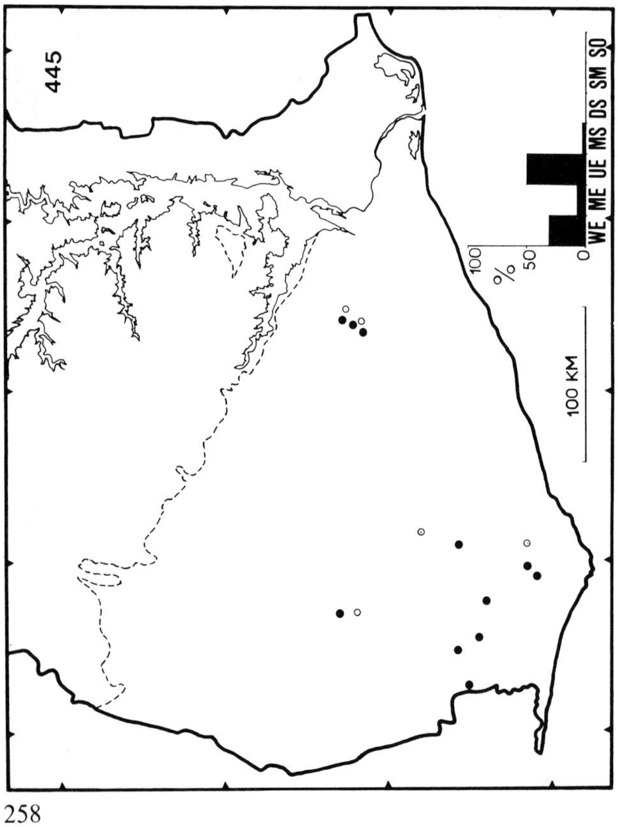
445
100 KM
100
%
50
0
WE ME UE MS DS SM SO

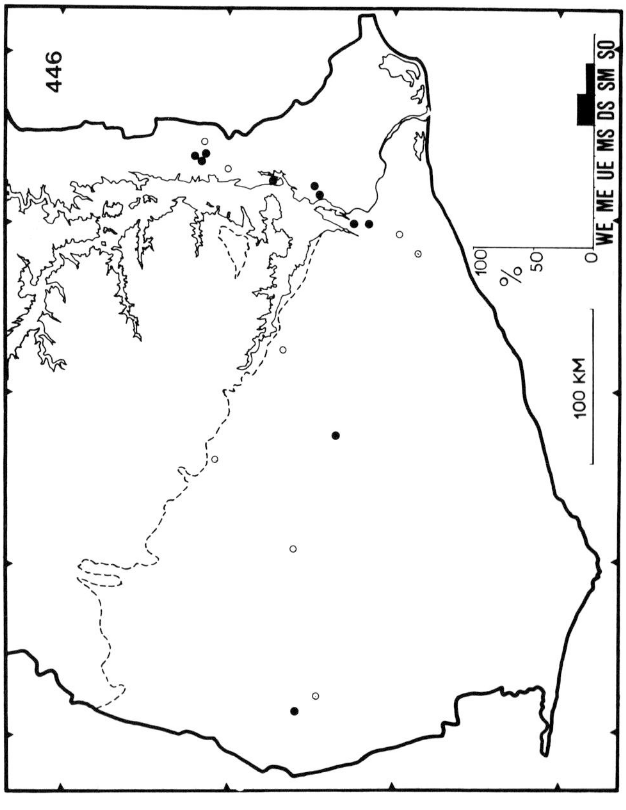
446
100 KM
100
%
50
0
WE ME UE MS DS SM SO

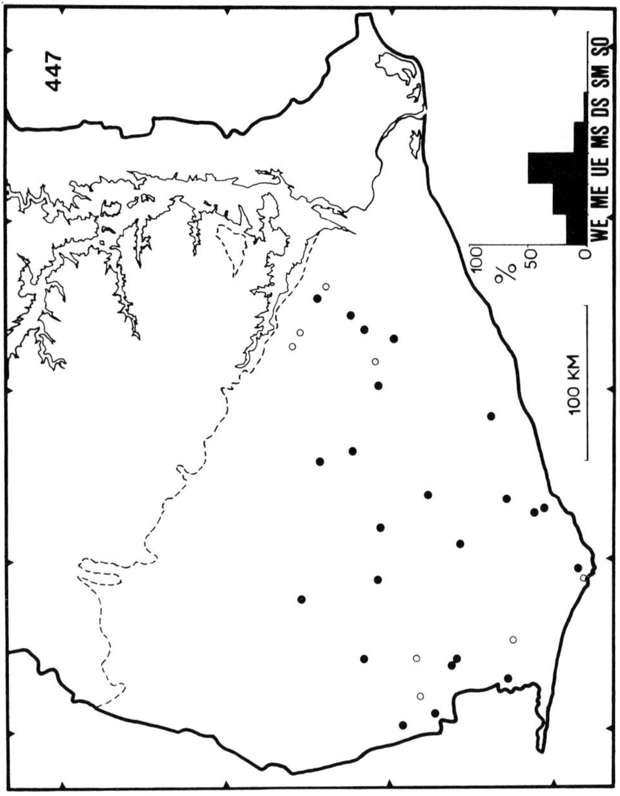
447
100 KM
100
%
50
0
WE ME UE MS DS SM SO

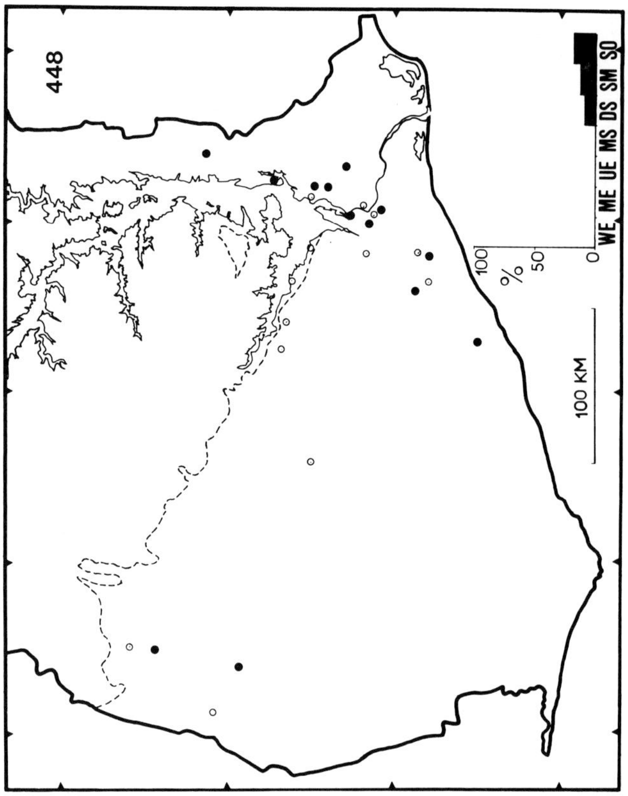
448
100 KM
100
%
50
0
WE ME UE MS DS SM SO

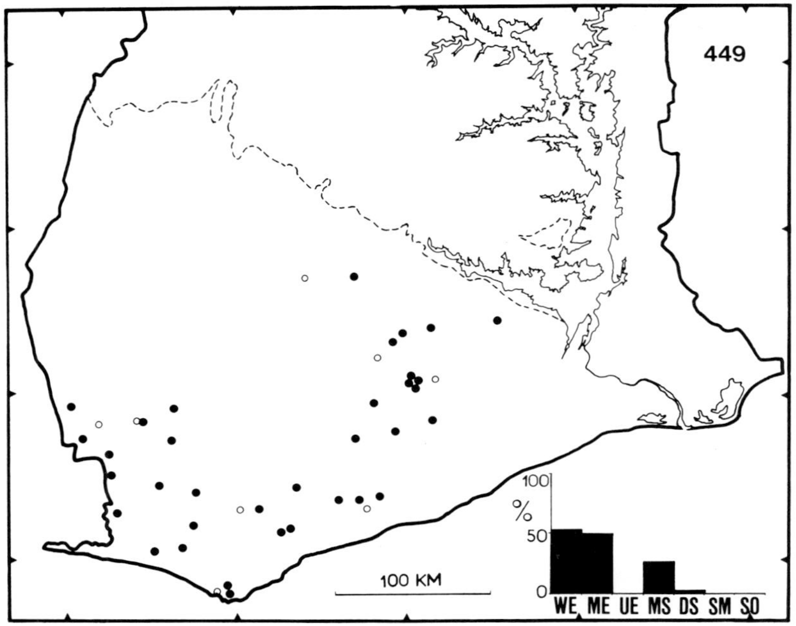

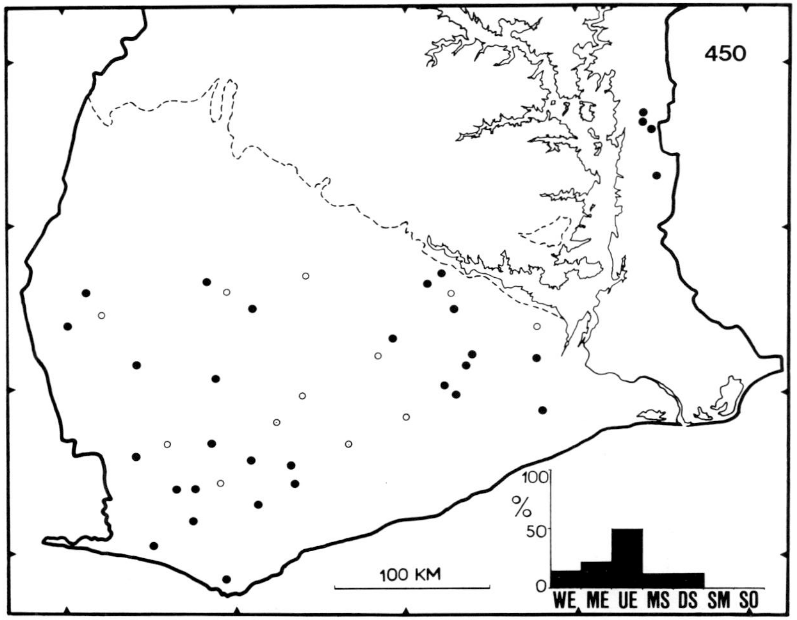

445 PAURIDIANTHA SYLVICOLA (Hutch. & Dalz.) Bremek., FWTA 2:169. Rubiaceae.

Pygmy tree; lvs notophyll. Fl whitish, small to medium, dish with tube, 8–1. Fr black berry; seeds small. Regenerates in shade. Plot records very significantly associated with base-poor soils. Liberia to Cameroun; (Guinea wide).

446 PAVETTA GENIPIFOLIA Schumach., FWTA 2:139. Rubiaceae. (Type from Akwapim).

Pygmy tree; lvs mesophyll, pubescent on undersurface and on twigs. Fl white, large, dish with tube, 3–6. Fr black berry; seed medium. Regenerates in shade. Ghana to Nigeria; (Guinea wide).

447 PAVETTA IXORIFOLIA Bremek., FWTA 2:139. Rubiaceae.
Pygmy tree; lvs mesophyll. Fl creamy-white, medium, dish with tube, 12–2. Fr berry; seeds medium. Regenerates in shade. Guinée to Nigeria; (Guinea wide).

448 PELLAEA DONIANA Hook., FWTA FF:43. Adiantaceae.
Ground fern; lvs pinnate, leaflets microphyll. Propagated by dust-like spores. Floor of dry forest, and on rock outcrops in moister forest. Widespread in Tropical Africa; (chorological transgressor).

449 PENIANTHUS ZENKERI (Engl.) Diels, FWTA 1:76. Menispermaceae. Akan: kramankote.

Pygmy tree less than 1 m high; lvs mesophyll. Corner's model (59). Fl small, yellowish, inconspicuous, dioecious, not clearly seasonal. Fr orange-red 1-seeded berry; seed large to very large. Regenerates in shade. Plot records very significantly correlated with acid, base-poor soils. S. Leone to Cameroun; (Guinea wide). Roots chewed as an aphrodisiac.

450 PENTACLETHRA MACROPHYLLA Benth., FWTA 1:487. Mimosaceae. Akan: ataa.

Medium-sized tree; lvs bipinnate, leaflets nanophyll, leaf axes and midrib pubescent. Troll's model (59). Fl creamy-yellow, small, clustered in dense brush inflorescence, 11–2. Fr large, woody, explosive pod, 9–12; seeds very large. Regenerates in shade. Senegal to Angola; (Guineo-Congolian wide). Seeds yield edible oil.

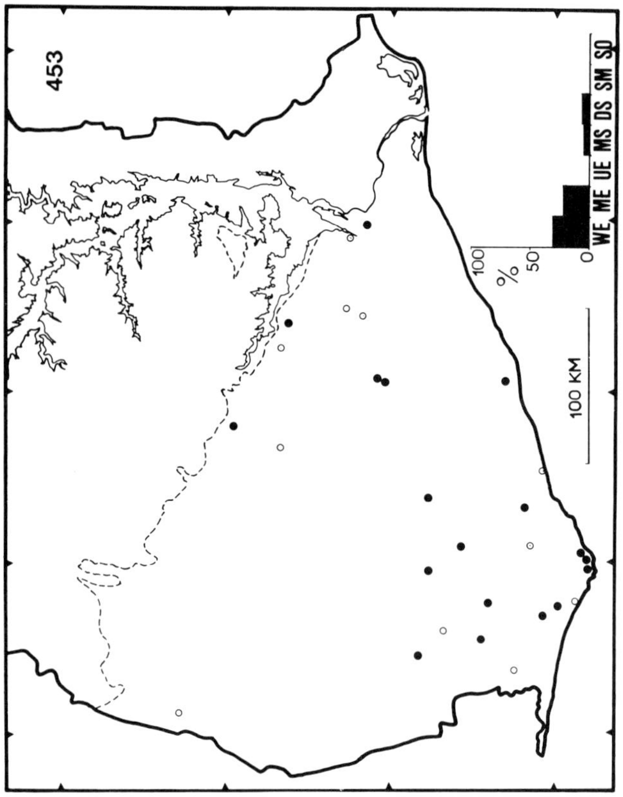
453
100 KM
%
100
50
0
WE ME UE MS DS SM SO

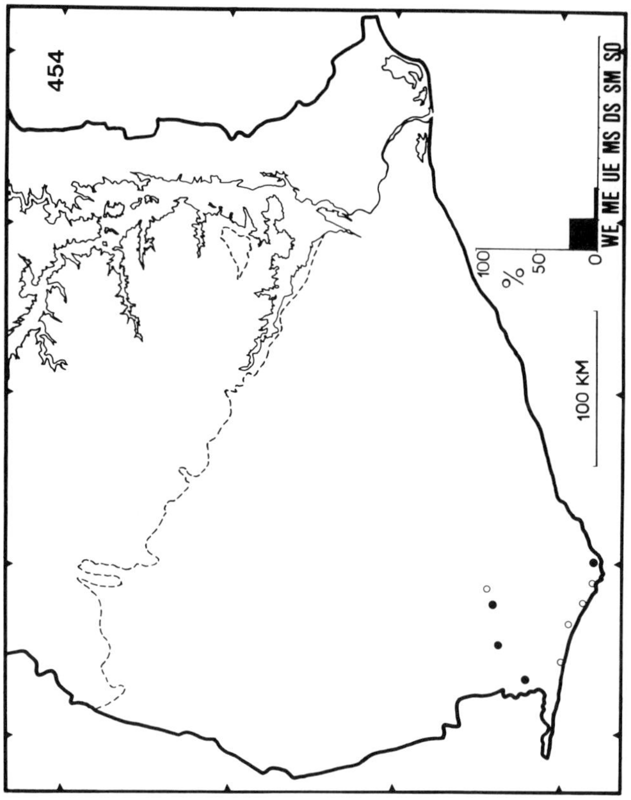
454
100 KM
%
100
50
0
WE ME UE MS DS SM SO

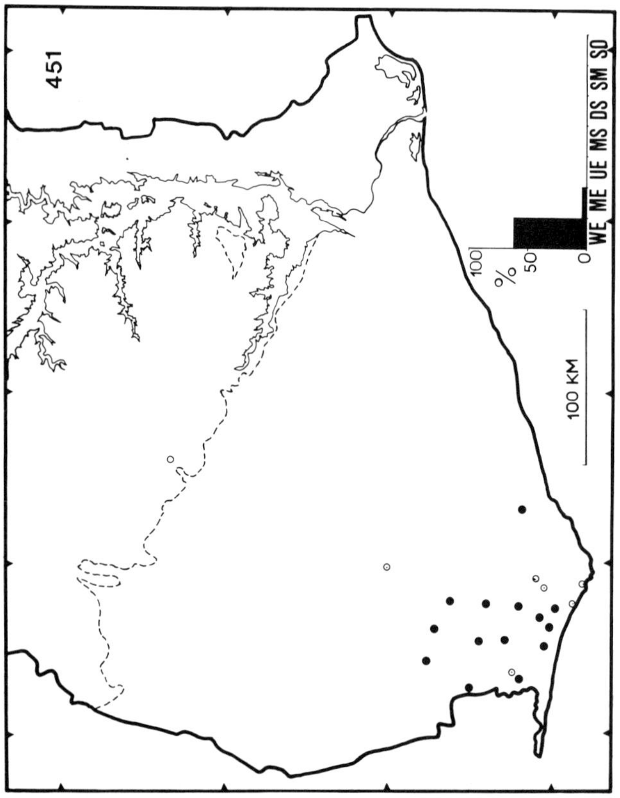
451
100 KM
%
100
50
0
WE ME UE MS DS SM SO

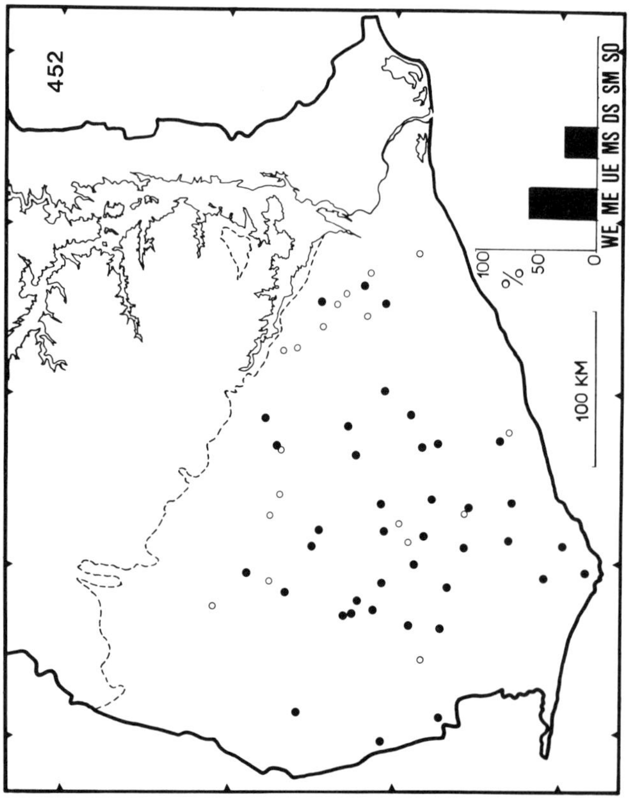
452
100 KM
%
100
50
0
WE ME UE MS DS SM SO

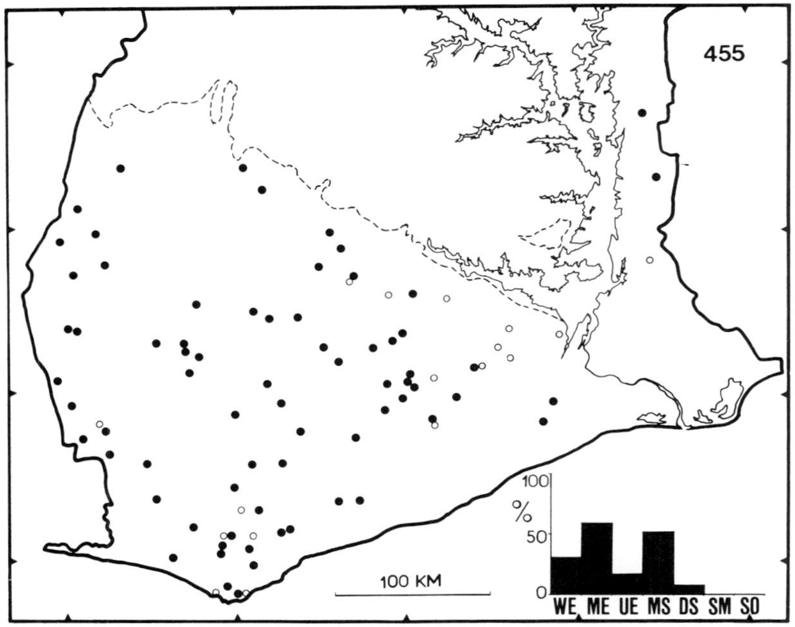

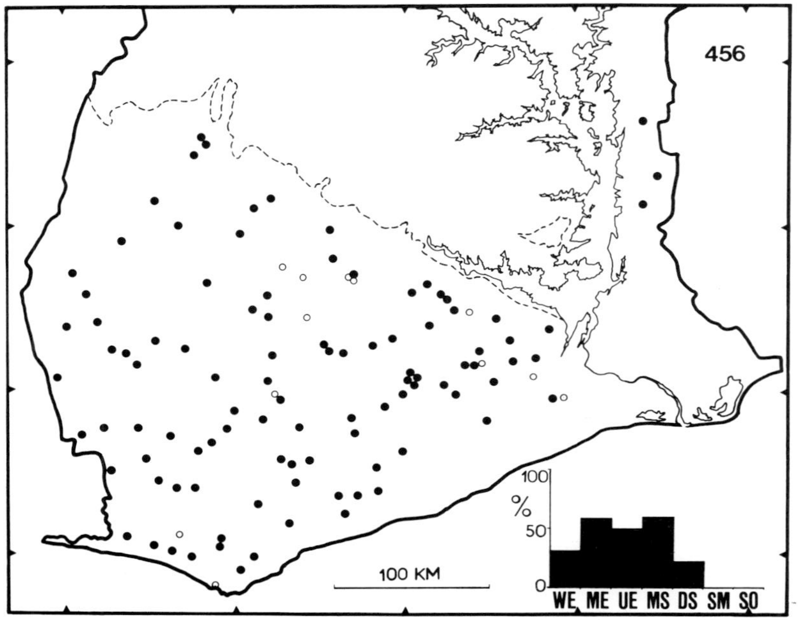

451 PENTADESMA BUTYRACEA Sabine, FWTA 1:291. Guttiferae (19). Akan: abotosebie.
Medium-sized tree, containing yellow latex; lvs mesophyll. Rauh's model (59). Fl white, strongly scented, large, bowl, mainly 12–2. Fr large, orange-brown berry, mainly 12–3; seeds very large, embedded in yellow pulp, dispersed by elephants (5). Regenerates in shade. Plot records very significantly correlated with base-poor soils. Some isolated populations occur in fringing forest near the savanna boundary. Seeds yield edible fat. Guinée to Zaire; (Guinea wide).

PERICOPSIS ELATA (Harms) van Meeuwen (107), (*Afrormosia elata* Harms, FWTA 1:510). Papilionaceae. Akan: kokrodua.
Large tree with buttresses; lvs pinnate, leaflets microphyll. Troll's model. Fl medium, white, flag, 4–5. Fr indehiscent, thin pod, wind-dispersed, 8–11; seeds large. Regenerates mainly in gaps and along roadsides, but also under light shade. Iv. Coast to Ghana, and also Cameroun and Zaire; (Guineo-Congolian wide). Class 1 timber; trade name: afrormosia (40). Probably because it regenerates mainly in gaps (6) small trees are scarce, and the species was recorded only twice in our survey; for distribution in Ghana as calculated from timber enumeration records see Fig. 11.6.

452 PETERSIANTHUS MACROCARPUS (P. Beauv.) Liben (79), (*Combretodendron macrocarpum* (P. Beauv.) Keay, FWTA 1:761; *Combretodendron africanum* (Welw. ex Benth. & Hook. f.) Exell, FWTA 1:242). Lecythidaceae. Akan: esia.
Large tree with deeply furrowed bark, sometimes buttressed; lvs notophyll, sometimes briefly deciduous. Koriba's model (59). Fl small to medium, greenish-white, dish, 1–6. Fr samara with 4 papery wings, wind-dispersed; seed large. Regenerates in shade or, more often, in small gaps. Guinée to Angola; (Guineo-Congolian wide). Class 3 timber.

453 PHYLLOCOSMUS AFRICANUS (Hook. f.) Klotzsch (42), (*Ochthocosmus africanus* Hook. f., FWTA 1:355). Ixonanthaceae. Akan: akokorabeditoa.
Small to medium tree, or shrub in thicket; lvs notophyll with curious deciduous teeth near apex. Fl white, small, bowl, not clearly seasonal, opening gregariously. Fr small capsule; seeds arillate, medium. Regenerates in shade. Guinea Bissau to Zaire; (Guineo-Congolian wide).

454 PHYLLOCOSMUS SESSILIFLORUS Oliv. (13, 42), (*Ochthocosmus chippii* Sprague & Hutch., FWTA 1:355). Ixonanthaceae.
Small tree; lvs notophyll, toothed. Fl pinky-white, small, dish, not clearly seasonal, opening gregariously. Fr small capsule; seeds arillate, medium. Occurs in forest understorey and secondary forest. Ghana, and Cameroun to Gabon; (Lower Guinea Satellite).

455 PIPER GUINEENSE Thonn., FWTA 1:84. Piperaceae. Akan: sorowisa. (Type from Akwapim).
Medium-sized woody climber; young plant with herbaceous stem attached to support by adventitious roots; old plant with woody stem to 3 cm thick, with concentric rings of xylem separated by phloem. Lvs notophyll, aromatic. Fl very small, yellow, crowded in spikes, not clearly seasonal. Fr red-brown 1-seeded berry; seed small to medium, peppery taste. Seeds appear to need exposure for germination; regeneration probably mainly vegetative. Guinea Bissau to Angola and Uganda; (Guineo-Congolian wide).

456 PIPTADENIASTRUM AFRICANUM (Hook. f.) Brenan, FWTA 1:489. Mimosaceae. Akan: dahoma.
Large tree with buttresses and enormous crown; lvs bipinnate, leaflets leptophyll, deciduous. Troll's model (59). Fl small, pale yellow, crowded in brush inflorescence, 4–7. Fr pod opening along one margin, 12–3; seeds very large, winged, wind-dispersed. Regenerates in shade. Senegal to Angola and Uganda; (Guineo-Congolian wide). Class 2 timber; trade name: dahoma (40).

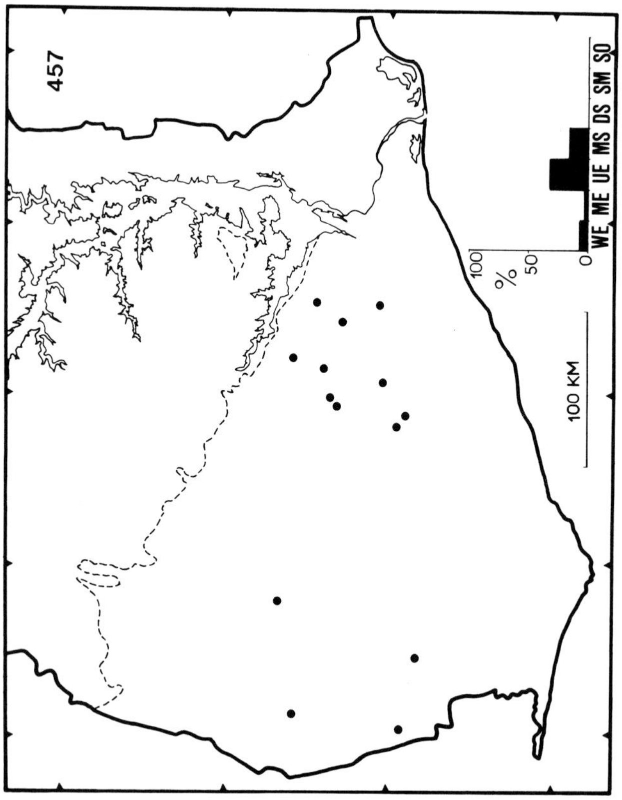
457
100 KM
100
%
50
0
WE ME UE MS DS SM SO

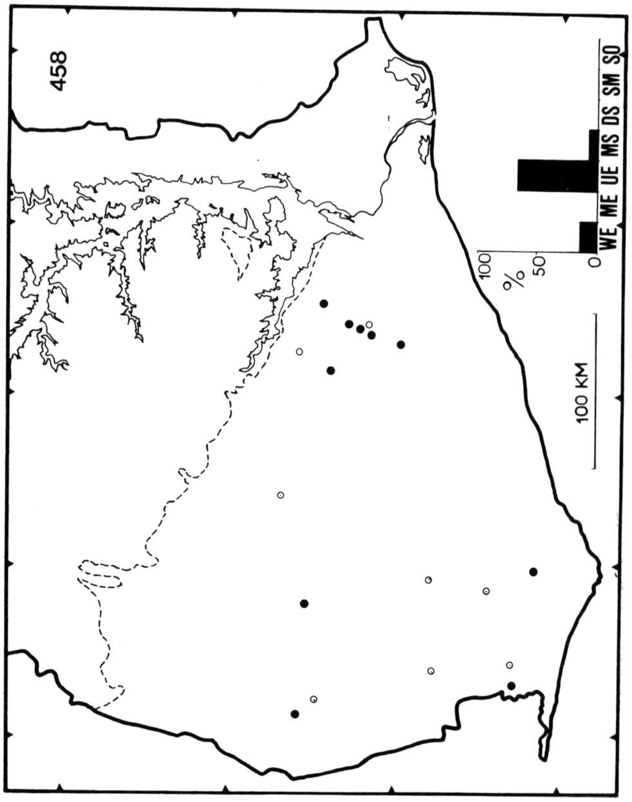
458
100 KM
100
%
50
0
WE ME UE MS DS SM SO

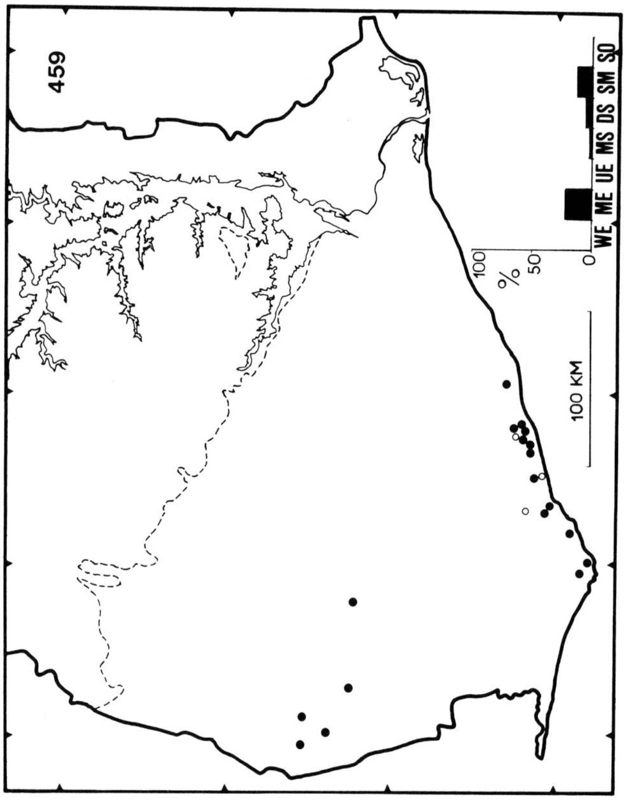
459
100 KM
100
%
50
0
WE ME UE MS DS SM SO

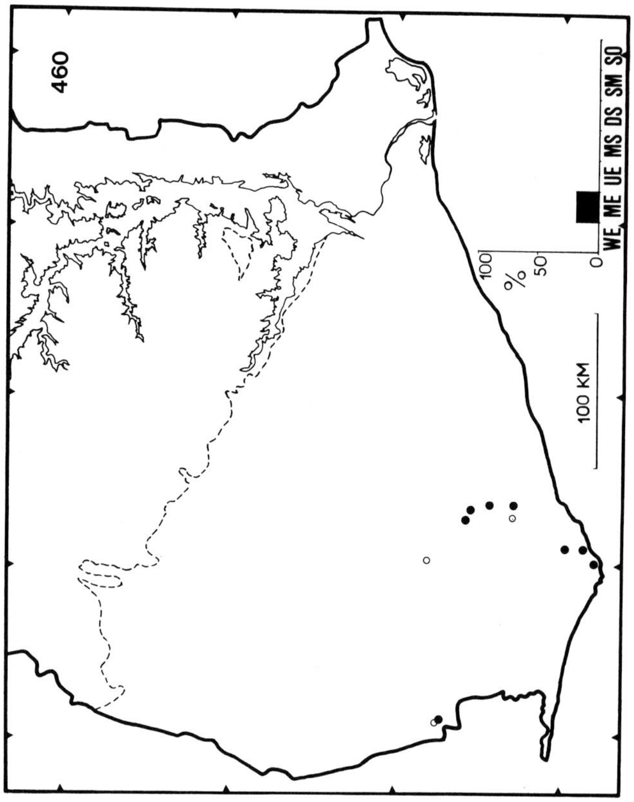
460
100 KM
100
%
50
0
WE ME UE MS DS SM SO

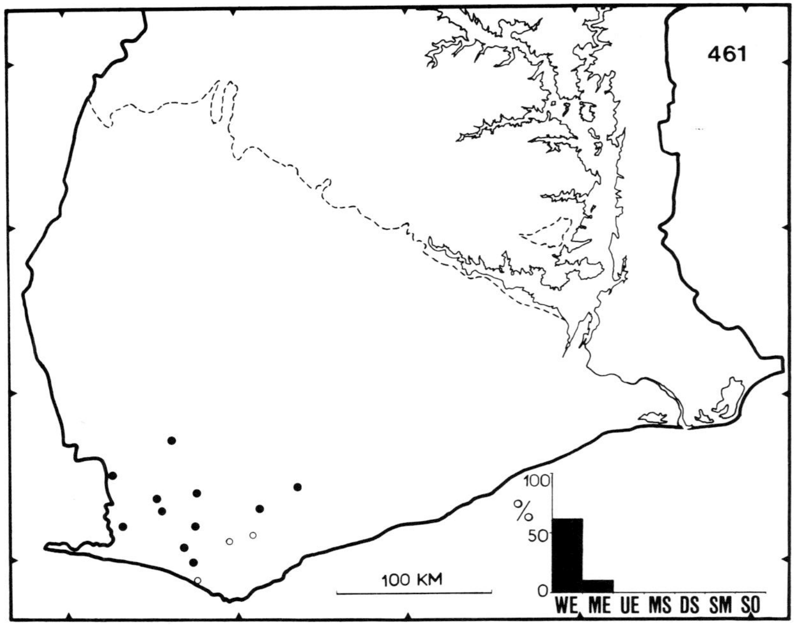

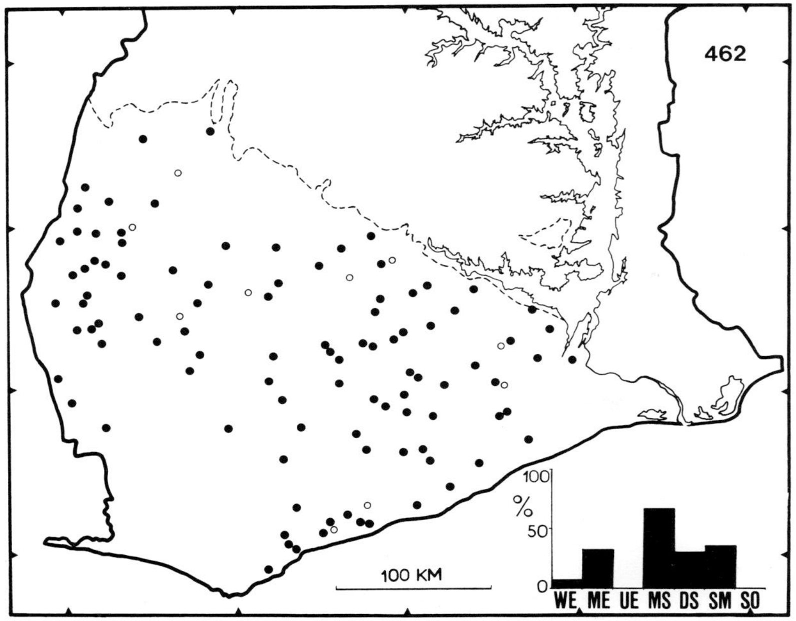

457 PIPTOSTIGMA FASCICULATUM (De Wild.) Paiva (88), (*Brieya fasciculata* De Wild., FWTA 1:39). Annonaceae.
Medium-sized tree; slash strongly scented. Lvs mesophyll; twigs and leaf underside puberulous. Roux's model. Fl large, yellowish, bowl, 5–7. Fr yellow-orange berry; seeds very large. Regenerates in shade. Iv. Coast to Angola; (Guineo-Congolian wide). The combination cannot be ascribed to Boutique, as he only suggested it, but did not accept it.

458 PIPTOSTIGMA FUGAX A. Chev. ex Hutch. & Dalz., FWTA 1:39. Annonaceae.
Small tree, slash aromatic; lvs mesophyll. Fl cream, large, bowl, cauliflorous, 5–7. Fr yellow berry; seeds very large. Regenerates in shade. Liberia to Ghana; (Upper Guinea endemic).

459 PLACODISCUS ATTENUATUS J. B. Hall (49). (*Placodiscus boya* of FWTA 1:720, partly (Aubréville 2017), not of Aubrév. & Pellegr.). Sapindaceae.
Small to medium-sized tree, or shrub in thicket. Lvs pinnate, leaflets notophyll to mesophyll. Fl small, white, bowl, probably dioecious, mainly cauliflorous, 7–9. Fr yellow berry, 10–12; seeds very large. Regenerates in shade. Iv Coast and Ghana; (Upper Guinea endemic).

460 PLACODISCUS BANCOENSIS Aubrév. & Pellegr., FWTA 1:720. Sapindaceae. Akan: kafuosonini.
Little-branched medium-sized tree; lvs pinnate, leaflets mesophyll to macrophyll. Corner's model. Fl small, white, bowl, probably dioecious, mainly cauliflorous, 11–12. Fr yellow berry; seeds very large. Regenerates in shade: sapling has simple leaves. Iv. Coast and Ghana; (Upper Guinea endemic).

461 PLACODISCUS OBLONGIFOLIUS J. B. Hall (49), (*Placodiscus leptostachys* of FWTA 1:720, partly, (Liberian specimens), not of Radlk.). Sapindaceae.
Small little-branched tree; lvs pinnate, leaflets mesophyll. Fl small, white or reddish, bowl, probably dioecious, mainly cauliflorous, mostly 7–8. Fr yellow berry; seeds very large. Regenerates in shade. Plot records very significantly associated with acid, base-poor soils. Liberia to Ghana; (Upper Guinea endemic).

462 PLATYSEPALUM HIRSUTUM (Dunn) Hepper, FWTA 1:524. Papilionaceae.
Medium-sized woody climber; stem 2-lobed in section; bark exuding red latex when cut. Lvs pinnate; leaflets notophyll, lower surface covered by fine, white, tomentellous indumentum, which is obscured in sun leaves by long brown hairs. Fl purplish, flag, medium, 8–11. Fr hairy pod, explosively dehiscent, 11–1; seeds medium to large. Regenerates in shade. Guinée to Ghana; (Upper Guinea endemic).

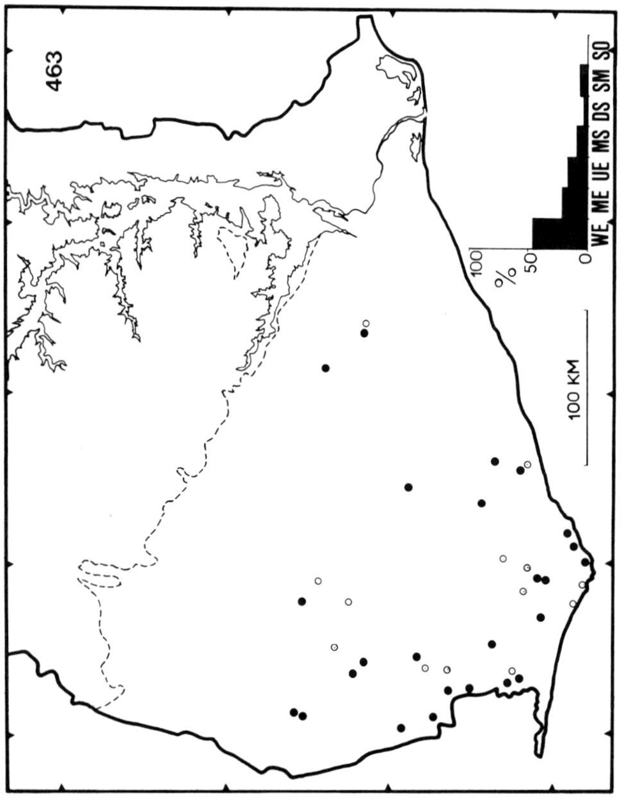
463
100 KM
%
100
50
0
WE ME UE MS DS SM SO

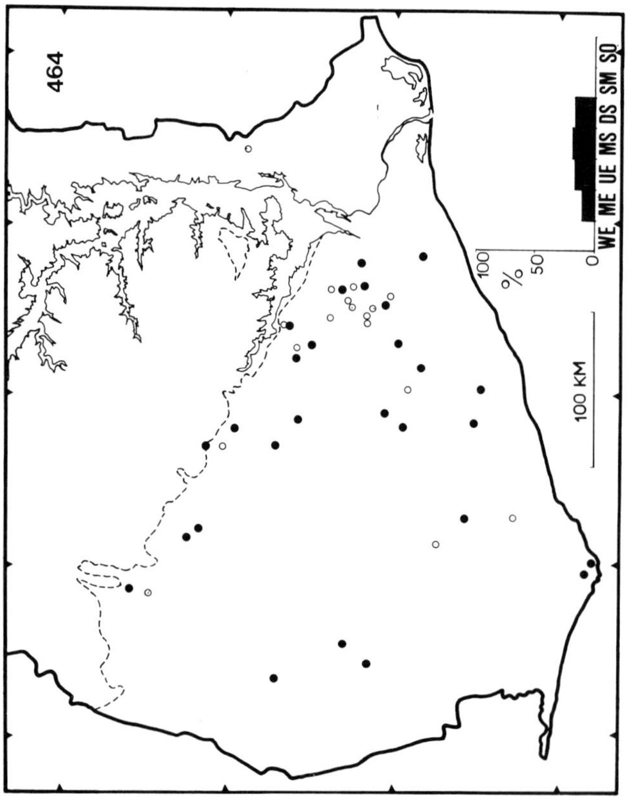
464
100 KM
%
100
50
0
WE ME UE MS DS SM SO

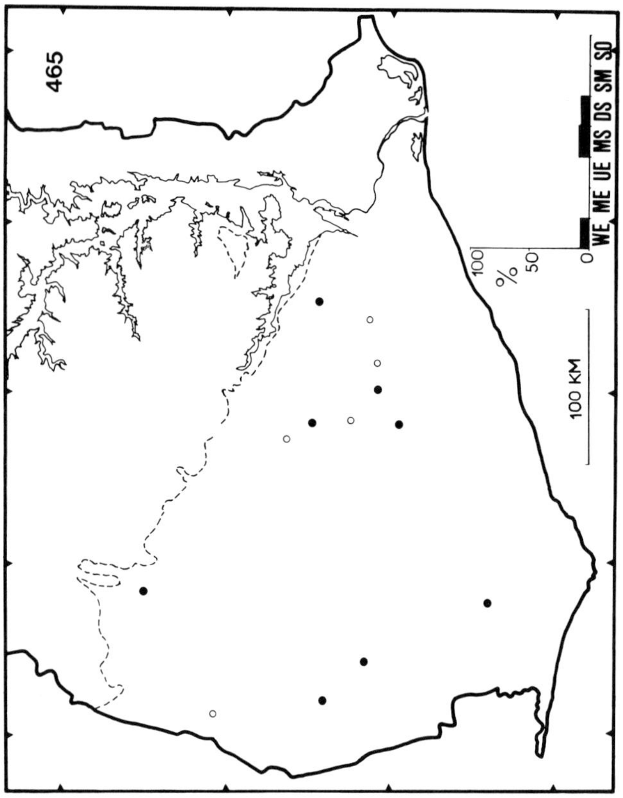
465
100 KM
%
100
50
0
WE ME UE MS DS SM SO

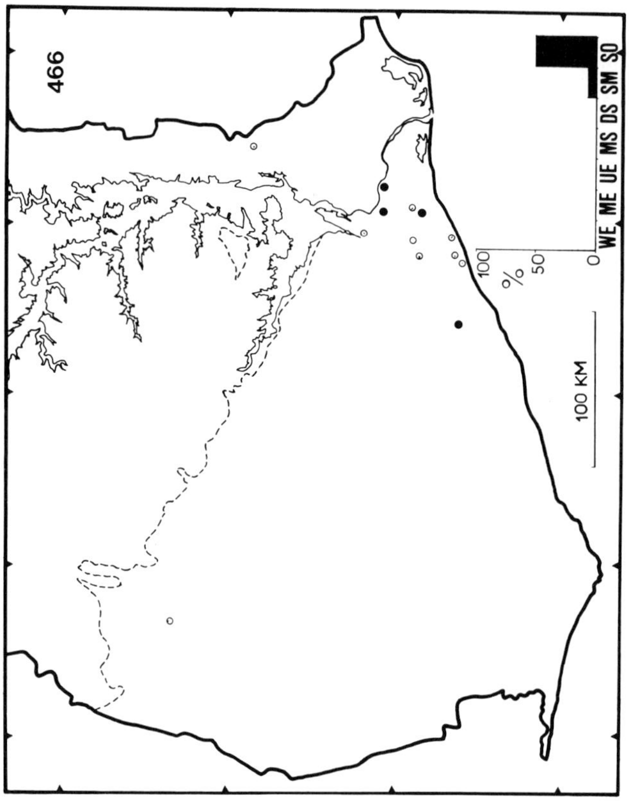
466
100 KM
%
100
50
0
WE ME UE MS DS SM SO

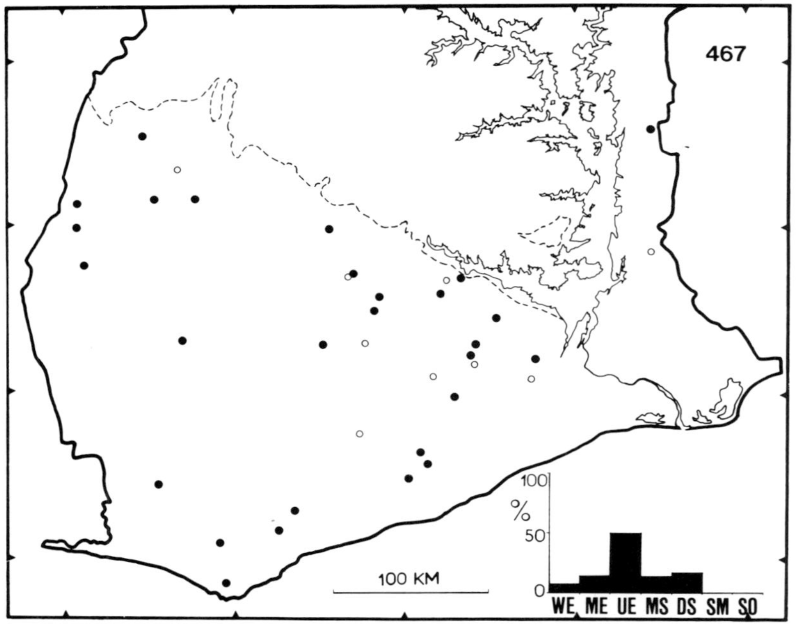

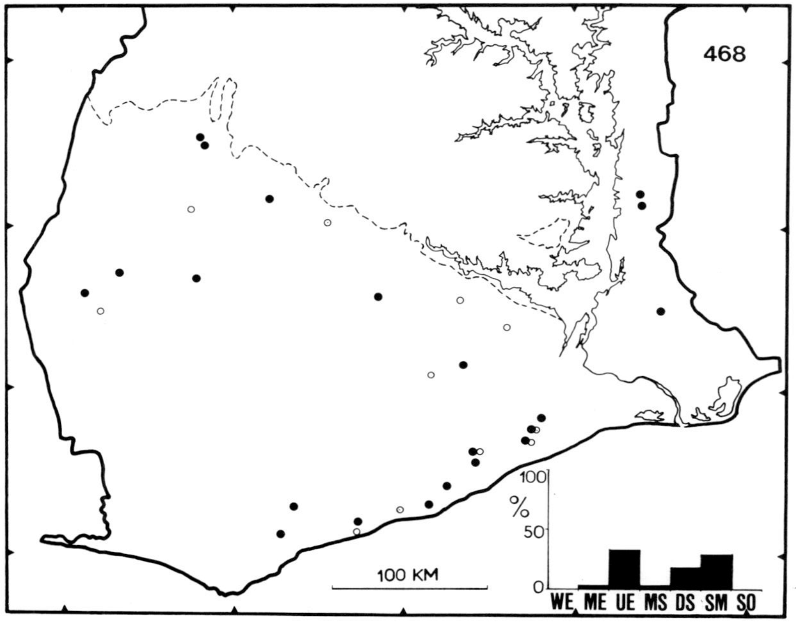

463 PLEIOCARPA MUTICA Benth., FWTA 2:63. Apocynaceae. Akan: onwenma.
Small tree or shrub, containing white latex; lvs notophyll to mesophyll. Fl white, medium, dish with tube, 12–4. Fr orange 1-seeded berry, 4–6; seed medium. Regenerates in shade. S. Leone to Zaire; (Guineo-Congolian wide).

464 POLYSPATHA PANICULATA Benth., FWTA 3:42. Commelinaceae.
Decumbent perennial ground herb; lvs mesophyll. Fl white, medium, dish, 10–2. Fr capsule; seeds medium. Regenerates, mainly vegetatively, in shade. Guinée to Zaire and Uganda; (Guineo-Congolian wide).

465 POLYSPHAERIA MACROPHYLLA K. Schum., FWTA 2:148. Rubiaceae.
Pygmy tree; lvs mesophyll. Fl white, medium, dish with tube, 3–5. Fr black berry, 10–12; seeds medium to large. Regenerates in shade. Ghana to Cameroun; (Guinea wide).

466 PREMNA QUADRIFOLIA Schum. & Thonn., FWTA 2:438. Verbenaceae. Ga: obosotso. (Type from near Accra).
Medium to large woody climber (or shrub in thicket), wood cylindrical; lvs notophyll, more or less toothed, aromatic. Fl white, small, tubular, in large inflorescence, not clearly seasonal. Fr orange 3–4-seeded drupe; seeds small. Regenerates in thin shade. Guinea Bissau to Nigeria; (Guinea wide).

467 PSEUDERANTHEMUM TUNICATUM (Afzel.) Milne-Redhead, FWTA 2:421. Acanthaceae.
Woody herb; lvs notophyll to mesophyll. Fl pale purplish, large, dish with tube, 12–4. Fr explosive capsule; seeds medium. Regenerates in shade. S. Leone to Angola; (Guineo-Congolian wide). There seems to be only one variable species of *Pseuderanthemum* in Ghana; we cannot follow FWTA in distinguishing *P. ludovicianum* (Buettner) Lindau in Ghana.

468 PSILANTHUS MANNII Hook. f., FWTA 2:157. Rubiaceae.
Pygmy tree; lvs notophyll. Fl large, white, dish with tube, mostly 2–3. Fr black berry, adorned by ribs and enlarged calyx lobes, mostly 4–6; seeds large. Regenerates in shade. Guinée to Angola; (Guineo-Congolian wide).

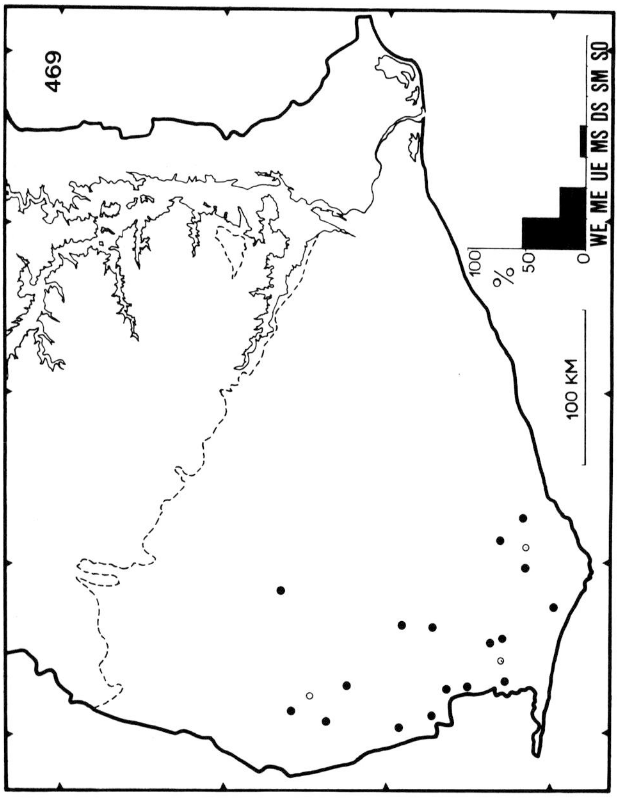
469
100 KM
%
100
50
0
WE ME UE MS DS SM SO

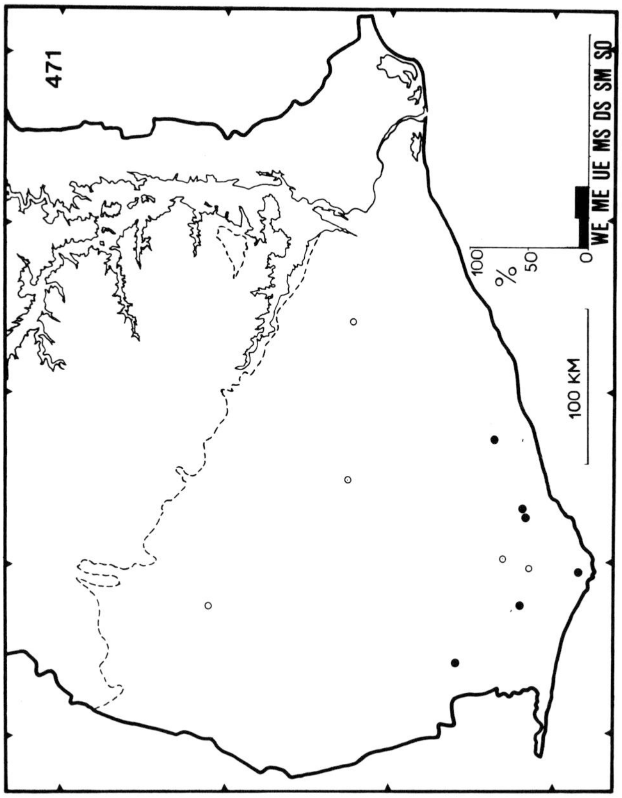
471
100 KM
%
100
50
0
WE ME UE MS DS SM SO

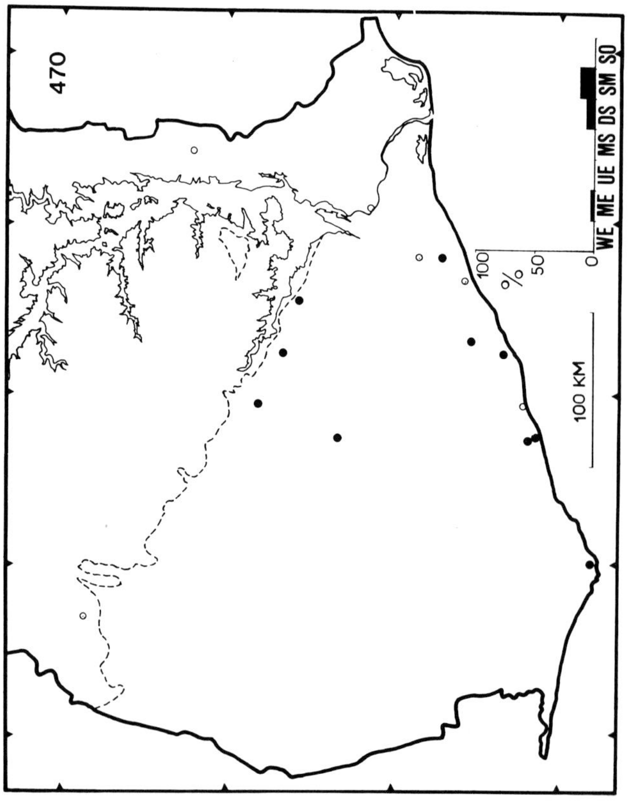
470
100 KM
%
100
50
0
WE ME UE MS DS SM SO

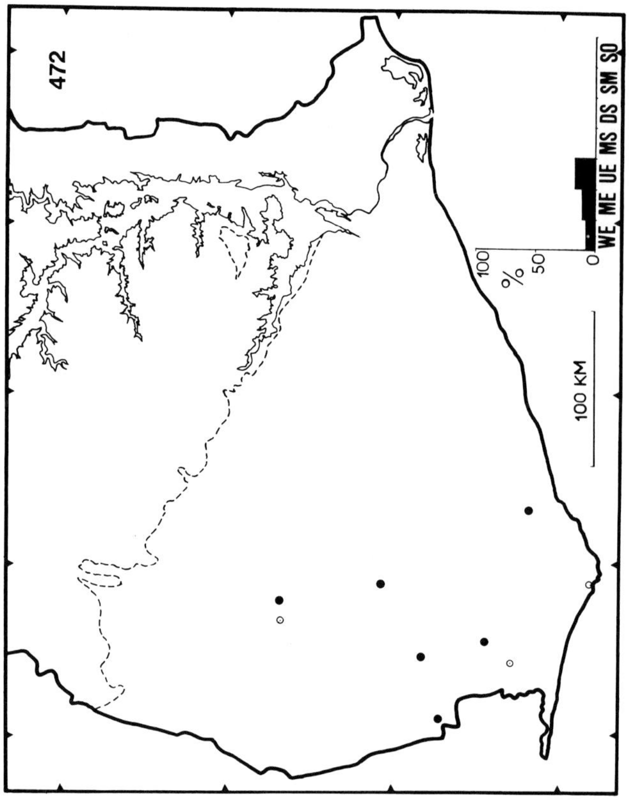
472
100 KM
%
100
50
0
WE ME UE MS DS SM SO

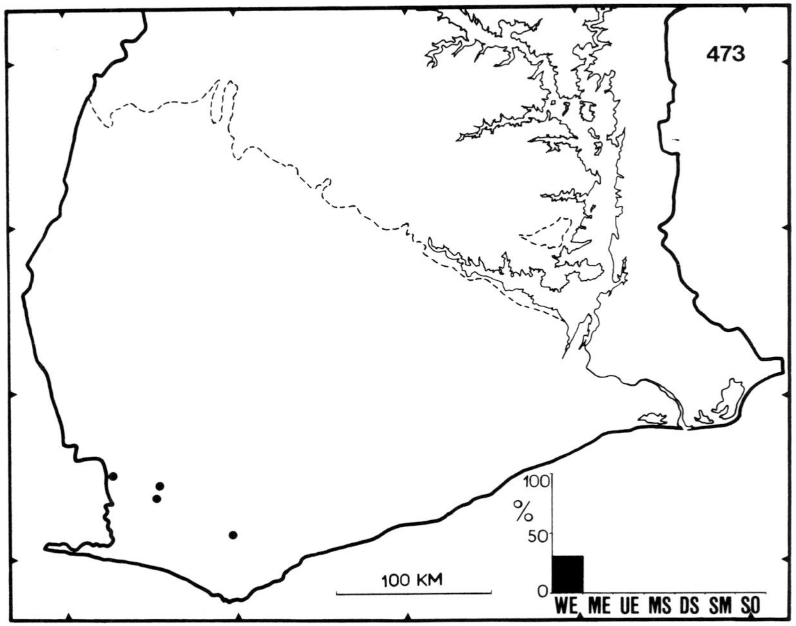

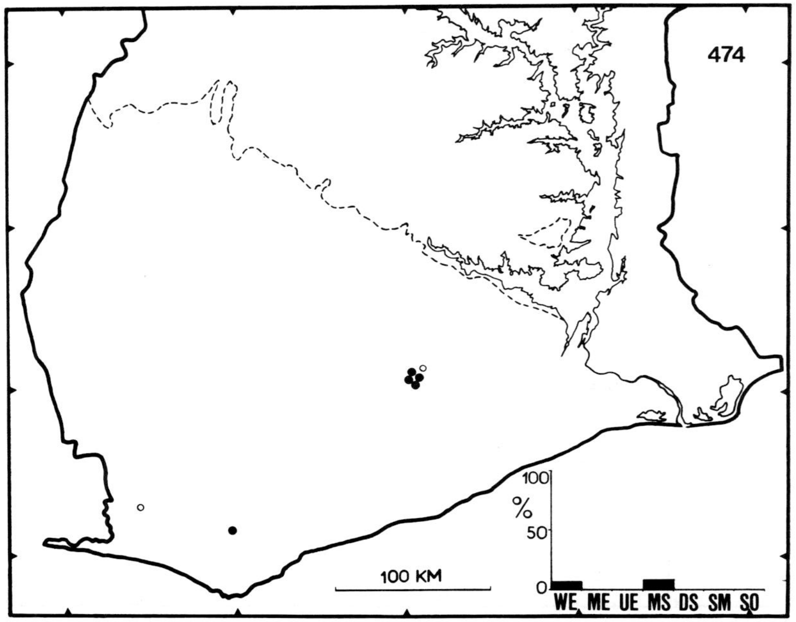

469 PSYCHOTRIA sp. aff. CALCEATA Petit (89), (*P. malchairei* of FWTA 2:201, partly (Burton s.n.), not of De Wild.). Rubiaceae.
Pygmy tree; lvs mesophyll to macrophyll. Fl white, small, bell, probably heterostyled, not clearly seasonal. Fr black berry; seeds medium. Regenerates in shade. Iv. Coast and Ghana; (Upper Guinea endemic). This species is rather uniform, and probably undescribed.

470 PSYCHOTRIA CALVA Hiern, FWTA 2:200. (incl. *P. huae* De Wild., FWTA 2:200). Rubiaceae.
Shrubs; lvs notophyll to mesophyll, with linear bacterial nodules along the main veins. Fl white, small to medium, bell, heterostyled, mostly 3–6. Fr red berry; seeds medium. Mostly in forest with rather thin canopy; extends to fringing forest in the savanna zone. Senegal to Zaire; (Guineo-Congolian wide).

471 PSYCHOTRIA IVORENSIS De Wild. (90), (*P. adafoana* of FWTA 2:201, not necessarily of K. Schum., as type is lost). Rubiaceae.
Pygmy tree; lvs notophyll, with linear bacterial nodules along midrib. Fl white, small, bell, heterostyled, mostly 3–5. Fr berry; seeds medium. Regenerates in shade. Iv. Coast and Ghana; (Upper Guinea endemic). Very similar to *P. brachyantha* Hiern, but differing in the linear, not punctate, bacterial nodules, and the glabrous, not puberulous, stipules.

472 PSYCHOTRIA LIBERICA Hepper, FWTA 2;201. Rubiaceae.
Small shrub; lvs notophyll, puberulous on midrib below and on twigs. Fl white, small, bell, probably heterostyled, 4–5. Fr red berry; seeds medium. Regenerates in shade. Liberia to Ghana; Upper Guinea endemic).

473 PSYCHOTRIA LONGITUBA A. Chev. ex De Wild., FWTA 2:200, partly, (48). Rubiaceae.
Small decumbent shrub; lvs mesophyll. Fl white, small, bell, 12–4. Fr red berry; seeds medium. Regenerates in shade. Iv. Coast and Ghana; (Upper Guinea endemic).

474 PSYCHOTRIA MANGENOTII (Aké Assi) Verdc. (110), (*Cephaelis mangenotii* Aké Assi, FWTA 2:204). Rubiaceae.
Creeping herb; lvs microphyll, leaves and stem hairy. Fl white, small, bell, 4–6. Fr berry; seeds medium. Regenerates in shade. Iv. Coast and Ghana; (Upper Guinea endemic).

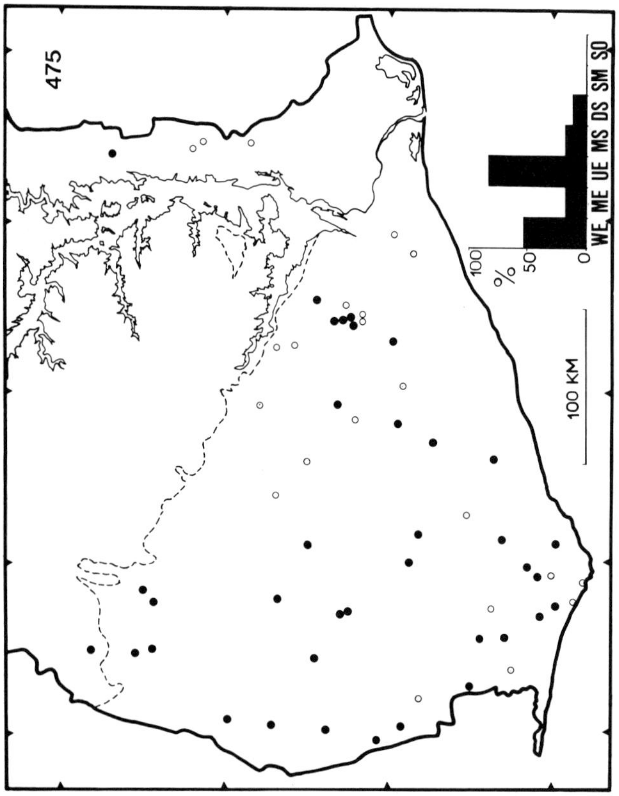
475
100 KM
%
100
50
0
WE ME UE MS DS SM SO

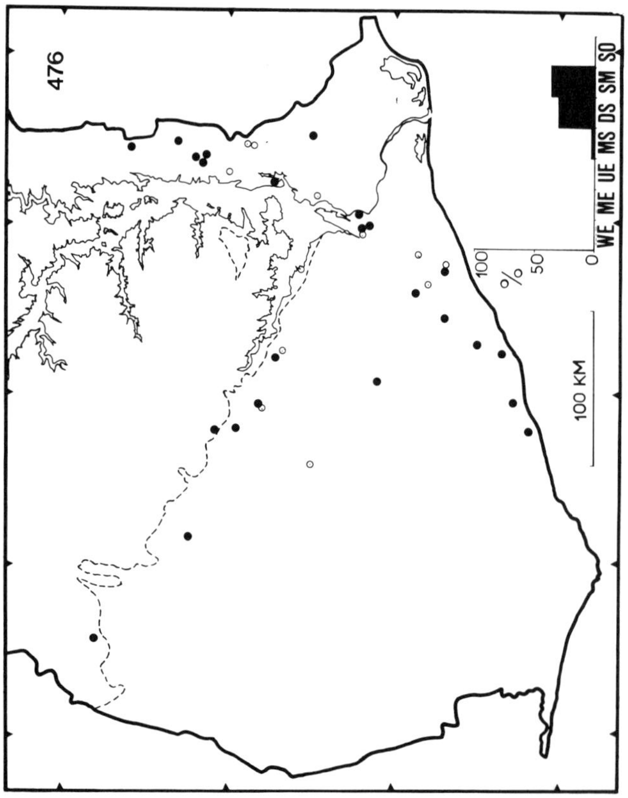
476
100 KM
%
100
50
0
WE ME UE MS DS SM SO

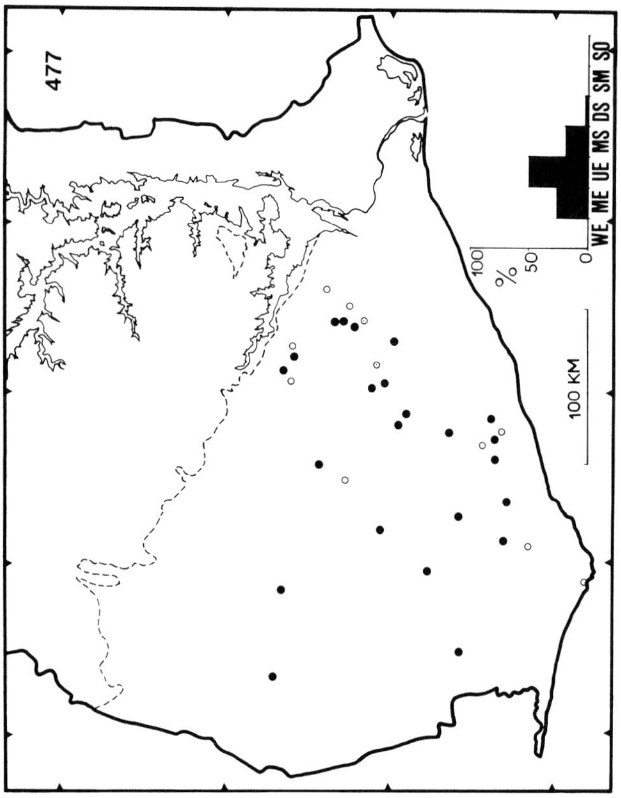
477
100 KM
%
100
50
0
WE ME UE MS DS SM SO

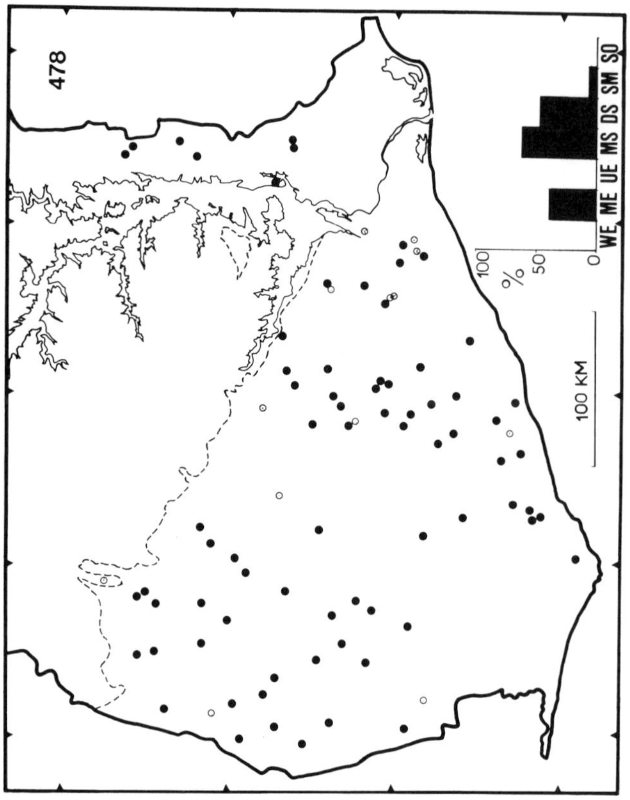
478
100 KM
%
100
50
0
WE ME UE MS DS SM SO

475 PSYCHOTRIA PEDUNCULARIS (Salisb.) Verdc. (110), (*Cephaelis peduncularis* Salisb., FWTA 2:204). Rubiaceae.

Shrub; lvs mesophyll, sometimes pubescent below. Leeuwenberg's model. Fl white, small, bell, clustered in bracteate head, not clearly seasonal. Fr blue-black berry; seeds medium. Widespread in Tropical Africa; (ecological and chorological transgressor). Several of the varieties dealt with in FWTA occur in Ghana, but they show little evidence of ecological discrimination.

476 PSYCHOTRIA SCHWEINFURTHII Hiern (89), (*P. obscura* Benth., FWTA 2:199). Rubiaceae.

Shrub; lvs notophyll to mesophyll, pubescent below. Fl white, small, bell, not clearly seasonal. Fr red berry; seeds medium. Regenerates in shade. Iv. Coast to Angola, Sudan and Uganda; (Guineo-Congolian wide).

477 PSYCHOTRIA SUBOBLIQUA Hiern, FWTA 2:200. Rubiaceae.
Shrub; lvs notophyll. Leeuwenberg's model. Fl white, small, bowl, clustered in head, 1–5. Fr red berry; seeds medium. Regenerates in shade. Iv. Coast to Zaire; (Guineo-Congolian wide).

478 PTERIS ACANTHONEURA Alston, FWTA FF:42. Adiantaceae.
Tufted ground fern; lvs pinnate, leaflets mesophyll, deeply lobed, spinulose on underside of midrib. Propagated by dust-like spores. High and low pH values very significantly under-represented in plot records. Iv. Coast to Zaire and Uganda; (Guineo-Congolian wide).

479 PTERIS ATROVIRENS Willd. FWTA FF:42. Adiantaceae.
Tufted ground fern; lvs pinnate, leaflets mesophyll to macrophyll, deeply lobed, spinulose on underside of midrib. Propagated by dust-like spores. Widespread in Tropical Africa; (chorological transgressor).

480 PTERIS BURTONII Bak. (*P. burtoni*, FWTA FF:42). Adiantaceae. (Type from near Tarkwa).

Tufted ground fern; lvs pinnate, leaflets mesophyll, very variable as to degree of lobing, even on one leaf. Propagated by dust-like spores, and by bulbils on the fronds. Plot records very significantly correlated with clay-rich, acid soils. Guinée to Angola; (Guineo-Congolian wide).

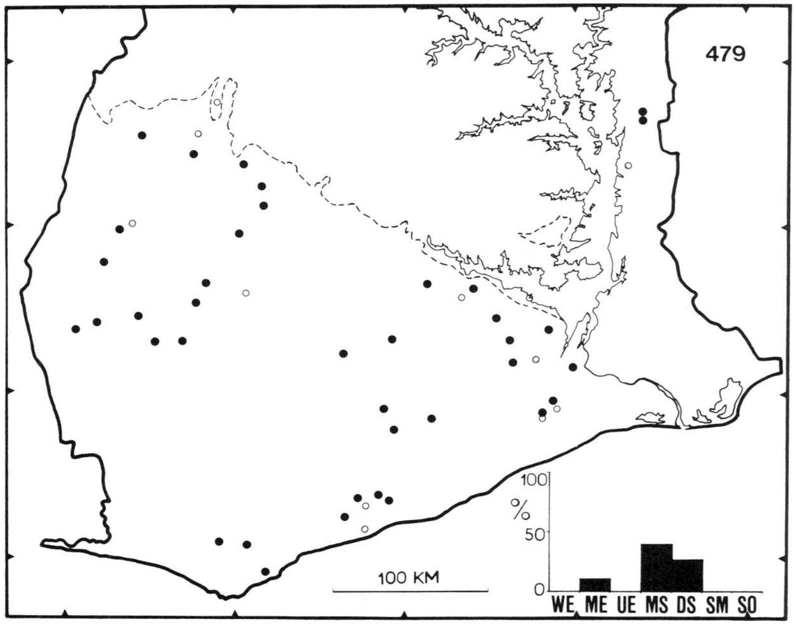

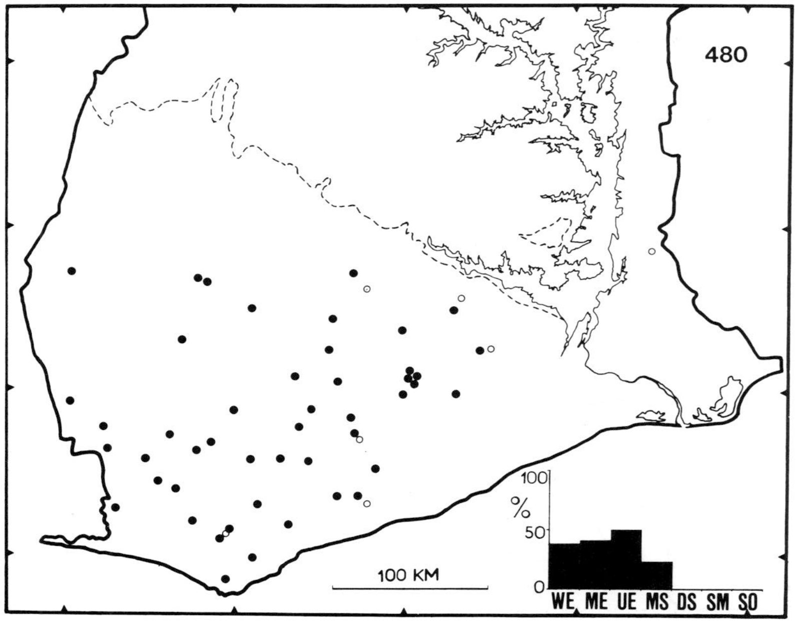

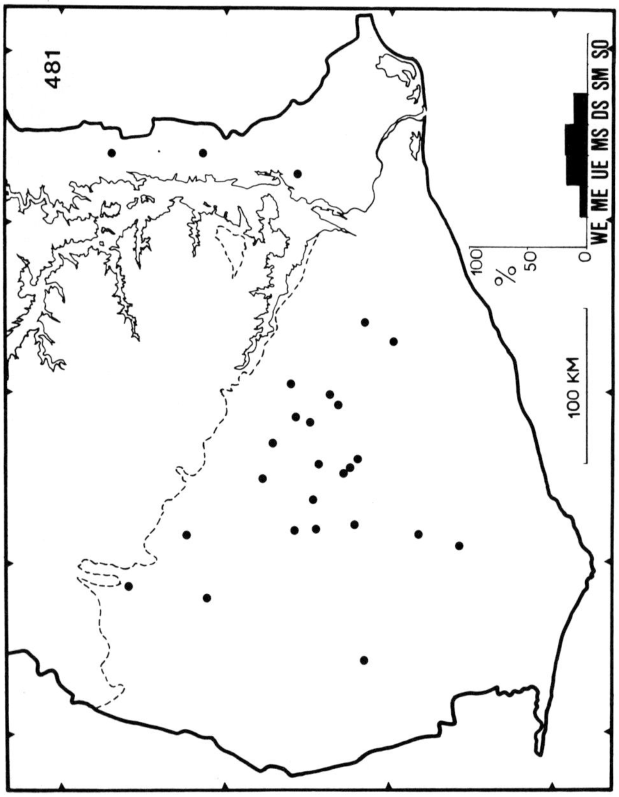
481
100 KM
%
100
50
0
WE ME UE MS DS SM SO

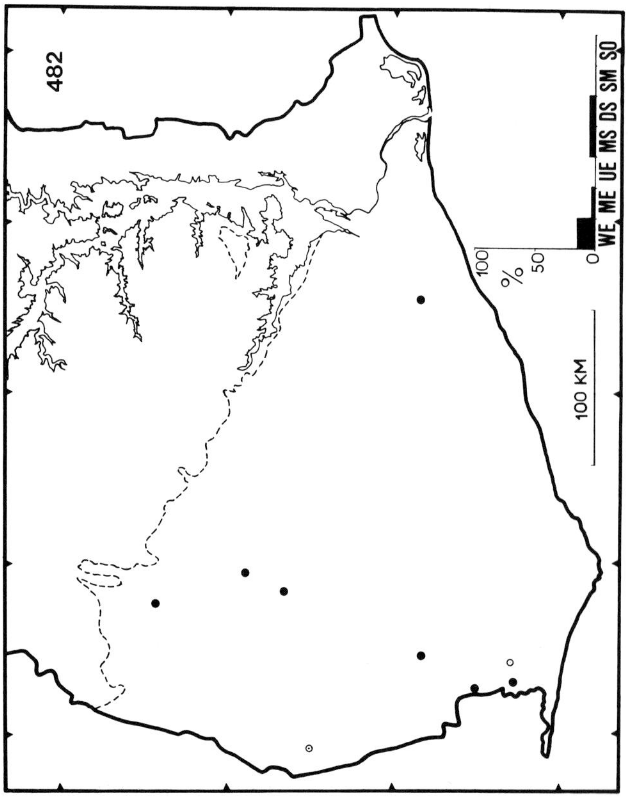
482
100 KM
%
100
50
0
WE ME UE MS DS SM SO

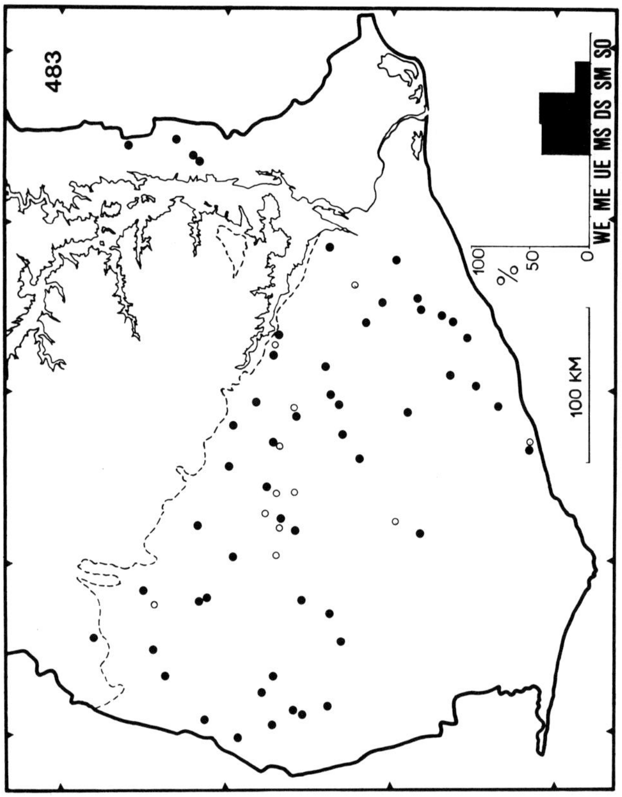
483
100 KM
%
100
50
0
WE ME UE MS DS SM SO

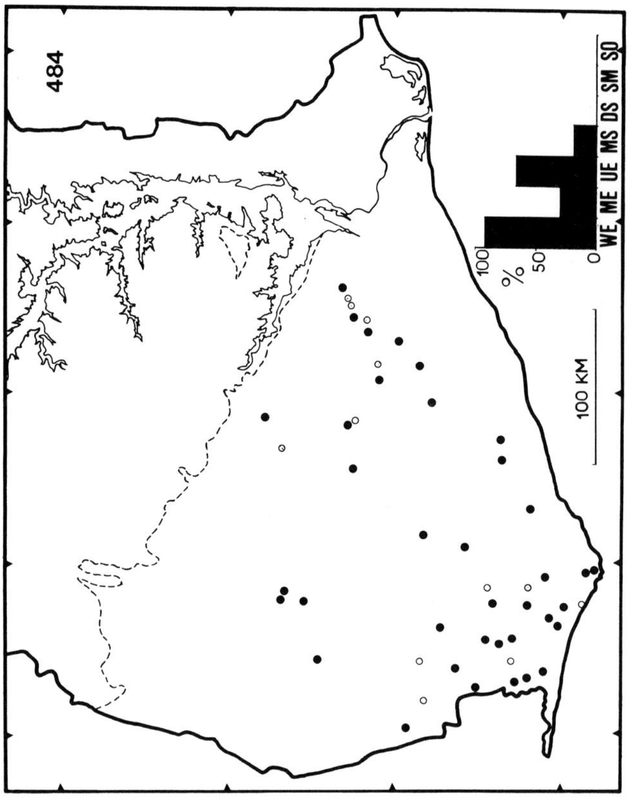
484
100 KM
%
100
50
0
WE ME UE MS DS SM SO

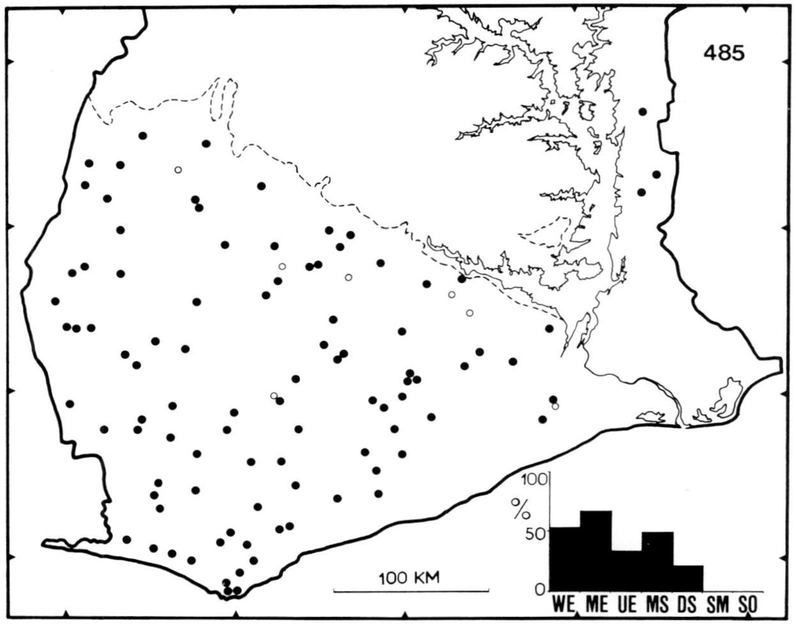

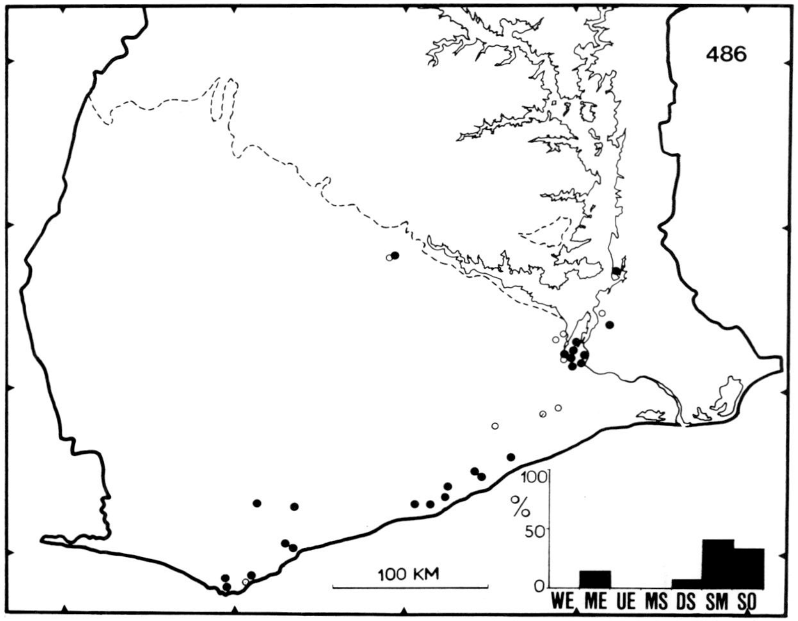

481 PTERIS TOGOENSIS Hieron., FWTA FF:40. Adiantaceae.
Tufted ground fern; lvs pinnate, leaflets microphyll, lobed almost to the midrib. Propagated by dust-like spores. Guinée to Cameroun; (Guinea wide).

482 PTERYGOTA BEQUAERTII De Wild., FWTA 1:320. Sterculiaceae. Akan: kyereyebere.
Large tree with buttresses; lvs mesophyll, deciduous; twigs puberulous. Fl brownish-green with red centre, medium, bowl, monoecious or polygamous, not clearly seasonal. Fr large follicle; seeds large, winged, wind-dispersed. Regenerates in shade. Iv. Coast to Zaire; (Guineo-Congolian wide).

483 PTERYGOTA MACROCARPA K. Schum, FWTA 1:320. Sterculiaceae. Akan: kyereye.
Large tree with buttresses; lvs mesophyll to macrophyll, deciduous. Fl brownish outside, pinkish inside, medium to large, bowl, monoecious, 10–1. Fr large follicle, 1–3; seeds very large, winged (Fig. 6.1g), wind-dispersed. Regenerates in shade. Plot records show a highly significant preference for base-rich, not-too-acid, soils. S. Leone to Cameroun; (Guinea wide). Timber is unclassified, but could be exploited (40).

484 PTYCHOPETALUM ANCEPS Oliv., FWTA 1:647. Olacaceae.
Small treelet or shrub; lvs notophyll. Fl white, small to medium, funnel, not clearly seasonal. Fr velvety red drupe, seed medium. Regenerates in shade. Plot records very significantly correlated with acid, base-poor soils. S. Leone to Ghana; (Upper Guinea endemic).

485 PYCNANTHUS ANGOLENSIS (Welw.) Warb., FWTA 1:61. Myristicaceae. Akan: otie.
Large tree, containing clear, pinkish latex; lvs mesophyll, puberulous on underside, characteristically much perforated by predator attack. Massart's model (59). Fl brownish, small, crowded in clusters on the inflorescences, monoecious (sometimes reported to be dioecious). Fr 1-seeded capsule; seed very large, with red aril. Regenerates in shade. Guinée to Angola und Uganda; (Guineo-Congolian wide). Class 3 timber; trade name: ilomba (40).

486 PYCNOCOMA CORNUTA Muell. Arg., FWTA 1:405. Euphorbiaceae. Akan: kafiekafie.
Unbranched or little-branched pygmy tree; lvs macrophyll, clustered at the ends of the stems. Corner's model. Fl whitish, medium, brush, monoecious, 11–4. Fr explosive capsule; seeds large. Regenerates in shade. Ghana to Nigeria; (Guinea wide). Very close to *P. macrophylla,* differing mainly in leaf shape and ecology; would better be regarded as a variety than as a full species.

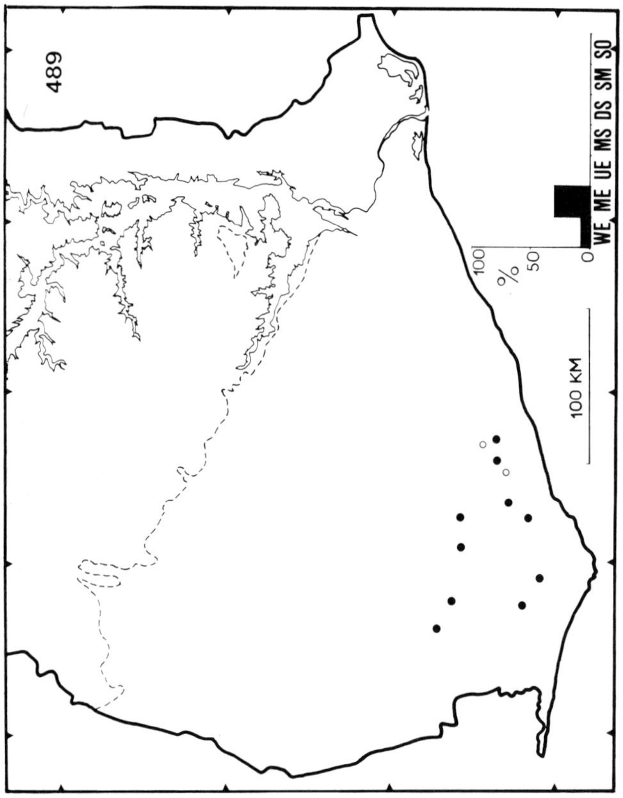
489
100 KM
%
100
50
0
WE ME UE MS DS SM SO

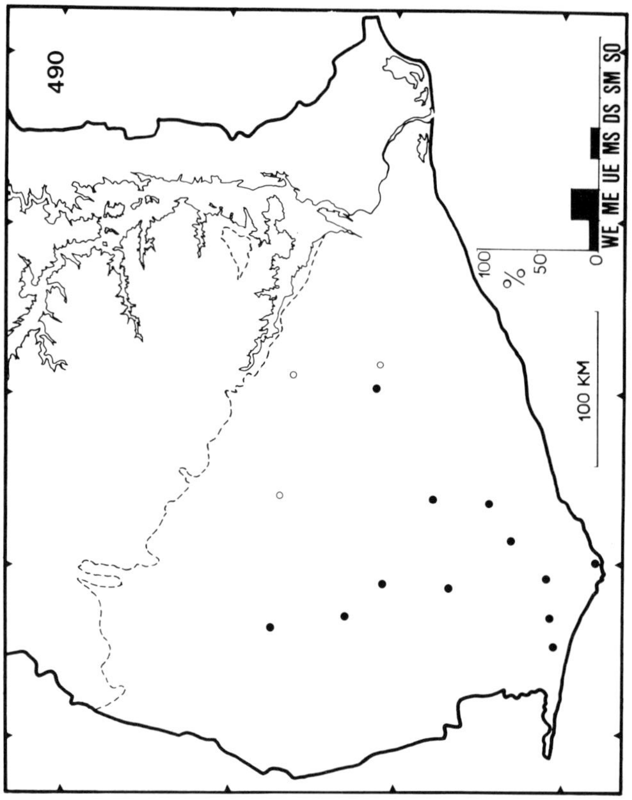
490
100 KM
%
100
50
0
WE ME UE MS DS SM SO

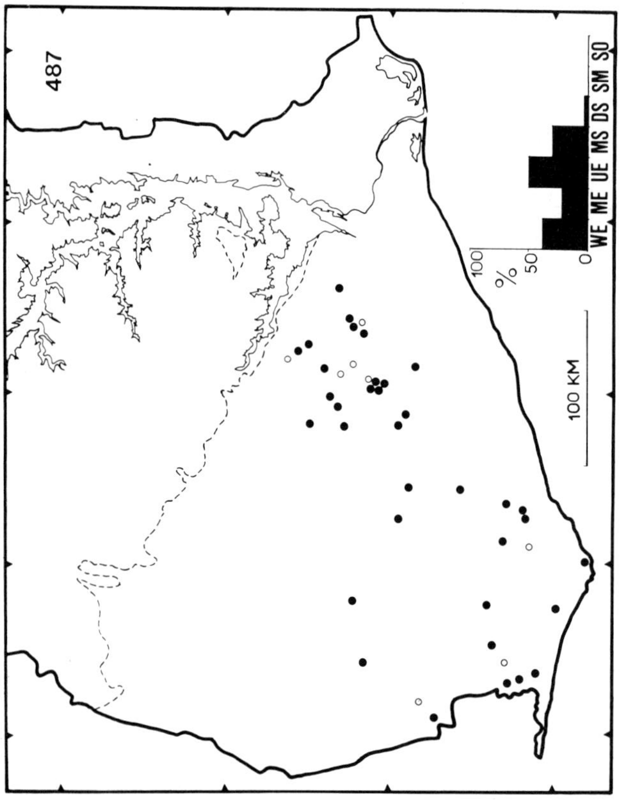
487
100 KM
%
100
50
0
WE ME UE MS DS SM SO

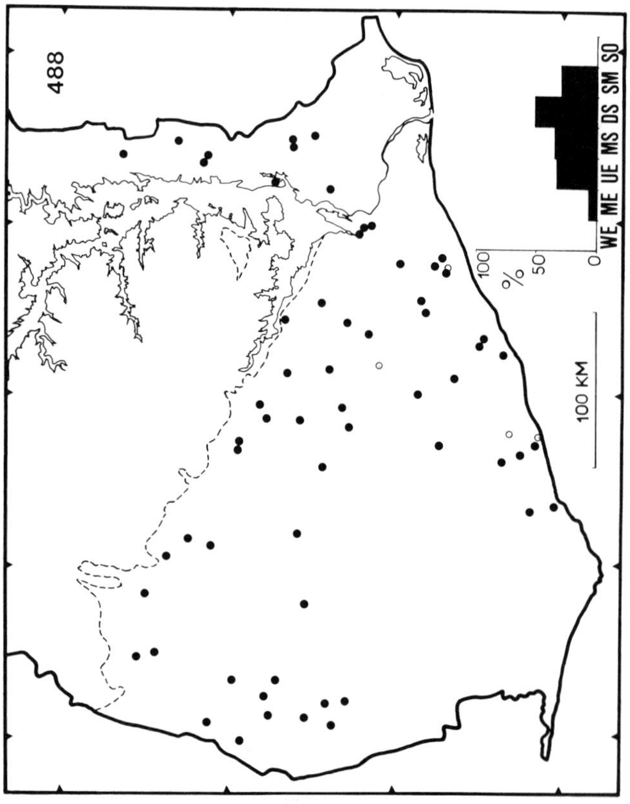
488
100 KM
%
100
50
0
WE ME UE MS DS SM SO

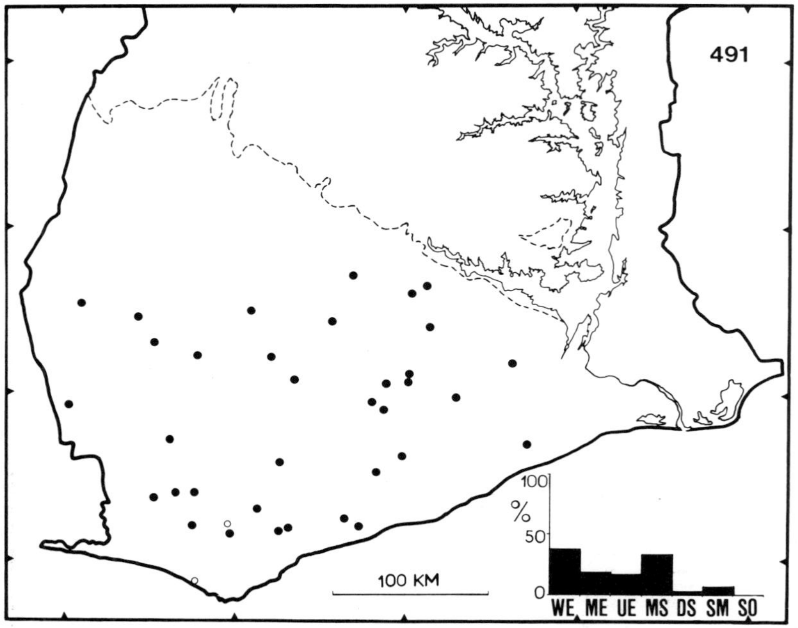

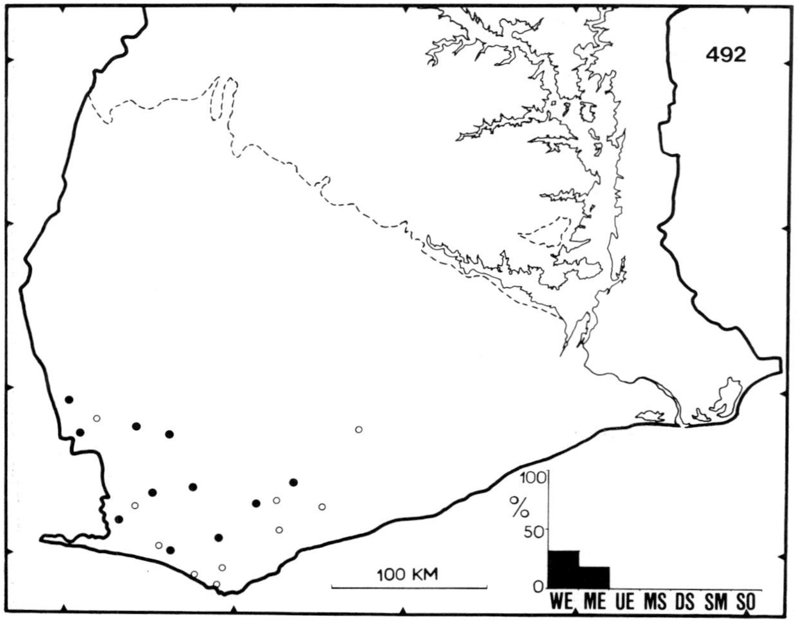

487 PYCNOCOMA MACROPHYLLA Benth., FWTA 1:405. Euphorbiaceae. Akan: kafiekafie.
Unbranched or little-branched pygmy tree; lvs macrophyll, clustered at the ends of the stems. Corner's model. Fl whitish, medium, brush, monoecious, not clearly seasonal. Regenerates in shade. Iv. Coast to Zaire; (Guineo-Congolian wide).

488 PYRENACANTHA ACUMINATA Engl., FWTA 1:642. (Incl. *P.* sp. *A*, FWTA 1:642). Icacinaceae.
Small woody climber; lvs notophyll to mesophyll, appressed pubescent on undersurface. Fl small, greenish, inconspicuous, dioecious, mostly 3–6. Fr orange drupe; seed medium. S. Leone to Zaire; (Guineo-Congolian wide).

489 PYRENACANTHA CORDICULA Villiers (112). Icacinaceae.
Small woody climber; lvs notophyll to mesophyll, appressed pubescent below. Fl small, greenish, inconspicuous, dioecious; (female flowers and fruits unknown). Regenerates in shade. Iv. Coast, Ghana and Cameroun; (Guinea wide).

490 PYRENACANTHA VOGELIANA Baill., FWTA 1:642. Icacinaceae.
Small woody climber; lvs notophyll to mesophyll, appressed puberulous below. Fl small, yellow, inconspicuous, dioecious. Fr orange drupe; seed medium. Regenerates in shade. S. Leone to Zaire and Tanzania; (Guineo-Congolian wide).

491 RAPHIA HOOKERI Mann & Wendl., FWTA 3:162. Palmae. Akan: adobe.
Medium-sized palm tree; lvs pinnate, leaflets megaphyll, prickly on the margin. Holttum's model. Fl yellowish, medium, funnel, monoecious, not seasonal. Fr drupe covered by hard, brown, shiny scales; seed very large, surrounded by orange flesh. Reaches maturity only in swamps, but seeds are widely dispersed and germinate under forest canopy on dry ground. Guinée to Gabon; (Guinea wide).

492 RAUVOLFIA CUMMINSII Stapf, FWTA 2:69. Apocynaceae. (Type from Assin Nyankumasi).
Pygmy tree, containing white latex; lvs mesophyll. Leeuwenberg's model. Fl small to medium, whitish, dish with tube, not clearly seasonal. Fr red 1-seeded berry; seed medium. Regenerates in shade. Liberia to Ghana; (Upper Guinea endemic).

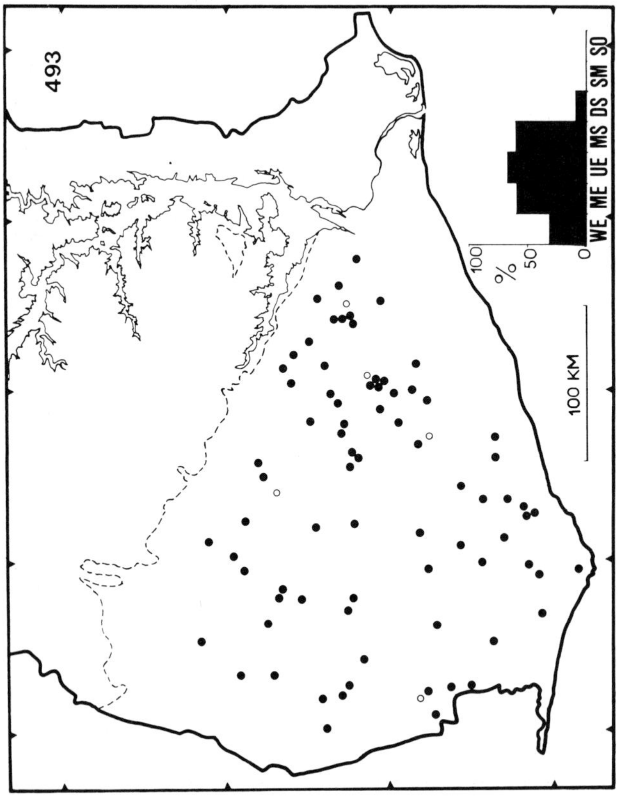

493
100 KM
%
100
50
0
WE ME UE MS DS SM SO

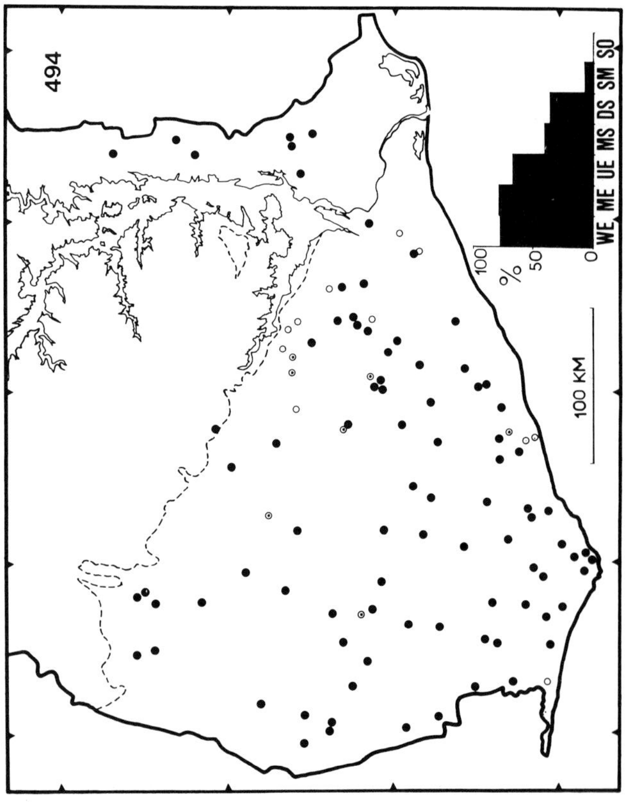

494
100 KM
%
100
50
0
WE ME UE MS DS SM SO

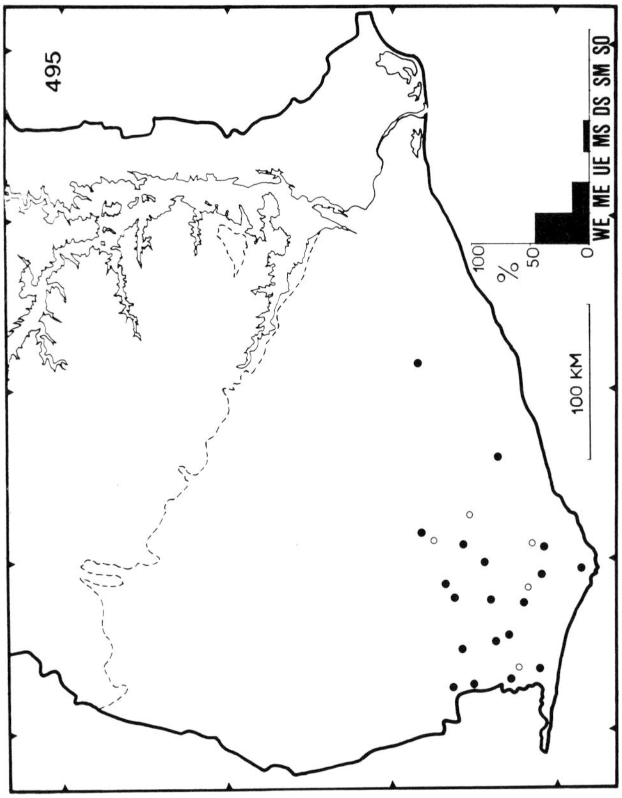

495
100 KM
%
100
50
0
WE ME UE MS DS SM SO

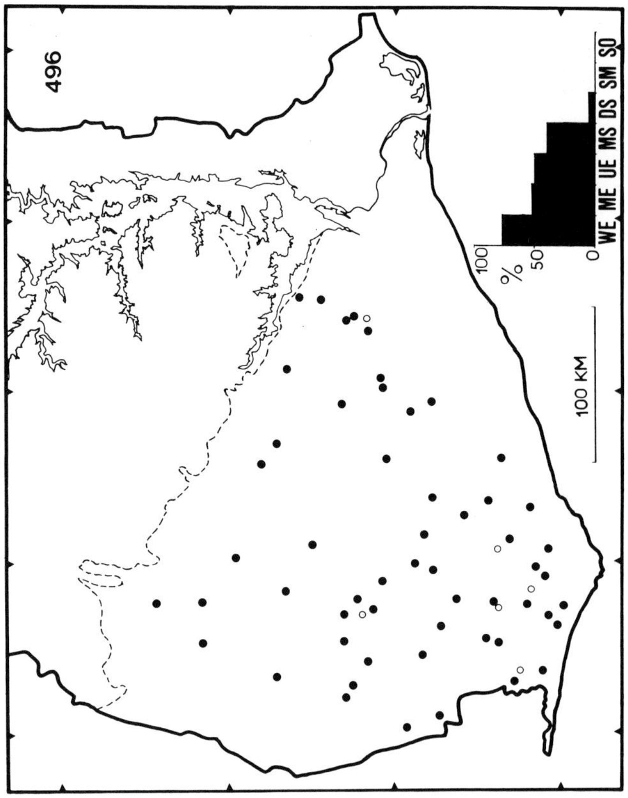

496
100 KM
%
100
50
0
WE ME UE MS DS SM SO

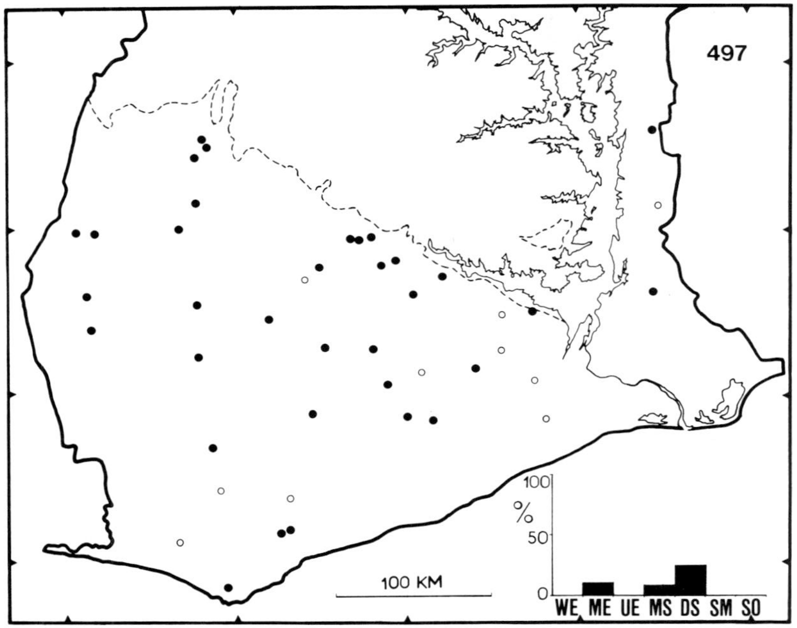

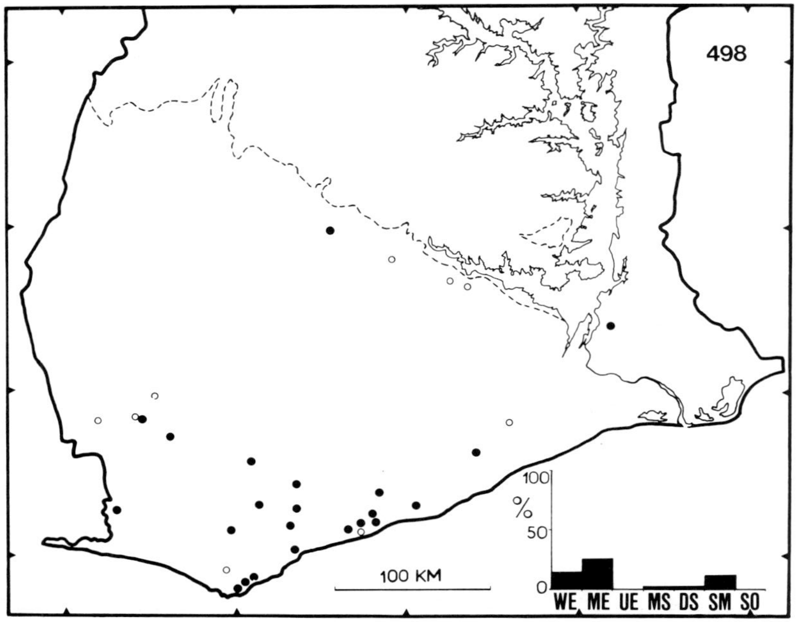

493 RHAPHIDOPHORA AFRICANA N.E. Br., FWTA 3:114. Araceae.
Large herbaceous root climber which may reach the forest canopy; lvs macrophyll. Fl white, small, crowded in spike with yellow spathe, 12–3. Fr compound berry; seeds small. S. Leone to Nigeria and Fernando Po; (Guinea wide).

494 RHAPHIOSTYLIS BENINENSIS (Hook. f. ex. Planch.) Planch. ex Benth., FWTA 1:638, and RHAPHIOSTYLIS PREUSSII Engl., FWTA 1:639. Icacinaceae.
Medium to large woody climbers (or shrubs in thicket), wood in concentric rings separated by phloem; lvs notophyll. Mangenot's model (59). Fl white, sweet-scented, medium, funnel, 9–1. Fr red drupe, 12–2; seed medium. Regenerate in shade. The two 'species' do not differ consistently except in ovary pubescence, and therefore could not be distinguished during the survey. Herbarium records are mapped separately: *R. beninensis* ○, *R. preussii* ⊙. The map shows that they do not appear to differ ecologically either, and it is very doubtful that they are good species. *R. beninensis* extends from Guinée to Angola and Zambia, (Guineo-Congolian wide); *R. preussii* from S. Leone to Gabon; (Guinea wide).

495 RHAPHIOSTYLIS CORDIFOLIA Hutch. & Dalz., FWTA 1:639. Icacinaceae.
Small woody climber; lvs notophyll, leaves and twigs spreading-hairy. Fl white, small to medium, funnel, 9–11. Fr red drupe; seed medium. Regenerates in shade. Liberia to Ghana; (Upper Guinea endemic).

496 RHAPHIOSTYLIS FERRUGINEA Engl., FWTA 1:639. Icacinaceae.
Large woody climber (or shrub in thicket) wood in concentric rings separated by phloem. Lvs notophyll, twigs and underside of midrib pubescent. Fl white, medium, funnel, not clearly seasonal. Fr large woody drupe; seed large. Regenerates in shade. Plot records very highly correlated with base-poor, acid soils. Iv. Coast to Zaire; (Guineo-Congolian wide).

497 RICINODENDRON HEUDELOTII (Baill.) Pierre ex Pax, FWTA 1:393. Euphorbiaceae. Akan: wama.
Large tree with buttresses; lvs digitate, leaflets mesophyll, slightly toothed, deciduous. Rauh's model (59). Fl white, small to medium, bell, dioecious, 12–3. Fr yellow 3-seeded drupe, 7–10; seeds large. Regenerates only in gaps. Guinea Bissau to Angola and Tanzania; (Guineo-Congolian wide). Timber very soft; trade name: erimado (40).

498 RINOREA BRACHYPETALA (Turcz.) O. Ktze., FWTA 1:104. (Incl. *R. poggei* Engl. (103, 104), FWTA 1:104). Violaceae.
Pygmy tree; lvs mesophyll, toothed. Fl small to medium, yellowish, bell, 12–5. Fr capsule; seeds medium. Regenerates in shade. S. Leone to Angola, Kenya and Zambia; (Guineo-Congolian wide).

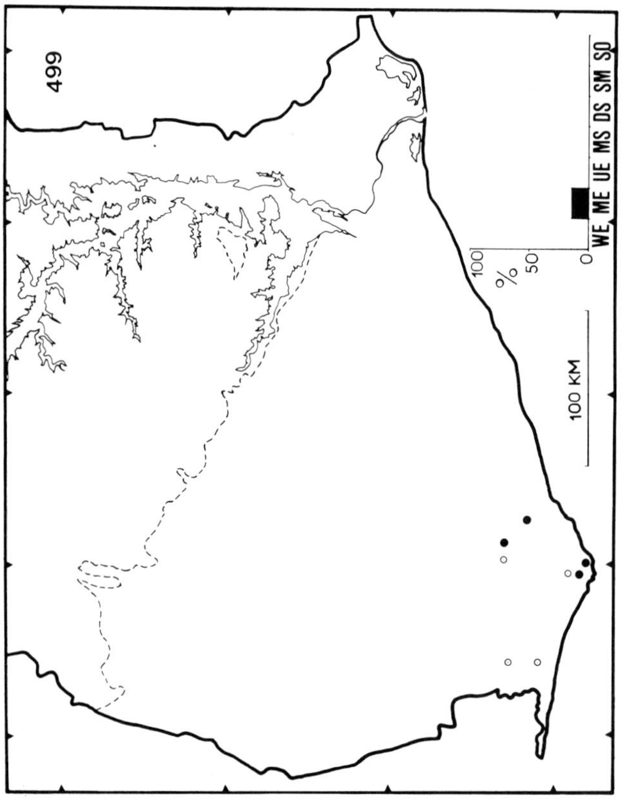
499
100 KM
100
%
50
0
WE ME UE MS DS SM SO

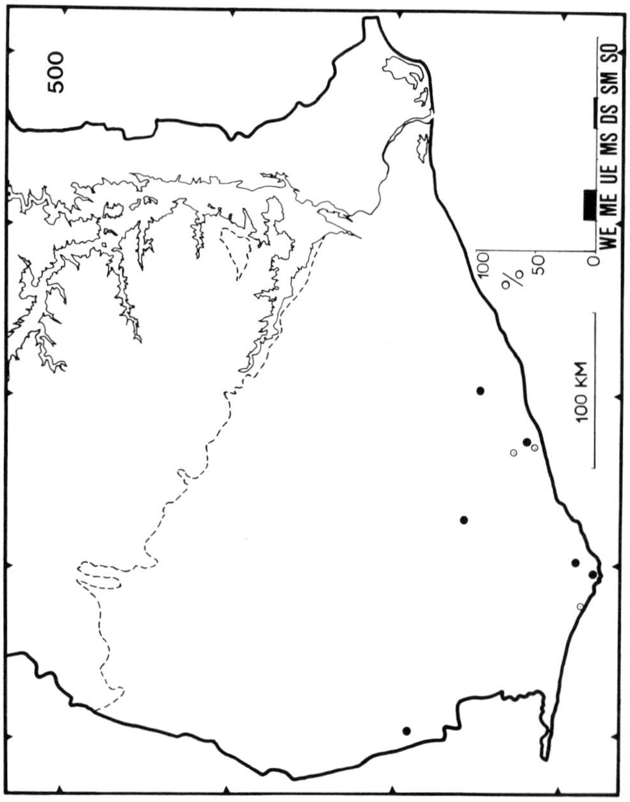
500
100 KM
100
%
50
0
WE ME UE MS DS SM SO

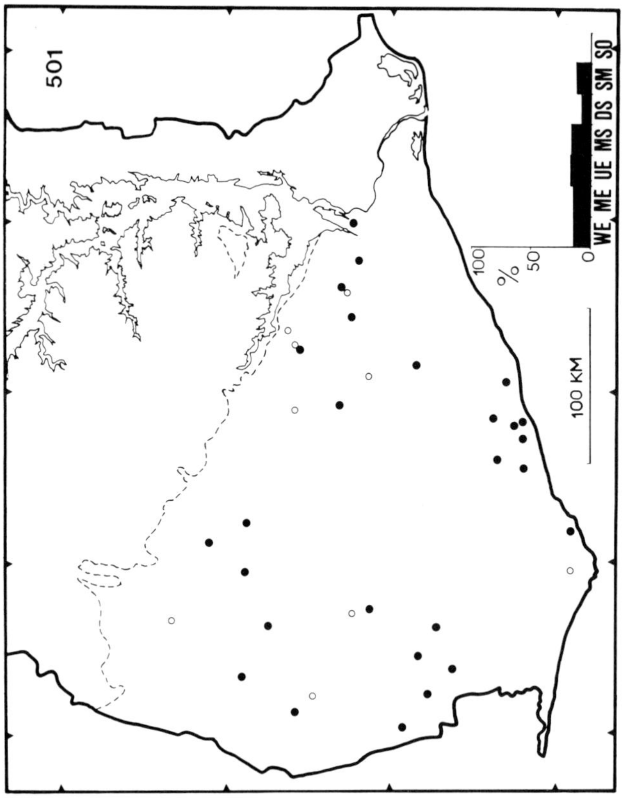
501
100 KM
100
%
50
0
WE ME UE MS DS SM SO

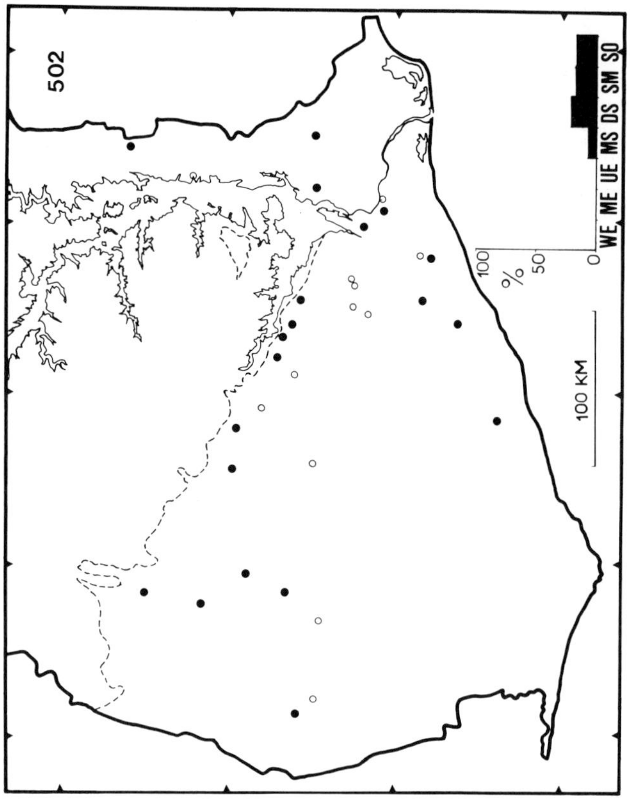
502
100 KM
100
%
50
0
WE ME UE MS DS SM SO

499 RINOREA BREVIRACEMOSA Chipp, FWTA 1:103. Violaceae.
Pygmy tree; lvs microphyll, slightly toothed. Fl small, cream, bell, 12–2. Fr capsule, seeds medium. Regenerates in shade. Often gregarious. S. Leone to Nigeria; (Guinea wide).

500 RINOREA DENTATA (P. Beauv.) O. Ktze., FWTA 1:104. Violaceae.
Pygmy to small tree; lvs notophyll to mesophyll, toothed; pubescent on underside of nerves, and on twigs. Fl greenish-yellow, small, bell, 9–12. Fr capsule, seeds medium. Regenerates in shade. Liberia to Angola and Tanzania; (Guineo-Congolian wide).

501 RINOREA ILICIFOLIA (Welw. ex. Oliv.) O. Ktze., FWTA 1:101. Violaceae.
Pygmy tree; lvs mesophyll with spiny teeth. Fl small, cream, bell, 12–3. Fr capsule; seeds medium. Regenerates in shade. Widespread in Tropical Africa and Madagascar; (chorological transgressor).

502 RINOREA KIBBIENSIS Chipp, FWTA 1:103. Violaceae. Akan: duaborobae. (Type from Kibi).
Small tree; lvs notophyll, toothed; nerves and twigs pubescent. Troll's model. Fl small, cream, bell, not clearly seasonal. Fr capsule; seeds medium. Regenerates in shade. Gregarious in forest understorey on rocky slopes. Iv. Coast to Zaire; (Guineo-Congolian wide, 103).

503 RINOREA OBLONGIFOLIA (C.H. Wright) Marquand ex Chipp, FWTA 1:104. Violaceae. Akan: mpawutuntum.
Small tree; lvs mesophyll, toothed. Troll's model. Fl cream, small, bell, not clearly seasonal. Fr capsule; seeds medium. Regenerates in shade. Tends to be shorter in evergreen than in semideciduous forest. Guinée to Angola and Uganda; (Guineo-Congolian wide).

504 RINOREA PRASINA (Stapf) Chipp, FWTA 1:103. Violaceae.
Pygmy to small tree; lvs mesophyll, toothed, with sessile glands on underside. Fl cream, small, bell, 10–1. Fr capsule; seeds medium. Regenerates in shade. Narrow-leaved form occurs on the banks of fast-flowing streams. Gambia to Ghana; (Upper Guinea endemic).

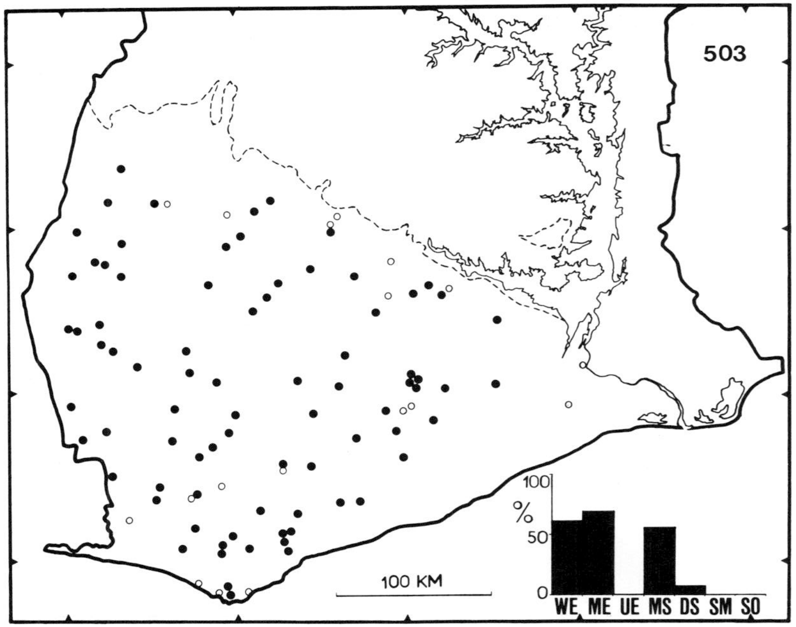

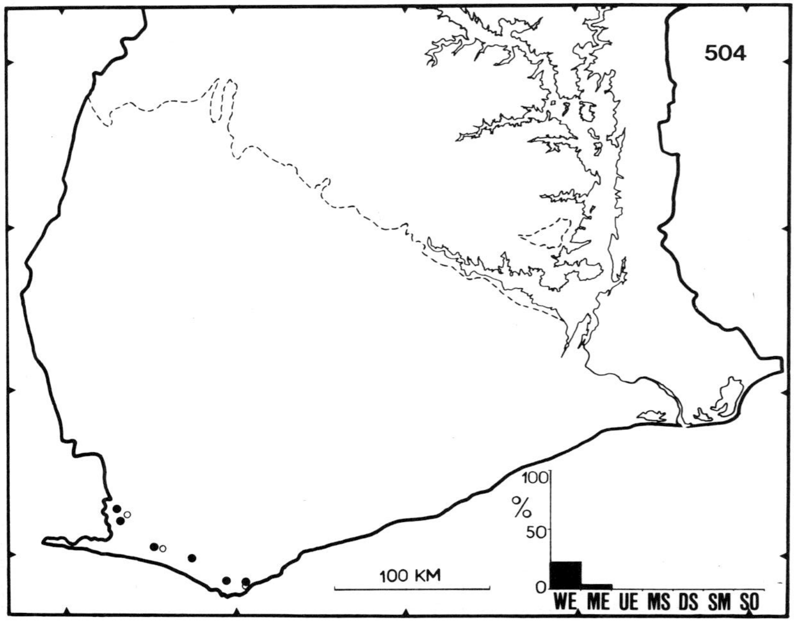

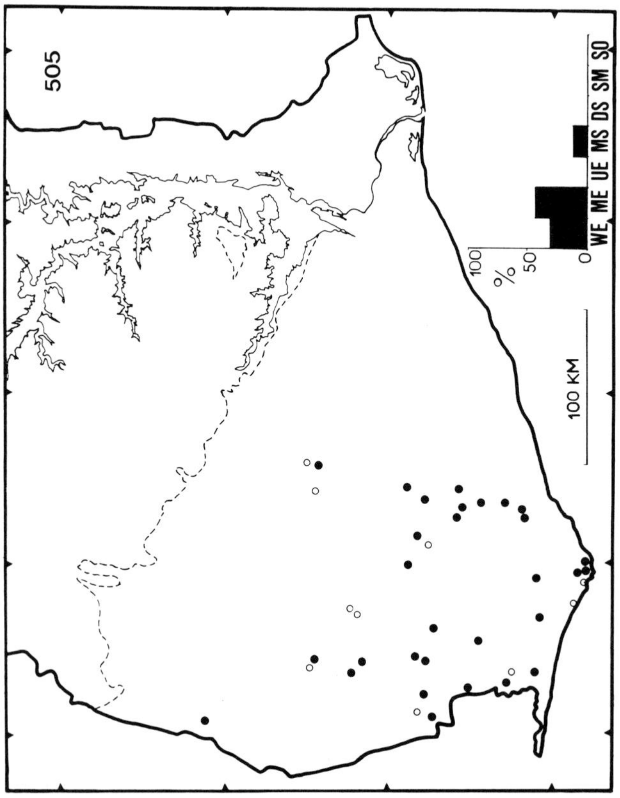
505
100 KM
%
100
50
0
WE ME UE MS DS SM SO

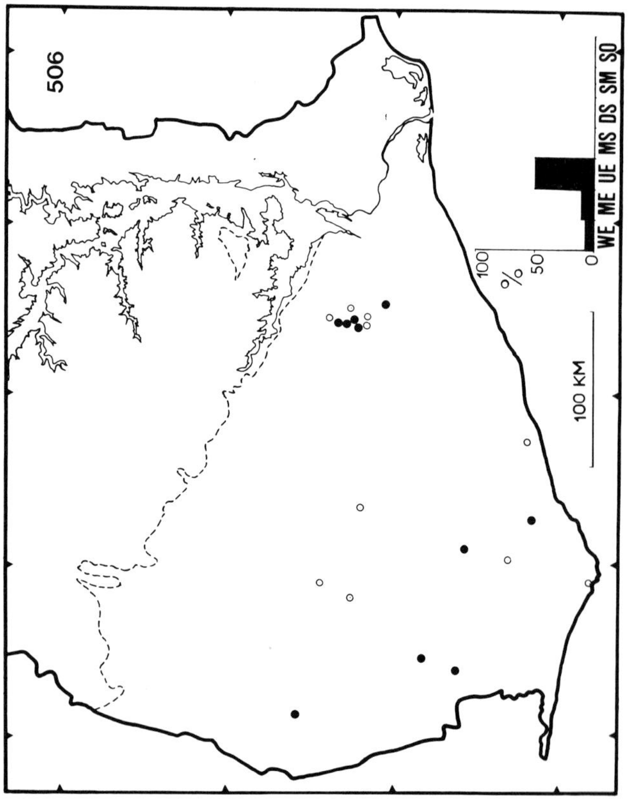
506
100 KM
%
100
50
0
WE ME UE MS DS SM SO

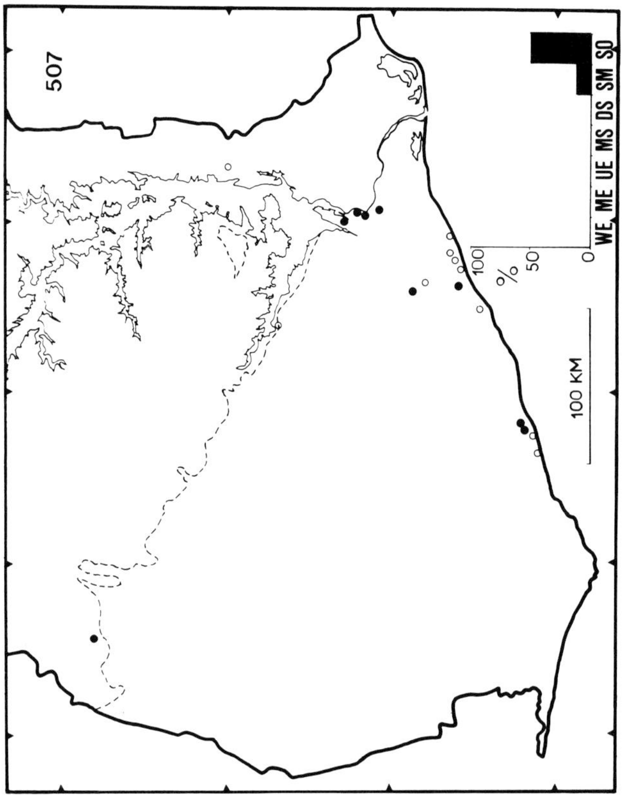
507
100 KM
%
100
50
0
WE ME UE MS DS SM SO

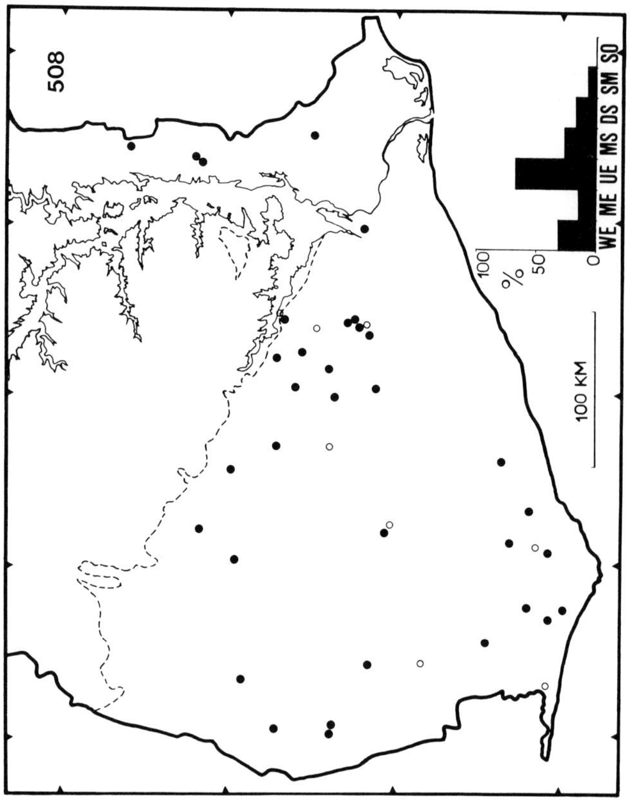
508
100 KM
%
100
50
0
WE ME UE MS DS SM SO

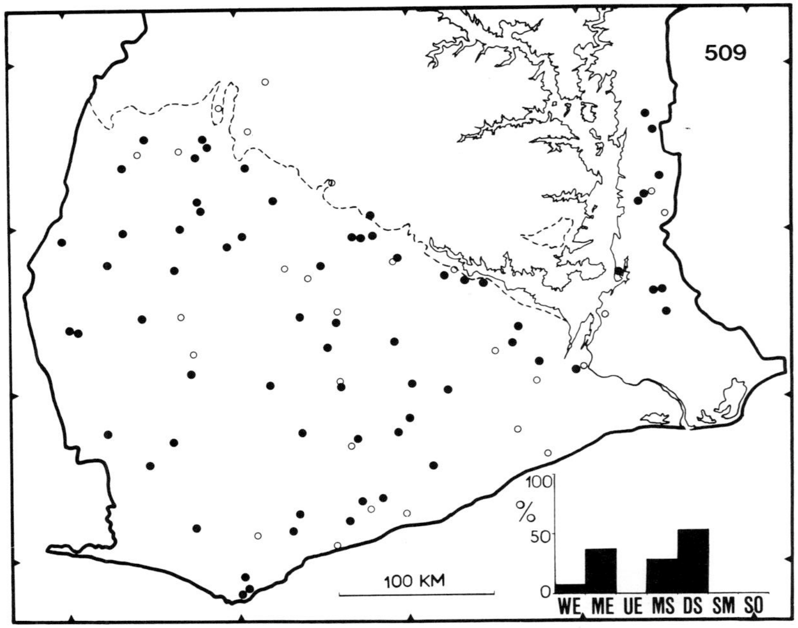

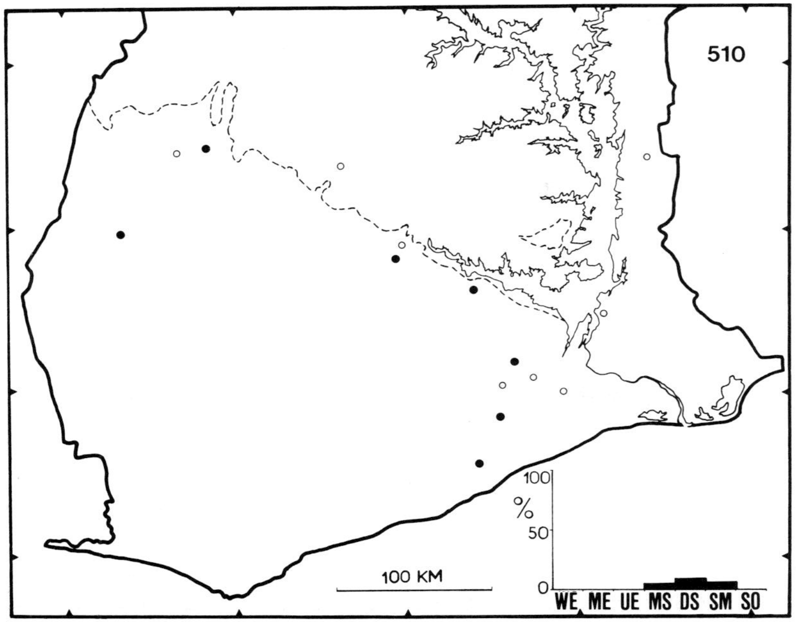

505 RINOREA SUBINTEGRIFOLIA (P. Beauv.) O. Ktze., FWTA 1:104. Violaceae.
Pygmy tree; lvs notophyll to mesophyll, more or less toothed. Troll's model. Fl small, cream, bell, not clearly seasonal. Fr capsule; seeds medium. Regenerates in shade. Guinée to Zaire and Tanzania; (Guineo-Congolian wide, 103).

506 RINOREA WELWITSCHII (Oliv.) O. Ktze. (Incl. *R. elliotii* Engl. (103, 104), FWTA 1:103, and *R. longicuspis* Engl. (104), FWTA 1:103). Violaceae. Akan: aposenini.
Small tree; lvs mesophyll, toothed, with sessile glands below, sometimes pubescent below. Troll's model. Fl cream, small, bell, 11–2. Fr capsule, seeds medium. Regenerates in shade. Appears to replace *R. oblongifolia* in Upland Evergreen forest. S. Leone to Angola and Zambia; (Guineo-Congolian wide).

507 RITCHIEA REFLEXA (Thonn.) Gilg & Benedict, FWTA 1:92. Capparaceae (Capparidaceae in FWTA). Ga: ayerebi. (Type from near Accra).
Small woody climber; lvs notophyll. Fl cream, large, brush, mostly 10–3. Fr yellow berry, 2–4; seeds medium. Regenerates in shade. Guinée to Togo; (Upper Guinea endemic).

508 ROTHMANNIA HISPIDA (K. Schum.) Fagerlind, FWTA 2:125. Rubiaceae. Akan: tukobo.
Small to medium tree; lvs notophyll to mesophyll, more or less spreading-hairy. Petit's model (59). Fl large, bell with long tube, white and purple, 11–1. Fr berry, 11–1; seeds medium. Regenerates in shade. Guinée to Zaire; (Guineo-Congolian wide).

509 ROTHMANNIA LONGIFLORA Salisb., FWTA 2:125. Rubiaceae. Akan: samankube.
Small tree or shrub; lvs notophyll. Fagerlind's model (59). Fl large, funnel with long tube (Fig. 9.1a), white and purple, mostly 11–1. Fr berry, c. 4; seeds medium. Regenerates in shade. Guinea Bissau to Angola and Uganda; (Guineo-Congolian wide).

510 ROTHMANNIA URCELLIFORMIS (Hiern) Bullock ex Robyns, FWTA 2:125. Rubiaceae.
Small tree; lvs notophyll. Fagerlind's model. Fl large, cream and purple, funnel with tube, 9–1. Fr berry, 11–1; seeds medium. Regenerates in shade. Widespread in Tropical Africa; (chorological transgressor).

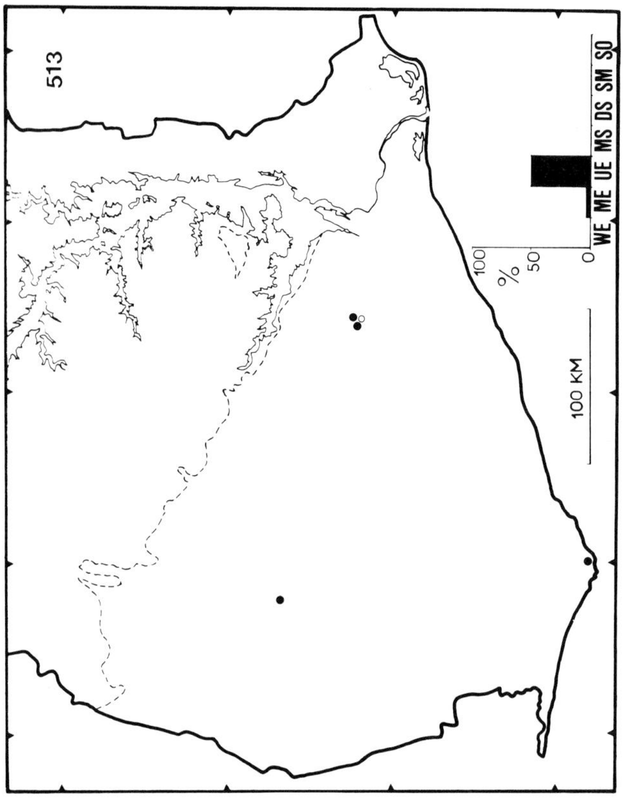
513
100 KM
% 100 50 0
WE ME UE MS DS SM SO

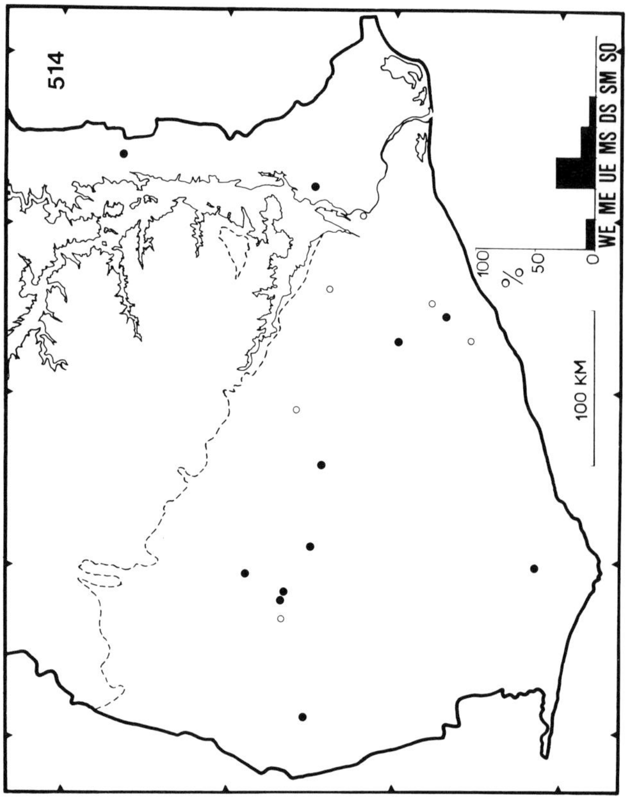
514
100 KM
% 100 50 0
WE ME UE MS DS SM SO

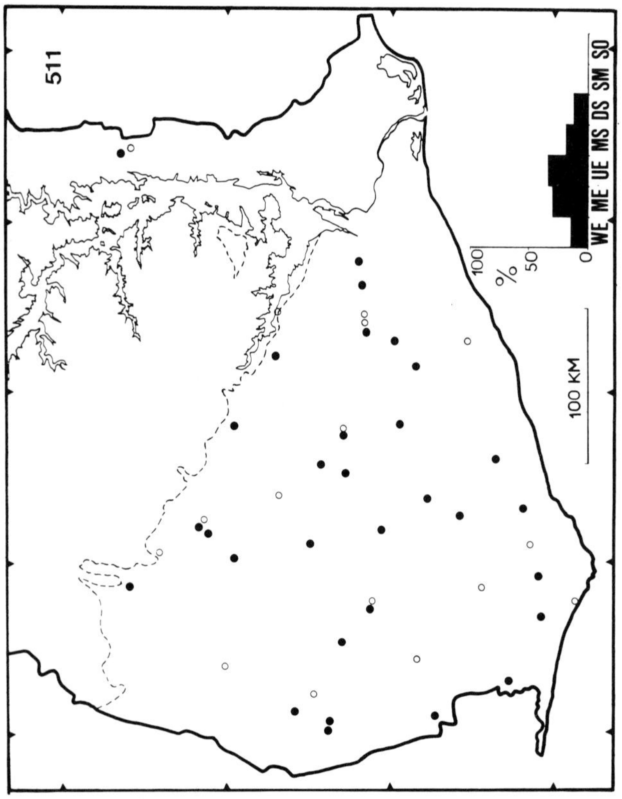
511
100 KM
% 100 50 0
WE ME UE MS DS SM SO

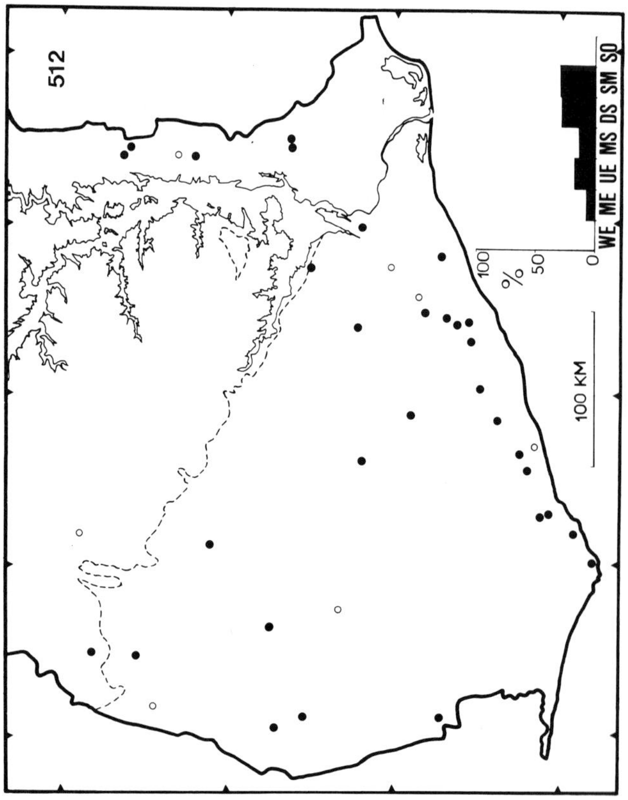
512
100 KM
% 100 50 0
WE ME UE MS DS SM SO

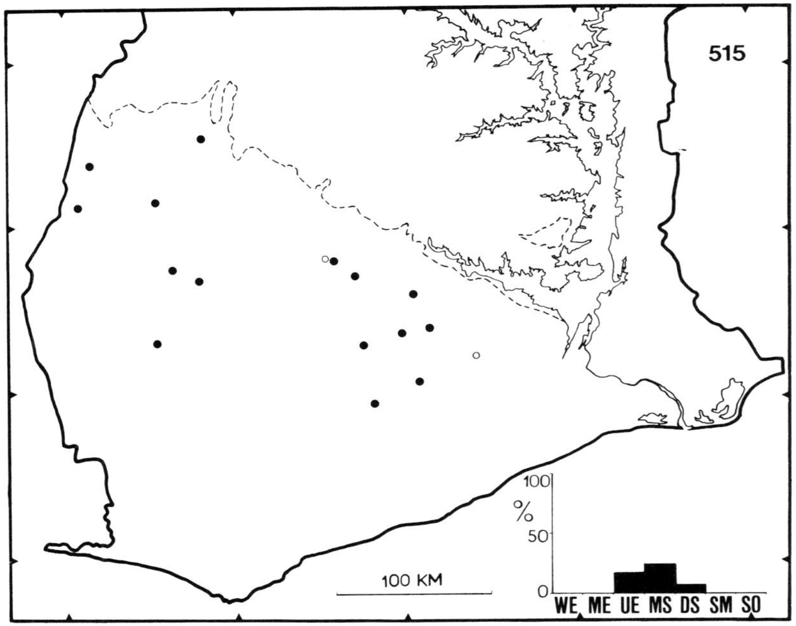

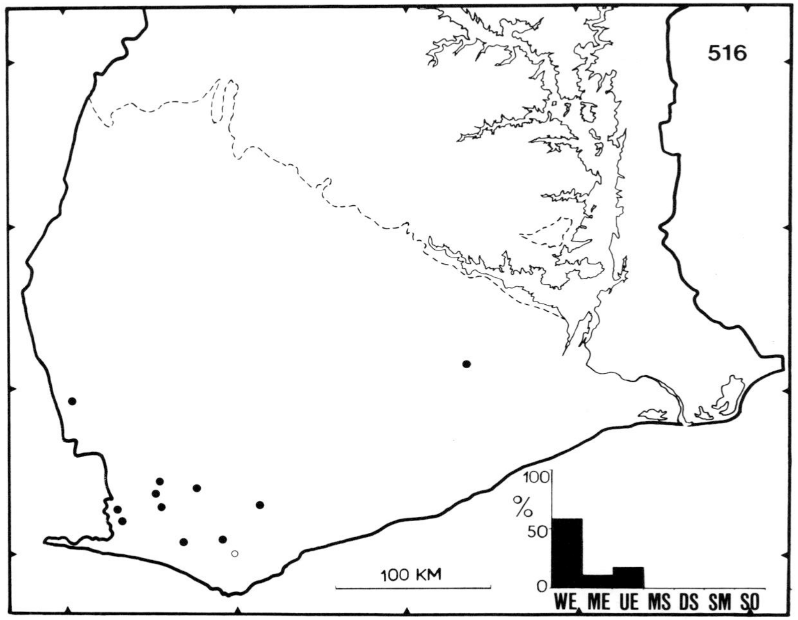

511 ROTHMANNIA WHITFIELDII (Lindl.) Dandy, FWTA 2:126. Rubiaceae. Akan: sabobe.
Small tree; lvs mesophyll. Petit's model. Fl large, funnel with tube, cream inside, brownish-hairy outside, not clearly seasonal. Fr woody berry; seeds medium to large. Regenerates in shade. Widespread in Tropical Africa; (chorological transgressor).

512 SABA THOMPSONII (A. Chev.) Pichon, FWTA 2:61. (Incl. *S. florida* of FWTA 2:61, partly, (Johnson 749) not of (Benth.) Bullock). Apocynaceae.
Large woody climber, containing white latex, woody tendrils, wood cylindrical, normal. Lvs notophyll to mesophyll. Fl white, large, dish with tube, 2–6. Fr yellow berry; seeds large. Regenerates in shade. Iv. Coast to Nigeria; (Guinea wide). Very close to *S. florida,* and perhaps not worthy of specific status.

513 SALACIA ALATA De Wild. var. *alata* (62), (*S. tuberculata* of FWTA 1:633, partly (Talbot 1705), not of Blakelock). Celastraceae.
Large woody climber; xylem separated into concentric rings by anomalous phloem (52). Lvs notophyll, slightly toothed. Roux's model. Fl small, orange, dish. Fr orange berry (not yet found in Ghana). Regenerates in shade. Ghana to Zaire; (Guineo-Congolian wide). The description and specimens cited in FWTA 1:631 for *S. alata* are *S. alata* var. *superba* N. Hallé which has not been found in Ghana.

514 SALACIA CAMERUNENSIS Loes., FWTA 1:633. Celastraceae.
Small to medium-sized woody climber; xylem with 1–2 anomalous rings (52). Lvs notophyll, toothed. Roux's model. Fl yellow-orange, small, dish, not clearly seasonal. Fr orange berry; seeds large. Ghana to Zaire; (Guineo-Congolian wide).

515 SALACIA CERASIFERA Welw. ex Oliv. (62), (*S. erecta* of FWTA 1:633, partly (Scott Elliot 4595), not of (G. Don) Walp.). Celastraceae.
Large woody climber; stem outline deeply fluted, network of anomalous phloem within the xylem (52). Lvs notophyll, slightly toothed. Roux's model. Fl yellowish, small, dish, not clearly seasonal. Fr orange berry, seeds large. Regenerates in shade. Guinée to Angola. Zambia and Kenya; (chorological transgressor).

516 SALACIA COLUMNA N. Hallé (62), (*S. senegalensis* of FWTA 1:631, partly, not of (Lam.) DC.). Celastraceae.
Small to medium woody climber; lvs notophyll, containing gutta threads. Roux's model. Fl small, whitish, dish. Fr orange berry; seeds large. Regenerates in shade. S. Leone to Ghana; (Upper Guinea endemic). The distinctness of *S. columna* from *S. senegalensis* requires study.

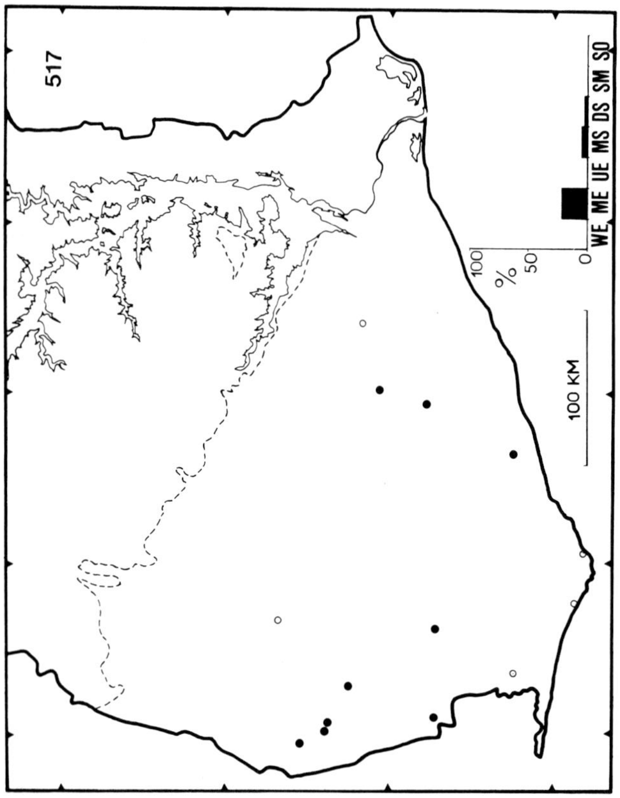
517
100 KM
%
100
50
0
WE ME UE MS DS SM SO

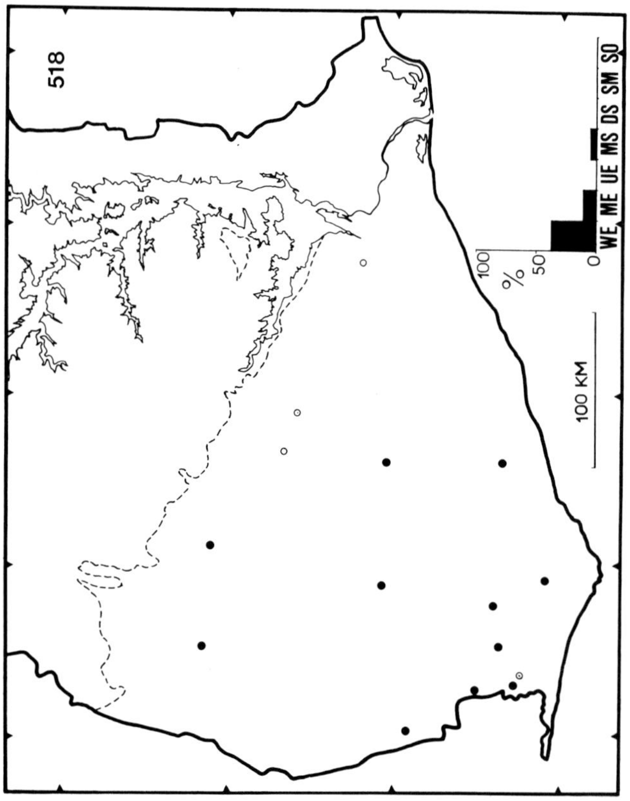
518
100 KM
%
100
50
0
WE ME UE MS DS SM SO

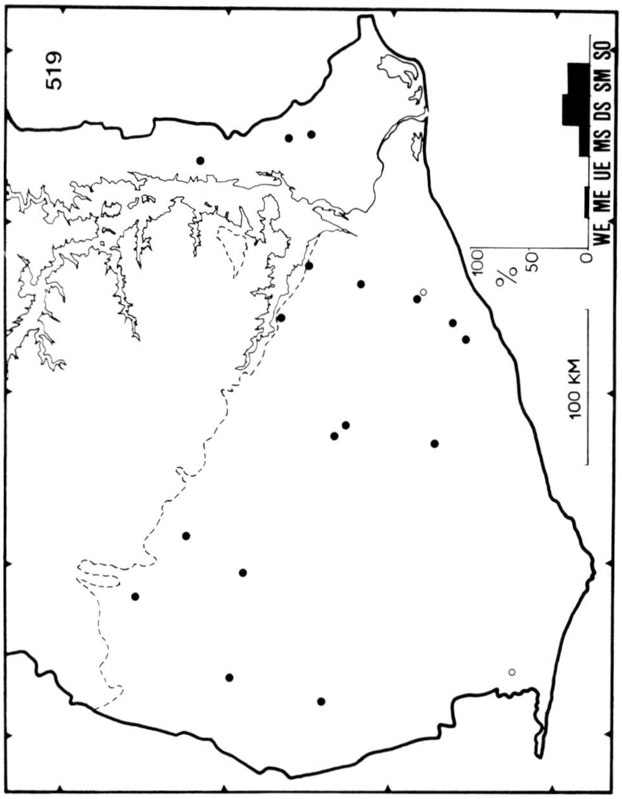
519
100 KM
%
100
50
0
WE ME UE MS DS SM SO

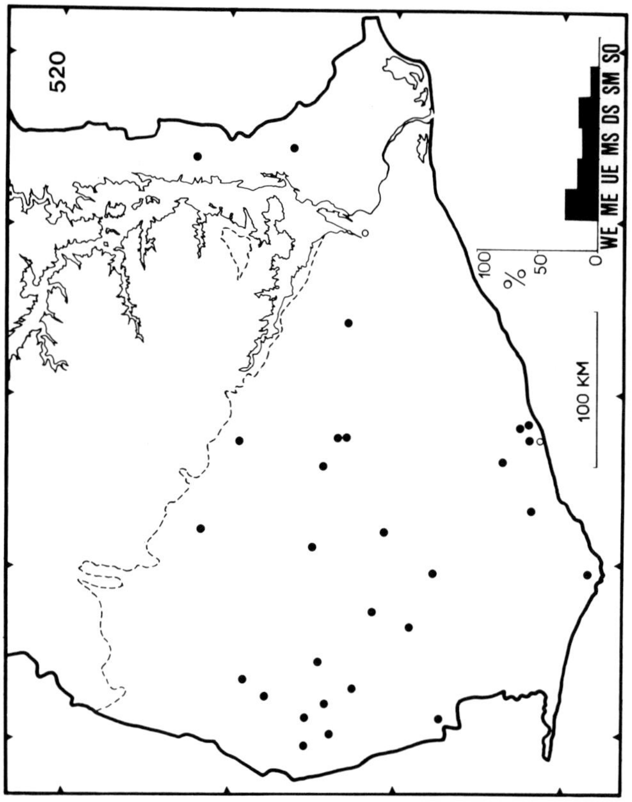
520
100 KM
%
100
50
0
WE ME UE MS DS SM SO

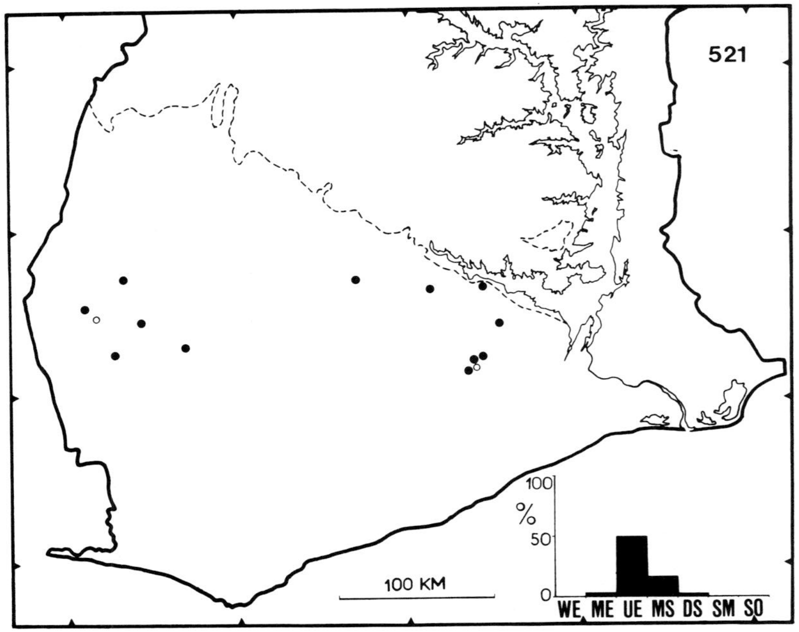

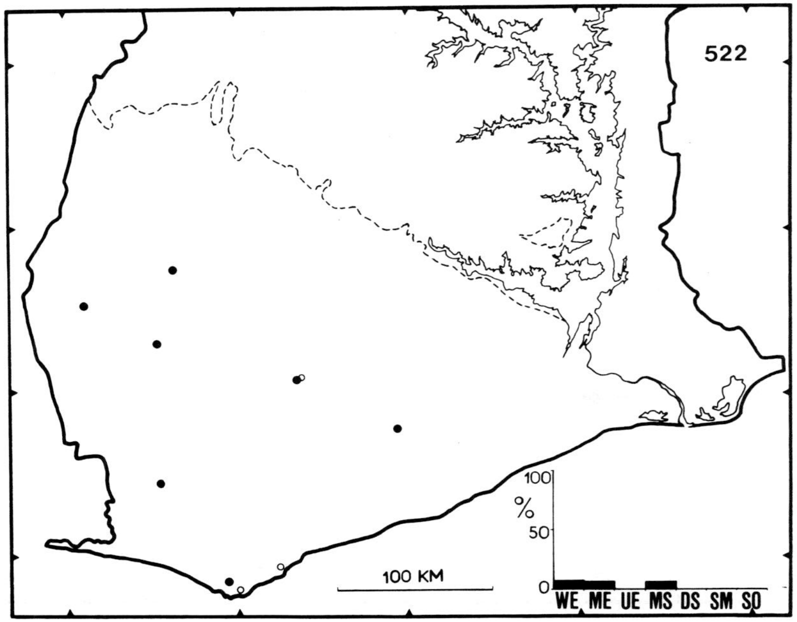

517 SALACIA CORNIFOLIA Hook. f. (62), (*S. erecta* of FWTA 1:633, partly, not of (G. Don) Walp.). Celastraceae.
Large woody climber; stem furrowed; xylem with network of included phloem (52). Lvs notophyll. Roux's model. Fl small, yellowish, dish. Fr orange berry; seeds large. Regenerates in shade. Sierra Leone to Central African Republic; (Guinea wide).

518 SALACIA DEBILIS (G. Don) Walp., FWTA 1:633. Celastraceae.
Large woody climber; stem deeply furrowed; xylem with included rings of anomalous phloem (52). Lvs notophyll, toothed. Roux's model. Fl small, yellowish, dish, 9–3. Fr reddish berry; seeds large. Regenerates in shade. Senegal to Zaire; (Guineo-Congolian wide).

519 SALACIA ELEGANS Welw. ex Oliv. (62), (*S. leptoclada* of FWTA 1:631, not of Tul.). Celastraceae.
Medium-sized woody climber; xylem with included discontinuous tangential bands of phloem (52). Lvs notophyll, toothed, containing gutta threads. Roux's model. Fl small, greenish-yellow, dish. Fr orange-red berry; seeds medium to large. Regenerates in shade. Guinée to Angola and Mozambique; (chorological transgressor). The herbarium specimen from Wet Evergreen forest is of the form with coriaceous, sub-entire leaves (62).

520 SALACIA ERECTA (G. Don) Walp. (98), FWTA 1:633, partly (excluding the synonyms). Celastraceae.
Large woody climber, or shrub in thicket; stem deeply furrowed; xylem with included tangential bands of anomalous phloem (52). Lvs microphyll, toothed. Roux's model. Fl small, yellowish, dish, not clearly seasonal. Fr orange berry; seeds large. Regenerates in shade. Guinée to Angola and Tanzania; (Guineo-Congolian wide).

521 SALACIA sp. aff. GABUNENSIS Loes. (122). Celastraceae.
Large woody climber; xylem in cross section showing network of included phloem (52). Lvs mesophyll. Roux's model. Fl small, yellowish, dish. Fr not known. Regenerates in shade. Specimens (Letouzey 4773) very similar to the Ghana material have been collected near Batouri in Cameroun; (probably Guinea wide).

522 SALACIA HOWESII Hutch. & M. B. Moss, FWTA 1:633. Celastraceae. (Type from Anaji near Sekondi).
Medium-sized woody climber; xylem in transverse section with included spots of phloem (52). Lvs notophyll to mesophyll; pubescent on underside of shade leaves, and on twigs. Roux's model. Fl yellowish, small, dish. Fr not known. Regenerates in shade. Iv. Coast to Nigeria; (Guinea wide).

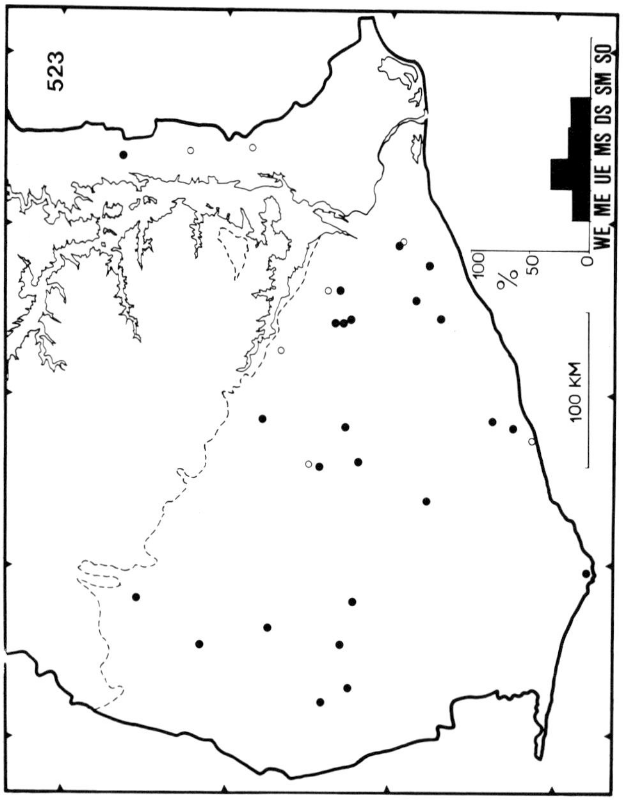
523
100 KM
%
100
50
0
WE ME UE MS DS SM SO

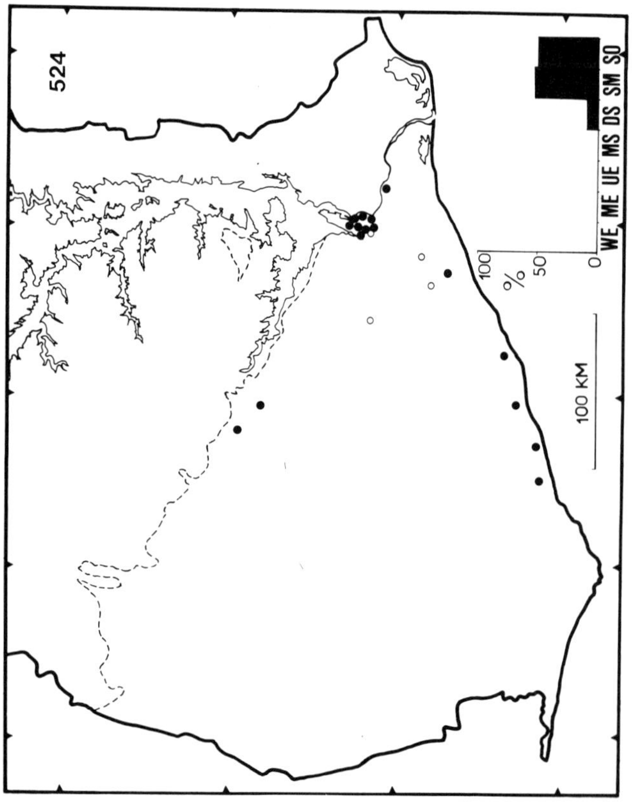
524
100 KM
%
100
50
0
WE ME UE MS DS SM SO

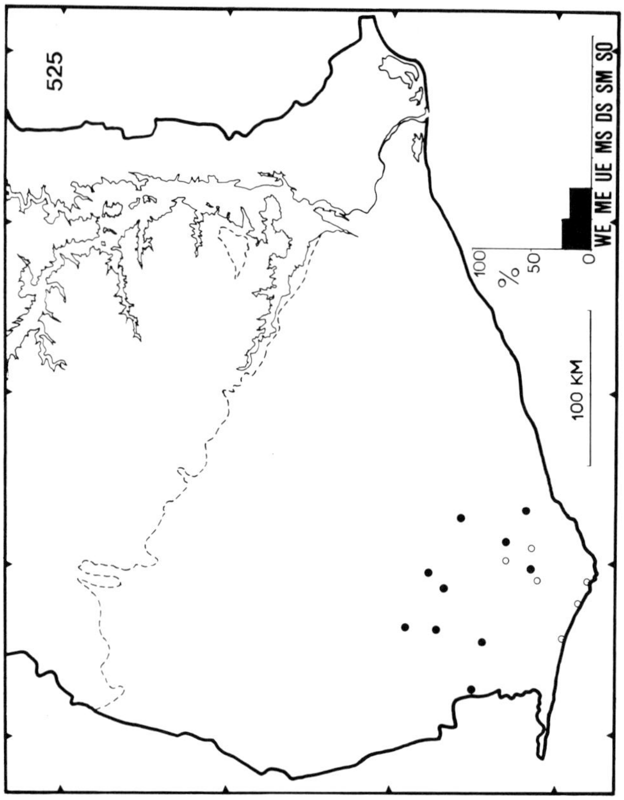
525
100 KM
%
100
50
0
WE ME UE MS DS SM SO

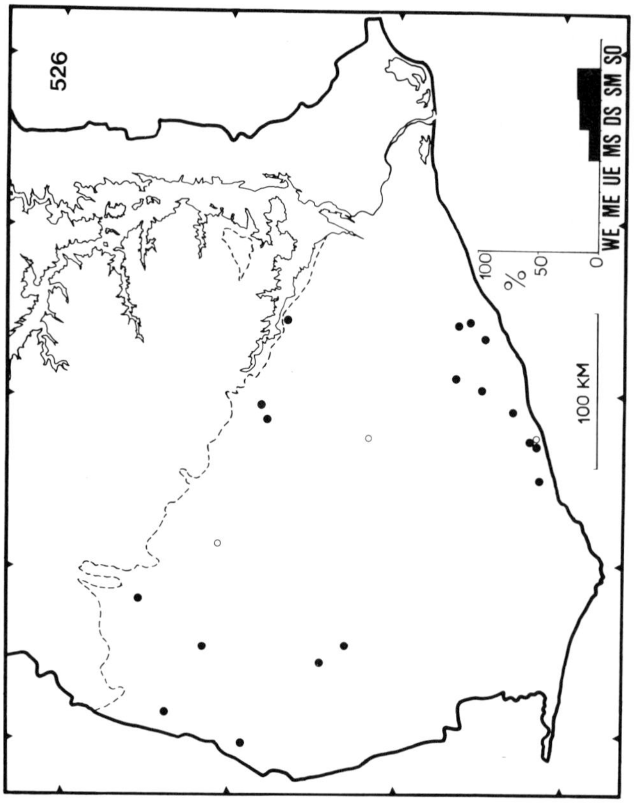
526
100 KM
%
100
50
0
WE ME UE MS DS SM SO

523 SALACIA ITURIENSIS Loes. (62), (*S. fimbrisepala* of FWTA 1:632, not of Loes.) Celastraceae.

Medium-sized to large woody climber, or shrub in thicket; network of included phloem appearing in transverse section of xylem (52). Lvs notophyll to mesophyll, toothed. Roux's model. Fl orange-brown, small, dish, 4–5. Fr large orange berry (size and shape of small mango); seeds large. Regenerates in shade. Iv. Coast to Zaire; (Guineo-Congolian wide).

524 SALACIA LEPTOCLADA Tul. (98), not of FWTA 1:631. (*S. erecta* of FWTA 1:633, partly, (*S. baumannii* Loes.) not of (G. Don) Walp.). Celastraceae.

Medium-sized woody climber, or shrub in thicket; network of included phloem in transverse section of xylem (52). Lvs microphyll to notophyll. Roux's model. Fl yellowish, small, dish not clearly seasonal. Fr orange berry; seeds medium to large. Regenerates in shade. Extends to forest patches on rocky scarps in the savanna zone. Iv. Coast to Togo and Kenya to S. Africa, and Madagascar; (chorological transgressor).

525 SALACIA NITIDA (Benth.) N. E. Br., FWTA 1:633. Celastraceae.

Medium-sized woody climber; stem with shallow grooves, xylem deeply sulcate with radial intrusion of phloem (52). Lvs notophyll. Roux's model. Fl yellowish, small, dish, not clearly seasonal. Fr orange berry; seeds large. Regenerates in shade. S. Leone to Zaire; (Guineo-Congolian wide).

526 SALACIA PALLESCENS Oliv., FWTA 1:632. Celastraceae.

Pygmy tree or shrub; lvs notophyll, toothed. Roux's model. Fl blackish, small, dish, not clearly seasonal. Fr orange berry with brown stripes; seeds large. Regenerates in shade. Guinée to Gabon; (Guinea wide).

527 SALACIA PYRIFORMIS (Sabine) Steud., FWTA 1:632. Celastraceae.

Medium to large woody climber, or shrub in thicket; phloem rings included in xylem (52). Lvs notophyll to mesophyll, containing gutta threads. Roux's model. Fl yellowish, small, dish, not clearly seasonal. Fr smooth or warted orange berry; seeds large. Regenerates in shade. Senegal to Zaire; (Guineo-Congolian wide).

528 SALACIA STAUDTIANA Loes. var. *leonensis* Loes. (62), (*S. caillei* A. Chev. ex Hutch. & M. B. Moss, FWTA 1:633). Celastraceae.

Medium-sized woody climber, or shrub in thicket; xylem sulcate with phloem in the radial clefts (52). Lvs microphyll to notophyll. Roux's model. Fl orange, small, dish, mostly 4–6. Fr orange berry; seeds medium. Regenerates in shade. Plot records very significantly associated with acid, base-poor soils. Guinée to Zaire; (Guineo-Congolian wide).

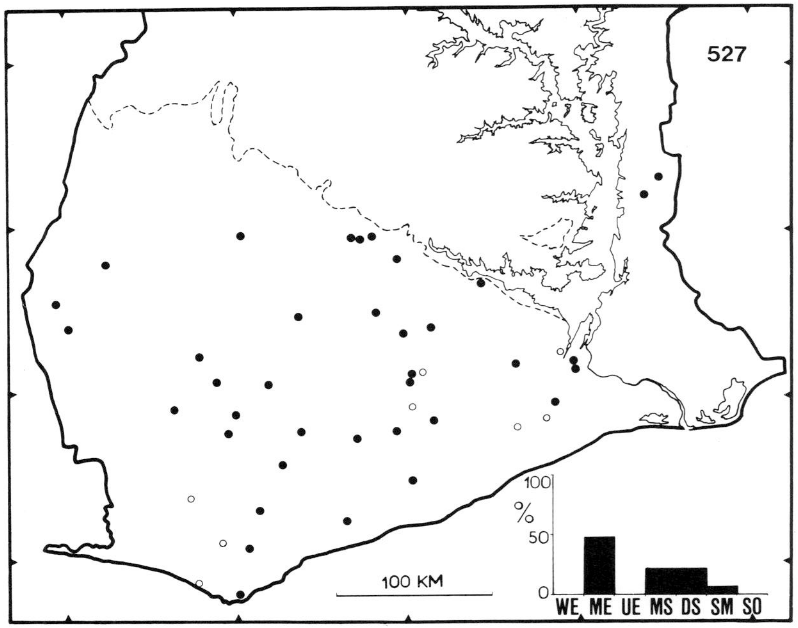

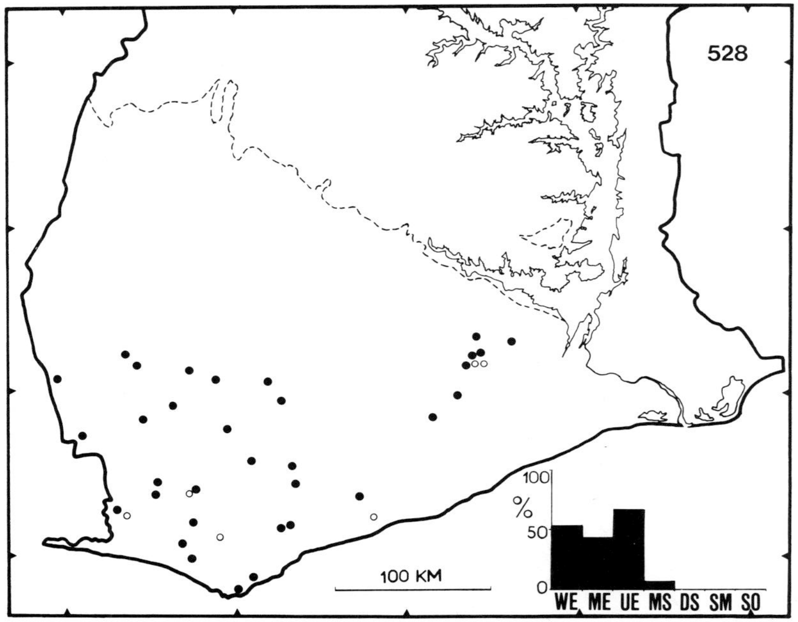

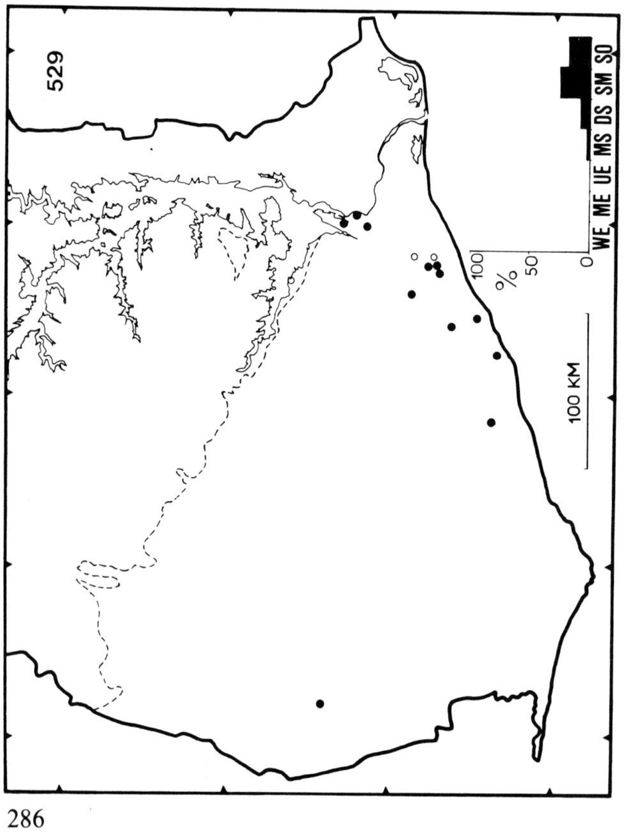
529
100 KM
%
100
50
0
WE ME UE MS DS SM SO

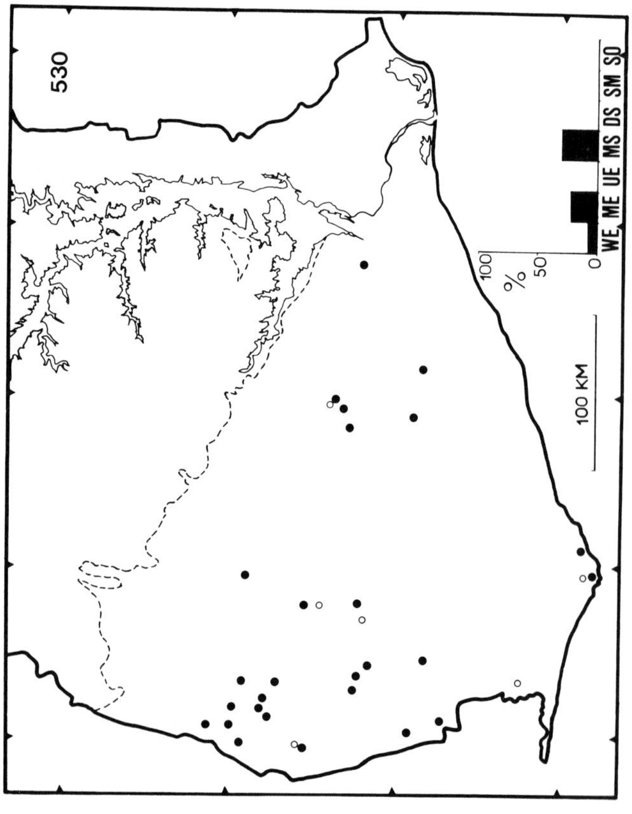
530
100 KM
%
100
50
0
WE ME UE MS DS SM SO

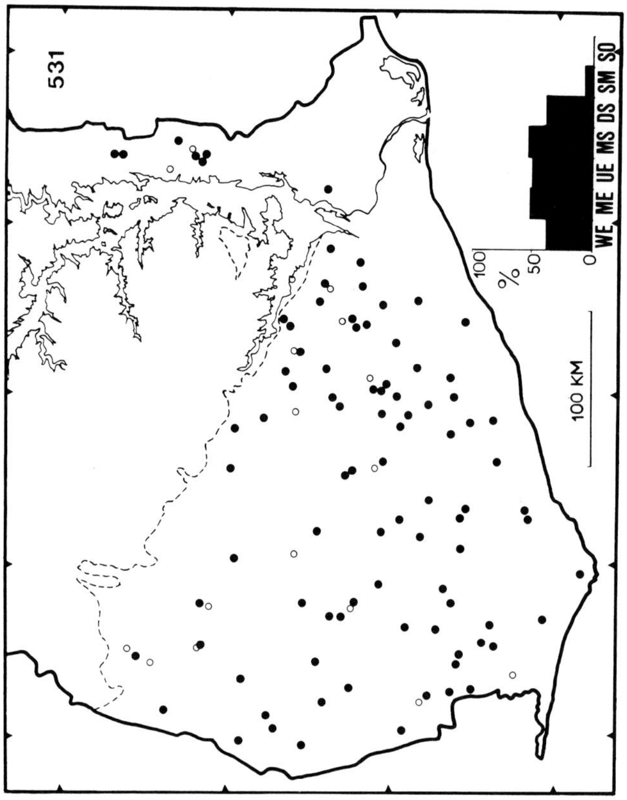
531
100 KM
%
100
50
0
WE ME UE MS DS SM SO

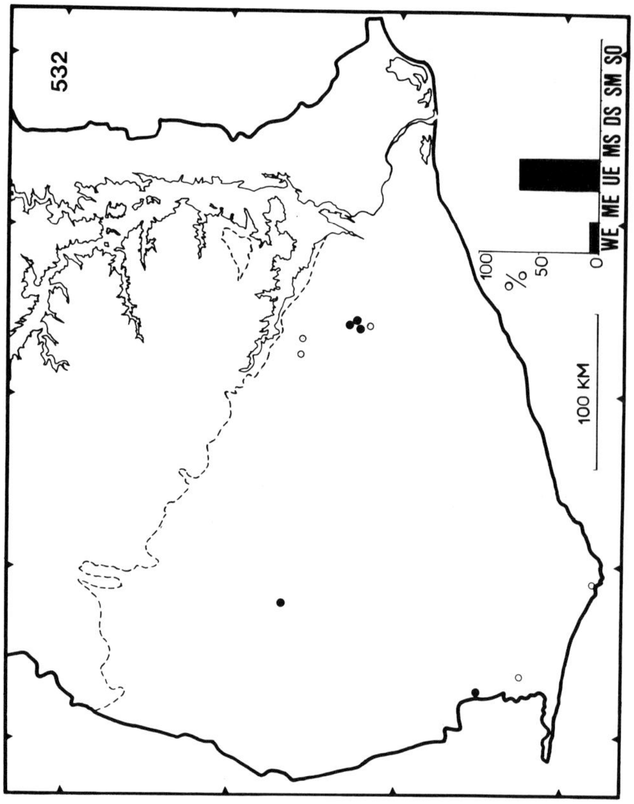
532
100 KM
%
100
50
0
WE ME UE MS DS SM SO

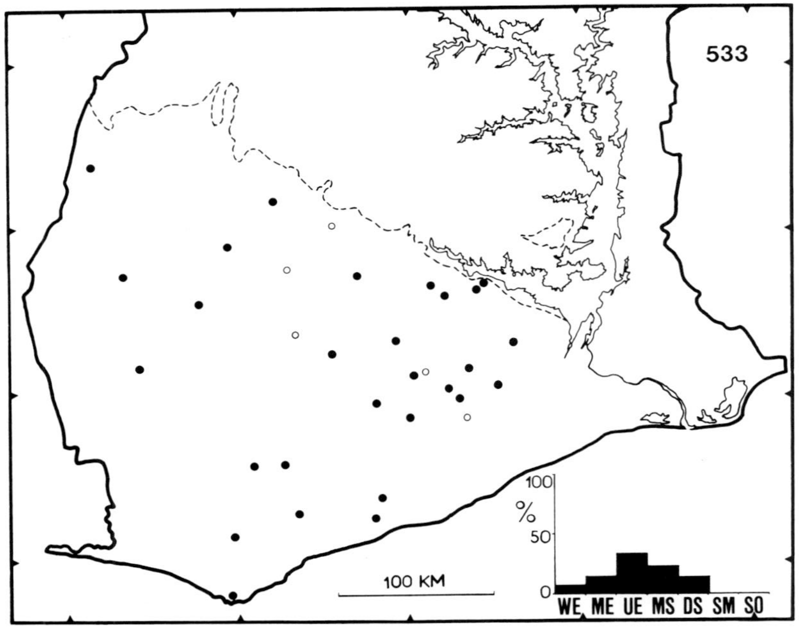

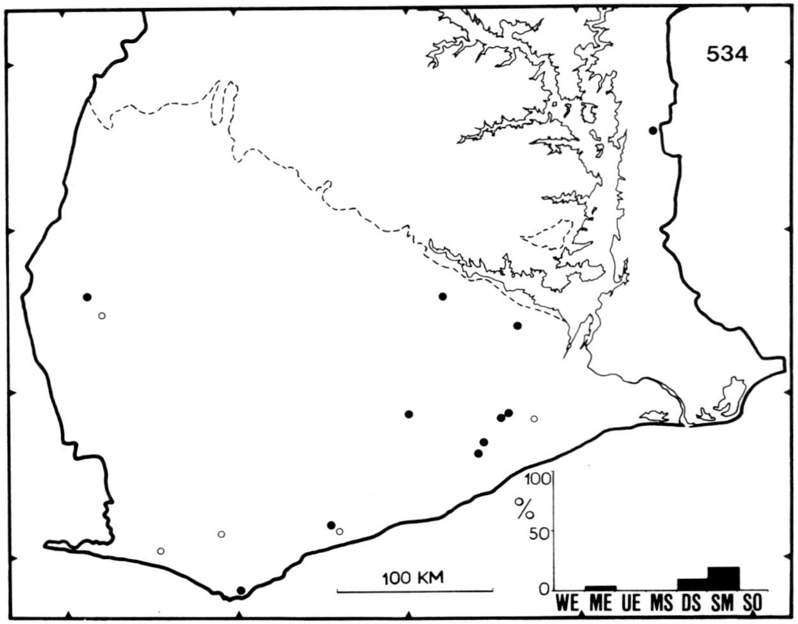

529 SALACIA TOGOICA Loes, FWTA 1:632. Celastraceae.
Medium-sized woody climber; xylem in transverse section showing discontinuous tangential lines of included phloem (52). Lvs notophyll. Roux's model. Fl orange-brown, small, dish, mostly 11–1. Fr red, seeds large. Regenerates in shade. S. Leone to Nigeria; (Guinea wide).

530 SALACIA UREGAENSIS R. Wilczek var. *aurantiaca* N. Hallé (62), (*S.* sp. aff. *leonensis* Hutch & M. B. Moss, FWTA 1:633). Celastraceae.
Pygmy tree or shrub; lvs notophyll to mesophyll. Roux's model. Fl orange, small, dish, not clearly seasonal. Fr orange berry, seed large. Regenerates in shade. S. Leone to Zaire; (Guineo-Congolian wide). Further study will be needed to show whether *S. uregaensis* is distinct from *S. leonensis* and *S. lehmbachii* Loes.

531 SARCOPHRYNIUM BRACHYSTACHYS (Benth.) K. Schum., FWTA 3:87. Marantaceae.
Perennial ground herb (Fig. 5.13c); lvs macrophyll. Fl whitish, medium, funnel, not clearly seasonal. Fr red berry; seeds medium to large. Regenerates in shade. Senegal to Zaire; (Guineo-Congolian wide).

532 SARCOPHRYNIUM PRIONOGONIUM (K. Schum.) K. Schum., FWTA 3:88. Marantaceae.
Perennial ground herb; lvs macrophyll, pubescent on sheaths and sometimes on underside. Fl whitish, medium, funnel, not clearly seasonal. Fr red berry; seeds medium to large. Regenerates in shade. Guinée to Zaire; (Guineo-Congolian wide).

533 SCADOXUS CINNABARINUS (Decne.) Friis & Nordal (46), (*Haemanthus cinnabarinus* Decne., FWTA 3:132). Amaryllidaceae.
Perennial bulbous ground herb; lvs mesophyll. Tomlinson's model. Fl red, large, brush, in dense umbel, 12–3. Fr orange berry, c. 5; seeds medium to large. Regenerates in shade from seed, and also from adventitious buds on midrib underside (57). S. Leone to Angola and Uganda; (Guineo-Congolian wide, 46).

534 SCADOXUS MULTIFLORUS (Martyn) Raf. subsp. *longitubus* (C. H. Wright) Friis & Nordal (46), (*Haemanthus longitubus* C. H. Wright, FWTA 3:312). Amaryllidaceae.
Perennial bulbous ground herb; lvs mesophyll. Tomlinson's model. Fl red, large, brush, in dense umbel, 1–4. Fr orange berry; seeds medium. Regenerates in shade. Guinée to Ghana (46). Species is a Guineo-Congolian wide.

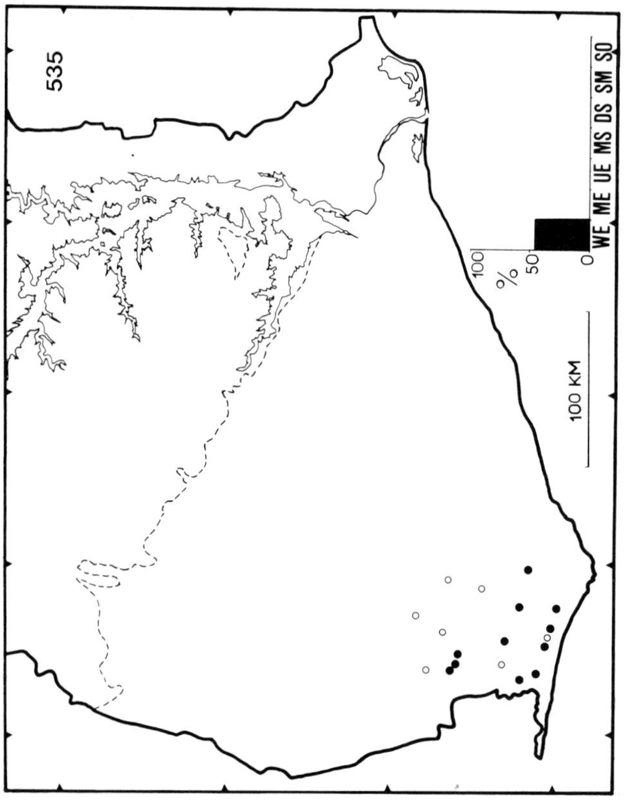
535
100
%
50
0
WE ME UE MS DS SM SO
100 KM

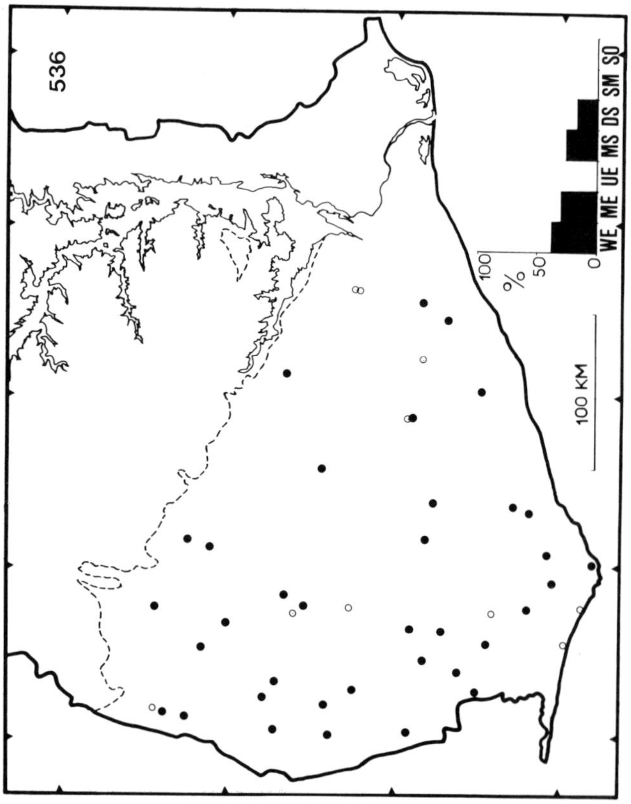
536
100
%
50
0
WE ME UE MS DS SM SO
100 KM

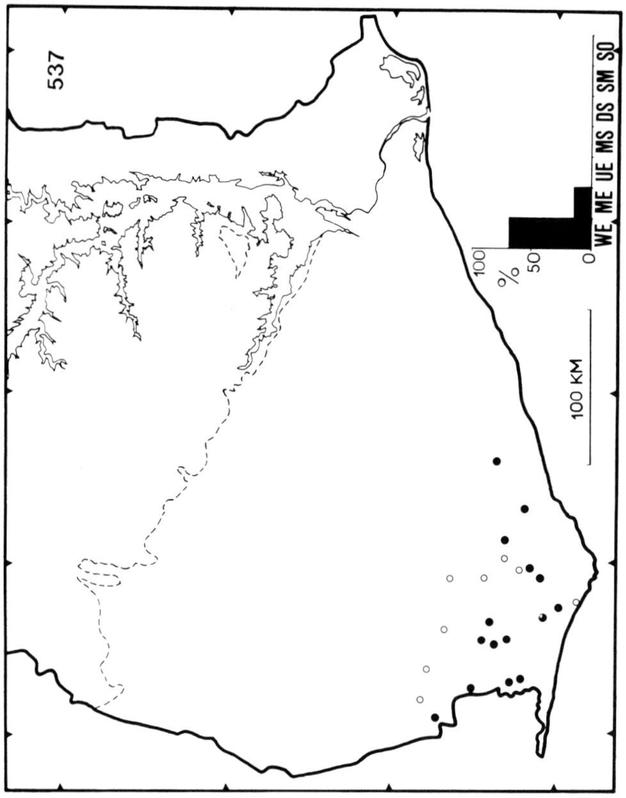
537
100
%
50
0
WE ME UE MS DS SM SO
100 KM

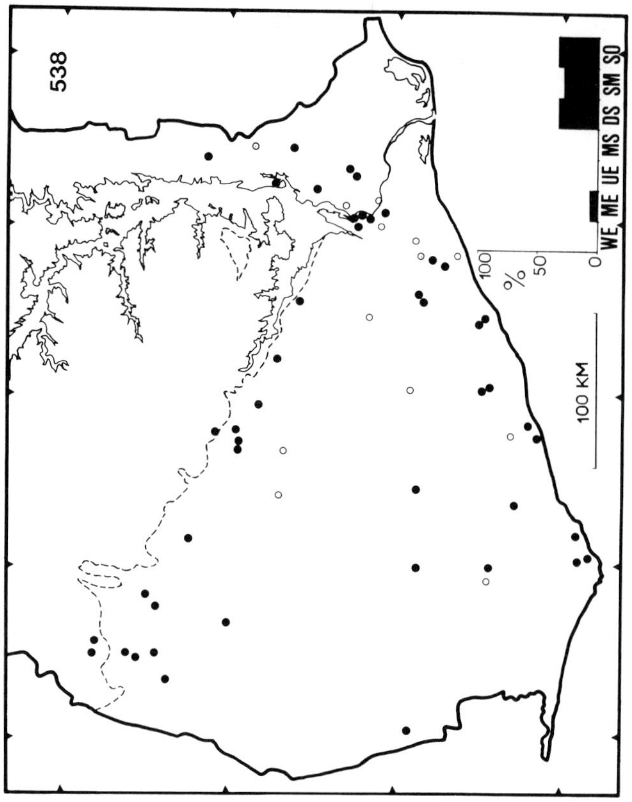
538
100
%
50
0
WE ME UE MS DS SM SO
100 KM

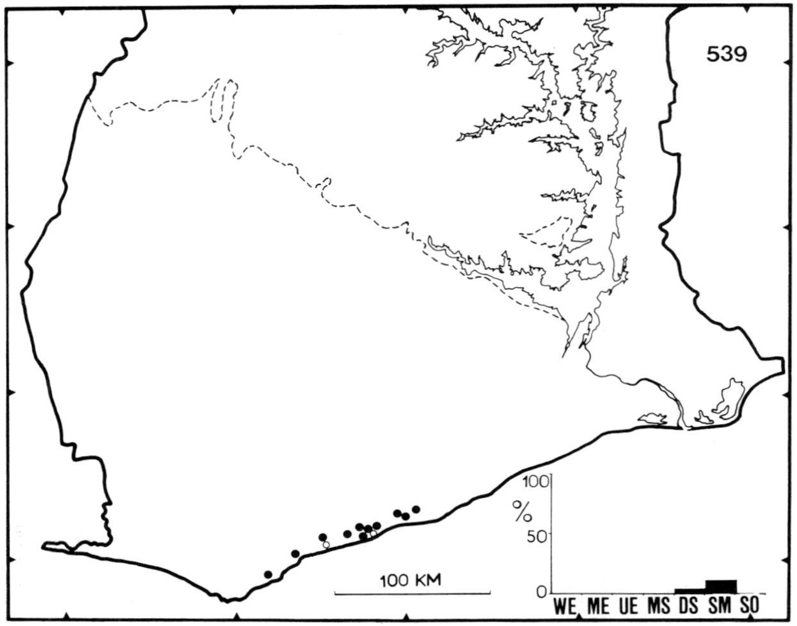

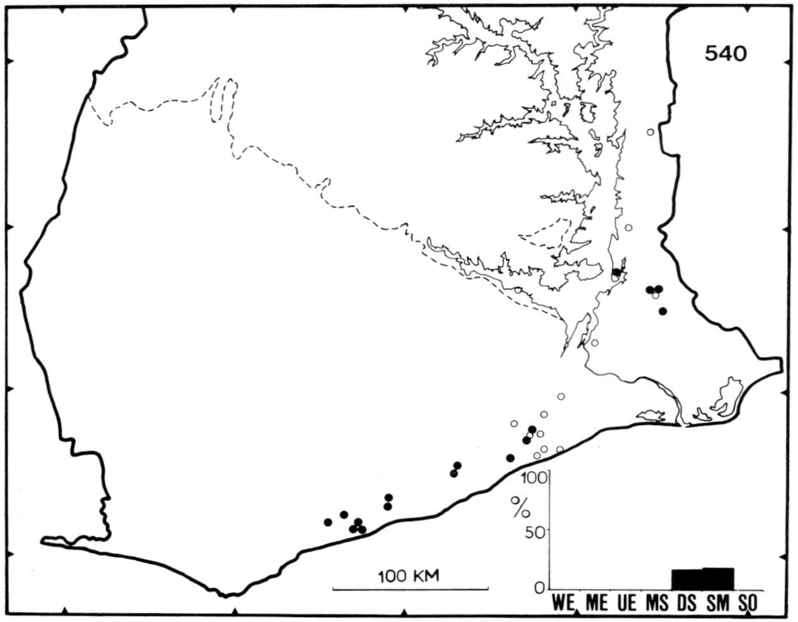

535 SCAPHOPETALUM AMOENUM A. Chev., FWTA 1:315. Sterculiaceae. Akan: nsoto.
Small tree or shrub with branches that take root when they touch the ground, so forming a dense monospecific thicket in the forest understorey (69). Champagnat's model (59). Lvs mesophyll to macrophyll. Fl yellowish with purple streaks, small to medium, bowl, not clearly seasonal. Fr capsule; seeds medium. Regenerates in shade both from seed, and vegetatively by layering of the branches. Liberia to Ghana; (Upper Guinea endemic). Possibly conspecific with *S. blackii* Mast. from Gabon (60).

536 SCOTTELLIA KLAINEANA Pierre (101), (Incl. *S. chevalieri* Chipp, FWTA 1:186, and *S. coriacea* A. Chev. ex Hutch. & Dalz., FWTA 1:187). Flacourtiaceae. Akan: tiabutuo.
Large tree with buttresses; lvs notophyll to mesophyll, more or less toothed. Fl white, small, bowl, hermaphrodite or polygamous, not clearly seasonal. Fr capsule; seeds small to medium. Regenerates in shade. S. Leone to Angola; (Guineo-Congolian wide). Larger-leaved specimens may be distinguished as var. *mimfiensis* (Gilg) Pellegr. from the smaller-leaved var. *klaineana,* but there are many intermediates, and no consistent ecological differences. Class 3 timber; trade name: odoko (40).

537 SCYTOPETALUM TIEGHEMII (A. Chev.) Hutch. & Dalz., FWTA 1:300. Scytopetalaceae. Akan: oprim.
Medium-sized tree; lvs microphyll to notophyll. Troll's model (59). Fl white, small to medium, bowl, 2–5. Fr red drupe, endocarp dehiscent, 7–8; seed large. Regenerates in shade. Plot records very significantly correlated with acid, base-poor soils. S. Leone to Ghana; (Upper Guinea endemic).

538 SECAMONE AFZELII (Schult.) K. Schum., FWTA 2:88. Asclepiadaceae.
Small to medium-sized woody climber, containing white latex; lvs microphyll. Fl yellowish, small, in crowded inflorescences, mostly 7–8. Fr follicle, 9–10; seeds medium, wind-dispersed with aid of tuft of long hairs. Senegal to Gabon; (Guinea wide).

539 SLOETIOPSIS USAMBARENSIS Engl. (22), (*Neosloetiopsis kamerunensis* Engl., FWTA 1:595). Moraceae.
Shrub or small tree; branches take root when they touch the ground, and a monospecific thicket may be formed in the understorey, as in *Scaphopetalum amoenum.* Troll's model (59). Lvs microphyll to notophyll. Fl dioecious, small, greenish-white, inconspicuous, not clearly seasonal. Fr white, fleshy, 1-seeded follicle, dispersing the medium-sized seed violently. Regenerates in shade by seed, and vegetatively. Guinée to Zaire, Tanzania and Mozambique; (chorological transgressor).

540 SORINDEIA WARNECKEI Engl., FWTA 1:738. Anacardiaceae.
Straggling shrub; lvs pinnate, leaflets generally notophyll. Fl pinkish, small, bowl, polygamous, 10–3. Fr orange drupe; seed large. Regenerates in shade. Sierra Leone to Nigeria; (Guinea wide).

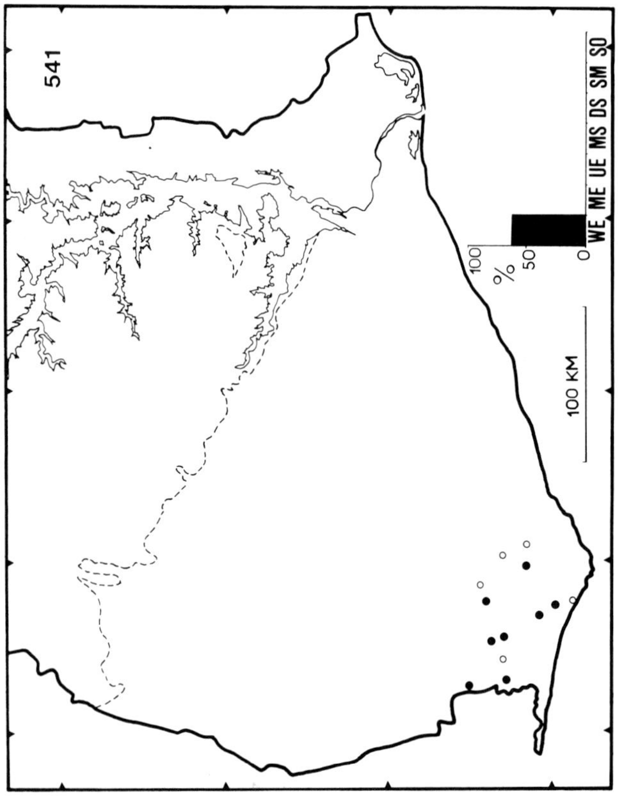
541
100 KM
100
%
50
0
WE ME UE MS DS SM SO

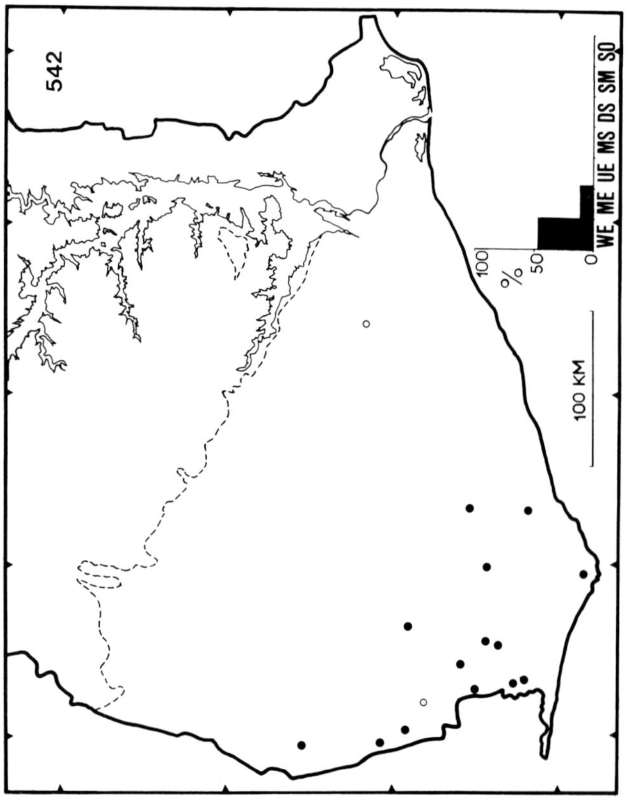
542
100 KM
100
%
50
0
WE ME UE MS DS SM SO

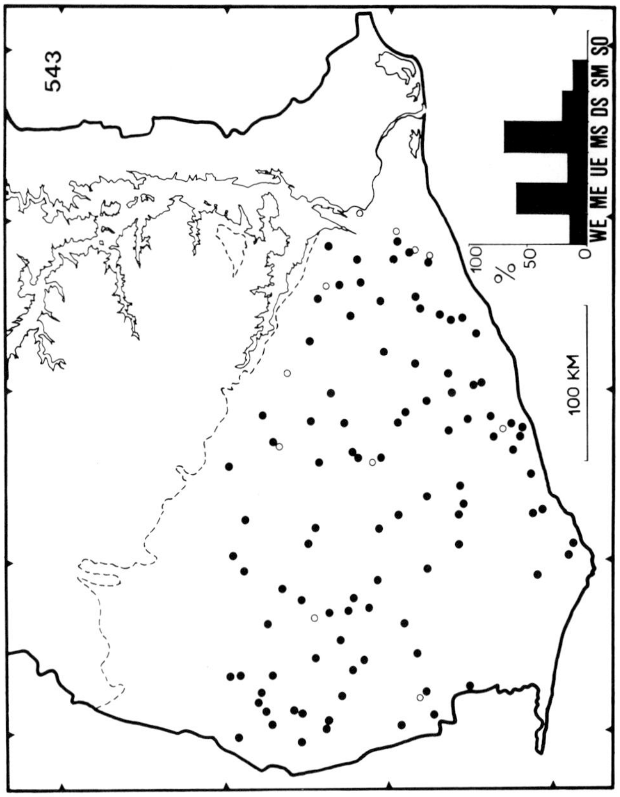
543
100 KM
100
%
50
0
WE ME UE MS DS SM SO

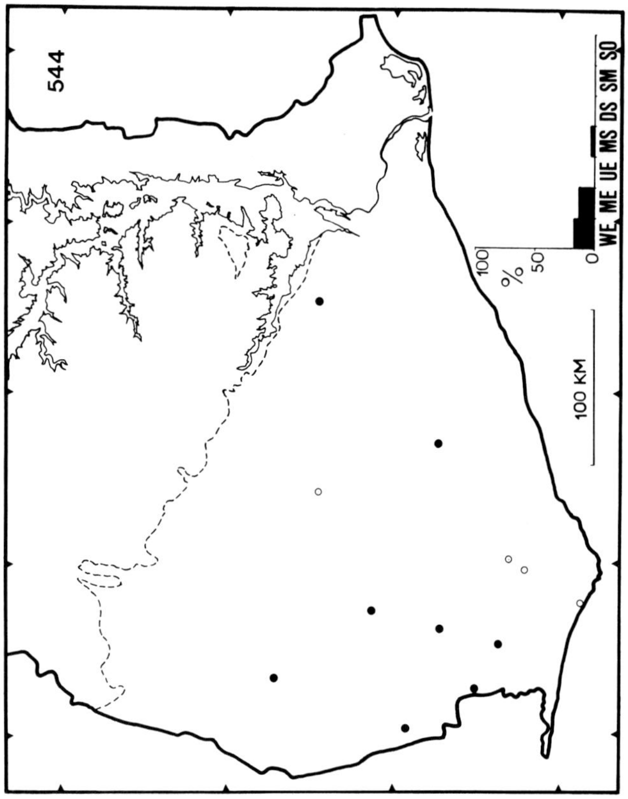
544
100 KM
100
%
50
0
WE ME UE MS DS SM SO

541 SOYAUXIA GRANDIFOLIA Gilg & Stapf, FWTA 1:653. Medusandraceae. Akan: abotesima.

Small to medium tree; lvs mesophyll. Fl white, small, brush, crowded in spikes, mostly 1–3. Fr 1-seeded reddish capsule; seeds large to very large. Regenerates in shade. Plot records very significantly correlated with base-poor soils. Liberia to Ghana; (Upper Guinea endemic).

542 SOYAUXIA VELUTINA Hutch. & Dalz., FWTA 1:653. Medusandraceae. Akan: abotesimanua.

Medium-sized tree; lvs notophyll to mesophyll, puberulous below and on twigs. Fl white, small, brush, crowded in spikes, mostly 5. Fr 1-seeded capsule; seed large. Regenerates in shade. Iv. Coast to Nigeria; (Guinea wide).

543 SPHENOCENTRUM JOLLYANUM Pierre, FWTA 1:76. Menispermaceae. Akan: kraman-kote.

Pygmy tree less than 1 m high; lvs mesophyll, usually deeply lobed. Corner's model. Fl dioecious, white, small, inconspicuous, cauliflorous, not clearly seasonal. Fr orange drupe; seed large. Regenerates in shade. Iv. Coast to Nigeria; (Guinea wide).

544 SPIROPETALUM HETEROPHYLLUM (Bak.) Gilg, FWTA 1:748. Connaraceae.

Large woody climber, or shrub in thicket; rings of anomalous phloem included in the xylem. Lvs pinnate, leaflets microphyll to notophyll. Fl white, small, clustered. Fr greenish 1-seeded follicle; seed large, black with orange aril. Regenerates in shade. S. Leone to Zaire; (Guineo-Congolian wide).

545 SPIROPETALUM SOLANDERI (Bak.) Gilg, FWTA 1:748. Connaraceae.
Woody climber, or shrub in thicket. Lvs pinnate; leaflets notophyll. Fl white, small, clustered. Fr a red 1-seeded follicle; seed large. S. Leone to Ghana (Upper Guinea endemic).

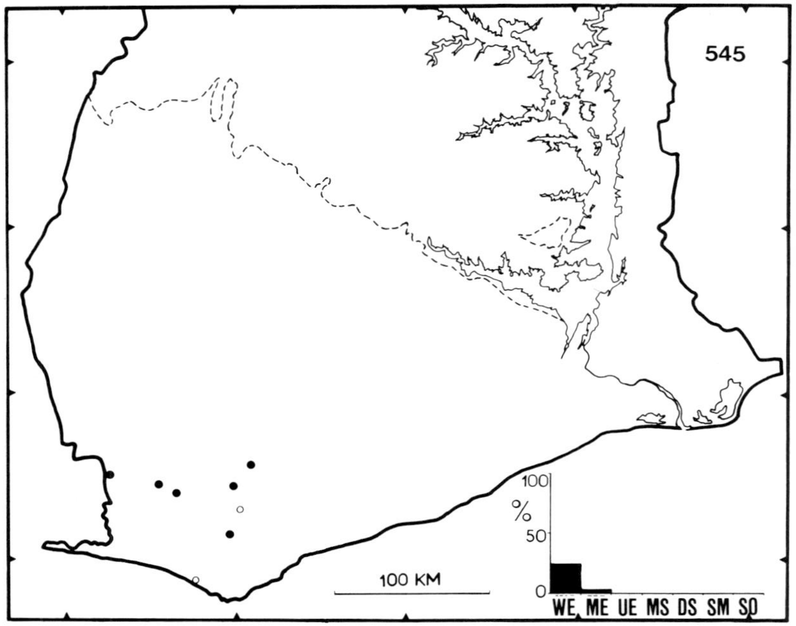

546 SPONDIAS MOMBIN L., FWTA 1:728. Anacardiaceae. Akan: atoa.
Medium-sized tree; lvs pinnate, leaflets notophyll, deciduous. Scarrone's model (59). Fl white, small, polygamous, in large panicles, 1–5. Fr yellow drupe with acid edible pulp, 7–9; seed large to very large. Widely planted for its fruit in villages throughout the forest zone; naturalized or possibly native in forest only in Volta Region. Senegal to Angola, and in Tropical America; (chorological transgressor, 7).

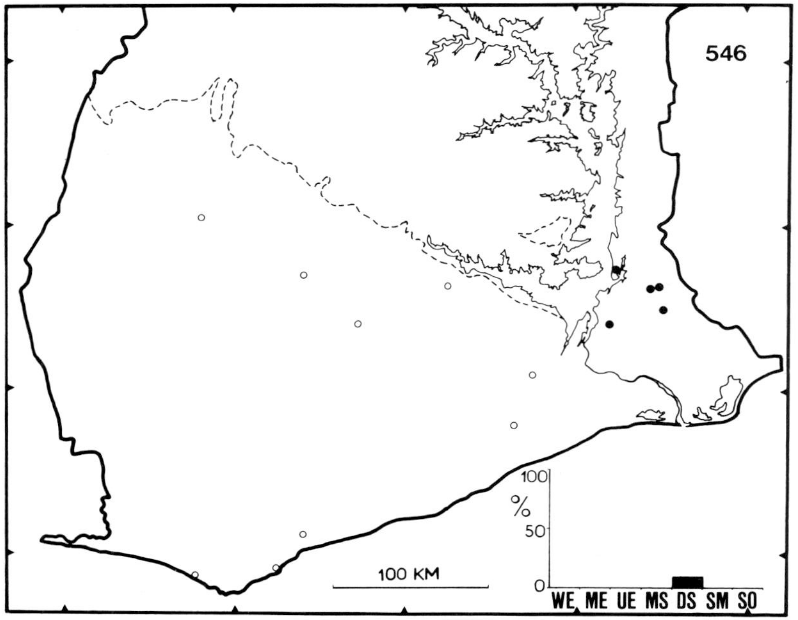

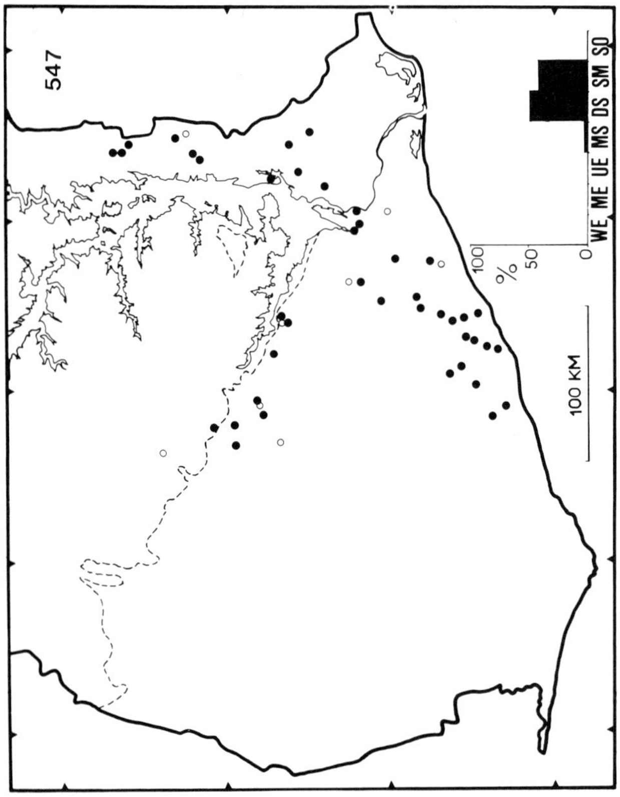
547
100 KM
%
100
50
0
WE ME UE MS DS SM SO

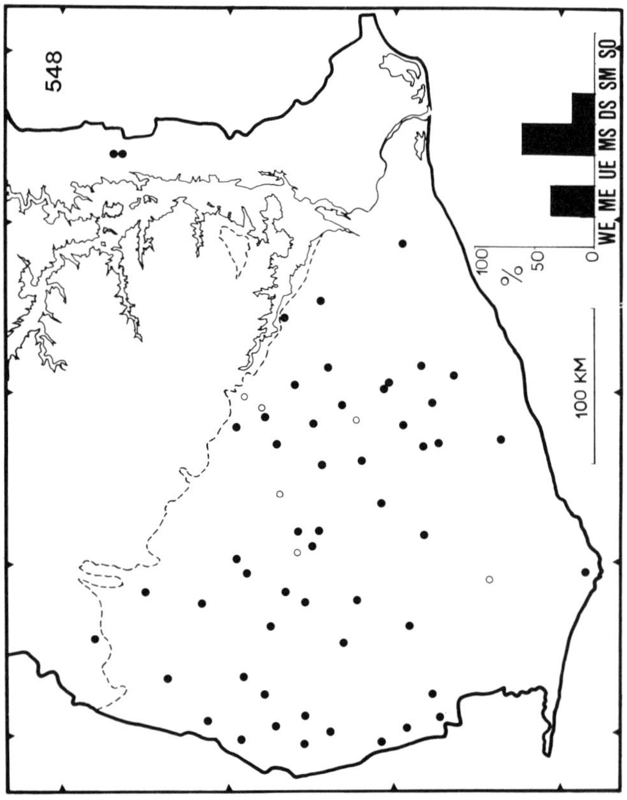
548
100 KM
%
100
50
0
WE ME UE MS DS SM SO

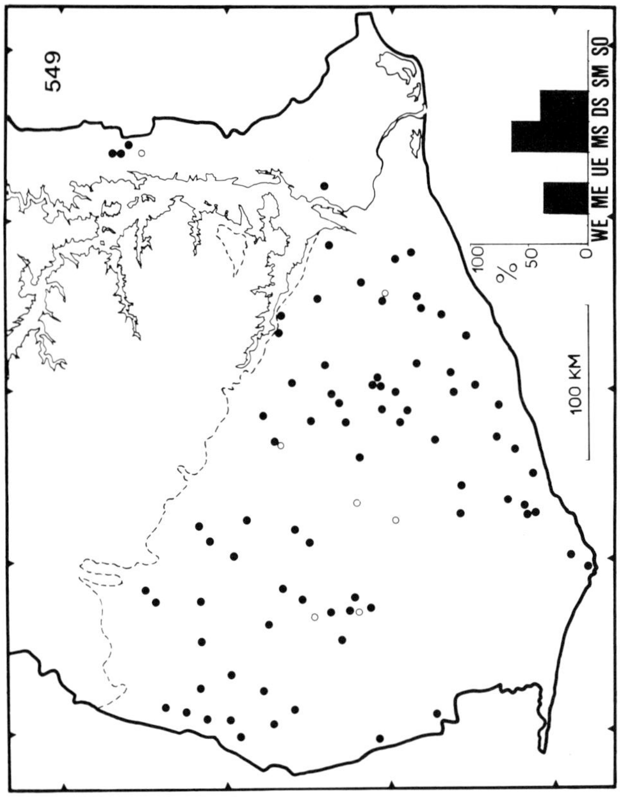
549
100 KM
%
100
50
0
WE ME UE MS DS SM SO

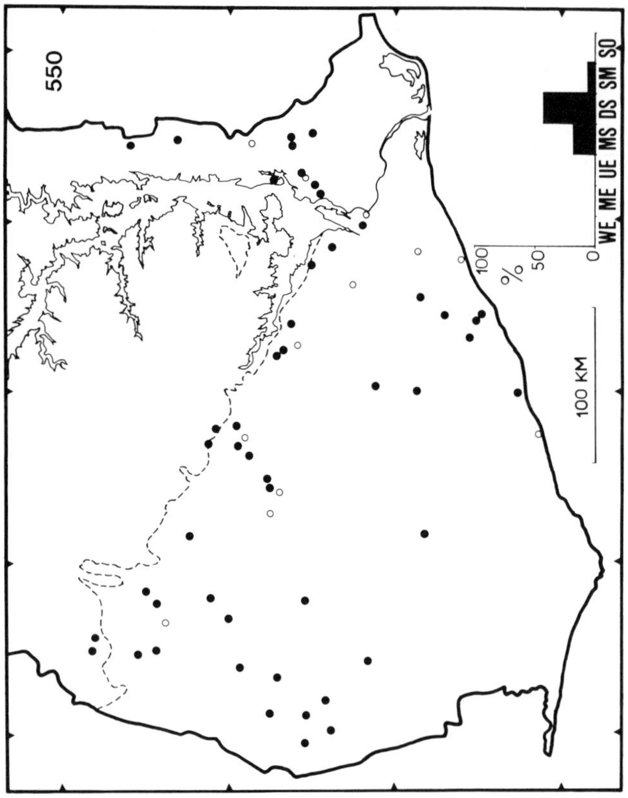
550
100 KM
%
100
50
0
WE ME UE MS DS SM SO

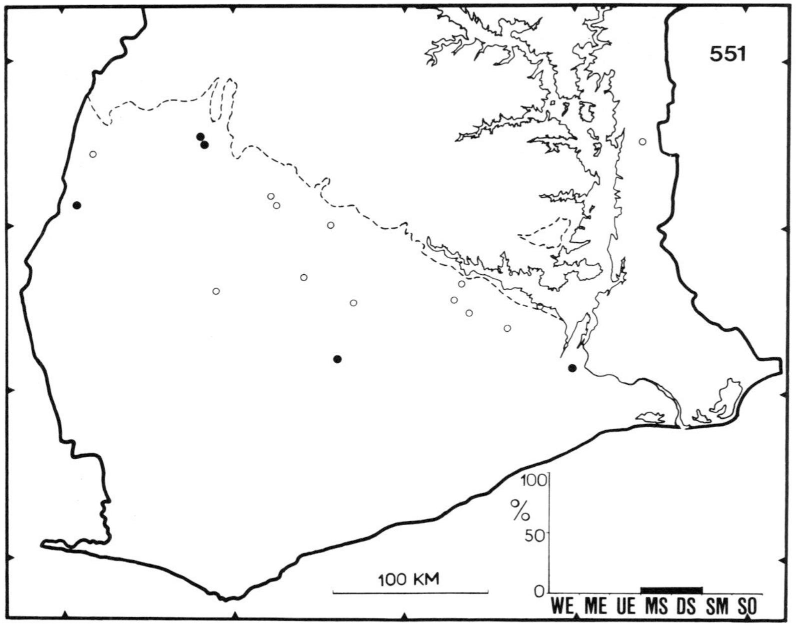

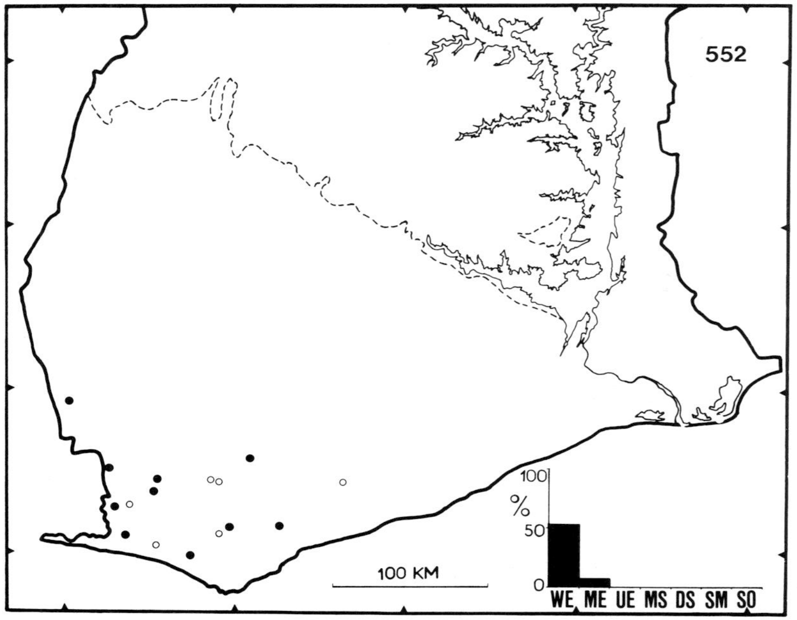

547 STACHYANTHUS OCCIDENTALIS (Keay & Miège) Boutique (27), (*Neostachyanthus occidentalis* Keay & Miège, FWTA 1:643). Icacinaceae.
Small slender woody climber; lvs notophyll to mesophyll. Undescribed model (59). Fl greenish-yellow, small, grouped in spikes, dioecious, not clearly seasonal. Fr orange drupe with edible flesh; seed medium. Regenerates in shade. Often gregarious in understorey. Plot records show highly significant correlation with the least acid soils. Iv. Coast to Nigeria; (Guinea wide).

548 STERCULIA OBLONGA Mast., FWTA 1:321. Sterculiaceae. Akan: ohaa.
Large tree with buttresses; lvs notophyll, puberulous below, deciduous. Fl cream, small to medium, bowl, polygamous, 8–12. Fr greenish-brown follicle, 11–2; seeds very large, covered by yellow aril. Regenerates in shade. Liberia to Gabon; (Guinea wide).

549 STERCULIA RHINOPETALA K. Schum., FWTA 1:321. Sterculiaceae. Akan: wawabima.
Large tree with buttresses; lvs mesophyll, deciduous. Fl cream, small, bowl, polygamous, 7–11. Fr brown follicle, 11–1; seeds large, covered with red aril. Regenerates in shade. Iv. Coast to Cameroun; (Guinea wide). Class 3 timber; trade name: sterculia brown (40).

550 STERCULIA TRAGACANTHA Lindl., FWTA 1:321. Sterculiaceae. Akan: sofo.
Medium-sized tree; lvs mesophyll, pubescent below, deciduous. Aubréville's model. Fl purplish, medium, bowl (or trap?), polygamous, 11–3. Fr red follicle, 10–12; seeds large, black. Regenerates in shade. Sometimes gregarious in forest near the savanna boundary, probably because it has some tolerance of fire. Guinea-Bissau to Angola and Sudan; (Guineo-Congolian wide).

551 STEREOSPERMUM ACUMINATISSIMUM K. Schum., FWTA 2:386. Bignoniaceae.
Medium-sized tree; lvs pinnate, leaflets notophyll. Scarrone's model. Fl large, pink, bell, 6–8. Fr long follicle-like capsule, 6–9; seeds medium, wind-dispersed with aid of papery wing. Occurs especially in forest on shallow soil at edge of rock outcrops. Guinée to Cameroun; (Guinea wide).

552 STREPHONEMA PSEUDOCOLA A. Chev., FWTA 1:264. Combretaceae. Akan: awuruku.
Medium to large tree; lvs mesophyll to macrophyll. Troll's model (59). Fl white, medium, brush, clustered in spikes, 8–11. Fr brown drupe; seed very large. Regenerates in shade. S. Leone to Ghana and Cameroun; (Guinea wide).

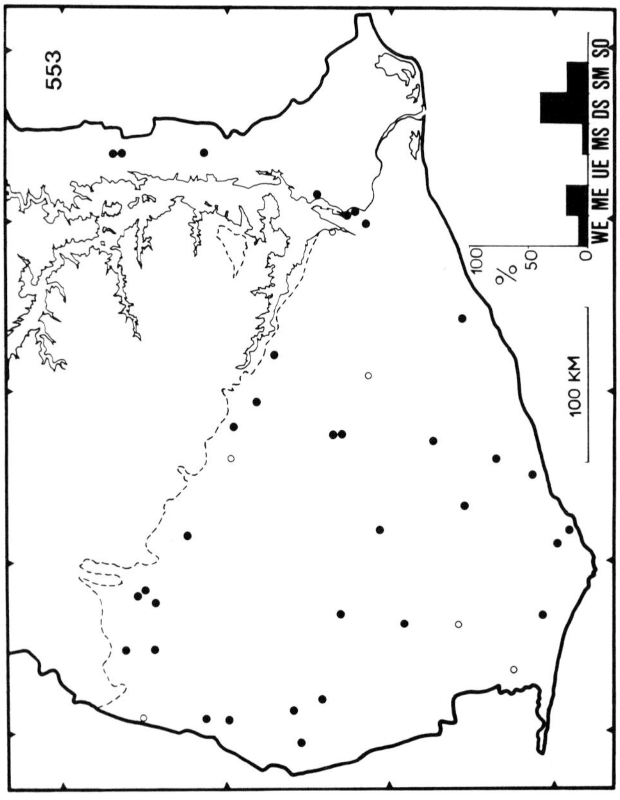
553
100 KM
100
%
50
0
WE ME UE MS DS SM SO

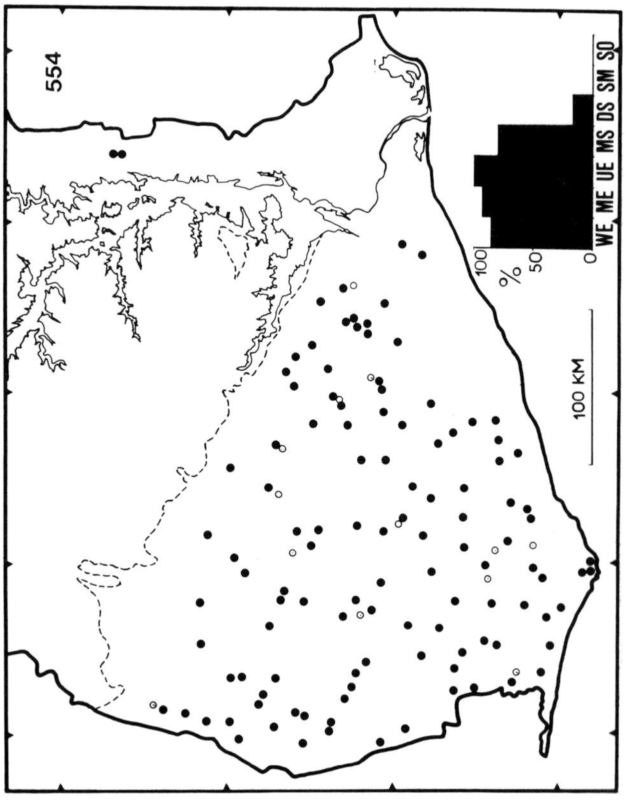
554
100 KM
100
%
50
0
WE ME UE MS DS SM SO

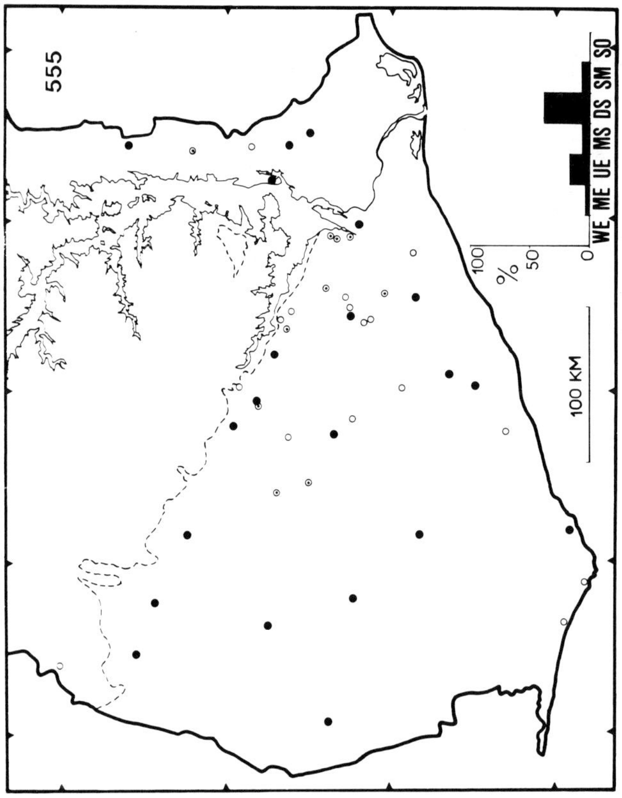
555
100 KM
100
%
50
0
WE ME UE MS DS SM SO

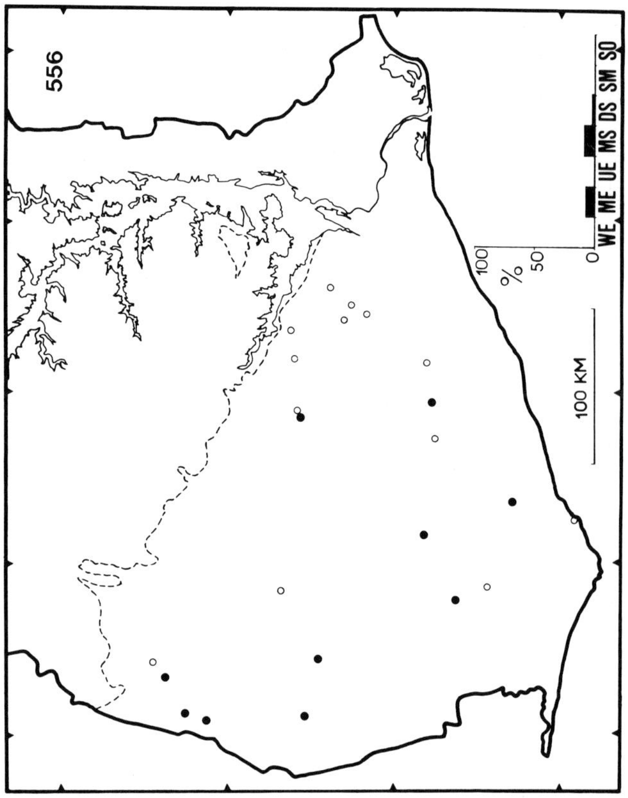
556
100 KM
100
%
50
0
WE ME UE MS DS SM SO

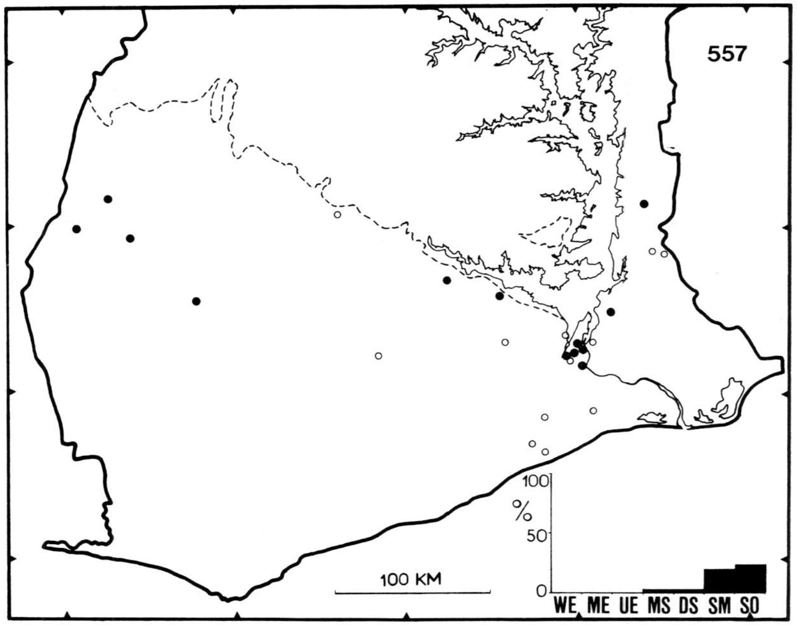

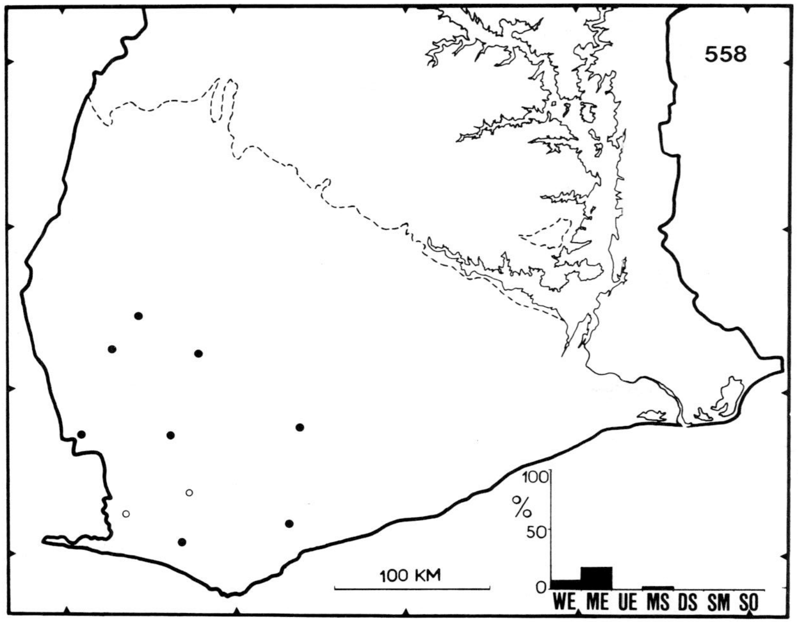

553 STREPTOGYNA CRINITA P. Beauv., FWTA 3:365. Gramineae. Akan: okyerotwe.
Perennial stoloniferous grass; lvs mesophyll. Tomlinson's model. Spikelets green, large, 11–3. Seed medium; dispersed by the attachment of spikelets to the hairs of animals. Regenerates at least vegetatively in shade. Senegal to Zaire and Uganda; also India and Ceylon; (chorological transgressor).

554 STROMBOSIA GLAUCESCENS Engl. var. *lucida* J. Léonard, FWTA 1:648. Olacaceae. Akan: afena.
Medium to large tree; lvs notophyll. Roux's model (59). Fl whitish, small, funnel, 8–9. Fr purplish drupe, 12–1; seed medium to large. Regenerates freely in shade. Plot records very significantly correlated with soils of low pH. S. Leone to Ghana, and Zaire; (Guineo-Congolian wide). Class 3 timber; trade name: afina (40). The type of var. *lucida* is from Zaire, and the description of its bark there (84) suggests a plant very different from the tree called var. *lucida* in West Africa (113). Further study may show that the two populations belong to different taxa, and thus that the apparent disjunction is merely a taxonomic artefact.

555 STROPHANTHUS BARTERI Franch., FWTA 2:72, and STROPHANTHUS PREUSSII Engl. & Pax, FWTA 2:72. Apocynaceae.
Large woody climbers with corky bark and clear latex; lvs microphyll. Fl yellowish and purple, large, funnel with tube and long 'tails' to the corolla lobes, mostly 12–4. Fr large follicle; seeds medium, wind-dispersed with the aid of a parachute of hairs. Leafy shoots of the two species cannot always be distinguished, and were not distinguished in the survey. Herbarium records are mapped separately: *S. barteri* ⊙, *S. preussii* ○. Distribution: *S. barteri*. Guinée to Cameroun, (Guinea wide); *S. preussi*, Guinée to Angola and Tanzania, (Guineo-Congolian wide).

556 STROPHANTHUS GRATUS (Hook.) Franch., FWTA 2:70. Apocynaceae.
Large woody climber; containing clear latex; lvs notophyll to mesophyll. Fl purplish, large, funnel, 11–2. Fr follicle; seeds medium to large, wind-dispersed with aid of a parachute of hairs. S. Leone to Zaire; (Guineo-Congolian wide). The seeds of this species are the main source of the cardiac glycoside strophanthin.

557 STROPHANTHUS HISPIDUS DC., FWTA 2:72. Apocynaceae.
Large woody climber containing clear latex; lvs notophyll, very hairy. Fl yellow and purple, large, funnel with tube, long tails to corolla lobes, 2–4. Fr long follicle, 5–11; seeds medium to large, wind-dispersed with aid of a parachute of hairs. Extends to woodland in the savanna zone. Senegal to Zaire and Uganda; (ecological and chorological transgressor). Seeds widely used in preparation of arrow poison.

558 STRYCHNOS ACULEATA Solered., FWTA 2:43. Loganiaceae (74).
Large woody climber, with tendrils in 1–3 pairs; bark thick, thorny; phloem strands in the xylem. Lvs notophyll. Mangenot's model. Fl white, small, funnel, 3–5. Fr large woody yellow-green berry, 2–6; seeds very large, dispersed by elephants. Regenerates in shade. Traces of alkaloid in shoots (25). S. Leone to Angola and Uganda; (Guineo-Congolian wide).

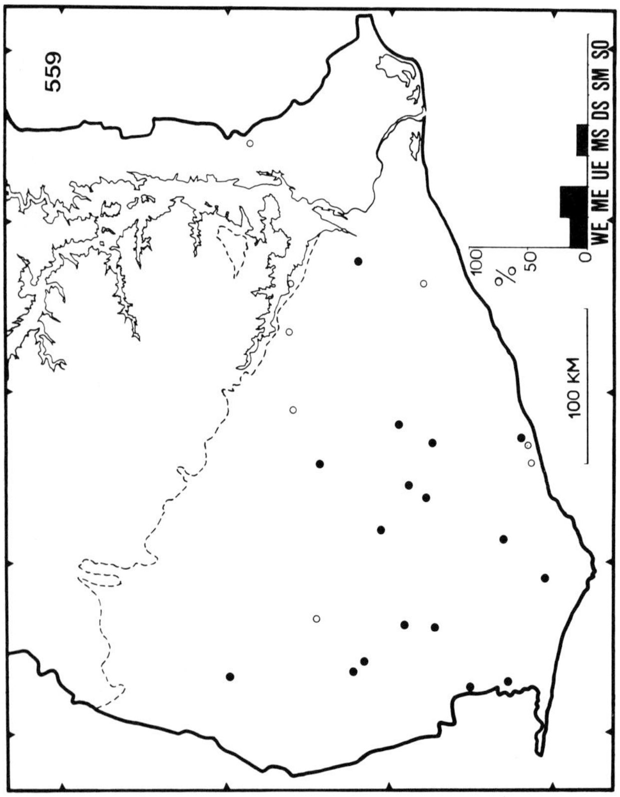
559
100
%
50
0
WE ME UE MS DS SM SO
100 KM

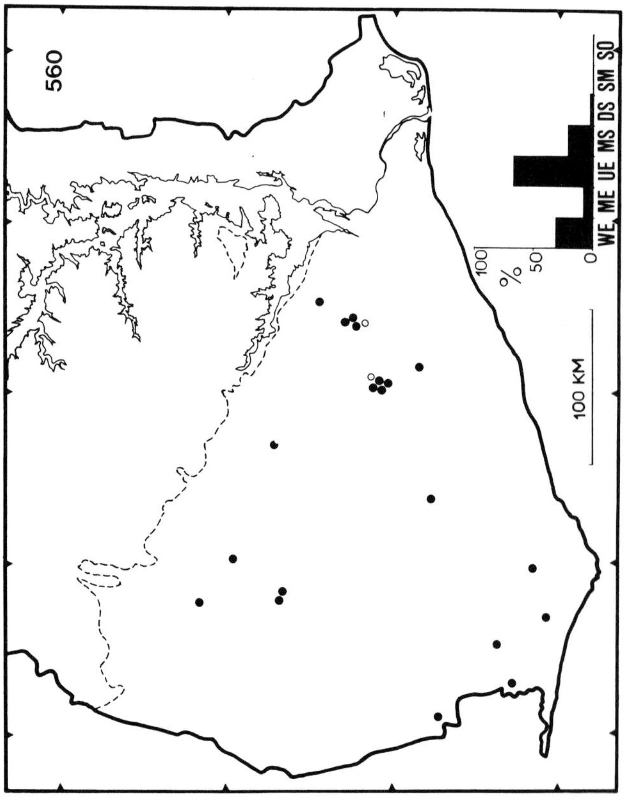
560
100
%
50
0
WE ME UE MS DS SM SO
100 KM

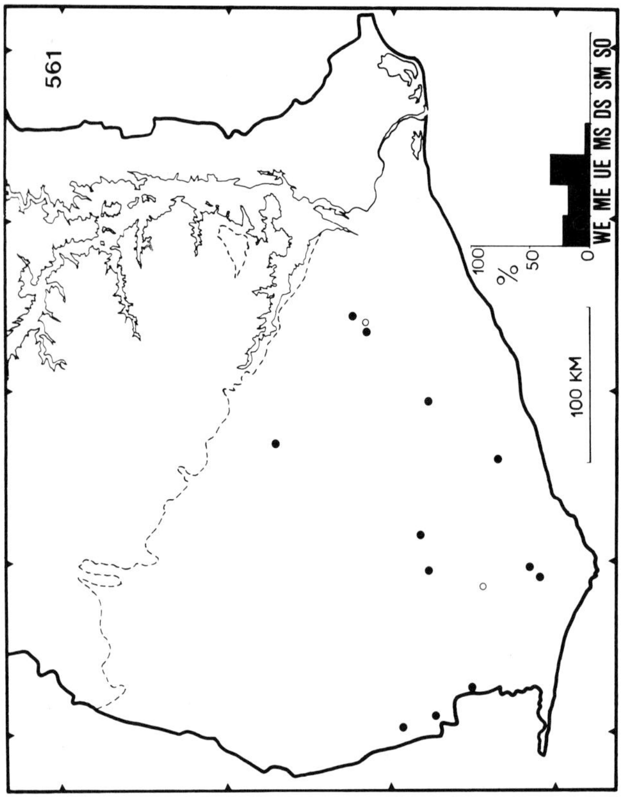
561
100
%
50
0
WE ME UE MS DS SM SO
100 KM

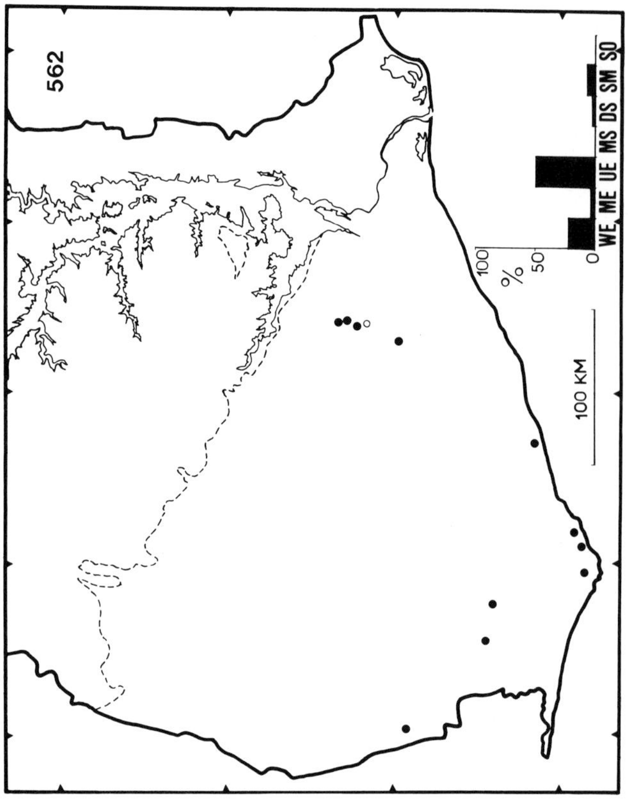
562
100
%
50
0
WE ME UE MS DS SM SO
100 KM

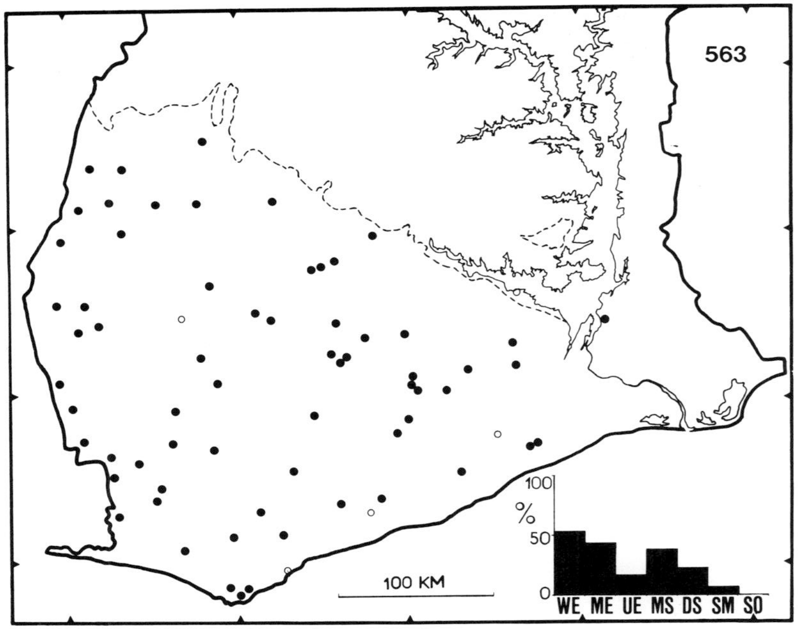

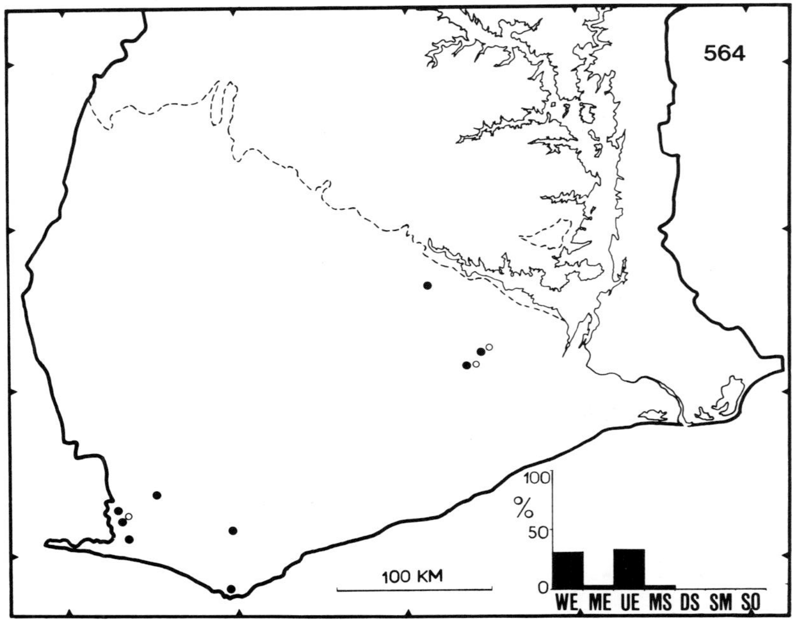

559 STRYCHNOS AFZELII Gilg, FWTA 2:43. Loganiaceae (74).
Large woody climber, or shrub in thicket, with solitary tendrils; bark with large lenticels; phloem strands in xylem. Lvs microphyll. Mangenot's model. Fl white, small, crowded, 3–7. Fr small 1-seeded orange berry; seed medium to large. Regenerates in shade. Alkaloid in shoots (25); but they are clove-scented and used as chewing sticks. Senegal to Nigeria; (Guinea wide).

560 STRYCHNOS ASTERANTHA Leeuwenberg (74). Loganiaceae.
Large woody climber, with tendrils in 1–3 pairs; bark with large lenticels; phloem strands in the xylem. Lvs microphyll to notophyll. Mangenot's model. Fl white, small, crowded. Fr 1–2-seeded berry; seed large. Regenerates in shade. Some alkaloid in shoots (25). Ghana to Cameroun; (Guinea wide).

561 STRYCHNOS CAMPTONEURA Gilg & Busse, FWTA 2:44. Loganiaceae (74).
Large woody climber, with tendrils in 1–3 pairs; bark with large lenticels; phloem strands in the yellowish xylem. Lvs mesophyll. Mangenot's model. Fl whitish, medium, dish with tube, not yet collected in Ghana. Fr large, woody, yellow berry; seeds very large. Regenerates in shade. Abundant alkaloids in the shoots (25). Liberia to Zaire; (Guineo-Congolian wide).

562 STRYCHNOS CUMINODORA Leeuwenberg (74). Loganiaceae.
Large woody climber, with paired tendrils; phloem strands in the yellowish wood. Lvs microphyll, pellucid-dotted. Mangenot's model. Fl pale yellow, small, inconspicuous. Fr small 1-seeded orange berry; seed medium. Regenerates in shade. Traces of alkaloid in shoots (25). Guinée to Ghana; (Upper Guinea endemic).

563 STRYCHNOS FLORIBUNDA Gilg, FWTA 2:44. Loganiaceae (74).
Large woody climber, with solitary tendrils; pale lenticels; phloem strands in xylem. Lvs microphyll. Mangenot's model. Fl whitish, small to medium, bell, not clearly seasonal. Fr 1-seeded orange berry, mostly 12–2; seed medium to large. Regenerates in shade. Alkaloid medium to abundant in shoots (25). S. Leone to Angola; (Guineo-Congolian wide).

564 STRYCHNOS ICAJA Baill., FWTA 2:44. Loganiaceae (74).
Large woody climber, with solitary tendrils; large lenticels; underbark red in specimens from north-east part of forest zone, brown in south-west; phloem strands in xylem. Lvs notophyll. Mangenot's model. Fl yellowish-white, small, inconspicuous. Fr 1-seeded, dark yellow berry; seed large to very large. Abundant alkaloid in shoots (25). Guinée to Angola; (Guineo-Congolian wide). Root bark used in Gabon and Zaire as arrow and ordeal poison (24).

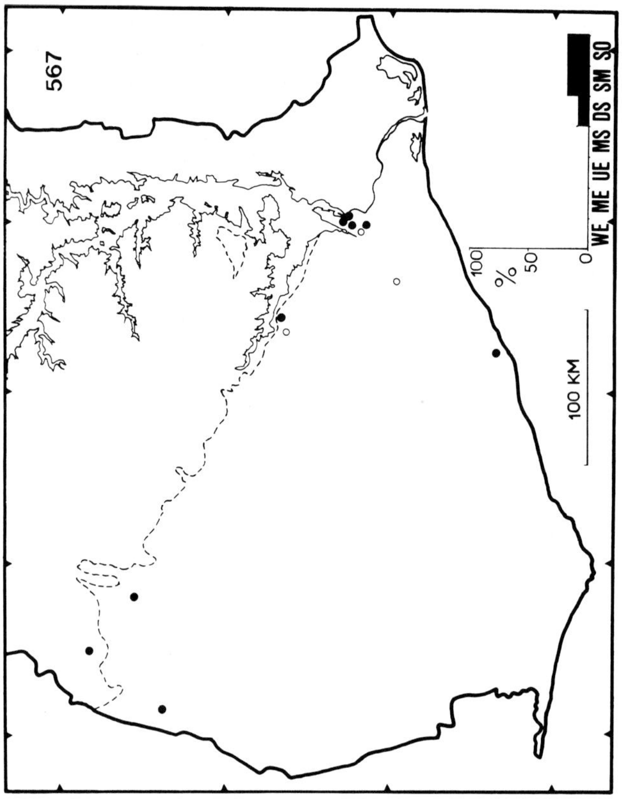
567
100 KM
100
%
50
0
WE ME UE MS DS SM SO

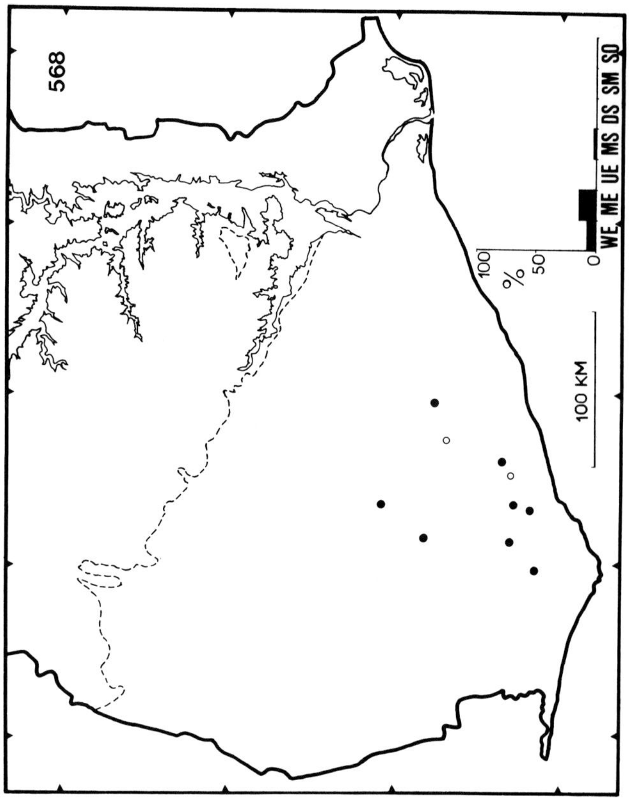
568
100 KM
100
%
50
0
WE ME UE MS DS SM SO

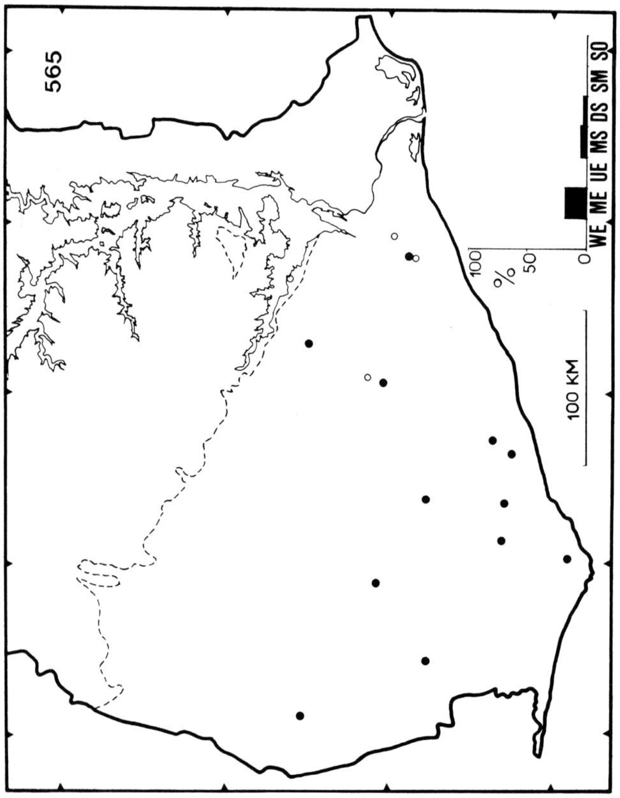
565
100 KM
100
%
50
0
WE ME UE MS DS SM SO

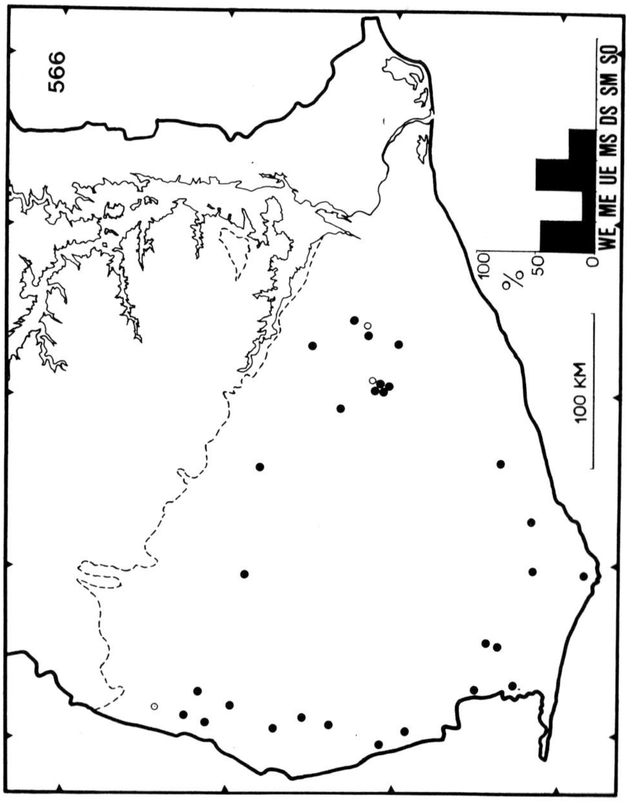
566
100 KM
100
%
50
0
WE ME UE MS DS SM SO

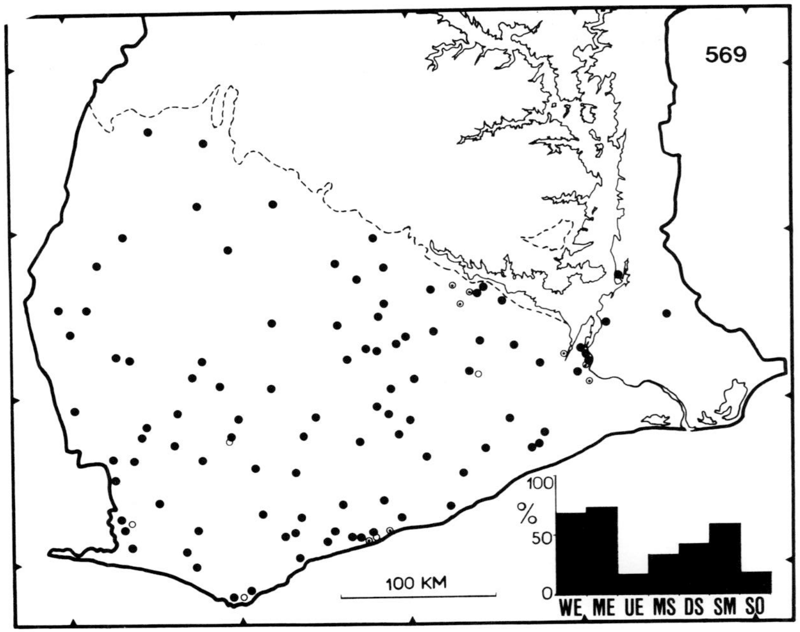

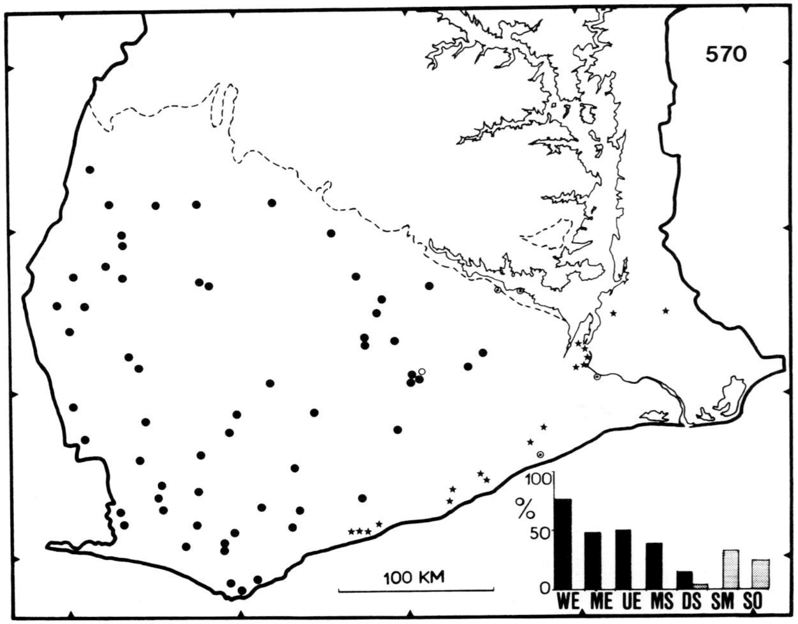

565 STRYCHNOS JOHNSONII Hutch. & M. B. Moss, FWTA 2:44 Loganiaceae. (74). (Type from Akropong-Akwapim).
Large woody climber, with paired tendrils; phloem strands in yellowish wood; twigs 4-angled. Lvs notophyll. Mangenot's model. Fl yellowish, small, bowl, 7–11. Fr small 1-seeded berry; seed medium to large. Regenerates in shade. Alkaloid at most trace in shoot (25). Guinée to Angola and Uganda; (Guineo-Congolian wide).

566 STRYCHNOS MALACOCLADOS C. H. Wright, FWTA 2:43. Loganiaceae (74).
Medium to large woody climber, with solitary tendrils; twigs pubescent; phloem strands in xylem. Lvs microphyll to notophyll. Mangenot's model. Fl yellow, small, inconspicuous. Fr 1-seeded orange berry; seed large. Regenerates in shade. Some alkaloid in shoot (25). Guinée to Gabon; (Guinea wide).

567 STRYCHNOS NIGRITANA Bak., FWTA 2:43 Loganiaceae (74).
Large woody climber, with paired tendrils; phloem strands in xylem. Lvs notophyll. Mangenot's model. Fl yellowish, medium, funnel, 3–5. Fr large orange berry with hard shell; seeds large. Extends to woodland and forest patches in the savanna zone. Alkaloid in shoot at most trace (25). Guinée to Zaire; (Guineo-Congolian wide).

568 STRYCHNOS PHAEOTRICHA Gilg, FWTA 2:41. Loganiaceae (74).
Large woody climber, with paired tendrils; pale lenticels; phloem strands in xylem inconspicuous. Lvs notophyll to mesophyll, spreading-hairy, as are the twigs. Mangenot's model. Fl whitish, small, bowl. Fr berry; seeds large. Regenerates in shade. No alkaloid in shoot. Ghana to Zaire; (Guineo-Congolian wide).

569 STRYCHNOS SOUBRENSIS Hutch. & Dalz. (74), (*S. ngouniensis* of FWTA 2:43, not of Pellegr.), and STRYCHNOS SPLENDENS Gilg, FWTA 2:41. Loganiaceae.
Large woody climbers, with paired tendrils; lenticels pale and conspicuous in *S. soubrensis*, less so in *S. splendens;* phloem strands in xylem. Lvs microphyll. Mangenot's model. Fl whitish, medium *(S. soubrensis)* or small *(S. splendens)*, mostly 4–5 *(S. soubrensis)* or 7–10 *(S. splendens)*. Fr 1–2-seeded orange berry; seed large. Regenerate in shade. Alkaloid in leafy shoots fairly plentiful, especially in *S. splendens* (25). Vegetative shoots of the two species are difficult to distinguish, and they were not separated during the survey. Herbarium records: *S. soubrensis* ○; *S. splendens* ⊙; *S. soubrensis* evidently tends to occur in forest under higher rainfall than does *S. splendens*. Distribution: *S. soubrensis:* Guinée to Nigeria, (Guinea wide); *S. splendens:* Guinea Bissau to Nigeria, (Guinea wide).

570 STRYCHNOS USAMBARENSIS Gilg, FWTA 2:44. Loganiaceae (74).
Large woody climber (at least in West Africa), with solitary tendrils; phloem strands in xylem. Lvs microphyll. Mangenot's model. Fl whitish, small, not clearly seasonal. Fr 1-seeded orange berry; seed large. Regenerates in shade. Abundant alkaloid in leafy shoot (25). Two rather well-marked forms occur in Ghana: (i) Leaves ovate, small, even in shade, main lateral nerves originating almost at leaf base, restricted to dry forest – survey records ★; herbarium records ⊙; histogram bars dotted; because this form grows as a shrub in thicket, flowers and fruits are readily collected. (ii) Leaves elliptic, shade leaves very much larger than sun leaves, main lateral nerves originating well above leaf base, found in moist forest – survey records ●; herbarium records ○; histogram bars solid black; flowers produced only in the canopy, so much less often collected than form (i), even though much more widely distributed. *S. usambarensis* occurs from Guinée to Zaire, Kenya, Rhodesia and S. Africa; (ecological and chorological transgressor).

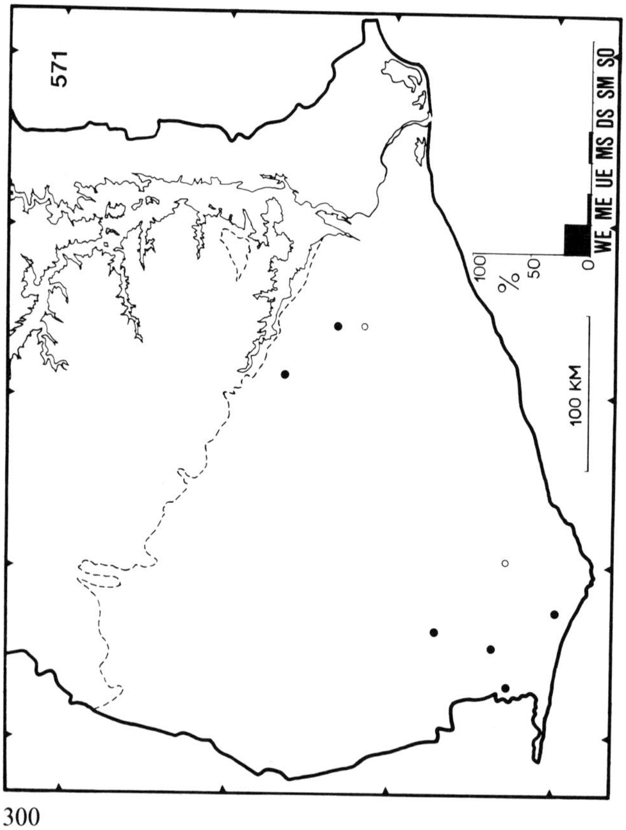
571
100 KM
100
%
50
0
WE ME UE MS DS SM SO

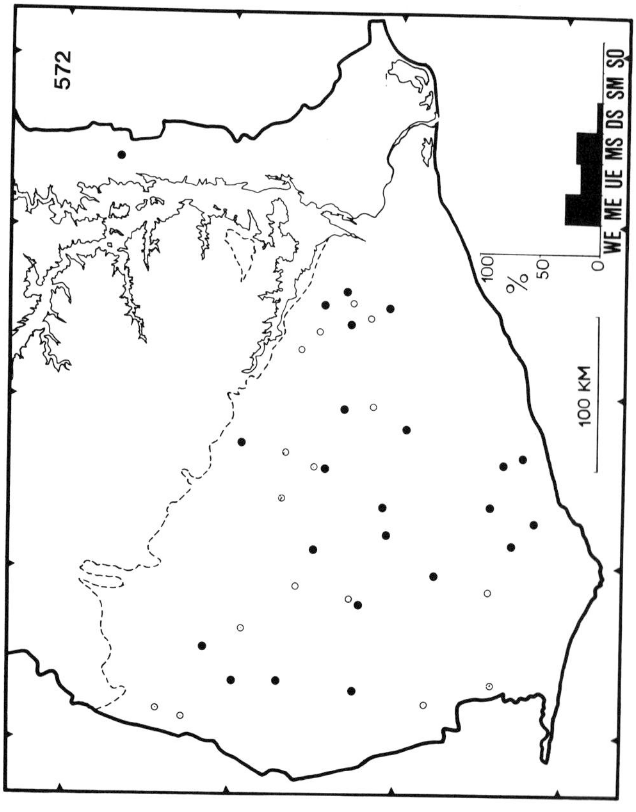
572
100 KM
100
%
50
0
WE ME UE MS DS SM SO

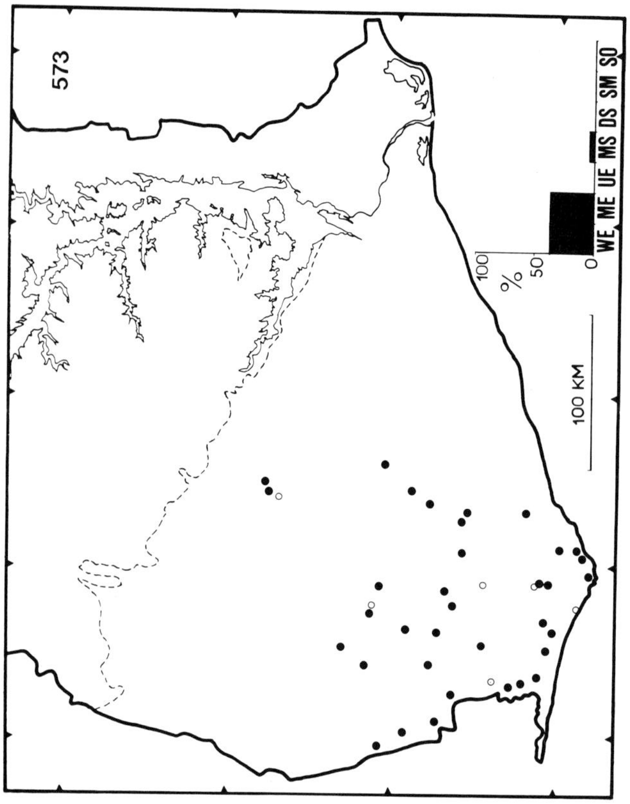
573
100 KM
100
%
50
0
WE ME UE MS DS SM SO

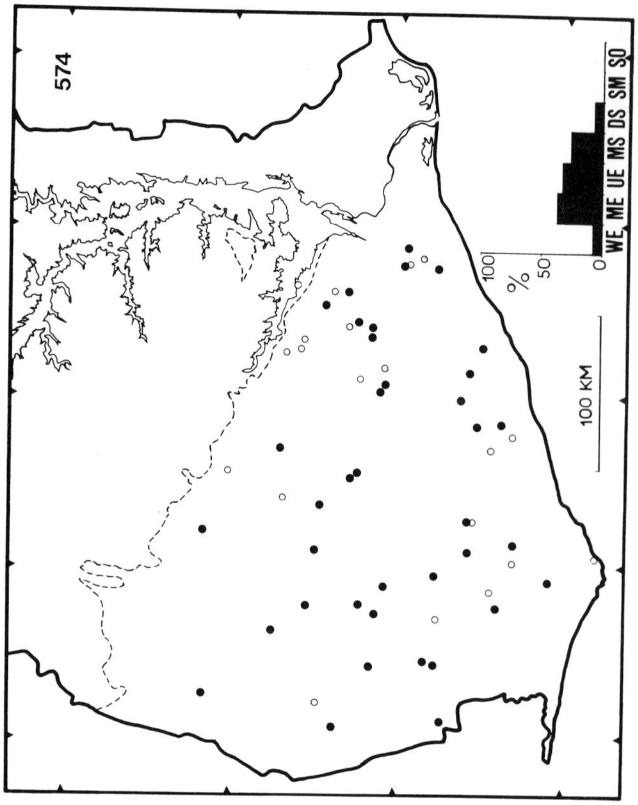
574
100 KM
100
%
50
0
WE ME UE MS DS SM SO

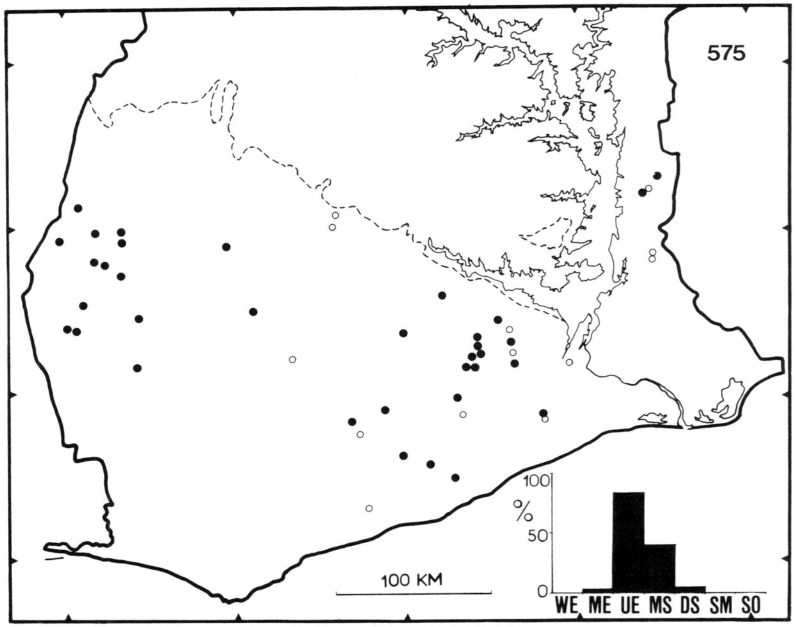

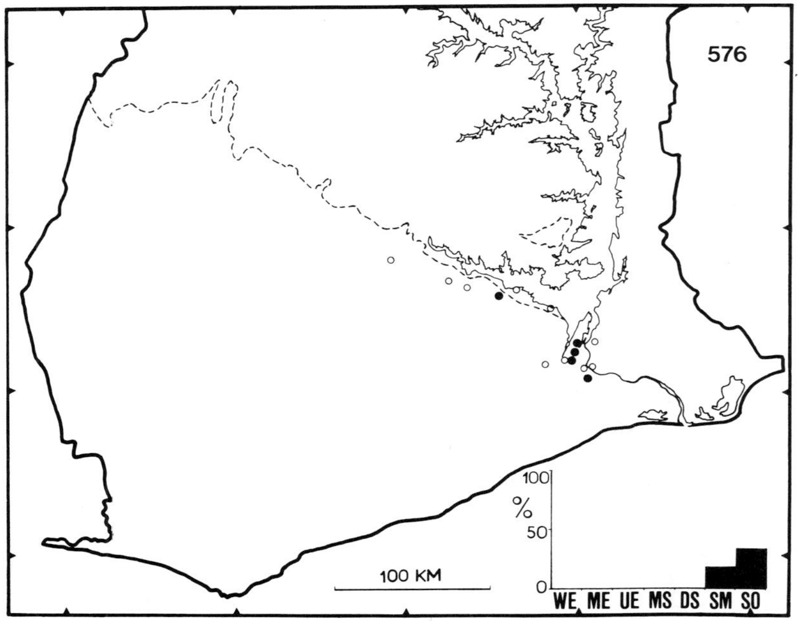

571 SYZYGIUM GUINEENSE (Willd.) DC. subsp. *occidentale* F. White (119), (*Syzygium staudtii* of FWTA 1:240, partly, not of (Engl.) Mildbr.). Myrtaceae. (Type of subspecies from Atewa Range F.R.).
Medium to large tree; trunk fluted and with adventitious branches. Lvs microphyll, with translucent glands. Fl cream, small, brush. Fr berry. Iv. Coast to Cameroun. Species is an ecological and chorological transgressor.

572 TABERNAEMONTANA CHIPPII (Stapf) Pichon, FWTA 2:66. Apocynaceae. Akan: obonawa. (Type from 'Fwirem', Ashanti).
Small tree, containing white latex; lvs mesophyll. Leeuwenberg's model. Fl white, large, dish with tube, not clearly seasonal. Fr yellowish berry; seeds medium to large. Regenerates in shade. Liberia to Ghana; (Upper Guinea endemic).

573 TABERNAEMONTANA CRASSA Benth., FWTA 2:66. Apocynaceae.
Small to medium tree, containing white latex; lvs mesophyll. Leeuwenberg's model (59). Fl white, large, dish with tube, not clearly seasonal. Fr yellowish berry; seeds medium to large. Regenerates in shade. Plot records very significantly correlated with acid, base-poor soils. S. Leone to Zaire; (Guineo-Congolian wide).

574 TABERNAEMONTANA GLANDULOSA (Stapf) Pichon, FWTA 2:66. Apocynaceae.
Small slender woody climber containing white latex; lvs notophyll to mesophyll. Fl large, white, dish with tube, not clearly seasonal. Fr orange berry; seeds medium to large. Regenerates in shade. S. Leone to Zaire; (Guineo-Congolian wide).

575 TABERNAEMONTANA PACHYSIPHON Stapf var. *cumminsii* (Stapf) H. Huber, FWTA 2:66. Apocynaceae. (Type of variety from Assin Nyankumasi).
Small to medium tree, with white latex; lvs mesophyll. Leeuwenberg's model. Fl white, large, dish with tube, not clearly seasonal. Fr yellowish-green berry; seeds medium to large. Regenerates in shade. Ghana to Cameroun; (Guinea wide).

576 TALBOTIELLA GENTII Hutch. & Greenway, FWTA 1:467. Caesalpiniaceae. Akan: takorowanua.
Medium-sized tree; lvs pinnate, leaflets nanophyll to microphyll. Troll's model. Fl pink and white, small, brush, 11–1. Fr explosive pod, c. 1; seeds large. Regenerates in shade. Forms pure stands in which most of the understorey consists of young plants of *Talbotiella* (102a). Endemic to Ghana; (Upper Guinea endemic).

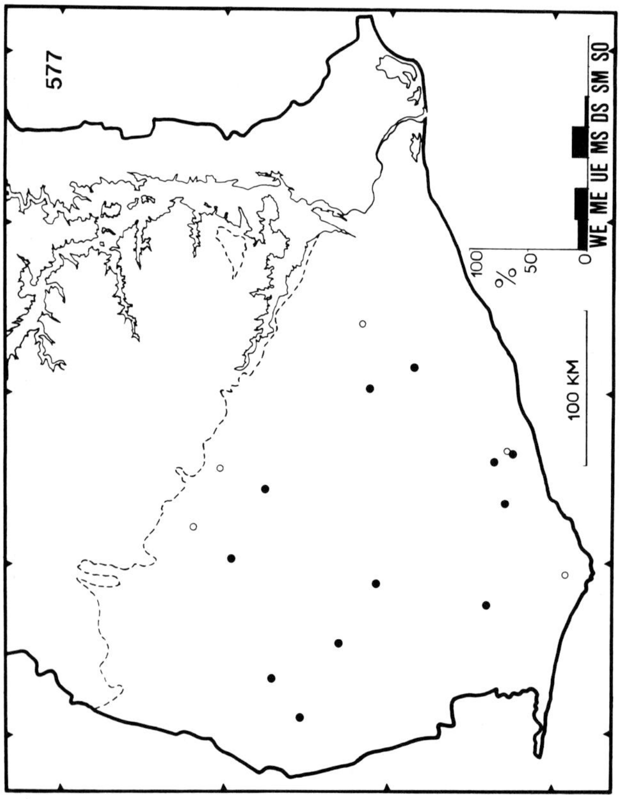
577
100 KM
%
100
50
0
WE ME UE MS DS SM SO

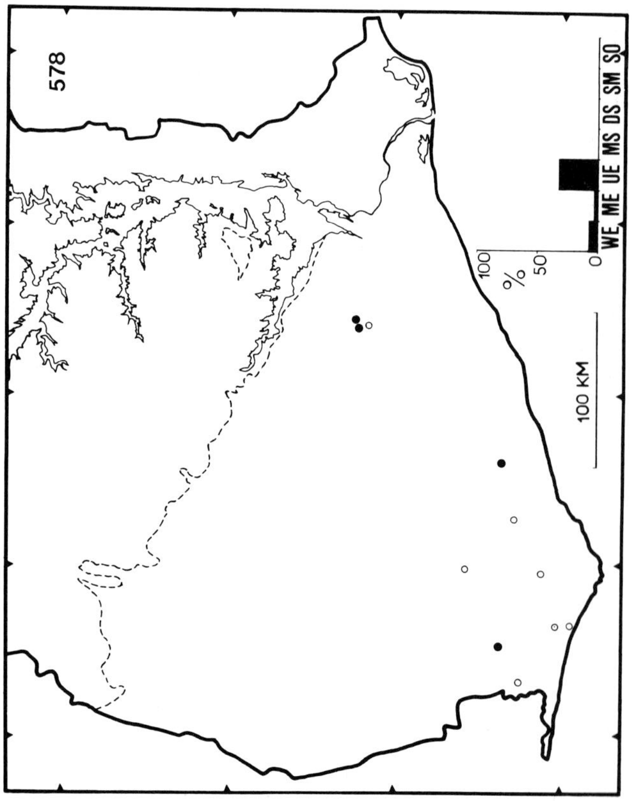
578
100 KM
%
100
50
0
WE ME UE MS DS SM SO

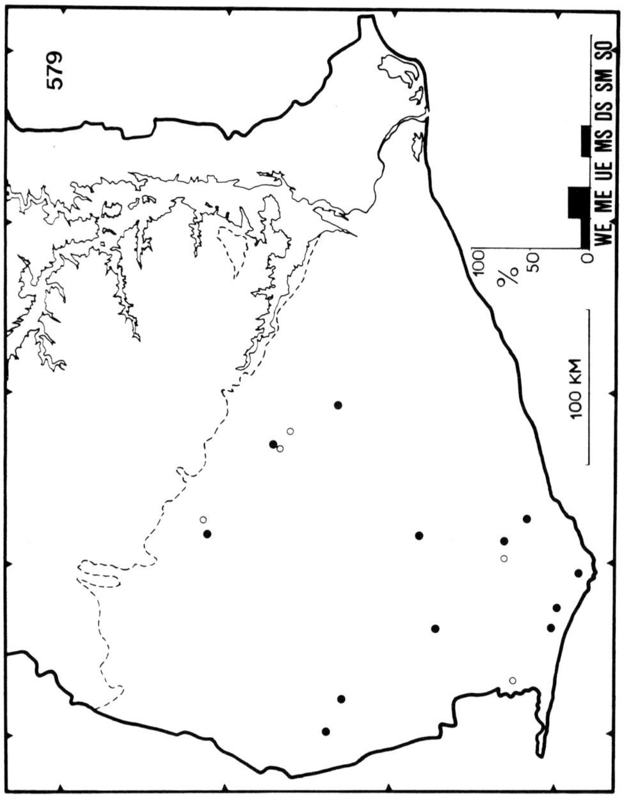
579
100 KM
%
100
50
0
WE ME UE MS DS SM SO

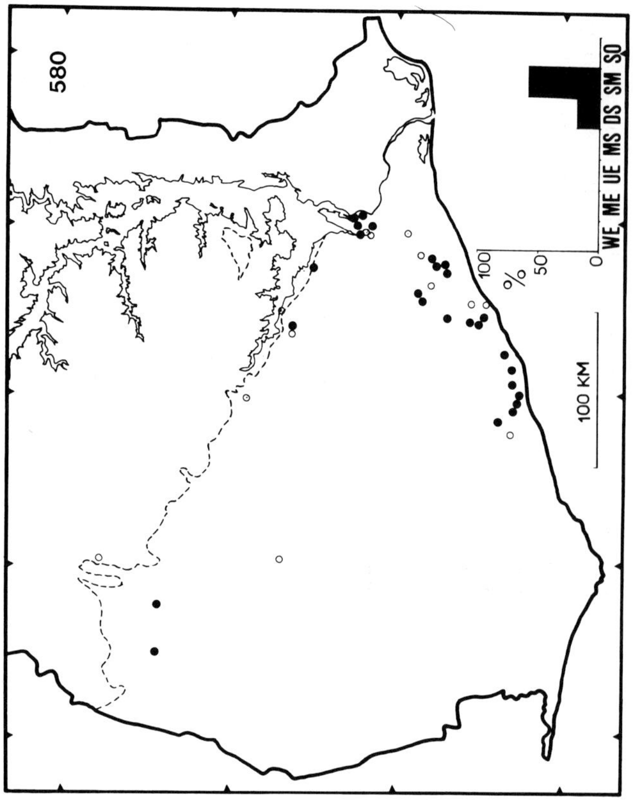
580
100 KM
%
100
50
0
WE ME UE MS DS SM SO

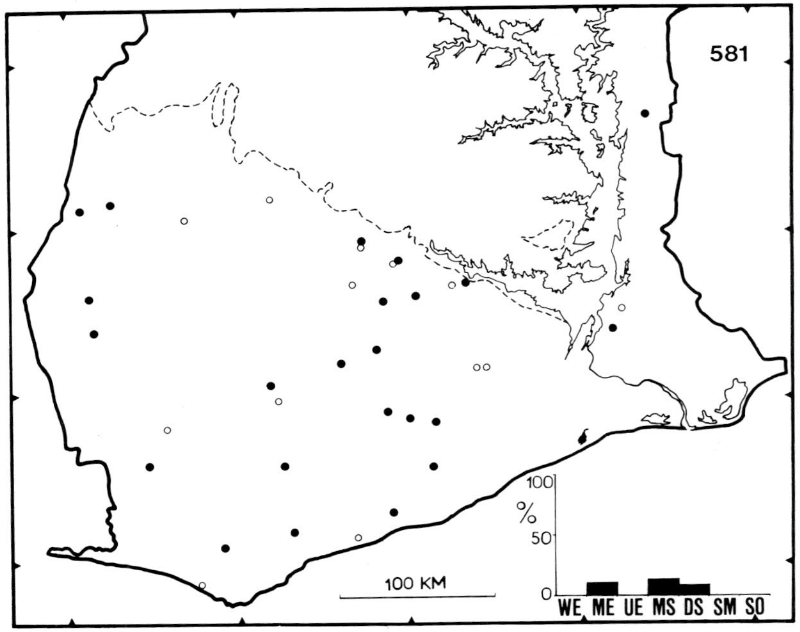

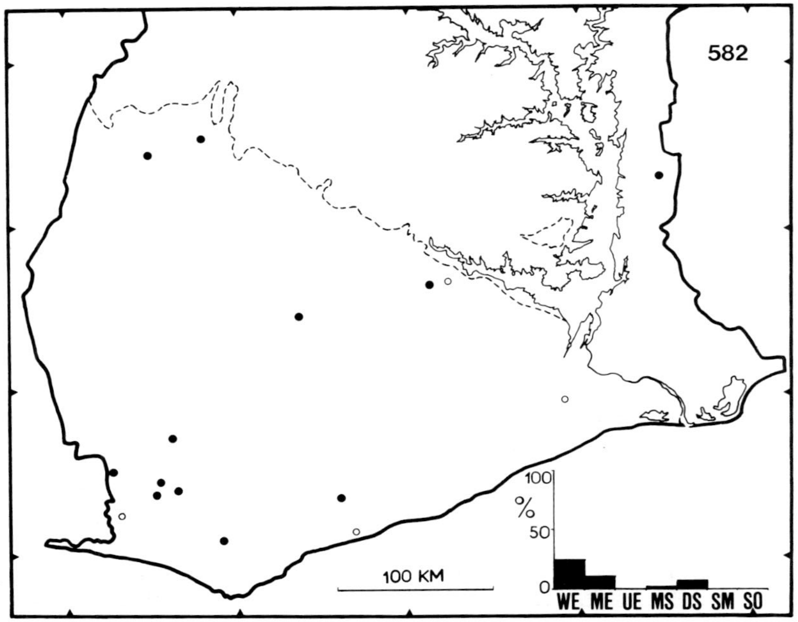

577 TARENNA BIPINDENSIS (K. Schum.) Bremek., FWTA 2:135. Rubiaceae.
Small woody climber; lvs notophyll. Fl pale green, medium, dish with tube, not clearly seasonal. Fr berry; seeds medium. Regenerates in shade. Guinée to Gabon; (Guinea wide).

578 TARENNA EKETENSIS Wernh., FWTA 2:135. Rubiaceae.
Small woody climber; lvs notophyll, pubescent on underside of midrib and lateral nerves, and on twigs. Fl white, medium, dish with tube, not clearly seasonal. Fr berry; seeds medium. Regenerates in shade. Liberia to Zaire; (Guineo-Congolian wide).

579 TARENNA FUSCO-FLAVA (K. Schum.) S. Moore (63), (*Tarenna flavo-fusca* in FWTA 2:134). Rubiaceae.
Small woody climber; lvs microphyll. Fl medium to large, greenish-yellow and purple, dish with tube, not clearly seasonal. Fr whitish berry; seeds medium. Regenerates in shade. Liberia to Angola and Uganda; (Guineo-Congolian wide).

580 TECLEA VERDOORNIANA Exell & Mendonça, FWTA 1:689. Rutaceae. Akan: owebiribi.
Small to medium tree; lvs trifoliolate, leaflets dotted with translucent glands, notophyll to mesophyll. Fl creamy-white, small, bowl, dioecious, 1–5. Fr red drupe, 3–6; seed medium. Regenerates in shade. Rich in alkaloids (115). S. Leone to Zaire (Guineo-Congolian wide).

TERMINALIA IVORENSIS A. Chev., FWTA 1:279. Combretaceae. Akan: emire.
Large tree with small buttresses; lvs notophyll, deciduous. Aubréville's model. Fl small, cream, brush, clustered in spikes, polygamous, mostly 1–5. Fr 1-seeded samara, wind-dispersed, 12–1; seed medium. Regenerates only in gaps; occurs in high forest as scattered large individuals and hence rarely recorded in the survey. Guinée to Cameroun; (Guinea wide). Class 1 timber; trade name; idigbo (40). Fig. 11.7 gives distribution in Ghana as calculated from timber enumeration records.

581 TERMINALIA SUPERBA Engl. & Diels, FWTA 1:277. Combretaceae. Akan: ofram.
Large tree with high buttresses; lvs notophyll to mesophyll, deciduous. Aubréville's model. Fl small, cream, brush, clustered in spikes, polygamous, mostly 1–2. Fr 1-seeded samara (Fig. 6.1e), wind-dispersed, 12–2; seed medium. Less intolerant of shade than *T. ivorensis*. Guinée to Zaire; (Guineo-Congolian wide). Class 3 timber; trade name: afara (40). See Fig. 11.8 for distribution in Ghana as calculated from timber enumeration records.

582 TETRACERA POTATORIA Afzel. ex G. Don, FWTA 1:180. (Incl. *T. affinis* Hutch., FWTA 1:181). Dilleniaceae.
Large woody climber; xylem cylindrical, yielding drinkable water when freshly cut. Lvs microphyll to notophyll, toothed, scabrid. Fl white, small to medium, bowl, 8–12. Fr follicle; seeds black, with red aril. Regenerates in shade. Senegal to Zaire and Uganda; (Guineo-Congolian wide).

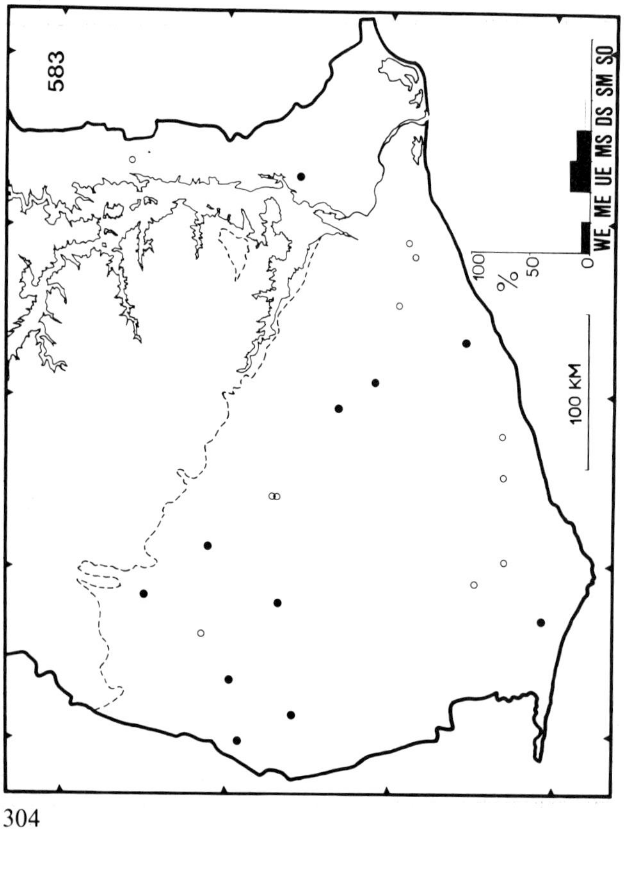
583
100 KM
%
100
50
0
WE ME UE MS DS SM SO

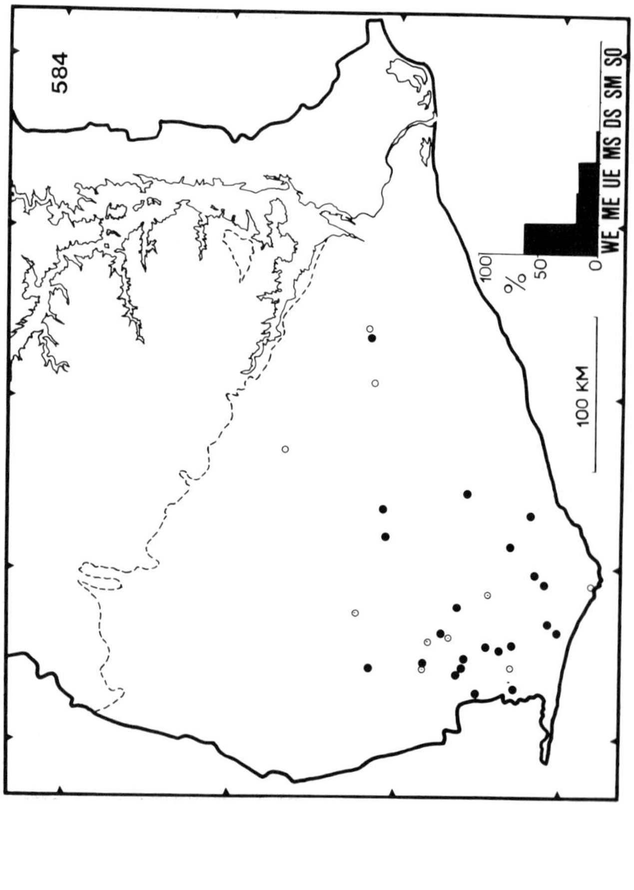
584
100 KM
%
100
50
0
WE ME UE MS DS SM SO

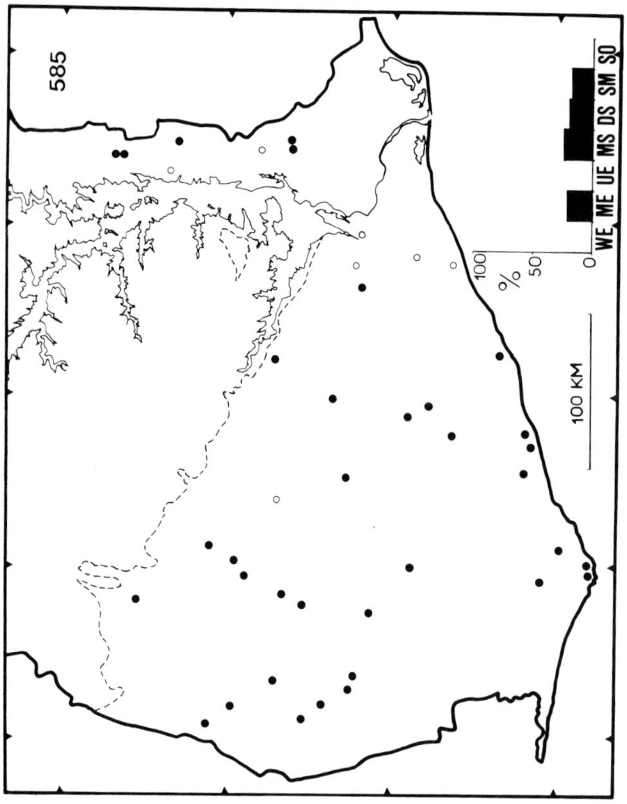
585
100 KM
%
100
50
0
WE ME UE MS DS SM SO

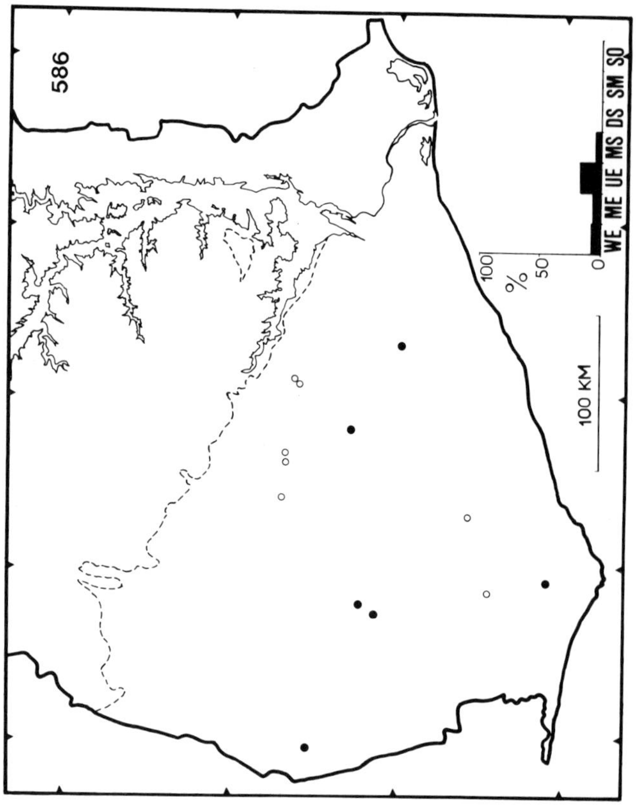
586
100 KM
%
100
50
0
WE ME UE MS DS SM SO

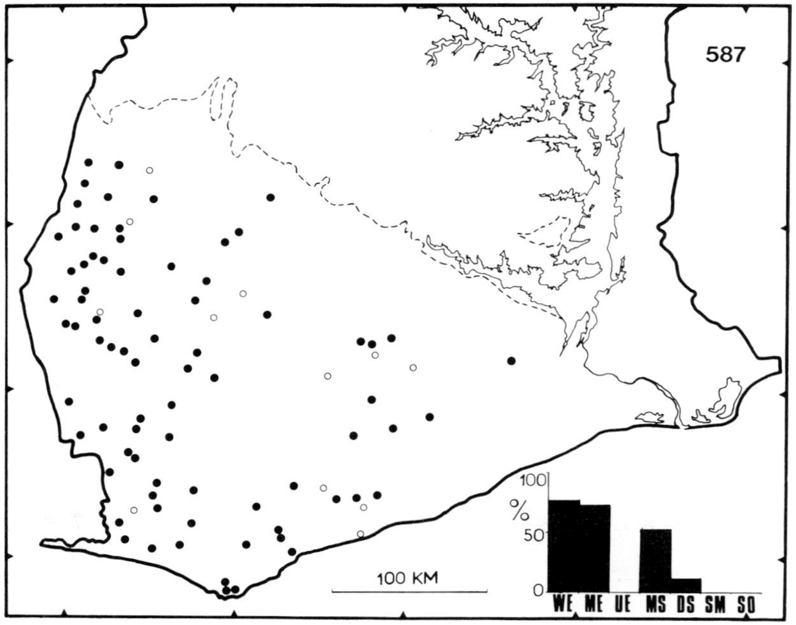

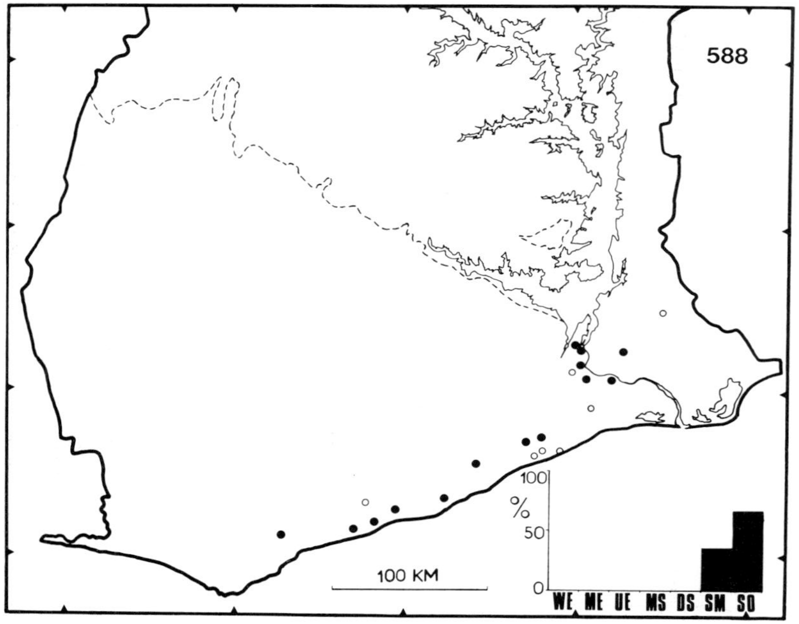

583 TETRAPLEURA TETRAPTERA (Schum. & Thonn.) Taub., FWTA 1:493. Mimosaceae. Akan: prekese. (Type from Akwapim).

Medium-sized tree with small buttresses; lvs bipinnate, leaflets nanophyll, deciduous. Fl small, yellowish, brush, crowded in spikes, 2–6. Fr winged fleshy indehiscent pod, 9–12, (aromatic and used as spice); seeds medium to large. Regenerates in shade. Senegal to Zaire and Uganda; (Guineo-Congolian wide).

584 THECACORIS STENOPETALA (Muell. Arg.) Muell. Arg., FWTA 1:372. Euphorbiaceae.

Pygmy tree; lvs notophyll to mesophyll. Fl small, greenish-yellow, crowded into spikes, dioecious, not clearly seasonal. Fr explosive capsule; seeds medium. Regenerates in shade. Plot records very significantly associated with acid, base-poor soils. S. Leone to Cameroun; (Guinea wide).

585 THONNINGIA SANGUINEA Vahl, FWTA 1:667. Balanophoraceae. Akan: kwabedwea. (Type from Akwapim).

Totally parasitic leafless herb on tree roots; only the inflorescence appears above ground (the only West African plant with this habit). Fl small, cream, monoecious, clustered within involucre of red, prickly bracts, not clearly seasonal. Fr 1-seeded nutlet; seed small. Regenerates in shade. Extends to the savanna zone. S. Leone to Angola, Kenya and Zambia; (ecological and chorological transgressor).

586 TIEGHEMELLA HECKELII Pierre ex A. Chev., FWTA 2:21. Sapotaceae. Akan: baku.

Large tree, containing white latex; lvs notophyll. Aubréville's model (59). Fl white, small, bowl, 2–5. Fr large yellow berry, 10–12; seeds very large with thick, woody testa, dispersed by elephants (5). Seedlings (26) occur in shade, but regeneration is rare – perhaps because of the lack of elephants (113). S. Leone to Nigeria; (Guinea wide). Class 1 timber; trade name: makore (40). The well-known 'Big Tree' in Esuboni F.R. is of this species. Distribution in Ghana as calculated from timber enumeration records is shown in Fig. 11.9.

587 TILIACORA DINKLAGEI Engl., FWTA 1:71. (incl. *T. dielsiana* Hutch. & Dalz., FWTA 1:71). Menispermaceae.

Large woody climber, stem somewhat lobed in section, xylem with rings of included phloem; lvs mesophyll. Roux's model. Fl yellow, small, bowl, dioecious, cauliflorous, 12–6. Fr not recorded. Regenerates in shade. Plot records very significantly correlated with acid, base-poor soils. S. Leone to Ghana; (Upper Guinea endemic). The characters used (106) to separate *T. dinklagei* from *T. dielsiana* are inadequate.

588 TILIACORA FUNIFERA (Miers) Oliv. (106), (*T. warneckei* Engl. & Diels and *T. johannis* Exell, FWTA 1:71). Menispermaceae.

Large woody climber (or shrub in thicket), stem somewhat lobed in section, xylem with rings of included phloem; lvs microphyll to notophyll. Roux's model. Fl yellow-green, small, bowl, dioecious, cauliflorous or axillary, not clearly seasonal. Fr yellow drupe; seed large. Regenerates in shade. Ghana to Angola, Kenya and Mozambique; (ecological and chorological transgressor, 105).

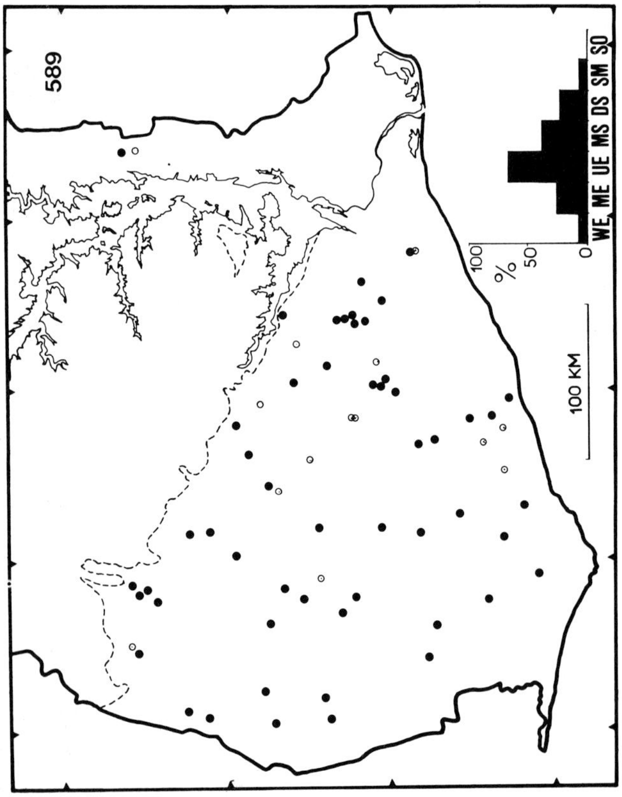
589
100 KM
100
%
50
0
WE ME UE MS DS SM SO

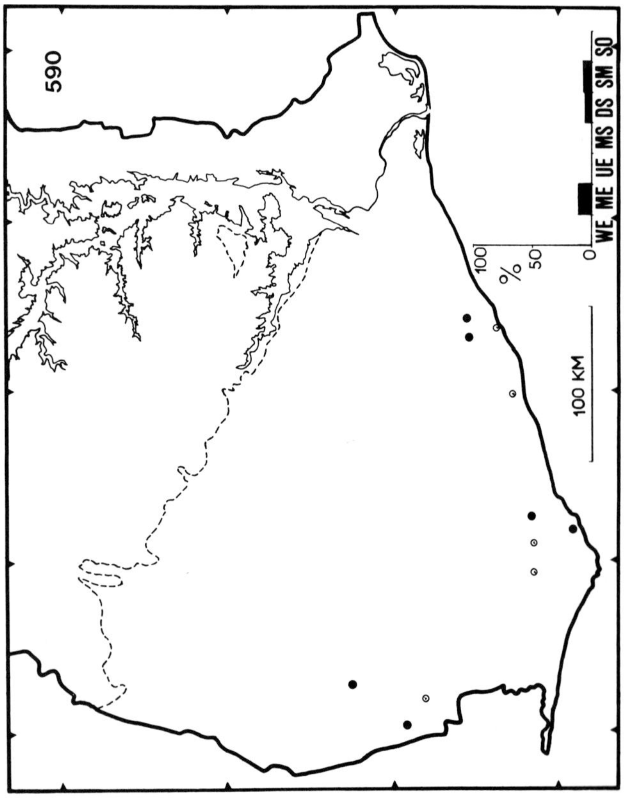
590
100 KM
100
%
50
0
WE ME UE MS DS SM SO

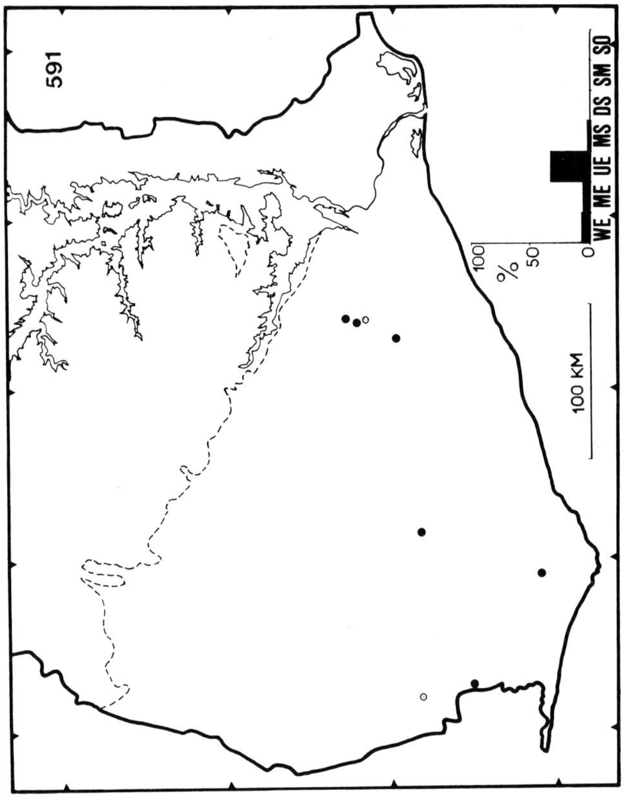
591
100 KM
100
%
50
0
WE ME UE MS DS SM SO

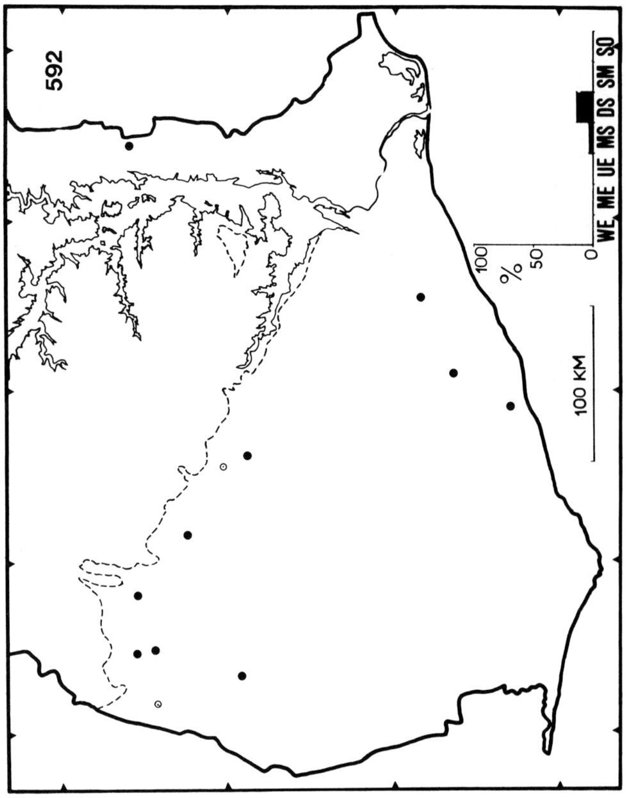
592
100 KM
100
%
50
0
WE ME UE MS DS SM SO

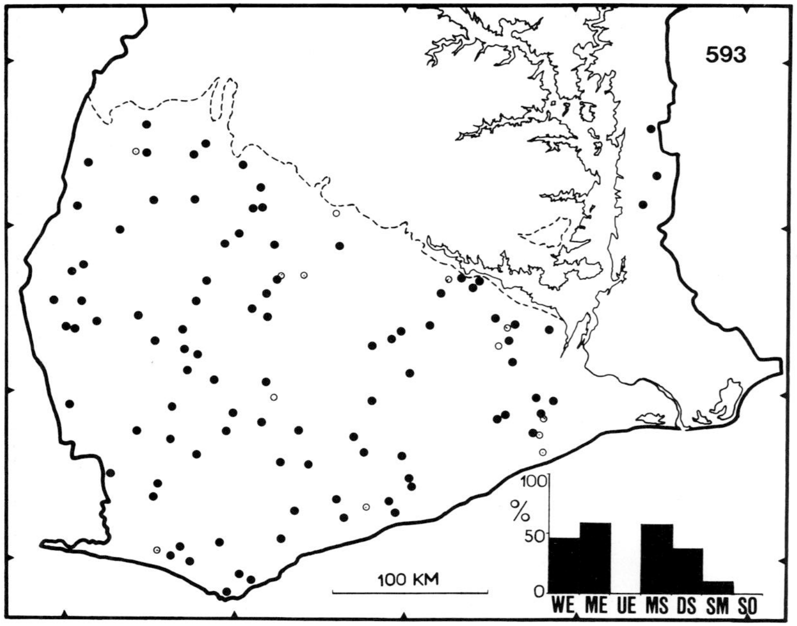

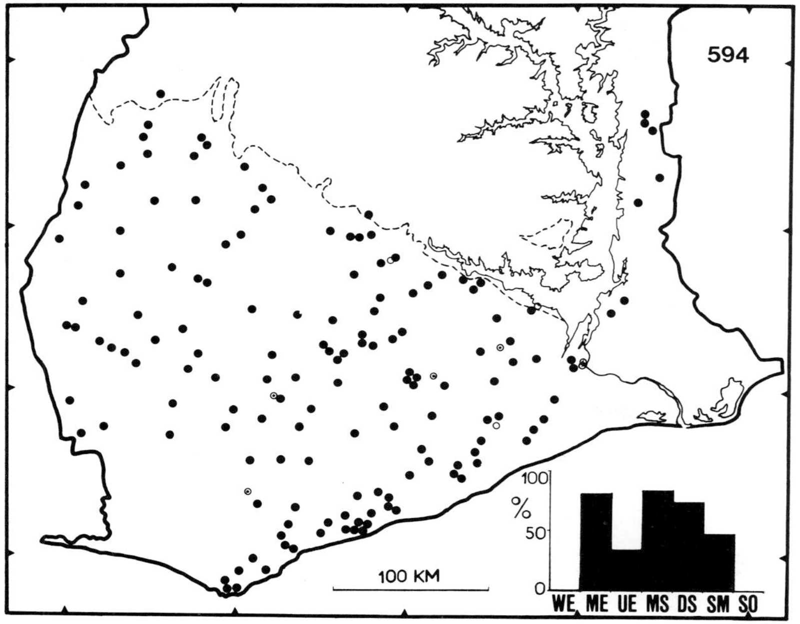

589 TRICALYSIA PALLENS Hiern, FWTA 2:152. Rubiaceae. Akan: turomdua.
Pygmy to small tree; lvs notophyll. Roux's model. Fl white, small to medium, dish with tube, 11–12. Fr red berry, 3–5; seeds small to medium. Regenerates in shade. Guinée to Angola; (Guineo-Congolian wide, 63).

590 TRICALYSIA REFLEXA Hutch., FWTA 2:151, (excluding Chipp 368). Rubiaceae.
Pygmy to small tree; lvs mesophyll. Roux's model. Fl white, medium, dish with tube, 8–1. Fr bluish-white berry, 10–4; seeds medium. Regenerates in shade. S. Leone to Cameroun; (Guinea wide). Chipp 368 (recorded in FWTA ed. 1 as *T. reticulata* (Benth.) Hiern, and in ed. 2 as *T. reflexa*), seems to represent a new species, known only from Ankasa G.P.R.

591 TRICALYSIA RETICULATA (Benth.) Hiern, FWTA 2:152. Rubiaceae.
Pygmy tree; lvs notophyll. Roux's model. Fl white, medium, dish with tube, 11–1. Fr orange berry, 11–5; seeds medium. Regenerates in shade. Guinea Bissau to Ghana; (Upper Guinea endemic).

592 TRICHILIA MEGALANTHA Harms, FWTA 1:705. Meliaceae.
Medium-sized tree; lvs pinnate, leaflets notophyll, pubescent below. Fl medium to large, cream, dish, dioecious, 4–5. Fr capsule, c. 9; seeds large, black with orange-red arillode. Iv. Coast to Nigeria; (Guinea wide).

593 TRICHILIA MONADELPHA (Thonn.) J. J. De Wilde (37), (*T. heudelotii* Planch. ex. Oliv., FWTA 1:704). Meliaceae. Akan: tanuro. (Type from Akwapim).
Medium-sized tree; lvs pinnate, leaflets notophyll. Rauh's model (59). Fl cream, medium, funnel, dioecious, mostly 2–4. Fr capsule, c. 7; seeds large, black with red arillode. Seeds germinate in shade, but regeneration usually under light canopy. Guinea Bissau to Angola; (Guineo-Congolian wide).

594 TRICHILIA PRIEURIANA A. Juss., (*T. prieureana* in FWTA 1:704), subsp. *prieuriana* and subsp. *vermoesenii* J. J. De Wilde (37). Meliaceae. Akan: kakadikuro.
Small to medium-sized trees with characteristic fluted trunk; lvs pinnate, leaflets notophyll. Fl cream, medium, funnel, dioecious, 10–1. Fr capsule, c. 4; seeds large, black, with orange-red arillode. Regenerates in shade. The varieties cannot always be identified in the field, and were not distinguished in the survey. In herbarium speciments subsp. *prieuriana* (○) is recognised by leaves drying greenish, and 3-lobed fruit; subsp. *vermoesenii* (⊙), which occurs in moister forest, by leaves drying brown, and 2-lobed fruit. Distribution: subsp. *prieuriana* – Senegal to Nigeria; subsp. *vermoesenii* – Iv. Coast to Angola and Uganda. The species is a Guineo-Congolian wide. The spelling 'prieureana' should be rejected as an orthographic error according to Article 73 of the International Code of Botanical Nomenclature.

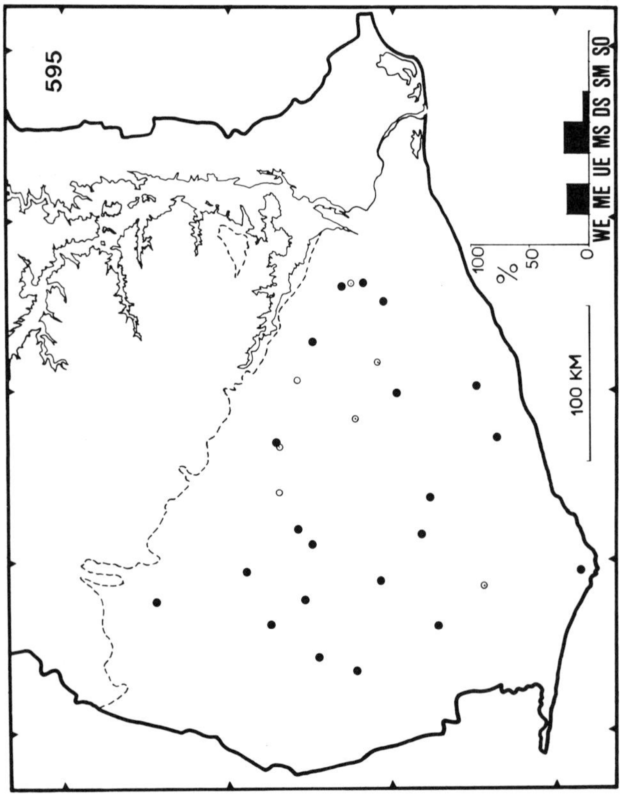
595
100 KM
%
100
50
0
WE ME UE MS DS SM SO

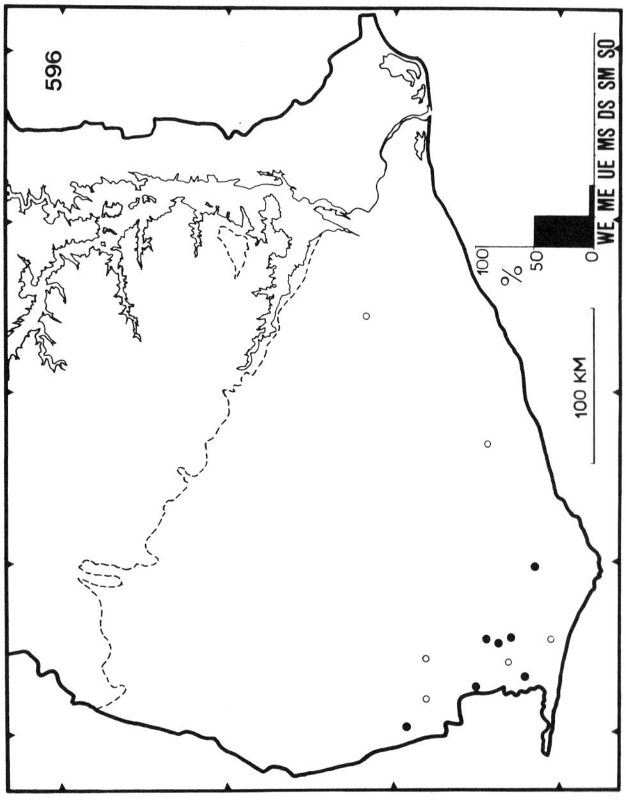
596
100 KM
%
100
50
0
WE ME UE MS DS SM SO

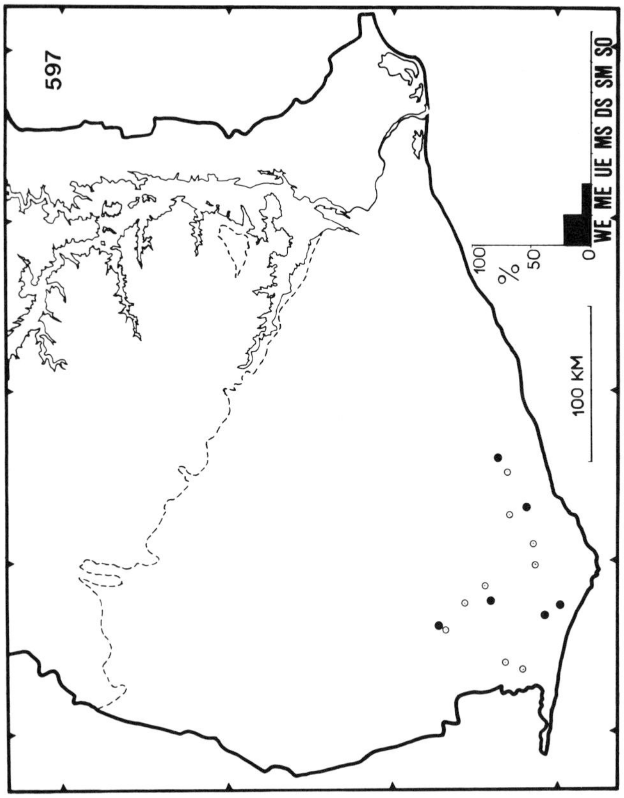
597
100 KM
%
100
50
0
WE ME UE MS DS SM SO

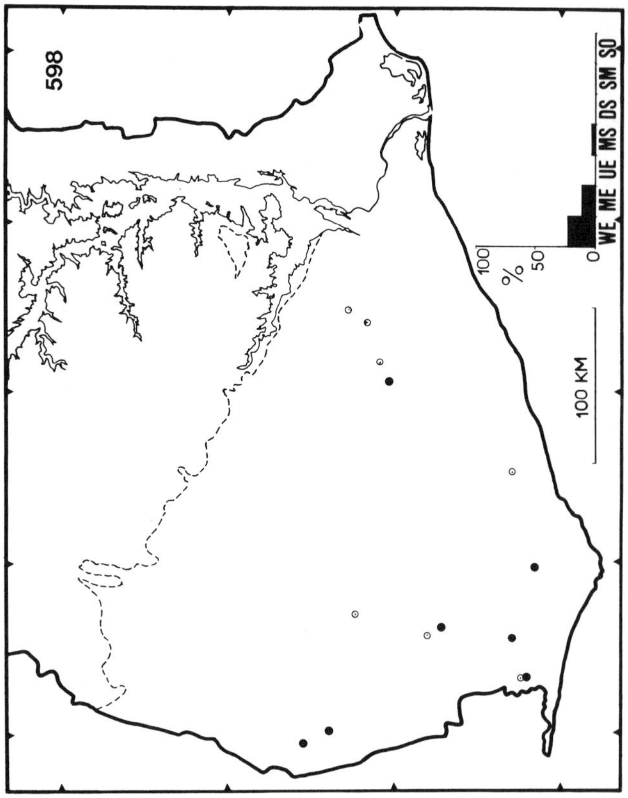
598
100 KM
%
100
50
0
WE ME UE MS DS SM SO

595 TRICHILIA TESSMANNII Harms (37), (*T. lanata* A. Chev., FWTA 1:705). Meliaceae. Akan: tanuronini.
Medium to large tree, with cream latex in bark; lvs pinnate, leaflets notophyll, densely tomentose below. Fl dish, greenish-white, medium to large, dioecious. Fr capsule, 1–2; seeds large to very large, black with orange-red arillode. Regenerates mainly under light canopy. Guinée to Angola; (Guineo-Congolian wide).

596 TRICHOMANES GUINEENSE Afzel. ex Sw., FWTA FF:31. Hymenophyllaceae.
Filmy fern growing on forest floor; lvs bipinnate, leaflets leptophyll. Regenerates in shade from dust-like spores. S. Leone to Cameroun; (Guinea wide).

597 TRICHOSCYPHA ARBOREA (A. Chev.) A. Chev., FWTA 1:736. Anacardiaceae. Akan: anaku.
Medium-sized tree, with some white latex in bark; lvs pinnate, leaflets mesophyll. Fl red, small, bowl, polygamous, 7–11. Fr red drupe; seed large. Regenerates in shade. S. Leone to Cameroun; (Guinea wide). A further eight or so species of *Trichoscypha* occur in Ghana; almost all are confined to evergreen forest.

598 TRICHOSTACHYS AUREA Hiern, FWTA 2:207. Rubiaceae.
Decumbent woody herb, or small shrub; lvs notophyll, hairy. Fl small, purple, bell, not clearly seasonal. Fr blue berry; seeds medium. Regenerates in shade. S. Leone to Gabon; (Guinea wide).

599 TRICLISIA DICTYOPHYLLA Diels (106), (*T. gilletii* (De Wild.) Staner, FWTA 1:71). Menispermaceae.
Large woody climber, stem outline lobed in section, xylem with included rings of anomalous phloem. Lvs macrophyll, pubescent on midrib and main nerves. Fl brownish-orange, small, bell, dioecious, cauliflorous. Fr orange drupe; seed large to very large. Generally found in small gaps or under light canopy. Iv. Coast to Zaire; (Guineo-Congolian wide).

600 TRICLISIA PATENS Oliv., FWTA 1:72. Menispermaceae.
Medium-sized woody climber; xylem with included rings of anomalous phloem; lvs notophyll to mesophyll. Fl yellow, small, dish, cauliflorous or axillary, 1–5. Fr yellow drupe; seed large. Regenerates in shade. Senegal to Ghana; (Upper Guinea endemic).

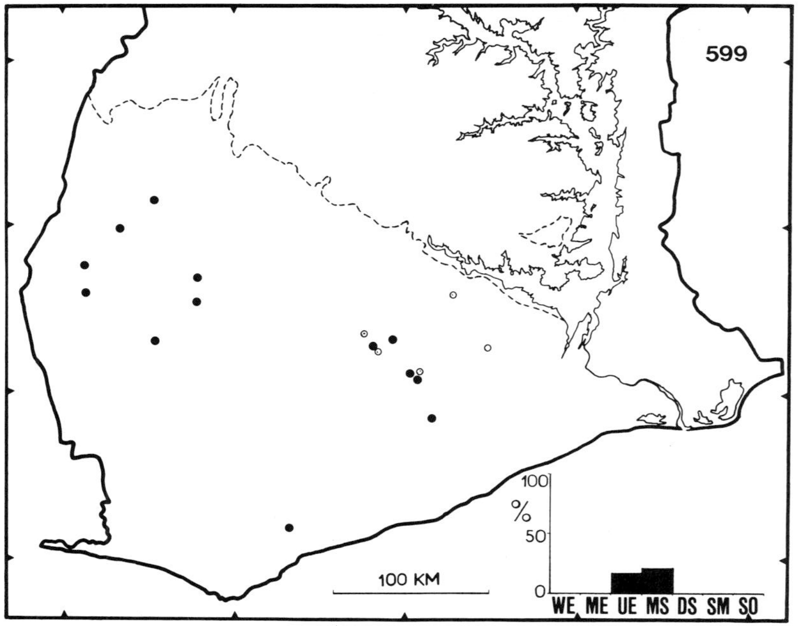

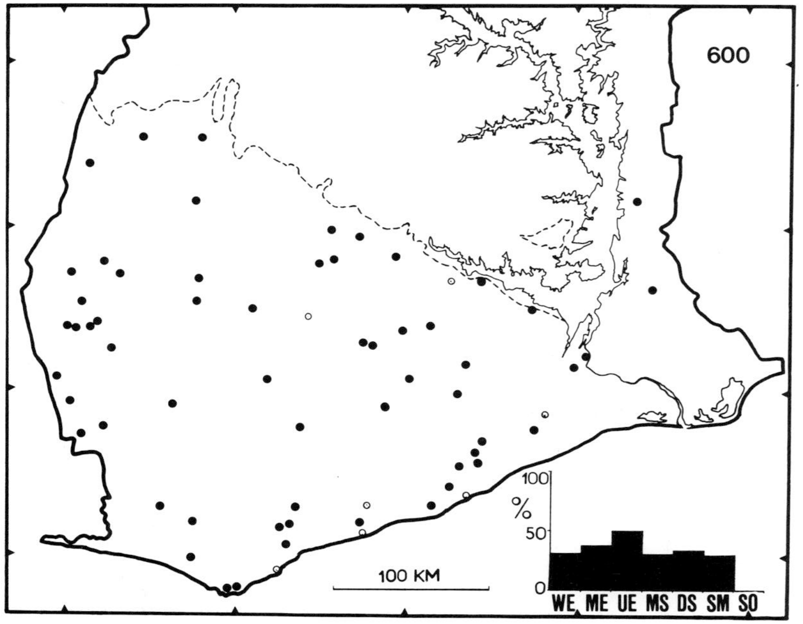

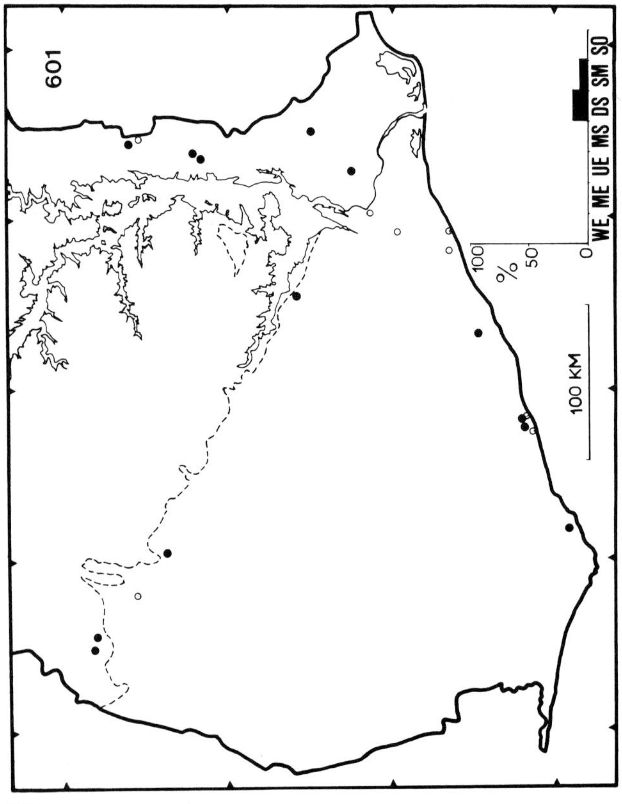
601
100 KM
%
100
50
0
WE ME UE MS DS SM SO

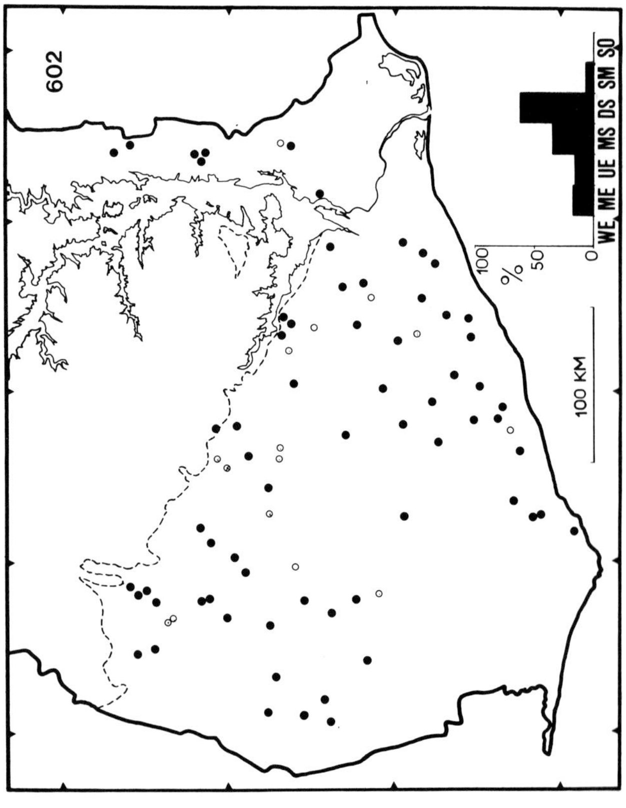
602
100 KM
%
100
50
0
WE ME UE MS DS SM SO

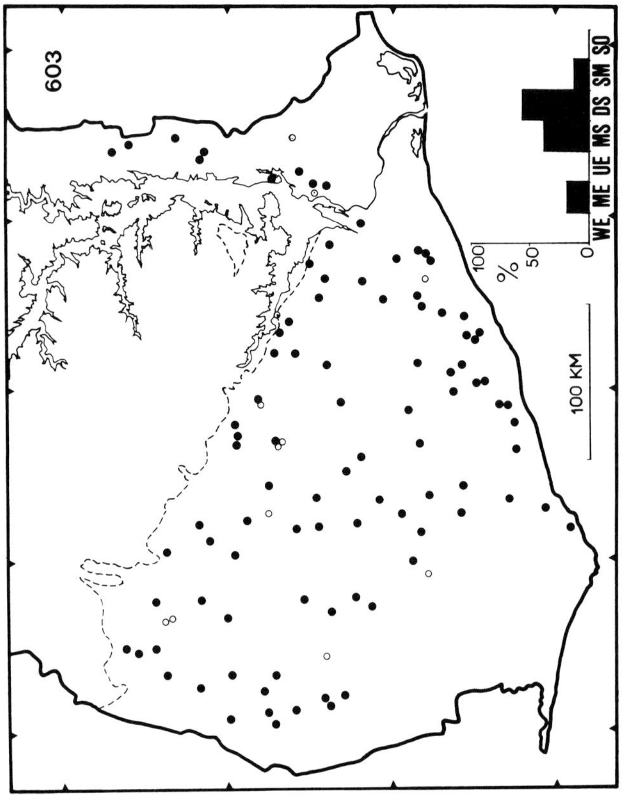
603
100 KM
%
100
50
0
WE ME UE MS DS SM SO

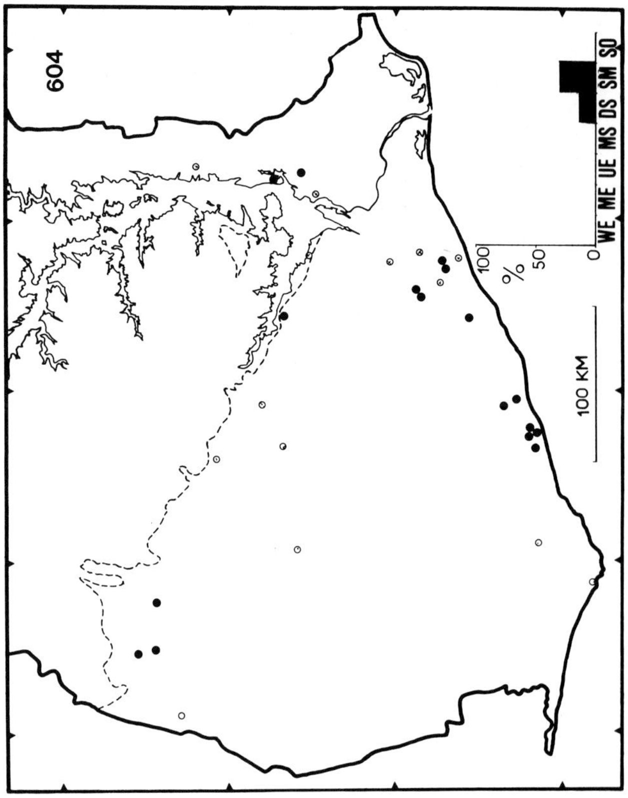
604
100 KM
%
100
50
0
WE ME UE MS DS SM SO

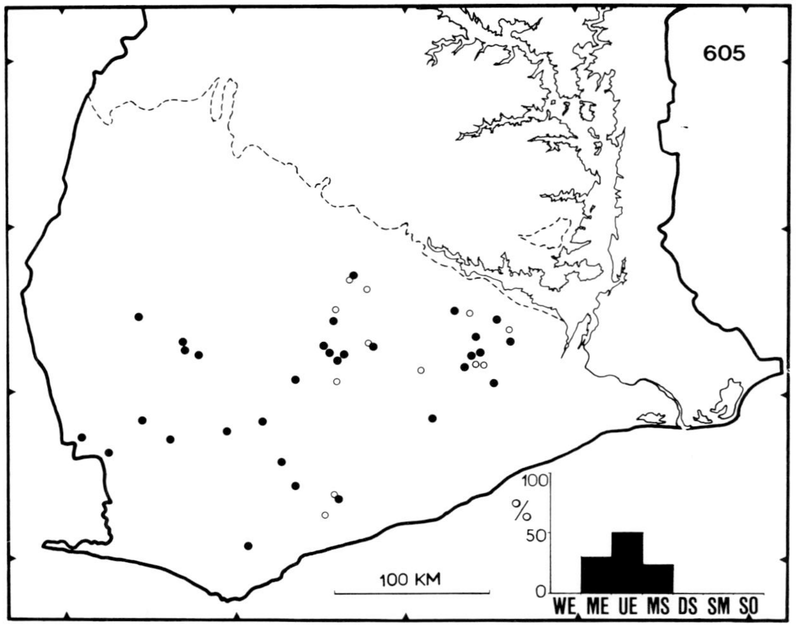

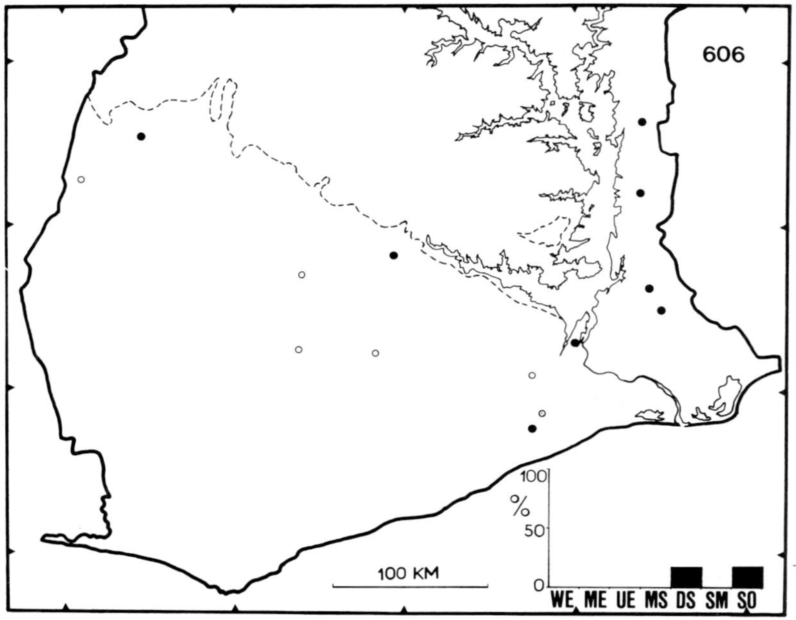

601 TRICLISIA SUBCORDATA Oliv., FWTA 1:72. Menispermaceae.
Slender woody climber with twining pubescent stems; lvs notophyll, pubescent below on nerves and midrib. Fl small, greenish, inconspicuous, mostly 1–4. Fr orange drupe; seed medium. Regenerates in light shade. Extends to woodland and fringing forest in the savanna zone. Iv. Coast to Nigeria; (ecological and chorological transgressor).

602 TRILEPISIUM MADAGASCARIENSE DC. (22), (*Bosqueia angolensis* Ficalho, FWTA 1:612). Moraceae. Akan: okure.
Medium to large tree with narrow buttresses and fluted trunk; brownish latex present; lvs notophyll. Troll's model (59). Fl small, red, brush, monoecious, clustered into a capitate inflorescence, 10–2. Fr black drupe, 11–5; seed medium to large. Regenerates in shade. Guinée to Angola, Kenya, S. Africa, Madagascar and Seychelles Is.; (chorological transgressor).

603 TRIPLOCHITON SCLEROXYLON K. Schum., FWTA 1:313. Sterculiaceae. Akan: wawa.
Large tree with buttresses; lvs mesophyll, palmately lobed, deciduous. Rauh's model (59). Fl white and purple, medium, dish, 11–1. Fr samara, wind-dispersed, 1–3; seed large. Regenerates in light shade. Plot records very significantly correlated with relatively base-rich soils of relatively high pH (57a). Guinée to Zaire; (Guineo-Congolian wide). Class 1 timber; trade name: obeche (40).

604 TURRAEA HETEROPHYLLA Sm., FWTA 1:708. Meliaceae. Akan: ehuna-nyankwa.
Pygmy tree or shrub; lvs microphyll to notophyll, variably lobed. Champagnat's model (59). Fl white, funnel, medium, not clearly seasonal. Fr capsule; seeds medium, black with orange arillode. Regenerates in shade. Ten out of eleven plot records were on soil with pH 6.2 or higher. S. Leone to Dahomey; (Guinea wide).

605 TURRAEANTHUS AFRICANUS (Welw. ex C. DC.) Pellegr., FWTA 1:707. Meliaceae. Akan: apapaye.
Medium-sized tree; cut bark strongly aromatic. Fl cream, large, dish with tube, dioecious. Fr orange capsule, 8–10; seeds large to very large, covered by yellow arillode. Regenerates in shade, seedlings near parent tree killed by nematodes (4). S. Leone to Angola; (Guineo-Congolian wide). Class 2 timber; trade name: avodire (40).

606 TYLOPHORA CONSPICUA N. E. Br., FWTA 2:96. Asclepiadaceae.
Slender twining woody climber, containing white latex; lvs notophyll. Fl deep red, medium, dish, 4–7. Fr follicle; seeds wind-dispersed with aid of tuft of long hairs. Regenerates under light shade and in gaps. Liberia to Zaire and East Africa; (Guineo-Congolian wide).

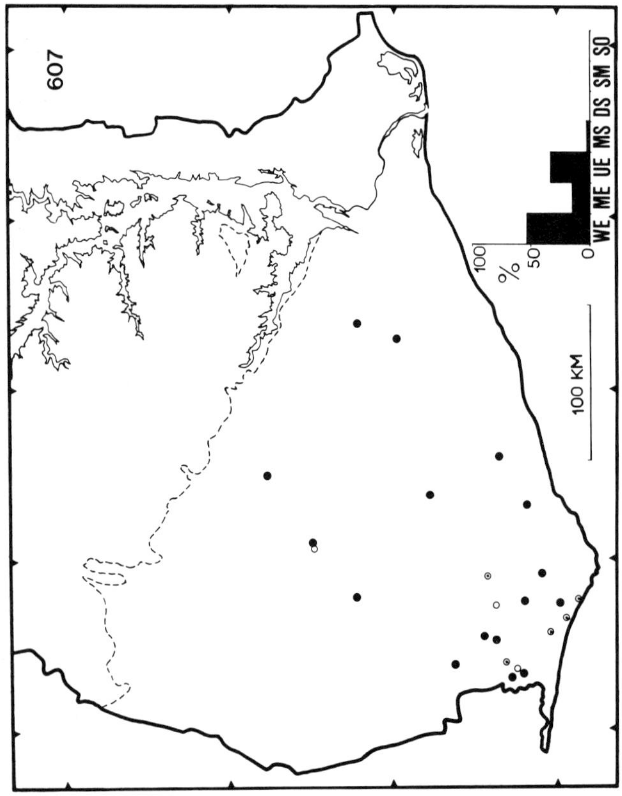
607
100 KM
%
100
50
0
WE ME UE MS DS SM SO

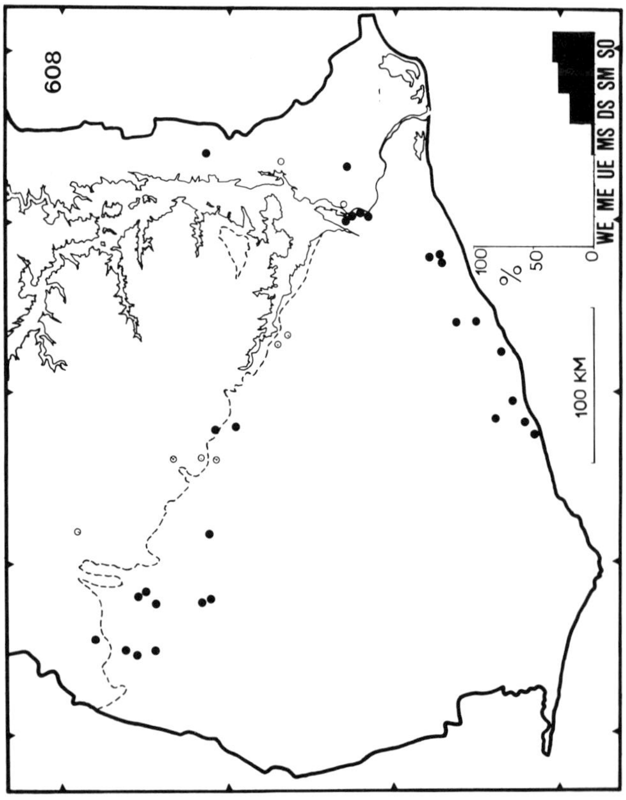
608
100 KM
%
100
50
0
WE ME UE MS DS SM SO

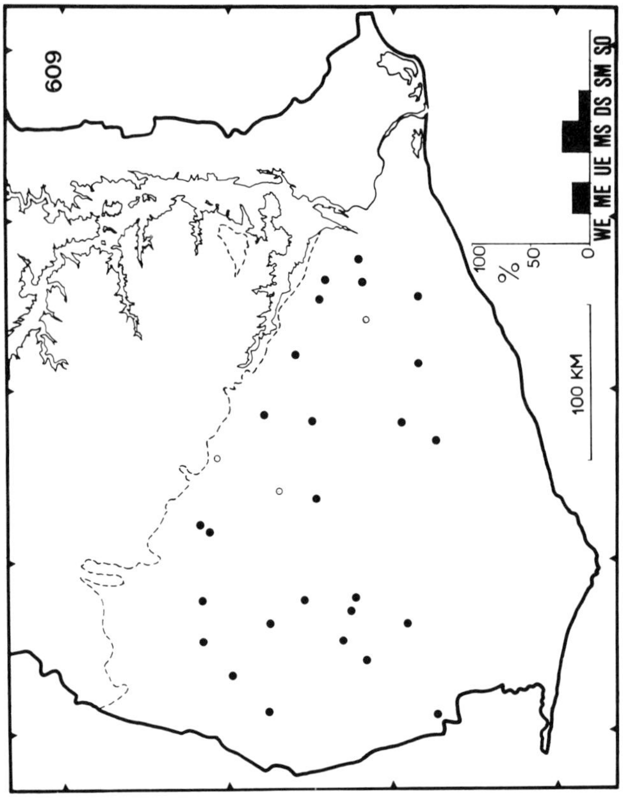
609
100 KM
%
100
50
0
WE ME UE MS DS SM SO

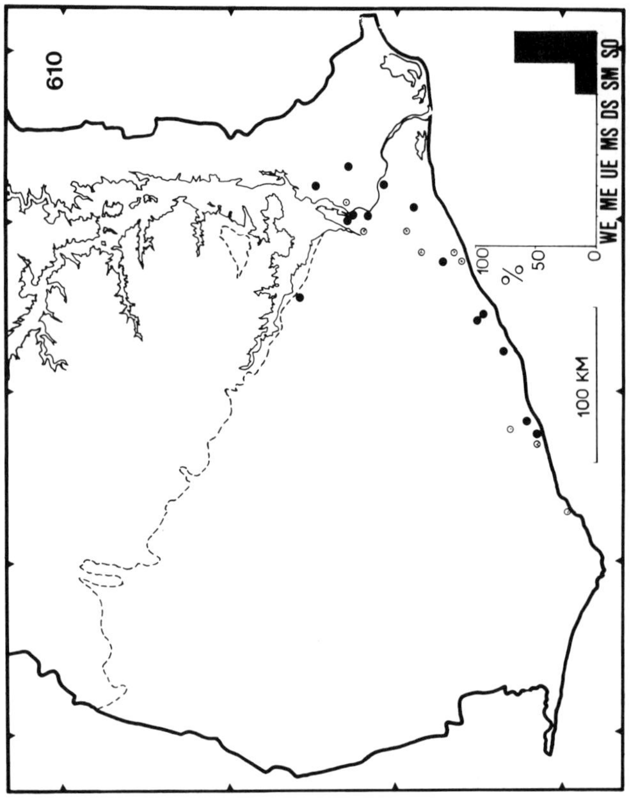
610
100 KM
%
100
50
0
WE ME UE MS DS SM SO

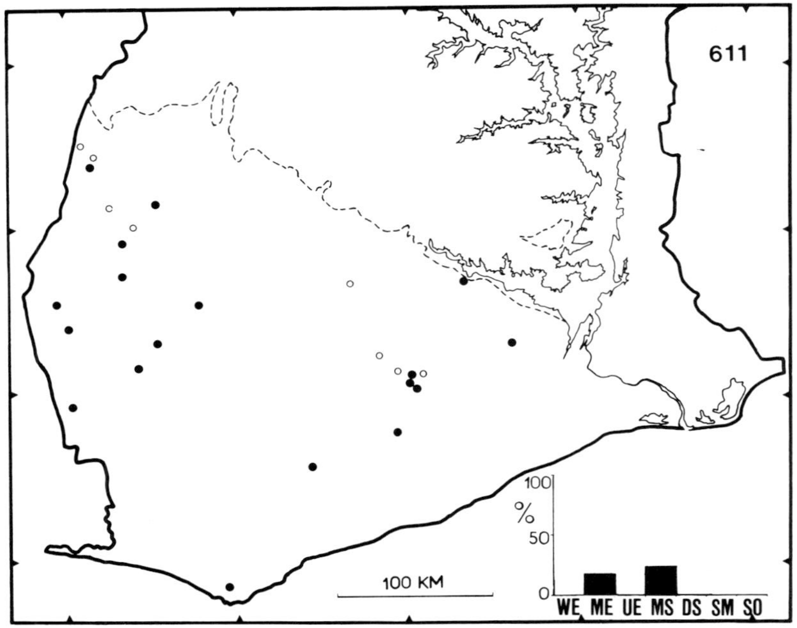

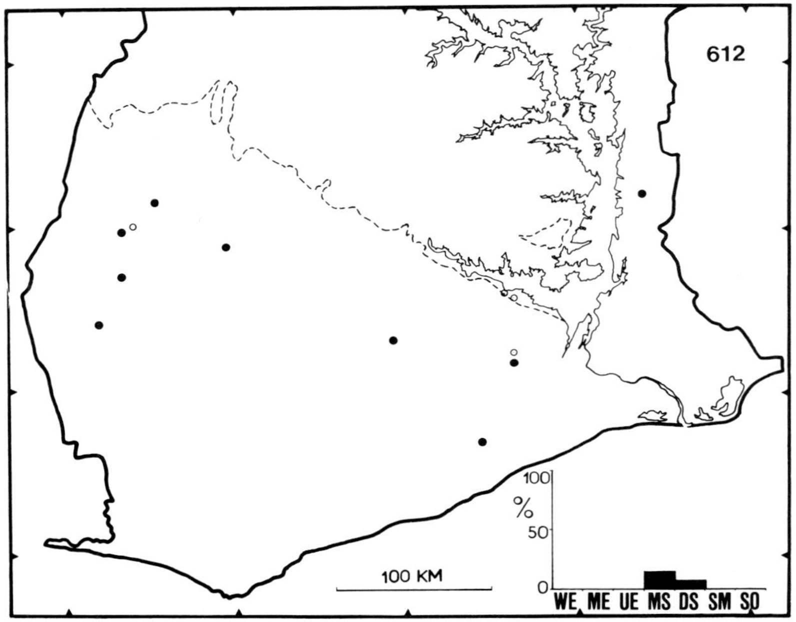

607 UAPACA GUINEENSIS Muell. Arg., FWTA 1:390, and UAPACA CORBISIERI De Wild. (113), (*U. esculenta* A. Chev. ex Aubrév. & Léandri, FWTA 1:392). Euphorbiaceae. Akan: kontan.

Medium to large trees, with conspicuous stilt roots; lvs mesophyll. Rauh's model (59). Fl small, dioecious, clustered in flower-like heads with yellow medium-sized petal-like bracts, not clearly seasonal. Fr berry; seeds large *(U. guineensis)*, large to very large *(U. corbisieri)*. Regenerates in shade. The two species are not readily distinguished in the field, and were not separated in the survey. Herbarium records: *U. guineensis* (⊙), *U. corbisieri* (○). Distribution: *U. guineensis* – S. Leone to Zaire; (Guineo-Congolian wide); *U. corbisieri* – S. Leone to Zaire; (Guineo-Congolian wide).

608 UVARIA DOERINGII Diels, FWTA 1:38. Annonaceae.

Large woody climber, with strong disagreeable smell; stem cylindrical, wood outline slightly angular. Lvs microphyll to notophyll, pubescent, as are the twigs. Roux's model. Fl pale yellow-green, medium, bowl, not clearly seasonal. Fr warty yellow berry; seeds medium to large. Regenerates in shade. Ghana to Dahomey; (Guinea wide).

609 UVARIA MOCOLI De Wild. & Th. Dur. (76), (*U. anonoides* of FWTA 1:38, partly not of Bak. f.). Annonaceae.

Large woody climber; leaves mesophyll, sparsely pubescent below. Roux's model. Fl orange, medium, dish. Fr berry; seeds medium to large. Regenerates in shade. S. Leone to Zaire; (Guineo-Congolian wide). *U. mocoli* differs from *U. anonoides* mainly in its mericarps which are sessile and tomentellous, not stipitate and glabrous. The only fruiting specimen known from Ghana has the characters of *U. mocoli,* and as the non-fruiting specimens appear to be conspecific, there is no evidence that *U. anonoides* occurs in Ghana.

610 UVARIA OVATA (Vahl ex Dunal) A. DC. subsp. *ovata* FWTA 1:36. Annonaceae. Akan: akotompo; Ga: aanyeli. (Type from Accra Plains).

Medium-sized woody climber, or scrambling shrub in thicket; lvs microphyll, pubescent below, as are the twigs. Roux's model. Fl orange-brown, medium, dish, not clearly seasonal. Fr orange berry; seeds medium. Regenerates in shade. Iv. Coast to Togo; (Upper Guinea endemic).

611 UVARIASTRUM PIERREANUM Engl., FWTA 1:47. Annonaceae. Akan: otwe-ehi.

Small to medium-sized tree; lvs notophyll, puberulous below and on twigs. Fl yellowish, large, dish, axillary or on twigs below leaves, mostly 2–3. Fr greenish-brown berry; seeds large to very large. Regenerates in shade. Iv. Coast to Zaire; (Guineo-Congolian wide).

612 UVARIODENDRON ANGUSTIFOLIUM (Engl. & Diels) R. E. Fries, FWTA 1:46. Annonaceae. Akan: bommofokwadu.

Small tree; lvs mesophyll, with strong smell of citronella when crushed. Fl large, whitish, dish, 1–3. Fr not known. Regenerates in shade. Iv. Coast to Cameroun; (Guinea wide).

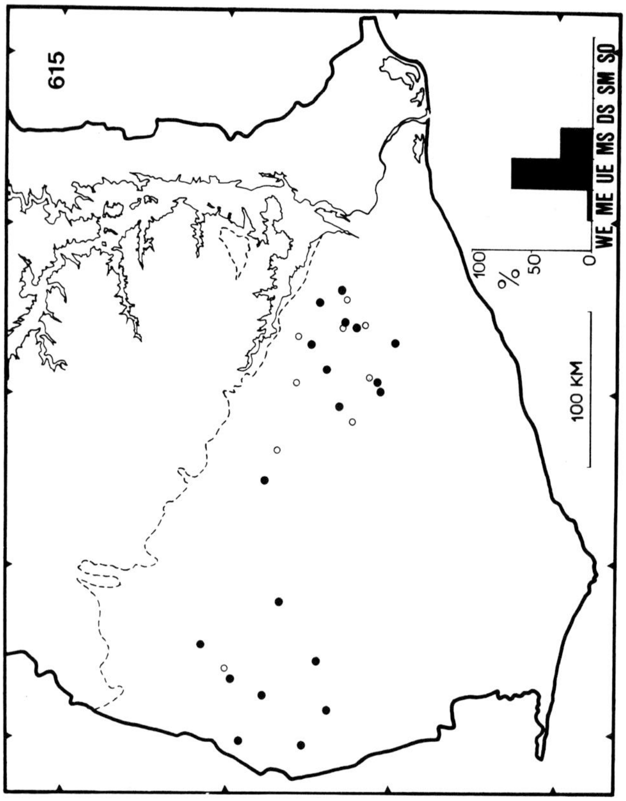
615
100 KM
100
50
0
%
WE ME UE MS DS SM SO

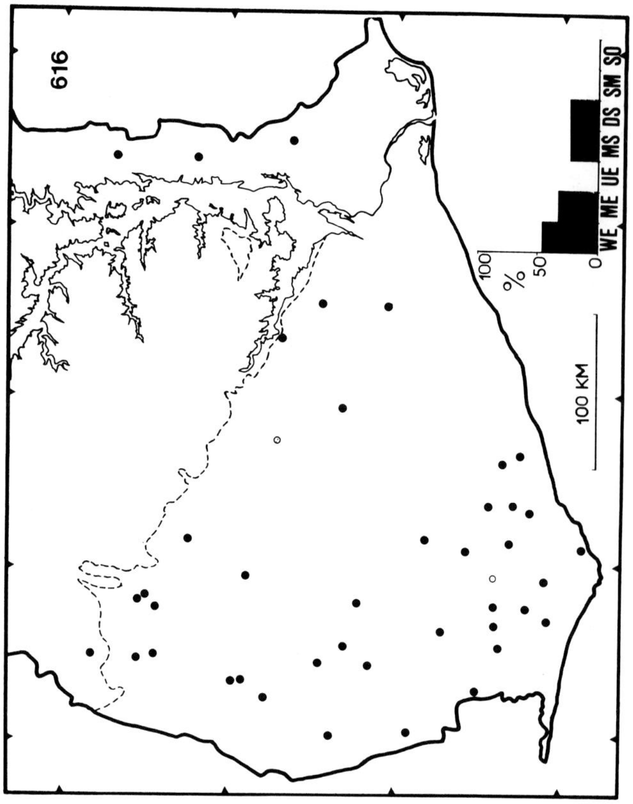
616
100 KM
100
50
0
%
WE ME UE MS DS SM SO

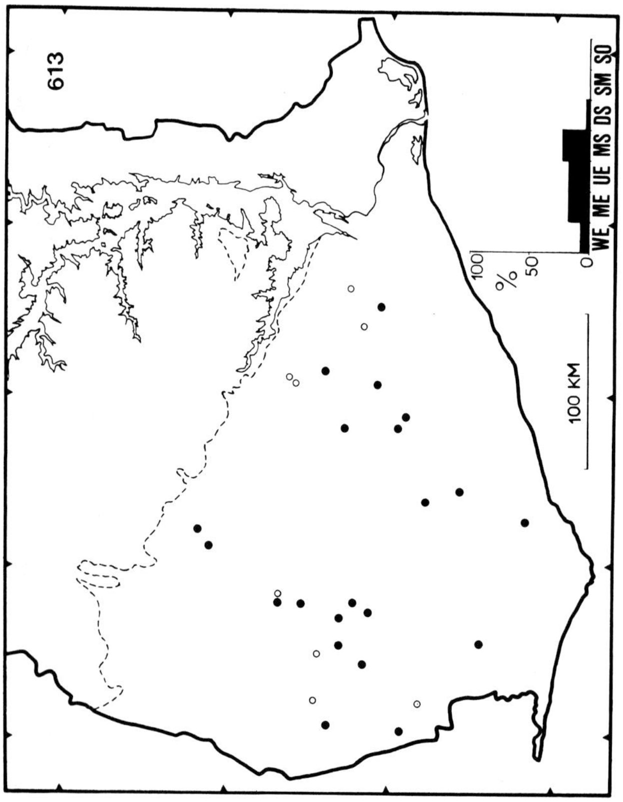
613
100 KM
100
50
0
%
WE ME UE MS DS SM SO

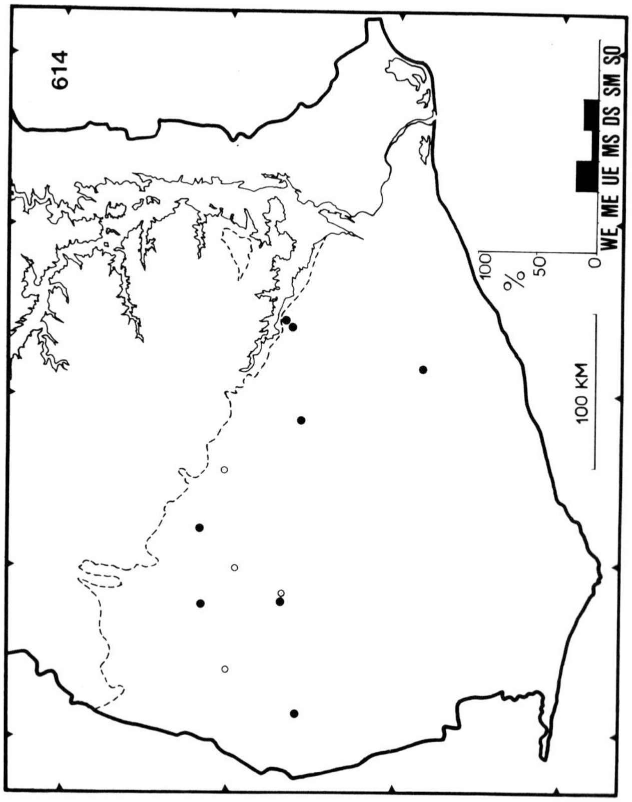
614
100 KM
100
50
0
%
WE ME UE MS DS SM SO

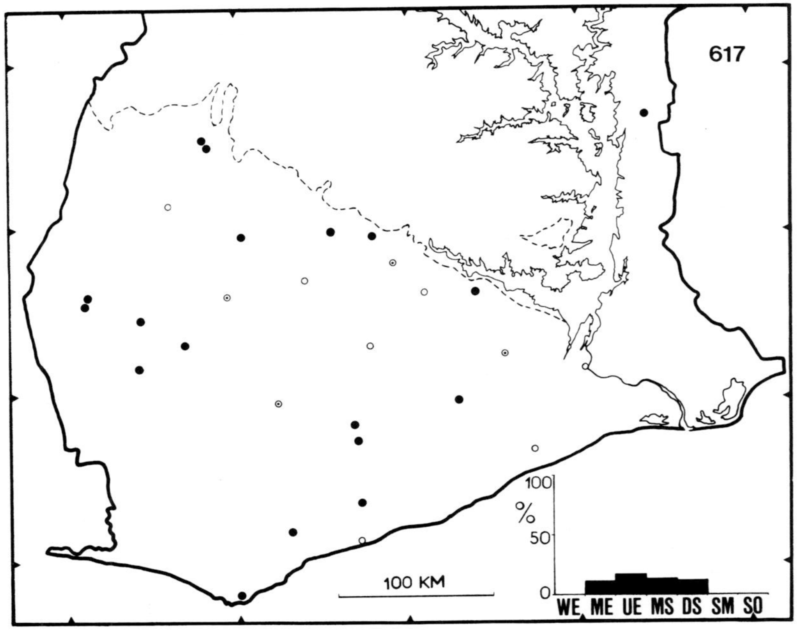

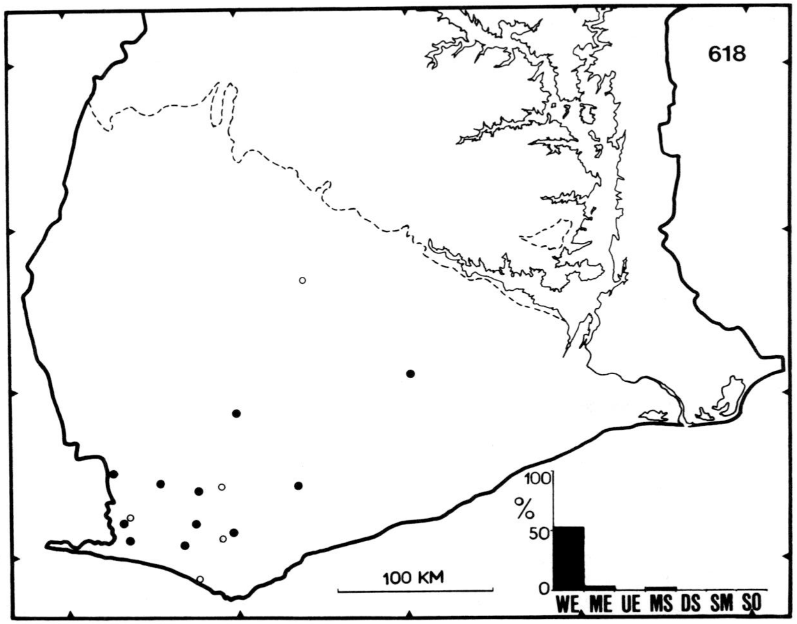

613 UVARIODENDRON CALOPHYLLUM R. E. Fries, FWTA 1:46. Annonaceae.
Small tree, with densely pubescent twigs. Lvs macrophyll, pubescent below. Fl large, red and yellow, dish, on main trunk, 1–2. Fr yellow-orange berry; seeds large. Regenerates in shade. Iv. Coast to Cameroun; (Guinea wide).

614 UVARIODENDRON OCCIDENTALE Le Thomas (75), (*U. mirabile* of FWTA 1:46, partly not of R. E. Fries). Annonaceae. Akan: esonokwadu.
Small tree; cut bark smelling of cloves; lvs mesophyll. Fl large, yellow and red, dish, on main trunk. Fr yellow-orange berry; seeds large. Regenerates in shade. Iv. Coast to Ghana; (Upper Guinea endemic).

615 UVARIOPSIS GLOBIFLORA Keay, FWTA 1:50. Annonaceae. (Type from Ashanti).
Small tree, with aromatic bark; lvs mesophyll. Fl yellowish with purple centre, medium to large, monoecious, axillary or cauliflorous, not clearly seasonal. Fr pinkish berry; seeds large. Regenerates in shade. Endemic to Ghana; (Upper Guinea endemic). *U. globiflora* was distinguished (70) from *U. guineensis* as having axillary rather than cauliflorous inflorescences, and by slight differences in the flowers. It has been shown (76) that cauliflory is not always a reliable character in *Uvariopsis;* further collecting in Ghana has shown that flowers occur in both positions on the same tree. Further collecting is neded to show whether these two species are really distinct.

616 VENTILAGO AFRICANA Exell, FWTA 1:670. Rhamnaceae.
Large woody climber; stem and wood cylindrical, normal. Roux's model. Lvs microphyll, toothed. Fl small, dish, greenish. Fr samara (Fig. 6.1m), wind-dispersed; seed medium. Regenerates in shade. Guinea Bissau to Angola and Uganda; (Guineo-Congolian wide).

617 VITEX FERRUGINEA Schum. & Thonn., FWTA 2:447, and VITEX RIVULARIS Gürke, FWTA 2:446. Verbenaceae. Akan: otwentorowa. (Type of *V. ferruginea* from Akwapim).
Small to medium-sized trees; lvs digitate, leaflets notophyll, puberulous and with sessile glands below. Fl bell, small and white *(V. rivularis)* or medium and blue *(V. ferruginea),* 3–5. Fr yellow 4-seeded drupe, 5–6; seed large. Regenerate in shade. These species are difficult to distinguish in the field (though this is possible from the indumentum on the midrib underside); they were not separated in the survey. Herbarium records: *V. ferruginea* (○); *V. rivularis* (⊙) Distribution: *V. ferruginea* – Guinea Bissau to Zaire, (Guineo-Congolian wide); *V. rivularis* – Liberia to Zaire; (Guineo-Congolian wide).

618 VITEX MICRANTHA Gürke, FWTA 2:446. Verbenaceae. Akan: otwentorowanini.
Medium-sized tree; lvs digitate, leaflets microphyll, somewhat toothed, with sessile glands on undersurface. Fl white, medium, bell, mostly 1–2. Fr drupe; seed medium to large. Regenerates in shade. Plot records very significantly associated with base-poor soils. Guinée to Ghana, and Cameroun; (Upper Guinea near-endemic).

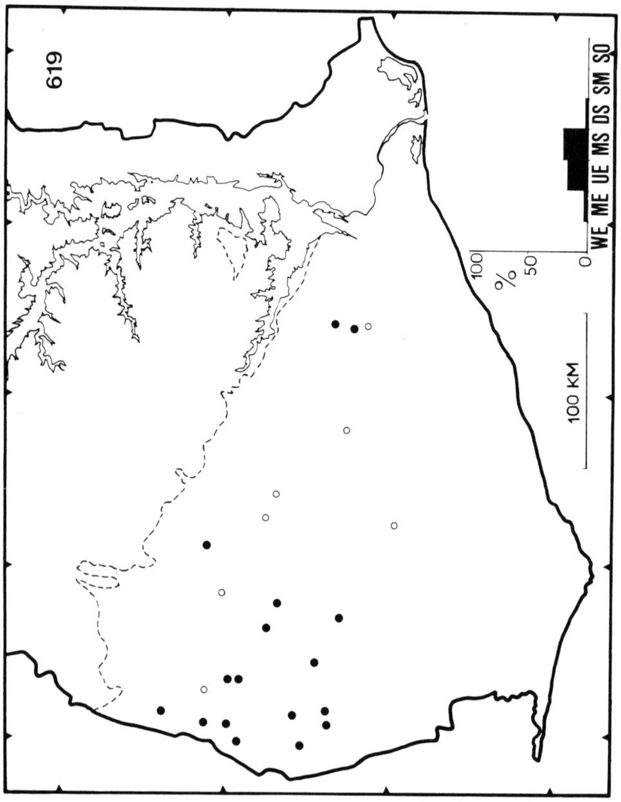
619
100 KM
100
%
50
0
WE ME UE MS DS SM SO

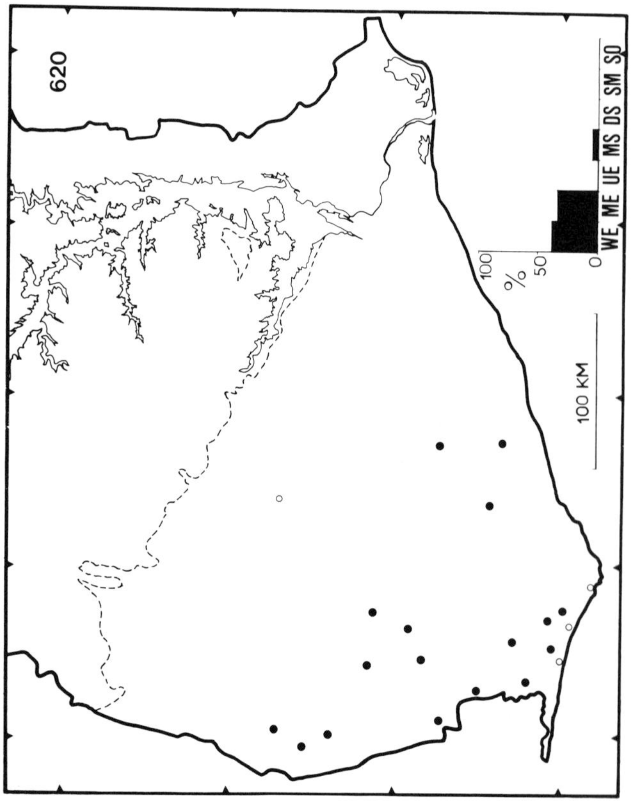
620
100 KM
100
%
50
0
WE ME UE MS DS SM SO

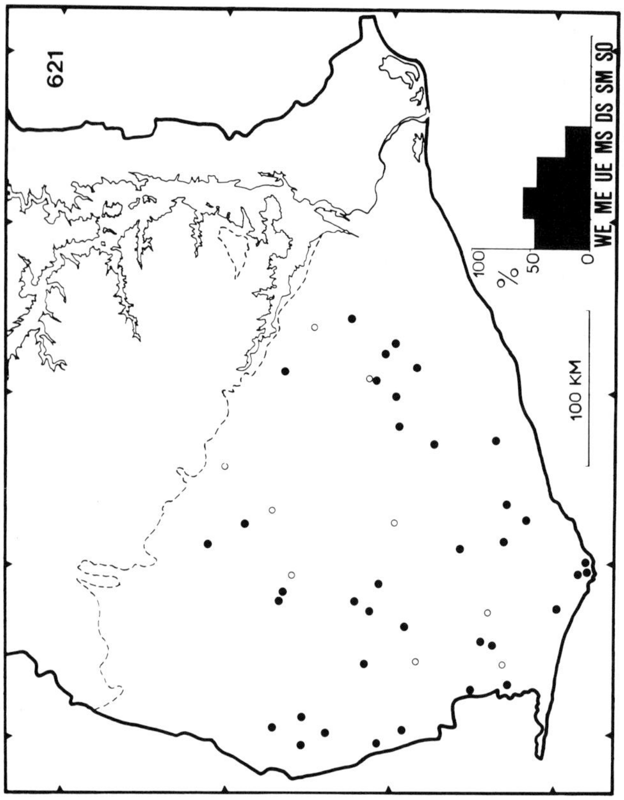
621
100 KM
100
%
50
0
WE ME UE MS DS SM SO

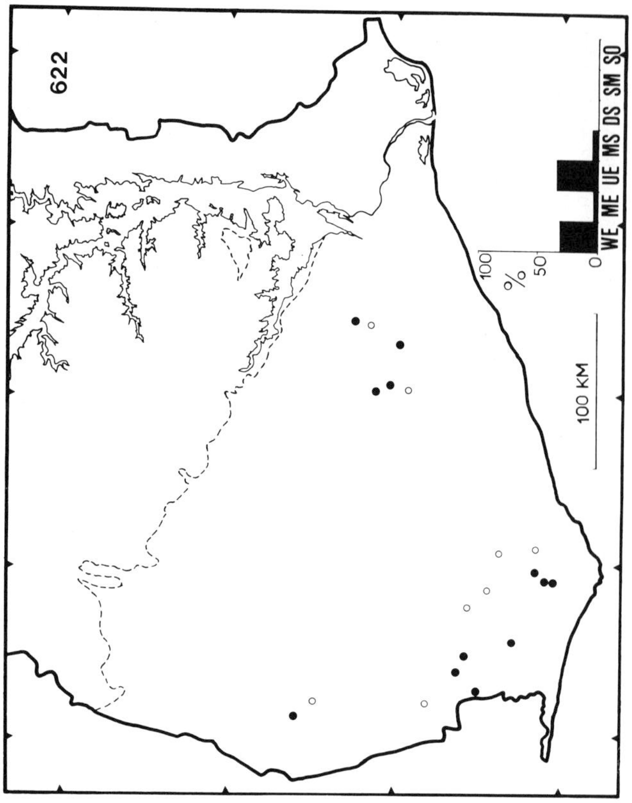
622
100 KM
100
%
50
0
WE ME UE MS DS SM SO

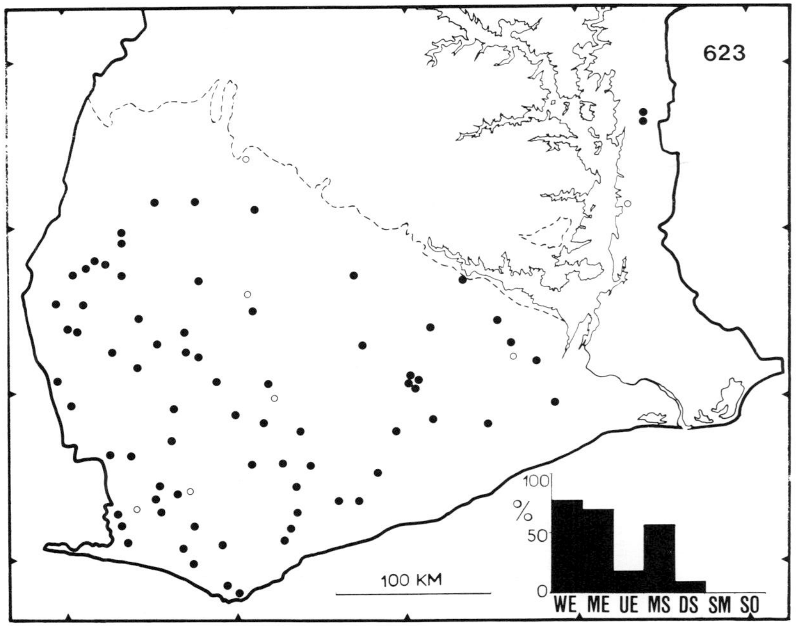

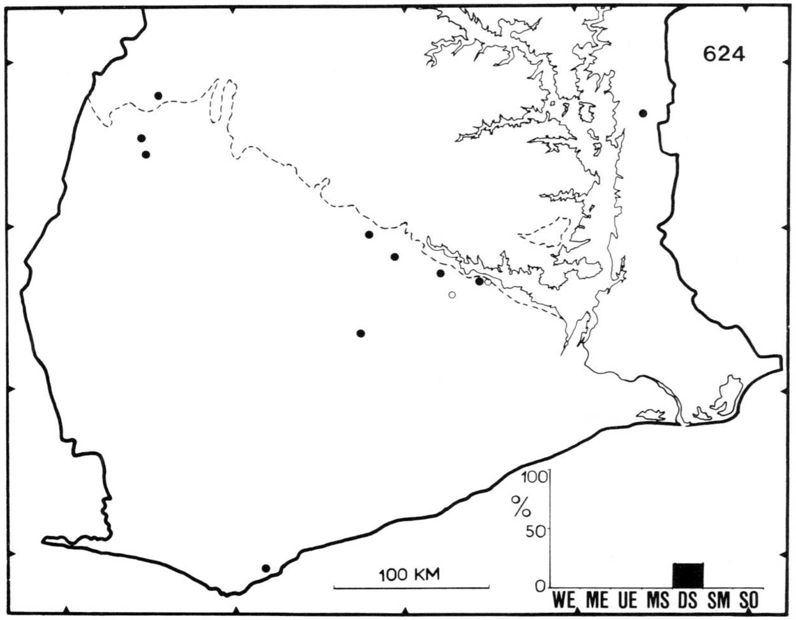

619 XYLIA EVANSII Hutch., FWTA 1-495. Mimosaceae. Akan: samantawa.
Medium to large tree with buttresses; lvs bipinnate, leaflets microphyll, pubescent below. Fl small, yellow, brush, clustered in heads, 2–4. Fr woody explosive pod, 11–1; seeds very large. Regenerates in shade. S. Leone to Ghana; (Upper Guinea endemic).

620 XYLOPIA ACUTIFLORA (Dunal) A. Rich., FWTA 1:42. Annonaceae.
Medium-sized woody climber, or small to medium tree, or shrub. Lvs microphyll to notophyll, appressed-pubescent below and on twigs. Roux's model. Fl whitish, large, funnel, mostly 12–3. Fr black berry with red flesh surrounding the medium to large red-brown seeds. Regenerates in shade. S. Leone to Angola and Zambia; (Guineo-Congolian wide). The usual life-form of Ghanaian plants referrable to *X. acutiflora* is a woody climber with leafy shoots of limited growth inserted approximately at right angles to the main stem, as in *Uvaria ovata*. Trees with similar leaves may be *X. elliotii* Engl. and Diels, if indeed this is distinct from *X. acutiflora*. Further study is needed to establish clear limits between these two species of *Xylopia*, and *X. parviflora* (A. Rich.) Benth.

621 XYLOPIA QUINTASII Engl. & Diels., FWTA 1:42. Annonaceae. Akan: obaa.
Medium-sized tree; lvs microphyll to notophyll. Roux's model. Fl white, medium, funnel, 12–3. Fr fleshy follicle, 7–10; seeds large, brown, almost entirely covered by red-brown aril. Regenerates in shade. S. Leone to Zaire; (Guineo-Congolian wide).

622 XYLOPIA STAUDTII Engl. & Diels, FWTA 1:41. Annonaceae. Akan: duanan.
Medium to large forest tree, with prominent stilt roots, and stilted pneumatophores when growing in swamps (68). Lvs microphyll to notophyll, appressed puberulous below. Roux's model. Fl yellow-orange, medium, funnel, 6–10. Fr black berry with red pulp, 1–3; seeds large, black, with orange aril. Regenerates in shade. S. Leone to Angola; (Guineo-Congolian wide).

623 XYLOPIA VILLOSA Chipp, FWTA 1:42. Annonaceae. Akan: obaafufuo.
Medium-sized tree with aromatic bark; lvs notophyll, pubescent below, as are twigs. Roux's model. Fl light yellow, large, funnel, 10–1. Fr berry, 6–7; seeds large, greenish-grey. Regenerates in shade. Iv. Coast to Nigeria; (Guinea wide).

624 ZANTHOXYLUM LEPRIEURII Guill. & Perr. (41), (*Fagara leprieurii* (Guill. & Perr.) Engl. in FWTA 1:686). Rutaceae. Akan: oyaa.
Small to medium thorny tree; lvs pinnate, leaflets notophyll, with pellucid glands. Scarrone's model. Fr small, cream, bowl, in large panicles, dioecious, 4–6. Fr small 1-seeded follicle; seed dark blue, small to medium. Senegal to Angola, Uganda, and South Africa; (ecological and chorological transgressor). Contains various alkaloids (41).

CHAPTER 11

Distribution of some important timber species

Species notes are provided for all the classified timber species in the preceding section. Four of them, however, lack the companion distribution map because they were so rarely recorded during our survèy.

Distribution maps for these species *(Mitragyna ciliata, Nauclea diderrichii, Pericopsis elata* and *Terminalia ivorensis)* are presented here. They were compiled from the Forestry Department enumeration records and thus include information on tree stock density within forest reserves. They are supplemented by maps of five other economic timber species (Table 11.1) which we felt would usefully complement the maps for these species given in Chap. 10.

The sampling unit for these maps is a forest reserve and those reserves for which stock density data are available are outlined on each map. Differential shading (Table 11.2) is used to denote classes of tree density. Not all reserves have been enumerated, and we have not been able to extract data for all enumerated reserves; consequently the sample distribution is rather uneven, especially as some forest types are virtually without reserves. Few of the reserves in the Volta Region seem to have been enumerated, and we have therefore omitted this region from the maps. The map for any species shows only those reserves in which it has been reported as present. We have not shown those reserves in which the species is reported as absent – in many of these cases the species must almost certainly have been there but over-looked, omitted in error from the record, or missed by chance by the sampling strips.

The maps should thus be interpreted with caution; apparent absences from particular areas may merely reflect a shortage of samples. Forest reserves in the west tend to be larger than those in the east. This can give a misleading visual impression about the abundance of certain species. *Terminalia superba* (Fig. 11.8), for instance, is probably at least as abundant around Oda and Juaso as it is in southern Brong-Ahafo.

The same conventions for indicating tree density are adopted for each species (Table 11.2), even though some species are much less common than others. Each map shows part of the political boundary of Ghana (solid line), and the northern limit of the forest zone (broken line).

Table 11.1 Distribution and density of nine important timber species in Ghanaian forest reserves.

Species	Notes on page	Map number	Figure
Chlorophora excelsa and *C. regia*	151	123	11.1
Khaya ivorensis	217	324	11.2
Lophira alata	227	–	11.3
Mitragyna ciliata and *M. stipulosa*	239	–	11.4
Nauclea diderrichii	243	–	11.5
Pericopsis elata	261	–	11.6
Terminalia ivorensis	303	–	11.7
Terminalia superba	303	583	11.8
Tieghemella heckelii	305	588	11.9

Table 11.2 Key to tree densities in Fig. 11.1 to 11.9.

Shading	Tree density class – all trees ⩾30 cm girth (trees km^{-2})
Solid black	⩾100
Vertical hatching	50–99
Stippled	10–49
Outline only	<10

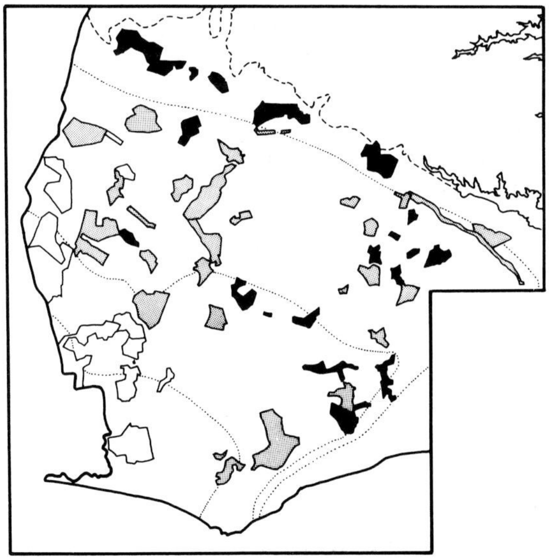

Fig. 11.1 Chlorophora spp.

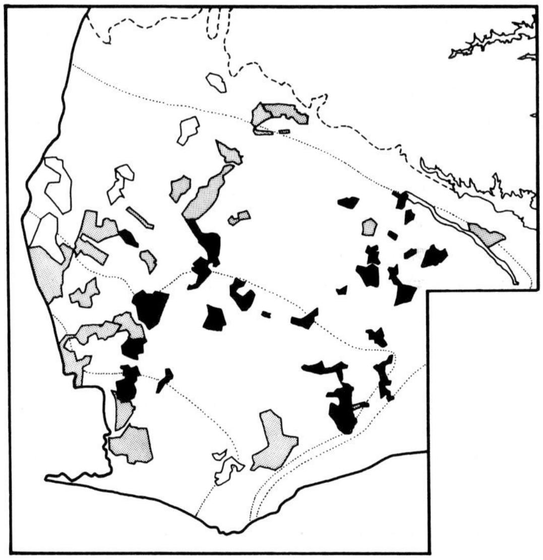

Fig. 11.2 Khaya ivorensis (for interpretation see Fig. 11.1).

Explanation Fig. 11.1 to 11.9.
Density of trees ⩾3 ft. (0.9 m) gbh in forest reserves. Data from Forestry Department enumerations. Solid black: ⩾100 trees per km^2; vertical shading: 50–99; open: <10. Broken line: forest-savanna boundary; dotted line: type boundaries.

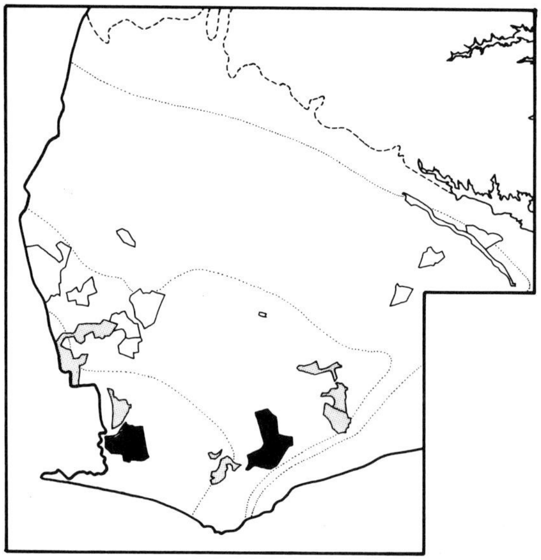

Fig. 11.3 Lophira alata.

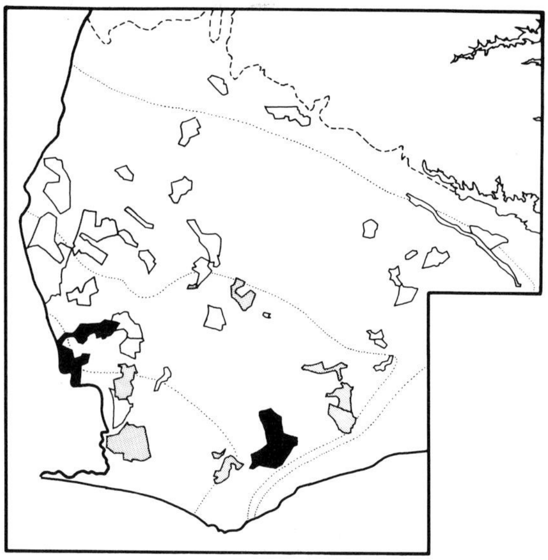

Fig. 11.4 Mitragyna spp.

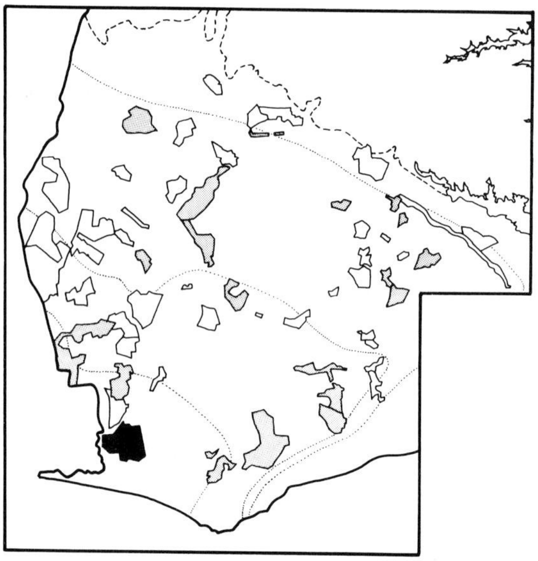

Fig. 11.5 Nauclea diderrichii.

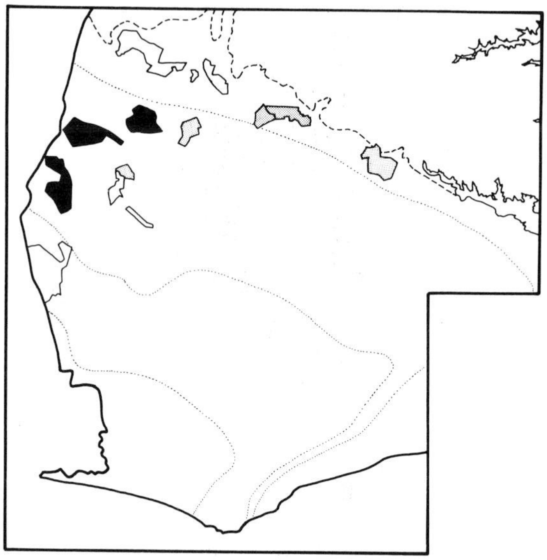

Fig. 11.6 Pericopsis elata.

Explanation to the figures:
Density of trees ⩾3 ft. (0.9 m) gbh in forest reserves. Data from Forestry Department enumerations. Solid black: ⩾100 trees per km^2; vertical shading: 50–99; open: <10. Broken line: forest-savanna boundary; dotted line: type boundaries.

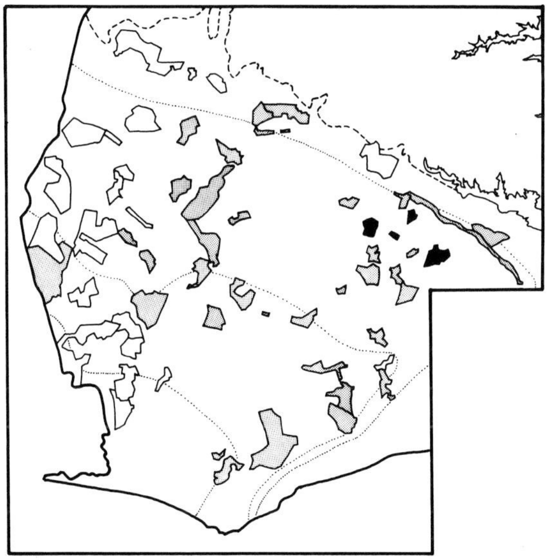

Fig. 11.7 Terminalia ivorensis.

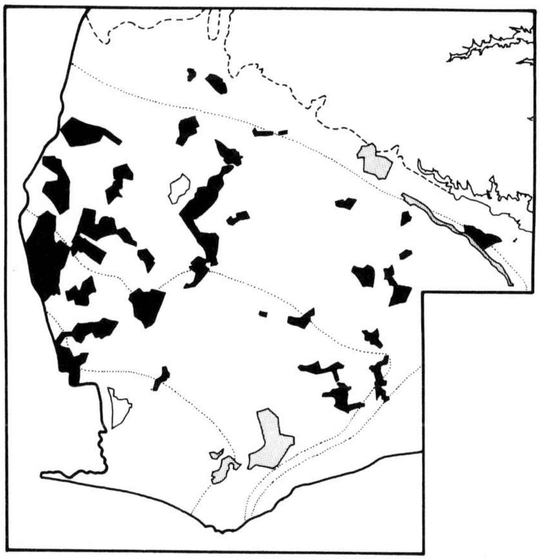

Fig. 11.8 Terminalia superba.

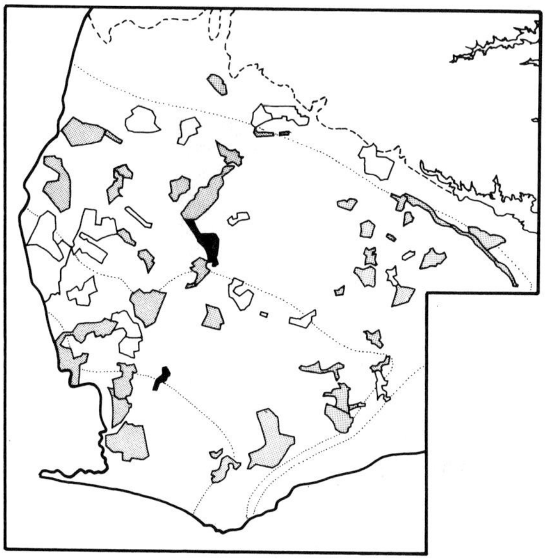

Fig. 11.9 Tieghemella heckelii.

Explanation to the figures:
Density of trees ⩾3 ft. (0.9 m) gbh in forest reserves. Data from Forestry Department enumerations. Solid black: ⩾100 trees per km^2; vertical shading: 50–99; open: <10. Broken line: forest-savanna boundary; dotted line: type boundaries.

Appendices

APPENDIX 1

Reserved forest in the forest zone of Ghana

1a. List of Forest Reserves, including former Forest Reserves which have been converted to Wildlife Reserves

Reserve	Lat.	Long.	Area (sq- km)	Area (sq. miles)	Forest-type	Date of first reservation
1. Abasumba	5 38 N	0 31 W	1.0	0.4	SM	1927
2. Abisu	6 37 N	0 32 W	9.1	3.5	DSIZ	1939
3. Aboben Hill	6 17 N	0 00 W	7.3	2.8	SM	1962
4. Aboma	7 11 N	1 29 W	45.6	17.6	DSFZ	1932
5. Aboniyere Shelterbelt	6 35 N	2 35 W	41.2	15.9	MSNW	1940
6. Abrimasu	7 17 N	1 27 W	26.2	10.1	DSFZ	1940
7. Afao Hills	6 14 N	2 19 W	34.7	13.4	MSSE	
8. Afia Shelterbelt	6 09 N	1 17 W	21.0	8.1	MSSE	1940
9. Afram Headwaters	7 10 N	1 40 W	201.2	77.7	DSFZ	1928
10. Afrensu-Brohuma	7 22 N	1 53 W	72.5	28.0	DSFZ	1934
11. Ahirasu	7 37 N	0 32 W	1.0	0.4	SM	1927
12. Aiyaola	6 09 N	0 57 W	34.7	13.4	MSSE	1929
13. Ajenjua Bepo	6 23 N	1 02 W	5.7	2.2	MSSE	1930
14. Ajuesu	5 30 N	1 17 W	9.6	3.7	ME	1943
15. Akrobong	5 38 N	0 33 W	2.6	1.0	SM	1930
16. Amama Shelterbelt	7 12 N	2 23 W	44.0	17.0	MSNW	1940
17. Angoben Shelterbelt	5 50 N	2 14 W	34.7	13.4	ME	1948
18. Anhwiaso East	6 18 N	2 10 W	124.3	48.0	MSSE	1926
19. Anhwiaso North	6 19 N	2 16 W	3.6	1.4	MSSE	1926
20. Anhwiaso South	6 12 N	2 23 W	22.3	8.6	ME	1926
21. Ankaful Fuelwood	5 09 N	1 19 W	2.1	0.8	SM	1954
22. Ankasa River	5 17 N	2 35 W	518.0	200.0	WE	1934
23. Anum Su	6 49 N	1 13 W	43.8	16.9	MSSE	1928
24. Anum Su Southern Sect.	6 48 N	1 09 W	12.7	4.9	MSSE	1950
25. Apamprama	6 19 N	1 53 W	34.7	13.4	MSSE	1952
26. Aparapi Shelterbelt	7 08 N	2 09 W	19.2	7.4	MSNW	1939
27. Apedwa	6 08 N	0 33 W	4.1	1.6	MS-UE	1926
28. Apepesu	7 50 N	0 35 E	60.6	23.4	DSFZ	1954
29. Asenanyo River	6 27 N	2 07 W	227.9	88.0	MSNW	1938
30. Asin Apimanim	5 31 N	1 14 W	11.4	4.4	ME	1927
31. Asin Atandaso	5 34 N	1 20 W	153.6	59.3	ME	1937
32. Asonari	6 51 N	1 19 W	1.6	0.6	DSIZ	1928
33. Asubima	7 27 N	1 52 W	78.7	30.4	DSFZ	1945
34. Asufu Shelterbelt East	7 05 N	1 40 W	11.4	4.4	DSIZ	1950

1a. List of Forest Reserves, including former Forest Reserves which have been converted to Wildlife Reserves

Reserve	Lat.	Long.	Area (sq- km)	Area (sq. miles)	Forest-type	Date of first reservation
35. Asufu Shelterbelt West	7 05 N	1 48 W	13.5	5.2	MS-DS	1951
36. Asukese	7 10 N	2 32 W	265.0	102.3	MSNW	1934
37. Asuokoko River	7 47 N	0 25 E	116.0	44.8	DSFZ	1939
38. Atewa Range	6 10 N	0 36 W	232.3	89.7	UE-MS	1926
39. Atewa Range Extension	5 59 N	0 40 W	26.4	10.2	UE-MS	1951
40. Auro River	6 18 N	1 07 W	8.5	3.3	MSSE	1948
41. Awura	7 20 N	1 22 W	133.9	51.7	DSFZ	1940
42. Ayum	6 45 N	2 41 W	112.9	43.6	MSNW	1940
43. Baku	5 33 N	1 04 W	13.0	5.0	MSSE	1929
44. Bandai Hills	6 48 N	0 55 W	160.8	62.1	DSFZ	1928
45. Bediako	6 10 N	0 58 W	7.0	2.7	MSSE	
46. Bemu	5 45 N	1 05 W	43.8	16.9	ME	1950
47. Ben East	5 35 N	1 43 W	25.4	9.8	ME	1954
48. Ben West	5 33 N	1 51 W	55.7	21.5	ME	1954
49. Bia Shelterbelt	7 03 N	2 42 W	29.5	11.4	MSNW	1940
50. Bia Tano	7 00 N	2 37 W	194.3	75.0	MSNW	1937
51. Bia Tawya	6 15 N	3 04 W	678.6	262.0	ME	1965
52. Bia Tributaries North	6 45 N	3 00 W	356.1	137.5	MSNW	1940
53. Bia Tributaries South	6 30 N	3 05 W	305.6	118.0	ME-MS	1940
54. Bimpong	5 43 N	1 28 W	104.1	40.2	ME	1937
55. Birim	5 55 N	1 09 W	39.1	15.1	MSSE	1927
56. Birim Extension	5 58 N	1 08 W	21.8	8.4	MSSE	1940
57. Bobiri	6 38 N	1 17 W	54.6	21.1	MSSE	1939
58. Bodi	6 09 N	2 52 W	175.3	67.7	ME	1967
59. Boi Tano	5 29 N	2 39 W	128.5	49.6	WE	1968
60. Boin River	5 44 N	2 56 W	277.6	107.2	WE-ME	1932
61. Bomfoum	7 01 N	1 10 W	294.7	113.8	DS	1928
62. Bonkoni	6 52 N	2 38 W	75.1	29.0	MSNW	1934
63. Bonsa Ben	5 40 N	1 45 W	155.4	60.0	ME	1939
64. Bonsa River	5 20 N	1 51 W	160.6	62.0	ME	1932
65. Bonsam Bepo	6 40 N	2 37 W	124.3	48.0	MSNW	1934
66. Bosomoa	7 55 N	1 49 W	170.9	66.0	DSFZ	1930
67. Bosumkese	7 06 N	2 15 W	138.3	53.4	MSNW	1937
68. Bosumtwi Range	6 30 N	1 25 W	78.7	30.4	MSSE	1931
69. Boti Falls	6 12 N	0 14 W	1.3	0.5	MSSE	1969
70. Bowiye Range	5 43 N	2 03 W	120.2	46.4	ME	1930
71. Brimso	5 12 N	1 18 W	10.6	4.1	SM-DS	1951
72. Bura River	5 50 N	2 18 W	103.1	39.8	ME	1932
73. Cape Three Points	4 50 N	2 03 W	51.0	19.7	ME	1950
74. Chai River	8 02 N	0 26 E	182.3	70.4	DSFZ	1962
75. Chiremoasi Bepo	6 14 N	1 28 W	6.0	2.3	MSSE	1931
76. Chirimfa	7 07 N	1 17 W	114.0	44.0	DSFZ	1932
77. Dadiaso	5 58 N	3 02 W	171.2	66.1	ME-WE	1977
78. Dampia Range	6 00 N	1 40 W	80.3	31.0	MS-ME	1937
79. Dede	6 34 N	0 16 W	51.0	19.7	DSFZ	1955
80. Denyau Shelterbelt	6 02 N	1 47 W	12.4	4.8	ME	1939
81. Desiri	6 40 N	2 20 W	151.0	58.3	MSNW	1954
82. Disue River	5 54 N	2 59 W	23.6	9.1	WE-ME	1943
83. Dome River	6 32 N	1 12 W	80.5	31.1	MSSE	1929
84. Draw River	5 12 N	2 20 W	235.4	90.9	WE	1937
85. Ebi River Shelterbelt	5 07 N	2 22 W	25.9	10.0	WE	1941

1a. List of Forest Reserves, including former Forest Reserves which have been converted to Wildlife Reserves

Reserve	Lat.	Long.	Area (sq- km)	Area (sq. miles)	Forest-type	Date of first reservation
86. Esen Epam	5 49 N	0 52 W	46.1	17.8	MSSE	1936
87. Esuboni	5 50 N	0 48 W	28.5	11.0	MSSE	1927
88. Esukawkaw	6 22 N	0 48 W	122.2	47.2	MSSE	1929
89. Fum Headwaters	6 15 N	1 21 W	72.5	28.0	MSSE	1932
90. Fure	5 22 N	2 17 W	158.2	61.1	WE	1939
91. Fure Headwaters	5 32 N	2 23 W	169.6	65.5	WE	1940
92. Gianima	7 04 N	1 35 W	17.1	6.6	DSIZ	1939
93. Goa Shelterbelt	6 58 N	2 28 W	23.8	9.2	MSNW	1940
94. Jade Bepo	6 28 N	0 51 W	5.2	2.0	MSSE	1932
95. Jade Bepo Extension	6 27 N	0 51 W	0.8	0.3	MSSE	1932
96. Jema Assamkrom	5 26 N	2 44 W	66.0	25.5	WE	>1977
97. Jeni River	6 30 N	1 55 W	21.5	8.3	MSNW	1937
98. Jimira	6 35 N	1 55 W	62.9	24.3	MSNW	1932
99. Kabakaba Hills East	6 39 N	0 29 E	7.3	2.8	DSFZ	1947
100. Kabakaba Hills North	6 42 N	0 30 E	1.3	0.5	DSFZ	1947
101. Kabakaba Hills West	6 40 N	0 27 E	5.4	2.1	DSFZ	1947
102. Kabo River	7 37 N	0 26 E	136.0	52.5	DSFZ	1931
103. Kade Bepo	6 29 N	0 48 W	16.8	6.5	MSSE	1930
104. Kajeasi	6 24 N	0 57 W	26.7	10.3	MSSE	1932
105. Kakum	5 26 N	1 19 W	212.4	82.1	ME	1931
106. Klemu Headwaters	6 46 N	0 31 E	10.9	4.2	DSFZ	1947
107. Kokotintin Shelterbelt	6 00 N	1 16 W	9.1	3.5	MSSE	1940
108. Komenda Fuelwood	5 04 N	1 30 W	2.1	0.8	SM	1946
109. Kpandu Range Dayi Block	5 45 N	0 17 E	30.3	11.7	DSFZ	1951
110. Kpandu Range West	6 50 N	0 14 E	35.5	13.7	DSFZ	1947
111. Krochua	5 41 N	1 11 W	10.6	4.1	ME	1932
112. Krokosua Hills	6 30 N	2 49 W	481.7	186.0	MSNW	1935
113. Kronwam	6 55 N	1 19 W	5.7	2.2	DSIZ	1928
114. Kunda	7 50 N	0 23 E	0.5	0.2	DSFZ	
115. Kunsimua Bepo	6 11 N	1 26 W	10.1	3.9	MSSE	1937
116. Kwamisa	7 08 N	1 54 W	82.9	32.0	DS-MS	1928
117. Kwekaru	6 27 N	0 59 W	11.7	4.5	MSSE	1929
118. Kwesi Anyinima	6 27 N	0 54 W	1.8	0.7	MSSE	1930
119. Mamang River	6 16 N	1 03 W	54.4	21.0	MSSE	1938
120. Mamiri	5 40 N	2 22 W	45.3	17.5	ME-WE	
121. Mankrang	7 21 N	2 04 W	85.5	33.0	DSFZ	1933
122. Manzan	6 50 N	3 09 W	305.6	118.0	MSNW	1972
123. Minta	5 48 N	1 40 W	21.8	8.4	ME	1938
124. Mirasa Hills	6 22 N	1 10 W	67.3	26.0	MSSE	1937
125. Mpameso	7 05 N	2 53 W	322.5	124.5	MSNW	1937
126. Muro	6 29 N	2 38 W	63.5	24.5	MSNW	1951
127. Ndumfri	5 10 N	2 09 W	72.5	28.0	WE	1937
128. Neung	5 07 N	2 02 W	157.7	60.9	WE-ME	1954
129. Nkawanda	6 29 N	0 43 W	8.0	3.1	MSSE	1939
130. Nkonto Ben	5 37 N	1 54 W	14.5	5.6	ME	1950
131. Nkrabia	6 00 N	1 35 W	100.2	38.7	ME-MS	1940
132. North Bandai Hills	6 52 N	0 53 W	72.8	28.1	DSFZ	1928
133. North Fomangsu	6 41 N	0 59 W	42.7	16.5	MSSE	1925
134. Northern Scarp East	6 40 N	0 33 W	49.2	19.0	DS	1935
135. Northern Scarp West	6 42 N	0 45 W	64.8	25.0	DS	1935
136. Nsemre	7 32 N	2 12 W	18.1	7.0	DSFZ	1939

1a. List of Forest Reserves, including former Forest Reserves which have been converted to Wildlife Reserves

Reserve	Lat.	Long.	Area (sq- km)	Area (sq. miles)	Forest-type	Date of first reser-vation
137. Nsuensa	6 09 N	0 59 W	62.7	24.2	MSSE	1929
138. Numia	6 02 N	1 24 W	50.2	19.4	MSSE	1938
139. Nyamibe Bepo	6 10 N	1 22 W	22.3	8.6	MSSE	1933
140. Obotumfo	5 36 N	0 34 W	1.6	0.6	SM	1930
141. Oboyow	5 46 N	0 55 W	63.7	24.6	MSSE	1927
142. Ochi Headwaters I	5 43 N	1 17 W	3.4	1.3	ME	1940
143. Ochi Headwaters II	5 42 N	1 13 W	9.3	3.6	ME	1940
144. Oda River	6 08 N	1 56 W	164.2	63.4	MS-ME	1939
145. Odomi River	7 19 N	0 29 E	16.1	6.2	DSFZ	1931
146. Ofin Headwaters	7 01 N	1 28 W	13.0	5.0	DSIZ	1927
147. Ofin Shelterbelt	6 40 N	2 03 W	60.3	23.3	MSNW	1951
148. Ongwam I, II & III	6 57 N	1 23 W	31.3	12.1	DSIZ	1951
149. Onuem Bepo	6 03 N	1 18 W	34.4	13.3	MSSE	1930
150. Onuem Nyamibe Shelterbelt	6 07 N	1 21 W	24.9	9.6	MSSE	1936
151. Onyimsu	6 43 N	1 01 W	8.5	3.3	MSSE	
152. Opimbo I & II	5 40 N	0 32 W	1.0	0.4	DSIZ	1927
153. Opon Mansi	5 45 N	1 53 W	116.6	45.0	ME	1930
154. Opro River	7 11 N	1 48 W	129.2	49.9	DS	1929
155. Pamu Berekum	7 25 N	2 56 W	189.1	73.0	DS-MS	1932
156. Pompo Headwaters	6 15 N	1 34 W	12.2	4.7	MSSE	1930
157. Pra Anum	6 15 N	1 10 W	123.3	47.6	MSSE	1908
158. Pra Birim	5 58 N	1 13 W	23.6	9.1	MSSE	1937
159. Pra Suhien I	5 16 N	1 32 W	82.1	31.7	ME	1928
160. Pra Suhien II	5 19 N	1 28 W	104.1	40.2	ME	1933
161. Prakaw	6 29 N	1 03 W	9.8	3.8	MSSE	1942
162. Santomang	6 00 N	2 42 W	21.2	8.2	ME	1944
163. Sapawsu	6 16 N	0 01 E	15.3	5.9	SM	1957
164. Sawsaw	7 36 N	2 10 W	62.9	24.3	DSFZ	1976
165. Sekondi Waterworks II & III	5 01 N	1 45 W	10.1	3.9	MSSE	1938
166. South Fomangsu	6 35 N	0 57 W	41.4	16.0	MSSE	1925
167. Southern Scarp	6 33 N	0 40 W	154.6	59.7	MSSE	1935
168. Subin	6 47 N	2 47 W	238.3	92.0	MSNW	1956
169. Subin Shelterbelt	6 03 N	1 52 W	22.5	8.7	ME	1940
170. Subri River	5 17 N	1 43 W	587.9	227.0	ME	1949
171. Suhuma	5 58 N	2 29 W	358.5	138.4	ME-MS	1935
172. Sukusuku	6 33 N	3 11 W	147.6	57.0	ME	1972
173. Sui River	6 10 N	2 44 W	333.9	128.9	ME-MS	1930
174. Sumtwitwi	6 20 N	2 10 W	3.6	1.4	MSSE	1939
175. Supong	5 36 N	1 07 W	35.7	13.8	MSSE	1954
176. Supuma Shelterbelt	6 00 N	1 42 W	25.0	9.7	ME	1938
177. Tain Tributaries I	7 25 N	2 14 W	30.6	11.8	DSFZ	1932
178. Tain Tributaries II	7 35 N	2 30 W	509.2	196.6	DSFZ	1934
179. Tano Anwia	5 50 N	2 35 W	153.1	59.1	ME	1935
180. Tano Ehuro	5 58 N	2 37 W	176.1	68.0	ME	1967
181. Tano Nimri	5 38 N	2 37 W	205.9	79.5	WE-ME	1935
182. Tano Ofin	6 40 N	2 10 W	402.2	155.3	MS-UE	1929
183. Tano Suhien	6 20 N	2 28 W	84.4	32.6	MS	1934
184. Tano Suraw	6 15 N	2 25 W	28.5	11.0	MSSE	1934
185. Tano Suraw Extension	6 21 N	2 23 W	75.1	29.0	MS	1935
186. Tinte Bepo	6 58 N	2 00 W	115.5	44.6	MSNW	1928
187. Togo Plateau	7 14 N	0 25 E	150.0	57.9	DSFZ	1929

APPENDIX 1 (Cont.)

1a. List of Forest Reserves, including former Forest Reserves which have been converted to Wildlife Reserves

Reserve	Lat.	Long.	Area (sq- km)	Area (sq. miles)	Forest-type	Date of first reser-vation
188. Tonton	6 00 N	2 05 W	146.3	56.5	ME	1936
189. Totua Shelterbelt	5 55 N	2 23 W	63.5	24.5	ME	1941
190. Upper Wasaw	6 07 N	2 13 W	100.8	38.9	ME	1925
191. Volta River	6 11 N	0 01 E	50.5	19.5	DS-SM	1928
192. Wawahi	5 40 N	1 06 W	138.9	15.0	MS-ME	1929
193. Worobong Kwahu	6 31 N	0 33 W	55.2	21.3	MS-DS	1930
194. Worobong North	6 37 N	0 26 W	13.2	5.1	DS	1927
195. Worobong South	6 28 N	0 28 W	106.2	41.0	MS-DS	1929
196. Yaya	7 27 N	2 08 W	51.3	19.8	DSFZ	1930
197. Yenku	5 24 N	0 43 W	21.2	8.2	SM	1937
198. Yogaga	6 11 N	0 03 E	0.8	0.3	SM	
199. Yongwa	6 13 N	0 03 W	7.8	3.0	SO	1957
200. Yoyo	5 55 N	2 48 W	235.7	91.0	ME	1932

Several former reserves which were abandoned when they were flooded by the Volta Lake have been omitted from the list. Areas given are total areas enclosed by the reserve boundary; no allowance has been made for permitted farms within the reserve. Some reserves or parts of reserves are now administered by the Department of Game and Wildlife (see Appendix 1b).

1b. National Parks (N.P.), Game Production Reserves (G.P.R.) and Wildlife Sanctuaries (W.S.) within the forest zone administered by the Department of Game and Wildlife

Wildlife reserve	Forest reserve*	Lat	Long.	Area (sq. km)	Forest. type	Date of first assignation to Dept. of Game & Wildlife
1. Ankasa G.P.R.	22	5 15 N	2 36 W	348.7	WE	1976
2. Bia G.P.R.	53	6 29 N	3 06 W	227.9	ME	1974
3. Bia N.P.	53	6 33 N	3 05 W	77.7	MSNW	1974
4. Bomfobiri W.S.	61	6 57 N	1 10 W	50.0	DSIZ	1976
5. Nini-Suhien N.P.	22	5 19 N	2 34 W	174.5	WE	1976
6. Owabi W.S.		6 44 N	1 42 W	14.5	MSNW	1971
7. Shai Hills G.P.R.		5 54 N	0 04 E	54.4	SO	1971

* The number is that of the forest reserve in Appendix 1(a) from which the Wildlife reserve has been formed.

Abbreviations for forest-types in Appendix 1a and 1b: WE = Wet Evergreen; ME = Moist Evergreen; UE = Upland Evergreen; MS = Moist Semi-deciduous; MSNW = Moist Semi-deciduous North-west subtype; MSSE = Moist Semi-deciduous South-east subtype: DS = Dry Semi-deciduous; DSFZ = Dry Semi-deciduous Fire Zone subtype; DSIZ = Dry Semi-deciduous Inner Zone subtype; SM = Southern Marginal; SO = South-east Outlier.

APPENDIX 2

Plots sampled in our survey of Ghana forest (Hall & Swaine 1976)

2a. Locations of A-sample plots (see Fig. 2.2)

Plot number	Latitude	Longitude	Notes
A1	6 08 30 N	0 55 50 W	Kade A.R.S.; Botany Department sample plot A (Lawson *et al.* 1970 Fig. 3).
A2	6 08 30 N	0 55 50 W	Kade A.R.S.; sample plot G (Lawson *et al.* 1970 Fig. 3).
A3	5 53 40 N	0 03 40 E	Shai Hills G.P.R., central valley.
A4	6 35 30 N	0 57 10 W	South Fomangsu F.R.; c. 1 km from main Kumasi road just E of road to Kade.
A5	6 42 10 N	1 18 30 W	N side of Bobiri F.R., near nature reserve.
A6	6 13 50 N	0 32 40 W	Atewa Range F.R. just off bauxite survey road from Sagyimase; 540 m asl.
A7	6 14 40 N	0 02 30 E	Akosombo, just south of the communications mast on hill.
A8	6 14 10 N	0 34 40 W	Atewa Range F.R., on summit at 760 m asl, on left-hand branch of Sagyimase survey road.
A9	6 05 10 N	0 03 20 E	Krobo Mountain, north slope near Akuse road.
A10	6 09 50 N	0 00 20 W	Volta River F.R.; just north of highest point on road from Odumase to Oterkpolu.
A11	5 24 00 N	0 43 40 W	Hill summit on W boundary of Yenku F.R.
A12	5 36 30 N	0 34 50 W	Obotumfo F.R. On steep slope above Asebu.
A13	5 33 00 N	0 32 40 W	On wooded slope of Osimpo Hill just W of Awutu to Obrachere road.
A14	6 13 50 N	0 04 10 W	Afram Bukunaw F.R. (dereserved). Just E of Pawmpawm arm of Volta Lake, near canoe landing to Yongwase.
A15	6 13 30 N	0 03 00 W	Yongwa F.R. On ridge at 450 m asl.
A16	6 13 20 N	0 03 30 W	Near NW boundary of Yongwa F.R.
A17	6 09 40 N	0 57 10 W	Beside logging road from Kade A.R.S. into Aiyaola F.R. about 2 km beyond boundary.
A18	6 08 40 N	0 56 30 W	Just off path from Kade A.R.S. dam into Aiyaola F.R.
A19	4 53 30 N	1 48 30 W	Grove N of road just E of Apowa.
A20	5 40 30 N	0 16 00 W	Sacred grove by main road, c. 2 km SE of Pokoase.
A21	6 39 30 N	0 50 50 W	Where Akwaseho to Abene road crosses the southern boundary of the Southern Scarp F.R.
A22	6 37 30 N	2 10 10 W	Tano Ofin F.R. On W side of road from Nyinahin to Geological Survey Camp.
A23	6 24 10 N	2 49 10 W	Krokosua Hills F.R., by road about half way from Asanwinso to Antobia.
A24	6 22 30 N	2 55 40 W	Extreme SW corner of Krokosua Hills F.R. 8 km W of Krokosua village.
A25	6 30 10 N	2 53 10 W	Krokosua Hills F.R. on hillside just N of Asempaneye.
A26	6 26 30 N	2 33 20 W	Just W of road N from Buaku, half way through Muro River F.R.
A27	6 30 10 N	2 12 50 W	N side of main road where it crosses SW boundary of Asenanyo F.R.
A28	6 39 40 N	2 11 50 W	Tano Ofin F.R. On summit plateau c. 3 km N of Geological Survey Camp.
A29	5 50 40 N	0 24 50 W	Steep NW slope of Koran Hill above Otoase.
A30	5 50 20 N	0 50 10 W	Just SW of Manso to Swedru road, along a path going into Esen Epam F.R. near NW boundary.

2a. Locations of A-sample plots (see Fig. 2.2)

Plot number	Latitude	Longitude	Notes
A31	6 29 40 N	0 29 40 E	Adaklu Mt.; on summit ridge at 570 m asl where path comes up from Kawdiabe village.
A32	6 38 20 N	0 27 50 E	About 1 km along NW boundary of Kabakaba Hills East F.R. SE of point where Ho to Have road cuts it.
A33	6 38 30 N	0 28 40 E	Kabakaba Hills East F.R. On very steep slope above Deme village.
A34	6 43 40 N	0 15 50 E	On W boundary of Kpandu Range Dayi Block F.R. to E of road 3 km S from Dzani.
A35	7 08 20 N	0 20 40 E	Togo Plateau F.R. about 2.5 km NE of Tayi on slope.
A36	7 08 50 N	0 21 00 E	Togo Plateau F.R. About 5 km NE of Tayi at 450 m asl, on steep slope.
A37	7 08 50 N	0 21 20 E	Togo Plateau F.R. About 5 km NE of Tayi at 630 m asl, in slight valley of summit plateau.
A38	7 18 50 N	0 28 40 E	Odomi River F.R.; steep slope at 480 m asl, near top of ridge.
A39	7 40 N	0 23 E	Kabo River F.R. 6 km NNW of Apesokubi; near reserve boundary.
A40	7 41 N	0 23 E	Kabo River F.R. 8 km NNW of Apesokubi; near reserve boundary.
A41	7 35 N	0 26 E	Kabo River F.R. 2 km E of Apesokubi, on slope.
A42	5 20 10 N	0 45 10 W	Grove on S side of main road, 2 km E of Apam junction.
A43	5 19 00 N	0 50 00 W	On N side of Amoawa Hill E of Legu Junction.
A44	5 21 10 N	1 23 50 W	Beside N boundary of Pra Suhien II F.R. just W of main Cape Coast to Dunkwa road.
A45	5 14 40 N	1 20 30 W	Grove c. 2 km S of Jukwa on W side of road.
A46	5 06 10 N	1 30 10 W	Konkondua grove; 1.5 km NE of Kissi.
A47	5 09 50 N	1 26 10 W	2 km W of Eguafo; summit of Dumpow Hill.
A48	5 09 40 N	1 20 10 W	About 1 km NNE of Ankaful Prison to the E of road to Simiw.
A49	5 03 30 N	1 18 20 W	Kwagya grove 1 km S of Ankaful village.
A50	5 16 10 N	1 01 50 W	2 km W of Mankesim on Kurum Hill S of road to Krofu.
A51	6 10 30 N	0 21 40 W	Nana Buo Powmu; sacred grove about 2 km SE of Kukurantumi, near main road from Koforidua.
A52	6 18 30 N	0 22 30 W	Southern Scarp F.R. 4 km NE of Bosuso, to E of road to Begoro, just below cliff at 480 m asl.
A53	6 26 30 N	0 26 30 W	10 km NW of Begoro, about 1.5 km across boundary of Worobong South F.R. on road to Enyobiesua; level ground to N of road, at 510 m asl.;
A54	6 38 20 N	0 24 50 W	Rocky slope just W of Worobong R c. 8 km W of Ampam.
A55	6 39 50 N	0 34 10 W	Northern Scarp East F.R. on path to Nkyenenkyene about 2 km NNE of Takwa; plateau near ridge crest at 450 m asl.
A56	6 39 10 N	0 34 40 W	About 0.5 km N of Takwa, steep slope near S boundary of Northern Scarp East F.R.
A57	6 41 50 N	0 46 50 W	About 5 km NW of Abetifi in Northern Scarp West F.R., on N side of road to Abene.
A58	6 50 20 N	1 02 40 W	About 7 km NE of Agogo on road to Ejan, in Bandai Hills F.R.
A59	5 56 30 N	0 07 00 W	In a sacred grove just N of Mamfe on road to Koforidua.
A60	6 18 40 N	0 03 10 E	In *Talbotiella* forest on S slope of Obotumfo Hill, 2 km WNW of Akosombo dam near road to Akosombo Port.
A61	6 17 50 N	0 02 50 E	Sapawsu F.R. In dry valley on W side of road from Akosombo to Port.
A62	6 19 20 N	0 51 20 W	Level ground at SW corner of Esukawkaw F.R. just inside boundary.
A63	6 21 50 N	1 01 10 W	NE corner of Ajenjua Bepo F.R.
A64	6 18 40 N	1 04 30 W	By road from Ntronang to Ofoase where it crosses N boundary of Auro River F.R.
A65	6 16 10 N	1 10 40 W	8 km N of Amentia, on E side of road through Pra Anum F.R. to Banka.
A66	6 18 30 N	1 15 00 W	Summit ridge of Gyedemoa Hill in NW corner of Pra Anum F.R.
A67	6 18 20 N	1 15 00 W	Gyedemoa Hill. Just S of A66.
A68	6 33 30 N	1 08 50 W	Just inside NE boundary of Dome River F.R. c. 3 km SW of Juaso.
A69	5 44 50 N	0 14 10 W	Sacred grove just S. of Berekuso.
A70	5 21 20 N	1 17 40 W	By boundary of Kakum F.R. 2 km WSW of Krua.

2a. Locations of A-sample plots (see Fig. 2.2)

Plot number	Latitude	Longitude	Notes
A71	5 23 20 N	1 09 20 W	On a hilly grove just W of Katakyiasi N of Fante Nyankomase.
A72	5 43 20 N	1 17 30 W	Ochi Headwaters I F.R. About 2 km N of Asin Foso on W side of road to Praso.
A73	6 12 40 N	1 23 00 W	Fum Headwaters F.R., c. 2 km WNW of Aboabo No. 1, on hillside near S boundary.
A74	6 25 20 N	1 25 00 W	Bosumtwi Range F.R. Near Dakabuoso c. 5 km from Kokobin, where road meets boundary.
A75	6 29 00 N	1 53 40 W	N of Manso Nkwanta on SE corner of Jeni River F.R.
A76	6 29 40 N	1 10 30 W	On S boundary of Dome River F.R. c. 7 km SW of Obogu.
A77	6 09 40 N	0 36 40 W	Atewa Range F.R. On summit ridge about 2 km NW of Apapam; 540 m asl.
A78	5 50 40 N	0 25 00 W	Foot of Koran Hill just W of A29.
A79	5 40 20 N	0 33 00 W	Obimpo F.R., c. 1 km S of Bawjiasi.
A80	5 31 10 N	0 41 50 W	About 2 km S of Agona Swedru; sacred grove on E side of road.
A81	4 46 20 N	2 02 20 W	Cape Three Points F.R. In SE corner of reserve c. 2.5 km NW of Akwida.
A82	4 51 40 N	2 03 20 W	Cape Three Points F.R. 150 m E from BP43. About 5 km S of Abra.
A83	4 46 10 N	1 58 30 W	Small hill just W of Achowa 2.5 km SW of Dixcove.
A84	6 57 50 N	1 12 20 W	Bomfoum F.R. 8 km E of Kwamang on road from Kumawu.
A85	7 06 50 N	1 17 50 W	Chirimfa F.R. 5 km NNE of Amoamang; E side of road.
A86	7 09 40 N	1 45 00 W	Opro River F.R. SE corner of reserve; c. 200 m along the boundary to NW of road c. 3 km from Abofuo.
A87	7 19 10 N	1 48 30 W	Opro River F.R. About 40 m W of road near N end of reserve, c. 2.5 km S of Asuasu.
A88	7 05 50 N	1 51 50 W	Asufu Shelterbelt West F.R. About half way along common boundary between Asufu F. R. and Kwamisa F.R., c. 2.5. km from Sabronomu.
A89	6 15 20 N	0 33 30 W	On summit plateau of Atewa Range F.R., alt. 770 m asl.
A90	5 51 00 N	0 03 10 E	Patch of forest on SE side of Mampong Hill, just S of Shai Hills G.P.R.
A91	6 03 30 N	0 11 20 E	By road from Asutsuare, just below summit of Osudoku Mountain.
A92	6 10 30 N	0 03 00 E	Yogaga Mountain. About 100 m up slope from where the road from Kpong Waterworks reaches the forest.
A93	6 56 40 N	1 56 40 W	Tinte Bepo F.R. 2 km along timber road off Sunyani road 1.5 km S of Fawumang.
A94	7 29 20 N	2 09 50 W	Yaya F.R. At Buoku junction on N side of Techiman road.
A95	7 31 10 N	1 10 00 W	Just inside NE corner of Nsemre F.R. c. 4 km from Buoku.
A96	7 25 10 N	2 12 50 W	At E boundary of Tain Tributaries Block I F.R. c. 5 km NW of Chiraa.
A97	7 10 10 N	2 12 30 W	Bosumkese F.R. On NE side of hill.
A98	7 09 50 N	2 27 30 W	About 1 km inside E edge of Asukese F.R. c. 5 km W of Atronie.
A99	7 26 30 N	2 30 05 W	At SW corner of Tain Tributaries Block II F.R. c. 2 km NNW of Suatre Amanfoso.
A100	7 35 00 N	2 36 00 W	About 15 km N of Berekum, just inside boundary of Tain Tributaries Block II F.R.
A101	7 08 00 N	2 55 10 W	3 km S of Mpameso F.R. boundary, Dormaa to Diabakrom road.
A102	7 21 40 N	2 51 40 W	3 km NNW of S boundary of Pamu-Berekum F.R. on Dormaa to Asunso road.
A103	6 57 25 N	2 39 30 W	Bia-Tano F.R. 6 km W of Bediakukrom.
A104	6 55 10 N	3 02 00 W	Bia Tributaries North F.R. 1.5 km S of reserve boundary on Okirakrom to Adowkrom road.
A105	6 54 20 N	2 39 15 W	Bia-Tano F.R. 9 km WNW of Mim.
A106	6 43 30 N	2 21 40 W	Desiri F.R. 5 km NE from Sienchem where road meets boundary.
A107	6 41 40 N	2 37 35 W	Bonsam Bepo F.R. 1.5 km inside reserve on road 9 km W of Akrodie.
A108	6 42 50 N	2 44 30 W	22 km E of Akrodie just inside Subin F.R. boundary.
A109	6 52 20 N	2 02 15 W	Tano Ofin F.R. 2.5 km E of Domiabra, near BP 27.
A110	6 59 20 N	1 26 40 W	Ofin Headwaters F.R. Half way up scarp, just E of the road.
A111	5 28 10 N	0 58 30 W	Grove c. 2.5 km N of Biseasi on E side of road to Asikuma.
A112	5 45 30 N	1 03 30 W	About 1 km inside Bemu River F.R. on road from Amanfopon to Aperade.
A113	5 54 40 N	1 07 40 W	Birim F.R. About 9 km on Aduasa to Apoli road.
A114	5 56 50 N	1 11 10 W	About 1 km inside Pra-Birim F.R. on path from Apoli to Praso.

2a. Locations of A-sample plots (see Fig. 2.2)

Plot number	Latitude	Longitude	Notes
A115	5 50 10 N	1 00 10 W	Grove on rock at Achiasi village.
A116	5 38 40 N	0 52 50 W	3 km N of Odoben just E of road to Brakwa.
A117	5 09 00 N	1 44 30 W	S corner of Subri River F.R. c. 10 km NW of Dompim.
A118	5 16 10 N	1 37 50 W	About 5 km NW of Krobo on small hill to the E side of Ateiku road, in Subri River F.R.
A119	5 25 30 N	1 38 10 W	Grove on outskirts of Mampanso village on Krobo to Ateiku road.
A120	5 34 10 N	1 42 40 W	About 2 km NW of Ateiku station, just inside Ben East F.R.
A121	5 45 30 N	1 36 50 W	SE corner of Minta F.R.
A122	5 49 20 N	1 50 10 W	About 2.5 km N of Atobiase just W of new road through Opon Mansi F.R.; foot of hill.
A123	6 02 00 N	1 47 30 W	Denyau Shelterbelt F.R. About 7 km NNW of Dunkwa on W side of road.
A124	6 03 50 N	2 05 20 W	Tonton F.R. About 2 km inside N edge of the reserve near Feiakobo.
A125	5 45 40 N	2 01 20 W	Bowiye Range F.R. On new road from Moseaso, just at foot of main hill inside W reserve boundary.
A126	5 34 30 N	1 54 30 W	4 km SE of Insu Junction on road to Ben West F.R.; steep hillside.
A127	5 19 30 N	1 51 30 W	About 1 km inside W boundary of Bonsa River F.R. on road from Aboso.
A128	5 09 50 N	2 01 40 W	Neung F.R. Just inside reserve boundary to S. of road from just S of Bonsaso to Benso; 1 km from junction.
A129	6 08 50 N	2 33 40 W	About 2 km SE of Anokokrom, c. 1 km inside boundary of Suhuma F.R.
A130	6 18 10 N	2 28 00 W	Tano Suhien F.R. About 5 km from Datano near SE boundary road.
A131	6 15 30 N	2 17 20 W	Afao Hills F.R. Hill summit just to N of road to new mine.
A132	6 12 40 N	2 11 30 W	Anwhiaso South F.R. On road from Anhwiaso to Dunkwa c. half way through reserve.
A133	6 07 00 N	2 16 00 W	Upper Wasaw F.R. About 3 km S of Humjibre on slope of small hill to W of road.
A134	6 14 00 N	2 42 20 W	Sui River F.R. About 6 km SW of Shi on S side of highest point of road to Atabuka.
A135	6 41 00 N	2 57 30 W	Bia Tributaries North F.R. About 1 km inside E boundary of reserve near Kutukrom.
A136	6 30 20 N	3 02 40 W	Near BP 19 on E boundary of Bia G.P.R.
A137	6 23 20 N	2 58 20 W	On Gliksten timber road about 7 km W of Abrokwofi about 1.5 km outside E boundary of Bia G.P.R.
A138	5 54 10 N	2 20 40 W	Totua Shelterbelt F.R. 6 km SSW of Aguna, just inside N boundary just NW of road.
A139	5 43 15 N	2 22 10 W	Mamiri F.R. Just inside NE boundary to S of Asankrangwa to Bawdia road.
A140	5 50 00 N	2 32 10 W	Tano Anwia F.R. 3 km WSW across river from Moseaso, on hill slope just inside E boundary.
A141	5 30 10 N	2 43 50 W	5 km S of Boinso along road to Hintimbo, just W of road at foot of slope.
A142	5 43 50 N	2 54 40 W	Boin River F.R. On old timber road c. 3 km W of Akunso; flat ground.
A143	5 55 10 N	2 57 40 W	Disue River F.R. On N edge of reserve to W of road from Enchi to Dadiaso.
A144	5 04 40 N	2 05 20 W	Neung F.R. About 3 km E of Nkroful junction on Takoradi to Tarkwa road; beside track through forest.
A145	5 24 30 N	2 14 00 W	Fure F.R. About 3 km inside E boundary of reserve to S side of Prestea to Samreboi road.
A146	5 27 10 N	2 26 35 W	About 37 km from Prestea on S side of road to Samreboi.
A147	5 22 25 N	2 28 40 W	Bronikrom; on W side of old timber road running S from Prestea to Samreboi road towards Draw River F.R.
A148	5 17 35 N	2 25 40 W	Kissikrom; c. 10 km S of plot 147.
A149	5 36 40 N	2 35 30 W	Tano Nimri F.R. 2.5 km NW of Samreboi, on top of small hill just S of road to Enchi.
A150	5 13 20 N	2 39 25 W	Ankasa River F.R. Just inside SW boundary 100 m W from old bridge over Ankasa R.
A151	5 16 40 N	2 40 50 W	Ankasa River F.R. Just N of road through reserve c. 10 km E of Elubo.

2a. Locations of A-sample plots (see Fig. 2.2)

Plot number	Latitude	Longitude	Notes
A152	5 03 30 N	2 18 00 W	Ebi Shelterbelt F.R. Just W of road c. 10 km N of Nkrofro just within SE boundary of reserve.
A153	4 59 10 N	2 15 10 W	Small hill just S of Nkroful to Bamiankaw road near Salman, c. 8 km from Nkroful.
A154	5 10 00 N	2 13 10 W	Draw River F.R. SE corner of reserve where it approaches R. Ancobra; side of hill near BP 37.
A155	5 58 10 N	0 41 30 W	Atewa Range Extension F.R. 8 km ENE of Osenasi; on path from Awaham at 480 m asl.

2b. Locations of mapped B-sample plots (see Fig. 2.2)

Plot number	Latitude	Longitude	Notes
B1	6 34 00 N	0 44 40 W	Southern Scarp F.R. below Atibie.
B8	5 08 50 N	1 18 40 W	Hillside E of Ankaful village.
B9	5 16 10 N	1 08 20 W	Otsiabo and Efefipow hills 9 km SSE of Abura Dunkwa.
B10	5 29 00 N	0 32 20 W	Granite hill near Ojobi.
B11	5 12 40 N	1 08 20 W	Fupow Hill 5 km NNW of Anomabu.
B12	5 15 40 N	1 00 10 W	Grove 2 km E of Ekumfi Swedru on S of main road.
B13	5 16 30 N	0 57 40 W	Hill by deep cutting in main road W of Nakwa junction.
B14	5 09 30 N	1 13 50 W	Sacred grove just S of Amosima.
B15	5 08 00 N	1 16 30 W	Grove just E of Aburankwanta.
B17	5 42 00 N	0 13 10 W	Sacred grove on flat ground W of Ashonman.
B18	5 41 30 N	0 14 50 W	Sacred grove just N of Kwabenya.
B19	6 16 10 N	0 17 00 E	Forest patch on grassy plain SW of Zondo hill.
B20	6 17 10 N	0 18 00 E	Summit of Zondo Hill.
B22	6 31 10 N	2 52 20 W	Krokosua Hills F.R., NE of Asempaneye.
B24	5 16 50 N	1 02 40 W	Secondary forest at Nkwanta 5 km WNW of Mankesim.
B25	5 20 30 N	1 05 20 W	Secondary forest at Opafi 12 km NW of Mankesim.
B26	5 30 20 N	1 09 50 W	Roadside thicket 1 km S of Manso.
B27	5 37 40 N	1 14 10 W	Abandoned cocoa farm at roadside 1 km NW of Nsuta.
B28	5 49 50 N	1 19 00 W	Secondary forest just NW of Akonfudi on roadside.
B29	6 02 20 N	1 23 30 W	Numia F.R., near main road.
B30	6 12 40 N	1 22 50 W	Fum Headwaters F.R. on roadside below A73.
B31	6 15 20 N	1 27 10 W	Steep slope in disturbed forest, Chiremoasi Bepo F.R.
B32	6 16 10 N	1 28 40 W	Secondary forest on hillside NE of Brofoyedru.
B33	6 27 40 N	1 38 00 W	Old secondary forest at Kumasi-Bekwai-Obuasi road junction.
B34	6 26 50 N	1 48 00 W	Sacred grove at Antoakrom.
B35	6 34 50 N	1 47 30 W	Good forest near road 1 km NE of Eduyaokrom.
B36	6 39 50 N	1 42 40 W	Abandoned cocoa farm ENE of Twedie.
B37	6 29 20 N	0 41 40 W	Good forest in Nkawanda F.R. S of Wawasi.
B38	6 02 40 N	0 27 10 W	Sacred grove at Neefio village just NW of Suhum.
B39	6 16 50 N	0 33 40 W	Near summit ridge of Atewa Range F.R.; 720 m asl.
B40	6 01 00 N	0 46 20 W	Old secondary forest by road 11 km SE of Kade.
B41	5 48 00 N	0 30 00 W	Old secondary forest just W of Adeiso.
B42	5 33 30 N	0 20 40 W	Rocky slope above small lake SW of Weija.
B43	6 09 30 N	0 35 20 W	Just outside Atewa Range F.R. by Takyiman road above Apapam.
B44	5 47 00 N	0 11 30 W	Rocky valley W of road from Ayimensa to Peduase.

2b. Locations of mapped B-sample plots (see Fig. 2.2)

Plot number	Latitude	Longitude	Notes
B45	5 51 20 N	0 10 30 W	Old secondary forest just W of Aburi Gardens.
B46	5 50 00 N	0 10 30 W	Sacred grove on SW outskirts of Aburi.
B47	6 12 40 N	0 34 40 W	Summit plateau of Atewa Range F.R.; 750 m asl.
B48	5 29 20 N	0 33 30 W	Granite slope to N of road 7 km W of Obutu towards Winneba.
B49	4 48 10 N	1 57 00 W	Slope near road 2 km N of Dixcove.
B50	4 51 00 N	1 54 30 W	Slope near road just E of Sese Oilmill.
B51	5 00 40 N	1 39 40 W	Hill just SE of Shama Junction.
B52	6 56 20 N	1 18 10 W	Secondary forest between Kronwam and Ongwam III forest reserves.
B53	6 56 20 N	1 18 20 W	Near B52.
B54	7 04 50 N	1 13 20 W	Fringing forest close to Jadiako in Chirimfa F.R.
B55	6 52 20 N	1 22 40 W	Secondary forest on roadside between Dadiase and Oyoko.
B56	6 48 10 N	1 26 10 W	Secondary forest 3 km SW of Juaben.
B57	6 45 50 N	1 30 20 W	Secondary forest near road just E of Antoa.
B58	6 44 30 N	1 33 20 W	Sacred grove at Kenyasi; good forest.
B59	6 53 50 N	1 44 30 W	Secondary forest just SW of Ebira.
B60	7 05 30 N	1 49 40 W	Flat ground by road from Abofuo to Sabronumu.
B61	6 03 30 N	1 38 10 W	Dampia F.R. by road from Akrofuom to Mensonso.
B62	7 36 20 N	2 07 10 W	Sawsaw F.R. by road from Tromiso to Nkunsia.
B63	7 09 00 N	2 12 30 W	Area in Bosumkese F.R. burnt 16 years earlier.
B64	7 38 40 N	2 33 20 W	Broken forest in Tain Tributaries Block II F.R. by road from Berekum to Sekwa.
B65	7 21 40 N	2 40 10 W	Old secondary forest by road 1.5 km SW of Amomaso.
B66	7 14 00 N	2 53 00 W	Secondary forest by road at Maso.
B67	7 09 10 N	2 45 00 W	By road from Kwamiesua to Tronang.
B68	6 56 30 N	2 55 20 W	Old secondary forest 13 km W of R. Bia by Bediakukrom to Okirakrom road.
B69	6 56 20 N	2 50 40 W	Good forest 1.5 km E of R. Bia by Bediakukrom to Okirakrom road.
B70	6 55 20 N	2 36 50 W	Ebaw Rock: granite outcrop 8 km NW Of Mim.
B71	6 46 10 N	2 48 30 W	Subin F.R.; good forest.
B72	6 44 50 N	2 52 00 W	Cut-over forest 3 km NW of Kojofosukrom.
B73	6 58 20 N	2 19 00 W	Poor secondary forest 1.5 km ENE of R. Tano by Hwidiem to Tepa road.
B74	6 40 10 N	0 44 40 W	Sacred grove at Abetifi.
B75	6 42 00 N	0 38 0C W	Secondary forest around Buruku rock.
B76	5 33 10 N	0 50 50 W	Secondary forest just E of Bobikuma.
B77	5 26 00 N	0 58 20 W	Secondary forest 2 km NNW of Ajumako.
B78	5 57 50 N	1 00 20 W	Abandoned cocoa farm near Oda Secondary School.
B79	6 02 50 N	1 06 00 W	Broken forest by road from Ayirebi to Abenasi.
B80	5 37 10 N	1 00 20 W	Old secondary forest by road from Benin to Jamra.
B81	5 05 40 N	1 43 40 W	Rocky hill just S of R. Nhwini.
B82	5 31 00 N	1 39 40 W	Near Adasi, SE of Ateiku.
B83	5 33 20 N	1 33 10 W	Rocky hillside just S of Ntafrewaso.
B84	5 53 00 N	1 32 40 W	By road from Twifu Praso to Dunkwa, 5 km NW Ewisam.
B85	5 56 10 N	1 43 00 W	Secondary forest on steep hillside by road from Kyekyewere to Dunkwa.
B86	6 12 40 N	1 45 20 W	Secondary forest just E of Hia.
B87	5 52 00 N	2 00 00 W	Abandoned cocoa farm on hillside 6 km NE of Tapa.
B88	5 40 00 N	2 07 00 W	S end of Bowiye Range F.R. by Dunkwa to Bogoso road.
B89	5 28 50 N	2 00 20 W	Poor secondary forest on steep hillside by road from Bogoso to Tarkwa.
B90	5 01 50 N	1 40 50 W	Grove on hill 2 km NW of Shama Junction.
B91	6 20 40 N	2 16 50 W	Secondary forest by road from Bibiani to Anwiawso.
B92	6 14 20 N	2 17 00 W	S boundary of Afao Hills F.R. near Awaso.
B93	6 12 00 N	2 37 00 W	5 km W of Jojokrom on Wiawso to Atabuka road.
B94	6 17 10 N	2 47 00 W	Steep slope by road from Suiano to Antobia.
B95	6 23 50 N	2 49 20 W	Krokosua Hills F.R., by road from Asanwinso to Antobia, near A23.
B96	5 46 20 N	2 33 20 W	Steep hillside by road 2 km E of R. Tano on Asankragwa to Enchi road.

2b. Locations of mapped B-sample plots (see Fig. 2.2)

Plot number	Latitude	Longitude	Notes
B97	5 45 40 N	2 45 00 W	Sacred grove by road from Asankragwa to Enchi.
B98	5 38 40 N	2 43 30 W	Steep slope just S of Kwawu by Enchi to Boinso road.
B99	6 04 00 N	3 01 20 W	Good forest 6 km S of Dadiaso by road to Enchi.
B100	5 37 20 N	2 12 30 W	Good forest on small hill near road 6 km ESE of Manso Amenfi.
B101	4 54 00 N	1 58 00 W	Grove 1 km N of Agona Junction.
B102	4 58 50 N	1 53 40 W	Abandoned cocoa farm by road from Mpoho to Domenase.
B103	5 02 50 N	1 55 30 W	Old secondary forest on small hill 1.5 km NNW of Edum Domenase.
B104	5 03 50 N	2 05 20 W	Neung F.R.
B105	5 24 30 N	2 20 20 W	Flat ground near main road at Fure Nkwanta.
B106	5 36 40 N	2 35 20 W	Good forest on level ground near road side below A149.
B107	5 34 30 N	2 31 00 W	Old secondary forest on small hill by main road at 'Woman-no-good' village.
B108	5 06 30 N	2 36 30 W	Good forest c. 2 km from Mpataba by main road to Elubo.
B110	5 02 40 N	2 28 20 W	Near stream in Crops Research Institute, Aiyinasi.
B111	5 00 20 N	2 20 20 W	Old secondary forest on small hill 1.5 km SE of Ebowu by Axim road.
B112	5 57 40 N	0 13 00 W	Secondary forest near road c. 1 km NW of Adawso.
B113	6 11 00 N	0 13 10 W	Steep slope in river gorge of Boti Falls F.R.
B114	6 23 50 N	0 08 30 W	Disturbed forest near stream bed just N of Asesewa.
B115	6 31 30 N	0 14 20 W	By road through Dede F.R.
B116	6 24 40 N	0 20 30 W	Poor secondary forest by road NE of Begoro; 450 m asl.
B117	6 24 40 N	0 10 10 E	Poor secondary forest at Asikuma, by main road junction.
B118	6 27 40 N	0 09 30 E	Poor secondary forest by roadside SSE of Anum.
B119	6 29 20 N	0 10 50 E	Rocky slope on Adansi Hill between Anum and Dzake.
B120	6 33 20 N	0 15 20 E	Secondary forest in valley near road NE of Peki Wudome.
B121	6 34 30 N	0 45 40 W	Secondary forest near summit of Mt. Ejuanema; 750 m asl.
B122	6 46 50 N	1 08 10 W	Boundary of Anum Su Southern Section F.R. just N of Amantama.
B123	7 21 20 N	1 54 10 W	Afrensu-Brohuma F.R.
B124	7 50 50 N	2 31 30 W	Forest along small seasonal stream at foot of hills W of Hani.
B125	7 49 40 N	2 30 20 W	Poor secondary forest near Chekru R. SW of Hani.
B126	7 48 00 N	2 26 50 W	Steep hillside above marsh NW of Degedege.
B127	7 48 10 N	2 26 20 W	Forest on level upland N of Degedege.
B128	5 10 00 N	1 42 40 W	Subri River F.R., by timber road from Daboase.
B129	5 27 50 N	0 40 40 W	Near stream by road 1 km N of Abasa.
B130	5 29 10 N	0 41 00 W	Sacred grove on S edge of Asamkrom.

Altitudes given only for plots ⩾450 m asl, (Appendix 2 (a) and 2 (b)).
asl – above sea level.

APPENDIX 3

Ordination scores of species found four or more times during our survey

Species ordination values for plot identification

Species name	Axis 1	Axis 6	Species name	Axis 1	Axis 6
Abrus pulchellus*	57	18	Albizia glaberrima	50	35
Acacia kamerunensis	36	44	Albizia zygia	34	43
Acacia pentagona	39	43	Alchornea floribunda	18	100
Acalypha neptunica	71	44	Allanblackia parviflora	11	35
Acridocarpus alternifolius	60	49	Allophylus africanus forma chrysothrix	43	46
Acridocarpus macrocalyx*	22	56	Alsodeiopsis staudtii	13	32
Acridocarpus smeathmannii	18	26	Alstonia boonei	30	36
Adenia cissampeloides*	39	26	Amorphophallus johnsonii	46	40
Adenia lobata*	53	47	Amphimas pterocarpoides	24	40
Adenia mannii*	21	44	Anchomanes difformis	62	29
(= A. tenuispira)			Ancistrophyllum opacum	18	33
Adenia rumicifolia*	37	25	Androsiphonia adenostegia	7	59
(= A. lobata p.p.)			Angylocalyx oligophyllus	18	27
Adiantum vogelii	26	38	Aningeria altissima	38	57
Aeglopsis chevalieri	54	40	Aningeria robusta	27	38
Aframomum chrysanthum	21	63	Anogeissus leiocarpus	64	8
Aframomum geocarpum	39	50	Anopyxis klaineana	14	44
Aframomum stanfieldii	44	47	Anthonotha fragrans	13	51
Afrobrunnichia erecta	25	35	Anthonotha macrophylla	21	37
Afrosersalisia afzelii	15	78	Antiaris toxicaria	39	40
Afzelia africana	56	34	Antidesma laciniatum	25	57
Afzelia bella	20	19	Antidesma membranaceum	60	26
Aganope leucobotrya	16	40	Aphanostylis mannii	21	43
Agelaea sp. aff. duchesnei	4	66	Aporrhiza urophylla	19	22
Agelaea nitida	17	34	Aptandra zenkeri	27	37
Agelaea obliqua	19	51	Argomuellera macrophylla	31	53
Agelaea cf. pilosa	25	37	Artabotrys insignis*	34	41
Agelaea trifolia	3	62	Artabotrys oliganthus	26	35
Aidia genipiflora	32	38	Artabotrys stenopetalus	35	36
Airyantha schweinfurthii	15	7	Artabotrys velutinus	62	48
Alafia barteri	36	48	Arthropteris palisotii	25	35
Alafia multiflora*	35	51	Asparagus warneckei	84	45
Alafia scandens	73	30	Asplenium africanum	18	44
Albertisia cuneata	70	54	Asplenium barteri	18	68
Albertisia scandens	22	42	Asplenium emarginatum	57	55
Albizia adianthifolia	37	39	Asplenium variabile	19	38
Albizia ferruginea	45	37	Ataenidia conferta	21	51

APPENDIX 3 (Cont.)

Species ordination values for plot identification

Species name	Axis 1	Axis 6	Species name	Axis 1	Axis 6
Atroxima liberica	11	38	Celtis wightii	51	49
Aubrevillea platycarpa	20	44	Celtis zenkeri	46	51
Aulacocalyx jasminiflora	20	45	Cercestis afzelii	19	39
Baissea breviloba	26	37	Cercestis ivorensis	5	58
Baissea leonensis*	28	42	Cercestis stigmaticus	15	47
Baissea multiflora	24	36	Chaetacme aristata	69	39
Baissea zygodioides	57	40	Chasmanthera dependens*	58	49
Baphia nitida	34	43	Chassalia afzelii*	3	60
Baphia pubescens	29	34	Chassalia corallifera	19	31
Beilschmiedia mannii	12	46	Chassalia kolly	54	40
Bequaertiodendron oblanceolatum	49	59	Chassalia subherbacea	8	52
Berlinia confusa	16	41	Chazaliella sciadephora (form i)	23	56
Berlinia tomentella	16	41	Chazaliella sciadephora (form ii)	10	35
Bersama abyssinica	25	41	Chidlowia sanguinea	22	80
Bertiera breviflora	28	45	Chlamydocarya macrocarpa	24	31
Bertiera racemosa*	14	46	Chlamydocarya thomsoniana	38	51
Blighia sapida	32	42	Chlorophora excelsa	46	18
Blighia unijugata	38	45	Chlorophora regia	46	18
Blighia welwitschii	32	42	Chlorophytum orchidastrum*	22	60
Bolbitis acrostichoides	38	52	Chrysophyllum giganteum	32	54
Bolbitis auriculata	23	42	Chrysophyllum perpulchrum	29	49
Bolbitis gemmifera	23	37	Chrysophyllum pruniforme	12	34
Bombax brevicuspe	12	20	Chrysophyllum subnudum	19	44
Bombax buonopozense	39	45	Chrysophyllum welwitschii	37	40
Bowringia mildbraedii	28	49	Chytranthus atroviolaceus*	21	25
Buchholzia coriacea	24	43	Chytranthus carneus	21	34
Bussea occidentalis	25	51	Chytranthus cauliflorus	9	33
Byrsocarpus coccineus	51	33	Chytranthus macrobotrys	29	39
Calamus deerratus	21	37	Chytranthus verecundus	13	27
Callichilia subsessilis	27	47	Cissus aralioides	35	38
Caloncoba echinata	13	37	Cissus petiolata	63	50
Caloncoba gilgiana	56	32	Cissus producta	23	35
Calpocalyx brevibracteatus	15	41	Clausena anisata	53	39
Calycobolus africanus	28	41	Cleidion gabonicum	33	60
Calycobolus heudelotii	52	46	Cleistopholis patens	40	50
Campylostemon angolense	50	41	Clerodendrum capitatum	42	30
Campylostemon mucronatum	31	53	Clerodendrum volubile*	36	36
Canarium schweinfurthii	27	31	Cnestis ferruginea	38	42
Canthium hispidum	48	43	Cnestis longiflora	16	18
Canthium horizontale	57	24	Cnestis racemosa	20	59
Canthium setosum	17	63	Coccinia barteri*	45	32
Canthium venosum	38	16	Coffea afzelii	27	45
Canthium vulgare	58	33	Coffea ebracteolata	41	41
Capparis erythrocarpos	74	66	Coffea rupestris	39	1
Carapa procera	17	36	Coffea spathicalyx*	32	59
Carpolobia lutea	35	36	Coffea togoensis	68	46
Cassipourea congoensis (form i)	78	28	Cola caricifolia	25	36
Cassipourea congoensis (form ii)	26	45	Cola chlamydantha	13	33
Cassipourea hiotou	1	82	Cola digitata*	6	78
Castanola paradoxa	24	41	Cola gigantea var. glabrescens	28	37
Ceiba pentandra	49	38	Cola heterophylla	11	29
Celtis adolfi-friderici	31	49	Cola lateritia var. maclaudii	28	37
Celtis mildbraedii	36	45	Cola millenii	53	44

APPENDIX 3 (Cont.)

Species ordination values for plot identification

Species name	Axis 1	Axis 6	Species name	Axis 1	Axis 6
Cola nitida	20	37	Dennettia tripetala	71	38
Cola reticulata	27	69	Desplatsia chrysochlamys	16	29
Cola umbratilis	1	79	Desplatsia subericarpa	29	42
Cola verticillata	20	37	Dialium aubrevillei	14	43
Combretum aphanopetalum*	20	19	Dialium dinklagei	23	32
Combretum dolichopetalum*	44	20	Dialium guineense	63	35
Combretum mucronatum	57	40	Diaphananthe pellucida*	36	48
Combretum oyemense	25	36	Dichapetalum albidum	22	37
Combretum paniculatum*	39	29	Dichapetalum angolense	9	40
Combretum platypterum*	27	49	Dichapetalum barteri	61	62
Combretum racemosum*	34	48	Dichapetalum crassifolium	28	42
Commelina capitata*	24	33	Dichapetalum filicaule	19	7
Commelina macrosperma	53	58	Dichapetalum guineense	49	46
Commelinidium gabunense	61	34	Dichapetalum heudelotii		
Connarus africanus	10	52	(sparsely hairy variety)	13	33
Copaifera salikounda	21	40	(tomentose variety)	9	37
Corymborkis corymbis	28	39	Dichapetalum pallidum	22	37
Corynanthe pachyceras	29	35	Dichapetalum toxicarium	15	27
Costus afer*	33	36	Dicranolepis grandiflora	43	27
Costus engleranus	19	53	Dicranolepis persei	21	49
Coula edulis	5	55	Dictyophleba leonensis	6	50
Craterispermum caudatum	19	44	Dioscorea minutiflora*	40	38
Craterispermum cerinanthum	25	35	Dioscorea praehensilis	44	40
Craterosiphon scandens*	24	40	Dioscorea smilacifolia	31	37
Cremaspora triflora	40	42	Dioscoreophyllum cumminsii	47	32
Croton penduliflorus*	44	50	Diospyros abyssinica	72	48
Crotonogyne manniana	12	39	Diospyros canaliculata	26	36
Crudia gabonensis	18	14	Diospyros chevalieri	5	57
Ctenitis jenseniae	11	40	Diospyros cooperi	19	29
Ctenitis lanigera*	20	46	Diospyros ferrea	36	47
Ctenitis pilosissima	18	30	Diospyros gabunensis	17	17
Ctenitis protensa	17	31	Diospyros heudelotii	27	40
Culcasia angolensis	23	39	Diospyros kamerunensis	16	39
Culcasia parviflora	23	39	Diospyros mannii	19	37
Culcasia scandens	19	41	Diospyros mespiliformis	59	44
Culcasia striolata	45	53	Diospyros monbuttensis	45	44
Cussonia bancoensis*	41	52	Diospyros piscatoria	22	22
Cylicodiscus gabunensis	22	26	Diospyros sanza-minika	14	42
Cynometra ananta	4	66	Diospyros soubreana	33	44
Cynometra megalophylla	62	63	Diospyros vignei	12	39
Cyphostemma adenopodum*	67	32	Diospyros viridicans	26	35
Dacryodes klaineana	18	36	Diphasia angolensis	34	49
Dalbergia afzeliana	31	38	Discoclaoxylon hexandrum*	20	54
Dalbergia hostilis	31	38	Distemonanthus benthamianus	25	35
Dalbergia oblongifolia	25	41	Dracaena arborea	55	53
Dalbergia saxatilis*	25	53	Dracaena camerooniana	21	37
Dalbergiella welwitschii	41	47	Dracaena elliotii	19	81
Daniellia ogea	17	39	Dracaena fragrans	39	45
Daniellia thurifera	17	39	Dracaena perrottetii	46	48
Dasylepis brevipedicellata	28	58	Dracaena phrynioides	23	42
Deinbollia grandifolia	22	32	Dracaena scoparia	6	46
Deinbollia pinnata	61	26	Dracaena surculosa	33	43
Delpydora gracilis	3	61	Dregea crinita*	39	28

APPENDIX 3 (Cont.)

Species ordination values for plot identification

Species name	Axis 1	Axis 6
Drypetes aframensis	34	54
Drypetes aubrevillei	7	73
Drypetes aylmeri	12	36
Drypetes chevalieri	27	42
Drypetes floribunda	78	34
Drypetes gilgiana	35	49
Drypetes ivorensis	11	36
Drypetes leonensis	27	62
Drypetes parvifolia	84	39
Drypetes pellegrinii	24	44
Drypetes principum	24	66
Drypetes singroboensis	55	73
Ehretia trachyphylla	20	13
Elaeis guineensis	43	36
Elaeophorbia grandifolia	63	53
Enantia polycarpa	21	38
Entandrophragma angolense	27	41
Entandrophragma candollei	24	49
Entandrophragma cylindricum	28	47
Entandrophragma utile	28	53
Eremospatha macrocarpa	23	38
Eriocoelum pungens*	19	44
Eriocoelum racemosum	7	45
Erythrococca africana*	29	43
Erythrococca anomala*	37	34
Erythroxylum emarginatum	85	54
Euadenia eminens	23	41
Euclinia longiflora	24	38
Eugenia calophylloides	3	79
Eugenia obanensis	50	46
Eulophia guineensis	60	24
Farquharia elliptica	28	33
Ficus exasperata	54	43
Ficus sagittifolia	47	55
Flabellaria paniculata*	62	24
Flagellaria guineensis	37	35
Friesodielsia enghiana	11	71
Friesodielsia velutina	16	44
Funtumia africana	24	32
Funtumia elastica	32	38
Gaertnera cooperi	1	79
Garcinia afzelii	65	28
Garcinia epunctata	2	76
Garcinia gnetoides	10	49
Garcinia smeathmannii	13	43
Geophila afzelii	13	48
Geophila obvallata subsp. obvallata	21	42
Geophila repens	41	48
Glyphaea brevis	26	37
Gongronema latifolium*	19	42
Greenwayodendron oliveri	50	38
Grewia carpinifolia	80	69
Grewia hookerana	16	34
Grewia malacocarpa	21	39
Griffonia simplicifolia	41	40
Grossera vignei*	25	72
Guaduella macrostachys	4	59
Guarea cedrata	23	36
Guarea thompsonii	24	38
Guibourtia ehie	25	43
Hannoa klaineana	22	33
Heisteria parvifolia	15	28
Heritiera utilis	7	46
Hexalobus crispiflorus	21	53
Hildegardia barteri	62	42
Hippocratea africana	48	43
Hippocratea clematoides	38	67
Hippocratea dewildemaniana	42	46
Hippocratea ectypetala*	49	43
Hippocratea apocynoides (H. guineensis)*	23	36
Hippocratea indica*	38	26
Hippocratea macrophylla	23	35
Hippocratea pallens*	26	43
Hippocratea plumbea	36	46
Hippocratea rowlandii	48	43
Hippocratea welwitschii	39	43
Holarrhena floribunda	57	44
Holoptelea grandis	50	49
Homalium letestui*	24	19
Homalium stipulaceum	17	21
Hugonia planchonii	44	35
Hugonia platysepala	17	47
Hugonia rufipilis	19	0
Hunteria eburnea	32	41
Hunteria ghanensis	32	41
Hunteria umbellata	32	41
Hymenocoleus hirsutus	10	46
Hymenocoleus neurodictyon		
var. neurodictyon	15	47
var. orientalis	18	95
Hymenostegia afzelii	43	56
Hypolytrum poecilolepis	12	35
Hypselodelphys violacea	39	40
Icacina mannii	15	29
Illigera pentaphylla	19	46
Iodes africana*	24	24
Iodes liberica	22	35
Irvingia gabonensis	27	36
Isolona campanulata	21	30
Isolona cooperi	17	20
Isolona deightonii	15	11
Ixora laxiflora	19	49
Jasminum pauciflorum	43	54
Jaundea pinnata	23	33
Justicia extensa	49	36
Khaya anthotheca	32	55

APPENDIX 3 (Cont.)

Species ordination values for plot identification

Species name	Axis 1	Axis 6	Species name	Axis 1	Axis 6
Khaya grandifoliola	49	50	Markhamia lutea	55	17
Khaya ivorensis	19	27	Massularia acuminata	15	42
Klainedoxa gabonensis	19	35	Memecylon afzelii	42	36
Kolobopetalum leonense	17	24	Memecylon aylmeri	21	30
Landolphia calabarica	41	27	Memecylon blakeoides	0	87
Landolphia dulcis var. barteri	20	42	Memecylon cinnamomoides	56	37
Landolphia hirsuta	42	33	Memecylon guineense	13	59
Landolphia micrantha	11	61	Memecylon lateriflorum	10	50
Landolphia owariensis var. owariensis	25	38	Memecylon membranifolium	29	44
Landolphia togolana	64	46	Memecylon normandii	18	53
Lankesteria brevior	15	39	Microdesmis puberula	28	39
Lankesteria elegans	37	52	Microsorium punctatum	26	40
Lannea nigritana	65	28	Mikaniopsis tedliei	19	24
Lannea welwitschii	37	51	Millettia barteri	25	45
Lasianthus batangensis	5	53	Millettia chrysophylla	24	52
Lasiodiscus mannii var. chevalieri	56	66	Millettia rhodantha	26	41
Lecaniodiscus cupanioides	52	44	Millettia thonningii	100	70
Leea guineensis*	20	45	Millettia zechiana	30	36
Leptactina densiflora	46	32	Mischogyne elliotiana*	26	49
Leptaspis cochleata	29	43	Momordica cabraei*	34	37
Leptaulus daphnoides	21	46	Momordica multiflora*	45	33
Leptoderris brachyptera	27	41	Monanthotaxis laurentii		
Leptoderris fasciculata	29	44	(Popowia congensis)*	35	41
Leptonychia pubescens	43	48	Monanthotaxis whytei	17	32
Lomariopsis guineensis	20	36	Monodora myristica	28	35
Lomariopsis rossii	11	37	Monodora tenuifolia	31	31
Lonchocarpus cyanescens	50	27	Morinda longiflora	26	47
Lophira alata	6	68	Morinda lucida*	52	35
Lovoa trichilioides	14	48	Morinda morindoides	25	28
Lychnodiscus dananensis	29	52	Morus mesozygia	39	52
Lychnodiscus reticulatus	25	44	Motandra guineensis	41	46
Macaranga heterophylla*	24	33	Myrianthus arboreus	26	31
Macrosphyra longistyla	57	31	Myrianthus libericus	21	29
Maerua duchesnei	62	58	Napoleonaea vogelii		
Maesobotrya barteri var. sparsiflora	13	33	(large-leaved variety)	53	44
Majidea fosteri	42	55	(small-leaved variety)	22	37
Malacantha alnifolia	63	24	Neostenanthera gabonensis	11	36
Mallotus oppositifolius	50	50	Nephthytis afzelii	16	40
Mammea africana	16	39	Nesogordonia papaverifera	38	48
Mangenotia eburnea	51	46	Neuropeltis acuminata	26	41
Manilkara obovata	32	44	Neuropeltis prevosteoides	9	41
Manniophyton fulvum	19	30	Neuropeltis velutina	24	28
Manotes longiflora	16	27	Newbouldia laevis	45	36
Mansonia altissima	45	61	Newtonia duparquetiana	10	99
Mapania baldwinii	8	54	Nichallea soyauxii (Tarenna soyauxii)	18	38
Maranthes aubrevillei	6	49	Ochna staudtii	41	54
Maranthes glabra	16	27	Octoknema borealis*	18	46
Maranthes robusta	28	47	Octolobus spectabilis	46	69
Marantochloa congensis	22	44	Oeceoclades maculata	59	51
Marantochloa cuspidata	42	52	Oeceoclades saundersiana	37	50
Marantochloa leucantha*	28	43	Olax gambecola	32	39
Maranthochloa purpurea*	42	54	Olax subscorpioidea	58	29
Mareya micrantha	24	54	Olyra latifolia	51	40

APPENDIX 3 (Cont.)

Species ordination values for plot identification

Species name	Axis 1	Axis 6	Species name	Axis 1	Axis 6
Oncinotis glabrata	26	38	Platysepalum hirsutum	35	45
Oncinotis gracilis*	46	38	Pleiocarpa mutica	20	45
Oncinotis pontyi	31	42	Pneumatopteris afra (Cyclosorus afer)*	23	46
Ongokea gore	33	28	Pollia condensata*	27	53
Oplismenus hirtellus	60	31	Polyspatha paniculata	30	47
Ostryocarpus riparius	23	28	Polysphaeria macrophylla	27	44
Ouratea affinis	26	39	Premna quadrifolia	93	80
Ouratea amplectens	5	56	Pseuderanthemum tunicatum	29	58
Ouratea calophylla	12	64	Psilanthus mannii	48	54
Ouratea duparquetiana	7	43	Psychotria sp. aff. calceata	12	40
Ouratea reticulata	10	56	Psychotria calva	62	31
Ouratea sulcata	21	72	Psychotria ivorensis	13	33
Oxyanthus formosus	23	44	Psychotria liberica	16	33
Oxyanthus pallidus	25	50	Psychotria longituba	5	59
Oxyanthus racemosus	78	19	Psychotria mangenotii	16	47
Oxyanthus speciosus	41	38	Psychotria peduncularis	22	50
Oxyanthus unilocularis	39	50	Psychotria schweinfurthii	60	41
Pachypodanthium staudtii	19	28	Psychotria subobliqua	23	42
Pachystela brevipes	37	59	Pteris acanthoneura	33	39
Palisota barteri	27	54	Pteris atrovirens	33	43
Palisota hirsuta	33	34	Pteris burtonii	17	37
Pancovia bijuga	76	28	Pteris togoensis	32	44
Pancovia sessiliflora	45	52	Pterocarpus mildbraedii*	37	53
Pancovia turbinata	13	50	Pterygota bequaertii	26	54
Panda oleosa	26	43	Pterygota macrocarpa	41	53
Pararistolochia goldieana	56	31	Ptychopetalum anceps	14	46
Pararistolochia macrocarpa	30	65	Pycnanthus angolensis	24	36
Parinari excelsa	12	44	Pycnocoma cornuta	61	22
Parkia bicolor	22	34	Pycnocoma macrophylla	19	47
Parquetina nigrescens*	51	39	Pyrenacantha acuminata	44	43
Paullinia pinnata	53	32	Pyrenacantha cordicula	16	12
Pauridiantha sylvicola	13	62	Pyrenacantha vogeliana	19	20
Pavetta corymbosa*	56	35	Raphia hookeri	23	34
Pavetta genipifolia	58	40	Rauvolfia cumminsii	10	29
Pavetta ixorifolia	19	39	Rauvolfia vomitoria*	32	31
Pellaea doniana	73	22	Rhaphidophora africana	23	43
Penianthus zenkeri	18	30	Rhaphiostylis beninensis	26	36
Pentaclethra macrophylla	24	39	Rhaphiostylis cordifolia	11	42
Pentadesma butyracea	6	44	Rhaphiostylis ferruginea	19	38
Petersianthus macrocarpus	22	24	Rhaphiostylis preussii	26	36
Phyllocosmus africanus	23	29	Ricinodendron heudelotii	42	42
Phyllocosmus sessiliflorus	9	52	Rinorea brachypetala	25	34
Phymatodes scolopendria*	24	30	Rinorea breviracemosa	19	8
Piper guineense	23	34	Rinorea dentata	25	23
Piptadeniastrum africanum	25	36	Rinorea ilicifolia	30	45
Piptostigma fasciculatum	22	53	Rinorea kibbiensis	54	52
Piptostigma fugax	17	82	Rinorea oblongifolia	22	34
Pisonia aculeata*	35	48	Rinorea prasina	5	63
Placodiscus attenuatus	37	34	Rinorea subintegrifolia	18	22
Placodiscus bancoensis	22	0	Rinorea welwitschii	15	66
Placodiscus oblongifolius	7	38	Ritchiea capparoides*	14	18
Platycerium elephantotis (P. angolense)*	29	53	Ritchiea reflexa	91	7
Platycerium stemaria*	32	43	Rothmannia hispida	29	50

Species ordination values for plot identification

Species name	Axis 1	Axis 6	Species name	Axis 1	Axis 6
Rothmannia longiflora	38	35	Strephonema pseudocola	8	47
Rothmannia urcelliformis	49	59	Streptogyna crinita	45	36
Rothmannia whitfieldii	26	38	Strombosia glaucescens var. lucida	22	41
Rutidea dupuisii*	14	48	Strophanthus barteri	49	41
Rutidea parviflora*	19	41	Strophanthus gratus	29	35
Rytigynia canthioides*	43	55	Strophanthus hispidus	75	17
Saba thompsonii	46	49	Strophanthus preussii	49	41
Salacia alata var. alata	21	77	Strychnos aculeata	18	16
Salacia camerunensis	26	52	Strychnos afzelii	19	26
Salacia cerasifera	32	38	Strychnos asterantha	18	60
Salacia columna var. akeassii	8	51	Strychnos camptoneura	16	38
Salacia columna var. columna	5	55	Strychnos cuminodora	25	66
Salacia cornifolia	15	10	Strychnos floribunda	26	40
Salacia debilis	13	45	Strychnos icaja	10	66
Salacia elegans	50	43	Strychnos johnsonii	23	17
Salacia erecta	32	34	Strychnos malacoclados	17	47
Salacia sp. aff. gabunensis	25	59	Strychnos nigritana	67	22
Salacia howesii	22	34	Strychnos phaeotricha	17	22
Salacia ituriensis	32	46	Strychnos soubrensis	36	40
Salacia leptoclada	76	29	Strychnos splendens	36	40
Salacia nitida	13	21	Strychnos usambarensis (form i)	76	22
Salacia pallescens	46	55	Strychnos usambarensis (form ii)	21	45
Salacia pyriformis	32	31	Suregada occidentalis		
Salacia staudtiana var. leonensis	16	36	(Gelonium occidentale)*	34	65
Salacia togoica	70	31	Syzygium guineense subsp. occidentale	9	50
Salacia uregaensis var. aurantiaca	24	36	Tabernaemontana chippii	25	33
Santaloides afzelii*	23	68	Tabernaemontana crassa	15	25
Sarcophrynium brachystachys	29	42	Tabernaemontana glandulosa	23	35
Sarcophrynium prionogonium	16	96	Tabernaemontana pachysiphon	27	56
Scadoxus cinnabarinus	28	47	Talbotiella gentii	87	7
Scadoxus multiflorus subsp. longitubus	57	61	Tarenna bipindensis	24	32
Scaphopetalum amoenum	3	63	Tarenna eketensis	12	81
Scottellia klaineana	27	38	Tarenna fusco-flava	18	25
Scytopetalum tieghemii	7	50	Teclea verdoorniana	65	48
Secamone afzelii	62	33	Tectaria angelicifolia*	27	27
Sherbournia bignoniiflora*	44	41	Tectaria fernandensis*	27	69
Sloetiopsis usambarensis	61	66	Terminalia superba	33	40
Smeathmannia pubescens*	25	28	Tetracera potatoria	20	38
Smilax kraussiana*	50	20	Tetrapleura tetraptera	30	55
Solenangis scandens*	16	24	Tetrorchidium didymostemon*	32	37
Sorindeia warneckei	61	35	Thecacoris stenopetala	10	45
Soyauxia grandifolia	4	59	Thonningia sanguinea	38	34
Soyauxia velutina	10	48	Tieghemella heckelii	18	47
Sphenocentrum jollyanum	29	40	Tiliacora dinklagei	22	37
Spiropetalum heterophyllum	17	35	Tiliacora funifera	84	27
Spiropetalum solanderi	8	43	Tragia cf. mildbraediana*	39	56
Spondias mombin	64	10	Tragia spathulata*	57	8
Stachyanthus occidentalis	56	50	Treculia africana*	21	19
Stanfieldiella imperforata*	35	39	Tricalysia discolor*	22	49
Sterculia oblonga	28	40	Tricalysia oligoneura*	24	33
Sterculia rhinopetala	32	43	Tricalysia pallens	31	46
Sterculia tragacantha	48	42	Tricalysia reflexa	37	46
Stereospermum acuminatissimum	44	38	Tricalysia reticulata	17	64

APPENDIX 3 (Cont.)

Species ordination values for plot identification

Species name	Axis 1	Axis 6
Trichilia megalantha	49	55
Trichilia monadelpha	29	39
Trichilia prieuriana	36	42
Trichilia tessmannii	27	33
Trichomanes guineense	4	64
Trichomanes liberiense*	23	43
Trichoscypha arborea	9	25
Trichostachys aurea	9	46
Triclisia dictyophylla	25	50
Triclisia patens	35	44
Triclisia subcordata	63	43
Trilepisium madagascariense	40	47
Triplochiton scleroxylon	41	45
Turraea heterophylla	62	55
Turraeanthus africanus	21	37
Tylophora conspicua	59	29
Uapaca corbisieri	11	54
Uapaca guineensis	11	54
Urera keayi (Urera repens)*	37	52
Urera oblongifolia*	27	48
Usteria guineensis*	34	18
Uvaria afzelii*	19	43
Uvaria doeringii	63	35
Uvaria mocoli	29	42
Uvaria ovata subsp. ovata	95	20
Uvariastrum pierreanum	23	39
Uvariodendron angustifolium	36	54
Uvariodendron calophyllum	24	38
Uvariodendron occidentale	35	68
Uvariopsis globiflora	24	63
Ventilago africana	27	32
Vitex ferruginea	33	43
Vitex micrantha	33	43
Vitex rivularis	8	40
Xylia evansii	28	52
Xylopia acutiflora	15	30
Xylopia quintasii	19	35
Xylopia staudtii	12	62
Xylopia villosa	21	37
Zanthoxylum gilletii*	22	33
Zanthoxylum leprieurii	54	53

Species marked with an asterisk (*) are not mapped.
FWTA names of unmapped species not listed in Appendix 5 given in brackets.

APPENDIX 4

Worked example of plot identification from ordination scores of species present

Identification of plot B119 from mean ordination values* of recorded species

Species	Species ordination values Axis 1	Axis 6	Species	Species ordination values Axis 1	Axis 6
Acacia kamerunensis	36	44	Griffonia simplicifolia	41	40
Adenia lobata	53	47	Hildegardia barteri	62	42
Adiantum confine	–	–	Hugonia planchonii	44	35
Afzelia africana	56	34	Hunteria ghanensis	32	41
Artabotrys velutinus	62	48	Hymenostegia afzelii	43	56
Asparagus warneckei	84	45	Hypselodelphys violacea	39	40
Asplenium africanum	18	44	Maerua duchesnei	62	58
A. buettneri	–	–	Millettia thonningii	100	70
Baissea zygodioides	57	40	Nesogordonia papaverifera	38	48
Calycobolus heudelotii	52	46	Oeceoclades maculata	59	51
Calyptrochilum emarginatum	–	–	Pavetta genipifolia	58	40
Cassipourea congoensis (form i)	78	28	Pellaea doniana	73	22
Ceiba pentandra	49	38	Psychotria albicaulis	–	–
Ceropegia yorubana	–	–	Rinorea kibbiensis	54	52
Coffea ebracteolata	41	41	Salacia camerunensis	26	52
Cola millenii	53	44	Sansevieria liberica	–	–
Commelinidium gabunense	61	34	Secamone afzelii	62	33
Diospyros abyssinica	72	48	Sterculia tragacantha	48	42
Dracaena arborea	55	53	Strophanthus hispidus	75	17
D. surculosa	33	43	Strychnos usambarensis (form i)	76	22
Drypetes parvifolia	84	39	Trichilia prieuriana	36	42
Elaeophorbia grandifolia	63	53	Triplochiton scleroxylon	41	45
Erythroxylum emarginatum	85	54	Uvaria ovata	95	20
Ficus polita	–	–	Mean values+	56.4	42.3

* Species ordination values are listed in Appendix 3 (p. 336). Of the 47 species recorded in B119, ordination values are available for 40.

+ Reference to Fig. 2.3 shows that mean species values of 56.4 (axis 1) and 42.3 (axis 6) place plot B119 (see Appendix 2 for location) in the Southern Marginal type.

APPENDIX 5

Unmapped Ghanaian forest trees

List of Ghanaian forest trees species for which distribution maps are not included in the present book+

ACANTHACEAE
Adhatoda robusta C.B. Cl.

ANACARDIACEAE
Antrocaryon micraster A. Chev. & Guillaum.
Pseudospondias microcarpa (A. Rich.) Engl.
Rhus longipes Engl.*
Sorindeia zenkeri Engl.*
Trichoscypha albiflora Engl.
T. atropurpurea Engl.
T. baldwinii Keay
T. beguei Aubrév. & Pellegr.*
T. cavalliensis Aubrév. & Pellegr.
T. chevalieri Aubrév. & Pellegr.
T. oba Aubrév. & Pellegr.*

ANNONACEAE
Anonidium mannii (Oliv.) Engl. & Diels
Isolona hexaloba (Pierre) Engl. & Diels* (= I. pleurocarpa Diels subsp. nigerica Keay)
Mischogyne elliotiana (Engl. & Diels) R.E. Fries var. glabra Keay (= Uvariastrum elliotianum)
Monocyclanthus vignei Keay
Monodora brevipes Benth.
Neostenanthera hamata (Benth.) Exell
Polyceratocarpus parviflorus (Bak.f.) Ghesq.,
Uvariastrum insculptum (Engl. & Diels) Sprague
Xylopia aethiopica (Dunal) A. Rich.
Xylopia elliotii Engl. & Diels*
X. rubescens Oliv.*
X. sp. aff. villosa Chipp*

APOCYNACEAE
Picralima nitida (Stapf) Th. & H. Dur.
Pleiocarpa pycnantha (K. Schum.) Stapf.
Rauvolfia vomitoria Afzel.
Voacanga africana Stapf
V. thouarsii Roem. & Schult.*

ARALIACEAE
Cussonia bancoensis Aubrév. & Pellegr.
Polyscias fulva (Hiern) Harms
Schefflera barteri (Seem.) Harms*

BALANITACEAE (not ZYGOPHYLLACEAE)
Balanites wilsoniana Dawe & Sprague

BIGNONIACEAE
Kigelia africana (Lam.) Benth.
Markhamia tomentosa (Benth.) K. Schum. ex Engl.
Spathodea campanulata P. Beauv.

BORAGINACEAE
Cordia millenii Bak.
C. platythyrsa Bak.
C. senegalensis Juss.
C. vignei Hutch. & Dalz.
Ehretia cymosa Thonn.

CAESALPINIACEAE
Anthonotha vignei (Hoyle) J. Léonard
Berlinia occidentalis Keay
Crudia senegalensis Planch. ex Benth.
Cryptosepalum tetraphyllum (Hook.f.) Benth.*
Detarium senegalense J.F. Gmel.*
Didelotia idae Oldeman De Wit & Léonard*
D. unifoliolata J. Léonard*
Erythrophleum ivorense A. Chev.
E. suaveolens (Guill. & Perr.) Brenan (= E. guineense)
Gilbertiodendron bilineatum (Hutch. & Dalz.) J. Léonard
G. limba (Sc. Elliot) J. Léonard
G. preussii (Harms) J. Léonard
G. splendidum (A. Chev. ex Hutch. & Dalz.) J. Léonard
Hymenostegia aubrevillei Pellegr.

List of Ghanaian forest trees species for which distribution maps are not included in the present book+

H. gracilipes Hutch. & Dalz.
Mildbraediodendron excelsum Harms
Pellegriniodendron diphyllum J. Léonard
Stemonocoleus micranthus Harms*
Swartzia fistuloides Harms*

CAPPARACEAE
Crateva adansonii DC. (= C. religiosa)
Euadenia trifoliolata (Schum. & Thonn.) Oliv.

CELASTRACEAE
Elaeodendron buchananii (Loes.) Loes.* (= Cassine buchananii)
Maytenus undata (Thunb.) Blakelock (= Maytenus undatus)

CHRYSOBALANACEAE (not ROSACEAE)
Acioa barteri (Hook.f. ex Oliv.) Engl.
A. dinklagei Engl.
A. cf. hirsuta A. Chev. ex De Wild.*
Chrysobalanus icaco L. subsp. atacorensis (A. Chev.) F. White* (= C. atacorensis)
Licania elaeosperma (Mildbr.) Prance & White (= Afrolicania elaeosperma)
Magnistipula butayei De Wild. subsp. sargosii* (= Hirtella butayei)
M. zenkeri Engl.* (= Hirtella fleuryana).
Maranthes chrysophylla (Oliv.) Prance (= Parinari chrysophylla)
M. kerstingii (Engl.) Prance* (= Parinari kerstingii)
Parinari congensis F. Didr.

COMBRETACEAE
Pteleopsis habeensis Aubrév. ex Keay*
P. hylodendron Mildbr.

COMPOSITAE
Vernonia amygdalina Del.
V. colorata (Willd.) Drake;
V. conferta Benth.
V. richardiana (O. Ktze.) P.-Sermolli
V. titanophylla Brenan*

DICHAPETALACEAE
Tapura fischeri Engl.
T. ivorensis Bretel.*

EBENACEAE
Diospyros barteri Hiern

ERYTHROXYLACEAE
Erythroxylum mannii Oliv.

EUPHORBIACEAE
Alchornea cordifolia (Schum. & Thonn.) Muell. Arg.
Amanoa bracteosa Planch.*
A. strobilacea Muell. Arg.
Anthostema aubryanum Baill.
Bridelia atroviridis Muell. Arg.
B. grandis Pierre ex Hutch.
B. micrantha (Hochst.) Baill.
Cleistanthus polystachyus Hook.f. ex Planch.
Croton aubrevillei J. Léonard*
C. penduliflorus Hutch.
C. sylvaticus Hochst. ex Krauss*
C. zambesicus Muell. Arg.
Crotonogyne chevalieri (Beille) Keay
Discoclaoxylon hexandrum (Mull. Arg.) Pax & K. Hoffm. (= Claoxylon hexandrum)
Discoglypremna caloneura (Pax) Prain
Drypetes afzelii (Pax) Hutch.
Erythrococca africana (Baill.) Prain
Euphorbia deightonii Croizat
Grossera vignei Hoyle
Hymenocardia lyrata Tul.
Keayodendron bridelioides (Mildbr. ex Hutch. & Dalz.) Léandri (= Casearia bridelioides)
Macaranga barteri Muell. Arg.
M. heterophylla (Muell. Arg.) Muell. Arg.
M. heudelotii Baill.
M. hurifolia Beille
M. spinosa Muell. Arg.*
Mallotus subulatus Muell. Arg.
Margaritaria discoidea (Baill.) Webster (= Phyllanthus discoideus)
Martretia quadricornis Beille*
Necepsia afzelii Prain*
Phyllanthus profusus N.E. Br.
Protomegabaria macrophylla (Pax) Hutch.
P. stapfiana (Beille) Hutch.
Pseudagrostistachys africana (Muell. Arg.) Pax & K Hoffm.*
Sapium aubrevillei Léandri*
S. ellipticum (Hochst. ex Krauss) Pax
Spondianthus preussii Engl.
Suregada occidentalis (Hoyle) Croizat (= Gelonium occidentale)
Tetrorchidium didymostemon (Baill.) Pax & Hoffm.
Uapaca heudelotii Baill.
U. paludosa Aubrév. & Léandri

FLACOURTIACEAE (incl. SAMYDACEAE)
Caloncoba glauca (P. Beauv.) Gilg*
Casearia barteri Mast. (= C. dinklagei)
C. calodendron Gilg* (= C. inaequalis)
Dovyalis zenkeri Gilg* (incl. D. afzelii)
Flacourtia flavescens Willd.
Homalium africanum (Hook.f.) Benth. (incl. H. molle)
H. dewevrei De Wild. & Th. Dur. (= H. angustistipulatum)
H. letestui Pellegr.
H. longistylum Mast.* (= H. aylmeri)

List of Ghanaian forest trees species for which distribution maps are not included in the present book[+]

Lindackeria dentata (Oliv.) Gilg
Oncoba brachyanthera Oliv.
O. spinosa Forsk.
Ophiobotrys zenkeri Gilg

GRAMINEAE
Bambusa vulgaris Schrad. ex Mendel.

GUTTIFERAE (incl. HYPERICACEAE)
Garcinia kola Heckel
Harungana madagascariensis Lam. ex Poir.
Symphonia globulifera L.f.
Vismia guineensis (L.) Choisy

HUACEAE (not STYRACACEAE)
Afrostyrax lepidophyllus Mildbr.

HUMIRIACEAE
Sacoglottis gabonensis (Baill.) Urb.

ICACINACEAE
Desmostachys vogelii (Miers) Stapf*

LOGANIACEAE
Anthocleista djalonensis A. Chev.
A. liebrechtsiana De Wild. & Th. Dur.*
A. microphylla Wernh.*
A. nobilis G. Don
A. vogelii Planch.
Nuxia congesta R. Br. ex Fresen.
Strychnos spinosa Lam.

MELASTOMATACEAE
Dichaetanthera africana (Hook.f.) Jac.-Fél.* (= Sakersia africana)
Memecylon barteri Hook.f.*
M. fasciculare (Planch. ex Benth.) Naud.*
M. memecyloides (Benth.) Exell

MELIACEAE
Ekebergia senegalensis A. Juss.
Trichilia martineaui Aubrév. & Pellegr.*
T. ornithothera J. de Wilde*
Turraea ghanensis J.B. Hall*

MIMOSACEAE
Albizia coriaria Welw. ex Oliv.
Aubrevillea kerstingii (Harms) Pellegr.*
Cathormion altissimum (Hook.f.) Hutch. & Dandy
Newtonia aubrevillei (Pellegr.) Keay*
Samanea dinklagei (Harms) Keay

MORACEAE
Dorstenia kameruniana Engl. (= Craterogyne kameruniana)
Ficus anomani Hutch.
F. barteri Sprague
F. bubu Warb.*
F. camptoneura Mildbr.*
F. capensis Thunb.
F. congensis Engl.
F. conraui Warb.*
F. cf. cyathistipula Warb.*
F. elasticoides De Wild.*
F. elegans (Miq.) Miq.
F. eriobotryoides Kunth & Bouché
F. kamerunensis Warb. ex Mildbr. & Burret
P. leprieurii Miq.
F. lingua Warb.*
F. lutea Vahl (= F. vogelii)
F. lyrata Warb.*
F. macrosperma Warb. ex Mildbr. & Burret*
F. mucuso Welw. ex Ficalho
F. natalensis Hochst.
F. ottoniifolia (Miq.) Miq.
F. ovata Vahl
F. polita Vahl
F. populifolia Vahl*
F. praticola Mildbr. & Hutch.*
F. recurvata De Wild.* (= F. goliath)
F. tessellata Warb.*
F. thonningii Blume
F. umbellata Vahl
F. variifolia Warb.
F. vogeliana (Miq.) Miq.*
Musanga cecropioides F. Br.
Treculia africana Decne.

MYRISTICACEAE
Coelocaryon oxycarpum Stapf

MYRSINACEAE
Maesa lanceolata Forsk.*

MYRTACEAE
Eugenia coronata Schum. & Thonn.
E. kalbreyeri Engl. & v. Brehm.*
E. leonensis Engl. & v. Brehm.
Syzygium rowlandii Sprague*

OCHNACEAE
Ochna afzelii R. Br. ex Oliv.*
O. membranacea Oliv.
O. ovata F. Hoffm.
Ouratea calantha Gilg*
O. congesta (Oliv.) Engl. ex Gilg
O. flava (Schumach.) Hutch. & Dalz. ex Stapf

OLACACEAE
Ximenia americana L.

List of Ghanaian forest trees species for which distribution maps are not included in the present book+

OLEACEAE
Chionanthus africanus (Knobl.) Stearn (= Linociera africana)
C. mannii (Solereder) Stearn var. congestus (Bak.) Stearn* (= Linociera congesta)
C. mannii (Solereder) Stearn var. mannii* (= Linociera mannii)
Schrebera arborea A. Chev.

PALMAE
Raphia palma-pinus (Gaertn.) Hutch.

PANDANACEAE
Pandanus candelabrum P. Beauv.

PAPILIONACEAE
Craibia atlantica Dunn*
Dalbergia sp. nr. pachycarpa (De Wild. & Th. Dur.) Uhlr. ex De Wild.
D. setifera Hutch. & Dalz.
Erythrina addisoniae Hutch. & Dalz.
E. mildbraedii Harms
E. vogelii Hook.f.
Lonchocarpus sericeus (Poir.) H.B. & K.
Millettia griffoniana Baill.* (= Lonchocarpus griffonianus)
Pterocarpus mildbraedii Harms
P. santalinoides L'Her. ex DC.

PASSIFLORACEAE
Paropsia guineensis Oliv.*
Smeathmannia pubescens Soland. ex. R. Br.

PTERIDOPHYTA
Alsophila manniana (Hook.) Tryon (= Cyathea manniana)

RHAMNACEAE
Lasiodiscus fasciculiflorus Engl.
Maesopsis eminii Engl.
Zizyphus cf. espinosa Buettn.*

RHIZOPHORACEAE
Anisophyllea meniaudii Aubrév. & Pellegr.*
Cassipourea afzelii (Oliv.) Alston*
C. gummiflua Tul. (= C. glabra)

RUBIACEAE
Belonophora hypoglauca (Welw. ex Hiern) A. Chev.
Bertiera racemosa (G. Don) K. Schum.
Canthium arnoldianum (De Wild. & Th. Dur.) Hepper*
Canthium pobeguinii Hutch. & Dalz.
C. subcordatum DC.
Coffea spathicalyx K. Schum.
Craterispermum laurinum (Poir.) Benth.
Cuviera acutiflora DC.
C. macroura K. Schum.
C. nigrescens (Sc. Elliot ex Oliv.) Wernb.
C. subuliflora Benth.
Dictyandra arborescens Welw. ex Hook.f.
D. involucrata (Hook.f.) Hiern
Gaertnera paniculata Benth.
Gardenia imperialis K. Schum.
G. nitida Hook.
G. vogelii Hook.f. ex Planch.
Heinsia crinita (Afzel.) G. Tayl.
Morinda lucida Benth.
Nauclea pobeguinii (Pobéguin ex Pellegr.) Petit
Pauridiantha hirtella (Benth.) Bremek.
Pausinystalia lane-poolei (Hutch.) Hutch. ex Lane-Poole
Pavetta corymbosa (DC.) F.N. Williams
P. lasioclada (K. Krause) Mildbr. ex Bremek.*
P. mollis Afzel. ex Hiern
P. mollissima Hutch. & Dalz.
P. owariensis P. Beauv.
P. sp. aff. mannioides Hutch. & Dalz.*
Psychotria gabonica Hiern
P. guineensis Petit
P. succulenta (Hutch.) Petit*
Robynsia glabrata Hutch.
Rothmannia megalostigma (Wernh.) Keay
Schumanniophyton problematicum (A. Chev.) Aubrév.
Sericanthe chevalieri (K. Krause) Robbrecht var. coffeoides (A. Chev.) Robbrecht (= Tricalysia chevalieri)
S. toupetou (Aubrév. & Pellegr.) Robbrecht* (= Tricalysia toupetou)
Tarenna gracilis (Stapf) Keay
T. sp. aff. laurentii (De Wild.) Garcia*
T. sp. aff. pavettoides (Garr.) Sim*
T. vignei Hutch. & Dalz.
Tricalysia biafrana Hiern
T. bracteata Hiern
T. coriacea (Benth.) Hiern
T. discolor Brenan
T. elliotii (K. Schum.) Hutch. & Dalz.
T. macrophylla K. Schum.
T. oligoneura K. Schum.*
T. vignei Aubrév. & Pellegr.
Vangueriopsis spinosa (Schum. & Thonn.) Hepper

RUTACEAE
Afraegle paniculata (Schum. & Thonn.) Engl.
Araliopsis soyauxii Engl. (= A. tabouensis)
Citropsis articulata (Willd. ex Spreng.) Swingle & Kellerman
C. gabunensis (Engl.) Swingle & Kellerman*
Oricia suaveolens (Engl.) Verdoorn*
Vepris heterophylla (Engl.) R. Letouzey (= Teclea sudanica)
Zanthoxylum chevalieri Waterman (= Fagara pubescens)
Z. gilletii (De Wild.) Waterman (= F. macrophylla)
Z. lemairei (De Wild.) Waterman (= F. lemairei, F. parvifoliola)
Z. psammophilum (Aké Assi) Waterman (= F. psammophila)*
Z. rubescens Planch. ex Hook.f. (= F. rubescens)
Z. viride (A. Chev.) Waterman (= F. viridis)

List of Ghanaian forest trees species for which distribution maps are not included in the present book[+]

Z. xanthoxyloides (Lam.) Waterman (= F. xanthoxyloides)

SANTALACEAE (not OCTOKNEMATACEAE)
Okoubaka aubrevillei Pellegr. & Normand

SAPINDACEAE
Allophylus spicatus (Poir.) Radlk.
Chytranthus atroviolaceus Bak.f. ex Hutch. & Dalz.
C. cf. ellipticus Hutch. & Dalz.*
Deinbollia molliuscula Radlk.*
Eriocoelum kerstingii Gilg ex Engl.*
E. pungens Radlk. ex Engl.
Lecaniodiscus punctatus J. B. Hall*
Lepisanthes senegalensis (Poir.) Leenhouts (= Aphania senegalensis)
Placodiscus boya Aubrév. & Pellegr.
P. bracteosus J. B. Hall*
P. pseudostipularis Radlk.
Zanha golungensis Hiern

SAPOTACEAE
Aubregrinia taiensis (Aubrév. & Pellegr.) Heine
Bequaertiodendron magalismontanum (Sond.) Heine & J. H. Hemsley
Breviea leptosperma (Baehni) Heine
Chrysophyllum albidum G. Don
C. azaguieanum Miège
C. beguei Aubrév. & Pellegr.
C. delevoyi De Wild.
C. pentagonocarpum Engl. & Krause
Gluema ivorensis Aubrév. & Pellegr.
Ituridendron bequaertii De Wild.
Neolemonniera clitandrifolia (A. Chev.) Heine*
Omphalocarpum ahia A. Chev.
O. elatum Miers
O. procerum P. Beauv.
Pachystela msolo (Engl.) Engl.
Synsepalum aubrevillei (Pellegr.) Aubrév. & Pellegr.*
S. dulcificum (Schum. & Thonn.) Daniell
Vincentella passargei (Engl.) Aubrév.

SCYTOPETALACEAE
Rhaptopetalum beguei Mangenot*

SIMAROUBACEAE
Brucea guineensis G. Don
Nothospondias staudtii Engl.* (not in Anacardiaceae)
Pierreodendron kerstingii (Engl.) Little

STERCULIACEAE
Cola boxiana Brenan & Keay
C. digitata Mast.
C. flavo-velutina K. Schum.
Dombeya buettneri K. Schum.

TILIACEAE
Christiana africana DC.
Desplatsia dewevrei (De Wild. & Th. Dur.) Burret
Duboscia viridiflora (K. Schum.) Mildbr.
Grewia mollis Juss. (G. pubescens sensu FWTA)

ULMACEAE
Celtis africana Burm.f.
Trema orientalis (L.) Blume (= T. guineensis)

VERBENACEAE
Premna angolensis Gürke
P. hispida Benth.
Vitex grandifolia Gürke

VIOLACEAE
Allexis cauliflora (Oliv.) Pierre*
Decorsella paradoxa A. Chev. (= Gymnorinorea abidjanensis)
Rinorea angustifolia (Thou.) Baill. (= R. ardisiiflora)
R. convallariiflora M. Brandt

[+] Distribution maps are not given for the species on this list either in Chap. 10 or 11; these trees are either too rare or restricted to secondary or swamp forest.
Names in brackets are used in second edition Flora of West Tropical Africa (FWTA) but are not currently accepted by us.
* Species not recorded for Ghana in the second edition of FWTA.

APPENDIX 6

Family representation in the Ghanaian forest flora

Physiognomy, ecology and distribution of vascular plant families in the Ghana forest flora; numbers of species

Family	Life-form categories					Successional preference			Zonal preference				Distribution		Total species
	Not climbing or epiphytic			Climbers	Epiphytes	Primary forest	Secondary forest & paths etc.	Swamp & riverain forest	'Wet' forest	'Dry' forest	Upland forest	Other forest	W. Afr. endemic	Wider distribution	
	Trees mega- & meso- phanerophytes	Trees & shrubs: micro. & nano- phanerophytes	Herbs	woody & herbaceous	woody & herbaceous										
Pteridophytes	0	2	50	31	46	102	13	11	19	14	32	61	19	107	126
Acanthaceae	0	7	40	8	0	18	35	4	9	18	4	26	27	30	57
Agavaceae	3	7	3	0	0	11	2	0	4	2	0	7	8	5	13
Amaranthaceae	0	0	15	1	0	2	13	1	0	9	1	6	0	16	16
Amaryllidaceae	0	0	6	0	0	2	1	3	2	1	0	3	3	3	6
Anacardiaceae	8	8	0	1	0	15	1	1	9	3	2	3	13	4	17
Ancistrocladaceae	0	0	0	1	0	1	0	0	1	0	0	0	1	0	1
Annonaceae	21	13	0	20	0	49	5	1	14	5	3	33	30	25	55
Apocynaceae	9	10	0	44	0	55	6	2	12	12	3	36	26	37	63
Araceae	0	0	2	6	10	16	0	3	5	3	1	10	13	6	19
Araliaceae	2	0	0	0	1	3	0	0	1	0	1	1	1	2	3
Aristolochiaceae	0	0	0	5	0	5	0	0	1	3	0	1	4	1	5
Asclepiadaceae	0	0	0	17	0	8	9	0	0	12	0	5	6	11	17
Balanitaceae	1	0	0	0	0	1	0	0	0	0	0	1	0	1	1
Balanophoraceae	0	0	1	0	0	1	0	0	0	0	0	1	0	1	1
Balsaminaceae	0	0	2	0	0	0	0	2	0	0	1	1	2	0	2
Begoniaceae	0	0	5	1	2	7	0	1	3	2	2	1	4	4	8
Bignoniaceae	3	3	0	0	0	4	1	1	0	3	0	3	4	2	6
Bombacaceae	3	0	0	0	0	2	1	0	1	0	0	2	1	2	3
Boraginaceae	4	2	1	1	0	3	5	0	1	2	0	5	4	4	8
Bromeliaceae	0	0	1	0	0	0	1	0	0	0	0	1	0	1	1
Burmanniaceae	0	0	2	0	0	2	0	0	2	0	0	0	0	2	2
Burseraceae	2	1	0	0	0	3	0	0	0	1	0	2	1	2	3
Buxaceae	0	1	0	0	0	1	0	0	0	1	0	0	0	1	1
Cactaceae	0	0	0	0	1	1	0	0	0	0	0	1	0	1	1
Caesalpiniaceae	39	6	4	3	0	35	9	8	19	11	1	21	19	33	52
Capparaceae	1	4	3	5	0	8	5	0	0	5	0	8	5	8	13
Caryophyllaceae	0	0	1	0	0	0	0	1	0	0	1	0	0	1	1
Celastraceae	0	4	0	45	0	48	1	0	12	6	1	30	19	30	49

Physiognomy, ecology and distribution of vascular plant families in the Ghana forest flora; numbers of species

Family	Life-form categories					Successional preference			Zonal preference				Distribution		Total species
	Not climbing or epiphytic			Climbers	Epiphytes	Primary forest	Secondary forest & paths etc.	Swamp & riverain forest	'Wet' forest	'Dry' forest	Upland forest	Other forest	W. Afr. endemic	Wider distribution	
	Trees mega- & meso-phanerophytes	Trees & shrubs: micro. & nano-phanerophytes	Herbs	woody & herbaceous	woody & herbaceous										
Chrysobalanaceae	11	1	0	0	0	10	0	2	6	2	0	4	4	8	12
Combretaceae	5	1	0	21	0	20	5	2	6	5	1	15	10	17	27
Commelinaceae	0	0	25	2	0	12	12	3	2	4	1	20	8	19	27
Compositae	0	7	24	9	0	2	36	2	1	9	3	27	7	33	40
Connaraceae	0	1	0	18	0	18	1	0	4	2	0	13	12	7	19
Convolvulaceae	0	0	0	26	0	8	16	2	3	3	1	19	7	19	26
Crassulaceae	0	0	1	0	0	0	1	0	0	0	0	1	0	1	1
Cruciferae	0	0	1	0	0	0	0	1	0	0	0	1	0	1	1
Cucurbitaceae	0	0	0	26	0	4	21	1	2	6	1	17	11	15	26
Cyperaceae	0	0	36	1	0	9	22	6	10	3	3	21	6	31	37
Dichapetalaceae	2	2	0	11	0	15	0	0	5	2	0	8	10	5	15
Dilleniaceae	0	0	0	3	0	3	0	0	1	0	0	2	1	2	3
Dioscoreaceae	0	0	0	9	0	5	4	0	0	4	0	5	4	5	9
Ebenaceae	12	5	0	0	0	17	0	0	3	3	0	11	6	11	17
Erythroxylaceae	1	1	0	0	0	1	1	0	0	1	0	1	1	1	2
Euphorbiaceae	36	43	10	8	0	47	40	10	20	22	0	55	38	59	97
Flacourtiaceae	8	11	0	0	0	12	6	1	5	5	1	8	5	14	19
Flagellariaceae	0	0	0	1	0	0	1	0	0	0	0	1	0	1	1
Gentianaceae	0	0	1	0	0	0	0	1	1	0	0	0	0	1	1
Gesneriaceae	0	0	1	0	0	0	0	1	0	1	0	0	1	0	1
Gramineae	1	0	54	0	0	7	39	9	4	11	2	38	4	51	55
Guttiferae	10	1	0	0	0	8	2	1	4	1	0	6	4	7	11
Hernandiaceae	0	0	0	2	0	2	0	0	1	0	0	1	1	1	2
Huaceae	1	0	0	0	0	1	0	0	1	0	0	0	1	0	1
Humiriaceae	1	0	0	0	0	1	0	0	1	0	0	0	0	1	1
Icacinaceae	1	3	1	16	0	20	0	0	7	3	0	10	9	11	20
Irvingiaceae	2	0	0	0	0	2	0	0	0	0	0	2	0	2	2
Ixonanthaceae	2	0	0	0	0	2	0	0	1	0	0	1	0	2	2
Labiatae	0	1	12	0	0	2	8	3	0	1	2	10	3	10	13
Lauraceae	1	0	0	0	0	1	0	0	1	0	0	0	0	1	1
Lecythidaceae	2	0	0	0	0	2	0	0	0	0	0	2	1	1	2
Leeaceae	0	1	0	0	0	0	1	0	0	0	0	1	0	1	1
Lentibulariaceae	0	0	1	0	0	0	0	1	0	1	0	0	0	1	1
Liliaceae	0	0	7	2	0	8	1	0	1	4	0	4	3	6	9
Linaceae	0	0	0	4	0	4	0	0	2	0	0	2	1	3	4

APPENDIX 6 (Cont.)

Physiognomy, ecology and distribution of vascular plant families in the Ghana forest flora; numbers of species

Family	Life-form categories: Not climbing or epiphytic: Trees mega- & meso- phanerophytes	Trees & shrubs: micro. & nano- phanerophytes	Herbs	Climbers: woody & herb-aceous	Epiphytes: woody & herb-aceous	Successional preference: Primary forest	Secondary forest & paths etc.	Swamp & riverain forest	Zonal preference: 'Wet' forest	'Dry' forest	Upland forest	Other forest	Distribution: W. Afr. endemic	Wider distribution	Total species
Loganiaceae	4	4	1	20	1	25	3	2	6	4	3	17	11	19	30
Loranthaceae	0	0	0	0	8	8	0	0	2	2	0	4	3	5	8
Malpighiaceae	0	0	0	8	0	7	0	1	0	4	1	3	6	2	8
Malvaceae	0	10	8	0	0	0	16	2	0	2	1	15	3	15	18
Marantaceae	0	0	13	3	0	9	5	2	0	2	2	12	2	14	16
Medusandraceae	2	0	0	0	0	2	0	0	2	0	0	0	2	0	2
Melastomataceae	3	9	8	1	3	12	7	5	9	1	3	11	19	5	24
Meliaceae	18	2	0	1	0	19	2	0	2	3	0	16	5	16	21
Melianthaceae	1	0	0	0	0	1	0	0	0	0	0	1	0	1	1
Menispermaceae	0	2	0	18	0	15	7	0	2	4	0	14	10	10	20
Mimosaceae	18	2	2	6	0	20	5	3	2	2	0	24	5	23	28
Molluginaceae	0	0	1	0	0	0	1	0	0	0	0	1	0	1	1
Moraceae	23	5	1	0	18	38	5	4	7	4	0	36	13	34	47
Myristicaceae	2	0	0	1	0	3	0	0	2	0	0	1	2	1	3
Myrsinaceae	0	1	0	1	0	2	0	0	0	0	2	0	1	1	2
Myrtaceae	2	7	0	0	0	8	0	1	2	1	3	3	8	1	9
Nyctaginaceae	0	0	2	1	0	1	2	0	0	0	0	3	0	3	3
Ochnaceae	1	17	0	0	0	18	0	0	7	3	0	8	10	8	18
Olacaceae	4	6	0	0	0	9	1	0	1	2	0	7	2	8	10
Oleaceae	3	1	0	3	0	6	1	0	1	3	0	3	1	6	7
Onagraceae	0	0	3	0	0	0	0	3	1	0	1	1	1	2	3
Orchidaceae	0	0	24	4	120	146	1	1	21	9	29	89	30	118	148
Palmae	3	1	0	5	0	3	1	5	3	0	0	6	3	6	9
Pandaceae	1	1	0	0	0	2	0	0	0	0	0	2	0	2	2
Pandanaceae	0	1	0	0	0	0	0	1	0	0	0	1	1	0	1
Papilionaceae	11	11	10	51	0	37	38	8	11	27	1	44	25	58	83
Passifloraceae	1	2	0	7	0	3	7	0	1	2	0	7	4	6	10
Pedaliaceae	0	0	3	0	0	0	3	0	0	1	0	2	0	3	3
Periplocaceae	0	0	0	7	1	2	4	1	1	4	0	2	4	3	7
Piperaceae	0	2	5	1	2	4	4	0	0	0	2	6	2	6	8
Plumbaginaceae	0	1	0	0	0	0	1	0	0	0	0	1	0	1	1
Podostemaceae	0	0	4	0	0	0	0	4	1	0	0	3	3	1	4
Portulacaceae	0	0	3	0	0	0	3	0	0	1	0	2	1	2	3
Polygalaceae	0	1	0	3	0	4	0	0	1	1	1	1	2	2	4
Polygonaceae	0	0	0	1	0	1	0	0	0	0	0	1	0	1	1

Physiognomy, ecology and distribution of vascular plant families in the Ghana forest flora; numbers of species

Family	Life-form categories					Successional preference			Zonal preference				Distribution		Total species
	Not climbing or epiphytic			Climbers	Epiphytes	Primary forest	Secondary forest & paths etc.	Swamp & riverain forest	'Wet' forest	'Dry' forest	Upland forest	Other forest	W. Afr. endemic	Wider distribution	
	Trees mega- & meso-phanerophytes	Trees & shrubs: micro. & nano-phanerophytes	Herbs	woody & herbaceous	woody & herbaceous										
Ranunculaceae	0	0	0	1	0	0	1	0	0	1	0	0	0	1	1
Rhamnaceae	2	1	0	2	0	3	2	0	0	1	0	4	0	5	5
Rhizophoraceae	5	1	0	0	0	6	0	0	4	1	0	1	3	3	6
Rosaceae	0	0	0	1	0	0	1	0	0	0	1	0	0	1	1
Rubiaceae	21	101	38	57	1	149	57	12	46	25	13	134	107	111	218
Rutaceae	8	11	0	2	0	16	5	0	4	4	0	13	11	10	21
Santalaceae	1	0	0	0	0	1	0	0	0	0	0	1	0	1	1
Sapindaceae	15	18	0	4	0	31	6	0	8	6	0	23	24	13	37
Sapotaceae	24	8	0	1	0	31	0	3	10	8	1	15	14	20	34
Scrophulariaceae	0	0	7	0	0	0	6	1	1	0	1	5	1	6	7
Scytopetalaceae	1	1	0	0	0	2	0	0	2	0	0	0	2	0	2
Simaroubaceae	3	1	0	1	0	1	4	0	0	1	0	4	2	3	5
Smilacaceae	0	0	0	1	0	0	1	0	0	0	0	1	0	1	1
Solanaceae	0	6	7	2	0	1	14	0	0	5	0	10	2	13	15
Sterculiaceae	17	9	3	1	0	25	4	1	6	2	1	21	18	12	30
Thymelaeaceae	0	3	0	1	0	4	0	0	0	1	1	2	3	1	4
Tiliaceae	2	7	1	4	0	10	3	1	1	4	0	9	4	10	14
Triuridaceae	0	0	1	0	0	1	0	0	1	0	0	0	1	0	1
Ulmaceae	7	1	0	0	0	7	1	0	0	3	0	5	0	8	8
Umbelliferae	0	0	1	0	0	0	1	0	0	0	0	1	0	1	1
Urticaceae	0	2	5	5	0	0	10	2	1	1	1	9	6	6	12
Verbenaceae	3	5	2	12	0	7	15	0	2	4	2	14	6	16	22
Violaceae	1	12	1	0	0	13	1	0	4	2	1	7	6	8	14
Vitaceae	0	0	0	19	0	5	14	0	0	7	1	11	7	12	19
Zingiberaceae	0	0	18	0	0	3	15	0	6	1	1	10	14	4	18
Zygophyllaceae	0	0	1	0	0	0	1	0	0	0	0	1	0	1	1
Total	399	419	483	677	214	1 365	604	148	388	365	141	1 223	778	1 339	2 117

Life-form categories are defined in Chap. 5. There is overlap between the categories of climbers and epiphytes (see p. 47); the row total may therefore exceed the family total.
The column totals for trees and herbs are smaller than those in the tables of Chap. 4 and 5 because epiphytes have been excluded. Other small discrepancies have arisen because the tables were computed at different times; percentages are not, however, significantly affected.

APPENDIX 7

Classification of Ghanaian timbers

Timber classification used by Department of Forestry, Government of Ghana, 1977

Class & subclass	Species	Local trade name*	Standard trade name*
1a	Chlorophora spp.	Odum	Iroko
	Entandrophragma angolense	Edinam	Gedu nohor
	E. cylindricum	Sapele	Sapele
	E. utile	Utile	Utile
	Khaya anthotheca	Anthotheca	Anthotheca
	K. grandifoliola	Odupon	African mahogany
	K. ivorensis	Mahogany	African mahogany
	Tieghemella heckelii	Baku	Makore
	Nauclea diderrichii	Kusia	Opepe
1b	Pericopsis elata	Kokrodua	Afrormosia
	Lovoa trichilioides	African walnut	African walnut
	Terminalia ivorensis	Emire	Idigbo
	Triplochiton scleroxylon	Wawa	Obeche
1c	Heritiera utilis	Nyankom	Niangon
2a	Entandrophragma candollei	Candollei	Omu
	Guarea cedrata	Kwabohoro	Scented guarea
	G. thompsonii	Kwadwuma	Black guarea
	Lophira alata	Kaku	Ekki
	Piptadeniastrum africanum	Dahoma	Dahoma
2b	Antiaris toxicaria	Kyenkyen	Antiaris
	Guibourtia ehie	Shedua	
	Mansonia altissima	Oprono	Mansonia
	Mitragyna ciliata	Subaha	Abura
	Nesogordonia papaverifera	Danta	Danta
	Turraeanthus africanus	Avodire	Avodire
3	Afzelia spp.	Papao	Afzelia
	Albizia adianthifolia	Pampena	Albizzia
	A. ferruginea	Awiemfo-samina	Albizzia
	A. zygia	Okuro	Okuro
	Anopyxis klaineana	Kokoti	Kokoti
	Canarium schweinfurthii	Bediwunua	Canarium
	Celtis adolfi-friderici	Esabiri	Celtis
	C. mildbraedii	Esa	Celtis
	C. zenkeri	Esakoko	Celtis
	Cylicodiscus gabunensis	Denya	Okan

APPENDIX 7 (Cont.)

Timber classification used by Department of Forestry, Government of Ghana, 1977

Class & subclass	Species	Local trade name*	Standard trade name*
	Cynometra ananta	Ananta	Apome
	Diospyros sanza-minika	Kisibiri	Kusibiri
	Distemonanthus benthamianus	Bonsamdua	Ayan
	Erythrophleum ivorense	Potrodom	Missanda
	Holoptelea grandis	Onakwa	Mumuli
	Mammea africana	Bompegya	African apple
	Morus mesozygia	Wonton	Difou
	Petersianthus macrocarpus	Esia	Essia
	Pyonanthus angolensis	Otie	Ilomba
	Scottellia klaineana	Koroko	Odoko
	Sterculia rhinopetala	Wawabima	Sterculia brown
	Strombosia glaucescens	Afina	Afina
	Terminalia superba	Ofram	Afara

* Names from Fergusson (1970) and Taylor (1960).

References

A. References in main text and appendices

Adams, C. D. & A. H. G. Alston (1955). A list of the Gold Coast Pteridophyta. *Bull. Br. Mus. Nat. Hist. Botany* 1: 145–185.

Adeyoju, S. K. (1975). *Forestry and the Nigerian Economy.* Ibadan Univ. Press, Ibadan.

Addicott, F. T. (1978). Abscission strategies in the behaviour of tropical trees. In: *Tropical Trees as Living Systems* (Ed.: P. B. Tomlinson & M. H. Zimmermann) pp. 381–398.

Adu-Tutu, M., Y. Afful, K. Asante-Appiah, D. Lieberman, J. B. Hall & M. Elvin-Lewis (1979). Chewing stick usage in Southern Ghana. *Econ. Bot.* 33: 320–328.

Ahn, P. M. (1958). Regrowth and swamp vegetation in the western forest areas of Ghana. *J. W. Afr. Sci. Ass.* 4: 163–173.

Ahn, P. M. (1959). The savanna patches of Nzima, south-western Ghana. *J. W. Afr. Sci. Ass.* 5: 10–25.

Ahn, P. M. (1961). Soils of the Lower Tano Basin, south-western Ghana. *Ghana Min. Food Agric. Soil Land-use Surv. Mem.* 2: 1–266.

Ahn, P. M. (1970). *West African Soils.* Oxford Univ. Press, London.

Alexandre, D. Y. (1977). *Rôle disseminateur des éléphants en forêt de Taï (Côte d'Ivoire).* Off. Rech. Scient. tech. Outre-mer, Adiopodoumé (cyclostyled).

Alexandre, D. Y. (1978). Observations sur l'écologie de *Trema guineensis* en basse Côte d'Ivoire. *Cah. Off. Rech. Scient. tech. Outre-mer* 13: 261–266.

Ampofo, S. T. & G. W. Lawson (1972). Growth of seedlings of *Afrormosia elata* Harms in relation to light intensity. *J. appl. Ecol.* 9: 301–306.

Ashton, P. S. (1969). Speciation among tropical forest trees: some deductions in the light of recent evidence. *Biol. J. Linn. Soc.* 1: 155–196.

Ashton, P. S. (1976). An approach to the study of breeding systems, population structure and taxonomy of tropical trees. *Linn. Soc. Symp.* 2: 35–42.

Asibey, E. O. A. (1970). The present status of Wildlife conservation in Ghana. In: *Wildlife Conservation in West Africa* (Ed.: D. C. D. Happold) pp. 15–21. Intern. Union Cons. Nat. Nat. Resources, Morges.

Asibey, E. O. A. (1972). Ghana's progress. *Oryx* 11: 470–475.

Assémien, P., J. C. Filleron, L. Martin & J. P. Tastet (1970). Le Quaternaire de la zone littorale de Côte d'Ivoire. *Bull. ASEQUA* 25: 65–78.

Aubert, G. (1963). La classification des sols utilisée par les pédologues français en zone tropicale ou aride. *Colloque C.C.T.A. sur la classification des sols des régions intertropicales, Léopoldville.*

Aubréville, A. (1938). La forêt coloniale. *Annls Acad. Sci. colon.* 9: 1–245.

Aubréville, A. (1949). *Contribution à la paléohistoire des forêts de l'Afrique tropicale.* Soc. Ed. géogr. marit. colon., Paris.

Aubréville, A. (1950–51). La concept d'association dans la forêt dense équatoriale de la basse Côte d'Ivoire. *Mém. Soc. Bot. Fr.* 98: 145–158.

Aubréville, A. (1962). Savanisation tropical et glaciations quaternaires. *Adansonia* sér. 2, 2: 16–84.

Aubréville, A. (1970). La flore tropicale tertiaire du Sahara. *Adansonia* sér. 2, 10: 9–14.

Aubréville, A. (1971). La flore saharo-libyéenne tropicale d'après Paul Louvet. *Adansonia* sér. 2, 11: 583–592.

Axelrod, D. I. & P. H. Raven (1978). Late Cretaceous and Tertiary vegetation history of Africa. In: *Biogeography and Ecology of Southern Africa* (Ed.: M. J. A. Werger) pp. 77–130. Junk, The Hague.

Baidoe, J. F. (1970). The selection system as practised in Ghana. *Comm. For. Rev.* 49: 159–165.

Baidoe, J. F. (1976). Yield regulation in the high forest of Ghana. *Ghana For. J.* 2: 22–27.

Baker, H. G. (1955). Self-compatibility and establishment after 'long-distance' dispersal. *Evolution* 9: 347–349.

Baker, H. G. (1958). Studies in the reproductive biology of West African Rubiaceae. *J. W. Afr. Sci. Ass.* 4: 9–24.

Baker, H. G. (1962). Heterostyly in the Connaraceae with special reference to *Byrsocarpus coccineus. Bot. Gaz.* 123: 206–211.

Baker, H. G., B. J. Harris (1959). Bat-pollination of the silk-cotton tree, *Ceiba pentandra,* in Ghana. *J. W. Afr. Sci. Ass.* 5: 1–9.

Barkman, J. J., J. Moravec & S. Rauschert (1976). Code of phytosociological nomenclature. *Vegetatio* 32: 131–185.

Bates, D. A. (1962). Geology. In: *Agriculture and Land Use in Ghana* (Ed.: J. B. Wills) pp. 51–61. Oxford Univ. Press, London.

Baur, G. N. (1964). *The Ecological Basis of Rainforest Management.* Forestry Commission, New South Wales.

Bawa, K. S. & P. A. Opler (1975). Dioecism in tropical forest trees. *Evolution* 29: 167–179.

Beer, Tom & M. D. Swaine (1977). On the theory of explosively dispersed seeds. *New Phytol.* 78: 681–694.

Bernhard-Reversat, F., C. Huttel & G. Lemée (1978). Structure and functioning of evergreen rainforest ecosystems of the Ivory Coast. In: *Tropical Forest Ecosystems* pp. 557–579, UNESCO, Paris.

Blanchot, A., J. P. Dumas & A. Papon (1972). *Carte géologique de la partie méridionale de l'Afrique de l'Ouest.* Bur. Rech. géol. min., Paris.

Boulet, R., R. Fauck, B. Kaloga, J. C. Leprun, J. Riquier & J. Vieillefon (1971). Soils. In: *International Atlas of West Africa.* Org. Afr. Unity Scient. Tech. Res. Commission.

Brammer, H. (1962). Soils: In: *Agriculture and Land Use in Ghana* (Ed.: J. B. Wills) pp. 88–126. Oxford Univ. Press, London.

Britwum, S. P. K. (1976). Natural and artificial regeneration practices in the high forest of Ghana. *Ghana For. J.* 2: 45–49.

Brunnschweiler, R. O., A. N. Dempster & I. Kusnir (1972). The 'Voltaian' ('Infracambrien') of western Niger and the Cambrian Pan-African Orogeny. In: *African Geology* (Eds.: T. F. Dessauvagie & A. J. Whiteman) pp. 35–43. Dept. of Geology, Univ. Ibadan, Ibadan.

Burrows, F. M. (1975). Wind-borne seed and fruit movement. *New Phytol.* 75: 405–418.

Burton, R. & V. L. Cameron (1883). *To the Gold Coast for Gold.* London.

Burtt Davy, J. & A. C. Hoyle (1937). *Draft of First Descriptive List of the Gold Coast.* Imp. For. Inst., Oxford.

Cagan, R. H. (1973). Chemostimulatory protein: a new type of taste stimulus. *Science* 181: 32–35.

Chakravarty, H. L. (1968). Cucurbitaceae of Ghana. *Bull. Inst. fond. Afr. noire* 30 sér. A: 400–468.

Chipp, T. F. (1913). *A list of trees, shrubs and climbers of the Gold Coast, Ashanti and the Northern Territories.* London.

Chipp, T. F. (1914). *A list of the herbaceous plants and undershrubs of the Gold Coast, Ashanti and the Northern Territories.* London.

Chipp, T. F. (1922). *The Forest Officer's Handbook of the Gold Coast, Ashanti and the Northern Territories.* Crown Agents, London.

Chipp, T. F. (1927). The Gold Coast forest. A study in synecology. *Oxf. For. Mem.* 7: 1–94.

Connell, J. H. (1978). Diversity in tropical rainforests and coral reefs. *Science* 199: 1302–1310.

Cooper, W. G. G. (1958). The bauxite deposits of the Gold Coast. *Bull. geol. Surv. Gold Cst.* 7: 1–33.

C.S.A. (1956). *Phytogeography.* Cons. Scient. Afr. Sud Sahara, Comm. Coop. tech. Afr. Sud Sahara. London.

Curtis, J. T. (1959). *The Vegetation of Wisconsin.* Univ. Wisconsin Press, Madison.

Dawkins, H. C. (1959). Volume increment in natural Tropical High Forest and limitations on its improvement. *Emp. For. Rev.* 38: 175–180.

De Rham, P. (1970). L'azote dans quelques forêts, savanes et terrains de cultures d'Afrique tropicale humide (Côte d'Ivoire). *Veröff. Geobot. Inst. ETH, Stift. Rübel, Zürich.* 45: 1–124.

D'Hoore, J. L. (1964). *Soil Map of Africa, Scale 1 to 5 000 000, Explanatory Monograph.* Comm. Coop. Tech. Afr. Sud Sahara. Lagos.

Dickson, K. B. (1971). *A Historical Geography of Ghana.* Cambridge Univ. Press, London.

Emberger, L., G. Mangenot & J. Miège (1950). Existence d'associations végétales typiques dans la forêt dense équatoriale. *C.r. hebd. Séanc. Acad. Sci., Paris* 231: 640–642.

Emiliani, C. (1966). Palaeotemperature analysis of Caribbean Cores P6304–8 and P6304–9 and a generalized temperature curve for the last 425 000 years. *J. Geol.* 74: 109–126.

Faegri, K. & L. van der Pijl (1971). *Principles of Pollination Ecology* (2nd edn). Pergamon Press, London.

Fedorov, A. A. (1966). The structure of the tropical rainforest and speciation in the humid tropics. *J. Ecol.* 54: 1–11.

Fergusson, K. (1960). *Ghana Hardwoods.* A. S. Worth, London.

Foggie, A. (1947). Some ecological observations on a tropical forest type in the Gold Coast. *J. Ecol.* 34: 88–106.

Foggie, A. (1957). Forestry problems in the closed forest zone of Ghana. *J. W. Afr. Sci. Ass.* 3: 131–147.

Foggie, A. & B. Piasecki (1962). Timber, fuel and minor forest produce. In: *Agriculture and Land Use in Ghana* (Ed.: J. B. Wills) pp. 236–251. Oxf. Univ. Press, London.

Frédoux, A. (1977). Étude palynologique de quelques sediments du quaternaire ivoirien. *Suppl. Bull. INQUA* 50: 181–186.

Frédoux, A. & J. P. Tastet (1976). Apport de la palynologie à la connaissance paléogéographique du littoral ivoirien entre 8 000 et 12 000 ans B.P. *Proc. 7th Afr. Micropaleontol. Colloq., Ile-Ife.*

Gaisie, S. K. & K. T. de Graft-Johnson (1976). *The Population of Ghana.* Comm. Intern. Coord. Nat. Res. Demogr. Paris.

Gan, Y. Y., F. W. Robertson, P. S. Ashton, E. Soepadmo & D. W. Lee (1977). Genetic variation in wild populations of rainforest trees. *Nature* 269: 323–325.

Gibbs, D. G. & D. Leston (1970). Insect phenology in a forest farm locality in West Africa. *J. appl. Ecol.* 7: 519–548.

Givnish, T. J. (1978). On the adaptive significance of compound leaves, with particular reference to tropical trees. In: *Tropical Trees as Living Systems* (Ed.: P. B. Tomlinson & M. H. Zimmermann) pp. 351–380. Cambridge Univ. Press, Cambridge.

Givnish, T. J. & G. J. Vermeij (1976). Sizes and shapes of liane leaves. *Amer. Natur.* 110: 743–778.

Goudet, J. P. (1975). Plantations experimentales d'espèces papetières en Côte d'Ivoire. *Bois Forêts Trop.* 159: 3–27.

Greenland, D. J. & J. M. L. Kowal (1960). Nutrient content of moist tropical forest of Ghana. *Pl. Soil* 12: 154–174.

Greig-Smith, P. (1964). *Quantitative Plant Ecology.* Butterworths, London.

Greig-Smith, P. (1969). Analysis of vegetation data: the user viewpoint. *Int. Symp. Statist. Ecol., New Haven* 3: 149–166.

Guillaumet, J. L. (1967). *Recherches sur la végétation et la flora de la région du Bas-Cavally (Côte d'Ivoire).* Off. Rech. Scient. tech. Outre-mer, Paris.

Guillaumet, J.-L. & E. Adjanohoun (1971). La végétation de la Côte d'Ivoire. *Mém. Off. Rech. Scient. tech. Outre-mer* 50: 157–263.
Hall, J. B. (1966). Some new plant records from Kwahu. *Ghana J. Sci.* 6: 149–152.
Hall, J. B. (1970). *Pteris vittata* Linn.; a gold-mine fern in Ghana. *Niger. Fld* 35: 29–37.
Hall, J. B. (1972). A new species of *Virectaria* (Rubiaceae) from Ghana. *Kew Bull.* 26: 569–571.
Hall, J. B. (1974a). African orchids XXXIV. *Kew Bull.* 29: 427–429.
Hall, J. B. (1974b). A new species of *Bowringia* (Leguminosae) from Ghana. *Kew Bull.* 29: 497–498.
Hall, J. B. (1976). A new species of *Turraea* (Meliaceae) from Ghana. *Adansonia* sér. 2, 15: 505–508.
Hall, J. B. (1980). Five new species of flowering plants from West Africa. *Bull. Jard. Bot Nat. Belg.* 50: 249–266.
Hall, J. B. & M. Bigger (1974). New and interesting records of pteridophytes from Ghana. *Bull. Inst. fond. Afr. noire* 36 sér. A: 880–885.
Hall, J. B. & J. C. Bowling (1969). Field key to the epiphytic orchids of Ghana based on characters of shoots and infructescences. *Adansonia* sér. 2, 9: 139–155.
Hall, J. B., J. Brookman-Amissah, D. Leston & R. Dodoo (1973). *Conservation and exploitation of Atewa Range Forest Reserve.* Council Scient. Indust. Res., Accra (cyclostyled).
Hall, J. B., R. Kumar & A. A. Enti (1972). The obnoxious weed *Eupatorium odoratum* (Compositae) in Ghana. *Ghana J. Agr. Sci.* 5: 75–78.
Hall, J. B. & A. J. M. Leeuwenberg (1979). A new species of *Hunteria* (Apocynaceae) from Ghana. *Bull. Jard. Bot. Nat. Belg.* 49: 421–425.
Hall, J. B. & J. M. Lock (1975). Use of vegetative characters in the identification of species of *Salacia* (Celastraceae). *Boissiera* 24: 331–338.
Hall, J. B., D. U. U. Okali & J. M. Lock (1975). Night illumination and flowering in an indigenous West African forest-tree, *Hildegardia barteri* (Mast.) Kosterm. *Bull. Inst. fond. Afr. noire* 37 sér. A: 282–287.
Hall, J. B. & M. D. Swaine (1976). Classification and ecology of closed-canopy forest in Ghana. *J. Ecol.* 64: 913–951.
Hall, J. B. & M. D. Swaine (1981). Seed stocks in Ghanaian forest soils. *Biotropica* (in press).
Hall, J. B., M. R. Talbot & M. D. Swaine (1978). An early holocene leaf flora from Lake Bosumtwi, Ghana. *Palaeogeogr., Palaeoclimatol., Palaeoecol.* 24: 247–261.
Hall, J. B. & P. B. Tomlinson (1973). Natural callus-like tissue produces vegetative buds on the leaves of the monocotyledon *Haemanthus cinnabarinus* Decaisne (Amaryllidaceae). *Nature* 243: 252–255.
Hall, John B. (1977). Forest-types in Nigeria: an analysis of pre-exploitation forest enumeration data. *J. Ecol.* 65: 187–199.
Hallé, F. & R. A. A. Oldeman (1970). *Essai sur l'architecture et la dynamique de croissance des arbres tropicaux.* Masson, Paris.
Hallé, F., R. A. A. Oldeman & P. B. Tomlinson (1978). *Tropical Trees and Forests.* Springer-Verlag, Berlin.
Hamilton, A. C. (1974). The history of the vegetation. In: *East African Vegetation* (Ed.: E. M. Lind & M. E. S. Morrison) pp. 188–209. Longman, London.
Hamilton, A. C. (1976). The significance of patterns of distribution shown by forest plants and animals in Tropical Africa for the reconstruction of Upper Pleistocene palaeoenvironments: a review. *Palaeoecol. Afr.* 9: 63–97.
Hedberg, I. (1979). Possibilities and needs for conservation of plant species and vegetation in Africa. In: *Systematic Botany, Plant Utilization and Biosphere Conservation* (Ed.: I. Hedberg) pp. 83–104. Almquist & Wiksell International, Stockholm.
Henwood, K. (1973). A structural model of forces in buttressed tropical rain forest trees. *Biotropica* 5: 83–93.
Hepper, F. N. (1976). *The West African Herbaria of Isert and Thonning.* Bentham-Moxon Trust, Kew.
Hill, M. O. (1973). Reciprocal averaging: an eigenvector method of ordination. *J. Ecol.* 61: 237–249.
Hill, Polly (1963). *Migrant Cocoa Farmers of Southern Ghana.* Cambridge Univ. Press, London.
Hutchinson, J. & J. M. Dalziel (1954–72). *Flora of West Tropical Africa* (2nd edn. by R. W. J. Keay & F. N. Hepper). Crown Agents, London.
Irvine, F. R. (1961). *Woody plants of Ghana.* Oxford Univ. Press, London.
Jack, W. H. (1959). *The structure of the Celtis-Triplochiton Forest Association in Ghana and its effect on management.* Ph.D. thesis, Univ. Edinburgh.
Janzen, D. H. (1975). *Ecology of Plants in the Tropics.* Edward Arnold, London.
Jeník, J. (1969). The life-form of *Scaphopetalum* A. Chev. *Preslia* 41: 109–112.
Jeník, J. & A. A. Enti (1968). Discontinuous distribution of *Allexis cauliflora* (Oliv.) L. Pierre in Equatorial Africa. *Novit. bot. Inst. bot. Univ. Carol. Prag.* 1968: 67–71.
Jeník, J. & J. B. Hall (1966). The ecological effects of the harmattan wind in Djebobo Massif, Togo. *J. Ecol.* 54: 767–779.
Jeník, J. & J. B. Hall (1976). Plant communities of the Accra Plains, Ghana. *Folia Geobot. Phytotax.* 11: 163–212.
Johansson, D. (1974). Ecology of vascular epiphytes in West African rainforest. *Acta Phytogeographica Suecica* 59: 1–129.
John, D. M. (1973). Accumulation and decay of litter and net production of forest in tropical West Africa. *Oikos* 24: 430–435.
Jones, E. W. (1955–56). Ecological studies on the rain forest of Southern Nigeria. 4. The plateau forest of the Okomu Forest Reserve. *J. Ecol.* 43: 564–594; 44: 83–117.
Jones, N. (1973). A note of some of the biological and technical problems affecting the harvesting of *Triplochiton scleroxylon* K. Schum. *Proc. Int. Union For. Res. Org. Symp., Bergen* 2.
Kaur, A., C. O. Ha, K. Jong, V. E. Sands, H. T. Chan, E. Soepadmo & P. S. Ashton (1978). Apomixis may be widespread among trees of the climax rain forest. *Nature* 271: 440–442.
Keay, R. W. J. (1957). Wind-dispersed species in a Nigerian forest. *J. Ecol.* 45: 471–478.
Keay, R. W. J. (1960). Seeds in forest soils. *Niger. For. Inf. Bull.* n.s. 4: 1–12.
Kio, P. R. O. (1979). Management strategies in the natural tropical high forest. *Forest Ecol. Manage.* 2: 207–220.

Kira, T. & H. Ogawa (1971). Assessment of primary production in tropical and equatorial forests. In: *Productivity of Forest Ecosystems* (Ed.: P. Duvigneaud) pp. 309–321. UNESCO, Paris.
Lawson, G. W., K. O. Armstrong-Mensah & J. B. Hall (1970). A catena in tropical moist semi-deciduous forest near Kade, Ghana. *J. Ecol.* 58: 371–398.
Lawton, R. M. (1978). The management and regeneration of some Nigerian high forest ecosystems. In: *Tropical Forest Ecosystems* pp. 580–588. UNESCO, Paris.
Lebrun, J. (1947). *La végétation de la plaine alluviale au sud du Lac Édouard.* Inst. Parcs natn. Congo Belge, Bruxelles.
Lebrun, J. & G. Gilbert (1954). Une classification écologique des forêts du Congo. *Publs Inst. natn. Étude agron. Congo belge série Scient.* 63: 1–89.
Lebrun, J.-P. (1976). Richesses spécifiques de la flore vasculaire des divers pays ou régions d'Afrique. *Candollea* 31: 11–15.
Lebrun, J.-P. & A. L. Stork (1977). *Index 1935–1976 des cartes de répartition des plantes vasculaires d'Afrique.* Inst. Élev. Méd. Vet. Pays trop., Paris.
Leenhouts, P. W. (1973). A revision of *Crossonephelis* (Sapindaceae). *Blumea* 21: 91–103.
Leeuwenberg, A. J. M. (1961). The Loganiaceae of Africa 2. A revision of *Mostuea* Didr. *Meded. Landb. Wageningen* 61–4: 1–31.
Léonard, J. (1965). Contribution à la subdivision phytogéographique de la région guinéo-congolaise d'après la répartition géographique d'Euphorbiacées d'Afrique tropicale. *Webbia* 19: 627–649.
Letouzey, R. (1968). *Étude Phytogéographique du Cameroun.* Éditions Paul Lechevalier, Paris.
Lieberman, D., J. B. Hall, M. D. Swaine & M. Lieberman (1979). Seed dispersal by baboons in the Shai Hills, Ghana. *Ecology* 60: 65–75.
Livingstone, D. A. (1975). Late quaternary climatic change in Africa. *Ann. Rev. Ecol. Syst.* 6: 249–280.
Lock, J. M. & J. B. Hall (1973). Three new species of *Aframomum* K. Schum. (Zingiberaceae) from Ghana. *Kew Bull.* 28: 441–449.
Lock, J. M. & J. B. Hall (1981). Floral biology of *Mallotus oppositifolius* (Euphorbiaceae). *Biotropica* (in press).
Logan, W. E. M. (1946). The Gold Coast Forestry Department, 1908–1945. *Emp. For. Rev.* 25: 52–59.
Logan, W. E. M. (1947). Timber exploitation in the Gold Coast. *Emp. For. Rev.* 26: 30–53.
Longman, K. A. (1969). The dormancy and survival of plants in the humid tropics *Symp. Soc. exp. Biol.* 23: 471–488.
Longman, K. A. & J. Jeník (1974). *Tropical forest and its environment.* Longman, London.
Lorougnon, G. (1973). Le vecteur pollinique chez les *Mapania* et les *Hypolytrum,* Cyperacées de sous-bois des forêts tropicales ombrophiles. *Bull. Jard. Bot. Nat. Belg.* 43: 33–36.
Mackenzie, J. A. (1959). Phenology of *Triplochiton. Dep. For. Res. Nigeria Tech. Note* 1.
Mc Key, D. (1975). The ecology of coevolved seed dispersal systems. In: *Coevolution of Animals and Plants* (Ed.: L. E. Gilbert & P. H. Raven) pp. 159–190. Univ. Texas Press, Austin.
Mc Key, D. (1978). Soils vegetation and seed-eating by Black Colobus monkeys. In: *The Ecology of Arboreal Folivores* (Ed.: G. G. Montgomery) pp. 423–437. Smithsonian Institution Press, Washington DC.
Mc Key, D., P. G. Waterman, C. N. Mbi, J. S. Gartlan & T. T. Struhsaker (1978). Phenolic content of vegetation in two African rain forests: ecological implications. *Science* 202: 61–64.
Mc Mahon, T. A. (1975). The mechanical design of trees. *Sci. Amer.* 233: 93–102.
Mangenot, G. (1950). Essai sur les forêts denses de la Côte d'Ivoire. *Bull. Soc. bot. Fr.* 97: 159–162.
Mangenot, G. (1955). Étude sur les forêts des plaines et plateaux de la Côte d'Ivoire. *Étud. éburn. Inst. fr. Afr. noire* 4: 5–61.
Mangenot, G. (1956). Les recherches sur la végétation dans les régions tropicales humides de l'Afrique occidentale. In: Study of Tropical Vegetation – *Proceedings of the Kandy Symposium* pp. 115–126. UNESCO, Paris.
Markham, R. H. & A. J. Babbedge (1979). Soil and vegetation catenas on the forest-savanna boundary in Ghana. *Biotropica* 11: 224–234.
May, R. M. (1976). Patterns in multi-species communities. In: *Theoretical Ecology – Principles and Applications* (Ed.: R. M. May) pp. 142–162. Blackwell, Oxford.
Mensah, K. O. A. (1966). An analysis of logging damage in compartment 4 of Tano Suhien Forest Reserve, Ghana. *Ghana J. Sci.* 6: 63–69.
Monk, C. D. (1966). An ecological significance of evergreenness. *Ecology* 47: 504–505.
Monod, T. (1957). Les grandes divisions chorologiques de l'Afrique. *Publs Cons. Scient. Afr. Sud Sahara Comm. Coop. tech. Afr. Sud Sahara* 24.
Mooney, J. W. C. (1961). Classification of the vegetation of the high forest zone of Ghana. In: *Tropical Soils and Vegetation – Proceedings of the Abidjan Symposium* pp. 85–86. UNESCO Paris.
Moreau, R. E. (1966). *The Bird Faunas of Africa and its Islands.* Academic Press, New York.
Morton, J. K. (1962). The upland floras of West Africa – their composition, distribution and significance in relation to climate changes. In: *Comptes rendus IVe Réunion AETFAT* (Ed.: A. Fernandes) pp. 391–409. Junta Investig. Ultramar, Lisbon.
Morton, J. K. (1963). Contributions towards a flora of Ghana. *Ghana J. Sci.* 3: 44–46.
Morton, J. K. (1964). Contributions towards a flora of Ghana. II. *Ghana J. Sci.* 4: 86–93.
Morton, J. K. (1966). Contributions towards a flora of Ghana III. *Ghana J. Sci.* 6: 1–8.
Obaton, M. (1960). Les lianes ligneuses à structure anormale des forêts denses d'Afrique occidentale: *Annls Sci. nat.* sér. 12 1: 1–219.
Okali, D. U. U., J. B. Hall & G. W. Lawson (1973). Root distribution under a thicket clump on the Accra Plains, Ghana. *J. Ecol.* 58: 89–101.
Organisation of African Unity (1970). *African Convention on the Conservation of Nature and Natural Resources.* OAU, Addis Ababa.
Osafo, E. D. (1969). Obstacles to natural regeneration in the high forest of Ghana. *For. Prod. Res. Inst. Ghana Tech. Newsl.* 3.
Papadakis, J. (1966). *Crop Ecologic Survey in West Africa.*

Food Agric. Org., Rome.
Pastouret, L., H. Chamley, G. Delibrias, J. C. Duplessy & J. Thiede (1978). Late quaternary climatic changes in Western Tropical Africa deduced from deep-sea sedimentation off the Niger Delta. *Oceanologica Acta* 1: 217–232.
Peltre, P. (1977). Le 'V' Baoulé (Côte d'Ivoire centrale) Heritage géomorphologique et paléoclimatique dans le tracé du contact forêt-savane. *Trav. et Docu. de l'ORSTOM* 80: 1–198.
Pennington, T. D. & B. T. Styles (1975). A generic monograph of the Meliaceae. *Blumea* 22: 419–540.
Perring, F. H. & S. M. Walters (1962). *Atlas of the British Flora*. Bot. Soc. Br. Is, London.
Pijl, L. van der (1972). *Principles of Dispersal in Higher Plants* (2nd edn.). Springer Verlag, Berlin.
Poore, M. E. D. (1955). The use of phytosociological methods in ecological investigations. I – The Braun-Blanquet system. *J. Ecol.* 43: 226–244.
Portères, R. (1950). Problèmes sur la végétation de la Basse Côte d'Ivoire. *Bull. Soc. bot. Fr.* 97: 153–156.
Posnansky, M. & R. McIntosh (1976). New radiocarbon dates for northern and western Africa. *J. Afr. Hist.* 17: 161–195.
Raunkiaer, C. (1934). *The Life-forms of Plants and Statistical Plant Geography*. Oxford Univ. Press, London.
Raven, P. H. & D. I. Axelrod (1974). Angiosperm biogeography and past continental movements. *Ann. Missouri Bot. Gard.* 61: 539–673.
Rees, A. R. (1964). Some observations on the flowering behaviour of *Coffea rupestris* in southern Nigeria. *J. Ecol.* 52: 1–7.
Richards, P. W. (1973). Africa, the 'odd man out'. In: *Tropical Forest Ecosystems in Africa and South America: a Comparative Review* (Eds.: B. J. Meggers, E. S. Ayensu & W. D. Duckworth) pp. 21–26. Smithsonian Institution, Washington.
Robbrecht, E. (1975). *Hymenocoleus*, a new genus of Psychotrieae (Rubiaceae) from Tropical Africa. *Bull. Jard. Bot. Nat. Belg.* 45: 273–300.
Robbrecht, E. (1977). The tropical African genus *Hymenocoleus* (Rubiaceae-Psychotriae): additions. *Bull. Jard. Bot. Nat. Belg.* 47: 3–29.
Rollet, B. (1974). *L'architecture des forêts denses humides sempervirentes de plaine*. Centre Tech. For. Trop., Nogent-sur-Marne.
Shackleton, N. J. & N. D. Opdyke (1977). Oxygen isotope and palaeomagnetic evidence for early Northern Hemisphere glaciation. *Nature* 270: 216–219.
Schnell, R. (1952). Contribution à une étude phytosociologique et phytogéographique de l'Afrique occidentale: les groupements et les unités géobotaniques de la région guinéenne. *Mém. Inst. fr. Afr. noire* 18: 45–234.
Schnell, R. (1976–77). *Flore et Végétation de l'Afrique Tropicale*. 2 vols. Gauthier-Villars, Paris.
Small, E. (1972). Photosynthetic rates in relation to nitrogen recycling as an adaptation to nutrient deficiency in bog plants. *Can. J. Bot.* 50: 2227–2233.
Smith, A. B. (1975). Radiocarbon dates from Bosumpra Cave, Ghana. *Proc. Prehist. Soc.* 41: 179–182.
Smith, A. P. (1972). Buttressing of tropical trees: a descriptive model and new hypotheses. *Amer. Natur.* 106: 32–46.
Smith, A. P. (1979). Buttressing of tropical trees in relation to bark thickness in Dominica, B.W.I. *Biotropica* 11: 159–160.
Sokal, R. R. & F. J. Rohlf (1969). *Biometry*. W. H. Freeman, San Francisco.
St. Clair Thompson, G. W. (1936). Forest conditions in the Gold Coast. *Imp. For. Inst. Paper* 1.
Styles, B. T. (1972). The flower biology of the Meliaceae and its bearing on tree breeding. *Silvae Genet.* 21: 175–182.
Styles, B. T. & P. K. Khosla (1976). Cytology and reproductive biology of Meliaceae. *Linn. Soc. Symp.* 2: 61–67.
Swaine, M. D. & Beer, Tom (1977). Explosive seed dispersal in *Hura crepitans* L. (Euphorbiaceae). *New Phytol.* 78: 695–708.
Swaine, M. D. & J. B. Hall (1976). An application of ordination to the identification of forest types. *Vegetatio* 32: 83–86.
Swaine, M. D. & J. B. Hall (1977). Ecology and conservation of upland forests in Ghana. In: *Proceedings of Ghana Scope's Conference on Environment and Development in West Africa* (Ed.: A. M. Laryea) pp. 151–158. Ghana Acad. Arts Sci., Accra.
Swaine, M. D. & J. B. Hall (1981). The monospecific tropical forest of the Ghanaian endemic tree *Talbotiella gentii*. In: *Proceedings of the International Conference on Biological Aspects of Rare Plant Conservation* (Ed.: H. Synge). Wiley, London.
Swaine, M. D., J. B. Hall & J. M. Lock (1976). The forest-savanna boundary in west-central Ghana. *Ghana J. Sci.* 16: 35–52.
Tansley, A. G. & T. F. Chipp (1926). *Aims and Methods in the Study of Vegetation*. Crown Agents, London.
Taylor, C. J. (1952). The vegetation zones of the Gold Coast. *Bull. Gold Cst. For. Dep.* 4: 1–12.
Taylor, C. J. (1960). *Synecology and Silviculture in Ghana*. Nelson, Edinburgh.
Thompson, H. N. (1910). Gold Coast: report on forests. *Colon. Rep. Miscell.* 66: 1–238.
Thorne, R. F. (1973). Floristic relationships between Tropical Africa and Tropical America. In: *Tropical Forest Ecosystems in Africa and South America: a Comparative Review* (Ed.: B. J. Meggers, E. S. Ayensu & D. W. Duckworth) pp. 27–47. Smithsonian Institution Press, Washington D.C.
Tufuor, K. (1976). Prospects for increasing forest productivity in West African high forest through forest genetics and tree improvement. *Ghana For. J.* 2: 15–19.
UNESCO (1973). *International Classification and Mapping of Vegetation*. UNESCO, Paris.
UNESCO/UNEP/FAO (1978). Palaeogeography and palaeoclimatology. In: *Tropical Forest Ecosystems*, pp. 61–90. UNESCO, Paris.
Waterman, P. G., I. A. Meshal, J. B. Hall & M. D. Swaine (1978). Biochemical systematics and ecology of the Toddalioideae in the central part of the West African forest zone. *Biochem. System. Ecol.* 6: 239–245.
Werger, M. J. A. (1974a). The place of the Zürich-Montpellier method in vegetation science. *Folia Geobot. Phytotax., Praha* 9: 99–109.
Werger, M. J. A. (1974b). On concepts and techniques applied in the Zürich-Montpellier method of vegetation survey. *Bothalia* 11: 309–323.
Westhoff, V. & E. van der Maarel (1973). The Braun-Blanquet approach. *Handb. Veg. Sci.* 5: 617–726.
White, F. (1962). Geographic variation and speciation in Africa

with particular reference to *Diospyros. Publs System. Assn* 4: 71–103.

White, F. (1965). The savanna woodlands of the Zambezian and Sudanian domains. *Webbia* 19: 651–681.

White, F. (1970). Floristics and plant geography. In: *The Evergreen Forests of Malawi* (J. D. Chapman & F. White) pp. 38–77. Commonw. For. Inst. Oxford.

White, F. (1976a). The vegetation map of Africa – the history of a completed project. *Boissiera* 24: 659–666.

White, F. (1976b). Chrysobalanaceae. *Distrib. Pl. Afr.* 281.

White, F. (1978). The taxonomy, ecology and chorology of African Ebenaceae. 1. The Guineo-Congolian species. *Bull. Jard. Bot. Nat. Belg.* 48: 245–358.

White, F. (1979). The Guineo-Congolian Region and its relationships to other phytochoria. *Bull. Jard. Bot. Nat. Belg.* 49: 11–55.

Whitmore, T. C. (1975). *The Tropical Rain Forests of the Far East.* Oxford Univ. Press, London.

Whitmore, T. C. (1978). Gaps in the forest canopy. In: *Tropical Trees as Living Systems* (Ed.: P. B. Tomlinson & M. H. Zimmermann) pp. 639–655. Cambridge Univ. Press, Cambridge.

Wickens, G. E. (1975). Changes in the climate and vegetation of the Sudan since 20 000 B.P. *Boissiera* 24: 43–65.

Williams, M. A. J. (1975). Late Pleistocene tropical aridity synchronous in both hemispheres? *Nature* 253: 617–618.

B. References, cited by number in the notes accompanying the maps

1. Adu-Tutu, M., Y. Afful, K. Asante-Appiah, D. Lieberman, J. B. Hall & M. Elvin-Lewis (1980). Chewing stick usage in Southern Ghana. *Econ. Bot.* 33: 320–328.
2. Aké Assi, L. (1960). Quatre espèces nouvelles de Côte d'Ivoire. *Bull. Jard. Bot. État Brux.* 30: 397–405.
3. Aké Assi, L. (1963). *Contribution à l'Étude Floristique de la Côte d'Ivoire et des Territoires Limitrophes.* Paul Lechevalier, Paris.
4. Alexandre, D. Y. (1977). Régéneration naturelle d'un arbre caractéristique de la forêt équatoriale de Côte d'Ivoire: *Turraeanthus africanus* Pellegr. *Oecologia Plantarum* 12: 241–262.
5. Alexandre, D. Y. (1978). La rôle disseminateur des éléphants en forêt de Tai, Côte d'Ivoire. *La Terre et la Vie* 32: 47–62.
6. Ampofo, S. T. & G. W. Lawson (1972). Growth of seedlings of *Afrormosia elata* Harms in relation to light intensity. *J. appl. Ecol.* 9: 301–306.
7. Anderson, E. (1954). *Plants, Man and Life.* Andrew Melrose, London.
8. Aubréville, A. (1959). *La Flore Forestière de la Côte d'Ivoire.* Centre Technique Forestier Tropical, Nogent sur Marne. 3 vols.
9. Aubréville, A. (1964). Sapotacées. In: *Flore du Cameroun* 2: 1–143. Mus. natl Hist. nat., Paris.
10. Aubréville, A. (1968). Legumineuses – Caesalpinioidées. In: *Flore du Gabon* 15: 1–362. Mus. natl Hist. nat., Paris.
11. Aubréville, A. (1971). Essai sur la géophylétique et l'écophylétisme des Manilkarées. *Adansonia,* sér. 2, 11: 251–265.
12. Badré, F. (1971). *Hugonia* africains et leurs fruits. *Adansonia,* sér. 2, 11: 95–106.
13. Badré, F. (1973). Ixonanthacées. In: *Flore du Gabon* 21: 55–62. Mus. natl Hist. nat., Paris.
14. Baker, H. G. (1958). Studies in the reproductive biology of West African Rubiaceae. *J. W. Afr. Sci. Ass.* 4: 9–24.
15. Baker, H. G. (1962). Heterostyly in the Connaraceae with special reference to *Byrsocarpus coccineus. Bot. Gaz.* 123: 206–211.
16. Baker, H. G. & B. J. Harris (1959). Bat-pollination of the silk-cotton tree, *Ceiba pentandra,* in Ghana. *J. W. Afr. Sci. Ass.* 5: 1–9.
17. Bamps, P. (1967). Ochnaceae. In: *Flore du Congo, du Rwanda et du Burundi.* Jard. Bot. nat. Belg., Bruxelles.
18. Bamps, P. (1969). Notes sur les Guttiferae d'Afrique tropicale. 3. Révision des Vismieae. *Bull. Jard. Bot. nat. Belg.* 39: 345–372.
19. Bamps, P. (1970). Guttiferae. In: *Flore du Congo, du Rwanda et du Burundi,* Jard. Bot. nat. Belg., Bruxelles.
20. Beentje, H. J. (1978). A revision of *Callichilia* Stapf (Apocynaceae). *Meded. Landb. Wageningen* 78–7: 1–32.
21. Bell, E. A. & D. H. Janzen (1971). Medical and ecological considerations of L-Dopa and 4-HTP in seeds. *Nature* 229: 136–137.
22. Berg, C. C. (1977). Revisions of African Moraceae. *Bull. Jard. Bot. nat. Belg.* 47: 267–407.
23. Berg, C. C. (1978). Revisions of African Moraceae (excl. *Dorstenia, Ficus, Musanga* and *Myrianthus*); corrections and additions. *Bull. Jard. Bot. Nat. Belg.* 48: 465–471.
24. Bisset, N. G. & A. J. M. Leeuwenberg (1968). The use of *Strychnos* species in Central African ordeal and arrow poisons. *Lloydia* 31: 208–222.
25. Bisset, N. G. & J. D. Philipson (1971). The African species of *Strychnos.* 2. The alkaloids. *Lloydia* 34: 1–60.
26. Bokdam, J. (1977). Seedling morphology of some African Sapotaceae and its taxonomical significance. *Meded. Landb. Wageningen* 77–20: 1–84.
27. Boutique, R. (1959). Icacinaceae novae congolanae. *Bull. Jard. Bot. État. Brux.* 29: 425–432.
28. Brenan, J. P. M. (1959). Leguminosae subfamily Mimosoideae. In: *Flora of Tropical East Africa.* Crown Agents, London.
29. Breteler, F. J. (1973). The African Dichapetalaceae; a taxonomical revision. *Meded. Landb. Wageningen* 73–13: 1–123.
30. Breteler, F. J. (1978). The African Dichapetalaceae 4. *Meded. Landb. Wageningen* 78–10: 1–84.
31. Breteler, F. J. (1979). The African Dichapetalaceae 5. *Meded. Landb. Wageningen* 79–16: 1–78.
32. Breteler, F. J. & A. A. S. Smissaert-Houwing (1977). Revision of *Atroxima* Stapf and *Carpolobia* G. Don (Polygalaceae). *Meded. Landb. Wageningen* 77–18: 1–45.
33. Brummitt, R. K. (1968). A new genus of the tribe Sophoreae (Leguminosae) from western Africa and Borneo. *Kew Bull.* 22: 375–386.
34. Cagan, R. H. (1973). Chemostimulatory protein: a new type of taste stimulus. *Science* 181: 32–35.
35. De Jong, B. H. J. (1979). A revision of the African species of *Alstonia* R. Br. (Apocynaceae). *Meded. Landb. Wageningen* 79–13: 1–16.

36. De Ruiter, G. (1976). Revision of the genera *Myrianthus* and *Musanga* (Moraceae). *Bull. Jard. Bot. Nat. Belg.* 46: 471–510.
37. De Wilde, J. J. F. E. (1968). A revision of the species of *Trichilia* P. Browne (Meliaceae) on the African continent. *Meded. Landb. Wageningen* 68–2: 1–207.
38. Enti, A. A. (1968). Distribution and ecology of *Hildegardia barteri* (Mast.) Kosterm. *Bull. Inst. fond. Afr. noire*, sér. A, 30: 881–895.
39. Fellows, L. E. & E. A. Bell (1970). 5-hydroxy-L-tryptophan, 5-hydroxytryptamine and L-tryptophan-5-hydroxylase in *Griffonia simplicifolia. Phytochemistry* 9: 2389–2396.
40. Fergusson, K. (1970). *Ghana Hardwoods.* A. S. Worth, London.
41. Fish, F. & P. G. Waterman (1973). Chemosystematics in the Rutaceae. 2. The chemosystematics of the *Zanthoxylum-Fagara* complex. *Taxon* 22: 177–203.
42. Forman, L. (1965). A new genus of Ixonanthaceae with notes on the family. *Kew Bull.* 19: 517–526.
43. Forman, L. (1966). The reinstatement of *Galearia* and *Microdesmis* in the Pandaceae. *Kew Bull.* 20: 309–321.
44. Forman, L. (1975). The tribe Triclisieae Diels in Asia, the Pacific and Australia. The Menispermaceae of Malesia and adjacent areas 8. *Kew Bull.* 30: 77–100.
45. Fouilloy, R. & N. Hallé (1973). Sapindacées. In: *Flore du Cameroun* 16: 1–202. Mus. natl Hist. nat., Paris.
46. Friis, I. & I. Nordal (1976). Studies on the genus; *Haemanthus* (Amaryllidaceae). 4. Division of the genus into *Haemanthus* s. str. and *Scadoxus* with notes on Haemanthus. s. str. *Norw. J. Bot.* 23: 63–77.
47. Garay, L. A. & P. Taylor (1976). The genus *Oeceoclades. Bot. Mus. Leafl. Harv. Univ.* 24: 249–274.
48. Hall, J. B. (1980). Five new species of flowering plants from West Africa. *Bull. Jard. Bot. Nat. Belg.* 50: 249–266.
49. Hall, J. B. (1980). New and little-known species of *Placodiscus* (Sapindaceae) in West Africa. *Andansonia,* sér. 2, 20: 287–295.
50. Hall, J. B. (1981). Three new combinations in tropical African Celastraceae and Ochnaceae. *Kew Bull.* (in press).
51. Hall, J. B. & A. J. M. Leeuwenberg (1979). A new species of Hunteria (Apocynaceae) from Ghana. *Bull. Jard. Bot. Nat. Belg.* 49: 421–425.
52. Hall, J. B. & J. M. Lock (1975). Use of vegetative characters in the identification of species of *Salacia* (Celastraceae). *Boissiera* 24: 331–338.
53. Hall, J. B., D. U. U. Okali & J. M. Lock (1975). Night illumination and flowering in an indigenous West African forest-tree, *Hildegardia barteri* (Mast.) Kosterm. *Bull. Inst. fond. Afr. noire,* sér. A, 37: 282–287.
54. Hall, J. B. & D. E. K. A. Siaw (1980). The varieties of *Grewia carpinifolia* Juss. (Tiliaceae) *Adansonia,* sér. 2, 20: 339–347.
55. Hall, J. B. & M. D. Swaine (1976). Classification and ecology of closed-canopy forest in Ghana. *J. Ecol.* 64: 913–951.
56. Hall, J. B., M. D. Swaine & M. R. Talbot (1978). An early holocene leaf flora from Lake Bosumtwi, Ghana. *Palaeogeogr., Palaeoclimatol., Palaeoecol.* 24: 247–261.
57. Hall, J. B. & P. B. Tomlinson (1973). Natural callus-like tissue produces vegetative buds on the leaves of the monocotyledon *Haemanthus cinnabarinus* Decne. *Nature New Biology* 243: 252–255.
57a. Hall, John B. & S. O. Baba (1979). The distribution and ecology of Obeche (*Triplochiton scleroxylon*). *J. Ecol.* 67: 543–564.
58. Hall, R. J. (1972). The distribution of organic fluorine in some toxic tropical plants. *New Phytol.* 71: 855–871.
59. Hallé, F., R. A. A. Oldeman & P. B. Tomlinson (1978). *Tropical Trees and Forests.* Springer-Verlag, Berlin.
60. Hallé, N. (1961). Sterculiacées. In: *Flore du Gabon* 2: 1–150. Mus. natl Hist. nat., Paris.
61. Hallé, N. (1962). Observations taxonomiques, morphologiques et écologiques sur deux *Lasiodiscus* d'Afrique (Rhamnacées). *Adansonia,* sér. 2, 2: 129–133.
62. Hallé, N. (1962). Monographie des Hippocrateacées d'Afrique Occidentale. *Mém. Inst. fr. Afr. noire* 64: 1–245.
63. Hallé, N. (1970). Rubiacées 2. In: *Flore du Gabon* 17: 1–335. Mus. natl Hist. nat., Paris.
64. Hallé, N. & L. Aké Assi (1962). Le genre *Chytranthus* en Côte d'Ivoire. *Adansonia,* sér. 2, 2: 291–299.
65. Hepper, F. N. (1976). *The West African Herbaria of Isert and Thonning.* Bentham-Moxon Trust, Kew.
66. Irvine, F. R. (1961). *Woody Plants of Ghana.* Oxford Univ. Press, London.
67. Jacques-Félix, H. (1979). Espèces nouvelles et peu connues du genre *Memecylon* (Melastomataceae) en Afrique. *Adansonia,* sér. 2, 18: 409–432.
68. Jeník, J. (1967). Root adaptations in West African trees. *J. Linn. Soc. (Bot.)* 60: 25–29.
69. Jeník, J. (1969). The life-form of *Scaphopetalum amoenum* A. Chev. *Preslia, Praha* 41: 109–112.
69a. Jeník, J. & K. O. A. Mensah (1967). Root systems of tropical trees. 1. Mycorrhiza of *Afzelia africana* Sm. *Preslia* 39: 59–65.
70. Keay, R. W. J. (1952). Revision of the 'Flora of West Tropical Africa' 1. *Kew Bull.* 1952: 149–165.
71. Keay, R. W. J., C. F. A. Onochie & D. P. Stanfield (1964). Meliaceae. In: *Nigerian Trees* 2: 257–284. Dept. of Forest Research, Ibadan.
72. Koechlin, J. (1965). Marantacées. In: *Flore du Cameroun* 4: 99–162. Mus. natl Hist. nat., Paris.
73. Kuijt, J. (1969). *The Biology of Parasitic Flowering Plants.* Univ. of California Press, Berkeley.
74. Leeuwenberg, A. J. M. (1969). The Loganiaceae of Africa: 8. *Strychnos:* 3. *Meded. Landb. Wageningen* 69–1: 1–316.
75. Le Thomas, A. (1967). À propos de l'*Uvariodendron mirabile* R. E. Fries. *Adansonia,* sér. 2, 7: 249–253.
76. Le Thomas, A. (1969). Annonacées. In: *Flore du Gabon* 16: 1–371. Mus. natl Hist. nat., Paris.
77. Letouzey, R. (1963). Rutacées. In: *Flore du Gabon* 6: 3–109. Mus. natl Hist. nat., Paris.
78. Letouzey, R. (1968). Ulmacées. In: *Flore du Cameroun* 8: 3–65. Mus. natl Hist. nat., Paris.
79. Liben, L. (1968). *Petersianthus* Merrill versus *Combretodendron* A. Chev. (Lecythidaceae). *Bull. Jard. Bot. Nat. Belg.* 38: 207–208.
80. Liben, L. (1971). Révision du genre africain *Napoleonaea* P. Beauv. (Lecythidaceae). *Bull. Jard. Bot. Nat. Belg.* 41: 363–382.

81. Lieberman, D., J. B. Hall, M. D. Swaine & M. Lieberman (1979). Seed dispersal by baboons in the Shai Hills, Ghana. *Ecology* 60: 65–75.
82. Lock, J. M. (1978). Notes on the genus *Aframomum* (Zingiberaceae). 3. Two new species from West Africa. *Bull. Jard. Bot. Nat. Belg.* 48: 393–398.
83. Lock, J. M. & J. B. Hall (1973). Three new species of *Aframomum* K. Schum. (Zingiberaceae) from Ghana. *Kew Bull.* 28: 441–449.
83a. Lock, J. M. & J. B. Hall (1981). Floral biology of *Mallotus oppositifolius* (Euphorbiaceae). *Biotropica* (in press).
84. Louis, J. & J. Léonard (1948). Olacaceae. In: *Flore du Congo Belge et du Raunda-Urundi* 1: 249–278. Inst. nat. Étude agron. Congo Belge, Bruxelles.
85. Morton, J. K. (1964). Contributions towards a flora of Ghana: 1. *Ghana J. Sci.* 4: 86–93.
86. Obaton, M. (1960). Les lianes ligneuses à structure anormale des forêts denses d'Afrique Occidentale. *Annls Sci. nat.*, sér. 12 Botanique 1: 1–220.
87. Okafor, J. C. (1975). Varietal delimitation in *Irvingia gabonensis* (Irvingiaceae). *Bull. Jard. Bot. Nat. Belg.* 45: 211–221.
88. Paiva, J. A. R. (1966). Revisão das Annonaceae de Angola. *Mem. Soc. Brot.* 19: 1–128.
89. Petit, E. (1964). Les espèces africaines du genre *Psychotria:* 1: *Bull. Jard. Bot. État Brux.* 34: 1–229.
90. Petit, E. (1966). Les espèces africaines du genre *Psychotria:* 2. *Bull. Jard. Bot. État Brux.* 36: 65–190.
91. Polhill, R. M. (1966). Ulmaceae. In: Flora of Tropical East Africa. Crown Agents, London.
92. Polhill, R. M. (1971). Some observations on generic limits in Dalbergieae-Lonchocarpineae. *Kew Bull.* 25: 259–273.
93. Poncy, O. (1978). Le genre *Pararistolochia.* Aristolochiaceae d'Afrique tropicale. *Adansonia*, sér. 2, 17: 465–494.
94. Quigley, F. & J. B. Hall (1979). The relation to cultigens of wild edible yams (*Dioscorea* sect. *Enantiophyllum*) in Ghana. *Proc. 9th Plen. Meeting* A.E.T.F.A.T., Las Palmas de Gran Canaria, pp. 75–79.
95. Rasmussen, F. N. (1977). The genus *Corymborkis* Thou. (Orchidaceae); a taxonomic revision. *Bot. Tidsskr.* 71: 161–192.
96. Robbrecht, E. (1975). *Hymenocoleus,* a new genus of Psychotrieae (Rubiaceae) from tropical Africa. *Bull. Jard. Bot. Nat. Belg.* 45: 273–300.
97. Robbrecht, E. (1977). The tropical African genus *Hymenocoleus* (Rubiaceae-Psychotrieae): additions. *Bull. Jard. Bot. Nat. Belg.* 47: 3–29.
98. Robson, N. K. B. (1966). *Celastraceae.* In: Flora Zambesiaca 2: 355–418. Crown Agents, London.
99. Schellenberg, G. (1938). *Connaraceae.* In: Das Pflanzenreich 4. 127: 65–93.
100. Sleumer, H. (1972). A taxonomic revision of the genus *Dasylepis* Oliv. (Flacourtiaceae). *Bot. Jahrb.* 92: 554–561.
101. Sleumer, H. (1972). A taxonomic revision of the genus *Scottellia* Oliv. (Flacourtiaceae). *Blumea* 20: 275–281.
102. Sleumer, H. (1973). Révision du genre *Homalium* Jacq. (Flacourtiacées) en Afrique (y compris Madagascar et les Mascareignes). *Bull. Jard. Bot. Nat. Belg.* 43: 239–328.
102a. Swaine, M. D. & J. B. Hall (1981). The monospecific tropical forest of the Ghanaian endemic tree *Talbotiella gentii.* In: *Proceedings of the International Conference on Biological Aspects of Rare Plant Conservation* (Ed.: H. Synge) pp. 355–363.
103. Taton, A. (1969). Violaceae. In: *Flore du Congo, du Rwanda et du Burundi.* Jard. Bot. nat. Belg, Bruxelles.
104. Tennant, J. (1963). Notes on Tropical African Violaceae. *Kew Bull.* 16: 409–435.
105. Troupin, G. (1956). Menispermaceae. In: *Flora of Tropical East Africa.* Crown Agents, London.
106. Troupin, G. (1962). Monographie des Menispermaceae africaines. *Mém. Acad. roy. Sci. outre-mer* 13: 1–312.
107. Van Meeuwen, M. S. K. (1962). Reduction of *Afrormosia* to *Pericopsis* (Papilionaceae). *Bull. Jard. Bot. État Brux.* 32: 213–219.
108. Verdcourt, B. (1969). The status of the genus *Polyalthia* Blume (Annonaceae) in Africa. *Adansonia,* sér. 2, 9: 87–94.
109. Verdcourt, B. (1971). Notes on East African Annonaceae. *Kew Bull.* 25: 1–34.
110. Verdcourt, B. (1975). Studies on the Rubiaceae-Rubioideae for the 'Flora of Tropical East Africa': 1. *Kew Bull.* 30: 247–326.
111. Verdcourt, B. (1977). A synopsis of the genus *Chazaliella* (Rubiaceae-Psychotrieae). *Kew Bull.* 31: 785–818.
112. Villiers, J.-F. (1973). Icacinacées. In: *Flore du Gabon* 15: 3–100. Mus. natl Hist. nat., Paris.
113. Voorhoeve, A. G. (1965). *Liberian High Forest Trees.* Centre for Agric. Publ. & Documentation, Wageningen.
114. Walker, J. W. (1972). Contribution to the pollen morphology and phylogeny of the Annonaceae: 2. *Bot. J. Linn. Soc.* 65: 173–178.
115. Waterman, P. G., I. A. Meshal, J. B. Hall & M. D. Swaine (1978). Biochemical systematics and ecology of the Toddalioideae in the central part of the West African forest zone. *Biochem. System. Ecol.* 6: 239–245.
116. White, F. (1958). Two new combinations in *Maerua* Forsk. *Bol. Soc. Brot.* 32: 33–35.
117. White, F. (1964). Sapotaceae. In: *Nigerian Trees* 2: 344–365. Dept. of Forest Research, Ibadan.
118. White, F. (1976). The taxonomy, ecology and chorology of African Chrysobalanaceae (excluding *Acioa*). *Bull. Jard. Bot. Nat. Belg.* 46: 265–350.
119. White, F. (1977). Some new taxa in African Myrtaceae. *Kirkia* 10: 401–404.
120. White, F. (1978). The taxonomy, ecology and chorology of African Ebenaceae. 1. The Guineo-Congolian species. *Bull. Jard. Bot. Nat. Belg.* 48: 245–358.
121. White, M. G. (1966). The problem of the *Phytolyma* gall bug in the establishment of *Chlorophora. Commonw. For. Inst. Oxford Institute Paper* 37: 1–52.
122. Wilczek, R. (1960). Hippocrateaceae. In: *Flore du Congo Belge et du Ruanda-Urundi* 9: 133–232. Inst. nat. Étude agron. Congo Belge, Bruxelles.
123. Willis, J. C. (1973). *A Dictionary of the Flowering Plants and Ferns.* 8th ed. revised by H. K. Airy Shaw. Cambridge Univ. Press, Cambridge.
124. Yakovlev, G. P., A. A. Yatsenko-Khmelevsky & I. G. Zoubkova (1978). Taxinomie et phylogenie du genre *Angylocalyx* et de la tribu des Angylocalyceae. *Adansonia,* sér. 2, 8: 317–335.

Index to scientific names of plants*

* Italics are used for synonyms and other names wrongly applied to Ghanaian plants.
Page numbers in italics for figures; bold face for map notes.

General index*

* Page numbers in italics for figures.